Lecture Notes in Computer Science 520

Edited by G. Goos and J. Hartmanis

Advisory Board: W. Brauer D. Gries J. Stoer

A. Tarlecki (Ed.)

Mathematical Foundations of Computer Science 1991

16th International Symposium
Kazimierz Dolny, Poland, September 9-13, 1991
Proceedings

Springer-Verlag
Berlin Heidelberg New York
London Paris Tokyo
Hong Kong Barcelona
Budapest

Series Editors

Gerhard Goos
GMD Forschungsstelle
Universität Karlsruhe
Vincenz-Priessnitz-Straße 1
W-7500 Karlsruhe, FRG

Juris Hartmanis
Department of Computer Science
Cornell University
Upson Hall
Ithaca, NY 14853, USA

Volume Editor

Andrzej Tarlecki
Institute of Computer Science, Polish Academy of Sciences
P. O. Box 22, PKiN, 00-901 Warsaw, Poland

CR Subject Classification (1991): D.3.3, F.1.1, F.2.2, F.3.1, F.4.1, F.4.3, D.2.1-2

ISBN 3-540-54345-7 Springer-Verlag Berlin Heidelberg New York
ISBN 0-387-54345-7 Springer-Verlag New York Berlin Heidelberg

Typesetting: Camera ready by author
Printing and binding: Druckhaus Beltz, Hemsbach/Bergstr.
2145/3140-543210 - Printed on acid-free paper

Preface

This volume contains the proceedings of the 16th International Symposium on Mathematical Foundations of Computer Science, MFCS'91, held in Kazimierz Dolny, Poland, 9–13 September 1991.

The series of MFCS symposia, organized alternately in Poland and Czechoslovakia since 1972, has a long and well-established tradition. The purpose of the series is to encourage high-quality research in all branches of theoretical computer science and to bring together specialists working actively in the area. Throughout the years, MFCS has served this purpose well, and we hope it will do so in the future.

Principal areas of interest of the symposium include: software specification and development, parallel and distributed computing, semantics and logics of programs, algorithms, complexity and computability theory (this is not an exclusive list).

The scientific programme of the symposium consisted of invited lectures by distinguished scientists and 38 presentations selected, strictly on the basis of the scientific merit, by the Programme Committee out of the total 109 submitted papers. This volume collects papers accompanying the presentations. Unfortunately, due to the usual space limitations, these are in much abridged form.

Thanks are due to the authors of all the submitted papers for their interest in MFCS'91 and for their valuable contribution to the symposium. I would also like to express my gratitude to all the members of the Programme Committee for the work they put into the careful evaluation of the papers. I am especially indebted to all the referees who assisted the members of the Programme Committee by reviewing the submissions in detail. Finally, let me thank the members of the Organizing Committee for their invaluable work in enabling the symposium to take place and for their assistance in putting the proceedings together, and Springer–Verlag for their efficient cooperation in the publication of this volume.

MFCS'91 is organized by the Institute of Computer Science of the Polish Academy of Sciences in cooperation with the Institute of Informatics of Warsaw University and the Polish Information Processing Society.

Warsaw, May 1991 Andrzej Tarlecki

Programme Committee

K. Apt, Amsterdam
J. Barzdin, Riga
A. Blikle, Warsaw
M. Broy, Munich
R. Cori, Bordeaux
R. De Nicola, Pisa
J. Gruska, Bratislava
Y. Gurevich, Ann Arbor
K. Jantke, Leipzig
C. Jones, Manchester
G. Mirkowska, Warsaw
V. Pratt, Stanford
B. Rovan, Bratislava
D. Sannella, Edinburgh
A. Tarlecki, Warsaw (chair)
J. Winkowski, Warsaw

Organizing Committee

P. Chrząstowski-Wachtel (co-chairman)
K. Diks
L. Holenderski
S. Jarominek
W. Penczek
A. Szałas
A. Tarlecki (co-chairman)

Referees

P.H.G. Aczel
A. Aceto
J. Albert
M. Alberti
H. Alt
S. Anderson
J. Andrews
A. Arnold
A. Asperti
A. Auguston
O. Bachmann
E. Badouel
B. Banieqbal
H. Barringer
E. Baudel
A. Bebják
M. Bednarczyk
J. Bergstra
D. Berry
J. Berstel
M. Bidoit
J. Bond
S. Bonnier
M. Bonuccelli
A. Borzyszkowski
C. Böhm
R. Breu
A. Bundy
G. Buntrock
H.-D. Burkhard
H.-J. Bürckert
R. Capocelli
Z. Chaochen
M. Chytil
J. Cleve
P. Clote
K. Compton
B. Courcelle
E. Csuhaj-Varju
L. Czaja
I. Černá
A. Černý
K. Čulik II
M. Dam
R. Danecki
J. Dassow
D. De Francesco
F. Dederichs
P. Degano
M. Delest
C. Dendorfer
J. Desel
Y. Deville
J. Díaz
M. Dietzfelbinger
Do Long Van
S. Dulucq
H. Ehler
H. Ehrig
M. Elvang
P. van Emde Boas
V. Engelson
A. Fantechi
G. Ferrari
R. Freivalds
M. Fuchs
U. Furbach
R. Gavaldà
T. Gecseg
R. Ghosh-Roy
S. Gnesi
J. Goguen

U. Goldammer
H.-J. Goltz
U. Goltz
R. Gorrieri
T. Gritzner
R. Grosu
C. Guerra
C. Gunter
C. Hankin
M. Hanus
R. Hartwig
M. Hennessy
R. Hennicker
U. Hertrampf
B. Hilken
J.J. Horning
J. Hromkovič
H. Hussman
O. Ibarra
P. Inverardi
B. Jay
M. Jantzen
C. Jones
A. Kalis
J. Kalninš
E. Kazmierczak
A. Kiehn
E. Kinber
B. Kirsig
H.C.M. Kleijn
J.W. Klop
R. Knast
B. Konikowska
I. Korec
A. Kreczmar
F. Kröger
A. Labella
S. Lange
F. Lesske
J. Leszczyłowski
C. Levcopoulos
A. Lingas
I. Litovski
Z. Luo
A. Maggiolo Schettini
T. Maibaum
A. Masini
U. Martin
S. Martini
G. Mauri
J. McKinna
Y. Metivier
J. Mitchell
K. Mitchell
W.P.R. Mitchell
E. Moggi
F. Moller
B.Q. Monahan
B. Monien
U. Montanari
A.W. Mostowski
B. Möller
J. Nawrocki
D. Nazareth
F. Nickl
T. Nipkow
D. Niwiński
L. Ong
P. Padawitz
D. Pardubska
H. Partsch
J. Penaud
U. Petermann
J.-E. Pin
E. Pippig
K. Podnieks
D. Pym
Z. Qian
A. Raspand
H. Reichel
J.M. Robson
F. Romani
K. Ross
B. Rounds
G. Ruhe
P. Ružička
D. Rydeheard
W. Rytter
A. Salwicki
S. Schönherr
L. Schrijver
G. Senizergues
A. Sinclair
A. Skowron
S. Smolka
S. Sokołowski
M. Srebrny
B. Steffen
C. Stirling
Q. Stout
T. Streicher
M. Sysło
A. Szałas
A. Szepietowski
J. Šturc
P. Taylor
J. Tiuryn
M. Tofte
Y. Toyama
T.H. Tse
P. Urzyczyn
M. Veldhorst
M. Venturini Zilli
W. Vogler
I. Vrto
P. Wadler
L. Wallen
E. Wanke
J.H.A. Warmerdam
R. Weber
J. Wiedermann
R. Wiehagen
R. Wilhelm
A. Wiweger
H. Wolter
S. Yoccoz
V. Zaťko
T. Zengmann
W. Zielonka

Table of contents

Invited lectures

Contributions

Elimination of Negation in Term Algebras

J-L. Lassez, M. Maher, K. Marriott
IBM T.J. Watson Research Center
P.O. Box 704
Yorktown Heights, NY 10598

Abstract

We give an informal review of the problem of eliminating negation in term algebras and its applications. The initial results appear to be very specialized with complex combinatorial proofs. Nevertheless they have applications and relevance to a number of important areas: unification, learning, abstract data types and rewriting systems, constraints and constructive negation in logic languages.

1 Initial Motivation: Learning

Plotkin [36] proposed a formal model for inductive inference which was based upon Popplestone's suggestion that

> Since unification is useful in automatic deduction, its dual might prove helpful for induction.

A similar formalism was independently introduced by Reynolds [38], who was more concerned with its algebraic properties than with its applications. The algebraic properties were further investigated by Huet [9, 10], who also studied the case of the infinitary Herbrand universe.

The key result in this theory is that, for any set of terms, there exists a unique (up to variable renaming) term that represents their most specific generalization. For instance the most specific generalization of $f(b,a)$, $f(a,b)$ and $f(a,c)$ is $f(x,y)$, where a,b,c are constant symbols, f is a function symbol and x,y are variables.

However, such generalizations may be too general and require counter examples to restrict them. For instance, if the above example also had counter examples $f(a,a)$ and $f(c,c)$, a reasonable generalization would be $f(x,y)/\{f(u,u)\}$ where this implicitly represents the ground instances of $f(x,y)$ such that x is different to y.

A generalization of the form

$$t/\{t_1 \vee ... \vee t_n\}$$

where a disjunction represents the union of the sets of the ground instances of the terms, and / represents set subtraction, is implicit because it does not provide direct access to the ground instances it represents.

Here we investigate when an implicit generalization has an equivalent finite disjunction of terms which can provide an explicit representation for the generalization. The two notations are equivalent if they have the same set of ground instances. For example, the generalization

$$f(x,y)/\{f(a,u) \vee f(u,a)\}$$

is equivalent to the disjunction of terms

$$f(b,b) \ \vee \ f(f(x,y),b) \ \vee \ f(b,f(x,y)) \ \vee \ f(f(x,y),f(x',y'))$$

when there are constant symbols a, b and function symbol f. However, not all implicit representations have an equivalent finite disjunction of terms. As for example $f(x,y)/\{f(u,u)\}$.

Transforming an implicit representation into an equivalent disjunction of terms is of importance to machine learning as it allows disjunctive concepts to be learnt from examples with counter examples.

The algorithm proposed in [21] can be used to learn disjunctive concepts from examples with counter examples. Consider the examples

$$[0,s(s(0)),s(0)], \ [s(0),s(s(0)),0], \ [0,s(0),0], \ [0,s(s(s(0))),0]$$

and the counter examples

$$[s(0),0,s(s(0))], \ [s(0),s(s(0)),s(0)], \ [0,0,0]$$

where s represents the successor function for integers. The examples generalize to give $[x,s(y),z]$. The counter examples can be generalized to give $[x,y,z]$. However, this is clearly too general as all of the examples are instances of it. A better generalization of the counter examples is the expression $[u,0,v] \vee [s(u),v,s(w)]$ which does not have any positive example as an instance of it. Thus a reasonable generalization taking account of both examples and counter examples is

$$[x,s(y),z]/\{[u,0,v] \vee [s(u),v,s(w)]\}$$

The equivalent disjunction

$$[0,s(x),y] \vee [x,s(y),0]$$

provides a natural way to express the concept learnt, which is that the list must contain a 0 in its first or last position but cannot contain a 0 in its middle position. The results we mention are to be found in [21], to which we refer for further details and proofs. Various cases are to be considered depending on the respective cardinalities of the Herbrand Universe and the set of function symbols. We present here only the most common case, where the Herbrand Universe is infinite and the set of function symbols is finite. The correctness of an algorithm **uncover**, described in [21], is established by the following theorem:

Theorem 1 *The algorithm* **uncover** *can be used to find an equivalent, finite disjunction of terms for*

$$t/\{t_1 \vee \ldots \vee t_n\}$$

if one exists. Otherwise **uncover** *will terminate with a canonical implicit representation.*

We can view the output of the algorithm as a canonical form, where the formula has been simplified by removing as much negation as possible. When all negation has been removed, the disjunction consists in a finite collection of terms, linear w.r.t. t, which represent disjoint sets of ground instances. We will see that this fact is useful in a later section.

The usefulness of this simplification is particularly evident when the positive part of the formula is in fact equivalent to the negative part. For example, consider the implicit generalization

$$f(x,y)/\{f(u,u) \vee f(f(u,v),w) \vee f(u,f(u,v)) \vee f(a,f(f(u,v),w))\}$$

when there is one function symbol f and one constant a. It is not immediately obvious that this is equivalent to the empty set, which is a more appropriate representation.

We now give a necessary and sufficient condition for an implicit generalization to have an equivalent finite disjunction.

Theorem 2 $t/\{t\theta_1 \vee \ldots \vee t\theta_n\}$ *has an equivalent finite disjunction iff* $\exists t\phi_1, \ldots, t\phi_m$ *s.t. each* $t\phi_i$ *is linear w.r.t.* t *and* $t\theta_1 \vee \ldots \vee t\theta_n \sim t\phi_1 \vee \ldots \vee t\phi_m$.

This theorem has a number of applications discussed later. An immediate consequence is that the sets of ground instances representable by finite sets of linear terms form a Boolean algebra.

Now, as the Herbrand Universe is a primitive structure, one could expect that most applications would be on higher level structures. So these notions will be more applicable to practical Learning problems when lifted to cater for more complex domains of computation. Several important extensions have been made that we briefly mention. In the Herbrand Universe two terms are equivalent only if syntactically identical. Kounalis has extended the algorithm for generalizing from examples and counter examples in the presence of equality theories that capture background knowledge not representable in the Herbrand Universe [14]. In particular the method is applied to induce concepts in elementary theories of arithmetic. Maher and Stuckey have examined the problem of inductive inference by lifting the notions and algorithm to the domain of pointer based data structures, through a rational trees formalization [31]. Two other papers, by Kounalis [15] and Kounalis and Rusinowitch [16], have a different main motivation but are still quite relevant to this study of generalization in the presence of counter examples. The first addresses the problem of understanding when a fact is new with respect to a given theory, and the second studies decidability issues for problems of finite presentation in Tree Languages generated by Rewrite Systems. Also relevant are the papers by Frisch and Page [7, 34].

2 Unification in the Presence of Negation

The problem of unification can be formulated this way: find if a system of equations has a solution, find a finite explicit representation of the set of solutions, or determine that no finite explicit representation exists. We will focus here on the problem of representation. We refer to [20] for background and formalization.

There are many different ways to explicitly represent the solutions of an equation set. These include the *most general unifier* (*mgu*) and the *most general solution set* (*mgss*). For systems which consist only of equations, these representations are equivalent. But when we introduce negative information, that is when we consider systems of equations and inequations, the equivalence is lost. Consider the system $\{x = f(a), y \neq x\}$. We know that it cannot be represented by a system consisting solely of equations. This means that we cannot use the standard definition of mgu to finitely represent its set of solutions, as this representation is equivalent to a system of equations. However the set of substitutions:

$$\{x \leftarrow f(a), y \leftarrow f(f(u))\}$$

and

$$\{x \leftarrow f(a), y \leftarrow a\}$$

represents finitely and explicitly the set of solutions of the system. We call such a set an mgss of the system of equations and inequations. We cannot, as in the case without negation, replace these substitutions by an equivalent set of equations. This is because an auxiliary variable u has been introduced, which cannot be eliminated. It is the use of this auxiliary variable which provides the mgss with more expressive power than the mgu. (In the mgu, auxiliary variables are not really needed as they can be eliminated).

We now show that it is decidable whether or not a given system has an mgss and compute it when it exists. This is done by transforming the problem into a problem of term subtraction. As a consequence we will be able to use the previously mentioned results and algorithms on term expressions.

To illustrate the transformation consider the system

$$y = f(x), x \neq f(f(a)), y \neq f(f(a))$$

This is transformed into the term expression

$$p(x,y)\{y \leftarrow f(x)\}/\{(p(x,y)\{y \leftarrow f(x)\})\{x \leftarrow f(f(a))\} \vee (p(x,y)\{y \leftarrow f(x)\})\{x \leftarrow f(a)\}\}$$

$$= p(x, f(x))/\{p(f(f(a)), f(f(f(a)))) \ \vee \ p(f(z), f(f(a)))\}$$

If Σ contains one function symbol f and one constant symbol a then this is equivalent to

$$p(a, f(a)) \ \vee \ p(f(f(f(u))), f(f(f(f(u)))))$$

and thus the set containing $\{x \leftarrow a, y \leftarrow f(a)\}$ and $\{x \leftarrow f(f(f(u))), y \leftarrow f(f(f(f(u))))\}$ is an mgss for the original system.

In general, consider a system S of equations E and inequations $I_1, \ldots, I_n$. Let μ_E be an idempotent mgu of E and let μ_i be an idempotent mgu of $E_i\mu_E$, where E_i is the equation obtained by negating I_i. If μ_E does not exist then the system has no solutions and if a μ_i does not exist then the inequation I_i can be removed, as it is redundant. The term expression corresponding to S is

$$p(v_1, \ldots, v_m)\mu_E/\{p(v_1, \ldots, v_m)\mu_E\mu_1 \vee \ldots \vee p(v_1, \ldots, v_m)\mu_E\mu_n\}$$

where p is a dummy function symbol and $v_1, v_2, \ldots, v_m$ are the variables appearing in S. From the results of the preceding section we have:

Proposition 1 *There exists an algorithm that will produce an mgss for a system of equations and inequations if one exists. Otherwise it will halt with failure.*

By taking the counterpart of Theorem 2, we obtain another characterization of systems that admit an mgss. This characterization is quite simple and furthermore algorithmic. We now give the theorem. Let a system contain equations E with idempotent mgu μ_E. The inequation I_j is *effectively ground* in this system if $E_j\mu_E$ has a grounding substitution as its idempotent mgu. For example, the inequation $x \neq y$ is effectively ground in the system with equations $\{z = f(y), x = a\}$.

Theorem 3 *A redundancy free system of equations and inequations has an mgss iff each inequation is effectively ground.*

In some applications, for example [33, 19], universally quantified variables appear naturally in the inequations.

For example, consider the logic program

$$\begin{array}{l} P(g(x)) \\ Q(g(y)) \leftarrow not\ P(y) \\ R(g(z)) \leftarrow not\ Q(z) \end{array}$$

The answer substitutions for $R(w)$ are just those assignments for w that satisfy the system $w = g(g(g(x)))$ or the system $w = g(z)$ and $\forall y, z \neq g(y)$.

We now consider the problems of solvability and representability within this context. We assume that there is a finite set U of the universally quantified variables that appear in any given system. This set is disjoint from the finite set of unquantified variables V. We will sometimes be dealing with equations on variables from $U \cup V$.

We now allow inequations of the form $I = \{\forall\, y_1, \ldots, y_k\ s_1 \neq t_1, \ldots, s_n \neq t_n\}$ to appear in a system S. A *solution* to a system S of equations E and inequations I is a grounding substitution α for V which solves E and is such that for all grounding substitutions β to U, $s_1(\alpha \cup \beta) \neq t_1(\alpha \cup \beta), \ldots, s_n(\alpha \cup \beta) \neq t_n(\alpha \cup \beta)$. For example, $\{w \leftarrow a, z \leftarrow a\}$ is a solution of the system $\{w = g(z), \forall y, z \neq g(y)\}$.

As for systems without universally quantified variables, determining whether a system has an mgss may be transformed into determining whether a term expression has an equivalent disjunction of terms. We first illustrate this process with an example.

The system $y = f(x), \forall z\ x \neq f(f(a)),\ y \neq f(f(z))$ may be translated into the term expression

$$p(x,y)\{y \leftarrow f(x)\}/\{(p(x,y)\{y \leftarrow f(x)\})\{x \leftarrow f(f(a))\} \vee (p(x,y)\{y \leftarrow f(x)\})\{x \leftarrow f(z)\}\}$$

$$= p(x, f(x))/\{(p(f(f(a)), f(f(f(a)))) \ \vee\ p(f(z), f(f(z))))\}$$

If the universe contains one function symbol f and one constant symbol a then this is equivalent to $p(a, f(a))$ and thus $\{x \leftarrow a, y \leftarrow f(a)\}$ is an mgss for the original system.

More precisely, consider a system S of equations E and inequations $\forall y_1, \ldots, y_k\ I_1, \ldots, I_n$. Let μ_E be an idempotent mgu of E and let μ_j be an idempotent mgu of $E_j\mu_E$. The corresponding term expression is

$$p(v_1, \ldots, v_m)\mu_E/\{p(v_1, \ldots, v_m)\mu_E\mu_1 \ \vee\ \ldots\ \vee\ p(v_1, \ldots, v_m)\mu_E\mu_n\}$$

where p is a dummy function symbol and $v_1, v_2, \ldots, v_m$ are the variables of V appearing in S.

As before, the transformation allows us to apply the uncover algorithm:

Theorem 4 *There exists an algorithm that will produce an mgss for a system of equations and universally quantified inequations if one exists. Otherwise it will halt with failure.*

Unlike the case when there are only unquantified variables, a redundancy free system may still have an mgss when it contains inequations that are not effectively ground. For example, if there is one constant symbol a and one binary function symbol f then the system

$$\{\forall u, v, u', v'\ f(x,y) \neq f(u,u),\ f(x,y) \neq f(f(u,v), f(u',v'))\}$$

has an mgss which consists of the substitutions $\{x \leftarrow a, y \leftarrow f(x',y')\}$ and $\{x \leftarrow f(x',y'), y \leftarrow a\}$.

In [5] Comon gives an authoritative survey of the theory of the algebra of finite trees, which is the appropriate context for the work presented in this section. Elements of an mgss have also been studied under the names basic sets [18], basic formulas [29], and parameterized substitutions [30], and the latter focuses on these objects.

Much of this work on unification in the presence of inequations was made easier by the drawing of an analogy between terms algebras and linear algebra. For instance we introduced in [20] the notion of dimension of the solution space of a system of equations, which was successfully exploited to derive interesting concepts and proofs. In [23] we studied linear constraints from an affine geometry point of view. Finding again analogies with the situation in term algebras led us to a systematic study of negative constraints in an abstract setting. The axiomatization proposed in [22] is sufficiently general to account for a variety of examples that come up in affine geometry, group theory, symbolic computation, term algebras and elsewhere.

3 Applications in Programming Languages and Term Rewriting

Results about term covering have two major applications in programming languages. The first is in languages which allow pattern-matching. The second is in algebraic specification and term rewriting as term covering is closely related to "ground reducibility". For notation and terminology that is not defined in this section, the reader is referred to Dershowitz and Jouannaud's presentation of Rewrite Systems [6]. We assume that the vocabulary of function symbols is divided into a set C of *constructors* and a set F of *defined* functions.

Many functional languages (and all logic programming languages) allow procedures to be defined using pattern-matching. The following example from [35] illustrates the idea.

Example: The function call ($mappairs\ f\ xs\ ys$) applies the function f to corresponding pairs from the lists xs and ys. The function is defined by:

$$\begin{aligned} mappairs\ f\ []\ ys &= [] \\ mappairs\ f\ (x : xs)\ [] &= [] \\ mappairs\ f\ (x : xs)\ (y : ys) &= f\ x\ y : mappairs\ f\ xs\ ys. \end{aligned}$$

In this example $C = \{:, []\}$ and $F = \{mappairs\}$. $\diamond$

The results in the previous section have two main implications in this context. The first, immediate implication, is that there is an algorithm to decide if the procedure definition covers all calling patterns, and if not, which patterns remain. Thus, in the above example, the definition of *mappairs* covers all calling patterns as:

$$mappairs(z, x, y)/\{mappairs(z, [], ys), mappairs(z, x : xs, []), mappairs(z, x : xs, y : ys)\} = \emptyset$$

where we consider only terms constructed from C and regard *mappairs* only as a grouping operator.

The second implication is more involved. In functional languages the function definition must be deterministic. Thus it is common to require that the patterns in the specification be disjoint. As the most general calling pattern is linear, it follows from Theorems 1 and 2 and the algorithm **uncover** that any disjoint cover for the calling patterns must consist only of linear terms, when each sort of the arguments is infinite. This provides a strong justification for the common requirement in functional languages that the patterns be, in fact, linear. A similar application of these results, pointed out by Comon [5], is that every ground-convergent constructor-based rewrite system is equivalent to a (ground-convergent constructor-based) rewrite system where all the patterns are linear. Term subtraction and the previous results are also relevant to the compilation of pattern-matching in functional languages, for example [24, 40].

Deciding whether a term is covered by a disjunction of terms is closely related to the concept of *ground reducibility* in term rewriting. Ground reducibility is important because it has application in proving that an algebraic specification is sufficiently complete and in proving inductive theorems from an algebraic specification.

Definition: An equality theory E is *sufficiently complete* for the set C of constructors if, for every ground term t, there is ground term s constructed from C such that $s =_E t$. $\diamond$

The notion of sufficient completeness was introduced in Guttag [8]. When a set of equations is both "consistent" and sufficiently complete it may be regarded as a reasonable specification of the defined functions. Thus it is important in algebraic specification to be able to prove that an equational theory is sufficiently complete. Unfortunately, in general this is undecidable. However, under certain reasonable assumptions, it can be shown that an equational theory E is sufficiently complete for constructors C if every term of the form $f(x_1, ..., x_n)$, where f is not a constructor, is ground reducible in the rewrite system corresponding to E.

Definition: A term s is *ground reducible* in rewriting system R, if all of its ground instances $s\sigma$ are rewritable by R. $\diamond$

The following example is taken from [6].

Example: (Stack)

The following rewrite system R is intended to capture the semantics of "stacks" equipped with the standard stack operations *top* and *pop* as well as an operation *alternate* that combines two stacks. Here C is $\{push, \Lambda\}$ and F is $\{tcp, pop, alternate\}$. We assume the functions have appropriate sorts.

$$\begin{array}{lcl} top(push(x, y)) & \rightarrow & x \\ pop(push(x, y)) & \rightarrow & y \\ alternate(\Lambda, z) & \rightarrow & z \\ alternate(push(x, y), z) & \rightarrow & push(x, alternate(z, y)) \end{array}$$

The terms $top(x)$ and $pop(x)$ are not ground reducible in R, since the instances $top(\Lambda)$ and $pop(\Lambda)$ cannot be rewritten by R. Thus the corresponding equations are not sufficiently complete for the constructors $\{\Lambda, push\}$. In order to extend the rewrite system so that the corresponding equation system is sufficiently complete for these constructors, it suffices to add the rules:

$$\begin{array}{lcl} top(\Lambda) & \rightarrow & \Lambda \\ pop(\Lambda) & \rightarrow & \Lambda \end{array}$$

It is then possible to show that $alternate(x, y)$, $top(x)$ and $pop(x)$ are ground reducible in this new system, R'. $\diamond$

Another important application of ground reducibility is in the "proof by consistency" method of inductive theorem proving, introduced by Musser [32]. The method relies on the observation that an equation $s = t$ is valid in the initial (or free) algebra of a ground convergent rewrite sytem R iff no additional equalities between ground terms follow from $R \cup \{s = t\}$. The method works as follows. First, a ground convergent system R' is obtained from $R \cup \{s = t\}$ by completing the definition using, for example, the Knuth-Bendix algorithm. By the nature of completion, every term reducible by R is reducible by R'. Then, to prove the converse, it is necessary and sufficient that the left hand side of each rule in R' is ground reducible by R. If R' can be constructed, and passes this test, then $s = t$ is valid in the initial algebra.

Example: Suppose that we wish to prove that $alternate(y, \Lambda) = y$ is an inductive theorem of the stack equations. Let R be the stack rewrite system. Completion ends with the system $R' = R \cup \{alternate(y, \Lambda) \rightarrow y\}$. Since the left-hand side of the new rule, $alternate(y, \Lambda)$, is ground reducible by R it follows that $alternate(y, \Lambda) = y$ holds in the initial algebra of the stack equations. ◇

Ground reducibility of a term rewriting system is closely related to term subtraction.

Proposition 2 *Let R be a rewrite system with left-hand sides $t_1, ..., t_n$. Term s is ground reducible for T if each term in $s/\{t_1, ..., t_n\}$ is ground reducible for T.*

This observation is the basis of the algorithm given by Thiel [42] to prove ground reducibility. In [25] a formal proof of correctness of the algorithm is given. It is the essential idea behind many other algorithms developed to test for ground reducibility, for example [3, 4, 17]. These algorithms iteratively use a simplification algorithm related to **uncover** to test for ground reducibility.

Note, however, that the condition given in the above proposition is a sufficient, but not a necessary condition for sufficient groundness to hold. Given that there are no equations between the constructors, it can be refined to give:

Proposition 3 *Let R be a rewrite system with left-hand sides $t_1, ..., t_n$. Term s is ground reducible for T iff each term in $s/\{t_1, ..., t_n\}$ is ground reducible for T.*

The completeness of the algorithm based on this proposition follows from the fact that a linear term cannot be covered by a disjunction of non-linear instances (which follows from the termination condition of the algorithm **uncover** and Theorem 1). The problem of ground-reducibility when there are no equations between the constructors is formulated as a logic program in [29].

Other algorithms, based on enumeration of test-cases have also been proposed for testing ground reducibility of a term, for example see Kapur et al. [13]. However, the approach based on term covering has the advantage that, when the term is not ground reducible, it suggests the "minimal" extension required in the rewrite system to make the term ground reducible.

4 Constructive Negation

In logic programming, considerable research has been devoted to the search for an appropriate notion of negation with which to extend definite clause logic programming ("pure PROLOG"). The problem is difficult because it seeks a notion of negation which is simultaneously semantically elegant and computationally feasible : in both execution and mathematical/logical semantics the extended language should cleanly extend the definite clause language.

The original notion of negation in PROLOG was negation-as-failure: if, for an atom A, the execution mechanism exhausted all possible proof attempts without success then execution of $\neg A$ would succeed; if a proof of A was found then execution of $\neg A$ would fail. In view of difficulties in giving a logical semantics when A has a proof which binds variables of A, an extra condition has generally been added to the negation-as-failure rule: all variables in A must be bound to variable-free terms before the negation-as-failure rule may be used on $\neg A$. But this introduces the possibility that execution might flounder [26], that is, be unable to proceed further, without the query being successful or failed. Clark proposed the use of the program completion (see later) as a logical semantics for programs and showed the soundness of negation-as-failure with respect to this semantics. Later, negation-as-failure was shown to be complete on some classes of programs.

Other notions of negation have been intended for deductive databases (function-free logic programs) or derived from an aim of common-sense or non-monotonic reasoning (see [37] for a survey). Although the logical semantics of these approaches is elegant, they are not computationally feasible when function symbols are allowed in programs. Indeed, these semantics are uncomputable, in general.

In this section we look at a notion of negation called "constructive negation". This form of negation is closely related to negation-as-failure and program completions, but avoids the problem of floundering. Its main feature is that negative subgoals are not simply tests, but can generate substitutions.

For a full definition of program completion we refer the reader to [26]. (This notion of completion is unrelated to the completion of a rewriting system mentioned earlier.) As an example, given the program P

$$\begin{array}{lll} append(nil, Y, Y). & & \\ append(U.X, Y, U.Z) & :- & append(X, Y, Z). \end{array}$$

which expresses that the result of appending the second argument to the first is the third argument, the program completion P^* is

$$\begin{array}{rl}\forall x, y, z\ append(x, y, z)\ \leftrightarrow & x = nil \wedge y = z \\ & \vee\ \exists u, v, w\ x = u.v \wedge z = u.w \wedge append(v, y, w)\end{array}$$

A central feature of logic programming as explained in [26] is the concept of substitution. The application of a substitution $\{x_1 \leftarrow t_1, \ldots, x_n \leftarrow t_n\}$ binds the variables x_i to terms t_i. All substitutions we need to consider are idempotent: no variable x_i occurs in a term t_j. Consequently we can associate to every such substitution a *substitution formula* $\exists u\ x_1 = t_1 \wedge \ldots x_n = t_n$ where u is the collection of variables appearing in the terms t_j. A disjunction of substitution formulas corresponds to a mgss. The equational part of each disjunct in a program completion can be expressed as a substitution formula. We will show that substitutions (and substitution formulas) are not sufficiently expressive to represent the negative information necessary in constructive negation.

We need some further definitions to restate previous results in this context. Let E_{FT} be the theory of the algebra of finite trees. Let s and s' be substitution formulas. We say that s' is *normalized* wrt s if $E_{FT} \models s' \rightarrow s$. In any formula $s \wedge \bigwedge_{i=1}^{n} \neg s_i$ we can assume, without loss of generality, that the s_i's are normalized wrt s, since $s \wedge s_i$ is equivalent to a substitution formula s_i' and $s \wedge \bigwedge_{i=1}^{n} \neg s_i$ is equivalent to $s \wedge \bigwedge_{i=1}^{n} \neg s_i'$. A substitution formula $\exists u\ x = t(u)$ is *linear* if every variable in u occurs only once in t. If s' is normalized wrt s we say that s' is *linear* wrt s if $E_{FT} \models s' \leftrightarrow s \wedge l$ for some linear substitution formula l. In the terminology of [21] this corresponds to a term (or tuple of terms) t' being unrestricted wrt t. Some results of earlier sections, and other results of [21], can be restated as follows (from the full version of [29]).

Theorem 5 *Let s be a substitution formula and $s_1, \ldots, s_n$ be substitution formulas normalized wrt s. Consider the following statement:*

There are substitution formulas $c_1, \ldots, c_k$ such that $E_{FT} \models s \wedge \bigwedge_{i=1}^{n} \neg s_i \ \leftrightarrow\ \bigvee_{j=1}^{k} c_j$

1. *The above statement is true iff there are substitution formulas $d_1, \ldots, d_m$ linear wrt s such that $E_{FT} \models \bigvee_{i=1}^{n} s_i \ \leftrightarrow\ \bigvee_{l=1}^{m} d_l$*
2. *The above statement is false if each s_i is non-linear wrt s*
3. *The algorithm* **uncover** *can be used to determine whether $s \wedge \bigwedge_{i=1}^{n} \neg s_i$ has an equivalent finite disjunction of substitution formulas, and to compute $c_1, \ldots, c_k$ if it does.*

Sato and Tamaki [39] first introduced a method where negated goals could be used to generate substitutions. Their method was a compile-time transformation which they referred to as "negation technique". Kunen [19, 18] (and, to a lesser extent, Lassez and Marriott [21]) outlined a runtime method and, independently, restricted versions of this method were studied in more detail by Maher [28] and Wallace [43]. Chan [1] introduced the name "constructive negation" and proved the soundness of a simple form of this method, and later [2] gave a more precise formulation of the method. The work of Lugiez [27] is closely related. A further generalization to constraint logic programs and a general soundness and completeness result are given by Stuckey [41].

In the Sato-Tamaki transformation, every predicate p which occurs negatively in some rule is compiled into rules for not_p, and every negative occurrence $\neg p(t_1, \ldots, t_n)$ is replaced by $not\_p(t_1, \ldots, t_n)$. For example, if the *append* predicate occurred negatively in a rule in a program then the method produces the rules

$$\begin{array}{lll} not\_append(x, y, z) & :- & \neg x = nil \wedge \neg \exists u, v, w\ x = u.v \wedge z = u.w \\ not\_append(nil, y, z) & :- & \neg \exists u\ y = u \wedge z = u \\ not\_append(u.v, y, u.w) & :- & not\_append(v, y, w) \end{array}$$

If Q is the result of transforming the append program P then

$$E_{FT}, P^* \models \neg append(x, y, z) \text{ iff } E_{FT}, Q^* \models not\_append(x, y, z)$$

This method is only applicable to some predicates and programs (for example, locally stratified programs and predicates defined with rules where every variable appearing in the body also appears in the head). Nevertheless, when it is applicable negation on predicates is compiled away and can be executed as a normal logic program *provided the remaining negations (of formulas involving only the predicate =) can be eliminated.*

In the example each formula which is negated is a substitution formula. (This is true in general.) By Theorem 5, since $\neg \exists u\ y = u \wedge z = u$ is non-linear, it has no equivalent finite disjunction of substitution formulas. Consequently, execution of this compiled program (and many others) requires an execution mechanism capable of handling inequalities as well as substitutions. Although it is difficult to characterize those cases where substitutions suffice, we can use Theorem 5 to show, for example, that if every maximally general head of a predicate is linear then the Sato-Tamaki transformation does not introduce inequalities.

The above method has the advantage that negated goals are not executed in one indivisible computation, but can generate substitutions and negated substitution formulas during execution. When combined with a sophisticated execution mechanism, as in MU- and NU-Prolog [33], this allows co-routining between positive and negative goals. The runtime form of constructive negation does not have this advantage, but applies to many more programs.

The simple form of the runtime method [1], given a negative goal $\neg A$, executes A; if A has a finite SLD-tree with answer substitutions (as substitution formulas) $s_1, \ldots, s_m$, then $\neg A$ gives as answer $\neg s_1 \wedge \ldots \wedge \neg s_m$. This answer can be presented as a finite collection of substitutions in cases as described by Theorem 5, but in general cannot be presented this way. For example, the goal $\neg append(x, y, z.nil)$ must return as an answer $\neg\exists v\ (x = nil \wedge y = v.nil \wedge z = v.nil) \wedge \neg\exists v\ (x = v.nil \wedge y = nil \wedge z = v.nil)$. Thus, again, Theorem 5 forces upon us an execution mechanism capable of handling inequalities. The general runtime method faces the same problems. Since the use of constraints has many advantages [11] and substitutions are too weak to represent the appropriate negative information, most approaches to constructive negation are based on constraints [28, 19, 43, 1, 41], rather than substitutions. The algorithm **uncover** is useful for reducing $\neg s_1 \wedge \ldots \wedge \neg s_m$, and the formulas produced by the Sato-Tamaki transformation, to a "simplest" form, with only inequalities that, as the theorem shows, are unavoidable.

Acknowledgements

Thanks to Peter Stuckey for comments on parts of this paper.

References

[1] D. Chan, Constructive Negation based on Completed Database, *Proc. 5th Symposium on Logic Programming*, 111–125, 1988.

[2] D. Chan, An Extension of Constructive Negation and its Application in Coroutining, *Proc. North American Conf. on Logic Programming*, Cleveland, 477–493, 1989.

[3] H. Comon, Sufficient Completeness, Term Rewriting Systems and Anti-Unification. *Proc. 8th Conference on Automated Deduction* (J. Siekmann, Ed.), Lecture Notes in Computer Science Vol. 230, Springer-Verlag, 128–140, 1986.

[4] H. Comon, An Effective Method for Handling Initial Algebras. *Proc. 1st International Workshop on Algebraic and Logic Programming*, Lecture Notes in Computer Science Vol. 343, Springer-Verlag, 108–118, 1988.

[5] H. Comon, Disunification: A Survey, in: *Computational Logic*, J-L. Lassez & G. Plotkin (Eds.), MIT Press, 1991.

[6] N. Dershowitz and J-P. Jouannaud, Rewrite Systems. *Handbook of Theoretical Computer Science*, (J. van Leeuwen Ed.), Elsevier Science Publishers, 243–320.

[7] A. Frisch & C.D. Page Jr., Generalization with Taxonomic Information, *Proc. AAAI-90*, Boston, 755–761, 1990.

[8] J.V. Guttag, Abstract Data Types and the Development of Data Structures. *Communications of the ACM*, **20**(6), 396–404, 1977.

[9] G. Huet, Résolution d'Equations Dans Des Langages D'Ordre 1,2, ... , ω (Thèse d'Etat), Université de Paris VII, 1976.

[10] G. Huet, Confluent Reductions: Abstract Properties and Applications to Term Rewriting Systems, *JACM*, Vol. 27, No. 4, Oct. 1980, 797-821.

[11] J. Jaffar & J-L. Lassez, Constraint Logic Programming, *Proc. Conf. on Principles of Programming Languages*, 111–119, 1987.

[12] J. Jaffar, J-L. Lassez & J.W. Lloyd, Completeness of the Negation as Failure Rule, *Proc. IJCAI*, Karlsruhe, 500–506, 1983.

[13] D. Kapur, P. Narendran and H. Zhang, On Sufficient Completeness and Related Properties of Term Rewriting Systems. *Acta Informatica* **24**, 395–415, 1987.

[14] E. Kounalis, An Algorithm For Learning from Examples and Counter Examples, manuscript, 1989.

[15] E. Kounalis, Pumping Lemmas For Tree Languages Generated by Rewrite Systems, *Mathematical Foundations of Computer Science*, 1990.

[16] E. Kounalis and M. Rusinowitch, Discovering New Facts For First Order Knowledge Based Systems, *4th International Symposium on AI*, Spain, 1990.

[17] E. Kounalis, Completeness in Data Type Specifications. *Proc. of EUROCAL '85 Conference*, Lecture Notes in Computer Science Vol. 204, Springer-Verlag, 348–362, 1985.

[18] K. Kunen, Answer Sets and Negation as Failure, *Proc. 4th. Int. Conf. on Logic Programming*, Melbourne, 219–228, 1987.

[19] K. Kunen, Negation in Logic Programming, *Journal of Logic Programming*, 4, 289–308, 1987.

[20] J-L. Lassez, M.J. Maher & K.G. Marriott, Unification Revisited in: *Foundations of Deductive Databases and Logic Programming*, J. Minker (Ed.), Kauffman, 587–625, 1988. Also in *Proc. Workshop on Foundations of Logic and Functional Programming*, LNCS 306, 1986.

[21] J-L. Lassez & K.G. Marriott, Explicit Representation of Terms Defined by Counter Examples, *Journal of Automated Reasoning*, 3, 301–317, 1987. Preliminary version in *Proc. Conf. on Foundations of Software Technology and Theoretical Computer Science*, LNCS 241, 1986.

[22] J-L Lassez, K. McAloon, A Constraint Sequent Calculus *LICS 90*. Philadelphia.

[23] J-L. Lassez and K. McAloon, A Canonical Form for Generalized Linear Constraints, *Journal of Symbolic Computation*, to appear.

[24] A. Laville, Lazy Pattern Matching in the ML Language, *Proc. Conf. on Foundations of Software Technology and Theoretical Computer Science*, Pune, LNCS 287, 400–419, 1987.

[25] A. Lazrek, P. Lescanne and J-J. Thiel, Tools for Proving Inductive Equalities, Relative Completeness, and ω-Completeness. *Information and Computation* **84**, 47–70, 1990.

[26] J.W. Lloyd, *Foundations of Logic Programming*, Springer-Verlag, 1987.

[27] D. Lugiez, A Deduction Procedure for First Order Programs, *Proc. ICLP-6*, 585–599, 1989.

[28] M.J. Maher, Logic Semantics for a Class of Committed-choice Programs, *Proc. 4th. Int. Conf. on Logic Programming*, Melbourne, 858–876, 1987.

[29] M.J. Maher, Complete Axiomatizations of the Algebras of Finite, Rational and Infinite Trees, *Proc. 3rd. Symp. Logic in Computer Science*, Edinburgh, 348–357, 1988. Full version IBM Research Report, T.J. Watson Research Center.

[30] M.J. Maher, On Parameterized Substitutions, IBM Research Report, T.J. Watson Research Center, 1990.

[31] M.J. Maher & P.J. Stuckey, On Inductive Inference of Cyclic Structures, *International Symposium on Artificial Intelligence and Mathematics*, Ft. Lauderdale, 1990.

[32] D.L. Musser, On Proving Inductive Properties of Abstract Data Types. *Proc. 7th ACM Symp. on Principles of Programming Languages*, ACM Press, 154–162, 1980.

[33] L. Naish, *Negation and Control in Prolog*, Lecture Notes in Computer Science 238, Springer-Verlag, 1986.

[34] C.D. Page Jr. & A. Frisch, Generalizing Atoms in Constraint Logic, *Proc. Conf. on Knowledge Representation and Reasoning*, San Mateo, 1991.

[35] S.L. Peyton Jones, *The Implementation of Functional Programming Languages*, Prentice Hall, 1987.

[36] G.D. Plotkin, A Note on Inductive Generalization, *Machine Intelligence 5*, (B. Meltzer & D. Michie Eds.), 1970, 153-163.

[37] H. Przymusinska & T. Przymusinski, Semantic Issues in Deductive Databases and Logic Programs, in: *Sourcebook on the Formal Approaches in Artificial Intelligence*, A. Banerji (Ed.), North-Holland, to appear.

[38] J.C. Reynolds, Transformational Systems and the Algebraic Structure of Atomic Formulas, *Machine Intelligence 5*, (B. Meltzer & D. Michie Eds.), 1970, 135-152.

[39] T. Sato & H. Tamaki, Transformational Logic Programming Synthesis, *Proc. FGCS'84*, Tokyo, 195–201, 1984.

[40] Ph. Schnoebelen, Refined Compilation of Pattern-matching for Functional Languages, *Proc. Workshop on Algebraic and Logic Programming*, LNCS 343, 233–243, 1988.

[41] P.J. Stuckey, Constructive Negation for Constraint Logic Programming, *Proc. 6th Symp. on Logic in Computer Science*, to appear.

[42] J-J. Thiel, Stop Losing Sleep Over Incomplete Data Type Specifications. *Proc. 11th ACM Symp. on Principles of Programming Languages*, ACM Press, 76–82, 1984.

[43] M. Wallace, Negation By Constraints: A Sound and Efficient Implementation of Negation in Deductive Databases, *Proc. 1987 Symposium on Logic Programming*, 253–263, 1987.

Rewrite orderings and termination of rewrite systems

Pierre LESCANNE*

Centre de Recherche en Informatique de Nancy (CNRS)

and

INRIA-Lorraine

Domaine Scientifique Victor Grignard, BP 239,

54506 Vandœuvre-lès-Nancy, France

email: lescanne@loria.crin.fr

Termination or halting is essential in the proof of properties of programs. In 1936, Turing proved its undecidability for the computation model he devised. In 1942, Newman [New42] noticed the importance of termination for the proof of confluence of what he called *equivalence* and what we would call today *rewrite rules*. In 1949, Turing proposes to use ordinals for proving termination of programs [Tur49, MJ84]. These days, the problem of proving termination of specific programs known as rewrite programs or equational programs is still the subject of much research and is covered by a very complete survey [Der87].

In this lecture, I am going to speak about recent results on proving uniform termination of term rewrite systems and rewrite systems modulo identities, like associativity and commutativity. Uniform termination is the termination of the rewrite process for any rewriting strategy. Proving uniform termination (later called just termination) of rewrite systems is known to be difficult, probably because it is undecidable [HL78]. Dauchet has proven that even if one restricts to one left-linear rewrite rule systems [Dau89] it is still undecidable. The problem is also interesting because there exist systems with hard proofs of termination like the following:

$$\begin{aligned} half(s(s(x))) &\to s(half(x)) \\ half(s(1)) &\to 1 \end{aligned}$$

$$mod2(s(s(x))) \to mod2(x)$$

$$three(s(x)) \to s(s(s(three(x))))$$

$$\begin{aligned} f(s(x), mod2(s(1))) &\to f(half(s(x)), mod2(half(s(x)))) \\ f(s(x), mod2(1)) &\to f(s(three(s(x))), mod2(s(three(s(x))))) \end{aligned}$$

*This research was partly supported by ESPRIT under Basic Research Working Group 3264, COMPASS

or its associative and commutative variant,

$$f(x+x) \to f(x)$$
$$f(x+x+1) \to f(x+x+x+1+1)$$

where + is an associative and commutative operator. This is known as the Syracuse problem, everybody suspects it terminates but nobody can provide a proof.

1 Well quasi-orderings and rewrite orderings

When we have a rewrite system R, we get a rewrite relation $\xrightarrow{+}_R$ and the proof of termination is just the proof of the fact that this relation is noetherian or well-founded. A natural approach is to take a known well-founded relation, say $\succ$ and to prove that the relation $\xrightarrow{+}_R$ is included in $\succ$. One of the most used ordering is the lexicographic path ordering $<_{lpo}$ that extends a quasi-ordering $\leq$ on the set of terms $\mathcal{T}(\mathcal{F})$, with the notations $<$ for $\leq$ & $\neq$ and $\simeq$ for $\leq$ & $\geq$. Often the ordering $<$ is called a semantic ordering and $<_{lpo}$ a semantic path ordering. It is defined as follows:

Definition 1 *Given two terms $s \equiv f\vec{s}$ and $t \equiv g\vec{t}$ in $\mathcal{T}(\mathcal{F})$, one says that*

$f\vec{s} <_{lpo} g\vec{t}$

iff

(ST) $(\forall i \in [1...m])$ $s_i <_{lpo} t$

and one of the following conditions is fulfilled

1. $s < t$,
2. $s \simeq t$ & $\vec{s} <_{lpo}^{lex} \vec{t}$,
3. $(\exists j \in [1...n])$ $s <_{lpo} t_j \ \vee \ s = t_j$.

Often the "semantic" ordering $<$ is a root ordering based on a precedence i.e., an ordering on symbols, this means that for comparing terms one compares just their root according to the precedence. Note that if one wants to implement this ordering it is wise not to use this definition directly and to massage it a little in order to avoid useless recursive calls, especially by factoring the condition *(ST)* that has not to be invoked in case 3 and can be simplified in case 2. Note also that in the part 2 of the previous definition one can replace a lexicographic comparison by a multiset one getting what is called the *multiset path ordering*, written $<_{mpo}$. One of the main features expected from a family of orderings when used in a completion procedure for instance is a property called *incrementality* [JLR82], it is the ability for an ordering of preserving the well-foundedness when the ordering is extended. For instance, suppose one is in the middle of a completion of groups using a lexicographic path ordering with the precedence $* > e$ and $i > e$. Suppose that the rewrite system

$$
\begin{aligned}
x * e &\longrightarrow x \\
e * x &\longrightarrow x \\
x * i(x) &\longrightarrow e \\
i(x) * x &\longrightarrow e \\
i(e) &\longrightarrow e \\
i(i(x)) &\longrightarrow x \\
(x * y) * z &\longrightarrow x * (y * z) \\
x * (i(x) * y) &\longrightarrow y \\
i(x) * (x * y) &\longrightarrow y
\end{aligned}
$$

is already generated, then the procedure computes the critical pair

$$i(x * y) = i(y) * i(y)$$

which cannot be oriented by the lexicographic path ordering, but if one extends the precedence by $i > *$ this identity can be oriented into

$$i(x * y) \longrightarrow i(y) * i(y).$$

It is known that any extension of a lexicographic path ordering is well-founded, but the question is whether this is always the case. In other words, can we characterize orderings that are such that all the orderings that contain them are well-founded. Actually this is just a definition of *well-quasi orderings* which turns to be a very natural concept in computer science. Notice that not all well-founded orderings are well-quasi orderings, as shown by the ordering on pairs of naturals that compares just the first component, i.e., $(n_1, n_2) < (m_1, m_2)$ *if and only if* $n_1 < m_1$. It is not a well quasi-ordering since it can be extended by a non well-founded ordering namely the ordering $(n_1, n_2) < (m_1, m_2)$ *if and only if* $n_1 < m_1$ & $n_2 > m_2$. Intuitively, a well quasi-ordering is a broad enough well-founded ordering which compares enough elements, say almost all of them. Now the question is to provide well-quasi orderings one can start with in order to build by incrementation a family of well quasi-orderings. A classical example is the *embedding*, but before giving its definition let us give few others.

Most notations are borrowed from [DJ90]. We say that a relation $\hookrightarrow$ satisfies the *replacement property* if

$$s \hookrightarrow t \Rightarrow f(\ldots, s, \ldots) \hookrightarrow f(\ldots, t, \ldots).$$

and that a relation $\hookrightarrow$ is *fully invariant* if for all substitution σ

$$s \hookrightarrow t \Rightarrow s\sigma \hookrightarrow t\sigma.$$

A fully invariant relation which has the replacement property is called a *rewrite relation*, it is written $\longrightarrow_R$ if it is generated by a set R of rewrite rules. If in addition it is transitive and reflexive it is called a *derivability* relation and written $\xrightarrow{*}_R$. If the derivability relation is well-founded it is called a *rewrite ordering*.

Now we define the embedding as follows: the *divisibility order* or *embedding* on $\mathcal{T}(\mathcal{F})$ is the rewrite relation $\xrightarrow{*}_{\mathcal{EMB}}$ where $\rightarrow_{\mathcal{EMB}}$ is given by the following set $\mathcal{EMB}$ of rewrite rules

$$f(x_1,\ldots,x_n) \underset{\mathcal{EMB}}{\longrightarrow} x_i \text{ for all } i \in [1..n]$$

where the f's are all the operators in $\mathcal{F}$. In other words $\xrightarrow{*}_{\mathcal{EMB}}$ is the smallest derivability relation, generated by $\mathcal{EMB}$. Higman has proven [Hig52] that the derivability relation $\xrightarrow{*}_{\mathcal{EMB}}$ is a well quasi-ordering. Suppose now $\mathcal{F} = \bigcup_{n=1}^{\infty} F_n$, where F is a given set of symbols and $\nu : F \longrightarrow F_n$ is a bijection $\nu : f \mapsto f_n$. Consider the set $\mathcal{K}$ of rules, called *erasing rules*:

$$f_n(x_1,\ldots,x_{k-1},x_k,x_{k+1},\ldots,x_n) \underset{\mathcal{K}}{\longrightarrow} f_{n-1}(x_1,\ldots,x_{k-1},x_{k+1},\ldots,x_n)$$

Kruskal theorem [Kru60] says that the systems generated $\mathcal{EMB} \cup \mathcal{K}$ is a well quasi-ordering on $\mathcal{T}(\bigcup_{n=1}^{\infty} F_n)$.

Since the lexicographic path ordering contains the embedding, it is well-founded by Higman theorem. But this does not help much in many situations. Let us consider for instance the one rule terminating rewrite system R defined by $f(f(x)) \rightarrow f(g(f(x)))$, its right-hand side and left-hand side satisfy $f(g(f(x))) \longrightarrow_{\mathcal{EMB}} f(f(x))$. A challenge is to extend Higman theorem to an ordering that contains the pair $\langle f(f(x)), f(g(f(x)))\rangle$. An idea is therefore to look for a rewrite system that contains the rule $f(f(x)) \rightarrow f(g(f(x)))$ plus some others similar to those of $\mathcal{EMB}$. For instance, adding rules like

$$\begin{aligned} g(g(x)) &\rightarrow x \\ f(g(x)) &\rightarrow x \\ g(f(x)) &\rightarrow x \end{aligned}$$

makes sense and indeed the rewrite ordering $\xrightarrow{+}_{\mathcal{FF}}$ defined by the four rules is a a well quasi-ordering. Is this method of using a rewrite ordering to define a well quasi-ordering general enough to give rise to an extension of both Higman divisibility order and of the previous ordering? First of all, the ordering has to be well-founded, i.e., the rewrite system has to terminate. This is obvious for $\mathcal{EMB}$ and a little less for $\mathcal{FF}$, but it is not an excessive constraint since after all we wish more than the well-foundedness, namely the fact that each ordering that contains it is well-founded. Moreover the well-foundedness plays an essential role in the proof of the theorem modeled after Nash-William's proof of Higman or Kruskal theorem; the proof is indeed by noetherian induction on the rewrite ordering, but when with Nash-William the noetherian induction is hidden behind a proof on the structure of the terms which is nothing more than an induction on the rewrite ordering $\longrightarrow^*_{\mathcal{EMB}}$, it is explicit here. Before looking at the other conditions, let me mention an interesting extension of Higman's theorem due to Laurence Puel [Pue89b]. Her idea is to replace the left-hand sides of $\mathcal{EMB}$ of the form $f(x_1,\ldots,x_n)$ by linear terms $p(x_1,\ldots,x_n)$.

Her theorem says that the rewrite ordering generated by the rewrite system

$$p(x_1,\ldots,x_n) \underset{\mathcal{P}}{\rightarrow} x_i \text{ for all } i \in [1..n]$$

is a well-quasi ordering if and only if the left-hand sides of $\mathcal{P}$ are unavoidable where *unavoidability* means that every term in $\mathcal{T}(\mathcal{F})$ but a finite number contains a subterm which is an instance of a left-hand side of $\mathcal{P}$. In our extension we are not able to keep unavoidability in its full generality, therefore we were led to loose the necessity of the condition and to replace it by a weaker condition that we call *basis property* which says that each term of $\mathcal{T}(\mathcal{F})$ but a finite number is an instance of a unique left-hand side, thus the basis property requires that the matching occurs at the top and is unique. The third property called *projectivity* is more technical, harder to prove and unfortunately undecidable in general. It is a generalization of the following remark: in the embedding if $s_1 \xrightarrow{*}_R t_1, \ldots, s_n \xrightarrow{*}_R t_n$ then $f(s_1,\ldots,s_n) \xrightarrow{*}_R f(t_1,\ldots,t_n)$. Projectivity says the same but replaces $f(s_1,\ldots,s_n)$ and $f(t_1,\ldots,t_n)$ by instances of left-hand sides and the s_i's and the t_i's by the corresponding right-hand sides. A formal definition is:

Definition 2 (Projectivity) *A set R of rewrite rules is* projective, *if for every left-hand side g of a rule in R and for every pair (σ,θ) of ground substitutions, one has*

$$[d_1\sigma \xrightarrow{*}_R d_1\theta \wedge \ldots \wedge d_i\sigma \xrightarrow{*}_R d_i\theta \wedge \ldots \wedge d_{n_g}\sigma \xrightarrow{*}_R d_{n_g}\theta]$$
$$\Longrightarrow$$
$$g\sigma \xrightarrow{*}_R g\theta$$

where $\{d_1,\ldots,d_{n_g}\}$ is the set of right-hand sides associated with g, i.e., R contains the rules $g \rightarrow d_1,\ldots,g \rightarrow d_n$.

The systems $\mathcal{EMB}$ and $\mathcal{FF}$ cited above are projective (it is trivial for $\mathcal{EMB}$, less for $\mathcal{FF}$.) The theorem [Les90] says that if R is a basic, projective and terminating set of rewrite rules, then the relation $\xrightarrow{+}_R$ is a well quasi-ordering.

2 Termination proofs and length of derivation

In this part I would like to examine the termination proof methods from the point of view of their complexity. A first approach consists of examining the complexity as ordinal. Another approach consists in evaluating the function that associates the size of a term with the length of the longest branches starting from this term which we call the *derivation length*. Finally one evaluates the complexity of the function computed by a rewrite system with respect to the size of the inputs.

Term orderings and ordinals

In [DO88], N. Dershowitz and M. Okada show the connection between ordering used in termination proofs of term rewrite systems and the Ackermann system for ordinal notations, more precisely they calculated the order type of $<_{lpo}$ and $<_{mpo}$. Cichon [Cic90] did the same for a subsystem of the Feferman-Aczel system of ordinal notations. Roughly speaking, they show that one can associated naturally with each ordering an ordinal that gives information on its "power". For $<_{mpo}$ on $\mathcal{T}(\mathcal{F})$, this ordinal is the ordinal noted $\theta_{|F|}0$, where $|\mathcal{F}|$ is the number of elements of $\mathcal{F}$ and for $<_{lpo}$ the ordinal noted $\theta_{\theta_0(\Omega.(M+1))}0$, where M is the maximum of the arities of the function symbols in $\mathcal{F}$.

Derivation length and termination orderings

Let us first study one of the most popular method for proving termination of rewrite systems, namely polynomial interpretations [Lan79]. It is based on the idea that a natural way for proving termination is to associate with each term a natural number. Actually since one works with term with variables it is wiser to associate a function. Therefore a natural way is to assign to each function symbol a function on the naturals and to compute by induction on the structure of the terms the function associated with compound terms. If one wants to have easy comparisons of functions and especially if one wants to mechanize the ordering one restricts the interpretations to polynomial only. If an operation is associative and commutative the interpretation has to satisfy condition that are easy to check, namely the polynomial has to be of the form $aXY + b(X + Y) + c$ with $ac + b - b^2 = 0$, this makes the polynomial interpretation method the only practical one in this case. Actually Narendran and Rusinowitch started from this feature to build a total ordering on associative and commutative terms providing a decision procedure for the words problems in theories with associative and commutative operators [NR91]. Methods based on polynomial interpretations have been implemented and included in softwares handling rewrite systems. The method [CL87] is naive and is basically a heuristic but works well on practical examples. Since proving that a polynomial is larger than another is equivalent to the tenth Hilbert problem and therefore undecidable, it is usual to restrict oneself to a comparison of polynomials over the reals where a decision procedure exists after Tarski [Tar51], however using such a brute force method is too inefficient to be practicable even if one uses the so called cylindric algebraic decomposition due to Collins [Col75], therefore specific approaches have to be devised [Rou91].

The connection between derivation length and ordering was first studied for polynomial orderings when people wanted to understand the constraint that a proof by polynomial interpretation makes on the derivation length [Les86]. It was proved by Hofbauer and Lautemann [HL89] and Oliver Geupel [Geu88] that derivation length is $O(2^{2^n})$ and $\Omega(2^{2^n})$. This is illustrated by the following rewrite system:

$$\begin{aligned} c(0,0) &\rightarrow 0 \\ c(s(x),0) &\rightarrow s(c(x,x)) \\ c(x,s(y)) &\rightarrow s(s(c(x,y))) \end{aligned}$$

Its termination can be proved by the polynomial interpretation:

$$[0] = 2$$
$$[s](X) = X + 1$$
$$[c](X, Y) = 2X^2 + 3Y.$$

The normal form of $c(s^m(0), s^n(0))$ is $s^{m^2+n}(0)$ with computation length $\frac{m(m+1)}{2} + n + 1$. Therefore the term

$$c(c(\dots c(c(s(s(0)), 0), 0) \dots, 0), 0)$$

with $k+1$ occurrences of c is of size $2k+5$ and reduces to $c(s^{2^{2^k}}(0), 0)$ which itself normalizes to $s^{2^{2^{k+1}}}(0)$ in $2^{2^{k+1}-1} + 2^{2^k-1} + 1$ steps that is larger than 2^{2^k} steps.

Similar results exist for other orderings, for instance, for the Knuth and Bendix orderings (KBO), there is no primitive recursive bound on the derivation length [HL89]. However with the multiset path ordering one gets a primitive recursive bound on the derivation length [Cic90, Hof90] and with the lexicographic path ordering one gets a multiple recursive function as bound [Cic90]. Drewes and Lautemann give examples of orderings that provide polynomial bounds, these orderings are "natural" and are built as combination of orderings that count components of terms occurring in rules [DL91]. Their measures are sharp in the sense they give the exact degree of the polynomial that reaches the bound.

Complexity of the functions computed by a rewrite system and termination orderings

Often rewrite systems are used for computing values of natural functions defined by a method called the *constructor discipline* [O'D85]. The constructor discipline partitions the function symbols into two classes: the *constructor symbols* 0 and s used to describe natural numbers and the *defined symbols* used to perform computations and associated with each computed functions. In addition, each left-hand side of a rule contains only constructors and variables except at the top where there is a defined symbol. For such rewrite systems the *derivation length* is not the adequate measure of the complexity of the function. For instance, suppose that a computation starts with

$$f(s^{n_1}(0), s^{n_2}(0), \dots, s^{n_p}(0))$$

and ends with

$$s^{\{f\}(n_1, \dots, n_p)}(0),$$

which is the value associated with the function f for the given input. It seems that the complexity of the rewrite system should be better seen as the complexity of the function

$$(n_1, \dots, n_p) \mapsto \{f\}(n_1, \dots, n_p)$$

since $\{f\}(n_1, \dots, n_p)$ is the intended value of the function described by the rewrite system. For instance, with the above function c, one gets $\{c\}(m, n) = \frac{m(m+1)}{2} + n + 1$

which is polynomial. Adam Cichon and I have proven [CL91] that if the function is described by a rewrite system with a polynomial interpretation based termination proof, then $\{f\}$ is polynomial in the size of the starting terms or equivalently is polynomial in $n_1, n_2, \ldots, n_p$ unlike its derivation length which is doubly exponential. This result holds even for functions with exponential computation length, like

$$g(0, y) \to y$$
$$g(s(x), y) \to g(s, g(x, y))$$

which is proved to terminate by the polynomial interpretation

$$[0] = 2$$
$$[s](X) = 3X$$
$$[g](X, Y) = X + Y$$

and computes $\{g\}(m, n) = n$ in 2^m steps.

3 New directions

I hope to have shown that termination of rewrite systems is an exciting and active field, the interested reader is also invited to consider the list of open problems published in [DJK91]. For those interested by general methods and their implementation let us mention methods for proving termination of specific systems, where there are still open research directions.

Self-embedding systems are systems that contain rules $l \to r$ with $r \longrightarrow^{+}_{\mathcal{EMB}} l$. Laurence Puel has proposed an extension of the recursive path ordering that can handle such systems [Pue89a].

Order-sorted rewrite systems. There are systems which do not terminate if sorts are forgotten but which do if sorts are taken into account. An extension of the recursive path ordering was suggested in this case by Isabelle Gnaedig [Gna91].

Associative and Commutative rewrite systems. Paliath Narendran and Michael Rusinowitch proposed an ordering compatible with associativity and commutativity and total on ground terms [NR91]. It seems that variants of this ordering can be devised and adapted to orderings on open terms.

I thank Françoise Bellegarde and Adam Cichon for comments on an early version of this paper.

References

[Cic90] E.A. Cichon. Bounds on derivation lenghts from termination proofs. Technical Report CSD-TR-622, Royal Holloway and Bedford New College, 1990.

[CL87] A. Ben Cherifa and P. Lescanne. Termination of rewriting systems by polynomial interpretations and its implementation. *Science of Computer Programming*, 9(2):137–160, October 1987.

[CL91] A. Cichon and P. Lescanne. Strict polynomial interpretation works only for termination of polynomial growth functions. in preparation, 1991.

[Col75] G. Collins. Quantifier elimination for real closed fields by cylindrical algebraic decomposition. In *Proceedings 2nd GI Conference on Automata and Formal Languages*, Lecture Notes in Computer Science. Springer-Verlag, 1975.

[Dau89] M. Dauchet. Simulation of Turing machines by a left-linear rewrite rule. In N. Dershowitz, editor, *Proceedings 3rd Conference on Rewriting Techniques and Applications, Chapel Hill, (North Carolina, USA)*, volume 355 of *Lecture Notes in Computer Science*, pages 109–120. Springer-Verlag, April 1989.

[Der87] N. Dershowitz. Termination of rewriting. *Journal of Symbolic Computation*, 3(1 & 2):69–116, 1987.

[DJ90] N. Dershowitz and J.-P. Jouannaud. Rewrite systems. In Van Leuven, editor, *Handbook of Theoretical Computer Science*. North-Holland, 1990.

[DJK91] N. Dershowitz, J.-P. Jouannaud, and J.W. Klop. Open problems in rewriting. In R.V. Book, editor, *Proceedings 4th Conference on Rewriting Techniques and Applications, Como, (Italy)*, volume 488 of *Lecture Notes in Computer Science*, pages 445–456. Springer-Verlag, April 1991.

[DL91] F. Drewes and C. Lauterman. Incremental termination proofs and the length of derivations. In R.V. Book, editor, *Proceedings 4th Conference on Rewriting Techniques and Applications, Como, (Italy)*, volume 488 of *Lecture Notes in Computer Science*, pages 49–61. Springer-Verlag, April 1991.

[DO88] N. Dershowitz and M. Okada. Proof-theoretic techniques and the theory of rewriting. In *Proceedings 3rd IEEE Symposium on Logic in Computer Science, Edinburgh (UK)*, pages 104–11. IEEE, 1988.

[Geu88] O. Geupel. Terminationbeweise bei termersetzungssytem, 1988. Diplomarbeit.

[Gna91] I. Gnaedig. Termination of rewriting in order-sorted algebras. draft, April 1991.

[Hig52] G. Higman. Ordering by divisibility in abstract algebra. *Proc. London Math. Soc.*, 3(2):326–336, 1952.

[HL78] G. Huet and D.S. Lankford. On the uniform halting problem for term rewriting systems. Technical Report 283, Laboria, France, 1978.

[HL89] D. Hofbauer and C. Lautemann. Termination proofs and the length of derivations. In N. Dershowitz, editor, *Proceedings 3rd Conference on Rewriting Techniques and Applications, Chapel Hill, (North Carolina, USA)*, volume 355 of *Lecture Notes in Computer Science*, pages 167–177. Springer-Verlag, April 1989.

[Hof90] D. Hofbauer. Termination proofs by multiset path orderings imply primitive recursive derivation lenghts. In H. Kirchner and W. Wechler, editors, *Proc. 2nd Int. workshop on Algebraic and Logic Programming*, volume 463 of *Lecture Notes in Computer Science*, pages 347–358, 1990.

[JLR82] J.-P. Jouannaud, P. Lescanne, and F. Reinig. Recursive decomposition ordering. In Bjørner D., editor, *Formal Description of Programming Concepts 2*, pages 331–348, Garmisch-Partenkirchen, Germany, 1982. North-Holland.

[Kru60] J.B. Kruskal. Well-quasi ordering, the tree theorem and Vazsonyi's conjecture. *Trans. Amer. Math. Soc.*, 95:210–225, 1960.

[Lan79] D.S. Lankford. On proving term rewriting systems are noetherian. Technical report, Louisiana Tech. University, Mathematics Dept., Ruston LA, 1979.

[Les86] P. Lescanne. Divergence of the Knuth-Bendix completion procedure and termination orderings. *Bulletin of European Association for Theoretical Computer Science*, 30:80–83, October 1986.

[Les90] P. Lescanne. Well rewrite orderings. In J. Mitchell, editor, *5th Symp. Logic in Computer Science*, pages 239–256. IEEE, 1990.

[MJ84] F. L. Morris and C. B. Jones. An early program proof by Alan Turing. *Annals of the History of Computing*, 6(2):139–143, April 1984.

[New42] M.H.A. Newman. On theories with a combinatorial definition of equivalence. In *Annals of Math*, volume 43, pages 223–243, 1942.

[NR91] P. Narendran and M. Rusinowitch. Any ground associative-commutative theory has a finite canonical system. In R.V. Book, editor, *Proceedings 4th Conference on Rewriting Techniques and Applications, Como, (Italy)*, volume 488 of *Lecture Notes in Computer Science*, pages 423–434. Springer-Verlag, April 1991.

[O'D85] M.J. O'Donnell. *Equational Logic as a Programming Language*. Foundation of Computing. MIT Press, 1985.

[Pue89a] L. Puel. Embedding with patterns and associated recursive path ordering. In N. Dershowitz, editor, *Proceedings 3rd Conference on Rewriting Techniques and Applications, Chapel Hill, (North Carolina, USA)*, volume

355 of *Lecture Notes in Computer Science*, pages 371–387. Springer-Verlag, April 1989.

[Pue89b] L. Puel. Using unavoidable set of trees to generalize Kruskal's theorem. *Journal of Symbolic Computation*, 8:335–382, 1989.

[Rou91] Jocelyne Rouyer. *Calcul formel en géométrie algébrique réelle appliqué à la terminaison des systèmes de réécriture.* PhD thesis, Université de Nancy 1, 1991.

[Tar51] A. Tarski. *A Decision Method for Elementary Algebra and Geometry.* University of California Press, Berkeley, second edition, 1951.

[Tur49] A.M. Turing. Checking a large routine. In *Report of a Conference on High Speed Automatic Calculing Machines*, pages 67–69, Cambridge, 1949. Univ. Math. Lab.

On the Faithfulness of Formal Models *

Zohar Manna
Stanford University[†]
and
Weizmann Institute of Science
zm@cs.stanford.edu

Amir Pnueli
Weizmann Institute of Science[‡]
amir@wisdom.weizmann.ac.il

Abstract

The paper presents a critical examination of the way certain central aspects of concurrent programs are formally modeled. The main formal model examined is the operational model of *fair transition system* which represents concurrency by interleaving of actions considered *atomic*. Several questions concerning the faithfulness of this representation naturally arise. The paper considers some of these questions and, while presenting and analyzing some of the alternatives, attempts to justify the design decisions actually taken in the construction of the model.

1 Introduction

There is an obvious fact which, self-evident as it is, omission of its clear statement has caused no end of confusion and led to overheated arguments about the value and possible contribution of formal methods to program construction.

Stated bluntly, this fact is that the scope of validity of each of the formal and analytical methods proposed for program construction, such as formal specification, verification, and systematic development, is always a certain *mathematical model* of the execution of a program, rather than the real thing itself. It follows that, while these methods have absolute validity (assuming they are mathematically sound) over the model, their relevance to the actual running of the program is only as good as the degree of faithfulness to which the model represents real executions of the program.

*This research was supported in part by the National Science Foundation under grants CCR-89-11512 and CCR-89-13641, by the Defense Advanced Research Projects Agency under contract N00039-84-C-0211, by the United States Air Force Office of Scientific Research under contract AFOSR-90-0057, and by the European Community ESPRIT Basic Research Action Project 3096 (SPEC).

[†]Department of Computer Science, Stanford University, Stanford, CA 94305

[‡]Department of Applied Mathematics, Weizmann Institute, Rehovot, Israel

The observation that mathematical methods are fully valid only over mathematical models is so well recognized by the philosophers of science, and considered to be one of the most elementary postulates of modern science, that it is seldom stated explicitly. However, every novice engineer who ever uses mathematical techniques for validating his design is continuously aware of the fact that his method applies to a certain approximation of reality. In practice, this is often observed by the fact that, after precisely computing what is a sufficient width of a pillar that is expected to support a building, the designer multiplies this result by a "safety factor", just to be sure.

In some fields, the inherent limitations of each individual model are overcome by constructing and using a *hierarchy* of models, each a refinement of it predecessor. A well known example is the design of integrated circuits where one designs and simulates first (even verifies) at a functional level, then successively tests and refines his design at gate level, transistor level, and even sometimes at the level of electron flows. An obvious tradeoff exists between the degree of detail (faithfulness) incorporated in the model and the complexity of using it. As a result, one tends to analyze the complete system within the least detailed model, and to apply the more accurate analysis only to small parts of the system which, according to judgment, may be problematic.

Somehow, these obvious facts of life have not been always consistently considered and respected in the area of program construction. Perhaps the fact that a program is nothing but a term in some formal language, have misled some people to forget that real executions do not involve programs alone but also the hardware executing them and the environment in which it is embedded. As a result we may have seen some careless declarations such as:

Formally verify your programs and you will never have to debug again!

There is no wonder that such overzealous (and clearly false) declarations caused some strong reactions. An example is [3], whose analysis has been interpreted by some to state that, since formal verification cannot guarantee absolute assurance, it is worthless. Such an extreme conclusion is obviously equally unacceptable. If we were to adopt it consistently in all areas of science and engineering, then no mathematical method would have survived.

As usual, the truth (as we see it) lies somewhere in between these two extreme positions. A conclusion about executions of programs, obtained by a formal method, is reliable only to the extent to which the underlying mathematical model is faithful. On the other hand, by developing more and more faithful models that account for a larger set of phenomena associated with real executions, we get closer (arbitrary close if so wished) to the thing itself.

Therefore, the activity of developing formal methodologies should consist not only of the construction of new models and the invention of new techniques for analyzing existing models, but also of a continuous assessment and comparison of models, both with one another and with the actual reality.

The domain of interest considered here is the family of *reactive programs* [4]. A reactive program is a program whose role is not to produce a final result, but to maintain some ongoing interaction with its environment. Examples of reactive programs are operating systems and programs controlling mechanical or chemical processes, such as a plane or a nuclear reactor. Some reactive programs are not expected to terminate. The concept of reactivity is strongly correlated with concurrency. To specify and analyze a reactive program we study the interaction between the program and its environment which run concurrently. Also, when specifying and analyzing a concurrent program, each process can be studies as a reactive component, where the other processes serve as its environment.

This paper presents an assessment and examination of the model of *fair transition system* proposed for modeling reactive and concurrent programs. This model has been formally introduced in [5], but has been used without formal definition in many other places. To examine the faithfulness of the model, we consider several phenomena that occur in real executions of programs, and inspect their representation in the model. In some of the cases, we consider several alternative options for representation, and attempt to justify the choice we have made.

An important and valuable guideline for the construction of useful models is the principle of *Occam's Razor*. In our context it can be formulated as a recommendation for choosing the simplest model that adequately represents the set of realistic phenomena one wishes to include. While this recommendation rarely identifies a unique model, it may considerably help in comparing several candidates.

2 A Model for Reactive Programs

Our model for concurrent programs is presented in two stages. The first stage presents the generic model of *fair transition system* [5], which is general enough to cover a multitude of concurrent systems with radically different syntactic presentations and communication mechanisms. In the second stage, we usually choose a particular syntax and a set of synchronization and communication mechanisms, to which we refer as a *concrete model*. The link between the generic and concrete models is established by mapping each abstract entity of the generic model to a corresponding construct in the concrete model.

The main concrete model we will study in detail is a programming language in which concurrent processes communicate by *shared variables*, and synchronize by *semaphores*. We will mention briefly additional concrete models, where communication is performed by *asynchronous* or *synchronous message-passing*, and refer the reader to [7] and [6] where some of these additional models are discussed in greater detail.

As we will see in the discussion, some issues of faithful representation are resolved by placing restrictions on the syntax of admissible programs in the concrete model. This is often an acceptable device for keeping the models simple.

2.1 Fair Transition System

To express the syntax of a basic transition system, we use an underlying first-order language. We refer to formulae in this language as *assertions*.

A *fair transition system* $\langle \Pi, \Sigma, \mathcal{T}, \Theta, \mathcal{J}, \mathcal{C} \rangle$, intended to represent a reactive program, is given by the following components:

- $\Pi = \{u_1, \ldots, u_n\}$: A finite set of typed *state variables*.

Some of the state variables represent *data variables*, which are explicitly declared and manipulated by statements in the program. These variables range over data domains used in programs, such as booleans, integers, or lists. By default, whenever we do not state otherwise, all our data variables are assumed to be integers. Other variables are *control variables*, which represent progress in the execution of the program; they may range over locations in the program. The *type* of each variable indicates the *domain* over which the variable ranges.

- Σ : A set of *states*.

Each state s in Σ is an interpretation of Π, assigning to each variable u in Π a value over its domain, which we denote by $s[u]$.

- $\mathcal{T}$: A finite set of *transitions*.

Each transition τ in $\mathcal{T}$ represents a state-transforming *atomic action* of the system and is defined as a function τ: $\Sigma \rightarrow 2^{\Sigma}$ that maps a state s in Σ into the (possibly empty) set of states $\tau(s)$ that can be obtained by applying action τ to state s. Each state s' in $\tau(s)$ is defined to be a τ-*successor* of s. If $\tau(s) \neq \phi$, we say that τ is *enabled* on s; otherwise, τ is *disabled* on s. It is required that, for each $s \in \Sigma$, there exists a transition $\tau \in \mathcal{T}$ that is enabled on s.

- Θ : An assertion called the *initial condition*.

Assertion Θ characterizes the states at which execution of the program can begin. A state s satisfying Θ, i.e., $s \models \Theta$, is called an *initial state*.

- $\mathcal{J} \subseteq \mathcal{T}$: A set of transitions called the *justice set*.

We refer to each transition $\tau \in \mathcal{J}$ as a *just transition*.

- $\mathcal{C} \subseteq \mathcal{T}$: A set of transitions called the *compassion set*.

We refer to each transition $\tau \in \mathcal{T}$ as a *compassionate transition*.

2.2 The Transition Relation ρ_τ

An effective representation of the state-transformation performed by each transition τ is provided by an assertion, called the *transition relation* for τ and denoted by $\rho_\tau(\Pi,\ \Pi')$.

Assertion ρ_τ relates the values of the state variables in a state s to their values in a successor state s' obtained by applying τ to s. The assertion refers to the state variables using two copies of $\Pi = \{u_1, \ldots, u_n\}$. An occurrence of u refers to the value of u in s, while an occurrence of u' refers to the value of u in s'. We call u' the *primed version* of u and denote the set of primed versions of all the state variables by $\Pi' = \{u'_1, \ldots, u'_n\}$.

For a transition τ associated with transition relation ρ_τ, we define state s' to be a τ-successor of state s if $\langle s, s'\rangle \models \rho_\tau$, where, for each $u \in \Pi$, the joint interpretation $\langle s, s'\rangle$ inerprets u as $s[u]$ and interprets u' as $s'[u]$. We restrict our attention to transitions whose state-transformation is defined by ρ_τ as described above.

Consider, for example, a transition τ whose transition relation is given by

$$\rho_\tau: \quad even(x) \;\wedge\; x+1 \leq x' \leq x+2,$$

where x is an integer state variable. This transition is enabled only on states in which x is even, and s' is asuccessor of x only if the value of x in s' equals to the value of $x+1$ or to the value of $x+2$ in s.

Normally, each transition modifies only few of the state variables and leaves the others unmodified. Consequently, a typical transition relation will have a conjunct of the form $u' = u$ for each unmodified variable. To simplify the presentation of transition relations we adopt a convention by which a variable u such that u' does not appear explicitly in ρ_τ is considered unmodified, and it is as though the conjunct $u' = u$ is automatically added to ρ_τ.

For example, if $\Pi = \{x, y, z\}$ and $\rho_\tau : z' = x + y$, then z is the only variable modified by τ, while x and y are preserved.

2.3 Computations

Consider an infinite sequence of states over Π

$$\sigma: \; s_0, \; s_1, \; s_2, \; \ldots \; .$$

For a given transition τ, we say that τ is *continually enabled beyond position j in σ* if τ is enabled on s_k, for every $k \geq j$. We say that τ is *enabled infinitely many times in σ* if τ is enabled on s_k for infinitely many k's.

Let P be the fair transition system $\langle \Pi, \Sigma, \mathcal{T}, \Theta, \mathcal{J}, \mathcal{C}\rangle$. A sequence σ is defined to be a *computation* of P if σ satisfies the following requirements:

- *Initiation*: The first state s_0 is initial, i.e., $s_0 \models \Theta$.
- *Consecution*: For each pair of consecutive states s_i, s_{i+1} in σ, $s_{i+1} \in \tau(s_i)$ for some transition τ in $\mathcal{T}$. That is, s_{i+1} is a *τ-successor* of s_i. In this case we say that τ is *taken* at position i. Note that it is possible for more than one transition to be considered as taken at a given position.
- *Justice* (also known as *weak fairness*): For each τ in $\mathcal{J}$ it is not the case that τ is *continually* enabled beyond some position in σ but taken only finitely many times.
- *Compassion* (also known as *srong fairness*): For each τ in $\mathcal{C}$ it is not the case that τ is enabled *infinitely many times* in σ but taken only finitely many times.

We often display computations as a sequence of states connected by arrows that are labeled by the transitions that cause the system to move to the next state. Thus, a presentation of a computation as

$$s_0 \xrightarrow{\tau_0} s_1 \xrightarrow{\tau_1} s_2 \longrightarrow \cdots$$

provides, in addition to the fact that $s_0, s_1, \ldots$ is a computation, also the identity of the transition $\tau_i \in \mathcal{T}$ leading from s_i to s_{i+1}.

Consider, for example, a (fair) transition system with state variables $\Pi = \{x\}$, transitions $\{\tau_1, \tau_2\}$, whose transition relations are given by

$$\rho_1: \ even(x) \ \wedge \ x+1 \leq x' \leq x+2 \quad \text{and} \quad \rho_2: \ odd(x) \ \wedge \ x' = x+1,$$

respectively. Assume that $\Theta : x = 0$, the justice set is $\mathcal{J} = \{\tau_1, \tau_2\}$, and the compassion set is $\mathcal{C} = \phi$.

Then, the following sequence which periodically takes τ_1 twice and then τ_2 once is an example of a computation:

$$\langle x:0\rangle \xrightarrow{\tau_1} \langle x:2\rangle \xrightarrow{\tau_1} \langle x:3\rangle \xrightarrow{\tau_2} \langle x:4\rangle \xrightarrow{\tau_1} \langle x:6\rangle \xrightarrow{\tau_1} \cdots.$$

3 A Programming Language: Syntax

As our main concrete model, we introduce a simple programming language in which processes communicate by shared variables. For each presented statement, we provide some intuitive explanation of its intended meaning.

- *Assignment*: For $y_1, \ldots, y_k$ a list of variables and $e_1, \ldots, e_k$ a list of expressions of corresponding types,

$$(y_1, \ldots, y_k) \ := \ (e_1, \ldots, e_k)$$

is an *assignment* statement. For the case that $k = 1$, we write simply $y := e$. The case $k = 0$ modifies no data variables, and is written as **skip**.

- *Await*: For c a boolean expression,

  await c

is an *await* statement. We refer to condition c as the *guard* of the statement.

Execution of **await** c changes no variables. Its sole purpose is to wait until c becomes true, at which point it terminates.

- *Conditional*: For S_1 and S_2 statements and c a boolean expression,

  if c **then** S_1 **else** S_2

is a *conditional* statement. Its intended meaning is that the boolean expression c is evaluated and tested. If the condition evaluates to T, statement S_1 is selected for subsequent execution; otherwise, S_2 is selected. Thus, the first step in an execution of the *conditional* statement is the evaluation of c and the selection of S_1 or S_2 for further execution. Subsequent steps continue to execute the selected substatement.

We abbreviate **if** c **then** S_1 **else skip** to **if** c **then** S_1.

- *Concatenation*: For statements $S_1, \ldots, S_k$,

  $S_1; \cdots; S_k$

is a *concatenation* statement. Its intended meaning is sequential composition. The first step in the execution of $S_1; \cdots; S_k$ is the first step in an execution of S_1. Subsequent steps continue to execute the rest of S_1, and when S_1 terminates, proceed to execute $S_2, S_3, \ldots, S_k$.

With *concatenation*, we can define the *when* statement **when** c **do** S as an abbreviation for the concatenation **await** c; S.

- *Selection*: For statements $S_1, \ldots, S_k$,

  S_1 **or** $\cdots$ **or** S_k

is a *selection* statement. Its intended meaning is that, as a first step, one of $S_1, \ldots, S_k$, which is currently enabled, is selected and the first step in the selected statement is executed. Subsequent steps proceed to execute the rest of the selected substatement. If more than one of $S_1, \ldots, S_k$ is enabled, the selection is non-deterministic.

The *selection* statement is often applied to *when* statements. This combination leads to *conditional selection*. For example, the general conditional command of the *guarded command* language (proposed by Dijkstra [1], [2]), of the form

  if $c_1 \rightarrow S_1 \;\square\; c_2 \rightarrow S_2 \;\square\; \ldots \;\square\; c_k \rightarrow S_k$ **fi**

can be represented in our language by a *selection* statement formed out of *when* statements:

  [**when** c_1 **do** S_1] **or** [**when** c_2 **do** S_2] **or** $\cdots$ **or** [**when** c_k **do** S_k].

- *While*: For c a boolean expression and S a statement,

  while c **do** S

is a *while* statement. Its execution begins by evaluating c. If c is found to be false, execution of the statement terminates. If c is found true, subsequent steps proceed to execute S. When S terminates, c is tested again.

- *Request*: For y an integer variable,

  request(y)

is a *request* statement. This is one of the two semaphore statements. It is enabled only when y is positive, and its execution subtracts 1 from y. As we will show below, the semaphore statements are used for synchronization between processes.

- *Release*: For y an integer variable,

  release(y)

is a *release* statement, which is the second semaphore statement. This statement is enabled whenever control is at its front, and its execution adds 1 to y. In that it is equivalent to the assignment $y := y + 1$.

3.1 Programs

A program P has the form

  $P :: \big[\text{declaration; } [P_1 :: S_1 \,\|\, \cdots \,\|\, P_m :: S_m]\big]$,

where $P_1 :: S_1, \ldots, P_m :: S_m$ are *named processes*. Each S_i is a statement, and P_i is a name for the process. The names of the program and of the processes are optional, and may be omitted.

A declaration consists of a sequence of *declaration statements* of the form

  mode variable, ..., variable: type **where** φ.

The mode of each declaration statement may be **in**, **local**, or **out** identifying variables that are inputs to the program, local, or outputs of the program. Declaration statements with no explicit mode specification retain the mode of the preceding statement.

The list

variable, ..., variable: type,

that appears in each declaration statement, lists several variables that share a common type and identifies their type, i.e., the domain over which the variables range. We use *basic types* such as **integer**, **character**, etc., as well as *structured types*, such as **array**, **list**, and **set**.

The assertion φ, which appears in a declaration statement, imposes constraints on the initial values of some of the variables declared in this statement. Assertion φ may be omitted from a declaration statement if no constraint is imposed on the variables declared in this statement.

Let $\varphi_1, \ldots, \varphi_n$ be the assertions appearing in the declaration statements of a program. We refer to the conjunction $\varphi : \varphi_1 \wedge \cdots \wedge \varphi_n$ as the *data-precondition* of the program.

3.2 Labels and Locations in Programs

Each statement in a program may be labelled. It is required that the labels appearing in the program are distinct. In Fig. 1 we present program GCD which computes the greatest common divisor of the inputs x_1 and x_2. All statements in this program are labelled. In principle, we consider each program to be fully labelled, and denote the label of statement S by *label*(S). In actual presentations of programs, we often omit many of the labels.

$$
\begin{array}{lll}
\textbf{in} & x_1,\ x_2: & \textbf{integer where } x_1 > 0,\ x_2 > 0 \\
\textbf{local} & y_1,\ y_2: & \textbf{integer where } y_1 = x_1,\ y_2 = x_2 \\
\textbf{out} & g & : \textbf{integer}
\end{array}
$$

$$
\ell_0: \left[\begin{array}{l}
\ell_1: \left[\begin{array}{l}\ell_2: \left[\begin{array}{l}\textbf{while } y_1 \neq y_2 \textbf{ do} \\
\quad \left[\begin{array}{l}\ell_3: \textbf{ when } y_1 > y_2 \textbf{ do } [\ell_4: y_1 := y_1 - y_2] \\
\quad \textbf{or} \\
\ell_5: \textbf{ when } y_2 > y_1 \textbf{ do } [\ell_6: y_2 := y_2 - y_1]\end{array}\right]\end{array}\right]\end{array}\right] \\
\ell_7: [g := y_1]
\end{array}\right]
$$

Fig. 1. A fully labeled GCD program.

Labels in programs have two important roles. The first is to provide a unique identification and reference to the statements. Thus we may talk about *when* statement ℓ_3 whose body is assignment ℓ_4.

The second important role of the labels is to serve as possible *sites* of control. Thus, we may consider a state in which control resides at label ℓ_2 in the program of Fig. 1 and, when finding that $y_1 > y_2$, execution moves to ℓ_4 where it pauses before subtracting y_2 from y_1.

The problem with interpreting labels as unique sites of control is that there are too many of them, and we do not necessarily wish to distinguish between them. For example, it is clear that when control is at ℓ_2 it is also at ℓ_3 and at ℓ_5, because any of these statments can be immediately executed if its guard is true. Thus, we define a *location* to be an *equivalence class* of labels, where the equivalence is defined by the rules:

$$\begin{array}{lll} label([S_1; \cdots ; S_k]) & \sim_L & label(S_1) \quad \text{and} \\ label([S_1 \textbf{ or } \cdots \textbf{ or } S_k]) & \sim_L & label(S_i) \quad \text{for each } i = 1, \ldots, k. \end{array}$$

For a label ℓ, we denote by $[\ell]$ the location corresponding to ℓ.

4 A Programming Language: Semantics

In order to establish the correspondence between the presented programming language and the generic model of fair transition systems, we have to map each of the components of a fair transition system to a construct in the programming language.

4.1 State Variables and States

For the programming language, the *state variables* Π consist of the *data variables* $Y = \{y_1, \ldots, y_n\}$ that are explicitly declared and manipulated by the program, and a single *control variable* π. The data variables Y include the input, output, and local variables, and range over their respectively declared data domains. The control variable π ranges over *sets of locations*. The value of π in a state denotes all the locations of the program in which control currently resides.

As states we take all possible interpretations that assign to the state variables values over their respective domains.

Consider, for example, the program

$$\left[\begin{array}{c} \textbf{out } x\text{: } \textbf{integer where } x = 0 \\ [\ell_1\text{: } x := x+1] \quad \| \quad [\ell_2\text{: } x := x+2] \end{array}\right].$$

The state variables for this program are $\Pi = \{\pi, x\}$. While there are many possible states, only four of them can ever arise in a computation of this program. These are:

$$\langle \pi : \{\ell_1, \ell_2\},\ x : 0\rangle \quad \langle \pi : \{\ell_2\},\ x : 1\rangle \quad \langle \pi : \{\ell_1\},\ x : 2\rangle \quad \langle \pi : \{\},\ x : 3\rangle$$

4.2 Transitions

For each statement S in a program, we define a set of labels, denotes by *tloc*(S), which represents the location at which execution resides after S is terminated. This set of labels is defined inductively in a top-down manner as follows:

For S a *process* of the program,	$tloc(S) = \phi$	
For $S = [S_1; \cdots; S_k]$,	$tloc(S_k) = tloc(S)$	
For $[\ell_1 : S_1; \cdots; \ell_k : S_k]$,	$tloc(S_i) = [\ell_{i+1}]$,	for each $i = 1, \ldots, k-1$
For $S =$ **if** c **then** S_1 **else** S_2,	$tloc(S_1) = tloc(S_2) = tloc(S)$	
For $S = [S_1$ **or** $\cdots$ **or** $S_k]$,	$tloc(S_i) = tloc(S)$,	for each $i = 1, \ldots, k$
For ℓ: **while** c **do** $\widetilde{S}$,	$tloc(\widetilde{S}) = [\ell]$	

For a statement $\ell : S$, we often denote $tloc(S)$ by $tloc(\ell)$.

To ensure that every state has some transition enabled on it, we standardly include in the transition system corresponding to each program the idling transition τ_I, whose transition relation is

$$\rho_I\colon \mathrm{T}.$$

We proceed to define the transition relations for the transitions associated with each of the statements.

- *Assignment*: With the statement $\ell : (y_1, \ldots, y_k) := (e_1, \ldots, e_k)$, we associate a transition τ_ℓ, with the transition relation

$$\rho_\ell\colon \quad ([\ell] \subseteq \pi) \ \wedge \ (\pi' = \pi - [\ell] \cup tloc(\ell)) \ \wedge \ (y_1' = e_1) \wedge \cdots \wedge (y_k' = e_k).$$

- *Await*: With the statement ℓ: **await** c, we associate a transition τ_ℓ, with the transition relation

$$\rho_\ell\colon \quad ([\ell] \subseteq \pi) \ \wedge \ c \ \wedge \ (\pi' = \pi - [\ell] \cup tloc(\ell)).$$

The transition τ_ℓ is enabled only when control is at ℓ and the condition c holds. When taken, it moves from ℓ to the location $tloc(\ell)$.

- *Conditional*: With the statement ℓ: **if** c **then** $[\ell_1\colon S_1]$ **else** $[\ell_2\colon S_2]$, we associate two transitions, τ_ℓ^{T} and τ_ℓ^{F} with transition relations

$$\rho_\ell^{\mathrm{T}}\colon \quad ([\ell] \subseteq \pi) \ \wedge \quad c \ \wedge \ (\pi' = \pi - [\ell] \cup [\ell_1])$$
$$\rho_\ell^{\mathrm{F}}\colon \quad ([\ell] \subseteq \pi) \ \wedge \ \neg c \ \wedge \ (\pi' = \pi - [\ell] \cup [\ell_2]).$$

Relation ρ_ℓ^{T} corresponds to the case that c evaluates to true and execution proceeds to ℓ_1, while ρ_ℓ^{F} corresponds to the case that c is false and execution proceeds to ℓ_2.

- *While*: With the statement ℓ: [**while** c **do** $[\widetilde{\ell}\colon \widetilde{S}\,]$], we associate two transitions, τ_ℓ^{T} and τ_ℓ^{F} with transition relations

$$\rho_\ell^{\mathrm{T}}\colon \quad ([\ell] \subseteq \pi) \ \wedge \quad c \ \wedge \ (\pi' = \pi - [\ell] \cup [\widetilde{\ell}])$$
$$\rho_\ell^{\mathrm{F}}\colon \quad ([\ell] \subseteq \pi) \ \wedge \ \neg c \ \wedge \ (\pi' = \pi - [\ell] \cup tloc(\ell)).$$

Note that in the case of a true c control moves, according to ρ_ℓ^{T}, from ℓ to $\widetilde{\ell}$, while in the case of a false c it moves, according to ρ_ℓ^{F}, from ℓ to $tloc(\ell)$.

- *Request*: With the statement ℓ: **request**(y), we associate a transition τ_ℓ, with the transition relation

$$\rho_\ell\colon \quad ([\ell] \subseteq \pi) \ \wedge \ x > 0 \ \wedge \ (\pi' = \pi - [\ell] \cup tloc(\ell)) \ \wedge \ y' = y - 1.$$

- *Release*: With the statement ℓ: **release**(y), we associate a transition τ_ℓ, with the transition relation

$$\rho_\ell\colon \quad ([\ell] \subseteq \pi) \ \wedge \ (\pi' = \pi - [\ell] \cup tloc(\ell)) \ \wedge \ y' = y + 1.$$

The transitions associated with the semaphore statements *request* and *release*, are called *semaphore transitions*. We refer to the transitions corresponding to statements of the program, i.e., all transitions except for τ_I, as the *diligent* transitions.

4.3 The Initial Condition

Consider a program

$$\big[\text{declaration; } [P_1 :: [\ell_1\text{: } S_1] \parallel \cdots \parallel P_m :: [\ell_m\text{: } S_m]]\big].$$

Let φ denote the *data precondition* of the program. We define the *initial condition* Θ for program P as

$$\Theta\text{:} \quad (\pi = \{\ell_1, \ldots, \ell_m\}) \land \varphi.$$

This implies that the first state in an execution of the program begins with the control set to the initial locations of the processes, *after* the initialization of the local and output variables has been performed.

4.4 Justice and Compassion

For the simple programming language we have presented, the justice and compassion requirements are straightforward.

- *Justice*: As the *justice* set, we take all the diligent transitions, i.e. $\mathcal{T} - \{\tau_I\}$.
- *Compassion*: As the *compassion* set, we take all the semaphore transitions.

5 Assessment of the Interleaving Model

An essential element of the generic model of transition systems is that concurrency is represented by *interleaving*. This means that, according to the formal model, two parallel processes never execute their statements at precisely the same instant, but take turns in executing atomic transitions. Formally, when one of them executes an atomic transition, the other is inactive. This model of computation is very convenient for the formalization, analysis, and manipulation of concurrent programs.

However, actual concurrent systems are usually composed of several independent processors, each of them executing a program of its own (a *process* in our terminology). In such systems, the execution of statements in different processors usually overlap rather than interleave. We refer to this behavior as *overlapped execution*.

A crucial question is how to reconcile the formal notion of *interleaved computations*, defined in our models, and the notion of *overlapped executions*, as realized on actual systems.

Two problems have to be resolved to achieve this reconciliation. The first problem is that of *interference*, and the second is *independent progress*.

- *Interference*

Interleaved execution provides a higher degree of protection from interference than is available in overlapped execution. This is because interleaved execution requires that when a transition is taken, all other transitions are inactive, so no interference *during* a transition is possible.

Consider, for example, the following statement:

when $y = y$ **do** S.

In interleaved computations, the condition $y = y$ is tested in one atomic step, so it always yields the value T. In comparison, overlapped execution (under naive implementation, with no optimization) of this statement will reference y twice. If, precisely between these two references, an overlapped statement performed by a concurrent processor changes y, the testing processor may find the condition to be false.

- *Independent Progress*

The problem of *independent progress* is that, in an overlapped execution, the computation of each process keeps advancing, since each processor is independently responsible for its own progress. In an interleaved computation, the only requirement is that enabled transitions be continuously chosen and executed. There is nothing to disallow a computation in which only transitions from one process are ever chosen. Such a computation ensures progress of the preferred process, but keeps all other processes stagnant.

Fair transition systems solve this problem by the two requirements of *justice* and *compassion*. We will discuss below the properties of each of these fariness requirements, and their implications.

5.1 Interleaving and Concurrency

Consider the following simple program, to which we refer as $A1$

$$\left[\begin{array}{c} \textbf{out } y\text{: } \textbf{integer where } y = 1 \\ P_1 :: \quad [\ell_0\text{: } y := y + 1] \quad \| \quad P_2 :: \quad [m_0\text{: } y := y - 1] \end{array}\right]$$

The overlapped execution of this program on a system consisting of two independent processors sharing a common memory yields $\{0, 1, 2\}$ as the set of possible outcomes for y. To see this, realize that execution of an *assignment* statement, such as $y := y + 1$, usually consists of three distinct steps:

- *Fetch Step*: The value in the shared-memory location corresponding to y is fetched and stored in a local register.
- *Compute Step*: The register is incremented, with the resulting value possibly being deposited in another local register.
- *Store Step*: The value in the result register is stored in the shared-memory location corresponding to y.

The partition of execution of an assignment into these three steps is common to many other statements. More complicated assignments may have several fetch steps, depending on the number of variables appearing in their right-hand sides.

When we consider program $A1$, there are such steps in the execution of assignments by both processes P_1 and P_2. We identify four of these steps as ***critical events***, in the sense that the relative ordering between them determines the final outcome of the execution. They are the events of fetching and storing values from and to the shared-memory location corresponding to y. We describe the critical events of this program as:

P_1 reads value m from y, denoted by $r_1(m)$

P_1 writes value m to y, denoted by $w_1(m)$

P_2 reads value m from y, denoted by $r_2(m)$

P_2 writes value m to y, denoted by $w_2(m)$.

Furthermore, we assume that these critical events are ***atomic***, in the sense that they cannot overlap. For simple data types, this is usually guaranteed by the underlying hardware: while one process is accessing a memory location, no other process may access the same location.

In an overlapped execution of program $A1$, these events can occur in several orders, some of which are listed below.

E_1: $r_1(1),\ w_1(2),\ r_2(2),\ w_2(1),$ yielding $y = 1$

E_2: $r_1(1),\ r_2(1),\ w_1(2),\ w_2(0),$ yielding $y = 0$

E_3: $r_1(1),\ r_2(1),\ w_2(0),\ w_1(2),$ yielding $y = 2$.

This shows how an overlapped execution of this program can yield the three possible results $0, 1, 2$, depending on the relative ordering of the critical events of the execution. Note that we do not consider the calculation step as critical, since it operates only on internal registers that are not accessible to the other process.

The situation illustrated here is general. In any concurrent program it is always possible to identify some events in the execution of the program as ***critical events***, such that their relative ordering in time uniquely determines the final result of the execution as well as its observable behavior, i.e., the sequence of values assumed by the observed variables. For programs that communicate by shared variables, these events are the reading and writing to the shared variables.

Consider now the possible results of computations of program $A1$, as defined in our model of fair transition system. The computational model assigns to program $A1$ two transitions ℓ_0 and m_0 (shorthand notation for τ_{ℓ_0} and τ_{m_0}) corresponding to the ***assignment*** statements in P_1 and P_2. Thus, the only possible computations of program $A1$ in our interleaved model are the following:

$$\langle\{\ell_0, m_0\},\ 1\rangle \xrightarrow{\ell_0} \langle\{m_0\},\ 2\rangle \xrightarrow{m_0} \langle\{\},\ 1\rangle \cdots$$

$$\langle\{\ell_0, m_0\},\ 1\rangle \xrightarrow{m_0} \langle\{\ell_0\},\ 0\rangle \xrightarrow{\ell_0} \langle\{\},\ 1\rangle \cdots.$$

In the interleaved model only one final result is possible: $y = 1$.

This seems to indicate that interleaved computation fails to capture the full range of behaviors exhibited by overlapped execution.

However, the problem is not in the interleaving model per se, but in the assignment of atomic transitions to statements in the above program. According to the rules we gave for our programming language, each *assignment* statement is represented as a single atomic transition. This representation leads to the undesirable effect that a transition such as ℓ_0 (corresponding to $y := y+1$) forces the two critical events r_1 and w_1 to happen in one step, thereby precluding the possibility of any of the critical events r_2, w_2 occurring after r_1 but before w_1. Indeed it is these possibilities that led to the outcomes 0 and 2, which the interleaved computation failed to produce.

One possible solution to this difficulty is that an assignment such as $y := y+1$ in program $A1$ should be associated with *two* atomic transitions, say τ'_{ℓ_0} and τ''_{ℓ_0}. The first transition τ'_{ℓ_0} should perform the fetch substep, while τ''_{ℓ_0} should perform the store substep. The calculation substep can be absorbed into either τ'_{ℓ_0} or τ''_{ℓ_0}.

5.2 Finer Granularity

There is another solution to the problem of discrepancy between interleaved computation and overlapped execution. This solution, which is the one we follow, is to modify program $A1$ in such a way that any atomic transition generated according to the rules as originally given, i.e., one transition per *assignment* statement, contains at most *one* critical event. Such a modification is presented by the following program $A2$:

$$\left[\begin{array}{c} \textbf{out } y\text{: } \textbf{integer where } y = 1 \\ P_1 :: \left[\begin{array}{l} \textbf{local } t_1\text{: } \textbf{integer} \\ \ell_0\text{: } t_1 := y \\ \ell_1\text{: } y := t_1 + 1 \end{array}\right] \quad \| \quad P_2 :: \left[\begin{array}{l} \textbf{local } t_2\text{: } \textbf{integer} \\ m_0\text{: } t_2 := y - 1 \\ m_1\text{: } y := t_2 \end{array}\right] \end{array}\right]$$

Each of the assignments of program $A1$ has been broken into two successive assignments in $A2$. The first assignment in each sequence performs the fetch substep, while the second assignment performs the store substep. To emphasize that the exact timing of the calculation substep is immaterial (as long as it comes after fetching and before storing), we have included it in the storing assignment in P_1 and in the fetching assignment in P_2. Note that local variables t_1 and t_2 represent the internal registers of the two independent processors.

The interleaved computations of $A2$ recreate all the overlapped executions of $A2$, which in turn are identical to the overlapped executions of program $A1$. To see this, we illustrate below three computations of $A2$ leading to the final results 1, 0, and 2, respectively. We represent the states occurring in these computations by listing the values assumed by the variables π, y, t_1 and t_2.

$$\begin{array}{ll} \sigma_1\text{:} & \langle\{\ell_0, m_0\},\ 1,\ -,\ -\rangle \xrightarrow{\ell_0} \langle\{\ell_1, m_0\},\ 1,\ 1,\ -\rangle \xrightarrow{\ell_1} \\ & \langle\{m_0\},\ 2,\ 1,\ -\rangle \xrightarrow{m_0} \langle\{m_1\},\ 2,\ 1,\ 1\rangle \xrightarrow{m_1} \langle\{\},\ 1,\ 1,\ 1\rangle \cdots, \end{array}$$

yielding $y = 1$.

$$\sigma_2\colon\ \langle\{\ell_0, m_0\},\ 1,\ -,\ -\rangle \xrightarrow{\ell_0} \langle\{\ell_1, m_0\},\ 1,\ 1,\ -\rangle \xrightarrow{m_0}$$
$$\langle\{\ell_1, m_1\},\ 1,\ 1,\ 0\rangle \xrightarrow{\ell_1} \langle\{m_1\},\ 2,\ 1,\ 0\rangle \xrightarrow{m_1} \langle\{\},\ 0,\ 1,\ 0\rangle \cdots,$$

yielding $y = 0$.

$$\sigma_3\colon\ \langle\{\ell_0, m_0\},\ 1,\ -,\ -\rangle \xrightarrow{\ell_0} \langle\{\ell_1, m_0\},\ 1,\ 1,\ -\rangle \xrightarrow{m_0}$$
$$\langle\{\ell_1, m_1\},\ 1,\ 1,\ 0\rangle \xrightarrow{m_1} \langle\{\ell_1\},\ 0,\ 1,\ 0\rangle \xrightarrow{\ell_1} \langle\{\},\ 2,\ 1,\ 0\rangle \cdots,$$

yielding $y = 2$.

This shows that, by proper restrictions on the syntax of programs we allow, it is possible to achieve a better matching between interleaved computations and overlapped executions.

Due to shortage of space, we will not present additional examples. However, the full version of this paper discusses many additional examples of discrepancies between the formal model and some phenomena arising in real executions, and how they can be resolved.

References

[1] E. Dijkstra. Guarded commands, nondeterminancy, and formal derivation of programs. *Comm. ACM*, 18(8):453–457, 1975.

[2] E. Dijkstra. *A Discipline of Programming.* Prentice-Hall, New Jersey, 1976.

[3] J. Fetzer. Program verification: The very idea. *Comm. ACM*, 31:420–422, 1988.

[4] D. Harel and A. Pnueli. On the development of reactive systems. In *Logics and Models of Concurrent Systems*, pages 477–498. Springer-Verlag, 1985.

[5] Z. Manna and A. Pnueli. How to cook a temporal proof system for your pet language. In *Proc. 10th ACM Symp. Princ. of Prog. Lang.*, pages 141–154, 1983.

[6] Z. Manna and A. Pnueli. The anchored version of the temporal framework. In J. de Bakker, W.-P. de Roever, and G. Rozenberg, editors, *Linear Time, Branching Time and Partial Order in Logics and Models for Concurrency*, pages 201–284. Lec. Notes in Comp. Sci. 354, Springer-Verlag, 1989.

[7] A. Pnueli. Applications of temporal logic to the specification and verification of reactive systems: A survey of current trends. In J. de Bakker, W.-P. de Roever, and G. Rozenberg, editors, *Current Trends in Concurrency*, pages 510–584. Lec. Notes in Comp. Sci. 224, Springer-Verlag, 1986.

MODELS FOR CONCURRENCY

(Invited talk – Abstract)

Mogens Nielsen
Computer Science Department, Aarhus University
DK-8000 Aarhus C, Denmark
mn@daimi.aau.dk

Background

The theory of concurrency has developed rapidly in recent years. Its goal is to establish a theoretical foundation for the understanding of and methods for reasoning about the complex behaviours of distributed computing systems. It is widely recognized that establishing such a theoretical foundation is one of the key issues in the practical use of the massive parallelism offered by the technological developments.

The models studied are mainly abstract mathematical models, each one applicable to a wide range of distributed systems, e.g. chips, networks of computing agents, computer based production systems etc.

Among the most developed models are Labeled Transition Systems, Petri Nets [Re, T], various term-based equivalences [H, Mi], Hoare Traces [H], Mazurkiewicz Traces [Ma], Synchronization Trees [W], and Event Structures [W].

All these models build on one common and important idea: they abstract away from details of real time, and use an abstract set of indivisible actions as atomic building blocks. This has proven to be an extremely useful idea in many different respects.

The models typically focus on behavioural aspects on the control level, e.g. the possible patterns of actions, capturing important aspects of distributed computations like deadlock, starvation, termination, etc.

The importance of a particular model is often a matter of balance between expressiveness and generality on one hand and simplicity and available proof techniques on the other.

The models are often informally classified in a number of different ways. Let us mention a few.

Some models aim at modeling at the so-called *behavioural* level (where an action is interpreted as having a unique occurrence in time and space, as in an Event Structure). Others aim at the so-called *system* level (where an action may have several occurrences in time and space, as in a Petri Net).

Models may be classified as either *interleaving* models (where the independent execution of different processes is modelled by the possible interleavings of their action patterns, as in Hoare Traces) or *noninterleaving* (where causal (in)dependencies between actions are represented directly in some formal sense, as in Mazurkiewicz Traces).

And finally, some models are classified as *branching-time* models (i.e. representing the branching structure of nondeterministic systems in some detail, as in Synchronization Trees), – others as *linear time* models (where nondeterministic systems are represented by their sets of deterministic computations, as in Hoare Traces).

Not surprisingly, the less expressive models are typically accompanied by a richer set of reasoning techniques (e.g. it is definitely the case that the interleaving models have been developed much further in this respect than noninterleaving models).

Relationships between models

It has proven fruitful to study the relationships between the various models formally in a categorical setting, where each model is equipped with a notion of behaviour preserving morphisms. This line of research was initiated by and has been pursued mainly by Winskel [W].

The purpose of such a study is to provide formal ways of expressing intuitive relationships between models, in terms of functors between the various categories of models.

As an example, one may prove the existence of certain kinds of adjunctions between say an interleaving model and its noninterleaving "counterpart" (e.g. a reflection between Hoare Traces and Synchronization Trees). The preservation of properties via such adjunctions is very helpful when transporting concepts and results from one model to another.

Also, one gets categorical characterizations of combinators of independent interest, e.g. the parallel combinators is often derived from categorical products.

We present and motivate some adjunctions along these lines in a survey form. Our choice of models will basically be the ones mentioned above, most of which have been related by a small and neat set of adjunctions ([W], [WN]). We present these adjunctions formally, with special emphasis on the associated motivation and intuitions. The goal of this work is partly to give formal underpinnings of the type of classifications mentioned above.

Net Systems and Transition Systems

We shall give special emphasis to the relationship between two of our models: (Standard) Labeled Transition Systems and Elementary Net Systems [T]. This is again one of the important relationships between an interleaving model and a noninterleaving "counterpart".

Labeled Transition Systems are equipped with behaviour preserving morphisms as in [W], and Elementary Net Systems [T] with a functional version of the net-morphisms suggested in [W].

There is a well-known functor from Elementary Net Systems to Labeled Transition Systems – the so-called case-graph construction [Ro]. We capture a few properties of these case-graphs in the form of axioms on Labeled Transition Systems, thus obtaining a full subcategory called Elementary Transition Systems (the objects satisfying our axioms). And in this setup we prove that the case-graph construction is the right adjoint of a coreflection between elementary Transition Systems and Elementary Net Systems ([NRT1]).

We shall elaborate on the consequences of such a categorical result. The most important consequence is that it allows you to work with a noninterleaving model of concurrency at the level of general Labeled Transition Systems, with its rich and well-established theory. Not by adding structure (as in Petri Nets) but by imposing restrictions in terms of our formal axioms. More concretely, when working with these restricted transition systems one keeps fundamental relations from Petri Nets like causality, conflict and concurrency as derived notions (but forget about how a particular case-graph may be programmed in nets using the extra structure of conditions).

Also, it turns out that by a smooth strengthening of our axiomatization of Elementary Transition Systems, we obtain a full subcategory, categorically equivalent to the well-known model of Prime Event Structures [W], called Occurrence Transition Systems.

It is widely recognized that one of the particularly attractive aspects of Net Theory is that it provides a uniform framework for talking about a distributed system *and* its behaviour – the behaviour obtained through a notion of unfolding, which has been categorically characterized by Winskel. We show that this idea transports to our abstract level of unfolding Elementary Transition Systems to Occurrence Transition Systems.

Summing up

To sum up, we shall attempt to motivate, present and interpret formal relationships between some of the well-known models for concurrency.

In particular, we shall provide characterizations in terms of standard Labeled Transition Systems of well-known noninterleaving models on the systems level (Nets) *and* on the behavioural level (Event Structures).

Also, we shall comment on advantages of some models over others with respect to particular applications.

Acknowledgements

The work reported here was done jointly with G. Winskel, G. Rozenberg, and P.S. Thiagarajan.

Some of the work reported was supported by Esprit Basic Research Action, CEDISYS.

References

[H] Hoare, C.A.R., *Communicating Sequential Processes*, Prentice Hall, (1985).

[Ma] Mazurkiewicz, A., Basic Notions of Trace Theory, Lecture Notes in Computer Science 354, 285-363, Springer-Verlag (1988).

[Mi] Milner, R., *Communication and Concurrency*, Prentice Hall, (1989).

[NRT1] Nielsen, M., Rozenberg, G., Thiagarajan, P.S., Elementary Transition Systems, DAIMI PB-310, Computer Science Department, Aarhus University (1990), to appear in Theoretical Computer Science.

[NRT2] Nielsen, M., Rozenberg, G., Thiagarajan, P.S., Transition Systems, Event Structures and Unfoldings, DAIMI PB-353, Computer Science Department, Aarhus University (1991).

[Re] Reisig, W., Petri Nets: an Introduction, EATCS Monographs on Theoretical Computer Science, Springer-Verlag, (1985).

[Ro] Rozenberg, G., Behaviour of elementary net systems, Lecture Notes in Computer Science 254, 60-94, Springer-Verlag (1987).

[T] Thiagarajan, P.S., Elementary net systems, Lecture Notes in Computer Science 254, 26-59, Springer-Verlag (1987).

[W] Winskel, G., Event Structures, Lecture Notes in Computer Science 235, 325-392, Springer-Verlag (1987).

[WN] Winskel, G., Nielsen, M., Models for Concurrency, in preparation.

ON A HIERARCHY OF FILE TYPES AND A TOWER OF THEIR THEORIES*

Andrzej Salwicki
Institute of Informatics, University of Warsaw
Departement d'Informatique, Université de Pau

Abstract

We present here a general method of software development which uses both: algorithmic logic and object programming. The approach will be presented in the form of a case study.
We discuss a rich system of files wich contains files of various types: text files, direct access files and binary sequential files. By introducing certain hierarchy one can gain a clear insight into the properties of files. A data structure of plain files is factorized out as a common factor of all mentioned types. All types of files are fully axiomatized in corresponding algorithmic theories. These theories form a tower since all of them are extensions of the theory of plain files.
Finally, one can construct a hierarchy of types which reflects the structure of the tower. In this way one obtains quite complicated software together with the clear specification of its functions. We hope that the reader will see the other applications of the general method presented here.

1. Introduction

We wish to convince the reader that the algorithmic approach to the specifications of abstract data types (or data structures) has many advantages. We are going to discuss the relationships between a hierarchy of data structures, a tower of its specifications (i.e. axiomatic theories) and another hierarchy of software modules which implement the models of the theories.
What are the goals to be achieved when one begins a work on the specification of an abstract data type? Surely, the specification should contain "truth, all truth and nothing but the truth" about the specified data structure. How to translate this legal term to the lan-

* The research was supported by Polish Ministry of National Education, grant RP.I.09 (1990) and by French Ministere d'Education Nationale grant for LITA Pau. (1991)

guage of computer science? "To say truth", this requirement is the easiest to fullfil. In nowadays practice many specifications are just collections of true statements. Much more difficult is to say "all truth", for, it requires certain degree of completeness i.e. that every true statement follows from given specification. We know from metamathematics that only a few structures can be given the complete axiomatization, in the language of first-oder logic. Here we propose to use an extended language which contains first-order formulas and algorithmic formulas. This enables to formulate the categorical axiomatization of many data structures of primary interest in computer science. As examples, we can quote the algorithmic theories of natural numbers, of references etc., cf. [4] Chapter IV. This permits also to formulate and prove representation theorems for many data structures. In the first case we prove that all the models of axioms of a theory are isomorphic. This settles the question. For all the truth about the axiomatized data structure follows logically from the axioms. On the other hand there are many cases in which we can not build a categorical axiomatization, for it was our intention to specify a class of data structures (like the class of groups in mathematics). In this case our goal will be achieved if we present a representation theorem which is also an evidence that all models of the specification have common structure. We are going to prove several theorems of this type.

"Nothing but the truth" what does it mean? One achieves this requirement by proving that his specification is consistent (is contradiction free).

Below we shall illustrate how to show that a specification says "truth, all truth and nothing but the truth". We shall apply our paradigm to a complicated case of a family of files data structures. This family occurs in Loglan'82 programming language where all of various kinds of files find applications. At first glance it seems that such collection of files is chaotic and difficult to understand. After a while we factorize out the common operations and common properties. This permits us to introduce certain classification and to tame the conceptual complexity of this rich system of files due to the classification.

It was quite astonishing for us to remark how nicely the structure of files: sequential, text and direct access is defined if a common part of plain files is factorized out. The difference between the idealized model as defined by our axioms and the present implementation is negligeable. Later we offer a class written in Loglan cf. section 6, which implements exactly the model described by axioms. To use it one has only to prefix his program with the name of this class.

The hierarchy is :

```
                PlainFiles
               /          \
  SequentialFiles    DirectAccessFiles
         /
  TextFiles
```

The hierarchy serves for two things, first, it explains our classification of algebraic data structures, secondly, this hierarchy is exactly the hierarchy of classes to be written. Each node in the above graph corresponds to a class which is an encapsulated data type and which is an extension of the class represented by the father of the current node. The extension beeing realized by means of inheritance mechanism. We would like to stress that both tools for modularization are needed: nesting to encapsulate types (or notions) and operations of a file structure and inheritance which permits to develop the file types from the basic type of plain files in an orderly and economic manner.

Our second goal is to convince the reader that the method proposed here has many more applications and to encourage him to follow the pattern presented here in his practice.

In most systems the documentation concerning files is long, obscure and complicated. Every programmer needs a short and clear information about the properties of operations on files. And he has difficulties in finding them. Our goal is therefore to provide a reasonably short and complete specification on abstract data type of files and to propose a corresponding implementation of it.

The studies on the notion of file has been done by many authors. The objective is always the same: to give a consistent and complete set of properties of files in such a way that the users can limit their knowledge of files to the formulas contained in the specification and that the implementors will arrive to the practically the same implementations (up to isomorphism). Among many approaches which can be taken (algebraic specification, denotational semantics etc.) we distinguish the algorithmic specification method.

There are two papers which use algorithmic logic to study the files. L.Stapp [7] wrote an axiomatization of sequential files, of catalogues and showed how to join them together. U.Petermann[5] in his voluminous doctoral dissertation studied the properties of hierarchical file systems i.e. the catalogues organized into a tree like in Unix, Dos and other systems

What differs our paper from the earlier ones? The difference from the papers of Stapp and Petermann lies in the exposition of the factorization method. We begin with 5 different types of files taken from the live programming language[3,7]. We factorize out the common features of them into a theory of plain files. On that base we construct the extensions: theory of direct access files and theory of binary sequential files. The text files are treated briefly as an variant of sequential files. In fact text files can be obtained from the sequential files of characters by extending the set of operations by read, write and similar operations and by forgetting the operations of sequential files: get and put.
Another problem of parametrized data types arrives when we consider three types of sequential files: real, integer,char. We have not enough space to discuss it here. Let us mention that due to the inheritance we have a liberty to choose between parametrized modules and extended (prefixed) modules.

We are going to present and study a tower of algorithmic theories which corresponds to the hierarchy of file structures.

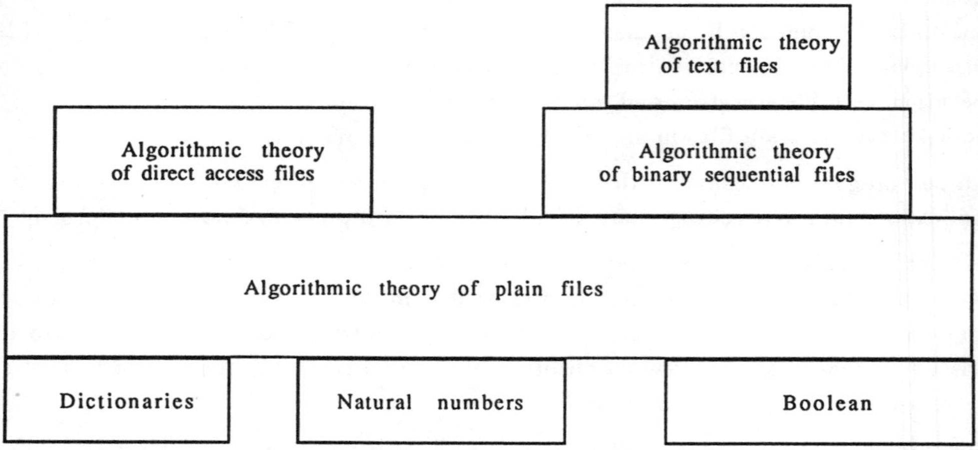

The theory of plain files is a common base for theories of binary sequential files and of direct access files. The theory contains and extends several other theories which concern the sorts of the algebraic structure of plain files: States, Types, Names, N - the sort of natural numbers and Boolean.
We do not explain here the theory of Boolean two-element algebra nor the algorithmic theory of natural numbers (see [4] Chapter 4). We assume that the intuitions of the reader will suffice. There are two sorts more which in our opinion need not an exagerated formalization. The sort Types has five elements

Types = {text, direct, real, integer, char(acter)}

i.e. just the names of kinds of files. The sort Names is the set of identifiers for files

Below we draw a diagram which shows both the signatures of data structures of files and the alphabets of corresponding formalized algorithmic theories. The structure of States is a formal counterpart of the catalogue and the mass storage. We abstract from the many details leaving only those things which are necessary to explain operations like create, open etc. We are viewing this structure as a dictionary structure with its usual operations cf [4] IV.2

in, del : Files x States $\rightarrow$ States
mb : Files x States $\rightarrow$ Boolean
amb : States $\rightarrow$ Files   empty: States $\rightarrow$ Boolean

An overview of the structures of files and their mutual relationships.

PLAIN FILES

Universe = States + Files + Types + Names + N + Bool

Operations and Relations:

```
create   : Names x Types x States → Files x States
open     : Names x States → Files x States
close    : Files x States → Files x States
unlink   : Files x States → Files x States
reset    : Files x States → Files x States
rewrite  : Files x States → Files x States

name     : Files → Names          isopen  : Files → Bool
type     : Files → Types          toread  : Files → Bool
pos      : Files → N              towrite : Files → Bool
len      : Files → N              eof     : Files → Bool
none     ∈ Files
```

SEQUENTIAL FILES

Universe of PlainFiles + Elem

Additional operations and relations

```
put : Elem x Files x States → Files x States
get : Files x States → Elem x Files x States
```

DIRECT ACCESS FILES

Universe of Plain Files + Arrays

Additional operations and relations

```
putrec: Arrays x Files x States → Files x States
getrec: Files x States → Files x States x Arrays
seek  : N x Files x States → Files x States
```

TEXT FILES

Operations

```
read :  Files x States → Files x States x Elem
write : Elem x Files x States → Files x States
eoln :  Files → Bool
```

In structure of Text Files one can usually find more operations We do not present all of them here. Note please, that operations get and put are hidden or forgotten. This kind of module construction is useful but it seems to be overlooked.

2. Plain Files; common operations on files

In this section we are going to describe the common part of different file types. Or to put it differently, we are presenting a theory T_{PF} such that all three theories of sequential file, of text files and of direct access files are the extensions of T_{PF}.

DENOTATIONS

In order to simplify the formulas, and to make them more readable, we shall assume the following denotations

f = f" iff name(f)=name(f") ∧ type(f)=type(f") ∧ pos(f)=pos(f") ∧ len(f)=len(f") ∧ isopen(f)=isopen(f") ∧ toread(f)=toread(f") ∧ towrite(f)=towrite(f")

f = f" mod(atr) means that all atributes of the file f are equal (as above) but the atribute or atributes mentioned, eg. f=f"mod(pos) means that files f,f" have the same names, types, len etc. and the values of pos(f), pos(f") need not to be equal .

member(n,s) iff P(¬ bool)

where P:
```
begin
    bool:=true;
    while (¬empty(s) ∧ bool )
    do
       f := amb(s);
       if  name(f)=n then bool := false else  s := del(f,s) fi
    od
end
```

The formula member(n,s) verifies if there is a file with the given name n in the state s.

s = s' iff (∀f) mb(f,s) ≡ mb(f,s')

The formulas below, denoted by OK, define the domaines of the operations on files.

OK(create(t,n,s)) $=^{df}$ ¬ member(n,s)

OK(open(n,s)) $=^{df}$ {member(n,s) ∧ (∀f)(mb(f) ∧ name(f)=n ⇒ ¬isopen(f))}

OK(close(f,s)) $=^{df}$ OK(unlink(f,s)) $=^{df}$ OK(reset(f,s)) $=^{df}$ OK(rewrite(f,s)) $=^{df}$ {mb(f,s) ∧ isopen(f) ∧ f≠none}

OK(name(f,s)) $=^{df}$ OK(type(f)) $=^{df}$ OK(pos(f)) $=^{df}$ OK(len(f)) $=^{df}$ f≠none

OK(in(s)) $=^{df}$ OK(del(s)) $=^{df}$ OK(amb(s)) $=^{df}$ ¬ empty(f))

AXIOMS OF PLAIN FILES PF

GROUP I All axioms of dictionaries for type States with elements in Files

Ax1 ¬ f=none ⇒ [s':=in(f,s)] { mb(f,s) ∧ (f≠f" ⇒ mb(f,s)= mb(f",s'))}

Ax2 ¬ f=none ⇒ [s':=del(f,s)] {¬ mb(f,s) ∧ (f≠f" ⇒ mb(f,s)= mb(f",s'))}

Ax3 mb(f,s) ≡ [bool:=false;
while (¬ empty(s) ∧ ¬ bool) **do**
f" := amb(s);
if f=f" **then** bool:= true **else** s := del(f,s) **fi**
od] bool

Ax4 **while** ¬ empty(s) **do** s := del(amb(s),s) **od** empty(s)

GROUP II Axioms which define the main operations on files

Ax1 OK(create(t,n,s)) ⇒
[(f,s') := create(t,n,s)] {name(f)=n ∧ type(f)=t ∧ pos(f)=0 ∧ len(f)=0 ∧
isopen(f) ∧ ¬toread(f)∧ ¬towrite(f)∧ s'=in(f,s)}

Ax2 {OK(open(n,s)) ∧ mb(f,s) ∧ name(f)=n} ⇒
[(f",s') := open(n,s)] {name(f")=n ∧ type(f")=type(f) ∧ pos(f")=0 ∧
len(f")=len(f) ∧ isopen(f") ∧ ¬toread(f") ∧ ¬towrite(f")∧ s'=in(f",del(f,s)) }

Ax3 OK(close(f,s)) ⇒
[(f",s') := close(f,s)] { ¬isopen(f") ∧ f=f"mod(isopen) ∧
s'=in(f",del(f,s)))}

Ax4 OK(unlink(f,s)) ⇒
[(f",s') := (unlink(f,s)] { f"=none ∧ s'=in(f",del(f,s)) }

Ax5 OK(reset(f,s)) ⇒
[(f",s') := reset(f,s)] {(toread(f") ∧ pos(f")=0 ∧
f=f"mod(pos,toread,towrite) ∧ s'=in(f",del(f,s)))}

Ax6 OK(rewrite(f,s)) ⇒
[(f",s') := rewrite(f,s)] {(name(f")=name(f) ∧ type(f")=type(f) ∧ pos(f")=0
∧ len(f")=0 ∧ towrite(f") ∧ s'=in(f",del(f,s)))}

Ax7 ¬ (toread(f) ∧ towrite(f))

Ax8 eof(f) ≡ pos(f)=len(f)

Ax9 pos(f) ≤ len(f)

GROUP III Auxiliary axioms which define the domaines of the operations

Ax1 OK($\varnothing$(x)) ≡ [y := $\varnothing$(x)] true for all operations $\varnothing$ in L_{PF}

e.g. OK(create(t,n,s)) ≡ [(f',s'):=create(t,n,s)] true

Ax2 f≠none ≡ [n:=name(f)] true ≡

[t:=type(f)] true ≡ [i:=pos(f)] true ≡ [j:=len(f)] true

Ax3 (f=none ⇒ ¬ mb(f,s) ∧ ¬ toread(f) ∧ ¬ towrite(f) ∧ ¬ isopen(f))

THEOREM 2.1

The theory of Plain Files is consistent.

PROOF.

To prove the theorem it is sufficient to present a data structure for L_{PF} which satisfies all axioms AxPF. Let us assume that the type Files is defined as

$$PFiles = \{<t,n,i,j,op,tr,tw> : (t,n)\in Types \times Names,\ i\leq j\in N,\ op,tr,tw\in Bool,\ \neg(tr\wedge tw)\}$$

and type States is defined as

$$\{\ s\in Files^{X} : (\ \forall x\in X)\ (s(x)=f \Rightarrow name(f)=x\ )\ and\ X\in Fin(Names)\ \}$$

Moreover, for arbitrary t∈Types, n∈Names, s∈States, f∈Files, we put:

mb, in, del - the set-theoretical operations of ∈, ∪, -. in the set States
amb(s) is a selector fuction such that amb(s)∈s

$$create(t,n,s)=^{df}\begin{cases}(f,s')\ \ where\ n\notin Dom(s),\ s'=s\cup\{(n,f)\}\ and\ f=<t,n,0,0,true,false,false>\\ undefined\ \ iff\ \ n\in Dom(s)\end{cases}$$

Let f=<t,n,i,j,op,tr,tw> then

$$open(n,s)=^{df}\begin{cases}(f'',s')\ \ where\ n\in Dom(s),\ s(n)=f,\ s'=s-\{(n,f)\}\cup\{(n,f'')\}\ and\\ \qquad op=false,\ f''=<t,n,0,j,true,false,false>\\ undefined\ \ iff\ \ n\notin Dom(s)\ or\ op=true\end{cases}$$

$$close(f,s)=^{df}\begin{cases}(f'',s')\ \ where\ s(n)=f,\ s'=s-\{(n,f)\}\cup\{(n,f'')\}\\ \qquad and\ f''=<t,n,i,j,false,tr,tw>\\ undefined\ \ iff\ \ (n,f)\notin s\ or\ f=none\end{cases}$$

$$unlink(f,s)=^{df}\begin{cases}(none,s')\ \ iff\ \ (n,f)\in s,\ s'=s-\{(n,f)\}\\ undefined\ \ iff\ \ (n,f)\notin s\ or\ f=none\end{cases}$$

$$reset(f,s)=^{df}\begin{cases}(f'',s')\ \ iff\ \ s(n)=f,\ s'=s-\{(n,f)\}\cup\{(n,f'')\},\ op=true\\ \qquad and\ f''=<t,n,0,j,true,true,false>\\ undef.\ \ iff\ \ (n,f)\notin s\ or\ f=none\ or\ op=false\end{cases}$$

$$\text{rewrite}(f,s) =^{df} \begin{cases} (f'',s') & \text{iff} \quad s(n)=f,\ s'=s-\{(n,f)\} \cup \{(n,f'')\}\ ,\ op=true \\ & \qquad\qquad \text{and} \quad f''=\langle t,n,0,0,true,false,true\rangle \\ \text{undefined} & \text{iff} \quad (n,f)\notin s \text{ or } f=none \text{ or } op=false \end{cases}$$

The operations name, type, pos, len are not defined for the argument f=none otherwise name(f) = n, type(f) = t, pos(f) = i, len(f) = j.

toread(f) iff f≠none and op=tr=true, tw=false

towrite(f) iff f≠none and op=tw=true, tr=false

isopen(f) iff f≠none and op=true

empty(s) iff s is empty set

By an easy verification we can prove that the above defined structure is a model of the theory of Plain Files, i.e. all axioms PF are valid in it. We shall denote this structure by M_{PF} and call it the standard structure of the theory of Plain Files determined by the given sets Names.

□

3. Binary sequential files

Sequential files realize certain policy of access to files. Position of a file increments by one or can be set to zero. No backward moves, no jumping ahead in a file are possible. Moreover, execution of any pair of put, get commands must be separated by a reset command and execution of any pair of get, put commands must be separated by a rewrite command. Therefore no mixed sequence of put and get commands is allowed. Binary sequential files BSF permit to store and read the binary information on the elements. There are three kinds of binary sequential files. Files of reals, files of integers and files of characters. For the simplicity we shall speak of binary sequential files of type Elem, where Elem can be defined as real, integer or char primitive type. Thus the sequential files form a family of three algebras. Their signatures admits all operations of plain files and additionally two operations put and get with the obvious meaning,

put : Elem x Files x States → File x States
get : File x States → Elem x Files x States

The binary representations of elements of different types are different. In fact for each type of BSF we ought to consider a constant which determines the size of elements. To simplify our considerations we assume here that the size of any element is 1 .

DENOTATIONS

Let us define as in the previous sections that s=s' iff for all f∈SFiles, mb(f,s) ≡ mb(f,s'). Assume that mb(f,s),mb(f'',s') and f=f'' (as in the sence defined for PlainFiles), then the formula eqval expresses the property: elements of the files f,f' are identical.

eqval(f,f") ≡ P {eof(f) ∧ eof(f")∧ bool}

where P:

```
begin
    if ¬isopen(f)
    then
        (f,s) := open(name(f),s);
        (f",s') := open(name(f"),s')
    fi;
    (f,s) := reset(f,s);
    (f",s) := reset(f",s');
    bool :=true;
    while ¬ eof(f)∧ bool
    do
        (e,f,s) := get(f,s);
        (e',f",s') := get(f,s');
        if e≠e' then bool :=false fi
    od
end
```

The following denotations are used to specify the domaines od the operations put and get

OK(put(e,f,s)) =^{df} (mb(f,s) ∧ isopen(f) ∧ towrite(f))
OK(get(f,s)) =^{df} (mb(f,s) ∧ isopen(f) ∧ toread(f) ∧ ¬ eof(f))

AXIOMS OF SEQUENTIAL FILES SF

The axioms of the binary sequential files contains all axioms of Plain Files with the obvious modifications according to the assumed above denotations and the following formulas defining the sence of the new operations.

OK(put(e,f,s)) ⇒ [(f",s') :=put(e,f,s)] {f=f"mod(pos,len) ∧ s'=in(f",del(f,s)) ∧ P(e*=e) ∧ pos(f")=pos(f)+1 ∧ len(f")=len(f) +1}

where

```
P: begin
       (f*,s*) := reset(f",s');
       while pos(f*) ≤ pos(f")
       do
           (e*,f*,g*) := get(f*,s*)
       od
   end;
```

OK(get(f,s)) ⇒ [(e',f",s') := get(f,s)] {f=f"mod(pos) ∧ s'=in(f",del(f,s)) ∧ pos(f")=pos(f)+1 }
OK(put(e,f,s)) ≡ [(f,s):=put(e,f,s)] true
OK(get(f,s)) ≡ [(e,f,s):=get(f,s)] true

DEFINITION 3.1

By a standard structure of Sequential Files M_{SF} determined by the set of Names we shall mean an extension of the standard structure defined for Plain Files by new sort of elements Elem and such that the sort Files is defined as follows:

SFiles = { (x,seq) : x $\in$PFiles, seq$\in$E* and if seq=<e_0,...,e_k> then len(x)=k+1}

The operation get and put are given as follows: for arbitrary f$\in$SFiles, if f=(<t,n,i,j,op,tr,tw>,<e_0,...,e_{j-1}>) then

$$\text{get(f,s)} =^{df} \begin{cases} \text{(e',f'',s')} \quad \text{if} \quad i<j,\ s(n)=f,\ s'=s-\{(n,f)\}\cup\{(n,f'')\},\ op=true \\ \qquad\qquad \text{and} \quad f''=(\langle t,n,i+1,j,true,false,true\rangle,\langle e_0,...,e_{j-1}\rangle) \\ \text{undefined} \quad \text{if} \quad (n,f)\notin s \text{ or } f=none \text{ or } op=false \end{cases}$$

$$\text{put(e,f,s)} =^{df} \begin{cases} \text{(f'',s')} \quad \text{if} \quad s(n)=f,\ s'=s-\{(n,f)\}\cup\{(n,f'')\},\ op=true \\ \qquad\qquad \text{and} \quad f''=\langle t,n,i+1,j+1,true,false,true\rangle,\langle e_0,...,e_{j-1},e\rangle)) \\ \text{undefined} \quad \text{if} \quad (n,f)\notin s \text{ or } f=none \text{ or } op=false \end{cases}$$

All other operations are defined as in the standard structure of Plain Files.

LEMMA 3.1

The standard structure of Binary Sequential Files M_{SF} is a model of axioms SF. i.e. The theory of BSF is consistent.

The proof is by easy verification.

Let M be an arbitrary data structure for the language L(PF). Let the universe of the corresponding sort be called as $Types_M$, $Names_M$, N, Bool, $Files_M$. $States_M$ and the corresponding operations and relations as o_M. We shall introduce an equivalence relation in the universe of M as follows

$$x \sim y \text{ iff } x,y \text{ are of the same sort and } x =_M y.$$

LEMMA 3.2

The relation $\sim$ is an equivalence relation and a congruence with respect to all the operations and relations in M but operation amb.

Let M* denote a reduct of the structure M obtained by removing the operation amb from the signature.

LEMMA 3.3

If M is a model of the set of axioms BSF then the quotient structure M*/$\sim$ is also a model provided we add an operation amb_M to M*/$\sim$ which satisfy the following condition: if x=amb_M([s]) then mb_M(x,[s]).

DEFINITION 3.2

We shall say that the structures M and W are weakly isomorphic iff there is an isomorphism between reducts M* and W* of M and W repectively.

The following theorem, called the theorem on represetation, states that the class of all models of the theory BSF is uniform.

THEOREM 3.4

Every model of algorithmic theory of binary sequential files is weakly isomorphic to the standard model determined by the same sets of Names and Elem.

PROOF

Let M be any model of the set of axioms BSF. Let us put

$$h: M^*/\sim \;\rightarrow\; M^*_{SF}$$

such that

$$h([f]) = (<name_M(f), type_M(f), pos_M(f), len_M(f), isopen_M(f), toread_M(f), towrite_M(f)>, <e_0,..,e_k>)$$

$$h([s]) = \{h([f]) : mb_M(f,s)\}$$

where $k=len_M(f)-1$ and the sequence $<e_0,\ldots,e_k>$ is determined by the following equalities

$$(e_0,f_0,s_0) = get_M(f,s)$$

$$(e_j,f_j,s_j) = get_M(f_{j-1},s_{j-1}) \text{ for } j\leq k.$$

It can be proved that the above function h preserves all operations and relations. Moreover it is a bijection.

□

4. Direct access files

Here we encounter an extension of the theory of plain files. The language of the extension contains additionally the sorts Elem and Arrays with auxiliary operations size, getel w

size : Arrays → N

getel : Arrays x N → Elem

and three operations of files: seek, getrec and putrec

putrec : Arrays x Files x States → Files x States

getrec : N x Files x States → Arrays x Files x States

seek : N x Files x States → Files x States

For direct access files we note two differences: firstly, a seqence of commands can contain both read and write commands intermixed, secondly, one can access any position of a file in a direct way. For it, it suffices to issue a seek command.

DENOTATIONS

For simplicity we use a[i] instead of getel(i,a). For arbitrary a,a' from the set Arrays we put

$a=a' \equiv (\forall i<size(a))\ a[i]=a'[i]$

The equality relations in the sorts of Files and of States are defined as in the case of sequential files.

The following formulas defines the domaines of the operations in DAF.

OK(putrec(a,f,s)) $=^{df}$ (mb(f,s) $\wedge$ isopen(f))
OK(getrec(k,f,s)) $=^{df}$ (mb(f,s) $\wedge$ isopen(f))
OK(seek(k,f,s)) $=^{df}$ (mb(f,s) $\wedge$ isopen(f))

AXIOMS OF DIRECT ACCESS FILES DAF

The axioms of the direct access files contains all axioms of Plain Files with the obvious modifications according to the denotations assumed above and the following formulas defining the meaning of the new operations.

OK(putrec(rec,f,s)) $\Rightarrow$
[(f",s') :=putrec(a,f,s)] {f=f"mod(pos,len) $\wedge$ s'=in(f",del(f,s)) $\wedge$ P(a*=a) $\wedge$
pos(f")=pos(f)+ size(a) $\wedge$ len(f")=max(len(f),pos(f")) }

where

```
P: begin
     (f*,s*) := seek(pos(f),f",s');
     (a*,f*,g*) := getrec(size(f),f*,s*)
   end
```

OK(getrec(k,f,s)) $\Rightarrow$ [(a,f",s') := getrec(k,f,s)] {f=f"mod(pos) $\wedge$ s'=in(f",del(f,s)) $\wedge$
pos(f")=pos(f)+ size(a) $\wedge$ size(a)=min(k,len(f)-pos(f)) }

OK(putrec(a,f,s)) $\equiv$ [(f,s):=putrec(a,f,s)] true
OK(getrec(k,f,s)) $\equiv$ [(a,f,s):=getrec(k,f,s)] true

DEFINITION 4.1

By a standard model of Direct Access Files we shall mean an extension of the standard structure for Plain Files by the set of elements Elem and a set Arrays and putting

DAFiles = { (x,seq) : x $\in$ PFiles, seq$\in$ Elem* and if seq=$<e_0,\ldots,e_k>$ then len(x)=k+1}

The operations getrec and putrec are defined as follows: for arbitrary $f \in$ DAFiles, if $f=(<t,n,i,j,op,tr,tw>,<e_0,\ldots,e_{j-1}>)$ then

$$getrec(k,f,s) =^{df} \begin{cases} (a,f'',s') & \text{if } s(n)=f,\ s'=s-\{(n,f)\} \cup \{(n,f'')\},\ op=true \\ & \quad \text{and } f''=(<t,n,i+p,j,true,false,true>,<e_0,\ldots,e_{j-1}>) \\ \text{undefined} & \text{if } (n,f) \notin s \text{ or } f=none \text{ or } op=false \end{cases}$$

where $p=\min(j-i,k)$, $a=<e_i,\ldots,e_{i+k-1}>$ if $k \leq j-i$, otherwise $a=<e_i,\ldots,e_{j-1}>$.

Assume that $a=<\underline{e}_0,\ldots,\underline{e}_{k-1}>$, then

$$putrec(re,f,s) =^{df} \begin{cases} (f'',s') & \text{if } s(n)=f,\ s'=s-\{(n,f)\} \cup \{(n,f'')\},\ op=true \\ & \quad \text{and } f''=(<t,n,i+l,p,true,false,true>,<e'_0,\ldots,e'_p>) \\ \text{undefined} & \text{if } (n,f) \notin s \text{ or } f=none \text{ or } op=false \end{cases}$$

where $p=j$ if $k \leq j-i$, otherwise $p=i+k$ and $e'_m = e_m$ for $k<i$ and $j-(i+l)<m<j-1$, $e'_m = \underline{e}_m$ for $i \leq k \leq m-(i+l)$

$$seek(k,f,s) =^{df} \begin{cases} (f'',s') & \text{if } s(n)=f,\ s'=s-\{(n,f)\} \cup \{(n,f'')\},\ op=true \\ & \quad \text{and } f''=(<t,n,i+p,j,true,false,true>,<e_0,\ldots,e_{j-1}>) \\ \text{undefined} & \text{if } (n,f) \notin s \text{ or } f=none \text{ or } op=false \end{cases}$$

where $p=\min(j-i,k)$.

All other operations are defined as in the standard structure of Plain Files.

The following lemma states that the theory of Direct Access Files is consistent.

LEMMA 4.1
The standard structure of Direct Access Files is a model of axioms DAF.

The following theorem is called the representation theorem for Direct Access Files.

THEOREM 4.2
Every model M of the theory DAF is weakly isomorphic to the standard model of DAF.

5. Text files

Text files can be constructed from binary sequential files of characters.They permit to either write or read. A write operation can be described as a superposition of two operations: first operation translates the argument (usually integer or real) into a sequence of characters and then the second operation sends the sequence to file by repeated put. Similarly one can view the read operation. The operation put and get of sequential files are usually hidden for the users and are not applicable to text files.

The theory of Text Files can be defined as an extension of Binary Sequential Files by the definition of the write and read operations. We omit further details in order to save the space.

6.Hierarchy of types

In this section we propose a sketch of an implementation of the hierarchy of file types discussed above. We are using the notions of class, object and of inheritance. We submit to the reader a tool which can be used in practical programs.

```
unit Files : class;

     unit States : class;
          ....
     end States;

     unit PlainFiles : class;
          unit pfile : class; ...
               var name: Names, type: Types,pos, len: integer, ...
          end pfile;
          unit create : function (n:Names,t:Types,s:States): pfile;
          end create; ...
          unit open : procedure ...
          end open;
          unit close : procedure ...
          end close;
          unit reset : procedure ...
          end reset;
          unit rewrite : procedure ...
          end rewrite;
           ...

     end PlainFiles;

     unit SequentialFiles : PlainFiles class;
          unit sfile: pfile class ...
          end sfile;
          unit put : procedure(e: Elem, f: sfile, s: states; output f': sfile, s': states); ...
          end put;
          unit get : procedure(f: sfile, s: states; output e: Elem, f': sfile, s': states); ...
          end get;

     end SequentialFiles;
```

```
unit DirectAccessFiles : PlainFiles class;
      unit dfile: pfile class ...
      unit putrec: procedure(a: Arrays, f: dfile, s: states; output f": dfile, s': states); ...
      unit getrec: procedure(f: dfile, s: states; output a: Arrays, f": dfile, s': states); ...
      unit seek: procedure(f: dfile, s: states, n: integer; output f": dfile, s': states); ...
end DirectAccessFiles ;

unit TextFiles : SequentialFiles class;
      unit tfile: sfile class ...
      unit write : procedure ...
      unit read : procedure ...
end TextFiles;

end Files;
```

The above sketch can be simplified in the case of applications to sequential programs by elimination of the class States. However, this class is useful for programs which use concurrent , distributed processes.

The reader will kindly note the simultaneous use of nesting of modules and of inheritance. A few programming languages only enjoy multileves inheritance [2,3].

7. Conclusions

Looking backward we see that the road we have passed gives an evidence of a methodology of software engineering. Our methodology is based on two ingredients: algorithmic specification and object programming.
It is algorithmic specification which brings a complete information about the data structure. In other words, without algorithmic properties we are not able to define the structure up to isomorphism. The algebraic or first-order specifications are much weaker: one can prove that given an algebraic axiomatization specification of files admits many unusual, undesired models. There are specifications of quality and just specifications. A first mark of quality for a specification is achieved when a specification is contradiction free. Only then the specification can have a commercial value. A second mark of quality is conferred to the specifications which are full i.e. which give a guarantee that they will never be replaced by better, fuller specifications. This guarantee comes from a proof of categoricity theorem or from representation theorem.

Another remark: the hierarchy of theories corresponds to a hierarchy of classes which implement the defined notions. This phenomenon pleases us and is a source of our belief that what we propose is a methodology of software construction. For it enables the development of software modules together with the proofs of their correctness.

Acknowledgments

I thank Grazyna Mirkowska for her help in research, writing and preparation of this paper. She is a coauthor of the paper, and we sign together a full version of it. We felt however it would inappropriate for these proceedings.

References

[1] Kerningham, B.W., Ritchie, D.M., The C programming language, Prentice Hall, Englewood Cliffs, 1978, pp1-237

[3] Kreczmar,A., et al., Report on Loglan'88 programming language, LNCS vol.414, Springer Verlag, Berlin Heidelberg, 1990, pp1-123

[2] Loglan'82 report, PWN Warsaw, 1983, pp 1-147

[4] Mirkowska,G.,Salwicki,A., Algorithmic Logic, PWN Warsaw and D.Reidel Dordrecht, 1987, pp. 1-358

[5] Petermann,U., File System - axiomatic specification and analysis of its realizability in the language Loglan, doctoral dissertation, Dep. Of Mathematics and Computer Science, University of Warsaw 1985

[6] Silvester, Peter,P., The Unix TM System Guidebook, Springer Verlag Inc, New York 1984, pp. 1-244

[7] Stapp,L., Axiomatic approach to the system of files, in Logics of programs and their applications, Proc. Poznan August 1980, LNCS v.148, Springer Verlag, Berlin Heidelberg 1983, pp. 270-294

[8] Szalas,A., Warpechowska,J., Loglan , WNT, Warsaw, 1991, pp. 1- 189

[9] Wirth,N., Programming in Modula 2, Springer Verlag, Berlin Heidelberg 1983, pp. 1-179

Strong Conjunction and Intersection Types

Fabio Alessi , Franco Barbanera
Dipartimento di Informatica. Corso Svizzera, 185 10149 Torino (Italy)

Abstract Provable realizability with untyped λ-terms as realizers of formulas of the propositional language with implication and the so called "strong conjunction" is proved to be equivalent to type inference in the system with intersection types of Coppo and Dezani. A similar equivalence is proved if realizers are terms of Combinatory Logic. These results are used to derive properties of models of λ-calculus and Combinatory Logic.

Introduction

In [BCD] and [CD] type assignment systems for λ-terms extending the inference system of Curry with intersection types have been introduced. The main feature of these systems is the introduction of the intersection type constructor ∧ and of the rules

$$(\wedge I)\frac{B \vdash^{\wedge} M : \sigma \quad B \vdash^{\wedge} M : \tau}{B \vdash^{\wedge} M : \sigma\wedge\tau} \qquad (\wedge E)\frac{B \vdash^{\wedge} M : \sigma\wedge\tau}{B \vdash^{\wedge} M : \sigma} \quad \frac{B \vdash^{\wedge} M : \sigma\wedge\tau}{B \vdash^{\wedge} M : \tau} .$$

While typed systems of λ-calculus closely correspond to various (constructive) logical systems, via the so called Curry-Howard isomorphism, problems arise if we consider type inference systems. This is the case for the system with intersection types, whose "logical meaning" is not straightforwardly clear and has been looked for since the first investigations on it.

Pottinger in the introduction of [Pot] hints that the type constructor ∧ suggests with its rules a new constructive logical connective,which he calls "strong conjunction", stating that

> *"The intuitive meaning of ∧ can be explained by saying that to assert $\alpha\wedge\beta$ is to assert that one has a reason for asserting α which is also a reason for asserting β".*

This is no doubt a good suggestion for a connective. One however could rigthly wonder if an investigation on strong conjunction is interesting. We think the answer is positive, since the definition of strong conjunction has certain peculiarities, the study of which can help to define, discuss and investigate new sorts of logical connectives. It was this which led other authors to get interested in strong conjunction. The first was Lopez-Escobar who, in his [Lop-Esc], investigates the strong conjunction connective. He refers to ∧ as *"..one of the first, if not the first, connective which is truly proof-functional".*

The study of strong conjunction can then be of help in the investigation of proof-functional connectives. These connectives are clearly intuitionistic.

Once a new connective is introduced one has to develop a formal concept of derivation for the sentences of a language containing the new connective. To focus the attention on strong conjunction the language considered by Lopez-Escobar was a propositional language containing only implication beside strong conjunction, a choice we make also in the present paper.

In order to give a formal system for strong conjunction the first thing to do is to restate its definition in terms of realizers. The definition of strong conjunction in a realizability context is : x realizes A∧B if x realizes both A and B. Now a good evidence that a formal system embodies some constructive concepts is the proof that it is possible to prove a formula in the formal system iff there exists a realizer such that it is possible to prove in (intuitionistic) first-order logic that the realizer found realizes the formula. The formal system defined in [Lop-Esc] however satisfied only one direction,i.e. it was not sound with respect to provable realizability.

The problem of defining a system which was a good formalization for the strong conjunction was dealt with also by Mints in [Min], who pointed out what lacked in Lopez-Escobar's system. Mints, unlike Lopez-Escobar who used CL-terms, uses untyped λ-terms as realizers. The sentences of his system contain typed λ-terms that, because of the presence of strong conjunction, are not well defined and thus the proof of completeness for this system fails.

What is proved in this paper is that one of the type inference systems with intersection types of [BCD] and [CD] is indeed the right formalization in a natural deduction style of the propositional logic for a language with implication and strong conjunction. It is interesting to note that indeed the system we propose "profits" from that which Mints pointed out was to be added to the system of [Lop-Esc]. Besides finding a system which is a good formalization for strong conjunction we then manage to give an answer to the problem of finding a "logical" caracterization of the systems of [BCD] and [CD]. Most of the proof methods for our system is borrowed from [Min].
Moreover, what we try to do in the paper is to face the problem when realizers are untyped terms of Combinatory Logic, a choice, among all the possible ones, quite reasonable, since CL is a good theory of function. The formal system for provable realizability, with CL-terms as realizers, for the propositional calculus with implication and strong conjunction will be found out to be the type assignment system for CL-terms with intersection types devised in [DH]. The results about provable realizability with CL-terms will be then used to derive some nice semantical equivalences in the field of models for λ-calculus and Combinatory Logic.

In what follows we refer to [Bar] or [HS] for the basic notions of λ-calculus and Combinatory Logic. To [HS] for the ones of type assignment systems.
The symbol "&", used in [Min] to denote the strong conjunction will be used here to denote the usual logical conjunction, while for strong conjunction we shall use the symbol of the intersection type constructor, namely "∧".

1. Intersection Types and Provable Realizability for Strong Conjunction.

As stated in the introduction, we wish to define a system which is a good formalization for the strong conjunction, i.e. a system such that it is possible to derive a formula in it iff there exists a realizer such that it is possible to prove in (intuitionistic) first-order logic that it realizes the formula (we shall specify better in the following the meaning of "to be a good formalization for the strong conjunction").
In this section *untyped* λ-terms are used as realizers.
The first thing to do is to formally define which formulas of first-order logic express the fact that a certain untyped λ-term realizes a proposition of the propositional calculus with implication and strong conjunction. This is done by associating in a very natural way to each propositional formula α a first-order predicate formula $r_\alpha[x]$ by induction on the structure of α. The definition of $r_\alpha[x]$ is taken from [Lop-Esc].

Definition 1.1 (*The first order formulas* $r_\alpha[x]$)
First we associate to any propositional variable φ a unary predicate symbol P_φ.

$$r_\varphi[x] \equiv P_\varphi(x)$$
$$r_{\sigma\to\tau}[x] \equiv \forall y\,(\, r_\sigma[y] \to r_\tau[xy]\,)$$
$$r_{\sigma\wedge\tau}[x] \equiv r_\sigma[x]\ \&\ r_\tau[x]\,.$$ ♪

<u>*Convention*</u> We shall abbreviate a term (or a subterm) of the form $PM_1..M_n$ by PM.

Definition 1.2 (*The systems* $\vdash_S$ and $\vdash_N$)
$\vdash_S$ is Gentzen's intuitionistic sequent calculus with equality where terms are λ-terms and the equality is the β-conversion among them.
$\vdash_N$ is the intuitionistic system of natural deduction with equality where the terms are λ-terms and the equality is the β-conversion among them. ♪

Let us give an example of derivation in $\vdash_S$.
<u>Example</u> 1.3

$$\dfrac{\dfrac{\dfrac{r_\alpha[y] \vdash_S r_\alpha[(\lambda x.x)y]}{\vdash_S r_\alpha[y]\to r_\alpha[(\lambda x.x)y]}}{\vdash_S \forall y(r_\alpha[y]\to r_\alpha[(\lambda x.x)y]) \equiv r_{\alpha\to\alpha}[\lambda x.x]} \qquad \dfrac{\dfrac{r_\beta[y] \vdash_S r_\beta[(\lambda x.x)y]}{\vdash_S r_\beta[y]\to r_\beta[(\lambda x.x)y]}}{\vdash_S \forall y(r_\beta[y]\to r_\beta[(\lambda x.x)y]) \equiv r_{\beta\to\beta}[\lambda x.x]}}{\vdash_S r_{\alpha\to\alpha}[\lambda x.x] \& r_{\beta\to\beta}[\lambda x.x] \equiv r_{(\alpha\to\alpha)\wedge(\beta\to\beta)}[\lambda x.x]}$$

In some proofs we shall use the equivalence between the system of Gentzen's sequents calculus and that of natural deduction. The following lemma states their equivalence for our case.

Lemma 1.4 $\Gamma \vdash_S \mathbf{r}_\alpha[M] \Leftrightarrow \Gamma \vdash_N \mathbf{r}_\alpha[M]$.

We can now formally define what we mean by "to be a good formalization for the strong conjunction" for a system :
we say that a system *S* is a *good formalization for the strong conjunction* iff
α is provable in *S* $\Leftrightarrow$ there exists a λ-term M s.t. $\Gamma \vdash_S \mathbf{r}_\alpha[M]$
i.e. iff *S* is correct and complete with respect to the notion of provable realizability.

The system defined by Lopez-Escobar in [Lop-Esc] was not a good formalization since it was not complete. It consisted essentially in the type assignment system for λ-terms $\vdash_\leq$ (see Def.1.6 below). The following counterexample to its completeness was given in [Min] : the formula $((\alpha\to\beta)\wedge\delta)\to(((\alpha\wedge\gamma)\to\beta)\wedge\delta)$ is realizable by the term $\lambda x.x$ but is unprovable in Lopez-Escobar's system. This counterexample relies on the fact that it lacks the correspondent of rule (η) (see Def.1.6 below) i.e. a rule of extensionality. The system for the provable realizability of strong conjunction defined in [Min] contains a rule of extentionality, but the use in its statements of *typed* λ-terms and not of *untyped* ones, prevents the proof of its completeness from working (Mints himself agrees on our remark [Min a]). This can be informally explained in the following way.
As shown in Example 1.3 above, the untyped λ-term $\lambda x.x$ is a realizer for the formula $(\alpha\to\alpha)\wedge(\beta\to\beta)$. In type assignment systems such as $\vdash_{\omega,\leq}$ (see Def.1.6 below) it is possible to infer the statement $\lambda x.x{:}(\alpha\to\alpha)\wedge(\beta\to\beta)$, while it is not possible to have, at least in Mints' typed system for λ-terms , a *typed* term t such that the statement $(\alpha\to\alpha)\wedge(\beta\to\beta)$ (t) is derivable, even if it is possible to derive the statements $\alpha\to\alpha$ (t_1) and $\beta\to\beta$ (t_2) for terms t_1 and t_2 such that the type erasures of t_1 and t_2 are both equal to $\lambda x.x$. This is because two typed terms with the same "structure" but with different types (even if the difference between them consists only in the names of type variables) are two *distinct* terms, a thing that prevents Mints' rule

$$(\wedge^+)\frac{\Gamma \vdash \gamma\,(t) \quad \Gamma \vdash \delta\,(t)}{\Gamma \vdash \gamma\wedge\delta\,(t)} ,$$

where t is a typed term, to be applied if we have $\alpha\to\alpha$ (t_1) and $\beta\to\beta$ (t_2), even if t_1 and t_2 have the same "erasure".
What we shall prove in the following is that in order to have a system which is a good formalization for the strong conjunction we have to remain in a type assignment context, using in such a context the rules suggested in [Min]. In particular the system we need is an already existing one, i.e. one of the type assignment systems with intersection types defined in [BCD][CD].

Type Assignment Systems with Intersection Types

Definition 1.5 (*Intersection types and the relation* $\leq$)
(i) The set of *intersection types* is the set of types built out of a set of type variables by means of two type constructors : $\to$ and $\wedge$.
(ii) A (*type assignment*) *statement* is of the form M:σ with σ a type and M a λ-term.
A *basis* B is a set of statements with only variables as terms. More than one occurrence of a variable can occur in a basis B.
(iii) The relation $\leq$ among types is the smallest relation satisfying

$$\tau \leq \tau \qquad\qquad \tau \leq \tau\wedge\tau$$
$$\sigma\wedge\tau \leq \sigma \qquad\qquad \sigma\wedge\tau \leq \tau$$
$$(\sigma\to\rho)\wedge(\sigma\to\tau) \leq \sigma\to(\rho\wedge\tau) \qquad\qquad \sigma\leq\tau\leq\rho \;\Rightarrow\; \sigma\leq\rho$$
$$\sigma\leq\sigma',\ \tau\leq\tau' \;\Rightarrow\; \sigma\wedge\tau \leq \sigma'\wedge\tau' \qquad\qquad \sigma'\leq\sigma,\ \tau\leq\tau' \;\Rightarrow\; (\sigma\to\tau) \leq (\sigma'\to\tau').$$ ♪

Definition 1.6 (*The system* $\vdash^\wedge$ *and its variants*) [BCD] [CD]
The type assignment system for λ-terms $\vdash^\wedge$ is defined by the following axioms and rules.

(Ax) $B, x{:}\sigma \vdash^{\wedge} x{:}\sigma$

$$(\rightarrow I)\ \frac{B, x{:}\sigma \vdash^{\wedge} M{:}\tau}{B \vdash^{\wedge} \lambda x.M : \sigma\rightarrow\tau}\ (*) \qquad (\rightarrow E)\ \frac{B \vdash^{\wedge} M : \sigma\rightarrow\tau \quad B \vdash^{\wedge} N : \sigma}{B \vdash^{\wedge} (MN) : \tau}$$

$$(\wedge I)\ \frac{B \vdash^{\wedge} M : \sigma \quad B \vdash^{\wedge} M : \tau}{B \vdash^{\wedge} M : \sigma\wedge\tau} \qquad (\wedge E)\ \frac{B \vdash^{\wedge} M : \sigma\wedge\tau}{B \vdash^{\wedge} M : \sigma} \quad \frac{B \vdash^{\wedge} M{:}\sigma\wedge\tau}{B \vdash^{\wedge} M : \tau}$$

(*) if x does not occur in B.

All the systems based on $\vdash^{\wedge}$ have the above rules. If other rules among the following are added, specifying a particular system, subscripts are used. When subscripts are present the superscript $\wedge$ is omitted.

$$(\leq)\ \frac{B \vdash^{\wedge} M : \sigma \quad \sigma\leq\tau}{B \vdash^{\wedge} M : \tau} \qquad (\omega)\ B \vdash^{\wedge} M : \omega \quad (\ast\ast)$$

$$(\eta)\ \frac{B \vdash^{\wedge} \lambda x.Mx : \sigma}{B \vdash^{\wedge} M : \sigma}\ x\notin FV(M).$$

(**) In systems containing rule (ω) it is assumed that in the definition of types the type constant ω is added and that the relation $\leq$ is extended by the following extra clauses :

$\sigma \leq \omega \qquad \omega \leq \omega\rightarrow\omega$. ♪

Remark 1.7 It is easy to see that if $B \vdash^{\wedge} M{:}\alpha$ then $B\cup B' \vdash^{\wedge} M{:}\alpha$ for any basis B'.

Lemma 1.8 [BCD] (η) is an admissible rule in $\vdash_{\leq}$ (and hence in $\vdash_{\omega,\leq}$).

Lemma 1.9 [CD] (i) $B \vdash^{\wedge} M{:}\alpha$ and $M \rightarrow_{\beta} M' \Rightarrow B \vdash^{\wedge} M'{:}\alpha$.
[BCD] (ii) $B\vdash_{\omega,\leq} M : \alpha$ and $M=_{\beta}N \Rightarrow B\vdash_{\omega,\leq} N : \alpha$.

It is beyond the scope of this paper to give in detail the motivations and the main achievements obtained by means of system $\vdash^{\wedge}$ and its variants. To show their relevance in the theory of λ-calculus it is however worth pointing out that they can be used to characterize important classes of pure λ-terms, namely the ones of strongly normalizable, normalizable and solvable terms ([BCD],[CD]). $\vdash_{\omega,\leq}$ is also used to build models of pure λ-calculus ([BCD],[CDHL])

$\vdash_{\omega,\leq}$ is a Good Formalization for Strong Conjunction

The rest of this section will be dedicated to proving that $\vdash_{\omega,\leq}$ (actually a slight restriction of it) is a good formalization for the strong conjunction. This will be done by first proving the equivalence of $\vdash_{\omega,\leq}$, when ω does not appear in the types of the assumptions and of the conclusion, and the type assignment system $\vdash_{\mathcal{M}}$ defined below and then by using this equivalence to prove that a type assignment statement $M{:}\alpha$ is derivable in $\vdash_{\omega,\leq}$ iff it is possible to prove in intuitionistic first order logic that M is a realizer for α.

Definition 1.10 (*The system* $\vdash_{\mathcal{M}}$)
The system $\vdash_{\mathcal{M}}$ is $\vdash^{\wedge}$ to which the following rules are added.

$$(\eta_1)\ \frac{B,x{:}\alpha \vdash_{\mathcal{M}} Mx : \beta}{B \vdash_{\mathcal{M}} M : \alpha\rightarrow\beta}\ x\notin FV(M) \qquad (eq\beta)\ \frac{B \vdash_{\mathcal{M}} N : \alpha \quad M=_{\beta}N}{B \vdash_{\mathcal{M}} M : \alpha}\ .$$

♪

Notice that $\vdash_{\mathcal{M}}$ is indeed an *untyped* version of the system for the provable realizability of formulas of the propositional language with implication and strong conjunction given by Mints in [Min] (the subscript $\mathcal{M}$ is used to recall this fact).

Lemma 1.11 Rule $(\leq)$ is admissible in $\vdash_{\mathcal{M}}$.
Proof
By induction on the derivation of $\sigma\leq\tau$. ∯

Lemma 1.12 Let $\omega\notin\alpha$,B. Then $B\vdash_{\omega,\leq} M:\alpha \;\Leftrightarrow\; B\vdash_{\mathcal{M}} M:\alpha$.
Proof
$\Leftarrow$) It is enough to prove that rules (eqβ) and (η_1) are admissible in $\vdash_{\omega,\leq}$.
• eqβ) By Lemma 1.9.
• η_1) By Lemma 1.8 the admissibility of rule (η_1) can be easily proved in the following way.

$$(\eta)\dfrac{(\to I)\dfrac{B,x:\alpha\vdash_{\omega,\leq} Mx:\beta}{B\vdash_{\omega,\leq}\lambda x.Mx:\alpha\to\beta}}{B\vdash_{\omega,\leq} M:\alpha\to\beta}$$

$\Rightarrow$) $B\vdash_{\omega,\leq} M:\alpha \;\Rightarrow\; B\vdash_{\omega,\leq} M^{\wedge}:\alpha$ by Lemma 1.9 (i) with $M^{\wedge}$ the normal form of M (M has normal form by Theorem 4.9 of [BCD])
$\Rightarrow\; B\vdash_{\leq} M^{\wedge}:\alpha$ by Lemma 4.5 of [BCD] (Subformula Principle)
$\Rightarrow\; B\vdash_{\mathcal{M}} M^{\wedge}:\alpha$ by Lemma 1.11 $\Rightarrow\; B\vdash_{\mathcal{M}} M:\alpha$ by rule (eqβ). ∯

As already stated in [Min] the cut rule can be proved admissible in $\vdash_S$ in the standard way, using the following obvious remark : if we have that M=N then we can transform the derivation of a sequent S[M] into a derivation of S[N] simply by replacing M by N. Moreover, since we are interested in derivations of formulas of the form $\mathbf{r}_\alpha[M]$ in $\vdash_S$, it is easy to check that by the subformula property of cut-free derivations they can contain only formulas of the form $\mathbf{r}_\alpha[M]$ and $\mathbf{r}_\gamma[N]\to\mathbf{r}_\delta[MN]$. The latter arises as a result of splitting $\mathbf{r}_{\gamma\to\delta}[M]$, i.e. $\forall y(\mathbf{r}_\gamma[y]\to\mathbf{r}_\delta[My])$.
Using the standard permutation of rules in Gentzen-type derivations [Kle] one can arrive at a situation where $\mathbf{r}_\gamma[N]\to\mathbf{r}_\delta[MN]$ is always split immediately above the formula $\mathbf{r}_{\gamma\to\delta}[M]$ from which $\mathbf{r}_\gamma[N]\to\mathbf{r}_\delta[MN]$ was generated. It can then be assumed that this is always done.
By using the above remarks it is possible to prove the following lemma .

Lemma 1.13 [Min] Any sequent occurring in the cut-free derivation of the sequent $\mathbf{r}_{\alpha1}[x_1M_1],..,\mathbf{r}_{\alpha k}[x_kM_k]\vdash_S\mathbf{r}_\beta[M]$ is of the form $\mathbf{r}_{\gamma1}[x_1N_1],..,\mathbf{r}_{\gamma h}[x_hN_h]\vdash_S\mathbf{r}_\delta[M]$.

From the above facts it is easy to see that if we are interested only in provably derivable formulas in $\vdash_S$ then it it is enough to use a restriction of $\vdash_S$ where contexts are of the shape $\mathbf{r}_{\alpha1}[x_1N_1],..,\mathbf{r}_{\alpha h}[x_hN_h]$ and the rules used are only the ones for conjunction ($(\&\vdash_S)$ and $(\vdash_S\&)$) and implication ($(\to\vdash_S)$ and $(\vdash_S\to)$). Since we shall deal with type assignment systems let us give now, in the following definition, a notational variation, in a type assignment style, of this restriction of $\vdash_S$.

Definition 1.14 (*The system* $\vdash$)
(Ax) $B, xN:\alpha\vdash M:\alpha$ where $xN=_\beta M$

$$(\wedge\vdash)\dfrac{B,\ N:\alpha,\ N:\beta\vdash M:\rho}{B,\ N:\alpha\wedge\beta\vdash M:\rho}\qquad\qquad(\vdash\wedge)\dfrac{B\vdash M:\alpha\quad B\vdash M:\beta}{B\vdash M:\alpha\wedge\beta}$$

$$(\to\vdash)\dfrac{B\vdash Q:\alpha\quad B,\ zPQ:\beta\vdash M:\rho}{B,\ zP:\alpha\to\beta\vdash M:\rho}\qquad\qquad(\vdash\to)\dfrac{B,\ y:\alpha\vdash My:\beta}{B\vdash M:\alpha\to\beta}\ (*)$$

(*) $y\notin FV(B)$, $y\notin FV(M)$ ♪

The restrictions on the applicability of rule $(\vdash\to)$ are not actually true restrictions. The restriction $y\notin FV(B)$ corresponds to the restriction on the applicability of right-introduction of $\forall$ in $\vdash_S$. About $y\notin FV(M)$ it is possible to show then that it is always possible from a derivation of $B,y:\alpha\vdash My:\alpha$ to get a derivation of $B,x:\alpha,y:\alpha\vdash M[x/y]y:\alpha$.

Notice that in the type assignment just defined the terms in a basis can be not only variables but whole λ-terms. To distinguish the two cases we shall call *small* a basis whose terms are only variables.

Definition 1.15 (i) A context Γ for $\vdash_S$ is a r-context iff each element is of the form $r_\alpha[M]$.
(ii) Let Γ be a r-context.
$\Gamma^\bullet$ is the basis for $\vdash$ obtained from Γ in the following way : we replace $r_\gamma[P]$ by $P{:}\gamma$.
$B^\vee$ is the r-context obtained in the opposite direction from a basis B for $\vdash$. ♪

Lemma 1.16 Let Γ be a r-context. Then $\Gamma \vdash_S r_\alpha[M] \Leftrightarrow \Gamma^\bullet \vdash M : \alpha$.

Claim 1.17 (i) Let $\omega \notin \alpha$,B. Then $B \vdash_{\omega,\leq} M : \alpha \Rightarrow B \vdash M : \alpha$.
(ii) $B \vdash M : \alpha$ with B small $\Rightarrow B \vdash_{\omega,\leq} M : \alpha$.

From Claim 1.17 , which we shall prove in the next sub-section, and Lemmas 1.16 it is straightforward to obtain the following theorem, which formally states what we wished, i.e. $\vdash_{\omega,\leq}$ is a good formalization for the strong conjunction when ω does not appear in the types of assumptions and conclusions.

Theorem 1.18 Let Γ be a r-context with variables as terms and let B be a basis for $\vdash_{\omega,\leq}$ such that $\omega \notin$ B.
(i) $\Gamma \vdash_S r_\alpha[M] \Rightarrow \Gamma^\bullet \vdash_{\omega,\leq} M : \alpha$
(ii) If $\omega \notin \alpha$ then $B \vdash_{\omega,\leq} M : \alpha \Rightarrow B^\vee \vdash_S r_\alpha[M]$

It is worth to pointing out that it is possible to prove in a more direct way that $\vdash_N$ is equivalent to $\vdash_{\omega,\leq}$, without using Gentzen's sequent calculus and without introducing systems $\vdash_{\mathcal{M}}$ and $\vdash$. The introduction of $\vdash_{\mathcal{M}}$ is justified by the aim of comparing what is done in the present paper with what is done in [Min]. System $\vdash$ is a nice notational variation for the restricted $\vdash_S$ and can be viewed as a type inference system equivalent to $\vdash_{\omega,\leq}$ and given in a real Gentzen style.

Proof of Claim 1.17 .

Definition 1.19 (*Main branches*)
Let Π be a derivation in $\vdash$. A *main branch* of Π is a list of sequents $<S_1,..,S_n>$ such that :
(i) S_1 is the bottom sequent of the derivation or a minor premise of an application of rule $(\rightarrow\vdash)$ i.e. the left hand premise if the rule is applied in the same form as in definition 1.14 .
(ii) If S_i is the result of application of a rule in Π, then S_{i+1} is the premise of the rule if the rule is $(\vdash\rightarrow)$ or $(\wedge\vdash)$, one of the two premises if the rule is $(\vdash\wedge)$ and the major premise if the rule is $(\rightarrow\vdash)$. ♪

If $<S_1,...,S_n>$ is a main branch, then S_n is an axiom. By axioms of the main branches of a sequent in a derivation we mean the sequents S_n's of all the main branches the sequent belongs to.

<u>*Convention*</u> In a node $B \vdash M : \alpha$ of a derivation the term M will be referred to as the *subject* of the sequent.

Lemma 1.20 Let $B \vdash M : \alpha$. Then the axioms in the main branches of $M{:}\alpha$ have the subjects of the shape My.
Proof
In a main branch the form of the subjects do not change but in rules $(\vdash\rightarrow)$. Then if the subject at the bottom of a main branch is M, necessarily the subject at the top of this branch has the form My. ∮

Lemma 1.21 $B \vdash M : \alpha$ and $M =_\beta N \Rightarrow B \vdash N : \alpha$.
Proof
By Lemma 1.20 the subjects of the main branches of $M : \alpha$ in the derivation of $B \vdash M : \alpha$ are of the shape My. Since $M =_\beta N$ we have also $My =_\beta Ny$. We can then replace N for M in these axioms and it is straightforward to check that replacing N for M in the whole derivation we get a correct derivation of $B \vdash N : \alpha$. ∮

Lemma 1.22 $B \vdash_{\mathcal{M}} M : \alpha \;\Rightarrow\; B \vdash M : \alpha$.

Proof

By induction on the derivation of $B \vdash_{\mathcal{M}} M:\alpha$. We shall give only the non-trivial cases.

• Ax) Since B is small, an axiom has to be of the shape $B',x:\alpha \vdash_{\mathcal{M}} x:\alpha$. Then $B',x:\alpha \vdash x:\alpha$.

• $\wedge$E) $\dfrac{B \vdash_{\mathcal{M}} M : \alpha\wedge\beta}{B \vdash_{\mathcal{M}} M : \alpha}$

By induction hypothesis $B \vdash M:\alpha\wedge\beta$. By Lemma 1.16 $B^{\smile} \vdash_S r_{\alpha\wedge\beta}[M]$ and then by Lemma 1.4 $B^{\smile} \vdash_N r_{\alpha\wedge\beta}[M] \equiv r_\alpha[M]\&r_\beta[M]$.

&E) $\dfrac{B^{\smile} \vdash_N r_\alpha[M]\&r_\beta[M]}{B^{\smile} \vdash_N r_\alpha[M]}$

It is easy now to obtain $B \vdash M:\alpha$ by Lemmas 1.4 and 1.16.

• $\to$I) $\dfrac{B,\, x:\alpha \vdash_{\mathcal{M}} N:\beta}{B \vdash_{\mathcal{M}} \lambda x.N : \alpha\to\beta}$

By induction hypothesis $B,x:\alpha \vdash N:\beta$ and hence by Lemma 1.21 $B,x:\alpha \vdash (\lambda x.N)x:\beta$. Using rule $(\vdash\to)$ we obtain $B \vdash \lambda x.N:\alpha\to\beta$.

• $\to$E) $\dfrac{B \vdash_{\mathcal{M}} P : \alpha\to\beta \quad B \vdash_{\mathcal{M}} Q : \alpha}{B \vdash_{\mathcal{M}} PQ : \beta}$

By induction hypothesis $B \vdash P:\alpha\to\beta$ and $B \vdash Q:\alpha$.
By Lemmas 1.16 and 1.4 $B^{\smile} \vdash_N \forall x(r_\alpha[x]\to r_\beta[Px])$ and $B^{\smile} \vdash_N r_\alpha[Q]$. Then by $(\forall E)_N$, $(\to E)_N$ (the rules for the elimination of $\forall$ and $\to$ in $\vdash_N$), Lemmas 1.4 and 1.16 $B \vdash PQ:\beta$.

• eqβ) $\dfrac{B \vdash_{\mathcal{M}} N:\alpha \quad N =_\beta M}{B \vdash_{\mathcal{M}} M:\alpha}$

By induction hypothesis $B \vdash N:\alpha$. Then by Lemma 1.21 $B \vdash M:\alpha$. ∲

Proof of Claim 1.17 (i) Immediate by Lemmas 1.12 and 1.22. ∲

Lemma 1.23 Let $B \vdash M : \alpha$ with $B = \{P_i:\gamma_i\}$ such that there exist bases B_i such that $B_i \vdash_{\omega,\leq} P_i : \gamma_i$. Then $\bigcup_i B_i \vdash_{\omega,\leq} M : \alpha$.

Proof

By induction on the derivation of $B \vdash M : \alpha$.

• Ax) $B',xN:\alpha \vdash M:\alpha$ where $M =_\beta xN$.
By hypothesis there exists B" such that $B'' \vdash_{\omega,\leq} xN:\alpha$. Then the thesis follows from Lemma 1.9.

• $\wedge\vdash$) $\dfrac{B \vdash M : \alpha \quad B \vdash M : \beta}{B \vdash M : \alpha\wedge\beta}$

By the ind.hyp. we have B_1, B_2 such that $B_1 \vdash_{\omega,\leq} M : \alpha$ and $B_2 \vdash_{\omega,\leq} M : \beta$. If $B'=B_1\cup B_2$ the thesis follows by Remark 1.7 using rule $\wedge$I).

• $\vdash\wedge$) $\dfrac{B,\, P:\delta,\, P:\beta \vdash M : \alpha}{B,\, P:\delta\wedge\beta \vdash M : \alpha}$

We have that there exists B' such that $B' \vdash_{\omega,\leq} P : \delta\wedge\beta$. By rule $(\wedge E)$ it is possible to apply the induction hypothesis which yields the thesis.

• $\to\vdash$) $\dfrac{B \vdash Q : \delta \quad B,\, zPQ:\beta \vdash M : \alpha}{B,\, zP:\delta\to\beta \vdash M : \alpha}$

We have that there exists B_1 such that $B_1 \vdash_{\omega,\leq} zP:\delta\to\beta$. As it is possible to apply the induction hypothesis to $\Gamma \vdash Q : \delta$, there exists B_2 such that $B_2 \vdash_{\omega,\leq} Q:\delta$. The thesis follows from the induction hypothesis on Γ, $zPQ:\beta \vdash M : \alpha$, which holds since by $(\to E)$ $B_1\cup B_2 \vdash_{\omega,\leq} zPQ : \beta$.

• $\vdash\to$) $\dfrac{B,\, y:\alpha \vdash My : \beta}{B \vdash M : \alpha\to\beta}$

We can apply the induction hypothesis since $y : \alpha \vdash_{\omega,\leq} y : \alpha$. Then $\bigcup_i B_i,y:\alpha \vdash_{\omega,\leq} My : \beta$ and by rule $(\to E)$ $\bigcup_i B_i \vdash_{\omega,\leq} \lambda y.My : \alpha\to\beta$. By the condition on applicability of $(\vdash\to)$ $y \notin FV(M)$. By rule (η), which is admissible in $\vdash_\leq$ (and hence in $\vdash_{\omega,\leq}$), we get $\bigcup_i B_i, \vdash_{\omega,\leq} M : \alpha\to\beta$. ∲

Proof of Claim 1.17 (ii) Immediate from Lemma 1.23. ∲

Because of lack of space we have dropped in this version of the paper the result about the characterization of those formulas realizable by using the λI-calculus : if we restrict realizers to be λI-terms a good formalization for strong conjunction is $\vdash_{\leq}$ where terms are λI-terms.

2. Combinatory Logic Terms as Realizers.

Up to now we have taken λ-terms as realizer for formulas of the implicational propositional logic with strong conjunction. This choice however is not compulsory, since also Combinatory Logic is a theory of functions and then a good choice could be also considering CL-terms as realizers. Indeed this latter choice is the one made in [Lop-Esc].
We have seen that if $\omega \notin \alpha$ then $\vdash_S \mathbf{r}_\alpha[M] \Leftrightarrow \vdash_{\omega,\leq} M{:}\alpha$. What we are asking now is what is the right system for provable realizability if we use CL-terms.
Speaking of CL-terms we have however to decide which notion of conversion it is better to take, since there exist two reasonable notions of conversion among CL-terms. The first, the weak equality, enables Combinatory Logic to have the properties of combinatory completeness and then to be considered a good theory of function, without having to deal with the notion of bound variables. The advantage of this theory is that it is first order axiomatizable by

$$\mathbf{K}xy = x \qquad \mathbf{S}xyz = xz(yz).$$

We shall use $=_w$ as symbol for weak equality and CL_w to denote its theory.
The second notion of conversion, the combinatory β-equality, whose theory shall be denoted by CL_β and whose symbol is $=_{c\beta}$, enables a perfect correspondence among CL-terms and β-terms with β-equality. It is possible to define translations between CL- and λ-terms ($(\text{-})_\lambda$ from CL-terms to λ-terms and $(\text{-})_H$ in the opposite direction), such that $X =_{c\beta} Y$ iff $X_\lambda =_\beta Y_\lambda$. In what follows we shall assume $(\text{-})_H$ to be one of the most powerful translations, for istance H^η, as defined in [CF].
The combinatory β-equality is first order axiomatizable (we refer to [Bar] par.7.3 for the rather long list of axioms needed). The class of its models is however not the same as the class of models of λ-calculus but a larger one : the class of λ-algebras, where rule (ξ) (a principle of weak extensionality) does not hold.

CL-terms with Combinatory β-equality.

We consider in what follows provable realizability with CL-terms (with **I**, **K** and **S** as basic combinators) and combinatory β-equality among them.

The definition of a realizability formula $\mathbf{r}_\alpha[X]$ with CL terms is obviously the same as the one with λ-terms.
We shall use superscripts to stress the fact that we are using systems with CL-terms.

Definition 2.1 (*The system* $\vdash_S^{CL}$)
The system $CL_\beta\vdash_S^{CL}$ is the intuitionistic sequent calculus with equality where terms are CL-terms and equality is the combinatory β-equality. ♪

It is quite staightforward to check that the following Lemma holds.

Lemma 2.2 $\qquad CL_\beta\vdash_S^{CL} \mathbf{r}_\alpha[X] \Leftrightarrow \vdash_S \mathbf{r}_\alpha[X_\lambda].$

We now give the definition of type assignment systems for CL-term with intersection types devised by Dezani & Hindley in [DH].

Definition 2.3 (*The System* $\vdash_{\omega,\leq}^{CL}$)
The axioms schemes of $\vdash_{\omega,\leq}^{CL}$ are

$$B\vdash_{\omega,\leq}^{CL} \mathbf{I} : \sigma\to\sigma$$

$B\vdash^{CL}_{\omega,\leq}$ **K** $: \sigma\to\tau\to\sigma$

$B\vdash^{CL}_{\omega,\leq}$ **S** $: (\sigma\to\tau\to\rho)\to(\sigma\to\tau)\to\sigma\to\rho$.

Its rules are $(\to E)$, $(\wedge I)$, $(\wedge E)$, (ω) and $(\leq)$. ♪

Lemma 2.4 [DH] $\vdash^{CL}_{\omega,\leq} X{:}\alpha \;\Leftrightarrow \vdash_{\omega,\leq} X_\lambda{:}\alpha$.

The presence of rule (ω) enables $\vdash^{CL}_{\omega,\leq}$ to have rule (eq_β) admissible even for CL-terms.

Lemma 2.5 [DH] $B\vdash^{CL}_{\omega,\leq} X{:}\alpha$, $X{=}_{c\beta}Y \;\Rightarrow\; B\vdash^{CL}_{\omega,\leq} Y{:}\alpha$.

It turns out that $\vdash^{CL}_{\omega,\leq}$ is the right system for provable realizability with CL-terms and combinatory β-equality. This fact is easily proved using the results obtained in the previous section and the results of [DH].

Theorem 2.6 $CL_\beta\vdash^{CL}_{S} \mathbf{r}_\alpha[X] \Leftrightarrow \vdash^{CL}_{\omega,\leq} X{:}\alpha$ (with $\omega\notin\alpha$).

Proof

$$\begin{array}{ccc} CL_\beta\vdash^{CL}_{S} \mathbf{r}_\alpha[X] & & \vdash^{CL}_{\omega,\leq} X{:}\alpha \\ \text{Lemma 2.2} \Updownarrow & \text{Theorem 1.18} & \Updownarrow \text{Lemma 2.4} \\ \vdash_S \mathbf{r}_\alpha[X_\lambda] & \Leftrightarrow & \vdash_{\omega,\leq} X_\lambda{:}\alpha . \end{array}$$

∮

It is possible to use what we have obtained above to derive some nice semantical equivalences. Let us define first the notion of validity for type assignment systems.

Definition 2.7 [BCD] Let $\mathcal{M}$ be a λ-model (λ-algebra) yielded by an applicative structure $\langle D, \cdot \rangle$.
(i) If ξ is a valuation of variables in D, then $[\![P]\!]^{\mathcal{M}}_{\xi}$ is the interpretation of the λ-term (CL-term) P in $\mathcal{M}$ via ξ.
(ii) Let V be a valuation of type variable in subsets of the domain D. Then the interpretation of a type σ in $\mathcal{M}$ via V is defined as follows.

$[\![\varphi]\!]^{\mathcal{M}}_{V} = V(\varphi)$ φ a type variable $\qquad [\![\omega]\!]^{\mathcal{M}}_{V} = D$

$[\![\sigma\to\tau]\!]^{\mathcal{M}}_{V} = \{d\in D \mid \forall e\in [\![\sigma]\!]^{\mathcal{M}}_{V}\ \ d\cdot e\in [\![\tau]\!]^{\mathcal{M}}_{V}\}$ $\qquad [\![\sigma\wedge\tau]\!]^{\mathcal{M}}_{V} = [\![\sigma]\!]^{\mathcal{M}}_{V} \cap [\![\tau]\!]^{\mathcal{M}}_{V}$.

(iii) $\mathcal{M},\xi,V \models P{:}\sigma$ iff $[\![P]\!]^{\mathcal{M}}_{\xi} \in [\![\sigma]\!]^{\mathcal{M}}_{V}$. $\mathcal{M},\xi,V \models B$ iff $\mathcal{M},\xi,V \models x{:}\sigma$ for all $x{:}\sigma\in B$

$B\models P{:}\sigma$ iff for all $\mathcal{M},\xi,V \models B$ $\mathcal{M},\xi,V \models P{:}\sigma$.

We shall denote by $CL_\beta\models X{:}\alpha$ (with X a CL-term) the semantic validity of $X{:}\alpha$ in all λ-algebras. ♪

System $\vdash_{\omega,\leq}$ is complete with respect to this notion of validity.

Lemma 2.8 [BCD] $B\models M : \alpha \;\Leftrightarrow\; B\vdash_{\omega,\leq} M : \alpha$.

By definition of realizability formula it is straightforward to see that the following fact holds :

$$CL_\beta\models \mathbf{r}_\alpha[X] \;\Leftrightarrow\; CL_\beta\models X{:}\alpha$$

where the symbol $\models$ to the left of the double implication denotes of course the validity in first order models.
Then by Gödel's completeness theorem and Theorem 2.6 the following theorem descends easily.

Theorem 2.9 $\vdash^{CL}_{\omega,\leq} X{:}\alpha \;\Leftrightarrow\; CL_\beta\models X{:}\alpha$.

This theorem can then be seen as a nice use of the fact that CL is first order.
Remark As pointed out by a referee the result above follows even by standard arguments.

It is possible to show now that for a CL-term X $X{:}\alpha$ is true in all λ-algebras iff its translation is true in all λ-models.

Corollary 2.10 $\quad CL_\beta \models X:\alpha \quad \Leftrightarrow \quad \models X_\lambda:\alpha$.
Proof

$$\begin{array}{ccccc} & CL_\beta \models X:\alpha & & \models X_\lambda:\alpha & \\ \text{Lemma 2.9} & \Uparrow\Downarrow & \text{Lemma 2.4} & \Uparrow\Downarrow & \text{Lemma 2.8} \\ & \vdash^{CL}_{\omega,\leq} X:\alpha & \Leftrightarrow & \vdash_{\omega,\leq} X_\lambda:\alpha\,. & \end{array}$$

∲

Let us see now that the same hold for λ-terms and for the translation $(\text{-})_H$.

Lemma 2.11 $\quad \vdash_S r_\alpha[M] \quad \Leftrightarrow \quad CL_\beta \vdash^{CL}_S r_\alpha[M_H]$.
Proof
$\Rightarrow$) We have $\vdash M:\alpha$ and then the axioms of the main branch containing $M:\alpha$ are of the form $\Gamma,P:\gamma \vdash My:\gamma$ with $P =_\beta My$. Then the following is a correct axiom : $CL_\beta + \Gamma_H, P_H:\gamma \vdash^{CL} (My)_H:\gamma$, since $P_H =_{c\beta} (My)_H$ and $(My)_H \equiv M_H y$. It is therefore quite staightforward to transform a derivation of $\vdash_S r_\alpha[M]$ in one of $CL_\beta \vdash^{CL}_S r_\alpha[M_H]$.
$\Leftarrow$) Similarly. ∲

Theorem 2.12 $\quad CL_\beta \models M_H:\alpha \quad \Leftrightarrow \quad \models M:\alpha$.
Proof

$$\begin{array}{ccccccc} & CL_\beta \models M_H:\alpha & & & & \models M:\alpha & \\ \text{Gödel's completeness} & \Uparrow\Downarrow & \text{Lemma 2.11} & & \text{Theorem 1.18} & \Uparrow\Downarrow & \text{Lemma 2.8} \\ & CL_\beta \vdash^{CL}_S r_\alpha[M_H] & \Leftrightarrow & \vdash_S r_\alpha[M] & \Leftrightarrow & \vdash_{\omega,\leq} M:\alpha\,. & \end{array}$$

∲

It is worth noting that the proof of Theorem 2.12 it is also a proof of the semantical completeness of $\vdash_S$, completeness which it is not an istance of Gödel's completeness, like for $\vdash^{CL}_S$, since terms are λ-terms and the equality is not first order.

We conclude by pointing out that it is not difficult to modify the systems described above in order to take into account also the case of weak equality on CL-term instead of combinatory β-equality.

Acknoledgments
We are grateful to Mariangiola Dezani for her gentle guidance, to Furio Honsell for useful suggestions and to an anonymous referee for helpful comments. We wish to thank also Pia Barbanera for her support.

References
[Bar] H.Barendregt. *Lambda Calculus : its Sintax and Semantics* . North Holland 1984.
[BCD] H.Barendregt, M.Coppo, M.Dezani. A Filter Lambda Model and the Completeness of Type Assignment. *Journal of Simbolic Logic*, 48, 931-940. 1983.
[CD] M.Coppo, M.Dezani. An extension of basic functionality theory for lambda-calculus. *Notre Dame Iournel of formal Logic* , 21, 685-693. 1980.
[CDHL] M.Coppo,M.Dezani,F.Honsell,G.Longo. Extended type structures and filter lambda models. *Proceedings of Logic Colloquium '82* (eds. G.Lolli, G.Longo, A.Marcja), 241-262. North-Holland. 1984.
[CF] H.B.Curry, R.Feys. *Combinatory Logic* . Vol.I. North Holland. 1958.
[DH] M.Dezani, R.Hindley. Intersection Types for Combinatory Logic. *J.W.de Bakker, 25 Jaar Semantiek* edited by J.W.Klop et al., published by Centrum voor Wiskunde, Amsterdam. 1989.
An expanded version to appear in *Teoretical Computer Science* .
[HS] R.Hindley, J.P.Seldin. *Introduction to Combinators and λ-calculus*. Cambridge University Press. 1986.
[Kle] S.C.Kleene. Permutability of Inferences in Gentzen's Calculi LK and LJ. *Memoirs of the American Mathematical Society*. 10, 1-26. 1952.
[Lop-Esc] E.G.K.Lopez-Escobar. Proof Functional Connectives. LNM, 1130,208-221. 1985.
[Min] G.E.Mints. The Completeness of Provable Realizability. *Notre Dame Journal of Formal Logic*. 30. 420-441. 1989.
[Min a] G.E.Mints. Personal communication, 1991.
[Pot] G.Pottinger. A Type Assignment for the Strongly Normalizable λ-terms. In *To H.B.Curry : Essays in Combinatory Logic, Lambda Calculus and Formalism* (ed. Seldin and Hindley). 561-577. Academic Press. 1980.

PARTIAL HIGHER-ORDER SPECIFICATIONS

Egidio Astesiano and Maura Cerioli
Dipartimento di Matematica - Università di Genova
Via L.B. Alberti 4 - 16132 Genova Italy
e-mail {Astes,Cerioli}@igecuniv.bitnet

Abstract. *In this paper we study the classes of extensional models of higher-order partial conditional specifications. After investigating the closure properties of these classes, we show that an inference system for partial higher-order conditional specifications, which is equationally complete w.r.t. the class of all extensional models, can be obtained from any equationally complete inference system for partial conditional specifications. Then, applying some previous results, we propose a deduction system, equationally complete for the class of extensional models of a partial conditional specification.*
Finally, turning the attention to the special important case of term-extensional models, we first show a sound and equationally complete inference system and then give necessary and sufficient conditions for the existence of free models, which are also free in the class of term-generated extensional models.

Introduction

Higher-order functions are now recognized as an important tool for the modular development of correct software systems. When the systems are developed by refinement from abstract specifications, it is rather natural to start with algebraic higher-order specifications. This higher-order approach, more or less explicit in the CIP method, is now at the basis of more recent projects, for example PROSPECTRA (see [K-B], also for further references).

Because of this solid practical motivation, a lot of work has been devoted recently to higher-order algebraic specifications (see, e.g. [M, MTW1, MTW2, Me, Q]). However little or nothing has been done for the algebraic specification of partial higher-order functions. Still in almost all real applications partial functions arise, especially in the abstract specification phase, when many details, including error messages, are not fixed.

This paper is an attempt at clarifying some basic hot points about partial higher-order specifications. We concentrate our attention on three classical issues: closure properties of the model classes, equational deduction and existence of free (and initial) models. It will turn out that in the partial case the situation is much more delicate then in the total one (see [Me] and [Q] for the same issues).

Following a classical approach (also adopted in [MTW1] and [Me], [Q]), we reduce higher-order specifications to special first-order specifications by introducing *apply* functions, one for each functional sort, and by restricting the class of models to *extensional* models, ie to the models where two functions giving the same result over any possible input are equal. So we are investigating classes of partial algebras satisfying not only the explicit axioms, but also the implicit *extensionality* axiom

$$* \qquad \forall\ f,\ g\ \ ((\forall\ x{:}\ apply(f,x) = apply(g,x)) \supset f = g).$$

The extensionality axiom differs from the usual positive conditional axioms (the most frequently used in algebraic approaches, being the widest class guaranteeing the existence of initial (free) models, see e.g. [T,W,BW,B,R]) because of the internal quantification, which only involves the premises, and since the equality $f(x) = g(x)$ in the premises is *strong*, ie it holds iff either both sides are undefined or both denote the same element of the algebra, and not *existential*, ie holding iff both sides are defined and denote the same element.

In section 1 we show that, as in the total case, *the model class is not closed w.r.t. subobjects*; but, differently from the total case, *the closure w.r.t. products is lost*, too, so that *in general a partial conditional higher-order specification does not admit free models for any family of variables of arbitrary cardinality.*

In the total case an equationally complete inference system may be obtained by adding to an equationally-complete first-order system some extensionality rules, one for each arity; for unary functions it takes the form

This work has been partially supported by Esprit-BRA-WG COMPASS and by MURST 40%

$$\frac{f(x) = g(x)}{f = g} \qquad \text{f, g terms of functional sort , x variable} \notin Var(f) \cup Var(g)$$

The proof of the completeness of the enriched system (see e.g.[Me]) relies on the existence of free objects for a sufficiently high cardinality of the family of variables.

Instead in the partial case *an equationally complete inference system enriched by the extensionality rules is not equationally complete* (see fact. 2.4); but we get (section 2) an equationally complete system for the extensional models of any partial higher-order conditional specification starting from an equationally complete system for (first-order) partial conditional specifications by using a skolemization procedure.

In computer science a special role is played by term-generated (or reachable) models; thus (section 3) we analyze the subclass of extensional term-generated models of a partial conditional higher-order specification. Since every element of a term-generated algebra A is the evaluation of a term, the extensionality axiom holds in A iff A satisfies the following axiom, which, note, is an infinitary (countable) conditional axiom:

$$** \qquad \{f(t) = g(t) \mid t \in T_\Sigma\} \supset f = g,$$

The axiom **, called "term-extensionality axiom", characterizes the class of term-extensional algebras, which is smaller than the class of extensional algebras but includes all term-generated extensional models so that the subclass of all term-generated extensional models coincides with the subclass of all term-generated term-extensional models.

Since the class of all term-extensional models is the model class of a partial conditional specification, on the basis of some results in [AC1,AC2], we may obtain *an equationally complete system* for this class, together with *necessary and sufficient conditions for the existence of free models in the class of term-extensional models.*

1 Extensional models

We reduce, as in [MTW1], higher-order specifications to particular classes of first-order specifications and consider the class of extensional models of a conditional specification Sp, which in general does not admit initial nor free models. Since the class of extensional models is not closed under sub-objects, the usual correspondence between the existence of an initial model in the whole model class and in the subclass of the term-generated models is missing.

1.1 Partial higher-order specifications

We assume known the notions of signature, partial algebra and evaluation of terms (see e.g. [B, R, AC1]).

Partial conditional specifications. Let $\Sigma = (S,F)$ be a signature; the homomorphisms, as quite standard in the initial approach (see e.g. [B] and total Σ-homomorphisms in [BW]), are chosen in a way that the initial model, if any, satisfies the conditions of *no-junk* and *no-confusion*; let us recall the definition.
Let A and B be two Σ-algebras and p be a family of total functions $p = \{p_s\}_{s\in S}$, s.t. $p_s: s^A \to s^B$. Then p is a *homomorphism* from A into B iff for any $op \in F_{(s_1\dots s_n, s_{n+1})}$, $n \geq 0$, and any $a_i \in s_i^A$, $i=1\dots n$,

$$op^A(a_1,\dots,a_n) \in s_{n+1}^A \text{ implies } p_{s_{n+1}}(op^A(a_1,\dots,a_n)) = op^B(p_{s_1}(a_1),\dots,p_{s_n}(a_n)).$$

Let X be a family of S-sorted variables.

- The set of all *elementary formulas* over Σ and X, denoted by $EForm(\Sigma,X)$, consists of
  $$EForm(\Sigma,X) = \{D(t) \mid t \in T_\Sigma(X)|_s, s \in S\} \cup \{t = t' \mid t,t' \in T_\Sigma(X)|_s, s \in S\}$$
  where D is called the *definedness predicate*.
- We denote by $D(X)$ the set $\{D(x) \mid x \in X\}$.
- The set of all *conditional formulas* over Σ and X, denoted by $CForm(\Sigma,X)$, consists of
  $$CForm(\Sigma,X) = \{\Delta \supset \varepsilon \mid \Delta \subseteq EForm(\Sigma,X), \varepsilon \in EForm(\Sigma,X)\}.$$
  If Δ is the empty set, then $\Delta \supset \varepsilon$ is an equivalent notation for the elementary formula ε.
- A *positive conditional formula* over Σ and X is a conditional formula $\Delta \supset \varepsilon \in CForm(\Sigma,X)$ s.t. for every $(t = t') \in \Delta$, $D(t) \in \Delta$ or $D(t') \in \Delta$.
- If $A \in PA(\Sigma)$, $\varphi \in CForm(\Sigma,X)$ and V is a (total) valuation for $Var(\varphi)$ (the S-sorted family of the variables occurring in φ) in A, then $A \models_V \varphi$ is defined by:
  - $A \models_V D(t)$ iff $t^{A,V} \in s^A$; $A \models_V t = t'$ iff either $t^{A,V}, t'^{A,V} \notin s^A$ or $t^{A,V}, t'^{A,V} \in s^A$ and $t^{A,V} = t'^{A,V}$;

- let φ be $\Delta \supset \varepsilon$; then $A \models_V \varphi$ iff $A \models_V \varepsilon$, or $A \not\models_V \delta$ for some $\delta \in \Delta$;

  We write $A \models \varphi$ for a formula φ and say that φ *holds* in (equivalently: is *satisfied by, is valid in*) A iff $A \models_V \varphi$ for all valuations V for $Var(\varphi)$ in A.
- If $C \subseteq PA(\Sigma)$ and $\varphi \in CForm(\Sigma,X)$, then $C \models \varphi$, iff $A \models \varphi$ for all $A \in C$. □

Remarks.

1 The equality that we consider is the so called *strong equality*, as opposed to the *existential equality* $t =_e t'$ ($A \models_V t =_e t'$ iff $t^{A,V}$ and $t'^{A,V}$ are both defined and equal). Note that $D(t)$ stands for $t =_e t$.

2 The above notion of validity is the usual one in the many-sorted case; however some comments can be helpful. If $Var(\varphi)_s \neq \varnothing$ and $s^A = \varnothing$, then $A \models \varphi$ holds; hence for any $C \subseteq PA(\Sigma)$, $C \models \varphi$ iff $A \models \varphi$ for all $A \in C$ s.t. $Var(\varphi)_s \neq \varnothing$ implies $s^A \neq \varnothing$. Thus if C contains an algebra with all supports non-empty (as it will always happens in the sequel), then the notion of validity for the class coincides with the classical one; for example we could not have both $C \models \varphi$ and $C \models \neg\varphi$ (but note that here we do not have negation). Finally it is also useful to emphasize that here we can stay within a two-valued logic, since a conditional formula for a total valuation of the variables is always either true or false.

3 Note that $A \models \varepsilon$ implies $A \models D(X) \supset \varepsilon$ for any X; if $X \subseteq Var(\varepsilon)$, then also the converse holds. Moreover, since $A \models D(x)$ for any variable x and any A, the presence of $D(x)$ in $D(X) \supset \varepsilon$ has the only effect of possibly increasing the variables of the formula and thus the domain of the variable valuations.

Def. 1.1.

- A *(positive) conditional specification* consists of a signature Σ and of a set Ax of (positive) conditional formulas over Σ. A generic conditional specification will be denoted by Sp; the formulas belonging to Ax are called the *axioms* of Sp and denoted by (possibly decorated) α.
- For any conditional specification $Sp = (\Sigma, Ax)$, $Mod(Sp) = \{A \mid A \in PA(\Sigma), A \models \alpha \ \forall\ \alpha \in Ax\}$; an algebra $A \in Mod(Sp)$ is called a *model* of Sp. The class $GMod(Sp)$ consists of all term-generated models of Sp. □

For any conditional specification Sp the class $Mod(Sp)$ is not empty, since the trivial (total) algebra, with singleton sets as carriers and the obvious (total) interpretation of function symbols, is always a model.

Higher-order specifications. We define higher-order specifications as a special class of first-order specifications.

Def. 1.2.

- If S is a set of *basic sorts*, then the set $S^{\rightarrow}$ of *functional sorts* over S is inductively defined by: $S \subseteq S^{\rightarrow}$ and if $s_1,\ldots,s_n,s_{n+1} \in S^{\rightarrow}$, then $s = (s_1 \times \ldots \times s_n \rightarrow s_{n+1}) \in S^{\rightarrow}$ for all $n \geq 1$.

  A subset $S' \subseteq S^{\rightarrow}$ is *downward-closed* iff $s_1,\ldots,s_n,s_{n+1} \in S'$ for all $(s_1 \times \ldots \times s_n \rightarrow s_{n+1}) \in S'$.
- A *higher-order signature* $F\Sigma$, from now on h.o. signature, is a signature (S,F), where S is a downward-closed set of functional sorts, s.t. for all $s = (s_1 \times \ldots \times s_n \rightarrow s_{n+1}) \in S$ with $n \geq 1$ there exists a distinguished operator $apply_s \in F_{(s\ s_1 \ldots s_n, s_{n+1})}$. We will often use the infix notation for the $apply_s$ operators, ie we will write $f(a_1,\ldots,a_n)$ for $apply_s(f,a_1,\ldots,a_n)$, dropping the sort indexes when there is no ambiguity. Moreover we will not explicitly mention the apply functions in the definitions of concrete functional signatures.
- Let $F\Sigma = (S,F)$ be a h.o. signature; then $A \in PA(F\Sigma)$ is an *extensional partial algebra* iff satisfies the following *extensionality condition*:

  for all $s = (s_1 \times \ldots \times s_n \rightarrow s_{n+1}) \in S$, with $n \geq 1$ and for all $f,g \in s^A$,

  if for all $a_i \in s_i^A$, $i=1,\ldots,n$, $f(a_1,\ldots,a_n) = g(a_1,\ldots,a_n)$, then $f = g$.

  An extensional partial algebra is called an E-algebra. We denote by $EPA(F\Sigma)$ the class of all E-algebras on $F\Sigma$.
- A *(positive) conditional higher-order specification* $(P)FSp = (F\Sigma, Ax)$ consists of a higher-order signature $F\Sigma$ and a set Ax of (positive) conditional axioms over $F\Sigma$.

  A generic (positive) higher-order specification will be denoted by $(P)FSp$. The class of *extensional models* of FSp, denoted by $EMod(FSp)$, is $Mod(FSp) \cap EPA(F\Sigma)$; while $EGMod(FSp)$ is the class of extensional term-generated models, i.e. $GMod(FSp) \cap EPA(F\Sigma)$. □

Note that for any h.o. signature $F\Sigma = (S,F)$, S is required to be downward closed in order that the operators $apply_s$ have arity in $S^* \times S$.

Remark. Any $A \in EPA(F\Sigma)$ is isomorphic to an algebra where the carriers of higher-order sort $(s_1 \times \ldots \times s_n \to s_{n+1})$ are subsets of the space of the partial functions from $s_1^A \times \ldots \times s_n^A$ into s_{n+1}^A and the $apply_s$ operators are interpreted in the standard way. Therefore in the following examples we assume that the higher-order carriers are function spaces and that the $apply_s$ functions are interpreted accordingly.

1.2 Counter-examples

Def. 1.3. Let A be a partial algebra on a signature $\Sigma = (S,F)$; then a Σ-algebra B is a *subalgebra* (regular subobject) of A iff $s^B \subseteq s^A\ \forall s \in S$ and op^B is the restriction of op^A to $s_1^B \times \ldots \times s_n^B\ \forall\ op \in F_{(s_1 \ldots s_n, s)}$. □

Analogously to the case of higher-order *total* algebras, the class of all extensional algebras is not closed w.r.t. subobjects so that, in particular, the class of all extensional algebras cannot be expressed as the model class of a conditional specification, because the model class of any conditional specification is closed under subobjects (see e.g. [AC2] prop.1.3). But while in the total case the extensional algebras are closed w.r.t. non-empty direct products (of course performed in the class of *all* algebras), as claimed for example by the theorem 5.3 in [Me], in the partial frame also this closure is missing.

Fact 1.4. Let $F\Sigma$ be a h.o. signature; in general $EPA(F\Sigma)$ *is not closed w.r.t. subobjects, nor w.r.t. non-empty direct products.*

Proof. Consider the following example.

Signature $F\Sigma$ Sorts: s, $(s \to s)$ Operations: $f,g: \to (s \to s)$

Consider the algebras A, B and C, defined by:

$s^A = \{\bullet\}$; $(s \to s)^A = \{\bot, Id\}$, $\bot(\bullet)$ is undefined, $Id(\bullet) = \bullet$, $f^A = Id$; $g^A = \bot$;
$s^B = s^A$; $(s \to s)^B = (s \to s)^A$ $f^B = \bot$; $g^B = Id$;
$s^C = \emptyset$; $(s \to s)^C = (s \to s)^A$ $f^C = f^A$; $g^C = g^A$.

Then obviously $A,B \in EPA(F\Sigma)$, while $C \notin EPA(F\Sigma)$ and C is a subalgebra of A, by definition. Therefore $EPA(F\Sigma)$ is not closed w.r.t. subobjects.

Let us define $A \times B$: $s^{A\times B} = \{(\bullet,\bullet)\}$; $(s \to s)^{A\times B} = (s \to s)^A \times (s \to s)^B$; $f^{A\times B} = (f^A, f^B)$; $g^{A\times B} = (g^A, g^B)$.

Thus $(s \to s)^{A\times B}$ has cardinality four, while there are just two distinct partial functions, the identity and the totally undefined function, from $s^{A\times B}$ into $s^{A\times B}$, because $s^{A\times B}$ has cardinality one. Therefore $A\times B \notin EPA(F\Sigma)$ and hence $EPA(F\Sigma)$ is not closed w.r.t. non-empty direct products. □

Def. 1.5. Let X be a family of variables and C be a class of Σ-algebras. A couple (Fr, m), where $Fr \in C$ and m is a valuation for X in Fr, is *free* over X in C iff

$\forall A \in C, \forall V: X \to A$ there exists a unique homomorphism p_V: $Fr \to A$ s.t. $p_V(m(x)) = V(x)\ \forall\ x \in X$.

An algebra I is initial in C iff it is free over the empty family of variables in C, ie iff $I \in C$ and for all $A \in C$ there exists a unique homomorphism from I into A. □

It is easy to see that initial and terminal algebras in $PA(F\Sigma)$ are also extensional and hence initial and terminal in $EPA(F\Sigma)$; but, although $EPA(F\Sigma)$ has an initial model, in general both the class of all extensional models and the class of all term-generated models for equational specifications have no initial model.

Fact 1.6. Let $F\Sigma = (S,F)$ be a higher-order signature and FSp be an equational specification $(F\Sigma, Ax)$. Then *in general there does not exist an E-algebra initial in* EMod(FSp) *nor in* EGMod(FSp).

Proof. Consider the following example.

Specification FSp_1

Sorts: s, $(s \to s)$ Operations: $e: \to s$ Axioms: α_1 D(f)
$f,g: \to (s \to s)$ α_2 D(g)

Then proceed by contradiction assuming that there exists I initial in $EMod(FSp_1)$ (resp. in $EGMod(FSp_1)$).

Let F and G be the E-algebras defined by:

$s^F = \{\bullet\}$; $(s \to s)^F = \{\bot, Id\}$, $\bot(\bullet)$ is undefined, $Id(\bullet) = \bullet$ $e^F = \bullet$ $f^F = Id$; $g^F = \bot$
$s^G = s^F$; $(s \to s)^G = (s \to s)^F$ $e^G = \bullet$ $f^G = \bot$; $g^G = Id$.

Both F and G belong obviously to $EGMod(FSp_1)$; thus, because of the initiality of I, there exist two homomorphisms $p^F: I \to F$ and $p^G: I \to G$. Then it is just routine to show that the existence of such p^F and p^G implies that for all $a \in s^I$ both $f^I(a)$ and $g^I(a)$ are undefined and hence that $f^I = g^I$, because of extensionality; thus we get $g^F = p^F(g^I) = p^F(f^I) = f^F$, in contradiction with the definition of f^F and g^F. □

The above example suggests that for the existence of the initial model, the minimal *definedness* may conflict with the minimal *equality*. Indeed if the elements in the domain are too few, then we cannot distinguish the functions and hence the minimal definedness (on the arguments) may force the *maximal* equality (on the functions). For the same reason we have that two functions having the same result over every tuple of terms because of the axioms, may differ on some *non-term-generated* argument-tuple, so that the equalities between ground terms holding in the term-generated models may be strictly more than the equalities holding in all models. In particular the equalities between ground terms holding in all the term-generated models may define an extensional algebra, so that there exists an initial model in EGMod(FSp), while the equalities between ground terms holding in all models are too few.

Fact 1.7. Let $F\Sigma = (S,F)$ be a h.o. signature and FSp be an equational specification $(F\Sigma,Ax)$ s.t. I is initial in EGMod(FSp). Then *in general* I *is not initial in* EMod(FSp) and the sets $\{\varepsilon \mid \varepsilon \in EForm(F\Sigma,\emptyset), EMod(FSp) \models \varepsilon\}$ and $\{\varepsilon \mid \varepsilon \in EForm(F\Sigma,\emptyset), EGMod(FSp)) \models \varepsilon\}$ are different.

Proof. Consider the following example.

Specification FSp_2

Sorts: $s_1, s_2, (s_1 \to s_2)$ Operations: $e: \to s_1$; $f,g: \to (s_1 \to s_2)$ Axioms: α_1 $D(f(e))$; α_2 $f(e) = g(e)$

Then all term-generated models are isomorphic to I, defined by:

$s_1{}^I = \{\bullet\}$; $s_2{}^I = \{\bullet\}$; $(s_1 \to s_2)^I = \{Id\}$, where $Id(\bullet) = \bullet$; $e^I = \bullet$; $f^I = g^I = Id$.

So that I is initial in $EGMod(FSp_2)$; however I is not initial in $EMod(FSp_2)$, since there are (no term-generated) models A for which $f^A \neq g^A$. Moreover $EGMod(FSp_2) \models f = g$, while $EMod(FSp_2) \not\models f = g$, because $A \not\models f = g$. □

In the total case if a family X of variables has a sufficiently high cardinality, then there exists the free model on X in the class of all extensional models of a conditional specification (see theorems 3.7 and 5.7 of [Me]). Instead in the partial case there are conditional specifications whose classes of extensional models do not admit free models for families of variables of arbitrary cardinality.

Fact 1.8. Let $F\Sigma = (S,F)$ be a h.o. signature, FSp be an equational specification $(F\Sigma,Ax)$ and X be a family of variables of arbitrary cardinality. Then *in general there does not exist a free model for* X *in* EMod(FSp).

Proof. Consider again the specification FSp_1 and the algebras F and G defined in fact 1.6; we show that there does not exist a free model for X in $EMod(FSp_1)$.

Assume by contradiction that (I,m) is free in $EMod(FSp_1)$ for a family X of variables. Let $V^F: X \to F$ and $V^G: X \to G$ be any valuations, which always exist, because F and G have all the carriers non-empty. Because of the freeness of I, there exist two homomorphisms $p^F: I \to F$ and $p^G: I \to G$ s.t. $p^F \cdot m = V^F$ and $p^G \cdot m = V^G$. Thus, as in in fact 1.6, we get $g^F =_e p^F(g^I) =_e p^F(f^I) =_e f^F$, in contradiction with the definition of f^F and g^F. □

Note that the above counter-example also applies to the subclass of extensional models generated by the family X of variables, $EGMod(FSp,X) = \{A \mid A \in EMod(FSp), \exists\, V:X \to A \text{ s.t. } eval^{A,V}(T_\Sigma(X)) = A\}$, because F and G, being term-generated, belong to EGMod(FSp,X).

2 Equational deduction

The focus of logic deduction in the total algebraic case is on *equational* deduction, because an inference system complete w.r.t. the equations gives the (initial) free model. In the partial case only the definedness and the equality between defined terms (ie existential equalities) are needed in order to characterize the (initial) free model, if any (see eg [BW]). Here we also need to consider conditional axioms with strong equalities in the premises and hence we deal also with strong equalities. Moreover our deduction is sound and complete not only w.r.t. equalities, but also w.r.t.

formulas of the form $D(X) \supset \varepsilon$ which corresponds to the formula $\forall\, (X \cup Var(\varepsilon)).\ \varepsilon$, with explicit quantification, considered in the many sorted total case (see [MG]) both for clarifying equational deduction and for obtaining models free w.r.t. a family of variables. Hence we give notions of soundness and completeness also dealing with such particular conditional formulas; our notions subsume the usual ones only dealing with equalities.

Def. 2.1. Let Sp be a conditional specification, C a subclass of Mod(Sp) and L an inference system.
L is *sound* for C iff $L \vdash \varphi$ implies $C \models \varphi$ for all conditional formulas φ.
L is *strongly complete* (for complete w.r.t. strong equalities) for C and a family of variables X iff for all $\varepsilon \in EForm(\Sigma,X)$ and all $Y \subseteq X$ $\quad C \models D(Y) \supset \varepsilon$ implies $L \vdash D(Y) \supset \varepsilon$. □

Note that if L is strongly complete for C and X, then in particular $C \models \varepsilon$ implies $L \vdash \varepsilon$.

In the total case (see [Me,Q]) a complete system for the class of extensional models may be obtained by enriching a complete system for the whole class of algebras by extensionality rules, one for each arity; for example for unary functions the rule takes the form

$$* \qquad \frac{f(x) = g(x)}{f = g} \qquad f, g \text{ terms of functional sort } s \to s',\ x \in X_s,\ x \notin Var(f) \cup Var(g)$$

Instead in the partial case the above rule * is insufficient to achieve a complete system. To propose an example of this claim and also for further use let us recall the definition of the system CL(Sp), from [AC2], which is sound and strongly complete (see theorem 3.11 in [AC2]).

Def. 2.2. The CL(Sp) system for a conditional specification $Sp = (\Sigma, Ax)$ consists of the axioms Ax and of the following axiom schemas and inference rules, where we assume that as usual $\varepsilon \in EForm(\Sigma,X)$, $\Delta, \Delta_\gamma, \Gamma, \Theta_j, \Gamma_j$ are countable subsets of $EForm(\Sigma,X)$, $x \in X$ and $t, t', t'', t_i, t'_i, t_x \in T_\Sigma(X)$.

0 $\quad D(x)$ — *Definedness of variables*

1 $\quad t = t$ — *Congruence*

2 $\quad t = t' \supset t' = t$

3 $\quad \{t = t', t' = t''\} \supset t = t''$

4 $\quad \{t_i = t'_i \mid i=1\ldots n\} \supset op(t_1,\ldots,t_n) = op(t'_1,\ldots,t'_n)$

5 $\quad D(op(t_1,\ldots,t_n)) \supset D(t_i)$ — *Strictness*

6 $\quad \{D(t), t = t'\} \supset D(t')$ — *Definedness and equality*

7 $\quad \dfrac{\Delta \cup \Gamma \supset \varepsilon,\ \{\Delta_\gamma \supset \gamma \mid \gamma \in \Gamma\}}{D(Var(\Gamma) - Var(\cup_{\gamma \in \Gamma} \Delta_\gamma \supset \varepsilon)) \cup \Delta \cup (\cup_{\gamma \in \Gamma} \Delta_\gamma) \supset \varepsilon}$ — *Modus Ponens*

8 $\quad \dfrac{\Delta \supset \varepsilon}{\{D(t_x) \mid x \in X\} \cup \{\delta[t_x/x \mid x \in X] \mid \delta \in \Delta\} \supset \varepsilon[t_x/x \mid x \in X]}$ — *Instantiation/Abstraction*

9 $\quad \dfrac{\{\Theta_j \cup \Gamma_j \supset \varepsilon \mid j \in J\}}{D(\cup_{j \in J} Var(\Gamma_j)) \cup (\cup_{j \in J} \Theta_j) \supset \varepsilon}$ — *Elimination* $\quad \forall\, \{\gamma_j\}_{j \in J}$ with $\gamma_j \in \Gamma_j\ \exists\, t, t'$ s.t. $D(t), D(t'), t = t' \in \{\gamma_j\}_{j \in J}$.

If the axioms of Sp are finitary (only a finite number of elementary formulas in the premises), then rules 7, 8 and 9 can be replaced by the rules

$7_f \quad \dfrac{\Delta \cup \{\gamma\} \supset \varepsilon,\ \Delta_\gamma \supset \gamma}{D(Var(\gamma) - Var(\Delta_\gamma)) \cup (\Delta \cup \Delta_\gamma) \supset \varepsilon} \qquad 8_f \quad \dfrac{\Delta \supset \varepsilon}{\{D(t)\} \cup \{\delta[t/x] \mid \delta \in \Delta\} \supset \varepsilon[t/x]}$

$9_f \quad \dfrac{\Delta_1 \cup \{D(t)\} \supset \varepsilon,\ \Delta_2 \cup \{D(t')\} \supset \varepsilon,\ \Delta_3 \cup \{t = t'\} \supset \varepsilon}{D(Var(t = t')) \cup (\Delta_1 \cup \Delta_2 \cup \Delta_3) \supset \varepsilon}$

where all sets of elementary formulas are finitary and in this case the system is called $CL_f(Sp)$. □

Remarks. Two comments are in order here.

1 Notice how the well-known empty-carrier problem is handled here (see [MG] and recall that an [MG]-like formula $\forall(X \cup Var(\Delta \supset \varepsilon)).\Delta \supset \varepsilon$ is represented by $D(X) \cup \Delta \supset \varepsilon$). We can eliminate $D(x)$ from the premises of a formula $\forall(X \cup Var(\Delta \supset \varepsilon)).\Delta \supset \varepsilon$ only if either $D(x)$ is redundant, because x already appears in $\Delta \supset \varepsilon$, or x does not appear in $\Delta \supset \varepsilon$ and there exists a *defined* ground term of the right sort to instantiate x. Indeed a premise may be eliminated only by rule 7 (modus ponens) by which the variables of a formula do not decrease; thus if x already appears in $\Delta \supset \varepsilon$, then applying rules 7 and 0 to $D(X) \cup \Delta \supset \varepsilon$ we can deduce $D(X-\{x\}) \cup \Delta \supset \varepsilon$; otherwise if x does not appear in $\Delta \supset \varepsilon$ and there exists a defined ground term t to instantiate x, then from rule 8 we get $D(X-\{x\}) \cup \{D(t)\} \cup \Delta \supset \varepsilon$ and hence from rule 7 and $D(t)$ we conclude $D(X-\{x\}) \cup \Delta \supset \varepsilon$. But if x does not appear in $\Delta \supset \varepsilon$ and there does not exist a defined ground term to instantiate x, then there is no way to eliminate $D(x)$ from the premises.

2 The elimination rule 9 is better understood as a generalization of the corresponding rule 9_f for the finitary case, which is rather simple and intuitive (though the proof that 9_f can replace 9 is quite difficult; see theorem 4.2 of [AC2]). Forgetting the definedness of variables, it is an inference rule which can be deduced in first-order logic with negation and disjunction (which we here not have): e.g. for $\Delta_1 = \Delta_2 = \Delta_3 = \emptyset$, the premises are $\neg D(t) \vee \varepsilon$, $\neg D(t') \vee \varepsilon$, $\neg t=t' \vee \varepsilon$ from which we get $(\neg D(t) \wedge \neg D(t') \wedge \neg t=t') \vee \varepsilon$, and since $\neg D(t) \wedge \neg D(t') \supset t=t'$, we get ε. Moreover it is straightforward to see 7_f and 8_f as particular cases of 7 and 8 in the finitary case. Note that in this section if we restrict ourselves to h.o. specifications with finitary axioms, then we can use the system for the finitary case; later on in section 3 it will be instead essential to use the system for the infinitary case, because of an implicit infinitary rule corresponding to the extensionality axiom for term-extensional models.

Theorem 2.3. Let $Sp = (\Sigma,Ax)$ be a conditional specification [s.t. all the axioms in Ax are finitary] and X be a [finitary] family of variables. Then $CL(Sp)$ $[CL_f(Sp)]$ is sound and strongly complete for $Mod(Sp)$ and X. □

Fact 2.4. Let FSp be a conditional higher-order specification. Then the system, from now on denoted by $FSp \vdash$, consisting of all the axiom schemas and inference rules of $CL(FSp)$ and of the following further inference rules:

$$10 \qquad \frac{f(x_1,\dots,x_n) = g(x_1,\dots,x_n)}{f = g} \qquad \begin{array}{l} x_i \in X_{s_i},\ x_i \notin Var(f) \cup Var(g),\ i=1\dots n \\ f,g \in T_\Sigma(X)_{|s_1 \times \dots \times s_n \to s} \end{array}$$

is not complete for $EMod(FSp)$ and the empty family of variables.

Proof. Consider the following specification $FSp = (F\Sigma,Ax)$, defined by:

Sorts	s, $(s \to s)$	Axioms	α_1: $f = g \supset D(e)$	
Operations	$f,g: \to (s \to s)$		α_2: $D(f(x)) \supset D(e)$	α_3: $D(g(x)) \supset D(e)$
	$e: \to s$		α_4: $D(f)$	α_5: $D(g)$

Then e is defined in each model A of FSp; indeed either there exists an element a s.t. $f^A(a)$ or $g^A(a)$ is defined, and in this case because of α_2 and α_3 also e^A is defined, or both f^A and g^A are defined (because of α_4 and α_5) and their result over any possible assignement is undefined so that, because of the extensionality, $f^A = g^A$ and hence $D(e)$ follows from α_1. But it easy to check that $FSp \nvdash D(e)$. □

Although rule 10 is insufficient to make the system CL complete, we can obtain a strongly complete inference system for the class of extensional models from any strongly complete inference system by applying a technique of skolemization; let us introduce the basic scheme of this translation before stating formally the result. To do this we informally use the full first-order language based on $EForm(F\Sigma,X)$, where the validity is defined in the obvious way.

Let $F\Sigma = (S,F)$ be a h.o. signature and FSp be a higher-order conditional specification $(F\Sigma,Ax)$.
Then $EMod(FSp)$ is the class of all (usual) models of FSp satisfying the non-conditional axioms $\alpha_s = \{\forall f,g: s.[\forall x_1:s_1,\dots,\forall x_n: s_n.\ f(x_1,\dots,x_n) = g(x_1,\dots,x_n)] \supset f = g\}$ for all $s = s_1 \times \dots \times s_n \to s_{n+1}$ in S.
In order to have a complete deduction system w.r.t. $EMod(FSp)$, we first reduce the α_s to conditional axioms, by a usual logical procedure of skolemization. Let us consider for simplicity unary functions, ie let s be $s' \to s''$; then α_s is logically equivalent to $\forall f,g:s.(\neg[\forall x:s'.f(x)=g(x)] \vee f=g)$ and then to $\forall f,g:s.([\exists x:s'.\neg f(x)=g(x)] \vee f=g)$. By using Skolem functions we reduce the last formula to $\forall f,g: s.[\neg f(x(f,g)) = g(x(f,g))] \vee f = g$ and finally this one is equivalent to $\beta_s = \{\forall f,g:s.f(x(f,g))=g(x(f,g)) \supset f=g\}$. Since skolemization preserves satisfiability, for

any conditional formula φ on $F\Sigma$ there exists $A\in$ EMod(FSp) which does not satisfy φ, ie $A \models Ax \wedge \{\alpha_s \mid s\in S\} \wedge \neg\varphi$, iff there exists $B\in$ Mod(FSp') which does not satisfy φ, where $FSp' = (F\Sigma,Ax) \cup (\cup_{s=(s_1\times\ldots\times s_n\to s_{n+1})\in S} (\{x_i: s\times s \to s_i \mid i=1,\ldots,n\},\{\beta_s\}))$.
Therefore any strongly complete deduction system for FSp' is a strongly complete deduction system for FSp, too.

Note that the axioms not in Ax are finitary.

Def. 2.5. Let $F\Sigma = (S,F)$ be a h.o. signature and FSp be the conditional specification $(F\Sigma,Ax)$. We denote by SK(FSp) the conditional specification $(SK(F\Sigma),SK(Ax))$, where

- $SK(F\Sigma) = (S,F\cup(\{x_i:s\times s\to s_i \mid i=1,\ldots,n\}_{s=(s_1\times\ldots\times s_n\to s_{n+1})\in S})$
- $SK(Ax) = Ax \cup \{f(x_1(f,g),\ldots,x_n(f,g)) = g(x_1(f,g),\ldots,x_n(f,g)) \supset f = g\}_{s=(s_1\times\ldots\times s_n\to s_{n+1})\in S}$. □

Theorem 2.6. Let FSp be a partial higher-order conditional specification and X be a family of variables. Every sound strongly complete system for Mod(SK(FSp)) and X, is a sound and strongly complete system for EMod(FSp) and X. □

Corollary 2.7. Let $FSp = (F\Sigma,Ax)$ be a higher-order conditional specification [s.t. all the axioms in Ax are finitary] and X be a [finitary] family of variables. Then CL(SK(FSp)) [CL_f(SK(FSp))] is sound and strongly complete for EMod(FSp) and X. □

Theorem 2.8. Let $FSp = (F\Sigma,Ax)$ be a higher-order conditional specification X be a family of variables and L be a sound and strongly complete system for EMod(FSp) and X. An $F\Sigma$-algebra F is free over X in EMod(FSp) iff it is isomorphic to $T_{F\Sigma}(X)/\equiv_L$, where $\equiv_L$ is the congruence

$$\{(t,t') \mid t,t'\in T_{F\Sigma}(X), \exists\, Y\subseteq X \text{ s.t. } L \vdash D(Y)\supset t=t', (L \vdash D(Y)\supset D(t) \text{ or } L \vdash D(Y)\supset D(t'))\}.$$ □

3 Term-extensional models

Although mathematical aspects may be more elegant if non-term-generated models are allowed and stepwise refinement is made easier, because extra-elements and structures may be added in a second moment, we cannot ignore that the computer science focus is on term-generated models; for example in [W] only (first-order) term-generated models, there called *computation structures*, are considered when defining abstract data types.

In the case of higher-order specifications, together with term-generated models we have also the interesting class of what we have called in [AC1] *term-extensional models*, i.e. the h.o. models where two functions are equal iff they give the same results when applied to tuples of term-generated arguments.

Partial specifications of term-extensional models can be seen as a special subclass of partial non-positive conditional specifications, which are studied in [AC1, AC2, C]. On the basis of these results we can obtain new results about equational deduction and existence of free models. For the class of term-extensional models it is possible to give directly, without skolemization, a strongly complete inference system. Moreover we can completely clarify the issue of free models and give necessary and sufficient conditions for their existence.

Def. 3.1. Let $F\Sigma = (S,F)$ be a h.o. signature.

- An $F\Sigma$-algebra A is *term–extensional* iff for any $f,g\in (s_1\times\ldots\times s_n \to s_{n+1})^A$, $f(t_1{}^A,\ldots,t_n{}^A) = g(t_1{}^A,\ldots,t_n{}^A)$ for all $t_i\in T_{F\Sigma|s_i}$ and $i=1\ldots n$ implies $f = g$.
- Let FSp be the conditional higher-order specification $(F\Sigma,Ax)$. Then the class TEMod(FSp) is the class of all *term-extensional* models of FSp, ie TEMod(FSp) = Mod(FSp^{ext}), where $FSp^{ext} = (F\Sigma, Ax \cup Ax^{ext})$ and Ax^{ext} is the set $\{\{f(t_1,\ldots,t_n)=g(t_1,\ldots,t_n) \mid t_i\in T_{\Sigma|s_i}, i=1,\ldots,n\} \supset f=g \mid (s_1\times\ldots\times s_n\to s_{n+1})\in S\}$ and f, g are variables of sort $(s_1\times\ldots\times s_n \to s_{n+1})$. □

Note that a term-generated algebra is extensional if and only if it is term-extensional, because every element is the evaluation of a ground term; so for any conditional higher-order specification FSp we have that in particular term-generated and term-extensional models are just the term-generated extensional models.

Equational deduction. Let us consider equational deduction in this special case of term-extensional models. First note that in order to get completeness it is not enough adding the rule

$$\frac{\{f(t) = g(t) \mid t \in T_{F\Sigma}\}}{f = g}$$

(see fact 2.4 for a motivation) which is a special case of the *infinitary induction ω-rule* we can use for making complete in GMod(Sp) any complete system for Mod(Sp) (see [W]). However since in this case higher-order specifications are reduced to particular non-positive conditional specifications, also inference systems for the higher-order case are particular inference systems for non-positive conditional specifications. Of course these systems are infinitary since Ax^{ext} contains infinitary conjunctions in the premises.

Def. 3.2. Let FSp be the conditional higher-order specification (FΣ,Ax) and FSp^{ext} be the conditional specification $(F\Sigma, Ax \cup Ax^{ext})$. The system FCL(FSp) is the system $CL(FSp^{ext})$. □

Theorem 3.3. Let FSp be the conditional higher-order specification (FΣ,Ax) and X be a family of variables. Then the conditional system FCL(FSp) is sound and strongly complete for TEMod(FSp) and X. □

Free models. The counter-example of fact. 1.6 shows that in general conditional higher-order specifications do not admit free and initial models in the class of all term-generated models.

The following result completely characterizes the existence of free models, giving necessary and sufficient conditions both in terms of semantic conditions and in terms of equational deduction.

Some informal comments may facilitate the understanding of the theorem. Condition 3 is a semantic condition, stating that if two functional terms are always defined, then either they are equal in all models or there exists a distinguishing tuple of arguments for them. Condition 4 is just the equivalent of condition 3 in terms of logical deduction and is an immediate consequence of the completeness of the system FL. Finally condition 5 is the specialization of the condition 4 to the complete system FLC(PFSp) that we have exhibited before.

It is interesting to note that starting from these conditions we can show that the existence of free models is undecidable (see [AC2]).

In the sequel by Gen(*C*, X), where *C* is a class of Σ-algebras and X a family of variables, we denote the subclass of *C* defined by $\{A \mid A \in C \text{ s.t. } \exists\ V: X \to A \text{ s.t. } eval^{A,V}(T_\Sigma(X)) = A\}$.

Theorem 3.4. Let PFSp = (FΣ,Ax) be a positive conditional higher-order specification, X be a family of variables and FL be a strongly complete system for TEMod(PFSp) and X.

Then the following conditions are equivalent.

1. there exists a free object for X in TEMod(PFSp);
2. there exists a free object for X in Gen(TEMod(PFSp),X);
3. $\forall\ f,g \in T_{F\Sigma}(X)|_{(s_1\times\ldots\times s_n\to s_{n+1})}$, $n\geq 1$, s.t. Gen(TEMod(PFSp),X)$\models$D(f) and Gen(TEMod(PFSp),X)$\models$D(g)
   - either Gen(TEMod(PFSp),X) $\models f = g$,
   - or there exist $t_i \in T_{F\Sigma}|_{s_i}$, $i=1,\ldots,n$, s.t. Gen(TEMod(PFSp),X) $\not\models f(t_1,\ldots,t_n) = g(t_1,\ldots,t_n)$ and (Gen(TEMod(PFSp),X) $\models D(f(t_1,\ldots,t_n))$ or Gen(TEMod(PFSp),X) $\models D(g(t_1,\ldots,t_n))$);
4. $\forall\ f,g \in T_{F\Sigma}(X)|_{(s_1\times\ldots\times s_n\to s_{n+1})}$, $n \geq 1$, s.t. FL $\vdash D(X) \supset D(f)$ and FL $\vdash D(X) \supset D(g)$
   - either FL $\vdash D(X) \supset f = g$,
   - or there exist $t_i \in T_{F\Sigma}|_{s_i}$, $i=1,\ldots,n$, s.t. FL $\not\vdash D(X) \supset f(t_1,\ldots,t_n) = g(t_1,\ldots,t_n)$ and (FL $\vdash D(X) \supset D(f(t_1,\ldots,t_n))$ or FL $\vdash D(X) \supset D(g(t_1,\ldots,t_n))$);
5. $\forall\ f,g \in T_{F\Sigma}(X)|_{(s_1\times\ldots\times s_n\to s_{n+1})}$, $n\geq 1$, s.t. FCL(PFSp) $\vdash D(X) \supset D(f)$ and FCL(PFSp) $\vdash D(X) \supset D(g)$
   - either FCL(PFSp) $\vdash D(X) \supset f = g$,
   - or $\exists\ t_i \in T_{F\Sigma}|_{s_i}$, $i=1,\ldots,n$, s.t. FCL(PFSp) $\not\vdash D(X) \supset f(t_1,\ldots,t_n) = g(t_1,\ldots,t_n)$ and (FCL(PFSp) $\vdash D(X) \supset D(f(t_1,\ldots,t_n))$ or FCL(PFSp) $\vdash D(X) \supset D(g(t_1,\ldots,t_n))$). □

In the particular case of *total* functions the above condition 3 in the theorem 3.4 is always satisfied, so that in the class of all term-extensional total models there exists a free model for each family of variables.

Corollary 3.5. Let X be a family of variables and PFSp = (FΣ,Ax) be a positive conditional higher-order specification. Then there exists a free model for X in the class of total term-extensional models of PFSp. □

Conclusion. This paper presents some basic results about partial h.o. specifications, which illustrate in particular the difference with the total h.o. case. Some of the results in the paper can be further enlightened when seen as particular applications of the technique of simulation of institutions (see [AC3] and also [Mes] for a similar notion). In particular this technique can be used to partly illustrate the relationship between total and partial specification of h.o. partial functions. We have the feeling that adopting a total approach to partial functions with explicit values for undefinedness does not change the nature of difficulties. We are currently working on this relationship and hope to come out with a complete picture, but we have found that things are less easy than thought and misbeliefs abound in the folklore.

References

AC1 Astesiano, E.; Cerioli, M. "On the Existence of Initial Models for Partial (Higher-Order) Conditional Specifications", Proc. TAPSOFT'89, vol.1, Lecture Notes in Computer Science n. 351, 1989.

AC2 Astesiano, E.; Cerioli, M. "Free Objects and Equational Deduction for Partial Conditional Specifications", Tecnhical Report n.3, Formal Methods Group, University of Genova, 1990.

AC3 Astesiano, E.; Cerioli, M. "Commuting between Institutions via Simulation", submitted, 1990.

B Burmeister, P. A Model Theoretic Oriented Approach to Partial Algebras, Berlin, Akademie-Verlag, 1986.

BW Broy, M.; Wirsing, M. "Partial abstract types", Acta Informatica 18, 1982.

C Cerioli, M. "A sound and equationally-complete deduction system for partial conditional (higher order) types", in Proc.3rd Italian Conference of Theoretical Computer Science,1989, Singapore, World Scientific.

GB Goguen J.A.; Burstall R.M. "Institutions: Abstract Model Theory for Specification and Programming". Technical Report of Computer Science Laboratory, SRI International, 1990.

K-B Krieg-Brückner B. "Algebraic Specification and Functionals for Transformational Program and Meta Program Development", in Proc.TAPSOFT'89, Lecture Notes in Computer Science n. 352, 1989.

M Möller, B. "Algebraic Specification with Higher-Order Operations", Proc. IFIP TC 2 Working Conference on Program Specification and Transformation, North-Holland, 1987.

Me Meinke, K. "Universal Algebra in Higher Types" to appear in Theoretical Computer Science, 1990.

Mes Meseguer J. "General logic" in *Proc. Logic Colloquium '87*, North-Holland, '89.

MG Meseguer, J.; Goguen, J.A. "Initiality, Induction and Computability", in Algebraic Methods in Semantics, Cambridge, Cambridge University Press, 1985.

MTW1 Möller B., Tarlecki A., Wirsing M. "Algebraic Specification with Built-in Domain Constructions", in Proc. of CAAP '88, Lecture Notes in Computer Science n.299, 1988.

MTW2 Möller B., Tarlecki A., Wirsing M. "Algebraic Specifications of Reachable Higher-Order Algebras", in Recent Trends in Data Type Specification, Lecture Notes in Computer Science n.332, 1988.

Q Qian Z. "Higher-Order Order-Sorted Algebras", Proc. 2nd International Conference on Algebraic and Logic Programming, Nancy Oct. 1990, Lecture Notes in Computer Science, Berlin, Springer-Verlag, 1990

R Reichel H. *Initial Computability, Algebraic Specifications, and Partial Algebras*, Berlin, Akademie-Verlag, 1986.

T Tarlecki A. "Quasi-varieties in Abstract Algebraic Institutions", *Journal of Computer and System Science*, n. 33, 1986.

W Wirsing, M. "Algebraic Specification", in Handbook of Theoretical Computer Science vol.B, North Holland, 1990.

Unification in Incompletely Specified Theories: A Case Study

Staffan Bonnier,
Department of Computer and Information Science,
Linköping University,
S-581 83 Linköping, Sweden,
E-mail: sbo@ida.liu.se

Abstract

Let $\mathcal{T}$ be an equational theory and let P be a unification procedure. In order to initiate P to do unification in $\mathcal{T}$, $\mathcal{T}$ must in some way be presented to P. Due to practical considerations, P can sometimes only obtain *approximations* of $\mathcal{T}$; each approximation specifies a *class* of theories in which $\mathcal{T}$ is just a single member. In this paper it is suggested to use *approximation frames* to formally characterize these classes. As a consequence, each approximation frame also specifies which complete sets of $\mathcal{T}$-unifiers that are *strongly complete*. Those are the only complete sets of $\mathcal{T}$-unifiers that can be found within the scope of accessing only available approximations of $\mathcal{T}$. A unification procedure which finds such sets whenever they exist is said to be *weakly complete*. These concepts are specialized to the study of so called *evaluation based unification algorithms*. An approximation frame characterizing the inherent incompleteness of such algorithms is given. Finally a weakly complete evaluation based algorithm is developed.

1 Introduction and Motivation

In recent years a number of methods for integrating logic and functional programming have been proposed (see e.g. [2,6] for surveys and motivation). What is common to most of these approaches is the use of *extended unification procedures* for the operational part of the integration [7]. Here "extended" means that the notion of equality is extended beyond identity among terms. Thus extended unification procedures are concerned with equation solving in some equational theory $\mathcal{T}$.

The question arises how $\mathcal{T}$ should be presented to the unification procedure. Some procedures are tailored for unification in one particular theory, and thus require no explicit presentation of $\mathcal{T}$. However, to be useful for combining logic and functional programming the procedure must be generic, i.e. it must allow $\mathcal{T}$ to vary. Most procedures of the latter kind require access to a complete axiomatization of $\mathcal{T}$ (see [15,11]). Among such procedures, narrowing [16] has probably attracted the most attention. Narrowing assumes the set of equations constituting the axiomatization to be a canonical term rewriting system. In EQLOG [10] narrowing was first suggested for the integration of relational and equational programming.

When taking practical considerations into account, there may be a discrepancy between the intended use of a unification procedure and the strive for supplying it with complete axiomatizations. This is for instance the case in [13,5], where the unification procedure constitutes the heart of the integration of Horn clause logic with external procedures; the more properties of $\mathcal{T}$ that are assumed to be observable, the fewer are the external procedures to which the system may be applied. In particular, if one assumes a complete axiomatization of $\mathcal{T}$ to be available one excludes the original objective of the work, since then only procedures coded in a fixed equational language may be used. Thus, in order to make the system useful, one may only assume certain restricted properties of $\mathcal{T}$ to be observable. Put another way, the presentation of $\mathcal{T}$ only supplies *approximations* of $\mathcal{T}$. Each approximation specifies a *class* of theories in which $\mathcal{T}$ is just a single member. With this view, it is important to ensure that all available information about $\mathcal{T}$ is properly exploited, i.e. that the unification procedure finds complete sets of unifiers as often as possible. This leads us to investigate the following two questions about sets U of equations to be unified (solved):

QUESTION 1: Is there a complete set C of $\mathcal{T}$-unifiers for U such that C can be seen to be complete by using only available approximations of $\mathcal{T}$?

QUESTION 2: Is there a unification algorithm which given U finds such C whenever the answer to question 1 is in the affirmative?

In order to formalize these questions this paper defines the notion of *approximation frames*. These essentially capture to what extent $\mathcal{T}$ can be distinguished from other theories. Consequently each approximation frame specifies which sets C that provide affirmative answers to question 1. Such sets are named *strongly* complete. A procedure which provides an affirmative answer to question 2 is then said to be *weakly* complete. Intuitively, a procedure is weakly complete if it is as complete as possible within the scope of accessing only available approximations of $\mathcal{T}$.

The main goal of the paper is however not to develop a general theory of approximation frames. Instead, so called *evaluation based unification algorithms* (*eba*'s) [7] are studied. Such algorithms assume $\mathcal{T}$ to be axiomatized by a functional program. The program is then used solely for reduction of ground terms. Consequently *eba*'s are inherently incomplete. Still, as suggested in e.g. Le Fun [1], Aflog [14] and in [9], *eba*'s are well-suited as the basis for integrating logic and functional programming. This is also the kind of algorithm used in [13,5].

When comparing narrowing with *eba*'s it is important to note that the former was developed with a declarative concept in mind; the aim of narrowing is to give a procedural counterpart to the declarative notion of complete sets of unifiers. Fay [8] and Hullot [12] provide completeness results which show that this aim is indeed accomplished. For evaluation based algorithms on the other hand, there is no such declarative prototype; it is clear *how* they compute but it is certainly not clear *what* they compute. It follows that the connection between the declarative semantics and the operational semantics of a program may be obscured when the interpreter employs an *eba*.

It is argued in this paper that a certain approximation frame, called EB, naturally reflects the restricted access to $\mathcal{T}$ that is characteristic for *eba*'s. The strongly complete sets of unifiers induced by this frame thus provides a declarative description of what ultimately should be computed by *eba*'s. The rest of the paper is then devoted to the development of an *eba* which is weakly complete for EB.

2 Preliminaries

The reader is assumed to be familiar with basic concepts of equational unification (see [11] and [15]). In order to clarify the notation used in the paper, this section reviews some of these concepts.

The set of substitutions will be denoted $\mathcal{S}ubst$. *Equations* are written as $u \doteq t$ where u and t are terms. For a set E of equations, $=_E$ is defined to be the least congruence relation on the termalgebra, subject to:

$$\forall \sigma \in \mathcal{S}ubst : u \doteq t \in E \quad \Rightarrow \quad u\sigma =_E t\sigma$$

This relation extends to substitutions; we write $\sigma =_E \theta[V]$ iff $x\sigma =_E x\theta$ for each x in the set V of variables. Furthermore, the preorder $\preceq_E$ is defined on substitutions by $\sigma \preceq_E \theta[V]$ iff $\sigma\gamma =_E \theta[V]$ for some substitution γ.

A set $\mathcal{T}$ of equations is an (equational) *theory* iff $u \doteq t \in \mathcal{T}$ whenever $u =_\mathcal{T} t$. The theory *axiomatized* (or *induced*) by a set E of equations is the least theory which includes E.

Let E and U be sets of equations. A substitution σ is an E-*unifier* for U iff $u\sigma =_E t\sigma$ for each $u \doteq t \in U$. A set C of E-unifiers for U is said to be *complete* iff for each E-unifier θ of U there is some σ in C such that $\sigma \preceq_E \theta[var(U)]$. Here $var(U)$ denotes the set of all variables that occur in U. A finite set of equations will be called a *unification problem*. The set of all such problems will be denoted $\mathcal{U}pr$.

3 Evaluation Based Algorithms

It was mentioned in the introduction that evaluation based unification algorithms assume the theory $\mathcal{T}$ to be axiomatized by a functional program. In this section, theories intended to be captured by this informal definition are given a formal characterization. This characterization is then used to give a more precise meaning to the notion "evaluation based unification algorithm" (*eba*).

As is customary when dealing with functional programs, the alphabet of function symbols is partitioned into two disjoint sets; the set $\mathcal{C}$ of *constructor symbols* and the set $\mathcal{D}$ of *defined symbols.* Terms which are built solely from constructor symbols and variables are called *structures.* Terms which are on the form $f(s_1 \ldots s_n)$ where f is a defined symbol and each s_i is a (not necessarily ground) structure are called *functional calls.* If d is a ground functional call and s is a ground structure, $d \doteq s$ will be called a *definitional equation.* Throughout this paper *the set of ground structures is assumed to be infinite.* Intuitively, we think of ground structures as canonical representations of values on which functions denoted by elements in $\mathcal{D}$ are defined. This leads us to consider equational theories which define $\mathcal{D}$ over $\mathcal{C}$ in the following precise sense:

Definition 3.1 Let $\mathcal{T}$ be an equational theory. $\mathcal{T}$ *defines* $\mathcal{D}$ *over* $\mathcal{C}$ iff the following three conditions are satisfied for all terms u_i and t_i:

C1: If $c(u_1 \ldots u_n) =_{\mathcal{T}} c(t_1 \ldots t_n)$ for some constructor c, then $u_i =_{\mathcal{T}} t_i$, $1 \leq i \leq n$.

C2: If c and c' are distinct constructors, then $c(u_1 \ldots u_n) \neq_{\mathcal{T}} c'(t_1 \ldots t_m)$.

C3: For each ground functional call d there exists a ground structure s such that $d =_{\mathcal{T}} s$.

The class of all theories which define $\mathcal{D}$ over $\mathcal{C}$ will be denoted DEF. □

Example 3.1 Let $\mathcal{C} = \{0, 1, 2 \ldots\}$ and let $\mathcal{D} = \{+, *\}$. The set $\mathcal{T}$ of all equations that hold in the algebra of nonnegative integers (with the standard interpretation of $\mathcal{C}$ and $\mathcal{D}$) then defines $\mathcal{D}$ over $\mathcal{C}$. □

Condition C3 postulates that each ground functional call $f(s_1 \ldots s_n)$ is equated with at least one ground structure s. Conditions C1 and C2 imply that this s is unique. Intuitively, s represents the value of the function denoted by f at $s_1 \ldots s_n$. Thus an "operation table" with definitional equations as "entries" is induced by $\mathcal{T}$:

Definition 3.2 Let $\mathcal{T} \in$ DEF. The *table* of $\mathcal{T}$, denoted $\mathsf{Tab}(\mathcal{T})$, is the set of all definitional equations $d \doteq s$ such that $d =_{\mathcal{T}} s$. We call a set E of equations a table iff E is the table of some $\mathcal{T} \in$ DEF. □

Example 3.2 Continuing example 3.1 yields:

$$\mathsf{Tab}(\mathcal{T}) = \{0 + 0 \doteq 0,\ 0 * 0 \doteq 0,\ 0 + 1 \doteq 1 \ldots\}$$

□

The table of a theory $\mathcal{T}$ reflects the input/output relationship defined by the underlying functional program. Thus the program and its interpreter supply a procedure for extracting "entries" from the table. Consequently a procedure which, when given any ground functional call d, terminates and returns the ground structure s uniquely determined by $d \doteq s \in \mathsf{Tab}(\mathcal{T})$ will be called a *table lookup* for $\mathcal{T}$.

We may now give an informal but sufficiently precise formulation of what should be satisfied by a procedure **P** to be classified as being an evaluation based unification algorithm:

> For any $\mathcal{T} \in$ DEF, if **P** is supplied with a table lookup for $\mathcal{T}$ and a set $U \in \mathcal{U}pr$ to be $\mathcal{T}$-unified, the following holds:
>
> (1) **P** terminates,
>
> (2) If **P** gives a set S of substitutions as output, S is a complete set of $\mathcal{T}$-unifiers for U.

Usually, *eba*'s are only required to be sound. However, if the algorithm is to be used for executing normal logic programs where negative literals are allowed in the clause-bodies, incompleteness may turn into unsoundness. Therefore we impose the stronger requirement (2).

Since **P** must function whichever $\mathcal{T}$ in DEF is chosen, **P** can retrieve information about $\mathcal{T}$ only by calling the table lookup. Due to the termination requirement, the number of such calls must be finite. This means that $\mathcal{T}$ can only be *approximated* by means of finite subsets of $\mathsf{Tab}(\mathcal{T})$.

Definition 3.3 Let $\mathcal{T} \in$ DEF. For each finite subset F of $\mathsf{Tab}(\mathcal{T})$, the set of all theories which include F is said to be an *approximation* of $\mathcal{T}$. The class of all approximations will be denoted $\mathsf{App}(\mathcal{T})$. □

To sum up this section, an *eba* **P** assumes $\mathcal{T}$ to be in DEF. Moreover, **P** can only distinguish $\mathcal{T}$ from other theories in DEF by selecting a finite subset of $\mathsf{Tab}(\mathcal{T})$.

4 Approximation Frames

This section defines the notion of approximation frames. An approximation frame formally characterizes the information that is passed to a unification procedure **P** when **P** is asked to do unification in a particular theory $\mathcal{T}$. As a special case, the information about theories in DEF that is passed to *eba*'s may be characterized in this way.

Definition 4.1 An *approximation frame* $\mathcal{F}$ is a pair $\langle \mathsf{T}, \mathsf{A} \rangle$, where:

- T is a class of theories, and
- A is a T-indexed family $(\mathsf{A}_\mathcal{T})_{\mathcal{T}\in\mathsf{T}}$ of classes of *approximations*, where, for each $\mathcal{T} \in \mathsf{T}$ and each $A \in \mathsf{A}_\mathcal{T}$, A is a set of theories with $\mathcal{T} \in A$.

□

Example 4.1 If $\mathcal{T}$ is any theory, then $\langle\{\mathcal{T}\}, (\{\{\mathcal{T}\}\})\rangle$ forms an approximation frame. A more interesting example is $\langle \mathsf{DEF}, (\mathsf{App}(\mathcal{T}))_{\mathcal{T}\in\mathsf{T}}\rangle$. □

The latter of these two approximation frames, i.e. the approximation frame for DEF, will in the sequel be denoted by EB.

Informally, an approximation frame $\langle \mathsf{T}, \mathsf{A} \rangle$ should be seen as a specification of a class of unification procedures. T specifies the degree of genericity to be supported by each procedure **P** in the class; when instantiated by any $\mathcal{T} \in \mathsf{T}$, **P** should specialize to a $\mathcal{T}$-unification procedure. $\mathsf{A}_\mathcal{T}$ then specifies which information **P** receives when instantiated by $\mathcal{T}$, and hence, to what extent **P** can distinguish $\mathcal{T}$ from other theories in T: When given a set U of equations to unify, **P** may select any (but at most one) approximation A in $\mathsf{A}_\mathcal{T}$ and use it to conclude that $\mathcal{T}$ is in the class $\mathsf{T} \cap A$ of theories. This, hence, puts definite limits to the performance of **P**. In order for **P** to generate a complete set S of $\mathcal{T}$-unifiers for U, there must be some A in $\mathsf{A}_\mathcal{T}$ such that S is a complete set of $\mathcal{T}'$-unifiers for *every* $\mathcal{T}' \in \mathsf{T} \cap A$. If this is not the case, **P** simply has too little information about $\mathcal{T}$ to determine that S is complete. This intuition is made precise in the following two definitions.

Definition 4.2 Let $\mathcal{F} = \langle \mathsf{T}, \mathsf{A} \rangle$ be an approximation frame, let $\mathcal{T} \in \mathsf{T}$ and let $U \in \mathcal{U}pr$. A set S of substitutions is said to be a *strongly complete* set of $\mathcal{T}_\mathcal{F}$-unifiers for U iff there exists $A \in \mathsf{A}_\mathcal{T}$ such that, for all $\mathcal{T}' \in \mathsf{T} \cap A$, S is a complete set of $\mathcal{T}'$-unifiers of U. □

Definition 4.3 Let $\mathcal{F} = \langle \mathsf{T}, \mathsf{A} \rangle$ be an approximation frame. A *unification function* for $\mathcal{F}$ is a mapping

$$\mathsf{U} : \mathsf{T} \to (\mathcal{U}pr \to (2^{Subst} \cup \{\bot\}))$$

such that, for each $\mathcal{T} \in \mathsf{T}$ and each $U \in \mathcal{U}pr$, if $(\mathsf{U}\mathcal{T})\,U = S$ and $S \neq \bot$, then S is a strongly complete set of $\mathcal{T}_\mathcal{F}$-unifiers for U. We will say that $\mathcal{F}$ *specifies* a unification procedure **P** if **P** computes a unification function for $\mathcal{F}$. This function will then be denoted by $uf(\mathbf{P})$. □

We shall not discuss the exact meaning of the word "computes" in the above definition. The intuitive idea, however, is that **P** is provided with some means to access $\mathcal{T}$, so that the approximations of $\mathcal{T}$ may be obtained. For instance, when $\mathcal{F} = \mathsf{EB}$ it is assumed that **P** is given some means to submit a finite number of ground functional calls and to obtain the corresponding ground structures (associated by $\mathcal{T}$) as result.

When the result of applying $uf(\mathbf{P})$ is a set of substitutions, **P** is assumed to generate a (possibly infinite) stream of (encodings of) the substitutions in this set. When $\bot$ is the result of applying $uf(\mathbf{P})$, this means that **P** is unable to generate a complete set of $\mathcal{T}$-unifiers for U. The only requirement put on the representation of $\bot$ is that it must not be confused with the representation of any set of substitutions.

Example 4.2 $\langle\{\mathcal{T}\}, (\{\{\mathcal{T}\}\})\rangle$ specifies procedures tailored for unification in $\mathcal{T}$. EB specifies all *eba*'s. Note however that EB additionally specifies some nonterminating procedures. □

Clearly, it may happen that no strongly complete set of unifiers exists. In such cases every unification procedure specified by the approximation frame in question is forced to give $\perp$ as output. The following example of this phenomenon is certainly not the simplest one, but has the advantage of giving a hint of the proof of proposition 5.1 in the next section.

Example 4.3 Let $\mathcal{C}, \mathcal{D}$ be defined as in 3.1. Moreover, let U be a unification function for EB and let $\mathcal{T}$ be any theory in DEF. Then:

$$(\mathsf{U}\mathcal{T})\{x+0 \doteq 0\} = \perp \tag{1}$$

Let F be a finite subset of some table, and let A be the set of all theories which include F. The following is then easily verified for any terms u, t:

$$(\forall \mathcal{T}' \in \mathsf{DEF} \cap A \;:\; u =_{\mathcal{T}'} t) \quad \Leftrightarrow \quad u =_F t \tag{2}$$

Now, to see why (1) holds, assume the opposite, i.e. that $(\mathsf{U}\mathcal{T})\{x+0 \doteq 0\} = S$ for some set S of substitutions. According to definitions 3.3, 4.2 and 4.3, there must be some finite $F \subseteq \mathsf{Tab}(\mathcal{T})$ such that S is a complete set of $\mathcal{T}'$-unifiers of $\{x+0 \doteq 0\}$ for each $\mathcal{T}'$ in $\mathsf{DEF} \cap A$, where A is the set of all theories which include F. For each $\sigma \in S$, (2) hence implies:

$$x\sigma + 0 =_F 0 \tag{3}$$

Note that, viewing the equations in F as rewrite-rules, F forms a canonical term-rewriting system. Since $F \subseteq \mathcal{T}'$ for each $\mathcal{T}' \in \mathsf{DEF} \cap A$, we may assume, without loss of generality, that $x\sigma$ is in normal form for each $\sigma \in S$. By (3) $x\sigma + 0$ must have 0 as normal form. Since the only possible redex in $x\sigma + 0$ is $x\sigma + 0$, we have:

$$\forall \sigma \in S \;:\; x\sigma + 0 \doteq 0 \in F \tag{4}$$

and hence:

$$\forall \sigma \in S \;:\; x\sigma \in \mathcal{C} \tag{5}$$

Since F is finite, there must be some $n \in \mathcal{C}$ which does not occur in any equation in F. Thus, for some $\mathcal{T}' \in \mathsf{DEF} \cap A$, $n+0 \doteq 0 \in \mathcal{T}'$. It follows that $\{x/n\}$ is a $\mathcal{T}'$-unifier of $\{x+0 \doteq 0\}$. Now (5) and the fact that S is complete implies that $x\sigma =_{\mathcal{T}'} n$ for some $\sigma \in S$. Again by (5), and by requirement C2 in definition 3.1, this means that $n = x\sigma$, which, by (4), is not possible because of the way n was selected. Thus the original assumption, that $(\mathsf{U}\mathcal{T})\{x+0 \doteq 0\} \neq \perp$, must be wrong. □

The previous example shows, in particular, that any *eba* must give $\perp$ as output when applied to $\{x+0 \doteq 0\}$.

For a given approximation frame $\mathcal{F}$, there is now a natural way to compare which one of two procedures specified by $\mathcal{F}$ that is better than the other, in the sense that it is the "least incomplete" of the two. This is done by introducing a preordering on the unification functions for $\mathcal{F}$.

Definition 4.4 Let $\mathcal{F} = \langle \mathsf{T}, \mathsf{A} \rangle$ be an approximation frame. The preordering $\sqsubseteq$ is defined as follows on the set of unification functions for $\mathcal{F}$:

$$\mathsf{U}_1 \sqsubseteq \mathsf{U}_2 \quad \Leftrightarrow \quad (\forall \mathcal{T} \in \mathsf{T})\,(\forall U \in \mathcal{U}pr) \;:\; (\mathsf{U}_2\mathcal{T})\,U = \perp \;\Rightarrow\; (\mathsf{U}_1\mathcal{T})\,U = \perp$$

When $\mathsf{U}_1 \sqsubseteq \mathsf{U}_2$ we also say that U_2 is *less incomplete* than U_1. □

When designing a unification procedure **P** specified by an approximation frame $\mathcal{F}$ one would like **P** to exploit the information provided about the underlying theories as efficiently as possible. Ultimately **P** should deliver a complete set of unifiers as output whenever this is done by *some* procedure among those specified by $\mathcal{F}$. According to definition 4.4 this will be the case if $uf(\mathbf{P})$ is $\sqsubseteq$-greatest.

Definition 4.5 Let $\mathcal{F} = \langle \mathsf{T}, \mathsf{A} \rangle$ be an approximation frame. A unification procedure **P** specified by $\mathcal{F}$ is said to be *weakly complete* for $\mathcal{F}$ iff, for each unification function U for $\mathcal{F}$, it holds that $\mathsf{U} \sqsubseteq uf(\mathbf{P})$. □

Note that if $\{\mathcal{T}\} \in \mathsf{A}_{\mathcal{T}}$ for each $\mathcal{T} \in \mathsf{T}$, then "weakly complete" means the same thing as "complete". This is, for instance, the case with narrowing procedures, where the given rewrite system is assumed to provide a complete axiomatization of the theory in which unification is to be done.

For a given approximation frame $\mathcal{F}$, it is natural to ask whether a weakly complete unification procedure for $\mathcal{F}$ exists. Obviously the answer need not be positive. However, the rest of this paper is devoted to give this question a constructive answer, in the affirmative, when $\mathcal{F} = \mathsf{EB}$.

5 A Unification Algorithm

In this section the concepts introduced in the previous two sections are tied together, and a weakly complete *eba* for EB is developed. The general strategy of this algorithm is to transform the set U of equations to be unified into a *normal form* N. A simple analysis of N then shows whether U has a strongly complete set of unifiers or not.

In the following definition, "u occurs in N" means "u is a subterm of some left hand or right hand side in N":

Definition 5.1 Let N be a finite set of equations. N is in *normal form* iff:

- $N = \{x_1 \doteq s_1 \ldots x_n \doteq s_n,\ r_1 \doteq d_1 \ldots r_m \doteq d_m\}$, where $n, m \geq 0$,
- each x_i is a variable which occurs only once in N,
- each d_i is a nonground functional call which occurs only once in N,
- all s_i and r_j are structures.

If, in addition, $r_1 \ldots r_m$ are all distinct variables and there exists a permutation π on $\{1 \ldots m\}$ such that $1 \leq i \leq j \leq m$ implies $r_{\pi i} \notin var(d_{\pi j})$, then N is in *solved form*. In this case the substitution $\mathsf{sol}(N)$ is defined by:

$$\mathsf{sol}(N) = \{x_1/s_1 \ldots x_n/s_n\}\{r_{\pi 1}/d_{\pi 1}\} \cdots \{r_{\pi m}/d_{\pi m}\}$$

□

Example 5.1 Let s and 0 be constructors and let f be a defined symbol. Then the sets:

$$N_1 = \{x \doteq s(y),\ y \doteq f(y),\ 0 \doteq f(s(y))\} \text{ and}$$
$$N_2 = \{x \doteq s(y),\ y \doteq f(s(w)),\ w \doteq f(z)\}$$

are both in normal form. In addition, N_2 is in solved form, and:

$$\mathsf{sol}(N_2) = \{x/s(f(s(f(z)))),\ y/f(s(f(z))),\ w/f(z)\}$$

□

For sets of equations in normal form there is a particularly simple way to decide whether strongly complete sets of $\mathcal{T}_{\mathsf{EB}}$-unifiers exists, and if so, to construct such sets:

Proposition 5.1 Let $\mathcal{T} \in \mathsf{DEF}$ and let N be a set of equations which is in normal form. Then the following holds:

- If N is not in solved form then N has no strongly complete set of $\mathcal{T}_{\mathsf{EB}}$-unifiers.
- If N is in solved form then $\{\mathsf{sol}(N)\}$ is a strongly complete set of $\mathcal{T}_{\mathsf{EB}}$-unifiers for N.

□

The following algorithm provides the core of the unification algorithm. Given a table lookup for any $\mathcal{T} \in \mathsf{DEF}$ it takes any $U \in \mathcal{U}pr$ and transforms it into normal form or into failure:

Normal Form Algorithm:
Initially, let $U_0 = U$ and let $i = 0$. Repeatedly do the following: Select any equation $u \doteq t$ in U_i such that one of rule 1 to 9 applies. If no such equation exists then stop with U_i as result. Otherwise perform the corresponding action, that is; stop with failure or construct U_{i+1} from U_i. Then increment i by 1.

1. $u = t$. Remove $u \doteq t$.

2. u is a variable which occurs elsewhere in U_i and t is a structure distinct from u. If u occurs in t then stop with failure, otherwise replace all other occurrences of u by t (i.e. leave $u \doteq t$ unchanged).
3. u is a nonvariable and t is a variable. Replace $u \doteq t$ by $t \doteq u$.
4. $u = c(u_1 \ldots u_n)$ and $t = c'(t_1 \ldots t_m)$ where c and c' are constructors. If $c \neq c'$ then stop with failure, otherwise replace $u \doteq t$ by $u_1 \doteq t_1 \ldots u_n \doteq t_n$.
5. u and t are both functional calls. Let z be a variable not in $var(U_i) \cup var(U)$. Replace $u \doteq t$ by the two equations $z \doteq u$ and $z \doteq t$.
6. u or t has a functional call d as proper subterm. Let z be a variable not in $var(U_i) \cup var(U)$. Replace all occurrences of d in U_i by z, then add the equation $z \doteq d$.
7. u is a functional call and t is a structure. Replace $u \doteq t$ by $t \doteq u$.
8. u is a structure and t is a functional call which occurs elsewhere in U_i. Replace all other occurrences of t by u (i.e. leave $u \doteq t$ unchanged).
9. t is a ground functional call. Replace $u \doteq t$ by $u \doteq s$, where $t \doteq s$ is in $\mathsf{Tab}(\mathcal{T})$.

□

The next proposition states that the normal form algorithm indeed has the properties one might expect:

Proposition 5.2 Let $\mathcal{T} \in \mathsf{DEF}$. Suppose the normal form algorithm is provided with a table lookup for $\mathcal{T}$ and a set $U \in \mathcal{U}pr$ as input. Then the algorithm terminates with one of the following two results: (1) failure, or (2) a set N of equations in normal form. Moreover:

(a) If the result is failure, then $\emptyset$ is a strongly complete set of $\mathcal{T}_{\mathsf{EB}}$-unifiers for U.

(b) If the result is a set N in solved form, then $\{\mathsf{sol}(N)\}$ is a strongly complete set of $\mathcal{T}_{\mathsf{EB}}$-unifiers for U.

(c) If none of conditions (a) or (b) above applies, then no strongly complete set of $\mathcal{T}_{\mathsf{EB}}$-unifiers for U exist.

□

What remains to be defined is the actual unification algorithm. When given a table lookup for a theory $\mathcal{T} \in \mathsf{DEF}$ and a finite set U of equations to unify, it performs as follows:

Unification Algorithm:
Use the table lookup and the normal form algorithm to transform U. Then choose, depending on the result of the transformation, the appropriate alternative below:

1. If the result is failure, then terminate and output $\emptyset$.
2. If the result is a set N in solved form, then terminate and output $\{\mathsf{sol}(N)\}$.
3. If the result is a set in normal form which is not in solved form, then terminate and output $\perp$.

□

The following proposition is an immediate corollary of proposition 5.2:

Proposition 5.3 The unification algorithm is weakly complete for EB. Moreover, when provided with any finite set U of equations as input the algorithm terminates. □

Thus, since the unification algorithm terminates it satisfies all requirements for being an *eba*.

All proofs of the propositions in this section may be found in [3]. These proofs do in fact show that stronger results hold: If T is taken to be the class of all theories axiomatized by tables (which is a proper subclass of DEF) and $\mathsf{A}_{\mathcal{T}} = \mathsf{App}(\mathcal{T})$ for each such theory $\mathcal{T}$, then all propositions still hold when EB is everywhere replaced by $\langle \mathsf{T}, \mathsf{A} \rangle$. Informally this means that decreasing the genericity specified by DEF to that specified by T does not pay off in an increased number of strongly complete sets of unifiers.

We close this section by giving some examples showing how the algorithm works:

Example 5.2 Let $\mathcal{T}$, $\mathcal{C}$ and $\mathcal{D}$ be as in example 3.1. Suppose we wish to solve the equations $U = \{x \doteq x + x, x + (x + x) \doteq 0\}$. The following is a possible trace of the normal form algorithm:

$$\begin{array}{ll} U_0 = \{x \doteq x + x,\ x + (x + x) \doteq 0\}, & \\ U_1 = \{x \doteq x + x, x + x \doteq 0\} & \text{by 8,} \\ U_2 = \{x \doteq x + x, x \doteq 0\} & \text{by 8,} \\ U_3 = \{0 \doteq 0 + 0, x \doteq 0\} & \text{by 2,} \\ U_4 = \{0 \doteq 0, x \doteq 0\} & \text{by 9.} \\ U_5 = \{x \doteq 0\} & \text{by 1.} \end{array}$$

Thus the answer would be $\{\{x/0\}\}$. When the unification algorithm is called with a set U of equations and outputs $\perp$, this does in general not mean that the work done by the normal form algorithm is wasted. Rather U is partially unified and put into a nice format, as in the following example:

Let $U = \{x \doteq c(f(x))\}$, where c is a constructor symbol and f is a defined symbol. This is a trace of the normal form algorithm when applied to U:

$$\begin{array}{ll} U_0 = \{x \doteq c(f(x))\}, & \\ U_1 = \{x \doteq c(z),\ z \doteq f(x)\} & \text{by 6,} \\ U_2 = \{x \doteq c(z),\ z \doteq f(c(z))\} & \text{by 2.} \end{array}$$

□

6 Discussion

The division of unification procedures into being either complete or incomplete sometimes seems too coarse. In particular this is the case when the underlying equational theory is underspecified, so that no (sound and) complete unification procedure can exist. One of the contributions of the paper is the introduction of approximation frames for handling this problem. Approximation frames extend the notion of completeness to that of weak completeness. Without being complete a procedure may be weakly complete, meaning that it finds complete sets of unifiers whenever possible. Moreover, here the use of "whenever possible" has a formally well-defined meaning, quite independent of operational considerations. It should be remarked though, that the definition of approximation frames given in this paper is by no means final. Many variations on the theme are possible in order to obtain appropriate notions of "weak completeness" in other contexts.

The paper is mainly devoted to the study of a particular approximation frame called EB. This study results in an *eba* which is weakly complete for EB. It may be argued that this algorithm is "too incomplete" to be of any practical use. It is certainly true that the class of equations that can be solved by the algorithm is quite small. This is however beside the point; the algorithm is weakly complete. Thus, to have a less incomplete algorithm one must use the functional program for more complex purposes than reducing ground functional calls. Moreover, the proper place for an *eba* is not as a pure equation solver, but rather as part of an interpreter for a combined logic and functional language. As such, its inherent incompleteness may be compensated by a proper integration with the resolution mechanism. For instance, in Le Fun [1] it is suggested to use *residuations* to delay unifications which can not be carried out immediately. Unification may then be resumed when more variables have become grounded.

Traditionally there is a drawback of using incomplete unification algorithms such as *eba*'s in an interpreter; the semantics of programs is only characterized operationally. Thus it is very hard to reason formally about programs. Since our unification algorithm is weakly complete, definitions 4.2 and 4.3 provide a purely declarative characterization of what is computed by the algorithm. Therefore it seems as if the notion of approximation frames may provide a way to overcome this problem: An interesting continuation of this work would be to study a logic language employing the *eba* of section 5, and to give its operational semantics a non-operational characterization in terms of strongly complete sets of unifiers.

As was mentioned in the introduction, this work has its origins in logic programming with external procedures. Put in this framework, proposition 5.3 shows that the assumptions made about the external procedures are fully exploited. In order to decrease the frequency of $\perp$, we are currently trying to develop a weakly complete unification algorithm when a stronger assumption is present: The external procedures are assumed to be written in a polymorphically typed language, and it is assumed that the unification algorithm can access these types. It turns out that with this information available, there are unification algorithms which terminate and give complete sets of unifiers as output when $\perp$ is given as output from the algorithm presented in this paper [4].

Acknowledgements

I would like to thank Jan Maluszynski and Andrzej Tarlecki for their helpful comments on an early draft of this paper.

References

[1] H. Ait-Kaci, P. Lincoln, and R. Nasr. Le Fun: Logic, Equations and Functions. In *Symp. on Logic Programming*, pages 17–23, 1987.

[2] M. Bellia and G. Levi. The Relation between Logic and Functional Languages. *J. Logic Programming*, (3):217–236, 1986.

[3] S. Bonnier. *Horn Clause Logic with External Procedures: Towards a Theoretical Framework.* Licentiate thesis 197, Dep. of Computer and Information Science, Linköping University, Linköping, Sweden, 1989.

[4] S. Bonnier. Type-driven Evaluation for Integrating FP and LP. Presented at the Integration workshop at Schloß Dagstuhl, March 1991. Draft.

[5] S. Bonnier and J. Maluszynski. Towards a Clean Amalgamation of Logic Programs with External Procedures. In *5:th Int. Conf. and Symp. on Logic Programming*, pages 311–326. MIT Press, 1988.

[6] D. DeGroot and G. Lindstrom, editors. *Logic Programming, Functions, Relations and Equations.* Prentice-Hall, 1986.

[7] M. Dincbas and P. van Hentenryck. Extended Unification Algorithms for the Integration of Functional Programming into Logic Programming. *J. Logic Programming*, (4):199–227, 1987.

[8] M.J. Fay. First-Order Unification in an Equational Theory. Master Thesis 78-5-002, University of California at Santa Cruz, 1978.

[9] U. Furbach and S. Hölldobler. Modelling the Combination of Functional and Logic Programming Languages. *J. Symbolic Computation*, (2):123–138, 1986.

[10] J. Goguen and J. Meseguer. EQLOG: Equality, Types and Generic Modules for Logic Programming. In D. DeGroot and G. Lindstrom, editors, *Logic Programming, Functions, Relations and Equations*, pages 295–363. Prentice-Hall, 1986.

[11] G. Huet and D.C. Oppen. Equations and Rewrite Rules a Survey. In R.V. Book, editor, *Formal Language Theory, Perspectives and Open Problems*, pages 349–405. Academic Press, 1980.

[12] J.M. Hullot. Canonical Forms and Unification. In *CADE-5*, pages 318–334, 1980.

[13] J. Leszczylowski, S. Bonnier, and J. Maluszynski. Logic Programming with External Procedures: Introducing S-Unification. *Information Processing Letters*, (27):159–165, 1988.

[14] D.W. Shin, J.H. Nang, S. Han, and S.R. Maeng. A Functional Logic Language based on Canonical Unification. In *Symp. on Logic Programming*, pages 328–333, 1987.

[15] J.H. Siekmann. Universal Unification. In *CADE-7*, pages 1–42, 1984.

[16] J.R. Slagle. Automated Theorem-Proving for Theories with Simplifiers, Commutativity and Associativity. *J. ACM*, 21(4):622–642, 1974.

Observing Localities *
(Extended Abstract)

G. Boudol, I. Castellani
INRIA, Sophia-Antipolis,
M. Hennessy
CSAI, University of Sussex,
A. Kiehn
TUM, Munich.

1 Introduction

There are by now a number of well-established semantic theories of processes in the research literature which are based on principles of observation. The main idea is that processes are deemed to be equivalent if there is no possible observation which can distinguish them. Different formalisations of this idea, which give rise to a number of semantic equivalences, may be found in [Mil89], [Hen88] and [Hoa85]. All of these formalisations are based on the same simple notion of observation, namely communication: one may observe a process by communicating with it via a communication channel. The resulting semantic theories are often called *interleaving theories*; they do not distinguish between concurrency and nondeterminism or more precisely they equate a parallel process with the purely nondeterministic one obtained by interleaving its primitive computation steps or actions.

Some attempts have been made to generalise this observation based approach in order to develop a semantic theory which does distinguish between these two phenomena, [CH89], [BC87], [DNM88], [BC89], [DD89]. Here we reexamine the approach taken in [CH89] and [Kie89] where the processes under observation are considered to be *distributed in nature.* So the observer can not only test the process by communicating with it but can also observe or distinguish that part of the distributed process which reacted to the test. A purely nondeterministic process is based at one site whereas in general a concurrent one may be distributed among a number of different locations. It follows that an observer will be able to distinguish them.

*This work has been supported by the ESPRIT/BRA CEDISYS project.

We use as a starting point the process algebra CCS, a process description language which describes processes in terms of the *actions* they can perform. So for example the process B_1 defined by

$$B_1 \Leftarrow in.out.B_1$$

is a simple process which repeatedly performs the actions *in, out.* If we run two copies of this in parallel we obtain a process which acts more or less like a two-place bag:

$$B_2 \Leftarrow B_1 \mid B_1.$$

Here | is the parallel operator of CCS which in this context defines a process which consists of two independent processes, two copies of B_1, running in parallel.

Processes running in parallel may also communicate or synchronise with each other. This is formalised by dividing the set of actions into two complementary subsets, the *input* actions and the *output* actions. Communication is then considered to be the simultaneous occurrence of complementary actions. Output actions are indicated by an overbar, such as $\overline{a}, \overline{in}$, etc., input actions by the absence of an overbar and there is a distinguished action τ to indicate a communication or more generally internal and unobservable computation. So if we define two processes *In, Out* by

$$\begin{aligned} In &\Leftarrow in.\alpha.In \\ Out &\Leftarrow \overline{\alpha}.out.Out \end{aligned}$$

then the process $In \mid Out$ acts somewhat like B_2. However *In, Out* are not obliged to synchronise via the action α. The actions α and $\overline{\alpha}$ may be performed independently, which corresponds to separate synchronisations with processes in their operating environments. To eliminate these possible communications with the environment and thereby force the synchronisation between the two processes we limit the scope of these actions using another operator of CCS, *restriction*, which in general is written as $\backslash A$ where A is a set of actions. So let NB_2 be defined by

$$NB_2 \Leftarrow (In \mid Out)\backslash\alpha.$$

These two processes, B_2 and NB_2, offer very similar behaviour to a user particularly as the synchronisation between *In* and *Out* is not supposed to be visible externally. According to the theory developed in [Mil89] they are *weak bisimulation equivalent*, denoted by $B_2 \approx NB_2$; in terms of the visible actions *in* and *out* they offer the same possible behaviour to any user of the systems.

However this reasoning is based on the assumption that the only property which can be observed of a process is its ability to perform particular actions. Now let us reinterpret the language by saying that $P \mid Q$ is a *distributed process* where the sub-process

P is at one site and Q is at another site; moreover let us suppose that an observer can distinguish between sites in the sense that when a distributed process performs an action the observer knows the site responsible for it. Thus one observer's view of the distributed process B_2 is

Here the observer has decided, out of personal choice, to call the site or location of the first subprocess l_1 and the second subprocess l_2. Now it is not possible to construct a similar view of NB_2. For example

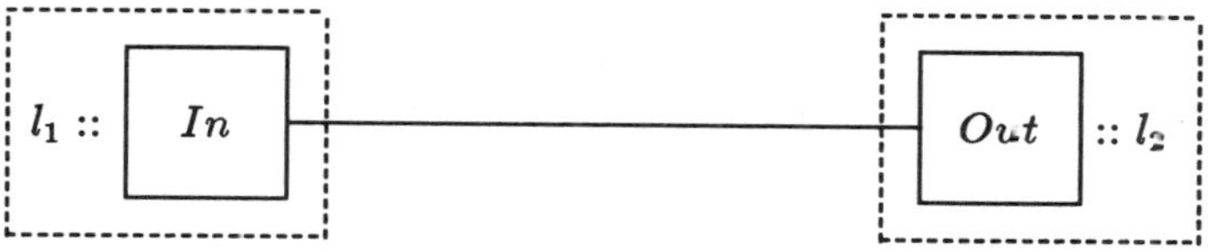

can easily be distinguished as here all *in* actions are seen to occur at the location l_1 and all *out* actions at location l_2. In contrast they are distributed between l_1 and l_2 in the distributed process B_2.

The basic difference between these two processes is that in NB_2 one site is responsible for the *in* actions and one for the *out* actions whereas B_2 has two equivalent sites each acting like a one place buffer. Viewing these as specifications this is a useful and meaningful distinction. To implement B_2 it is necessary to have independent locations each acting like buffers whereas an implementation of NB_2 would always have to localise the responsibility for the *in* actions and the *out* actions in independent locations.

Let us now address the question of how exactly an observer should be allowed to perceive the distributed nature of a process. In this respect we are guided by principles of extensionality; we would like the resulting equivalence to be as extensional as possible in that the semantics of a process should only be determined by its external manifestations rather than its internal structure or behaviour. It is reasonable to argue that at least some aspect of the distribution of subprocesses of a distributed system is a part of its extensional behaviour and therefore if we are to view CCS as a language for describing distributed systems an observer should be able to view in $P \mid Q$ a distributed system which potentially has two distinct sites; any externally visible action performed by P should be recognizable as emanating from one location and any performed by Q should be recognizable as coming from a different location.

In this paper we develop the idea that location names are assigned dynamically as part and parcel of the act of observation; when an observer sees an action it will naturally be seen as emanating from a particular location and the observer may then choose to allocate a name to that location. All subsequent actions performed at that location will then be recognised as emanating from this same location. Technically this involves developing an operational semantics for the language by replacing the usual statements of the form $P \xrightarrow{a} Q$ with new ones of the form

$$P \xrightarrow[u]{a} Q$$

which carry information about the location names which the observer has assigned to particular locations in the system. This is very similar to the approach taken in [CH89], [Cas88] and [Kie89] where *distributed bisimulations* are defined and in the full version of the paper we offer a detailed comparison.

2 Location Equivalence

The language we use to formalize our approach is a slight extension of Milner's pure CCS. The main extension is an additional operator called *location prefixing* representing the allocation of locations to processes. A process p prefixed by a location u will be denoted by $u :: p$. Intuitively, this means that process p is at a location called u. However, in general we will assume these locations are introduced via the observation of visible actions. That is, initially, before any experiment has been performed the process under investigation does not contain any location. With the observation of an action the location of the action is also perceived and assigned a name. Thus the observers we assume here are more powerful than those usually considered for CCS or other process algebras. So we will have a transition rule

$$a.p \xrightarrow[u]{a} u :: p \quad \text{for any location name } u$$

which means that a is performed at a location to which the observer permanently assigns the name u. If further experiments are performed on $u :: p$ then the location u will always be observed. Moreover the location called u may contain sub-locations which in turn may also be observed. For example

$$u :: (a.nil \mid b.nil) \xrightarrow[uv]{a} u :: (v :: nil \mid b.nil)\,.$$

Here the location which has been called u by a previous observation contains two sub-locations and at one of them a is performed. The name v is allocated to this subsite via the observation of a.

Formally we assume an infinite set of basic location names or site names Loc, ranged over by $l, k, etc.$, and general locations will be sequences from Loc^*, ranged over by

$u, v, w, \ldots$. So we work with the following abstract syntax:

$$\begin{array}{rcl} t & = & nil \mid \mu.t \mid t+t \mid t \mid t \mid t[f] \mid t\backslash b \\ & & \mid x \mid rec\,x.\,t \mid t \lfloor\!\lfloor t \mid t|_c t \mid u :: t \end{array}$$

where μ ranges over $Act \cup \{\tau\}$, Act is a basic set of actions and f is a bijective relabelling function which preserves complementation. Most of these operators are taken directly from CCS but $\lfloor\!\lfloor$ and $|_c$ are the leftmerge and communication merge operators from [BK85], [Hen88].

We will give two operational semantics to the closed terms of this language. The first one generalizes bisimulation equivalence, $\approx$, in a straightforward way and is omitted in this extended abstract. This equivalence considers the ability of performing visible actions and only in this respect bisimilar processes exhibit the same behaviour. The second semantics we give to the language additionally takes the distribution in space into account. As already discussed above there will be two points in which it differs from bisimulation equivalence. The first point is that locations may be introduced via the observation of actions. Secondly processes may contain locations and actions in the scope of locations will be observed at those locations. Formally these two points are reflected by the following two rules in the location transition system.

$$\begin{array}{lll} (\mathrm{L1}) & a.p \xrightarrow[u]{a} u :: p & u \in Loc^* \\ (\mathrm{L2}) & p \xrightarrow[u]{a} p' \quad \text{implies} & v :: p \xrightarrow[vu]{a} v :: p' \end{array}$$

Here u is an access path representing a location where the action a is performed and in the second rule this is extended by v to give the new location vu.

The τ–transitions are considered, as usual, to be invisible, so no location is observed when they are performed. They are of the form $p \xrightarrow{\tau} p'$ and are defined through the standard transition system for CCS on which bisimulation equivalence is based. Visible transitions are defined by the location transition system given in Figure 1. They always have the form $p \xrightarrow[u]{a} p'$ where we call u the location where the action a is performed. Weak transitions are defined in the usual manner:

$$p \overset{\varepsilon}{\Rightarrow} p_1,\ p_1 \xrightarrow[u]{a} p_2,\ p_2 \overset{\varepsilon}{\Rightarrow} p' \quad \text{implies} \quad p \overset{a}{\underset{u}{\Longrightarrow}} p'.$$

We can now see how parallelism is differentiated from nondeterminism. For the process $a.nil \mid b.nil$ we can derive $a.nil \mid b.nil \xrightarrow[l]{a} l :: nil \mid b.nil$ while its nondeterministic counterpart would perform the transition $a.b.nil + b.a.nil \xrightarrow[l]{a} l :: b.nil$. Now with the observation of the action b different locations would be perceived. In $l :: nil \mid b.nil \xrightarrow[k]{b} l :: nil \mid k :: nil$ the b is performed at the location k which is independent of l whereas in $l :: b.nil \xrightarrow[lk]{b} lk :: nil$ it is performed at a sublocation of l, namely lk.

For each $a \in Act$ let $\xrightarrow[u]{a} \subseteq (\mathbb{P} \times Loc^* \times \mathbb{P})$ be the least binary relation satisfying the following axioms and rules.

(L1)	$a.p \xrightarrow[u]{a} u :: p$	$u \in Loc^*$	
(L2)	$p \xrightarrow[u]{a} p'$	implies	$v :: p \xrightarrow[vu]{a} v :: p'$
(L3)	$p \xrightarrow[u]{a} p'$	implies	$p + q \xrightarrow[u]{a} p'$ $q + p \xrightarrow[u]{a} p'$
(L4)	$p \xrightarrow[u]{a} p'$	implies	$p \mid q \xrightarrow[u]{a} p' \mid q$ $q \mid p \xrightarrow[u]{a} q \mid p'$
(L5)	$p \xrightarrow[u]{a} p'$	implies	$p \not\mid q \xrightarrow[u]{a} p' \mid q$
(L6)	$p \xrightarrow[u]{a} p'$	implies	$p[f] \xrightarrow[u]{f(a)} p'[f]$
(L7)	$p \xrightarrow[u]{a} p'$	implies	$p\backslash b \xrightarrow[u]{a} p'\backslash b,\ a \notin \{b, \bar{b}\}$
(L8)	$t[rec\, x.\ t/x] \xrightarrow[u]{a} p'$	implies	$rec\, x.\ t \xrightarrow[u]{a} p'$

Figure 1: Location Transitions

Based on this transition system we now define location equivalence. Two processes p and q are location equivalent if every move of one of them is matched by a similar move of the other and in particular if for every visible transition $p \overset{a}{\underset{u}{\Longrightarrow}} p'$ the matching transition $q \overset{a}{\underset{u}{\Longrightarrow}} q'$ has the same location.

Definition 2.1 [Location Equivalence]
A symmetric relation $R \subseteq \mathbb{P} \times \mathbb{P}$ is called a *location bisimulation* iff $R \subseteq C(R)$ where $(p, q) \in C(R)$ iff

(i) $p \overset{\varepsilon}{\Rightarrow} p'$ implies $q \overset{\varepsilon}{\Rightarrow} q'$ for some $q' \in \mathbb{P}$ such that $(p', q') \in R$

(ii) $p \overset{a}{\underset{u}{\Longrightarrow}} p', a \in Act, u \in Loc^*$

implies $q \overset{a}{\underset{u}{\Longrightarrow}} q'$ for some $q' \in \mathbb{P}$ such that $(p', q') \in R$.

Two processes p and q are said to be *location equivalent*, $p \approx_l q$, iff there is a location bisimulation R such that $(p, q) \in R$. □

One may now easily check that the two processes B_2 and NB_2 defined in the introduction, are distinguished by $\approx_l$.

3 Properties of Location Equivalence

In this section we briefly outline some properties of $\approx_l$. The details may be found in the full version of the paper.

Proposition 3.1

1. $p \approx_l q$ *implies* $p \approx q$.

2. *If* p, q *contain no occurrence of* $|$ *then* $p \approx q$ *implies* $p \approx_l q$.

3. *Let* $p, q \in \mathbb{P}$, *and* $\pi\colon Loc \longmapsto Loc^*$ *be an arbitrary relabelling of location names. Then* $p \approx_l q$ *implies* $\pi(p) \approx_l \pi(q)$.

4. $\approx_l$ *is preserved by all the operators of* CCS *except* $+$ *and* $\not|$.

Like weak bisimulation $\approx$, the equivalence $\approx_l$ is not preserved by $+$. Let $\approx_l^c$ be the largest equivalence contained in $\approx_l$ which is preserved by all operators. This congruence satisfies the usual static laws of CCS [Mil89] and may be characterised equationally on the subset of CCS which does not include restriction and relabelling. The required equations, which involve the auxiliary operators $\not|$, $|_c$ are given in Figure 2, where the notation $x \sqsubset y$ means that x is absorbed by y, i. e. $y + x = y$. In the full paper we show that on the same sub-language location equivalence $\approx_l$ coincides with *distributed bisimulation* as defined in [CH89], [Kie89].

It is well known that bisimulation equivalence may be characterized using a simple modal language called *HML*, in the sense that two processes are bisimulation equivalent if and only if they satisfy exactly the same set of formulae, [Mil89]. *HML* is a simple modal logic based on two modalities $\langle a\rangle$ and $[a]$ where a is an arbitrary action. So an obvious extension to cope with location equivalence is to parameterise these modalities by locations. However, we will introduce a slightly more general modal language, which we feel is somewhat more natural. If a process contains no locations then it should be unnecessary for the formulae which characterise it to contain locations. In such processes, elements of CCS, locations are potential rather than actual and it should also be thus in their characterising formulae. For this reason we introduce location variables and quantification over these variables. One can imagine an expressive term language for locations but here we consider only a very simple language given by

$$t ::= l, l \in Loc \quad | \quad x, x \in LVar \quad | \quad \varepsilon \quad | \quad t.t.$$

$$
\begin{array}{lrcl}
\text{(A1)} & x+(y+z) & = & (x+y)+z \\
\text{(A2)} & x+y & = & y+x \\
\text{(A3)} & x+nil & = & x \\
\text{(A4)} & x+x & = & x \\
 & & & \\
\text{(LP1)} & (x+y) \lfloor\!\lfloor z & = & x \lfloor\!\lfloor z + y \lfloor\!\lfloor z \\
\text{(LP2)} & (x \lfloor\!\lfloor y) \lfloor\!\lfloor z & = & x \lfloor\!\lfloor (y \mid z) \\
\text{(LP3)} & x \lfloor\!\lfloor nil & = & x \\
\text{(LP4)} & nil \lfloor\!\lfloor x & = & nil \\
 & & & \\
\text{(I1)} & x + \tau x & = & \tau.x \\
\text{(I2)} & \mu.\tau x & = & \mu.x \\
\text{(I3)} & \mu.(x+\tau.y)+\mu.y & = & \mu.(x+\tau.y) \\
 & & & \\
\text{(NI1)} & \tau.x \lfloor\!\lfloor y & = & \tau.(x \mid y) \\
\text{(NI2)} & x \lfloor\!\lfloor y & = & x \lfloor\!\lfloor \tau.y \\
\text{(NI3)} & x \lfloor\!\lfloor (y+\tau.z) + x \lfloor\!\lfloor z & = & x \lfloor\!\lfloor (y+\tau.z) \\
 & & & \\
\text{(CPE)} & x \mid y & = & x \lfloor\!\lfloor y + y \lfloor\!\lfloor x + x \mid_c y \\
 & & & \\
\text{(CP0)} & \tau.x \mid_c y & = & x \mid_c y \\
\text{(CP1)} & (x+y) \mid_c z & = & (x \mid_c z) + (y \mid_c z) \\
\text{(CP2)} & x \mid_c y & = & y \mid_c x \\
\text{(CP3)} & x \mid_c nil & = & nil \\
\text{(CP4)} & (a.x \lfloor\!\lfloor x') \mid_c (b.y \lfloor\!\lfloor y') & = & \begin{cases} \tau.(x \mid y) \lfloor\!\lfloor (x' \mid y') & \text{if } a = \bar{b} \\ nil & \text{otherwise} \end{cases} \\
\text{(CP5)} & a.(x \mid x') \lfloor\!\lfloor (y \mid y') & \sqsubset & a.(c.x \lfloor\!\lfloor x' + v) \lfloor\!\lfloor (\bar{c}y \lfloor\!\lfloor y' + w)
\end{array}
$$

Figure 2: Equations G: Standard Non-Interleaving Laws.

Here *LVar* is a set of location variables, ε represents the degenerate location and is sequence concatenation. The language for property formulae is then defined by the following abstract syntax

$$\begin{array}{rcll} \Phi & ::= & \bigwedge\{\Phi \mid \Phi \in I\} & \mid \neg\Phi \\ & & \langle a\rangle_t\Phi \quad \mid & \langle\varepsilon\rangle\Phi \\ & & t = t' \quad \mid & \exists x.\Phi \end{array}$$

In the formula $\langle a\rangle_t\Phi$ the term t represents a location and intuitively the formula is satisfied by a process which can perform an action a at a location specified by t and in so doing reaches a state which satisfies Φ. Both $\forall$ and $\exists$ bind location variables and this leads to the usual definition of free and bound variables. We are only interested in formulae which are closed , i.e. which contain no free occurrences of location variables, which we denote by $\mathcal{L}$. More generally, for $L \subseteq Loc$ we let $\mathcal{L}_L$ denote the set of closed formulae which only use locations from L. So in particular formulae in $\mathcal{L}_\emptyset$ use no locations and all location variables are bound.

The satisfaction relation between processes and formulae, $\models \subseteq \mathbb{P} \times \mathcal{L}$, is a straightforward extension of the standard one, [Mil89], and is defined by structural induction on formulae. For example

$$\begin{array}{ll} p \models \langle a\rangle_u\Phi & \text{if } p \overset{a}{\underset{u}{\Longrightarrow}} p' \text{ and } p' \models \Phi \\ p \models \exists x.\Phi & \text{if for some } u \in Loc^* \;\; p \models \Phi[x \rightarrow u]. \end{array}$$

Theorem 3.2 *$p \approx_l q$ iff $\mathcal{L}_L(p) = \mathcal{L}_L(q)$ where $L = loc(p) \cup loc(q)$.*

As an immediate corollary we have

Corollary 3.3 *For $p, q \in$ CCS, $p \approx_l q$ if and only if $\mathcal{L}_\emptyset(p) = \mathcal{L}_\emptyset(q)$.* □

These technical results, only sketched here, show that location equivalence is a natural refinement of bisimulation equivalence which inherits many of its interesting properties but distinguishes between nondeterminism and concurrency.

References

[BC87] G. Boudol and I. Castellani. On the semantics of concurrency: partial orders and transition systems. In *Proceedings of TAPSOFT 87*, number 249 in Lecture Notes in Computer Science, pages 123–137, 1987.

[BC89] G. Boudol and I. Castellani. Permutation of transitions: an event structure semantics for CCS and SCCS. In *Proceedings of Linear Time, Branching Time and Partial Order in Logics and Models for Concurrency*, number 354 in Lecture Notes in Computer Science, pages 411–427, 1989.

[BK85] J. Bergstra and J.W. Klop. Algebra of communicating processes with abstraction. *Theoretical Computer Science*, (37):77–121, 1985.

[Cas88] I. Castellani. *Bisimulations for Concurrency*. Ph.d. thesis, University of Edinburgh, 1988.

[CH89] I. Castellani and M. Hennessy. Distributed bisimulations. *JACM*, 10(4):887–911, 1989.

[DD89] P. Degano and P. Darondeau. Causal trees. In *Proceedings of ICALP 88*, number 372 in Lecture Notes in Computer Science, pages 234–248, 1989.

[DNM88] P. Degano, R. De Nicola, and U. Montanari. A distributed operational semantics for CCS based on condition/event systems. *Acta Informatica*, 26:59–91, 1988.

[Hen88] M. Hennessy. Axiomatising finite concurrent processes. *SIAM Journal of Computing*, 17(5):997–1017, 1988.

[Hoa85] C.A.R. Hoare. *Communicating Sequential Processes*. Prentice-Hall, 1985.

[Kie89] A. Kiehn. Distributed bisimulations for finite CCS. Report 7/89, University of Sussex, 1989.

[Mil89] R. Milner. *Communication and Concurrency*. Prentice-Hall, 1989.

ABSTRACT DYNAMIC DATA TYPES: A TEMPORAL LOGIC APPROACH

GERARDO COSTA - GIANNA REGGIO
DIPARTIMENTO DI MATEMATICA - UNIVERSITÀ DI GENOVA
VIA L.B.ALBERTI 4 - 16132 GENOVA (ITALY)

INTRODUCTION

Dynamic data types are (modelled by) dynamic algebras, which are a particular kind of partial algebras with predicates. These, in turn, are just the familiar algebraic structures that are needed to interpret (many sorted) 1st order logic: a family of sets (the carriers) together with a set of operations and predicates on the carriers [GM]. The operations are partial in order to model situations like trying to get the first element of an empty list.

The distinguishing feature of dynamic algebras is that for some of the carriers there are special ternary predicates $\longrightarrow$; $d \xrightarrow{l} d'$ means that element d can perform a transition labelled by l into element d'. If we use a dynamic algebra to model (some kind of) processes, then we may have transition predicates corresponding to send and receive actions. A dynamic algebra for lists, may have a transition predicate corresponding to the tail operation; thus: $list \xrightarrow{l} list'$ is true, for some appropriate label l, whenever list' = tail(list). Of course we are not forced to have such predicates: when modelling processes it is natural to use them (one could even say that we need them); in the case of lists we have a choice: we can use the (classical) static view, or a dynamic one (closer to the way we regard lists when programming within the imperative paradigm).

The basic idea behind dynamic algebras is very simple. There are some technical problems; but they are orthogonal w.r.t. the dynamic features, as they concern handling partial operations, and have been dealt with in the literature, see [BW, AC] for instance. The name seems appropriate, even though it has already been used to denote structures for interpreting dynamic logic.

The question is whether dynamic algebras are of any use; we think the answer is yes. Indeed, they are, in disguise, a basic tool in the SMoLCS methodology, which has been used in practice, and for large projects, with success (see eg [AR2, AFD]). SMoLCS is a specification methodology, for specifying concurrent systems, that provides a framework for handling both ordinary (static) data types and data types with dynamic features (process types). The logical language used in SMoLCS is many sorted 1st order logic with equality and transition predicates. Such a language allows reasonable specifications for many properties of concurrent systems, however it becomes cumbersome when dealing with properties involving the transitive closure of the transition relations such as (some) liveness or safety properties [L]. A really significant example would take up too much space here; a simple, but still interesting, example can be cooked up using buffers (we shall use it also in the following sections). So let us consider a set B of buffers together with the operations Put, Get and Remove: Put(e, b) adds element e to buffer b; Get(b) yields the "first" element in b and leaves b unchanged; Remove(b) removes this first value from b producing a new buffer. As example of constraints on B, Get, Put and Remove we can consider:

This work has been supported by CNR-PF-Sistemi Informatici e Calcolo Parallelo (Italy), COMPASS-Esprit-BRA-W.G. n 3264 and by a grant ENEL/CRA (Milano Italy).

(i) The buffers in B follow a LIFO policy, ie: Get(Put(e, b)) = e and Remove(Put(e, b)) = b.

(ii) If b is non-empty, b' = Remove(b) and e = Get(b), then there is an elementary transition from b to b' corresponding to "output e". Using a transition predicate, we can phrase this by saying that $b \xrightarrow{O(e)} b'$ is true (here O(e) is a label meaning "output of e"; see Sec. 2 for a more precise formulation).

(iii) The buffers in B are such that they have the capability of returning any element, say e, that they receive and mantain this capability until e is actually delivered.

Condition (i) is standard and does not need comments. (iii) is a liveness constraint: once a buffer b inputs e it will evolve (through input/output transitions) in such a way that at any "state" (or moment) either e can be output or another state can be reached in which e can be output. Notice one important difference between (i) and (iii): the first specifies the structure of our buffers, while the other specifies their behaviour, without constraining the internal structure. Finally, one way of reading (ii) is: if b is nonempty then it can always output a (stored) value; thus we have an example of a simple safety property. Being a simple one it can be easily expressed in 1st order logic (with transition predicates). Of course, with the same language one can also express properties such as (iii); but the corresponding formulae are almost unreadable. This difficulty provides the principal motivation for the present work: our aim is to find a logic which is well suited for expressing in a (reasonably) natural way properties such as (iii), but also (i) and (ii).

Various temporal logics have been proposed as an adequate tool for specifying the behaviour of concurrent systems, see eg [P, Em, K]. They allow to write concise formulae corresponding to (iii), but are not well suited for properties such as (i) or even (ii). Therefore we have been brought to a language which is, more or less, a 1st order version of CTL* with explicit transition predicates (notice that with these predicates the accessibility relations used in Kripke frames are - or can be referred to - within the language). Our choices, both for syntax and semantics, have been motivated by previous experience with the problems involved in the specification of concurrent systems [AR1, AR2]. We are now experimenting with our framework (as part of a larger project aimed at extending the algebraic approach to the specification of dynamic systems) and comparing it with similar ones, such as those reported in [FL, FM, SFSE]. We hope that a better insight into the problems will either allow us to switch to a more established setting, or provide strong arguments in favour of ours. Presently we can at least say that our framework is a sound one, indeed it forms an institution [GB].

One part of the experiment consists in testing our approach against "real" problems, in an industrial environment; the other concerns the theoretical side. In the first place we are going through the well known concepts and results concerning specifications for abstract data types (see eg [W90]) and replacing 1st order logic with ours, algebras with dynamic algebras, and so on: it appears that many concepts and results can be extended to our setting and in a natural way. Here we give a first, and concise, account of what happens in connection with existence of initial models and related proof systems. In a more complete paper [CR] we consider also structured and hierarchical specifications and implementations of specifications, in the style of [W89].

Acknowledgements. The present work is strictly connected with other researches carried out in Genova. Thanks to Egidio Astesiano for many fruitful discussions on this and related topics. Thanks also to the referees for helpful comments.

1 PARTIAL ALGEBRAS WITH PREDICATES

Here we summarize the main definitions and facts about *partial algebras with predicates,* which are derived from the partial algebras of Broy and Wirsing (see [BW]) and from the algebras with predicates of Goguen and Meseguer (see [GM]).

A *predicate signature* (shortly, a *signature)* is a triple $\Sigma = (SRT, OP, PR)$, where: SRT is a set (the set of the *sorts*); OP is a family of sets: $\{OP_{w,srt}\}_{w \in SRT^*, srt \in SRT}$ and PR is a family of sets: $\{PR_w\}_{w \in SRT^*}$.

We shall write $Op: srt_1 \times \ldots \times srt_n \rightarrow srt$ for $Op \in OP_{srt_1 \ldots srt_n, srt}$, $Pr: srt_1 \times \ldots \times srt_n$ for $Pr \in PR_{srt_1 \ldots srt_n}$ and also, when sorts are irrelevant, $Op \in OP$, $Pr \in PR$.

A *partial Σ-algebra with predicates* (shortly a *Σ-algebra*) is a triple

$A = (\{A_{srt}\}_{srt \in SRT}, \{Op^A\}_{Op \in OP}, \{Pr^A\}_{Pr \in PR})$

consisting of the *carriers*, the *interpretation of the operation symbols* and the *interpretation of the predicate symbols*; ie:

- if $srt \in SRT$, then A_{srt} is a set;
- if $Op: srt_1 \times \ldots \times srt_n \rightarrow srt$, then $Op^A: A_{srt_1} \times \ldots \times A_{srt_n} \rightarrow A_{srt}$ is a (partial) function;
- if $Pr: srt_1 \times \ldots \times srt_n$, then $Pr^A \subseteq A_{srt_1} \times \ldots \times A_{srt_n}$.

Usually we write $Pr^A(a_1, \ldots, a_n)$ instead of $(a_1, \ldots, a_n) \in Pr^A$.

Given an SRT-indexed family of sets of variables X, the *term algebra* $T_\Sigma(X)$ is the Σ-algebra defined as as usual, with the condition that $Pr^{T_\Sigma(X)} = \emptyset$ for all $Pr \in PR$. If $X_{srt} = \emptyset$ for all $srt \in SRT$, then $T_\Sigma(X)$ is simply written T_Σ and its elements are called *ground terms*.

If A is an algebra, $t \in T_\Sigma(X)$ and $V: X \rightarrow A$ is a *variable evaluation*, ie a sort-respecting assignment of values in A to *all* the variables in X, then the *interpretation of* t *in* A *w.r.t.* V, denoted by $t^{A,V}$, is given as usual, but note that now it may be undefined; if t is a ground term then we use the notation t^A.

In what follows we assume that sorts and arities are respected and also that our algebras have *nonempty carriers*.

If A and B are Σ-algebras, a *homomorphism* h from A into B (written $h: A \rightarrow B$) is a family of *total* functions $h = \{ h_{srt} \}_{srt \in SRT}$ where for all $srt \in SRT$ $h_{srt}: A_{srt} \rightarrow B_{srt}$ and

- for all $Op \in OP$: if $Op^A(a_1, \ldots, a_n)$ is defined, then so is $Op^B(h_{srt_1}(a_1), \ldots, h_{srt_n}(a_n))$ and $h_{srt}(Op^A(a_1, \ldots, a_n)) = Op^B(h_{srt_1}(a_1), \ldots, h_{srt_n}(a_n))$;
- for all $Pr \in PR$: if $Pr^A(a_1, \ldots, a_n)$, then $Pr^B(h_{srt_1}(a_1), \ldots, h_{srt_n}(a_n))$.

The interpretation of a formula of (many sorted) 1st order logic with equality (with operation and predicate symbols belonging to Σ) in a Σ-algebra A is given as usual, but: for t_1, t_2 of the same sort, $t_1 = t_2$ is *true in* A *w.r.t. a variable evaluation* V iff $t_1^{A,V}$ and $t_2^{A,V}$ are both defined and equal in A (we say that = denotes "existential equality").

We write $A, V \models \theta$ when the interpretation of the formula θ in A w.r.t. V yields true; then θ is *valid* in A (written $A \models \theta$) whenever $A, V \models \theta$ for all evaluations V.
Usually we simply write D(t) for $t = t$ and use it to require that the interpretation of t is defined.

Given a class of Σ-algebras *C*, an algebra I is *initial* in *C*
iff $I \in C$ and for all $A \in C$ there exists a unique homomorphism $h: I \rightarrow A$. The following holds:
if I is initial in *C*, then for all ground terms $t_1, \ldots, t_n$ and all predicates $Pr \in PR$:

- $I \models t_1 = t_2$ iff for all $A \in C$: $A \models t_1 = t_2$; thus $I \models D(t_1)$ iff for all $A \in C$: $A \models D(t_1)$; therefore, in general, the term algebra T_Σ is not initial in the class of all Σ-algebras;
- $I \models Pr(t_1, \ldots, t_n)$ iff for all $A \in C$: $A \models Pr(t_1, \ldots, t_n)$.

2 DYNAMIC ALGEBRAS

A *dynamic signature* DΣ is a couple (Σ, STATE) where:

- Σ = (SRT, OP, PR) is a predicate signature,
- STATE ⊆ SRT (the elements in STATE are the *dynamic sorts*, ie the sorts of dynamic elements),
- for all st ∈ STATE
  there exist a sort lab(st) ∈ SRT - STATE such that lab(st') = lab(st") iff st' = st"
  and a predicate $\_ \overset{-}{\longrightarrow} \_$: st × lab(st) × st ∈ PR.

A (*dynamic*) DΣ-algebra is just a Σ-algebra; the term algebra $T_{D\Sigma}(X)$ is just $T_\Sigma(X)$.

Notation: in this paper, for some of the operation and predicate symbols, we use a mixfix notation. This is explicit in the definition of the signatures: for instance, $\_ \overset{-}{\longrightarrow} \_$: st × lab(st) × st ∈ PR means that we shall write $t \overset{t'}{\longrightarrow} t''$ instead of ⟶(t, t', t"); ie terms of appropriate sorts replace underscores.

If DA is a DΣ-algebra and st ∈ STATE, then the elements of sort st, the elements of sort lab(st) and the interpretation of the predicate $\_ \overset{-}{\longrightarrow} \_$ are respectively the states, the labels and the transitions of a labelled transition system, describing the activity of the dynamic elements of sort st. The whole activity of the dynamic elements is represented by the *maximal* labelled paths, such as

$s_0 \overset{l_0}{\longrightarrow}_{DA} s_1 \overset{l_1}{\longrightarrow}_{DA} s_2 \ldots$ (either finite, and non-extendable, or infinite).

We denote by PATH(DA, st) the set of such paths for the dynamic elements of sort st.

If σ is the path above, then: $S(\sigma)$ denotes s_0, $L(\sigma)$ denotes l_0, $\sigma|_n$ denotes the subpath from s_n onwards (if it exists).

In what follows DΣ will denote a generic dynamic signature (Σ, STATE), where Σ = (SRT, OP, PR); moreover we often write: **sorts** S **dsorts** STATE **opns** OP **preds** PR for the dynamic signature (Σ, STATE), where Σ is:

(S ∪ STATE ∪ { lab(st) | st ∈ STATE }, OP, PR ∪ {$\_ \overset{-}{\longrightarrow} \_$: st × lab(st) × st | st ∈ STATE}).

Example: buffers containing natural values organized in a LIFO way

Consider the dynamic signature BUFΣ

```
sorts   nat
dsorts  buf
opns    0:  → nat
        Succ: nat → nat
        Empty:  → buf
        Put: nat × buf → buf
        Get: buf → nat
        Remove: buf → buf
        I, O: nat → lab(buf)
```

The elements built by the two operations I and O label the transitions corresponding to the actions of receiving and returning a value, respectively.

The buffers are modelled by the BUFΣ- algebra STACKBUF, where:

- $STACKBUF_{nat} = \mathbb{N}$; $STACKBUF_{buf}$ and the interpretation of the operations Empty, Put, Get and Remove are respectively the set of stacks of natural numbers and the usual operations EmptyStack, Push, Top and Pop.
- If we assume that the buffers are bounded and can contain k elements at most, then the interpretation of ⟶ in STACKBUF is the relation consisting of the following triples (here and below the inter-

pretation of a [predicate / operation] symbol Symb in STACKBUF, $Symb^{STACKBUF}$, is simply denoted by Symb):

$b \xrightarrow{I(n)} Put(n, b)$ for all n and all b having k-1 elements at most,

$b \xrightarrow{O(Get(b))} Remove(b)$ for all b s.t. Get(b) is defined.

– If we assume that buffers are unbounded, then $\longrightarrow$ consists of the triples:

$b \xrightarrow{I(n)} Put(n, b)$ for all n and all b,

$b \xrightarrow{O(Get(b))} Remove(b)$ for all b s.t. Get(b) is defined.

The activity of a bounded buffer, with k = 2, which is initially empty (represented by the term Empty) is given by the set of paths starting from the root of the following tree; notice that they are exactly the elements of PATH(STACKBUF, buf) with initial element the empty buffer.

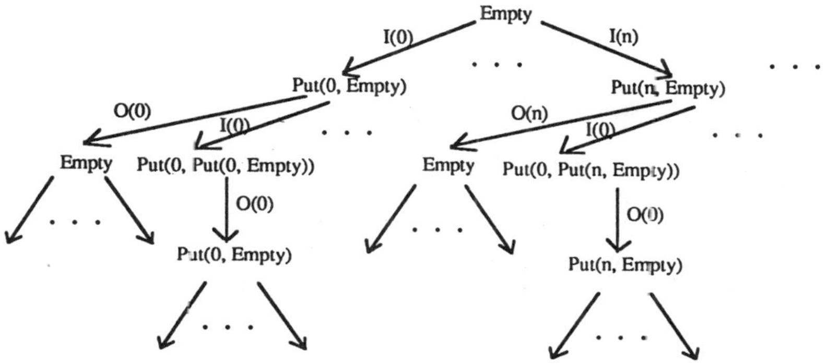

End of Example

Let DA and DA' be DΣ-algebras; a *(dynamic) homomorphism* h: DA → DA' is just a homomorphism from DA into DA' as Σ-algebras. It is easy to see that, for each signature DΣ, DΣ-algebras and dynamic homomorphisms form a category, that we denote by $DAlg_{D\Sigma}$.

Homomorphisms between dynamic algebras preserve the activity of the dynamic elements; formally:

if h is as above, for all s, l, s' ∈ DA: if $s \xrightarrow{l}{}^{DA} s'$, then $h(s) \xrightarrow{h(l)}{}^{DA'} h(s')$.

If DI is initial in a class $\mathcal{D}$ of DΣ-algebras then its element have the minimum amount of activity:

$DI \models t \xrightarrow{t'} t''$ iff for all $DA \in \mathcal{D}$: $DA \models t \xrightarrow{t'} t''$.

3 A Logic for Specifying Dynamic Data Types

Following a widely accepted idea (see eg [W90]) a (static) *abstract data type* (shortly ADT) is an isomorphism class of Σ-algebras and it is usually given by a *specification*, ie a couple sp = (Σ, AX), where Σ is a signature and AX a set of 1st order formulae on Σ (the *axioms* of sp) representing the properties of the ADT. The *models* of sp are precisely the Σ-algebras which satisfy the axioms in AX; more precisely:

$Mod(sp) = \{ A \mid A$ is a Σ-algebra and for all $\theta \in AX$: $A \models \theta \}$.

In the *initial algebra approach* sp defines the ADT consisting of the (isomorphism class of the) initial elements of the class Mod(sp). In the *loose* approach, instead, sp is viewed as a description of the main properties of an ADT; thus it represents a class, consisting of all the ADT's satisfying the properties expressed by the axioms (more formally: the class of all isomorphism classes included in Mod(sp)).

The above definition of ADT can be easily adapted to the dynamic case: an *abstract dynamic data type* (shortly ADDT) is an isomorphism class of DΣ-algebras. In order to extend the definition of specification, the problem is choosing the appropriate logical framework. We have already discussed some of the problems in the introduction, therefore we first define our logic and then comment on it.

Recall that $D\Sigma = (\Sigma, STATE)$ and $\Sigma = (SRT, OP, PR)$; moreover let X be a fixed SRT-sorted family of variables s.t. for each sort srt X_{srt} is a denumerable set.

The sets of *dynamic formulae* and of *path formulae* of sort $st \in STATE$ on $D\Sigma$ and X, denoted respectively by $F_{D\Sigma}(X)$ and $P_{D\Sigma}(X, st)$, are inductively defined as follows (where $t_1, \ldots, t_n$ denote terms of appropriate sort and we assume that sorts are respected):

dynamic formulae

- $Pr(t_1, \ldots, t_n)$, $t_1 = t_2 \in F_{D\Sigma}(X)$ if $Pr \in PR$
- $\neg \phi$, $\phi_1 \supset \phi_2$, $\forall x . \phi \in F_{D\Sigma}(X)$ if $\phi, \phi_1, \phi_2 \in F_{D\Sigma}(X)$, $x \in X$
- $\Delta(t, \pi) \in F_{D\Sigma}(X)$ if $t \in T_{D\Sigma}(X)_{st}$, $\pi \in P_{D\Sigma}(X, st)$

path formulae

- $[\lambda x . \phi]$, $<\lambda y . \phi> \in P_{D\Sigma}(X, st)$ if $x \in X_{st}$, $y \in X_{lab(st)}$, $\phi \in F_{D\Sigma}(X)$
- $\neg \pi$, $\pi_1 \supset \pi_2$, $\forall x . \pi$, $\pi_1 \mathcal{U} \pi_2 \in P_{D\Sigma}(X, st)$ if $\pi, \pi_1, \pi_2 \in P_{D\Sigma}(X, st)$, $x \in X$.

Let DA be a $D\Sigma$-dynamic algebra and $V: X \to DA$ be a variable evaluation (ie an SRT-family of total functions). We now define by multiple induction when a formula $\phi \in F_{D\Sigma}(X)$ *holds in* DA *under* V (written $DA, V \models \phi$) and when a formula $\pi \in P_{D\Sigma}(X, st)$ *holds on a path* $\sigma \in PATH(DA, st)$ *under* V (written $DA, \sigma, V \models \phi$). Recall that the interpretation of a term t in DA w.r.t. V is denoted by $t^{DA,V}$ and that, for a path σ, $S(\sigma)$ and $L(\sigma)$ have been defined in Sec. 2.

dynamic formulae

- $DA, V \models Pr(t_1, \ldots, t_n)$ iff $(t_1^{DA,V}, \ldots, t_n^{DA,V}) \in Pr^{DA}$;
- $DA, V \models t_1 = t_2$ iff $t_1^{DA,V} = t_2^{DA,V}$ (both sides must be defined and equal);
- $DA, V \models \neg \phi$ iff $DA, V \not\models \phi$;
- $DA, V \models \phi_1 \supset \phi_2$ iff either $DA, V \not\models \phi_1$ or $DA, V \models \phi_2$;
- $DA, V \models \forall x . \phi$ iff for all $v \in DA_{srt}$, with srt sort of x, $DA, V[v/x] \models \phi$;
- $DA, V \models \Delta(t, \pi)$ iff for all $\sigma \in PATH(DA, st)$, with st sort of t, if $S(\sigma) = t^{DA,V}$ then $DA, \sigma, V \models \pi$;

path formulae

- $DA, \sigma, V \models [\lambda x . \phi]$ iff $DA, V[S(\sigma)/x] \models \phi$;
- $DA, \sigma, V \models <\lambda x . \phi>$ iff either $DA, V[L(\sigma)/x] \models \phi$ or $L(\sigma)$ is not defined;
- $DA, \sigma, V \models \neg \pi$ iff $DA, \sigma, V \not\models \pi$;
- $DA, \sigma, V \models \pi_1 \supset \pi_2$ iff either $DA, \sigma, V \not\models \pi_1$ or $DA, \sigma, V \models \pi_2$;
- $DA, \sigma, V \models \forall x . \pi$ iff for all $v \in DA_{srt}$, with srt sort of x, $DA, \sigma, V[v/x] \models \pi$;
- $DA, \sigma, V \models \pi_1 \mathcal{U} \pi_2$ iff there exists $j > 0$ s.t. $\sigma|_j$ is defined and $DA, \sigma|_j, V \models \pi_2$ and for all i s.t. $0 < i < j$ $DA, \sigma|_i, V \models \pi_1$.

A formula $\phi \in F_{D\Sigma}(X)$ is *valid* in DA (written $DA \models \phi$) iff $DA, V \models \phi$ for all evaluations V. Validity is preserved under isomorphisms.

Remarks. Dynamic formulae include the usual (hence static) many-sorted 1st order logic with equality; if $D\Sigma$ contains state-sorts, they include also formulae built with the transition predicates.

The formula $\Delta(t, \pi)$ can be read as "for every path σ starting from the state denoted by t, (the path formula) π holds on σ". We have borrowed Δ and ∇ below from [S]. We anchor those formulae to states because we do not refer to a single transition system but to a whole set of them.

The formula $[\lambda\, x\, .\, \phi]$ holds on a path σ whenever ϕ holds of the first state of σ; similarly the formula $<\lambda\, x\, .\, \phi>$ holds on σ whenever ϕ holds of the first label of σ. The need for both state and edge formulae has been already discussed in [L]. Finally, $\mathcal{U}$ is the so called strong until operator. **End remarks**

In the above definitions we have used a minimal set of combinators; in practice, however, it is convenient to use other, derived, combinators; the most common are:

true, **false**, $\vee, \wedge, \exists$, defined in the usual way; $\nabla(t, \pi) =_{def} \neg\, \Delta(t, \neg\, \pi)$;

$\Diamond\pi =_{def}$ **true** $\mathcal{U}\ \pi$ (eventually π); $\Box\, \pi =_{def} \neg \Diamond \neg\, \pi$ (always π);

$\pi_1\ \mathcal{WU}\ \pi_2 =_{def} \pi_1\ \mathcal{U}\ \pi_2 \vee \Box\, \pi_1$ (π_1 weak until π_2); $\bigcirc\, \pi =_{def}$ **false** $\mathcal{WU}\ \pi$ (next π).

A few examples should clarify the meaning of the non-standard constructs in our language; in particular, example 3) should explain the role of the binders λ x. We assume that: Cs is a constant symbol of state-sort st; Ps and Pl are unary predicate symbols of sort st and lab(st), respectively; x, x' and y are variables of sort st and lab(st) respectively. Moreover, for simplicity, we do not distinguish between the symbols Cs, Ps, Pl, ... and their interpretations.

1) $\Delta(Cs, \Diamond<\lambda\, y.Pl(y)>)$ can be read: on each path from the state Cs there exists a label satisfying Pl;

2) $\nabla(Cs, \Box\, \Diamond[\,\lambda\, x\, .\, Ps(x)])$ can be read: there exists a path from the state Cs that has infinitely many states satisfying Ps;

3) $\Delta(Cs, \Box\, [\lambda\, x.\nabla(x, \Diamond[\,\lambda\, x'.\ Ps(x')])])$ can be read: for every path σ from Cs, for every state x on σ, there is a path from x such that along this path there is a state x' satisfying Ps.

Our framework corresponds to an institution [GB]; here we just outline the basic definitions; full details will appear in [CR]. ***dyn*** = (**DSign**, DSen, DAlg, $\models$) is an institution, where:

DSign is the category whose objects are dynamic signatures and whose morphisms are the subclass of the morphisms of predicate signatures respecting the dynamic features (ie: dynamic sorts are mapped into dynamic sorts, special sorts and predicates are mapped into the corresponding special sorts and predicates); DSen is the sentence functor: $DSen(D\Sigma)$ is the set of formulae in $F_{D\Sigma}(X)$; DAlg is the algebra functor: $DAlg(D\Sigma)$ is the category $DAlg_{D\Sigma}$; $\models$ is our validity relation.

4 Dynamic Specifications

A *dynamic specification* is a couple sp = $(D\Sigma, AX)$, where $AX \subseteq F_{D\Sigma}(X)$. The loose semantics for sp is the class of all isomorphism classes in Mod(sp); its initial semantics is the isomorphism class of the initial elements of Mod(sp).

Notation: usually the dynamic specification $(D\Sigma, AX)$ will be written as $D\Sigma$ **axioms** AX.

Examples: we use the signature BUFΣ defined in Sec. 2; in ex. 1 we refer to the initial semantics and in ex. 2, 3 to the loose one.

Example 1: unbounded buffers with a LIFO policy
BUF = (BUFΣ, BUF-AX1), where BUF-AX1 consists of the following axioms:

-- properties of the data contained into the buffers (the terms 0 and Succ(n) are always defined):
D(0) D(Succ(n))
-- static properties (LIFO organization of the buffers)
D(Put(n, b)) ¬ D(Get(Empty)) ¬ D(Remove(Empty))
Get(Put(n, b)) = n Remove(Put(n, b)) = b
-- definition of the dynamic activity of the buffers

$D(Get(b)) \supset b \xrightarrow{O(Get(b))} Remove(b)$ a buffer can always return its first element (if it exists)

$b \xrightarrow{I(n)} Put(n, b)$ a buffer can always receive a value.

This specification admits initial models (see Propostion below): algebras where the carrier of sort buf is the set of unbounded stacks of natural numbers.

Example 2: a very abstract specification of buffers containing natural values; we only require the essential properties.
BUF2 = (BUFΣ, BUF-AX2) where BUF-AX2 consists of the following axioms.

-- properties of the data contained into the buffers (natural numbers):
D(0) D(Succ(n)) ¬ 0 = Succ(n) Succ(n) = Succ(m) ⊃ n = m
-- static properties (the operations Get and Remove are not defined on the empty buffer):
¬ D(Get(Empty)) ¬ D(Remove(Empty))
-- dynamic properties (safety properties):

$b \xrightarrow{O(n)} b' \supset n = Get(b) \wedge b' = Remove(b)$ specifies the action of returning a value

$b \xrightarrow{I(n)} b' \supset b' = Put(b, n)$ specifies the action of receiving a value

Here the operations Get, Put and Remove are not defined as in BUF1: we only specify some of their properties; clearly such a specification is oriented to a loose semantics. If A is an algebra which is a model of BUF2, the set A_{buf} may contain, for instance, unbounded and bounded buffers, FIFO and LIFO buffers.

Example 3: a very abstract specification of buffers containing natural values where the received values are always returned.
BUF3 is given by adding the following axioms to BUF2.

$b \xrightarrow{I(n)} b' \supset \Delta(b', (\neg <\lambda\ x.\ x = I(n)>)\ \mathcal{WU}\ <\lambda\ x.\ x = O(n)>)$
(a safety property) a buffer that has received a value n must return it before it can receive another copy of n (this ensures that the buffer contains distinct values)

$b \xrightarrow{I(n)} b' \supset \Delta(b', \Diamond <\lambda\ x.\ x = O(n)>)$
(a liveness property) eventually, a received value will be returned (recall that the elements in a buffer are distinct, so we know that it is the same n which appears in I(n) and O(n)).

If a buffer interacts with its users in a synchronous way, the last axiom is not very appropriate: indeed in case no one wants to accept the returned value, this axiom prescribes that the whole system, including the buffer and its users, will eventually deadlock. This problem can be avoided by replacing this last axiom with the two formulae below. They just require that a buffer will have the capability of returning any value it receives (*) and that such capability remains until the value is actually returned (**).

(*) $b \xrightarrow{I(n)} b' \supset \Delta(b', \Diamond\ [\lambda\ b''\ .\ Out\_Cap(b'', n)\ ])$

(**) $Out\_Cap(b, n) \supset \Delta(b, [\ \lambda\ b'\ .\ Out\_Cap(b', n)]\ \mathcal{WU} <\lambda\ x.\ x = O(n)>)$

where Out_Cap(x, y) stands for $\exists\ z\ .\ x \xrightarrow{O(y)} z$.

BUF3 specifies a subclass of the buffers defined by BUF2: the buffers where received values will, eventually, be returned (if someone requests them). It includes bounded FIFO buffers, but also, say, unbounded LIFO buffers where the "fair behaviour" is obtained by using auxiliary structures. On the other hand, the intial models of BUF1 are not included, since there each buffer admits an infinite path composed of input (push) actions only. *End of examples*

Not all dynamic specifications admit initial models. Classical (static) specifications are a particular case and it is well known that axioms like $t_1 = t_2 \vee t_3 = t_4$ or $\exists x . Pr(x)$ do not allow initial models. One can show that the same happens with formulae including existential temporal operators; for instance: $\nabla(t, \pi)$ or $\Delta(t, \Diamond\pi)$. However, as in the case of classical specifications, we can guarantee the existence of initial models by restricting the form of the axioms.

A formula $\phi \in F_{D\Sigma}(X)$ is *dynamic positive conditional* iff it has the form $\wedge_{i=1,\dots,n} \alpha_i \supset \psi$, where: $n \geq 0$, α_i is an atom (ie of one of the forms: $t_1 = t_2$; $Pr(t_1, \dots, t_n)$) and ψ is either an atom or has the form $\Delta(t, \pi)$ with π built using $[\dots]$, $<\dots>$, $\Box, \bigcirc$ only, and the formulae inside $[\dots]$ and $<\dots>$ are themselves dynamic positive conditional. The properties that can be specified using axioms of this kind include "usual" static properties and *safety properties*.

Proposition. Let dsp = $(D\Sigma, AX)$ be a dynamic specification; if the formulae in AX are dynamic positive conditional, then Mod(dsp) has initial models. □

Under the hypotheses of the above proposition, we have also a deductive system which is sound and complete with respect Mod(dsp). The first step is to consider a deductive system for equational logic with partially defined terms (but recall that algebras have nonempty carriers). This can be obtained, as in [C], by considering a system which is sound and complete for the total case and modifying it as follows: suppress reflexivity of equality; allow substitution of t for x only when t is defined (rule SUB below); add rules to assert that operations and predicates are strict (rules STR below).

One such system is given by the following rules:

$$\frac{t = t'}{t' = t} \qquad \frac{t = t' \quad t' = t''}{t = t''} \qquad \text{(SUB)} \ \frac{\phi[x] \quad D(t)}{\phi[t]}$$

$$\frac{D(Op(t_1,\dots,t_n)) \qquad t_i = t_i' \ (i=1,\dots,n)}{Op(t_1,\dots,t_n) = Op(t_1',\dots,t_n')} \qquad \frac{Pr(t_1,\dots,t_n) \qquad t_i = t_i' \ (i=1,\dots,n)}{Pr(t_1',\dots,t_n')}$$

$$\frac{(\wedge_{i=1,\dots,n} \phi_i) \supset \phi \qquad \phi_i \ (i=1,\dots,n)}{\phi} \qquad \text{(STR)} \ \frac{D(Op(t_1, \dots, t_n))}{D(t_i)} \qquad \frac{Pr(t_1, \dots, t_n)}{D(t_i)}$$

The second step is to extend the system by adding rules for the temporal operators (in the context of dynamic positive conditional formulae). Let us consider for each dynamic sort, st, of $D\Sigma$ a (new) predicate symbol $Trans_{st}$: st × st and the set of axioms TRANS = $\cup$ { $TRANS_{st}$ | st $\in$ STATE }, where, if x, z, w are variables of sort st and y is a variable of sort lab(st):

$$TRANS_{st} = \{\ x \xrightarrow{y} z \supset Trans_{st}(x, z),\ Trans_{st}(x, z) \wedge Trans_{st}(z, w) \supset Trans_{st}(x, w)\ \}.$$

Then, we consider the 4 rules below plus the 4 rules obtained by reversing them (ie exchanging premise with consequence):

$$\frac{\Delta(t, [\lambda\ x\ .\ \phi])}{\phi[t/x]} \qquad \frac{\Delta(t, <\lambda\ x\ .\ \phi>)}{t \xrightarrow{y} z \supset \phi[y/x]}$$

$$\frac{\Delta(t, \Box\ \pi)}{Trans(t, z) \supset \Delta(z, \pi)} \qquad \frac{\Delta(t, \bigcirc\ \pi)}{t \xrightarrow{y} z \supset \Delta(z, \pi)}$$

Proposition. Let dsp = (DΣ, AX) be a dynamic specification and $\vdash$ the deduction relation associated with the full system. If the formulae in AX are dynamic positive conditional, then: Mod(dsp)$\models\phi$ iff AX $\cup$ TRANS $\vdash$ ϕ. □

References ("LNCS x" stands for Lectures Notes in Computer Science vol. x)

[AC] Astesiano E.- Cerioli M. "On the existence of initial models for partial (higher order) conditional specifications", TAPSOFT'89, LNCS 351, 1989.

[AFD] Astesiano E. - Bendix Nielsen C. - Botta N. - Fantechi A. - Giovini A. - Inverardi P. - Karlsen E. - Mazzanti F. - Storbank Pedersen J. - Zucca E. "The Draft Formal Definition of Ada" Deliverable of the CEC MAP project: The Draft Formal Definition of ANSI/STD 1815A Ada, 1987.

[AR1] Astesiano E. - Reggio G. "On the Specification of the Firing Squad Problem" Workshop on The Analysis of Concurrent Systems, Cambridge, 1983, LNCS 207, 1985.

[AR2] Astesiano E. - Reggio G. "An outline of the SMoLCS methodology" Advanced School on Mathematical Models of Parallelism 1987, LNCS 280, 1987.

[BW] Broy M. - Wirsing M. "Partial abstract data types" Acta Informatica **18** (1982).

[C] Cerioli M. "A sound and equationally-complete deduction system for partial conditional (higher order) types" 3rd Italian Conf. on Theoret. Comp. Sci. Mantova 1989, World Scientific Pub., 1989.

[CR] Costa G.- Reggio G. "Specification and implementation of abstract dynamic data types: a temporal logic approach" Technical Report, University of Genova 1991.

[Em] Emerson A.E. "Temporal and modal logic" in Handbook of Theoret. Comp. Sci. Elsevier 1990.

[FL] Feng Y. - Liu J. "Temporal approach to algebraic specifications" Concur 90, LNCS 485, 1990.

[FM] Fiadeiro J. - Maibaum T. "Describing, Structuring and Implementing Objects", Draft, presented at the REX School/Workshop on Foundations of Object-Oriented Languages, May, 1990.

[GB] Goguen J.A. - Burstall R. "Institutions: abstract model theory for specification and programming" LFCS Report Jan.'90, Dept. of Comp. Sci. Univ. of Edinburgh.

[GM] Goguen J.A. - Meseguer J. "Models and equality for logic programming" TAPSOFT'87, LNCS 250, 1987.

[K] Kröger F. "Temporal logic of programs" EATCS Monographs, Springer 1987.

[L] Lamport L. "Specifying concurrent program modules" ACM TOPLAS **5** (1983).

[P] Pnueli A. "Applications of temporal logic to the specification and verification of reactive systems: a survey of current trends" in Current Trends in Concurrency, LNCS 224 (1986).

[S] Stirling C. "Comparing linear and branching time temporal logics" in Temporal Logic of Specification, LNCS 398 (1989).

[SFSH] Sernadas A. - Fiadeiro J. - Sernadas C. - Ehrich H.D. "Abstract object types: a temporal perspective" in Temporal Logic of Specification, LNCS 398 (1989).

[W89] Wirsing M. "Proofs in structured specifications" Preprint 1989.

[W90] Wirsing M. "Algebraic specifications" in Handbook of Theoret. Comput. Sci. vol.B, Elsevier 1990.

GENERATING WORDS BY CELLULAR AUTOMATA

Anton Černý
Department of Informatics, Comenius University,
Bratislava, Czechoslovakia[1]

ABSTRACT. Linear cellular automata with left-to-right communication are used for generating words. The terminal state of each cell corresponds to the symbol generated in the given position. In nondeterministic case such an automaton describes a language. The class obtained is exactly the class of context-sensitive languages. This fact provides a possible approach to defining context-sensitivity in the 2-dimensional case.

In the deterministic case generating infinite words rather than finite ones seems to be of interest. We show that all fixed points of (productive) substitutions and — in the 2-dimensional case — both regular and modular trellises can be generated.

1. Introduction

One interesting model for investigation of the computing power of parallel automata is the systolic trellis automaton [CGS 1982]. This kind of automaton is a recognizing device consisting of an infinite rectangular network of processors called *trellis structure* (Fig. 1).

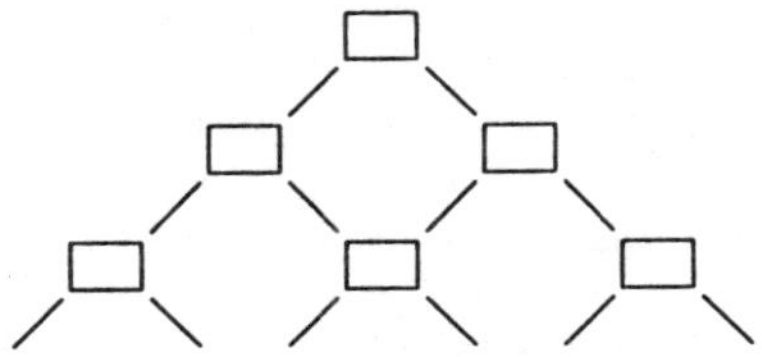

Fig. 1

The processors need not be all of the same type. If we denote each type of a processor by a distinct symbol (the processor's name), then a prticular placement of the symbols in the trellis structure will be called *trellis*. It is quite naturel to require some kind of regularity in the distribution of symbols in a trellis. One way how to achieve this goal was described in the original paper [CGS 198] where regular trellises were introduced. This approach was profoundly investigated in [K 1984] where regular trellises are called generalized Pascal triangles. Another approach — modular trellises — was considered in [CG 1986]. The classes of modular and regular trellises are proved to be incomparable.

Regular [modular] trellises can be obtained from so called strictly regular [strictly modular] trellises by alphabetical renaming of symbols. In a strictly regular trellis the topmost symbol is

[1]This work was partially supported by the research grant SM-058 of Kuwait University.

chosen arbitrarily and each pair of neighbor symbols (parents) in a horizontal row (one symbol only at the margin) uniquely determines the symbol inbetween in the next row (their son — see Fig. 1). A strictly modular trellis is obtained as a result of infinite iteration (or equivalently as a fixed point) of a uniform morphism.

In an equivalent approach a modular trellis can be completely described by a finite automaton which for any position in the trellis, given by its coordinates, determines the corresponding symbol. As shown in [CG 1986] modular trellises can be created in a top-down way very similar to that used for the regular ones if each node in the trellis behaves like the above finite automaton. Hence a trellis structure as in Fig. 1 with its nodes being finite state automata and the signals flowing top-down seems to be an interesting model for description of infinite trellises.

In the present paper, we investigate this kind of model being in principle a kind of cellular automaton [G 1984], however we will use it for generating rather than recognizing purpose. We call our model *generating cellular automaton*. Moreover, because of simplicity, we concentrate our attention to the one-dimensional case of our model where the trellis structure is replaced by a one-way infinite series of nodes generating finite or infinite sequences. Depending on whether the automata in the computational nodes are deterministic we obtain either a single word or a language of words as the result of generation. We prove that in the case of finitary languages the model describes exactly the class of context-sensitive languages. In the deterministic case the class of infinite sequences generated by the model includes both regular (i.e. quasiperiodic) and modular sequences and moreover all fixed points of productive uniform substitutions. The analogical results hold in the two-dimensional case as well.

2. Basic notions

Throughout this text we will identify a singleton set with its only element.

We denote $\mathbb{N} = \{0, 1, \dots\}$ — the set of all natural numbers. For $p, q \in \mathbb{N}$ let $[p] = \{0, 1, \dots, p-1\}$ (in particular $[0] = \emptyset$), $[p, q] = [p] \times [q]$.

For a finite alphabet Σ let Σ^*, Σ^ω denote the set of all finite and infinite words, respectively. A finite word (of length p) over Σ is a maping $w\colon [p] \to \Sigma$ (we denote $|w| = p$). An infinite word over Σ is a mapping $w\colon \mathbb{N} \to \Sigma$. The empty word (of length 0) is denoted by ε. We will use the usual operation of concatenation of finite words and of a finite and an infinite word.

Analogically, in the two-dimensional case we denote by Σ^{**}, $\Sigma^{\omega\omega}$ the sets of all array words (of type (p, q)) and trellises over Σ, respectively. Array words are mappings $w\colon [p, q] \to \Sigma$ (we denote $|w| = (p, q)$), trellises are mappings $w\colon \mathbb{N} \times \mathbb{N} \to \Sigma$. We depict a two-dimensional word as in Fig. 2.

$$\begin{array}{ccccc} & & w(0,0) & & \\ & w(1,0) & & w(0,1) & \\ w(2,0) & & w(1,1) & & w(0,2) \\ & & \dots & & \end{array}$$

Fig. 2

Let Σ, Γ be two alphabets. We will use the usual notion of a morphism from Σ^* to Γ^*. An (m, p)-substitution from Σ^* to Γ^* is a mapping h assigning to each word of length m over Σ a word of length p over Γ. h can be naturaly extended to all words over Σ of length being a multiple of m, and to all infinite words over Σ. In the case $m = 1$ h is a morphism called p-uniform morphism. In two-dimensional case we will use the analogical notion of $((m, n), (p, q))$-substitution.

3. Generating cellular automata

In our approach we will concentrate on the one-dimensional generating device because of its better transparency and simpler notation.

Informally, a one-dimensional generating cellular automaton (1-GCA, or if there is no danger of confusion, GCA) is a left-to-right infinite sequence of identical finite state automata — cells — working in synchronized steps (Fig. 3). The state of each cell depends on its state and on the state of its left neighbour in the previous step. Originally, all the cells are in a fixed "sleeping" state s. A cell can be "woken up" by its nonsleeping left neighbour only. The leftmost cell can wake up by itself. Once being woken up, a cell cannot sleep anew.

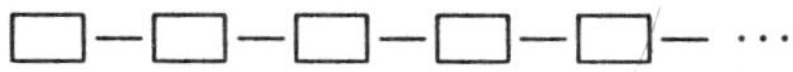

Fig. 3

The states of the cells are divided into two disjoint sets — nonterminals and terminals. A terminal state once reached cannot be changed. Moreover there is a nonterminal "dead" state \$ with the same property. We are interested in generating words over the alphabet of terminal states.

We will often use a dual view of a CGA as an infinite tape divided to fields, each containing a symbol of a given fixed alphabet (Fig. 4). In this view a field of the tape corresponds to a cell of the CGA and the symbol in the field to the state of the cell.

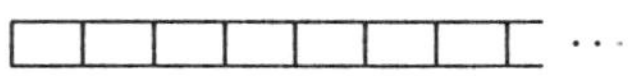

Fig. 4

Definition 1. A *generating cellular automaton* is a triple $\mathcal{A} = (N, T, \delta)$ where,

N is a finite set (alphabet) of nonterminal states (symbols) containing two different special elements s (sleeping) and \$ (dead);

T is a finite set (alphabet) of terminal states (symbols), $N \cap T = \emptyset$, we denote $K = N \cup T$;

δ: $K \times K \rightarrow (2^K - \emptyset)$ is the transition function of $\mathcal{A}$ such that for all $q \in K$, $a \in T$ $\delta(q, a) = a$, $\delta(q, \$) = \$$, $\delta(s, s) = s$.

A is *deterministic* if for all $a, b \in K$ $\delta(a, b)$ is a singleton set.

Definition 2. Any element of K^*s^ω is a *configuration* of $\mathcal{A}$. s^ω is the *initial configuration.*

A *computational step* of A is the relation $\succ$ on the set of all configurations defined for configurations x, y as

$$x \succ y \qquad \text{iff} \qquad y(i) \in \delta(x(i-1), x(i)) \quad \text{for all } i \in \mathbb{N}$$

(by default, $x(-1) = \$$).

A configuration x is *accessible* if $s^\omega \succ {}^*x$.

$w \in T^*$, $w \neq \varepsilon$, is *generated* by A if ws^ω is an accessible configuration; ε is generated by A if $s \in \delta(\$, s)$.

$w \in T^\omega$ is *generated* by A if each its prefix is a prefix of some accessible configuration of A not containing \$.

The set of all finite (infinite) words over T generated by A is denoted by $L(A)$, $(S(A))$, the class of all languages of finite (infinite) words generated by GCA-s is denoted by $\mathcal{L}(\text{GCA})$, $\mathcal{S}(\text{GCA})$.

4. Generating finite words by GCA

Obviously, if we allow the communication in a linear cellular automaton to proceed in both directions, the cells can simulate the Turing tape. We will show that the same is possible even in the case of one-way communication, if combined with nondeterminism, and if we allow each cell to decide to "veto" the current computation.

Suppose we want to simulate a Turing machine computation by the GCA. The GCA will use multiple symbols as nonterminals. One GCA field will correspond to one field of the Turing tape. The position of the Turing head will be denoted by the name of the current state of the Turing machine in the corresponding field of the GCA. The simulation of a right move of the head is straightforward. To simulate a left move each cell can any time decide that the head has moved to its position. In the following step the right neighbour has to check whether the decision was correct. (Hence if a cell contains the head which is to be moved left, this cell has to remember it still in the next step, to be able to approve the correct decision of the left neighbour.) The cell having discovered an error enters the "dead" $ state to "veto" the further computation.

Example: Simulation of the Turing step $p, b \longrightarrow q, c, \text{left}$:
(The subscript $_{\text{left}}$ serves for distinguishing from the case when the head has moved to right.):

a) a correct left move:

$$\begin{bmatrix} a \\ \ \end{bmatrix} \begin{bmatrix} b \\ p \end{bmatrix} \longrightarrow \begin{bmatrix} a \\ q_{\text{left}} \end{bmatrix} \begin{bmatrix} c \\ \text{check } q \end{bmatrix} \longrightarrow \begin{bmatrix} \ \\ \ \end{bmatrix} \begin{bmatrix} c \\ \ \end{bmatrix}$$

b) an incorrect left move:

$$\begin{bmatrix} a \\ \ \end{bmatrix} \begin{bmatrix} b \\ \ \end{bmatrix} \longrightarrow \begin{bmatrix} a \\ q_{\text{left}} \end{bmatrix} \begin{bmatrix} b \\ \ \end{bmatrix} \longrightarrow \begin{bmatrix} \ \\ \ \end{bmatrix} \begin{bmatrix} \$ \\ \ \end{bmatrix}$$

According to the definition of generating finite words on GCA the space of the GCA "tape" which can be used for simulating a Turing machine is exactly the length of the resulting generated word, thus GCA is able to simulate a linear bounded automaton (LBA) only.

The simulation of a LBA computation on a GCA proceeds then as follows. Since GCA is a generating device and LBA an accepting device, the GCA first nondeterministically generates (guesses) the terminal word to be generated (as the first and second parts of the symbols) and then simulates the LBA on the second parts of the symbols, while the first part is kept untouched. The rightmost symbol is not allowed to simulate receiving the head from right. If the word is accepted, then the simulated head is "moved" to the leftmost cell and all nonterminals are changed to terminals corresponding to the first parts of the symbols.

Since a CGA can be easily simulated by a LBA, we have proved

Theorem 1. $\mathcal{L}$(CGA) *is exactly the class of all (extended) context-sensitive languages.*

Theorem 1 can be proved even by simulation of a generative process of a nonshortening grammar. Basically, each cell of the CGA simulates one symbol. In each time step a signal starts to propagate from left to right. A cell can either let it go through or "swallow" it and start rewriting according to some rule of the grammar, the first symbol of the left-hand side of which is the current content of the cell. The new content is the first symbol of the right-hand side and information sent to the right neighbour on the rest of both sides of the rule (to be compared and rewritten). In this way several rewriting rules are simulated in parallel.

The technique of LBA simulation can be almost directly transformed to a two-dimensional case using 2-GCA — a trellis analogue of the 1-GCA generating words from T^{**}. Just a slight problem appears with the margin cells, that can be solved by simulation of four neighbour positions in one margin cell. Then a LBA with a 2-dimensional working tape can be simulated.

5. Generating infinite words by deterministic GCA-s

Considering a deterministic GCA (DGCA for short) A, there is just one its possible computation waking up either a finite or an infinite number of cells. An infinite terminal word can be generated in the latter case if the terminal prefixes of the configurations are strictly increasing. The language $S(A)$ is either empty or contains a single element.

We can characterize the following properties of $\mathcal{S}$(DGCA):

Lemma 1. $\mathcal{S}$(DCGA) *is closed under inverse of an injective uniform morphism.*

Proof (idea): If the morphism under consideration is p-uniform, then we can simulate p cells of the original DGCA by one cell of the new one. □

Theorem 2. $\mathcal{S}$(DCGA) *is closed under deterministic nonerasing finite-state single-input transductions.*

A nonerasing finite-state transducer is a finite-state device which in one step reads (on-line) a nonempty portion from the input word, outputs a nonempty word (not necessarily of a uniform length), depending on the current state and input portion, and changes the internal state. The device is single-input if the input portion is always a letter. We do not consider accepting states here since the transducer works on infinite words. (See e.g. [LT 1990] for an equivalent algebraic definition.)

Proof (sketch): Assume a DGCA A and a nonerasing finite state transducer T working on the terminal alphabet of A. We want to construct a DGCA A' generating the T-transduction of $S(A)$.

A' will work exactly like A, but whenever a cell has to reach a terminal state a, its next state will be $[a,-,-,0]$ in general, and $[a,q_0,w_0,0]$ for the leftmost symbol, where q_0 is the initial state of T and w_0 is the output of T on input a in state q_0. The transition function of A' will map $[a,q,w,k][b,-,-,0]$ to $[a,q',w',0]$ whenever T being in the state q and reading a enters the state q', and being in q' and readin b outputs w'. Thus the resulting transduction is formed by the third components of the compound symbols.

Now the third components have to be unpacked and pushed to right. This will be done one symbol per step. However, by this speed the propagating letters will soon reach a point where no bracketed symbol was created yet. To slow down the process, the fourth compnents of the bracketed symbols will be used as counters counting cyclically up to $n+1$, where n is the total number of states of A. A cell starts counting just after its left neighbour's counter has reached $n+1$ for the first time. This slowing down will be sufficient, since a cell in DGCA A having its left neighbour in a terminal state will reach the terminal state at most after n steps. Whenever the third component consists of a single symbol, the cell of A' can enter the corresponding terminal state. □

Corollary 1. $\mathcal{S}$(DCGA) *is closed under* (m,p)*-substitutions and nonerasing morphisms.*

$\mathcal{S}$(DCGA) contains all regular sequences (analogues of regular trellises) being actually ultimately periodic sequences. The results from [CG 1986] imply, that $\mathcal{S}$(DCGA) contains all modular sequences. Modular sequences are fixed points of (m,p)-substitutions such that p is a multiple of m. What about the substitutions where this is not the case? If $p \leq m$ then no finite initial part of a fixed point sequence determines the whole sequence, and according to [K 1990] the fixed points can be very strange, even nonrecursive in any reasonable way. If $p > m$, then the substitution is "productive", i.e. the initial part of length m determines the whole fixed point sequence. We are able to show that such sequences belong to $\mathcal{S}$(DCGA) .

Theorem 3. *Let m, p; $m < p$ be integers. Assume a (m,p)-substitution on a single alphabet. Then all its fixed points belong to $\mathcal{S}$(DCGA) .*

Proof (sketch): Assume a (m,p)-substitution σ on an alphabet T and its fixed point $w \in T^\omega$. We will give an idea how to construct a DGCA A such that $w = S(A)$.

Suppose we are given a position $t \in \mathbb{N}$ and we want to know the value $w(t)$. $w(t)$ is the initial letter of a subword of w of length mp contained in the image $\sigma(x)$ of some subword x' of w of length mp, starting at some position $t' < t$ being a multiple of m. The distance j in w between t and the first letter of $\sigma(x')$ can be chosen to be less than p. x is fully determined by x', σ, and j. In a similar way x' is fully determined by σ, some $j' < p$, and a subword x'' of w of length mp starting at some positon $t'' < t$ being a multiple of m. By iterating this process we will finish at the initial subword of w of length mp. By summarizing the above facts we obtain

Lemma 2. *For each $t \in \mathbb{N}$ there is a sequence of words $x_0, x_1, \ldots, x_k$ and corresponding sequences of integers $j_0, j_1, \ldots, j_k$; $t_0, t_1, \ldots, t_k$ such that for each $i \in [k+1]$*

(1) *x_i is a subword of w of length mp starting at the position t, in particular $t_0 = 0$, $t_k = t$*
(2) *for $i < k$ t_i is a multiple of m*
(3) $j_i \in [p]$; $j_i = t_i \bmod p$
(4) *for $i < k$ x_{i+1} is a subword of $\sigma(x_i)$ starting at the position j_i*
(5) $t_{i+1} = t_i * p/m + j_{i+1}$

The sequence $j_0 j_1 \ldots j_k$ is by (3) and (5) fully determined by t and forms a kind of notation for t expressed in base p/m. Indeed,

$$t = j_0(p/m)^k + j_1(p/m)^{k-1} + \cdots + j_k$$

e.g. 17 in base 3/2 can be expressed as 21012. This notation is independent on the leading zeroes and arithmetic is performed in the usual way, just each carry has to be multiplied by m.[2]

Based on Lemma 2 we can construct a finite automaton F for determining the value $w(t)$. The states of the automaton are the words over the terminal alphabet of A of length pm. The input alphabet is $[p]$. The initial state is x_0. This automaton for the input sequence being a (p/m)-notation of t will reach the state w_k with the initial letter being $w(t)$, as seen from (4) of Lemma 2.

To describe the desired DGCA A we would like its cells to behave like F. In this case we have to supply the t-th cell (digit-by-digit from its left neighbour) by the (p/m)-notation of t. If we supply it in the reversed order, then the t-th cell can easily increment the obtained notation by 1 and send it on to the right neighbour. To be able to use the reversed notation we have to rebuild F to follow the reversed input. This is a technical exercise on finite automata using some modification of standard techniques of reversing automata and constructing deterministic ones. □

The above results stated in Theorem 2, Corollary 1, and Theorem 3 can be directly transformed to the two-dimensional case. Thus $\mathcal{S}$(2-DCGA) contains all fixed points of $((m,n),(p,q))$-substitutions over a single alphabet, such that $m < p$ and $n < q$.

The proof of Theorem 3 provides an algorithm for determining for a fixed point of a (m,p)-substitution its value at some position without generating the preceding initial part of the sequence. We can achieve a more general result by using even a simpler proof technique. Theorem 3 in the case $m = 1$ deals with fixed points of uniform morphisms. The assertion can be generalized, up to some restrictions, to an arbitrary morphism as stated in the following Theorem 4, yielding, combined with Lemma 1, a generalization of Theorem 3. The restriction consists in the fact that a general morphism can have finite fixed points as well. Any infinite combination of such finite fixed points is a fixed point again. Obviously, if we form some kind of nonrecursive combination of

[2] A more authentical notation in base p/m should use digits j_i/p rather than j_i but the above one is more suitable for our purpose.

at least two different finite fixed points, the resulting sequence cannot be generated by a DGCA. That is why the notion of fixed point from Theorem 3 is replaced in Theorem 4 by the notion of the result of an infinite iteration of a morphism on a letter.

Given a morphism $h\colon \Sigma^* \longrightarrow \Sigma^*$, h is called ***prolongable*** on a letter $a \in \Sigma$, if the initial letter of $h(a)$ is a. In this case for any natural i $h^i(a)$ is an initial part of $h^{i+1}(a)$ and the process of infinite iteration of h starting on a has a limit (finite or infinite word) denoted by $h^\omega(a)$.

Theorem 4. *Let h be a morphism prolongable on a. Then $h^\omega(a)$ (whether being finite or infinite) can be generated by a DGCA.*

Proof (idea): Since $h^i(a) = h^{i-1}(a)w$ for some word w, $h^{i+1}(a)$ can be obtained from $h^i(a)$ by adding $h(w)$. This can be achieved in a DGCA if each of the letters of w (starting from left to right) produces its h-image, which starts to be propagated to the right. Having reached the right end of the current word, the h-image is joined to it. Technically this can be achieved by dividing the working tape of the DGCA to two parts, the upper part holding the terminal symbol to be the resulting state of the cell, and the lower part being used for propagating the strings coming from left. Each cell enters the terminal state after it has sent out its h-image. The cell produces its h-image, when its left neighbour enters the terminal state. If the h-image is empty, the cell enters the terminal state directly. □

Nondeterministic GCA-s can generate languages of infinite words (ω-languages). If we define generating of ω-languages in an analogical way on (deterministic) Turing machines, then using the idea of the simulation of a Turing machine on GCA from the previous section and define $\mathcal{S}$(TM) ($\mathcal{S}$(DTM)) to be the generated class of ω-languages, we obtain the following theorem:

Theorem 5.

$$\mathcal{S}(\mathrm{CGA}) = \mathcal{S}(\mathrm{TM})$$
$$\mathcal{S}(\mathrm{DCGA}) \subsetneqq \mathcal{S}(\mathrm{DTM})$$

Proof (of the second assertion — sketch): The inequality can be obtained by diagonalization. The i-th cell of a DGCA has to reach a terminal state after at most k^{i+1} steps, k being the total number of its states. Hence a deterministic Turing machine can find the i-th symbol generated by the i-th DGCA. □

Finally, we provide a nondeterministic analogue of Theorem 2:

Theorem 6. $\mathcal{S}$(CGA) *is closed under nonerasing finite-state transductions.*

Proof : Assume a GCA A and a transducer described by numbered quadruples of the form

$$m\colon \quad p, x, y, q$$

where during the corresponding step the tranducer changes state from p to q by reading $x = x_m^1 x_m^2 \dots x_m^{c(m)}$, and outputs $y = y_m^1 y_m^2 \dots y_m^{d(m)}$, m being the number of the quadruple. We describe a GCA A' generating all transductions of sequences from $\mathcal{S}$(CGA) .

A' first nondeterministically generates multiple-symbol contents of the cells, cells being groupped to blocks (Fig. 5).

$$\left[\begin{array}{llll|l|l|l}
y^1_{m1} & & & & y^2_{m1} & \cdots & y^{d(m1)}_{m1} \\
m1 & & & & m1 & \cdots & m1 \\
1 & & & & 2 & \cdots & d(m1) \\
x^1_{m1} & x^2_{m1} & \cdots & x^{c(m1)}_{m1} & \rightarrow & \rightarrow & \rightarrow \\
s & s & \cdots & s & s & s & s
\end{array}\right]
\left[\begin{array}{llll|l|l|l}
y^1_{m2} & & & & y^2_{m2} & \cdots & y^{d(m2)}_{m2} \\
m2 & & & & m2 & \cdots & m2 \\
1 & & & & 2 & \cdots & d(m2) \\
x^1_{m2} & x^2_{m2} & \cdots & x^{c(m2)}_{m2} & \rightarrow & \rightarrow & \rightarrow \\
s & s & \cdots & s & s & s & s
\end{array}\right]$$

Fig. 5

Fig. 5 shows two blocks of cells. The bold vertical lines separate blocks, the normal vertical lines separate cells and the short vertical lines separate symbols in the first cell of a block. The first row of the cells contains the guessed result — one letter per cell. The second contains the quadruple number — the same in one block. The word in the first row of the block must correspond to the y-part of the quadruple. The neighbour blocks can contain only such quadruple numbers, that the transducer states bind together. The ordering numbers in the third row are used in the process of building blocks. The first cell of each block contains in the fourth row the input portion of the quadruple. The last row, containing originaly s, is used for the simulation of A. Only the first cells of blocks are actually used for the simulation (of several cells of A), the others denoted by arrows are used just for propagating the rightmost simulated symbol to the next block.

After generating the blocks the simulation of A starts in the fifth row. Whenever all the symbols in the fifth row of the first cell of a block are terminals, they are compared to the fourth row. If a mismatch is indicated the whole process is aborted by generating \$. If matching is perfect, then the contents of all the cells in the block are replaced from left to right by the terminal symbols kept in the first row. □

References:

[CG 1986] Černý, A. – Gruska, J.: Modular trellises. In: Rosenberg, I. – Salomaa, A. (Eds.): The book of L. Springer, Berlin, 1986, 45–61

[CGS 1982] Culik, K. II – Gruska, J. – Salomaa, A.: Systolic trellis automata (for VLSI). University of Waterloo, 1982

[G 1984] Gruska, J.: Systolic automata – power, characterizations, nonhomogeneity. In: Chytil, M. – Koubek, V. (Eds.): Proceedings MFCS'84, Prague, Springer LNCS 176, Berlin, 1984

[K 1984] Korec, I.: Generalized Pascal Triangles. (in Slovak), doctoral thesis, Comenius University, 1984

[K 1990] Korec, I.: Personal communication.

[LT 1990] Latteaux, M. – Timmerman, E.: Rational ω-transductions. In: Rovan, B. (Ed.): Proceedings MFCS'90, B. Bystrica, Springer LNCS 452, Berlin, 1990

Atomic Refinement in Process Description Languages

PIERPAOLO DEGANO† and ROBERTO GORRIERI

Dipartimento di Informatica, Università degli Studi di Pisa
Corso Italia 40, I-56100 Pisa, Italy. e_mail uucp: {degano, gorrieri}@dipisa.di.unipi.it.

The hierarchical development of concurrent systems is investigated in a linguistic approach, by introducing a a new combinator for action refinement that substitutes a process for an action. In this way, the classic horizontal modularity is amalgamated with the new vertical one. The semantic definitions have been driven by two methodological criteria arising from a quest for compositionality that enforce to consider as *atomic* the behaviour of the processes refining actions. The first criterion requires that refinement must preserve the structure of the semantic object to be refined; the second one calls for a compositional refinement operation at the semantic level. Thus, *refinement is not syntactic substitution*, rather it is a compositional operation, which results to be context-free graph replacement of transition systems for transitions. The operational semantics implements this operation through states which are sort of stacks, used to make atomic the behaviour of processes refining actions; the denotational one uses tags expressing the start and the end of atomic sequences. Moreover, we define equivalences on both semantics, based on strong and rooted branching bisimulations, and we prove them *congruences* with respect to all the combinators of the language, and coincident.

1. INTRODUCTION

There is a general agreement on the fact that complex systems can be better understood by imposing structure on them, in order to build their parts, and reason on them, separately. This approach urged for the definition of a set of combinators suitable for constructing new systems from existing ones. *Compositionality* is the main achievement of this approach to the semantics and specification of concurrent systems. Once chosen a particular set of combinators, we obtain a *Process Description Language* as the free algebra generated by them. Possibly, congruences that identify systems with indistinguishable behaviour according to some notion of observation may be defined. Among these languages, we mention CCS [21], TCSP [18] and ACP [5], which can be equipped with various notions of observational semantics, such as the bisimulation-based strong and weak ones [21], based on bisimulation.

Another orthogonal direction for adding structure to concurrent systems consists of defining processes at conceptually different levels of description. Since describing a complex system is hard, we can formulate a simpler high level description of the system, and then implement its high level primitives in terms of a lower level system. This well known approach is sometimes referred to as the *hierarchical specification methodology*. It has been successfully developed for sequential systems, yielding, for instance, the technique known as *top-down system specification and analysis*, where a high level primitive stands for a lower level procedure. We advocate for Process Description Languages a notion playing a similar rôle in the specification and development of concurrent systems.

The two ways sketched above of giving structure to systems can be named as *horizontal* and *vertical modularity*, respectively. Besides imposing a structure to system specification, they provide the user with a facility for managing the complexity of verifying their properties.

We propose a uniform process description language that permits describing systems at different levels of detail, by freely exploiting both the horizontal and vertical development modularities. In the vain of [1, 23], we introduce a combinator for *action refinement*, understood as the replacement of a whole process, defined at a lower level, for an elementary action. Syntactically, our operator is $E[\lambda \rightsquigarrow E']$ in which E and E' are processes and λ is an activity that may be performed by E. The intuition behind its operational semantics is that whenever E will be engaged in λ, E' has to be executed in its place. An analogy between the refinement operator and (parameterless) procedure declaration in sequential languages can be drawn. Indeed, $E[\lambda \rightsquigarrow E']$ can be read as `let λ=E' in E`, which introduces the name λ, its body E', and its scope. In the definitions of its operational semantics we will be driven by this analogy, as we aim at conservatively extending the mechanism provided by sequential languages.

Research partially supported by Hewlett-Packard, Pisa Science Center, Corso Italia 115, I-56100 Pisa, Italy.
† Present address: Dipartimento di Matematica, Università di Parma, Via M. d'Azeglio, 85, I-43100 Parma, Italy.

We think that the sole justification for introducing a refinement operator is to enable hierarchical development of systems. Thus, in giving the semantics to our refinement operator, we will follow two methodological criteria, that enforce a compositional hierarchical system design.

The first major methodological point is that the overall structure of the system can be further detailed by refinement, but not altered at all. We claim that

> DEVELOPPING CONCURRENT PROCESSES AND REASONING ON THEM BY LEVELS OF ABSTRACTION REQUIRES THAT THE STRUCTURE OF CONTROL AT A GIVEN LEVEL IS PRESERVED AFTER REFINING SOME ACTIONS OCCURRING IN IT.

The second criterion stems from the fact that we are working in a linguistic framework. Process Description Languages should come equipped with an *operational semantics*, usually in terms of labelled transition systems in Plotkin's style. This definition is the first of the so-called "two-step approach" [19]. The second one consists of defining equivalences that abstract from those details considered immaterial, according to a notion of observation; in this way, processes are equated whenever no observation differentiates them. To be useful in a compositional approach, these behavioural equivalences need to be congruences. In our issue of refinement, this means that whichever two congruent processes are refined in congruent ways, two processes are obtained which are still congruent. Process Description Languages should also be equipped with a *denotational semantics* which maps syntactic terms to elements of the semantic domain, *via* an algebra with suitable operations. With respect to our problem, the semantic operations of refinement that we are going to define, must obey the following criterion

> SEMANTICALLY REFINING THE DENOTATION OF A PROCESS GIVES THE *SAME* DENOTATION OF THE SYNTACTIC REFINEMENT OF THE PROCESS ITSELF.

In other words, the diagram in Fig. 1 commutes.

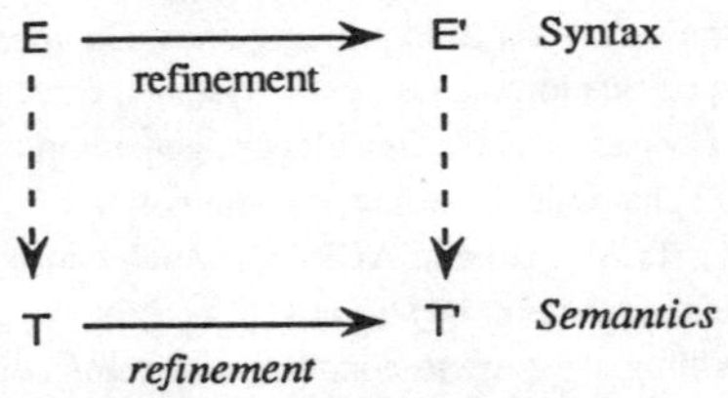

Figure 1. Process term E has T as semantic denotation; the refining operation (refinement) at the syntactic level has *refinement* as semantic counterpart.

We want to clarify the relationship between the two criteria. The second criterion essentially requires the existence of a compositional semantics, where syntactic refinement has a corresponding semantic operation. In a linguistic framework where the second criterion is satisifed, the first one calls for a more demanding property: it requires that the semantic operation of refinement be *context-free*, i.e., it must have a *local* effect only on the component to be refined, without affecting the overall structure of the system. This requirement seems very natural in a pragmatical perspective, where a concurrent system should be designed in a strictly hierarchical and modular fashion.

The contribution of our paper is the following. We consider a process description language with a set of combinators borrowed from ACP, CCS and TCSP, plus the new combinator of action refinement $E[\lambda \rightsquigarrow E']$. Note that this set is expressive enough to describe the usual process combinators [20]. The set $\mathcal{A}$ of *agents* contains the *closed, guarded* terms generated by the following syntax over the set of *basic* actions $M = \Lambda \cup \{\tau\}$ (ranged over by μ), the set of variables X (ranged over by x) and the set of communication sequences $\wp_{fin}(\Lambda^+)$ (ranged over by A):

$$E \rightarrow nil \mid \mu \mid E;E \mid E{+}E \mid E \|_A E \mid x \mid rec\ x.E \mid E[\lambda \rightsquigarrow E],$$

Sometimes refinements will be denoted by Ψ. First, we define a *denotational semantics* in terms of (a slight generalization of) Synchronization Trees. The meaning of the construct $E[\lambda \rightsquigarrow E']$ is intuitively simple: an arc labelled by λ of the tree which is the meaning of E is replaced by the tree of E'. Equivalences are then studied that are based on strong [19] and rooted branching [16] bisimulations, and proved them to be *congruences*. Our next step is giving an *operational semantics* to the language by defining in a structural way a transition system with the obvious goal of implementing the denotational definitions. In doing this, we follow the analogy with procedure call, respecting the two methodological criteria put forward above. To this aim, the states of the transition system are (almost) stacks, the elements of which are (almost) terms of the language. Stacks are exploited for storing the nested structure of refinements. A term E' is pushed on the stack with top $E[\lambda \rightsquigarrow E']$ when E may immediately execute λ, and it will be popped when terminated. The resulting operational semantics, up to strong and rooted branching bisimulations, *coincides* with the denotational one, and is also *fully abstract* for finite processes. Furthermore, a complete proof system for finite processes can be easily obtained from the denotational semantics, at the price of having axioms relating also some auxiliary terms that are not process terms.

2. INSTANTIATING THE METHODOLOGICAL CRITERIA

The two criteria introduced in the previous section reflect our intuition about what the meaning of refinements should be, and naturally discipline the semantics of systems developed in a horizontal / vertical methodology. Accordingly, constraints are to be imposed on how a concurrent system should be hierarchically specified and developed, so that all aspects of its semantic structure are preserved when step-wise passing from one level of description to a more detailed one. Also, these criteria imply that refinement is *not*, in general, syntactic substitution of a process for an action, rather it is a disciplined version of it. Indeed, the refining process is constrained to be executed *atomically*, i.e., it cannot be interrupted by, or interleaved with the actions of other concurrent processes. Moreover, the control structure of the system must be preserved under refinement. In other words, when two actions are ment to be executed in sequence or concurrently or are mutally exclusive, so must be their refinements; additionally, when processes do (or do not) communicate *via* actions, they must (or must not) communicate also after refining those actions. A more detailed, and hopefully convincing rationale of the impact of our choices follows.

2.1 Relationships between Refinement and the Horizontal Combinators

a) The actions at *any* level of abstraction are intrinsically *atomic*, and this property has to be preserved when they are refined. Thus, a process refining an action has to be executed *atomically*. Moreover, since only visible actions can be refined, the start and the end of the process refining an action must be "observable" to preserve such a visibility.
MOTTO: *action refinement can change neither the overall control structure of the system, nor the granularity of atomicity* — a well-established paradigm in top-down development of sequential systems — *nor the visibility structure.*

b) Since $E \parallel_A E'$ denotes two processes E and E' which can communicate *via* actions in the set A at a given level of description, the communication should be still possible when these actions have been refined *via* refinement Ψ. Of course, also the reverse should be true: if two processes cannot communicate, their refined processes cannot communicate. Thus, passing from process $E \parallel_A E'$ to the refined process $(E \parallel_A E')[\Psi]$ should not alter the structure of the communication channels. New communications can be established only *within* the refining processes. Since establishing the communication structure is the *major activity in designing concurrent systems,* we claim that these semantic-driven constraints enforce cleaner system specification and permit easier reasoning. Actually, also the communication structure is hierarchically organized and developed.
MOTTO: *action refinement can change neither the structure of communication nor of asynchrony.*

c) At a given level of abstraction, an action represents an activity which takes a *finite* amount of time to successfully terminate; the unsuccessful behaviour is represented simply by the non-occurrence of the action itself. Therefore, a process refining an action cannot definitively be empty. Additionally, a process refining an action may turn out specifying also an unsuccessful behaviour of the activity, e.g., deadlocks may arise. Moreover, a refining process can also be recursively defined, but it may infinitely often terminate. Summing up, refining an action consists of specifying its successful behaviour and, if divergence or deadlock are introduced, these must be reflected as divergence or deadlocks of the whole system (deadlocks will be dealt with in the full paper [10]). In this way, the sequential structure of control is not altered by successful behaviours, and erroneous situations are reported.
MOTTO: *action refinement should respect the structure of sequential composition.*

d) The final motto is:
MOTTO: *action refinement cannot change the structure of non-determinism.*

The only feature of the language that is affected by the atomicity constraint is parallel composition. As an example, the process $\alpha\|_\alpha\alpha$ may only synchronize with visible effect α; if we refine it to $(\alpha\|_\alpha\alpha)[\alpha\rightsquigarrow\alpha';\alpha'']$, the visible effect is the *atomic sequence* $\alpha'\alpha''$: the communication action itself has been refined. The key point is that also the visible effect of a communication is *always* considered as atomic. Since we aim at liberally permitting to intertwine vertical and horizontal development steps, we are compelled to accept that communication may take place on *atomic sequences* of actions. Thus, in the family of operators $\|_A$, the communication set A may contain also strings of actions. A further decision concerns how an atomic sequence of communications occurs. Motto a enforces that the synchronization of two processes on a sequence of actions must begin and end with an explicit, visible communication. The initial and final communications model a sort of *commitment* on the beginning and the end of the atomic sequence and "keep track" of the visibility of the refined communication.

2.2 Impact on the Denotational Semantics

The denotational model of Synchronization Trees [19] must be slightly refined in order to express atomicity which is crucial in guaranteing compositionality. These trees – called *Atomic Trees*, or *A-trees*– are labelled on a richer set of labels, also expressing the start and the end of an atomic sequence. When a visible action λ is the beginning of one of these sequences, it will be prefixed by tag "‹" and appear in the tree as $‹\lambda$; analogously, when λ is the last action of an atomic sequence, it will appear as $\lambda›$. The A-Trees we will use have actually a restricted format, due to the fact that no nesting of atomic sequences can be observed. A nested refinement can be syntactically defined which corresponds to a nested atomic sequence; however this will be seen as a single atomic sequence, for we are in an interleaving model. Indeed, all the processes which are concurrent with one performing an atomic sequence are frozen, and no change is observable by them if that sequence is in turn made of nested atomic sequences. As an example, consider the term $\alpha;\beta[\alpha\rightsquigarrow\gamma;\rho[\gamma\rightsquigarrow\theta;\zeta]]$. To perform action α, it should atomically execute $\gamma;\rho$. Since γ is in turn refined to $\theta;\zeta$, the overall atomic sequence is $\theta\zeta\rho$, which turns out to be the same observable behaviour of $\alpha;\beta[\alpha\rightsquigarrow\theta;\zeta;\rho]$. The consideration above suggests that the special labels for starting and ending actions come always in pair: any tag "‹" will be eventually followed by "›", without any other special labels in the middle.

We define an algebra on A-Trees, the operations of which roughly correspond to the syntactic operations of the language. Besides the obvious nondeterministic sum operator + (which is assumed associative, commutative, idempotent and with nil as neutral element), the other five operations on A-Trees are $;$, $\rightsquigarrow$, $\|_A$, ${}^{B}\rfloor_A$, ${}^{B}|_A$, where ${}^{B}\rfloor_A$ and ${}^{B}|_A$ are the generalizations of the auxiliary operators of left and communication merge of [5] to the case of atomic sequences. The semantic operation of sequential composition ; is trivial (see Appendix A for its definition). The atomicity constraint of refinement is easily achieved in the denotational model, because *action refinement is simply substitution of a tree for those arcs labelled by that action.* The only non trivial case is parallel composition, for which a more general version of the expansion law holds, dealing also with atomic sequences. For example, if we consider the synchronization tree $‹\alpha'\cdot\alpha''›\cdot nil$, and we put it in parallel with the synchronization tree $\beta\cdot nil$, we will obtain $‹\alpha'\cdot\alpha''›\cdot\beta\cdot nil + \beta\cdot‹\alpha'\cdot\alpha''›\cdot nil$, which is the semantics of $(\alpha\|_\emptyset\beta)[\alpha\rightsquigarrow\alpha';\alpha'']$. Of course, the new expansion law will also synchronize paths, starting with a $‹\lambda$, and ending with a $\lambda'›$.

- $$\Big(\sum_i ‹\lambda_i\cdot T_i + \sum_j \mu_j\cdot T_j\Big) \|_A \Big(\sum_h ‹\lambda_h\cdot U_h + \sum_k \mu_k\cdot U_k\Big) =$$
$$\sum_{\mu_j\notin A} \mu_j\cdot(T_j \|_A U) + \sum_{\mu_k\notin A} \mu_k\cdot(T \|_A U_k) + \sum_{\mu_j=\mu_k\in A} \mu_j\cdot(T_j \|_A U_k) + \sum_i ‹\lambda_i\cdot(T_i\, {}^{A_i}\rfloor_A U) +$$
$$\sum_h ‹\lambda_h\cdot(U_h\, {}^{A_h}\rfloor_A T) + \sum_{\lambda_i=\lambda_h\in Pref(A\setminus\{\lambda_i\})} ‹\lambda_i\cdot(T_i\, {}^{A_i}|_A U_h)$$

where $A_i=\{w \mid \lambda_i w\in A\}$, $Pref(A) = \{\lambda \mid \lambda w\in A\}$ and the two auxiliary operators are defined as follows:

- $$\Big(\sum_i \lambda_i›.T_i + \sum_j \mu_j\cdot T_j\Big)\, {}^{B}\rfloor_A U = \sum_{\lambda_i\notin B} \lambda_i›.(T_i \|_A U) + \sum_j \mu_j\cdot(T_j\, {}^{B_j}\rfloor_A U) \qquad \text{with } B_j=\{w \mid \mu_j w\in A\}$$
- $$\Big(\sum_i \lambda_i›.T_i + \sum_j \mu_j\cdot T_j\Big)\, {}^{B}|_A \Big(\sum_h \lambda_h›.U_h + \sum_k \mu_k\cdot U_k\Big) = \sum_{\lambda_i=\lambda_h\in B} \lambda_i›.(T_i \|_A U_h) +$$
$$\sum_{\mu_j=\mu_k\in Pref(B)} \mu_j\cdot(T_j\, {}^{B_j}|_A U_k) + \sum_{\lambda_i=\lambda_k\in Pref(B)} \lambda_i + \sum_{\lambda_j=\lambda_h\in Pref(B)} \mu_j + \sum_{\mu_j=\tau} \tau.(T_j\, {}^{B}|_A U) + \sum_{\mu_k=\tau} \tau.(U_k\, {}^{B}|_A T)$$

The first, second and third summands of this generalized expansion law come directly from the usual law. The fourth (symmetrically the fifth) one expresses that an atomic sequence is being asynchronously taken from T and must be completed before taking any action from U. This will be ruled by operation ${}^{B}\rfloor_{A}$ which extends to strings the left merge of [5]. The upper set B is used to keep track of the rest of those communication sequences starting with the performed action, and A is kept until the completion of the atomic sequence. The sixth summand enforces communication until the completion of a sequence in A. The auxiliary operator is a generalization to strings of the communication merge of [5]. Note that the start action of a communication sequence cannot be the internal action τ.

The aim of the auxiliary operator ${}^{B}\rfloor_{A}$ is to permit its left operand to complete an atomic sequence of actions not belonging to B. An atomic sequence succesfully terminates when its last action is not in B (first summand) and the ${}^{B}\rfloor_{A}$ operator is transformed back to the $\|_{A}$ operator. The operator for communication merge ${}^{B}|_{A}$ enforces its operands to communicate completing a sequence in B. The first summand of the equation says that a communication sequence can be completed provided that the ending action belongs to B; the second summand states that communication can be further carried on if the action to be synchronized is in the prefixes of B; the third and fourth addends consider the case when the last action of a sequence is synchronized with an intermediate action of the other sequence, which results in a synchronization (followed by a deadlock, [10]); the fifth summand permits τ-actions to be always performed in the middle of a communication sequence.

Let $\mathcal{Den}_{\langle\rangle}$: $\mathcal{A}\to$AT, the interpretation function mapping agents into A-Trees, be defined as follows:

- $\mathcal{Den}_{\langle\rangle}(\mathrm{nil}) = \mathrm{nil}$
- $\mathcal{Den}_{\langle\rangle}(\mu) = \mu$
- $\mathcal{Den}_{\langle\rangle}(E_1;E_2) = \mathcal{Den}_{\langle\rangle}(E_1) ; \mathcal{Den}_{\langle\rangle}(E_2)$
- $\mathcal{Den}_{\langle\rangle}(E_1+E_2) = \mathcal{Den}_{\langle\rangle}(E_1) + \mathcal{Den}_{\langle\rangle}(E_2)$
- $\mathcal{Den}_{\langle\rangle}(E_1 \|_A E_2) = \mathcal{Den}_{\langle\rangle}(E_1) \|_A \mathcal{Den}_{\langle\rangle}(E_2)$
- $\mathcal{Den}_{\langle\rangle}(E_1[\lambda\rightsquigarrow E_2]) = \mathcal{Den}_{\langle\rangle}(E_1)\ [\lambda\rightsquigarrow \mathcal{Den}_{\langle\rangle}(E_2)]$
- $\mathcal{Den}_{\langle\rangle}(\mathrm{rec}\ x.E) = \mathrm{fix}.\Gamma\ (\mathcal{Den}_{\langle\rangle}(E[\Gamma/x]))$

(Note that a partial ordering on A-trees can be easily defined — see, e.g., [30] — such that all the operators of the language are continuous with respect to it, and therefore fix is well-defined.)

2.3 Impact on the Operational Semantics

In this sub-section we will sketch how the denotational semantics and the above criteria have influenced the operational definitions of the language we are proposing.

For we are in an interleaving setting, the operational semantics is based on labelled transition systems, and is the usual one except for refinement. In this case, the basic idea is defining a single transition system for a process out of a hierarchy of transition systems, each defining its refinements at a specific level of description. We will illustrate this intuition through an example. Suppose we are given a process term E_1, with associated a transition system T_1, and a refinement of an action λ to a process E'_1, with transition system T'_1. The transition system of $E_1[\lambda\rightsquigarrow E'_1]$ is defined as T_1, where T'_1 replaces all the transitions labelled by λ. Of course, if $E_1 \!-\!\lambda\!\to E_2$ is a transition of T_1, then all the states reachable from E'_1 *via* an action μ are made reachable from E_1 (rather, from $E_1[\lambda\rightsquigarrow E'_1]$), *via* μ, as well. Analogously, all the transitions of T'_1 leading to a final state will reach the state E_2 (rather, $E_2[\lambda\rightsquigarrow E'_1]$). In this way, the operational semantics guarantees atomicity of refinement, according to mottos a and c. Indeed, if we see a transition system as a graph, our first criterion can be rephrased in graph-theoretical terms as follows: the refinement of an arc by a graph must be a context-free replacement operation.

More in detail, our operational semantics will implement this idea following the analogy with procedure call in sequential languages. To this aim, the states of the transition system are a sort of stacks, which grow letfwards by pushing a state s onto another state. The *states* are generated by the following syntax:

$$s ::= E \mid s{\cdot}s \mid s\,|_A s \mid s\backslash A$$

An agent is a state; $s\,|_A s'$ represents a situation in which both the components of the parallel composition $\|_A$ are performing an atomic sequence of basic actions in a synchronous way; $s\backslash A$ stands for a situation in which one of the components in $\|_A$ is performing asynchronously an atomic sequence, which cannot be in the communication set A.

Typically, a transition will have the following shape $E{\cdot}s \!-\!\lambda\!\to s'{\cdot}E'{\cdot}s$. The intuition behind such a transition is that the process E will evolve to E' by performing an atomic sequence having λ as its first step. The agents stacked in s' reflect the rest of that atomic sequence. Since refinements may be nested recursively, refining an action of E, say λ', may lead to store several nested continuations in s' before reaching the first action λ actually executable. In other words, an agent is pushed on the stack when it is the refinement of an immediately executable action λ'; recursively, λ is the first action of the innermost refining process. The state s represents the fact that the atomic action λ' of E is simply an intermediate step of the atomic action of the agent at the bottom of the stack.

Coming back to the above example, when the active process is $E_1[\lambda \rightsquigarrow E'_1]$ and action λ has to be executed, one of the possible first moves of E'_1 is performed in its place. The state reached in this way is pushed in a (sort of) stack, as it represents the active process, on top of $E_2[\lambda \rightsquigarrow E'_1]$, that represents the return point from which the computation will continue after the refinement is completed. An example of this construction is shown in Fig. 2, where $E_1 = \lambda;\varepsilon+\sigma$, $E_2 = nil;\varepsilon$, $E_3 = E'_3 = nil$, $E'_1 = rec\ x.\beta;\alpha;x+\gamma$, $E'_2 = nil;\alpha;(rec\ x.\ \beta;\alpha;x+\gamma)$, $E'_4 = nil;(rec\ x.\ \beta;\alpha;x+\gamma)$. Atomicity is indeed guaranteed by this mechanism: either an action λ is executed in a single step, if not refined, or the actions of its refining process are executed atomically, for those actions concurrent with λ are hidden by the new top of the stack, and thus frozen. As a final remark, we want to stress the fact that *our construction gives the linguistic counterpart to the semantic operation of graph replacement on labelled transition systems.*

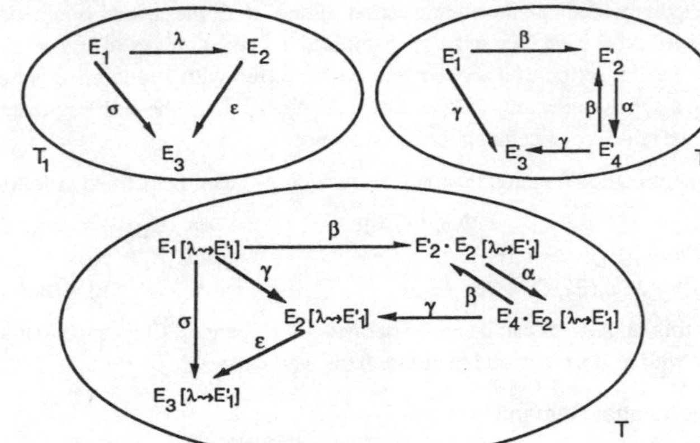

Figure 2. A transition system T_1 for E_1, T'_1 for E'_1 and T for $E_1[\lambda \rightsquigarrow E'_1]$, assuming that E'_3 is a successfully terminated state. Stacks grow to the left, and the push operation on them is denoted by a dot. The effect of refining λ with the one-step computation $E'_1 \text{—}\gamma\rightarrow E'_3$ is the transition $E_1[\lambda \rightsquigarrow E'_1] \text{—}\gamma\rightarrow E_2[\lambda \rightsquigarrow E'_1]$, i.e., this is a case of *relabelling*.

Operationally, refinement is defined by the following inference rules, where $\surd(E)$ denotes proper termination of E (roughly, E is made of nils, only), homomorphically extended to states. (The other inference rules for the operational semantics are listed in Appendix B.)

$$\text{Ref-1} \quad \frac{E_1 \text{—}\mu\rightarrow \{s\cdot\}E_1'}{E_1[\lambda \rightsquigarrow E_2] \text{—}\mu\rightarrow (\{s\cdot\}E_1')[\lambda \rightsquigarrow E_2]} \quad \mu \neq \lambda$$

$$\text{Ref-2} \quad \frac{E_1 \text{—}\lambda\rightarrow \{s_1\cdot\}E_1' \text{ and } E_2 \text{—}\mu\rightarrow E_2' \text{ and } \surd(E_2')}{E_1[\lambda \rightsquigarrow E_2] \text{—}\mu\rightarrow (\{s_1\cdot\}E_1')[\lambda \rightsquigarrow E_2]}$$

$$\text{Ref-3} \quad \frac{E_1 \text{—}\lambda\rightarrow \{s_1\cdot\}E_1' \text{ and } E_2 \text{—}\lambda'\rightarrow \{s_2\cdot\}E_2' \text{ and not } \surd(\{s_2\cdot\}E_2')}{E_1[\lambda \rightsquigarrow E_2] \text{—}\lambda'\rightarrow (\{s_2\cdot\}E_2')\cdot((\{s_1\cdot\}E_1')[\lambda \rightsquigarrow E_2])}$$

Rule Ref-1 states that if E_1 can perform an action μ, then $E_1[\lambda \rightsquigarrow E_2]$ can do so provided that μ is not under refinement. (The expression $\{s\cdot\}$ stands for optional presence of stack s, hence Ref-1 is a shorthand for two different rules; if present, s is componentwise affected by the application of $[\lambda \rightsquigarrow E_2]$.) Rule Ref-2 is for relabelling. If the refining process can perform a computation that successfully terminates in a single step, then it is not pushed onto the stack, because relabelling should not change the level of description (this is not standard relabelling, because it permits to relabel an action through any agent, the computations of which consist of a single action, e.g., $[\lambda \rightsquigarrow \alpha+(\beta[\beta \rightsquigarrow \gamma])]$). Rule Ref-3 deals with the general case. If the refining process may evolve to the state $\{s_2\cdot\}E_2'$, this state is put on top of the stack, becoming the active component. It can be easily proved by induction that the top element of stack $\{s_2\cdot\}E_2'$ is not properly terminated; nonetheless, some of its elements can be in the relation $\surd$. Note that only a visible action is permitted as the start action of an atomic sequence, according to motto a. Note also that only the (old) part of the stack $\{s_1\cdot\}E_1'$ is updated with the refinement function.

The rules for handling stacks follow. If a state can perform an action, it may do so when it is on top of a stack (S-1); state s' is popped as soon as it successfully terminates with a visible action (S-2), according to motto **a**. The top of the stack can be properly terminated only when it has been pushed by Ref-3. Therefore, we also need a rule that allow to remove from stacks terminated states (S-3).

$$\text{S-1}\ \frac{s \xrightarrow{\mu} s' \text{ and not } \surd(s')}{s \cdot s'' \xrightarrow{\mu} s' \cdot s''} \qquad \text{S-2}\ \frac{s \xrightarrow{\lambda} s' \text{ and } \surd(s')}{s \cdot s'' \xrightarrow{\lambda} s''} \qquad \text{S-3}\ \frac{\surd(s) \text{ and } s'' \xrightarrow{\mu} s'}{s \cdot s'' \xrightarrow{\mu} s'}$$

Few comments are in order about parallel composition. In order to deduce a transition of communication between two refined processes, they may do before being refined. Also, the refining processes will not communicate each other or with actions at a different level, because they will appear at different depth in the stack. Therefore, the structure of communication and asynchrony of the system is hierarchically defined according to motto **b**. Indeed, our proposal of atomic refinement *does preserve* the notion of level of abstraction [3].

In order to relate the operational and the denotational semantics, the former must be extended to explicitly represent atomicity, call it $Op_{\diamond}$. Note that this information is already available in the transition system given above. Indeed, an atomic sequence starts when a state is pushed on an agent, and ends when the stack is popped to become again an agent. We first slightly modify transitions with the enriched labels. An A-tree is then obtained as $Op_{\diamond}$ by unfolding the resulting transition system, i.e., by taking its computations ordered by prefix. (Details about the isomorphism of the two transition systems in the full paper [10].)

Our main result is that the operational and the denotational semantics coincide (up to strong bisimulation [21], already enforced in the algebra of A-Trees), and moreover that they coincide for any context.

Theorem (*full abstractness*) Given an agent E, the following properties hold:

i) $Op_{\diamond}(E) = \mathcal{D}en_{\diamond}(E)$

ii) $\mathcal{D}en_{\diamond}(E) = \mathcal{D}en_{\diamond}(E')$ if and only if, for every finite context $C[]$, $Op_{\diamond}(C[E]) = Op_{\diamond}(C[E'])$.

2.4 Impact on Equivalences

Various notions of behavioural congruence have been defined which can be adapted to support abstraction from unwanted details. However, not all of them are resistant to action refinement. First, we stress the fact that strong congruence [19] is not preserved under syntactic action refinement, understood as *non-atomic* syntactic substitution (see [25]). An example, presented in [6], makes the point more clear and considers the following equivalent processes $\alpha \|_{\emptyset} \beta$ and $\alpha;\beta + \beta;\alpha$. The two refined processes $(\alpha \|_{\emptyset} \beta)[\alpha';\alpha''/\alpha]$ and $(\alpha;\beta + \beta;\alpha)[\alpha';\alpha''/\alpha]$ are not equivalent, not even for trace equivalence, for a β-action can occur in between α' and α''. Thence, the naïve notion of refinement as syntactic substitution does not preserve the key property illustrated in Figure 1. According to our atomicity constraint, the refined processes are still equivalent, for the β-action cannot interrupt the atomic sequence $\alpha';\alpha''$. Also weak congruence [21] is not preserved under action refinement (e.g., its axiom $\mu(x + \tau y) = \mu(x + \tau y) + \mu y$ does not hold when μ is refined to $\mu'\mu''$). We will use the more demanding notion of rooted branching bisimulation [16], that requires choices to be maintained under τ-moves, and prove that it is preserved by all the combinators of the calculus, namely it is a congruence.

Theorem The equivalences induced by strong and rooted branching bisimulations, respectively, are congruences.

The next step is concerned with the definition of a complete system of axioms characterizing strong and rooted branching congruences for finite processes. As noted in the Introduction, in order to guarantee the atomicity requirement, extra states are needed that do not correspond to process terms. Therefore, the denotational semantics is not sufficient as it is to induce a complete axiomatization on process terms, only. If we incorporate all the (semantic and syntactic) operators used to define the (denotational and operational) semantics of the calculus within its syntax, we obviously have a sound and complete proof system for it. This is a further step generalizing the use of a left, right and communication merge, as done in [5]. Some interesting properties on process terms hold involving refinement, besides the usual ones about the other combinators. As a matter of fact, all combinators commute with refinement, except for $\|_A$ because pushing refinements inside its arguments may introduce new communications, thus violating motto **b**. This shows that our notion of refinement cannot be interpreted as syntactic substitution, instead it is a first class operator of the language.

3. RELATED WORK

In spite of its theoretical and pragmatic relevance, only recently the issue of refinement in concurrency models has received the attention it deserves. The first attempts arose in the area of Petri Nets, thus considering a semantic model rather than a Process Description Language. Among the various proposals, refinements on nets are defined in [28, 27, 13, 29] by considering transitions, seldom places, as abstractions of entire sub-nets. Restrictions are imposed on the proposed replacement operations and a general solution is not available yet. An appealing proposal comes from the categorical treatment of Petri Nets [22], where refinements are morphisms which may map transitions to computations. Still in the semantic side, but in the interleaving approach, semantic refinement in Synchronization Trees has been defined as replacement of a tree for an arc, thus respecting our motto a. Notably, this operation has been proved to respect strong and rooted branching congruences [17, 9]. Refinement has been studied also in Causal Trees [7]. Both atomic [9] and non-atomic [8] refinements are defined, that respect the causal counterparts of strong and rooted branching congruences. In the domain of Event Structures [24], refinement turns out to be replacement of an event structure for an event in [13] and [8].

Refinement from a linguistic point of view has been studied in [15, 1, 23], where simple finite calculi without either autoconcurrency, or communication or restriction (or combinations of these three ingredients) are considered. In [2], Aceto and Hennessy extend their proposal to a language with communication and restriction. However, their notion of refinement as syntactic substitution may lead to some problems. Consider the two CCS agents $\alpha | \alpha^-$ and $\tau + \alpha | \alpha^-$ which are equivalent according to all the so far proposed equivalences; then, replacing syntactically $\alpha_1.\alpha_2$ for α (dually for α^-), we obtain the agents $\alpha_1.\alpha_2 \,|\, \alpha^-_1.\alpha^-_2$ and $\tau + \alpha_1.\alpha_2 \,|\, \alpha^-_1.\alpha^-_2$ which are not even maximal trace equivalent, for only the second agent can perform a maximal trace τ. Therefore, no strong bisimulation equivalence may be a congruence w.r.t. their operator. They prove that there exists a weak bisimulation, weak timed-bisimulation equivalence, which is a congruence, provided that refinement functions satisfy a restrictive condition. Although successfull in defining a notion of weak congruence, we share a strong criticism against defining refinement as syntactic substitution, expressed in [3]. Indeed, a confusion of the abstraction levels is made, because refined actions (those of E' in $E[\lambda \rightsquigarrow E']$) and unrefined ones (those of E) can now synchronize, possibly changing in unexpected ways the overall structure of the system.

The dual notion of process abstraction is also considered in the literature. It is studied in [14] where CCS is extended with an explicit combinator of strong prefixing, expressing the beginning of an atomic sequence. Atomic actions can be defined in this way, but only at two levels of description; an abstraction homomorphism is defined between them that preserves strong congruence. A different notion of process abstraction is presented in [4], for a calculus with an explicit operation on data. Atomic actions are obtained by abstracting to a single event a whole operation on data, i.e., a terminated computation.

For a wider presentation of the various approaches to refinement, the reader is referred to [26, 12].

4. CONCLUSIONS

The semantics of our language takes as basic principle that the execution of processes refining actions is atomic, thus a combinator for atomicity of a sequence of actions is immediately expressed in terms of refinements. As an example, if $\langle E \rangle$ denotes the operator of making atomic the execution of process E, a the process $\langle \alpha';\alpha'' \rangle \|_{\varnothing} \beta$ can be implemented by $(\alpha[\alpha \rightsquigarrow \alpha';\alpha''] \|_{\varnothing} \beta)$. Atomicity is a crucial notion in many areas; for instance, in distributed data bases, it is exploited for implementing transactions, that model possibly complex, non interruptable activities. Moreover, the effects of a transaction are visible only if it has been successfully terminated, namely, transactions enjoy the *all-or-nothing* property. Note that, for defining *all-or-nothing* behaviours, only the rules for asynchrony need to be slightly changed. Actually, a simple check has to be performed on the whole atomic sequence an agent is starting, and only the successfully completed ones should be kept. Thus, a language with atomicity and transactions can be implemented in ours.

We have chosen here an interleaving model of concurrency to experiment on our ideas in a firm framework, even if the methodological requirement of preserving system structure during refinement can be naturally expressed also in "truly concurrent" models. In such a case, the notion of atomicity itself seems to disappear, because processes lay in distinct places and there is *no* global state. As a matter of fact, extending our approach to the truly concurrent case is an interesting issue, and a first attempt in this direction is reported in [11, 12].

ACKNOWLEDGEMENTS. We would like to thank Ugo Montanari for encouragement. The first author acknowledges Jaco de Bakker for an extremely pleasent stay at CWI, where the first ideas of this paper spring out.

REFERENCES

[1] Aceto L., Hennessy M., Towards Action Refinement in Process Algebra, Proc. LICS'89, 138-145, (1989).

[2] Aceto L., Hennessy M., Adding Action Refinement to a Finite Process Algebra, Proc. ICALP'91, LNCS, (1991).

[3] Boudol G., Atomic Actions, Bulletin of the EATCS **38**, 136-144, (1989).

[4] Boudol G., Castellani I., Concurrency and Atomicity, *Theoretical Computer Science*, **59**, 25-84, (1988).

[5] Bergstra J.A., Klop J.-W., Process Algebra for Synchronous Communication, *Info and Co* **61**, 109-137, (1984).

[6] Castellano L., De Michelis G., Pomello L., Concurrency vs Interleaving: an Instructive Example, Bullettin of the EATCS **31**, 12-15, (1987).

[7] Darondeau Ph., Degano P., Causal Trees, in Proc. 11th ICALP, LNCS 372, 234-248, (1989).

[8] Darondeau Ph., Degano P., Event Structures, Causal Trees, and Refinements, Proc. MFCS'90, LNCS 452, (1990).

[9] Darondeau Ph., Degano P., About Semantic Action Refinement, *Fundamenta Informaticae* XIV, 221-234, (1991).

[10] Degano, P., Gorrieri, R., Atomic Refinement for Process Description Languages, TR 17-91 HP Pisa Center, 1991.

[11] Degano, P., Gorrieri, R., Parallel Procedure Call as Event Structure Substitution, preliminary report, March1991.

[12] Gorrieri R., *Refinement, Atomicity and Transactions*, Ph.D. thesis, Uni. di Pisa, Dip. di Informatica, TD-2/91, 1991.

[13] van Glabbeek R.J. and Goltz U., Refinement of Actions in Causality Based Models, in [26], 267-300.

[14] Gorrieri R., Marchetti S. and Montanari U., A^{2}CCS: Atomic Actions for CCS, *T.C.S*, **72**, 203-223 (1990).

[15] van Glabbeek R.J., Vaandrager F.W., Petri Nets Models for Algebraic Theories of Concurrency, Proc. PARLE, LNCS 259, 224-242, (1987).

[16] van Glabbeek R.J., Weijland W.P., Branching Time and Abstraction in Bisimulation Semantics, Proc. IFIP'89, 613-618.

[17] van Glabbeek R.J., Weijland W.P., Refinement in Branching Time Semantics, Proc. AMAST'89, 197-201, (1989).

[18] Hoare C.A.R., *Communicating Sequential Processes*, Prentice Hall International, (1985).

[19] Milner R., A Calculus of Communicating Systems, LNCS 92, (1980).

[20] Milner R., Process Constructors and Interpretations, Proc. IFIP'86, (North-Holland), 507-514, (1986).

[21] Milner R., Communication and Concurrency, Prentice Hall International, (1989).

[22] Meseguer J., Montanari U., Petri Nets are Monoids, *Information and Computation*, **88** (2), 105-155, 1990.

[23] Nielsen M., Engberg U., Larsen K.S., Fully Abstract Models for a Process Language with Refinement, Proc. REX School on *Linear Time, Branching Time and Partial Order in Concurrency*, LNCS 354, 523-548, (1989).

[24] Nielsen M., Plotkin G., Winskel G., Petri Nets, Event Structures and Domains part I, *TCS*, **13**, 85-108, (1981).

[25] Pratt V.R., Modelling Concurrency with Partial Orders, *Int. Journal of Parallel Prog.*, **15** (1), 33-71, 1986.

[26] REX Workshop on *Stepwise Refinement of Distributed Systems*, Mook, 1989, LNCS 430, (1990).

[27] Suzuki I, Murata T, A Method for Stepwise Refinement and Abstraction of Petri Nets, *JCSS* **27** (1), 51-76, (1983)

[28] Valette R., Analysis of Petri Nets by Stepwise Refinements, *J.C.S.S.*, **18**, 47-64, (1979).

[29] Vogler W., Failure Semantics based on interval semiwords is a congruence for refinement, *Distributed Computing* **4**, 139-162, (1991).

[30] Winskel G., Synchronization Trees, *Theoretical Computer Science*, **34**, 33-82, 1984.

APPENDIX A: The Algebra of Atomic Trees

Function *norm*, transforming an a-tree into a synchronisation tree, is defined on labels as follows:

- $norm(\langle\lambda) = \lambda$
- $norm(\lambda\rangle) = \lambda$
- $norm(\mu) = \mu$,

and is homomorphically extended to atomic trees as follows:

- $norm(\mathrm{nil}) = \mathrm{nil}$
- $\mathrm{norm}(\sum_h \rho_h . T_h) = \sum_h norm(\rho_h).norm(T_h)$

Functions $\langle\cdot\rangle$, $\cdot\rangle$ and $\langle\cdot$ are defined as follows:

- $\langle \mathrm{nil} \rangle = \mathrm{nil}$
- $\langle(\sum \mu_i . T_i + \sum \mu_j .\mathrm{nil})\rangle = \sum_{\mu_i \neq \tau} \langle \mu_i . T_i \rangle + \sum \mu_j .\mathrm{nil}$
- $(\sum \mu_i . T_i + \sum \mu_i .\mathrm{nil})\rangle = \sum \mu_i . T_i\rangle + \sum_{\mu_j \neq \tau} \mu_j\rangle.\mathrm{nil}$
- $\langle\sum \mu_i . T_i = \sum_{\mu_i \neq \tau} \langle \mu_i . T_i$

The algebra of Atomic Trees is defined as follows:

+) $\bullet\ T + U = U + T \qquad \bullet\ T + (U + T') = (T + U) + T' \qquad \bullet\ nil + T = T \qquad \bullet\ T + T = T$

;) $\bullet\ nil\,;T = T \qquad \bullet\ (\sum_h \rho_h.T_h)\,;U = \sum_h \rho_h.(T_h;U)$

$\rightsquigarrow$) $\bullet\ nil[\lambda \rightsquigarrow T] = nil$

$$\bullet\ (\sum_h \rho_h.T_h)[\lambda \rightsquigarrow U] = \sum_{norm(\rho_h) \neq \lambda} \rho_h.(T_h[\lambda \rightsquigarrow U]) + \sum_{\rho_h = \langle\lambda} (\langle norm(U));(T_h[\lambda \rightsquigarrow U]) +$$

$$\sum_{\rho_h = \lambda, \rangle(T_h)} (norm(U));(T_h[\lambda \rightsquigarrow U]) + \sum_{\rho_h = \lambda, \langle\rangle(T_h)} (\langle norm(U)\rangle);(T_h[\lambda \rightsquigarrow U]) + \sum_{\rho_h = \lambda\rangle} (norm(U)\rangle);(T_h[\lambda \rightsquigarrow U])$$

$\|_A$ and auxiliary operators ${}^{B}\rfloor_A$ and ${}^{B}|_A$ are given in the text.

APPENDIX B: The Inference Rules for the Operational Semantics

Act $\quad \mu \xrightarrow{\mu} nil$

Seq-1 $\dfrac{E_1 \xrightarrow{\mu} \{s_1\cdot\}E_1'}{E_1;E_2 \xrightarrow{\mu} \{s_1\cdot\}(E_1';E_2)}$ $\qquad$ Seq-2 $\dfrac{\surd(E_1) \text{ and } E_2 \xrightarrow{\mu} s}{E_1;E_2 \xrightarrow{\mu} s}$

Nd-1 $\dfrac{E_1 \xrightarrow{\mu} \{s_1\cdot\}E_1'}{E_1+E_2 \xrightarrow{\mu} \{s_1\cdot\}E_1' \text{ and } E_2+E_1 \xrightarrow{\mu} \{s_1\cdot\}E_1'}$

Asy-1 $\dfrac{E_1 \xrightarrow{\mu} E_1'}{E_1 \|_A E_2 \xrightarrow{\mu} E_1' \|_A E_2 \text{ and } E_2 \|_A E_1 \xrightarrow{\mu} E_2 \|_A E_1'} \quad \mu \notin A$

Asy-2 $\dfrac{E_1 \xrightarrow{\lambda} s_1\cdot E_1'}{E_1 \|_A E_2 \xrightarrow{\lambda} s_1\backslash A'\cdot E_1' \|_A E_2 \text{ and } E_2 \|_A E_1 \xrightarrow{\lambda} s_1\backslash A'\cdot E_2 \|_A E_1'} \quad A'=A\otimes\lambda$

Res-1 $\dfrac{s \xrightarrow{\lambda} s' \text{ and } \surd(s')}{s\backslash A \xrightarrow{\lambda} s'} \quad \lambda \notin A$ $\qquad$ Res-2 $\dfrac{s \xrightarrow{\mu} s' \text{ and not } \surd(s')}{s\backslash A \xrightarrow{\mu} s'\backslash A'} \quad A'=A\otimes\mu$

Syn-1 $\dfrac{E_1 \xrightarrow{\lambda} E_1' \text{ and } E_2 \xrightarrow{\lambda} E_2'}{E_1 \|_A E_2 \xrightarrow{\lambda} E_1' \|_A E_2' \text{ and } E_2 \|_A E_1 \xrightarrow{\lambda} E_2' \|_A E_1'} \quad \lambda \in A$

Syn-2 $\dfrac{E_1 \xrightarrow{\lambda} s_1\cdot E_1' \text{ and } E_2 \xrightarrow{\lambda} s_2\cdot E_2' \qquad \text{with } \lambda \in Pref(A\backslash\{\lambda\}),\ A'=A\otimes\lambda}{E_1 \|_A E_2 \xrightarrow{\lambda} (s_1 |_{A'} s_2)\cdot E_1' \|_A E_2' \text{ and } E_2 \|_A E_1 \xrightarrow{\lambda} (s_2 |_{A'} s_1)\cdot E_2' \|_A E_1'}$

Com-1 $\dfrac{s_1 \xrightarrow{\lambda} s_1' \text{ and } s_2 \xrightarrow{\lambda} s_2' \text{ and } \surd(s_1') \text{ and } \surd(s_2')}{s_1|_A s_2 \xrightarrow{\lambda} s_1'|_\varnothing s_2'} \quad \lambda \in A$

Com-2 $\dfrac{s_1 \xrightarrow{\lambda} s_1' \text{ and } s_2 \xrightarrow{\lambda} s_2' \text{ and } \surd(s_1'), \text{ not } \surd(s_2')}{s_1|_A s_2 \xrightarrow{\lambda} s_1'|_\varnothing s_2' \text{ and } s_2|_A s_1 \xrightarrow{\lambda} s_2'|_\varnothing s_1'} \quad \lambda \in Pref(A)$

Com-3 $\dfrac{s_1 \xrightarrow{\lambda} s_1' \text{ and } s_2 \xrightarrow{\lambda} s_2' \text{ and not } \surd(s_1'), \text{ not } \surd(s_2')}{s_1|_A s_2 \xrightarrow{\lambda} s_1'|_{A'} s_2' \text{ and } s_2|_A s_1 \xrightarrow{\lambda} s_2'|_{A'} s_1'} \quad \lambda \in Pref(A),\ A'=A\otimes\lambda$

Com-4 $\dfrac{s_1 \xrightarrow{\tau} s_1'}{s_1|_A s_2 \xrightarrow{\tau} s_1'|_A s_2' \text{ and } s_2|_A s_1 \xrightarrow{\tau} s_2'|_A s_1'} \quad A \neq \varnothing$

Rec-1 $\dfrac{E[rec\ x.E / x] \xrightarrow{\mu} s}{rec\ x.E \xrightarrow{\mu} s}$

where $A\otimes\lambda = \{w \in A^+ \mid \lambda w \in A\}$ and $A\otimes\tau = A$.

The rules for refinement and stacks, and the definition of Pref(A) are given in the text.

RECOGNIZABLE COMPLEX TRACE LANGUAGES* (ABSTRACT)

Volker DIEKERT
Technische Universität München
Institut für Informatik
Arcisstr. 21
D-8000 MÜNCHEN 2

Paul GASTIN
Université PARIS 6
LITP, Institut Blaise Pascal
4, place Jussieu
F-75 252 PARIS CEDEX 05

Antoine PETIT
Université PARIS SUD
LRI, URA CNRS 410
Bât. 490
F-91 405 ORSAY CEDEX

Abstract: A. Mazurkiewicz defined traces in order to modelize non-sequential processes. Complex traces have been recently introduced as a generalization of both traces and infinite words. This paper studies the family of recognizable complex trace languages. It is proved that this family is closed under boolean operations, concatenation, left and right quotients. Then sufficient conditions ensuring the recognizability of the finite and infinite iterations of a recognizable complex trace language are given. The notion of co-iteration is defined and the Kleene-Ochmanski theorem is generalized to complex traces.

Introduction

The concept of traces has been introduced by A. Mazurkiewicz as a suitable semantics for non-sequential processes. Let us refer, for instance, surveys [Maz86], [AR88], [Per89] or the monograph [Die90] where also extensive bibliographies on the subject are given.
Recognizable languages describe the behaviour of finite state systems and hence form one of the basic families of a monoid. For trace monoids, this family is closed under boolean operations and concatenation [Fli74], [CP85], [Och85] but it turns out that T may be recognizable whereas T^* is not. Several papers discussed the recognizability of T^* [FR85], [Sak87], [Roz90]. One of the most interesting result has been found by Métivier [Met86] who showed that if a recognizable language T is connected then the language T^* is recognizable too. Independently Ochmanski [Och85] defined a concurrent version of the star iteration, the concurrent iteration, and proved the equality of the families of recognizable trace languages and co-rational trace languages. The latter family is obtained as the rational one by simply replacing the star iteration by the concurrent one.
In order to describe non-sequential processes which never terminate, e.g. distributed operating systems, the notion of infinite trace was introduced. An infinite (real) trace is an infinite dependence graph where every vertex has only finitely many predecessors. The theory of infinite real traces has been the subject of several papers [FR85], [Maz86], [BMP89], [Kwi90], [Gas90], [GR91], [GPZ91]. It turned out that there is no convenient notion of concatenation and Gastin [Gas90] proposed to solve this problem by adding a new element, called the error, to the set of real traces in order to get a monoid. In this monoid, the notion of recognizable languages has been introduced in a natural way as an extension of recognizable (infinite) word languages [Gas90]. This family is closed under boolean operations and concatenation, but similarly to the finite case, this family is closed neither under Kleene's iteration nor infinite

* This work has been partly supported by the ESPRIT Basic Research Actions N° 3166 (ASMICS) and N° 3148 (DEMON) and by the PRC Math-Info.

iteration [Gas91]. Gastin, Petit and Zielonka [GPZ91] proved recently that any sufficient condition ensuring the recognizability of T^* ensures that T^ω is also recognizable. Then, extending the ideas of Ochmanski, they defined a concurrent version of the ω-iteration. They generalized both Ochmanski's result on finite traces and Büchi's result on infinite words and obtained a Kleene like theorem for infinite real traces.
One of the main drawback of this concatenation on infinite real traces is that $a^\omega a \neq a^\omega$ contrary to the word case. Diekert [Die91] proposed a new solution to the problem of concatenation and defined the notion of complex trace. A complex trace is simply a real trace together with a second component which is some alphabetic information.
The aim of this paper is to define in a proper way the family of recognizable complex trace languages and to study its properties. We prove that this family is closed under boolean operations, concatenation, left and right quotients. Then we give sufficient conditions ensuring the recognizability of T^* and T^ω. These conditions are more general than those proposed in [GPZ91]. We define concurrent iterations on complex trace languages which leads to the family of co-rational languages. We show that this family is equal to the family of recognizable languages. Hence we obtain a generalization of Kleene's theorem to complex trace languages. At last we propose a canonical representation for recognizable languages.

For lack of space almost all proofs are refered to the full version of this abstract, which is available as Technical Report [DGP91] and which will be published elsewhere.

1. Preliminaries

In the following (X,D) denotes a finite dependence alphabet, i.e., X is a finite alphabet together with a reflexive and symmetric dependence relation $D \subseteq X \times X$. We denote by I the independence relation $X \times X \setminus D$ which is the complement of D. The monoid of finite traces is the quotient $M(X,D) = X^* / \{ab = ba / (a,b) \in I\}$. It is well-known that every trace can be identified with its dependence graph.
A dependence graph (over (X,D)) is (an isomorphism class of) a labelled acyclic graph $[V,E,\lambda]$ where V is a countable set of vertices, $E \subseteq V \times V$ is the set of arcs, $\lambda: V \to X$ is the labelling and it holds that $E \cup E^{-1} \cup id_V = \lambda^{-1}(D)$. We put the restriction on these graphs that the induced partial order $(V, \leq)$ is well-founded. Then, for any a in X the restriction to the set of vertices with label a is well-ordered. This allows to think of dependence graphs by standard representations where the vertices are pairs (a,i) with $a \in A$ and i is a countable ordinal.
The set of dependence graphs is denoted by G(X,D). It is a monoid by the operation $[V_1,E_1,\lambda_1][V_2,E_2,\lambda_2] = [V,E,\lambda]$ where $[V,E,\lambda]$ is the disjoint union of $[V_1,E_1,\lambda_1]$ and $[V_2,E_2,\lambda_2]$ together with new arcs (p_1,p_2) for all vertices $p_1 \in V_1$, $p_2 \in V_2$ such that $(\lambda_1(p_1),\lambda_2(p_2)) \in D$. The neutral element is the empty graph $1 = [\emptyset,\emptyset,\emptyset]$. In the following the monoid M(X,D) of finite traces will be identified with the submonoid of G(X,D) of finite dependence graphs.
Each element $s = [V,E,\lambda] \in G(X,D)$ splits into its real part Re(s) which is (the restriction of the dependence graph s to) $\{p \in V / p{\downarrow} \text{ is finite}\}$ and its transfinite part $Tr(s) = \{p \in V / p{\downarrow} \text{ is infinite}\}$. Here $p{\downarrow}$ is the downward closed subset of s with maximal vertex p, i.e.,

$p\downarrow = \{q \in V / q \leq p\}$. Obviously, Re(s) and Tr(s) are dependence graphs and, in G(X,D), we have the equation: s = Re(s)Tr(s).
The set of dependence graphs where the transfinite part is empty is called the set of **real traces** and is denoted by $\mathbb{R}$(X,D). It is clear that $\mathbb{R}$(X,D) is the set of dependence graphs which are representable by (finite or infinite) words from $A^\infty = A^\omega \cup A^*$. It follows that a real trace is exactly the same as a finite or infinite trace in the notion of [FR85], [Maz86]...
For a dependence graph s, the length of s is the number of vertices, the alphabet is alph(s) = {a ∈ X / a occurs in s at least one} and the alphabet at infinity is alphinf(s) = {a ∈ X / a occurs in s infinitely often} ∪ alph(Tr(s)).
For a subset $B \subseteq A$, define D(B) = {a ∈ X / (a,b) ∈ D for some b ∈ B}and I(B) = X \ D(B). This is extended to any dependence graph s by D(s) = D(alph(s)) and I(s) = I(alph(s)). For s,t ∈ G(X,D) we also write (r,s) ∈ I if alph(s) × alph(t) ⊆ I.
The elements of G(X,D) are good candidates to describe the behaviour of concurrent processes, but they are not abstract enough. Following the terminology suggested by A. Tarlecki, let us say that two dependence graphs s,t ∈ G(X,D) are **practically distinguishable** if there exists a (finite) real trace u such that su and tu have different real parts, i.e., if Re(su) ≠ Re(tu). In particular, two traces with different real parts are practically distinguishable. It turns out [Die91] that practical-undistinguishability is the largest congruence of G(X,D) which respects real parts. The quotient of G(X,D) by this congruence yields the monoid of **complex traces** $\mathbb{C}$**(X,D)**.
It is not difficult to see that two dependence graphs s,t in G(X,D) are not practically distinguishable if and only if Re(s) = Re(t) and D(alphinf(s)) = D(alphinf(t)). This follows from Re(sa) = Re(s)a ≠ Re(s) if a ∉ D(alphinf(s)) and Re(sa) = Re(s) otherwise. To simplify the notations, we define the imaginary part of s by Im(s) = D(alphinf(s)). Hence a complex trace is a pair (u,D(A)) = (Re(s),Im(s)) for some s ∈ G(X,D) and the multiplication in $\mathbb{C}$(X,D) is given by:

$$(u,D(A))(v,D(B)) = (u\mu,D(A \cup B \cup alph(\mu^{-1}v))).$$

Here μ is an abbreviation for the real trace $\mu_A(v)$. This μ-notation $\mu_A(s)$ is defined for any subset $A \subseteq X$ and dependence graph s ∈ G(X,D) by the maximal real prefix of s containing letters from I(A) only. Hence for s = [V,E,λ] we have $\mu_A(s) = \{n \in V / n\downarrow$ is finite & $alph(n\downarrow) \subseteq I(A)\} \in \mathbb{R}(X,D)$. Since G(X,D) is left-cancellative [Die91], we can write $s' = (\mu_A(s))^{-1}s$ for the unique s' such that $s = \mu_A(s)s'$ and we have $min(s') \cap I(A) \subseteq D(alphinf(\mu_A(s)))$. Here and in the following, min(s') denotes the set of labels of the minimal elements in the dependence graph s'.

Let (s_i) be any sequence of dependence graphs. Then $s = s_1s_2...s_i... \in G(X,D)$ is defined as the disjoint union of the s_i with the obvious new arcs from s_i to s_j for i<j whenever the vertices have dependent labels. Let (t_i) be another sequence of G(X,D) such that for all i, $Re(s_i) = Re(t_i)$ and $Im(s_i) = Im(t_i)$. Let $t = t_1t_2...t_i...$. One can verify that Re(s) = Re(t) and Im(s) = Im(t). Hence, for any language $L \subseteq \mathbb{C}(X,D)$, we can define the infinite iteration L^ω as follows: $L^\omega = \{\chi(s_1s_2...s_i...) / \forall i\ s_i \in \chi^{-1}(L)\}$ where χ is the surjective homomorphism from

$G(X,D)$ in $\mathbb{C}(X,D)$, $s \mapsto (Re(s),Im(s))$. Furthermore, the ω-iteration commutes with the projection χ.

To summarize, we have the following situation. There is a canonical mapping $\varphi: X^\infty \to G(X,D)$ with the image $\varphi(X^\infty) = \mathbb{R}(X,D) \subseteq G(X,D)$. We have a surjective homomorphism $\chi: G(X,D) \to \mathbb{C}(X,D)$ and a surjective mapping Re: $\mathbb{C}(X,D) \to \mathbb{R}(X,D)$, $(u,D(A)) \mapsto u$. For a complex trace x, the image Re(x) is also called the real part of x.
Since $\mathbb{R}(X,D) \to G(X,D) \to \mathbb{C}(X,D) \to \mathbb{R}(X,D)$ is the identity we may view $\mathbb{R}(X,D)$ also as a subset of $\mathbb{C}(X,D)$. More precisely, a real trace $u \in \mathbb{R}(X,D)$ is identified with its image $(u,Im(u))$ in $\mathbb{C}(X,D)$.
Note that the set of real traces $\mathbb{R}(X,D)$ is not a submonoid of $G(X,D)$. However, the complement $\perp = G(X,D) \setminus \mathbb{R}(X,D)$, which is the set of dependence graphs with non-empty transfinite part, forms an ideal. Taking the Rees quotient we obtain the monoid $\mathbb{R}(X,D) \cup \{\perp\}$ which has been studied in [Gas90], [Gas91], [GR91], [GPZ91]. In fact, the computations with respect to $\mathbb{R}(X,D) \cup \{\perp\}$ are also the basic for our results on complex traces.
Let $u,v \in \mathbb{R}(X,D)$ be real traces. Then $uv \neq \perp$ means that the dependence graph uv has no transfinite part, hence we can split u in wu' where $w \in M(X,D)$ is finite, $(u',v) \in I$ and we have $uv = wu'v = wvu'$. Of course, when $uv \neq \perp$, this concatenation commutes with the embedding of real traces in $\mathbb{C}(X,D)$.
Note however that the image by χ of the ideal $\perp \subseteq G(X,D)$ covers all infinite complex traces. It follows that if we start with real traces $u,v,w \in \mathbb{R}(X,D)$ such that the equation $uv = w$ holds in $\mathbb{C}(X,D)$, then we cannot infer that $uv \neq \perp$.

Example: Let $(X,D) = a — b — c — d — e — f$. Then we have:

In $\mathbb{R}(X,D) \cup \{\perp\}$		In $\mathbb{C}(X,D)$	
$a^\omega a$	$= \perp$	$a^\omega a$	$= (a^\omega, D(a))\ (a, \emptyset) = (a^\omega, D(a)) = a^\omega$
$a^\omega b$	$= \perp$	$a^\omega b$	$= (a^\omega, D(a))\ (b, \emptyset) = (a^\omega, D(a,b)) \neq a^\omega$
$b^\omega d^\omega f^n$	$= (bdf)^n (bd)^\omega = f^n (bd)^\omega$	$b^\omega d^\omega f^n$	$= ((bdf)^n (bd)^\omega, X\setminus\{f\}) = f^n (bd)^\omega$
$b^\omega d^\omega e$	$= (bd)^\omega e = \perp$	$b^\omega d^\omega e$	$= ((bd)^\omega, X) = (bd)^\omega e \neq (bd)^\omega$
$a^\omega b^\omega d^\omega$	$= \perp\ d^\omega = \perp$	$a^\omega b^\omega d^\omega$	$= ((ad)^\omega, X\setminus\{f\}) = (ad)^\omega$
$a^\omega c^\omega e^\omega f$	$= (ace)^\omega f = \perp$	$a^\omega c^\omega e^\omega f$	$= ((ace)^\omega, X) = (ace)^\omega$

Note in particular the difference in the first and last two lines. In $\mathbb{C}(X,D)$, $a^\omega a$, $a^\omega b^\omega d^\omega$ and $a^\omega c^\omega e^\omega f$ are representable by real traces, although they yield error in the monoid $\mathbb{R}(X,D) \cup \{\perp\}$.

Finally let us note that for $(X,D) = (X, X \times X)$ the monoid $\mathbb{C}(X,D)$ is just the usual construct X^∞. This follows because in this case the second component is redundant. It is $\emptyset$ if the real part is finite and X otherwise. Thus all our considerations about complex traces cover, as a special case, the classical situation of finite and infinite words.

Remarks. 1) Beside $\mathbb{C}(X,D)$, it could be also convenient to consider the monoid $\mathbb{C}_\alpha(X,D)$ of α-complex traces. It is defined by the set of pairs (Re(s),alphinf(s)) for some $s \in G(X,D)$. All

results proved here for $\mathbb{C}(X,D)$ hold analogously for $\mathbb{C}_\alpha(X,D)$ too. However for lack of space of this abstract we deal with complex traces only.
2) For a complex trace x, we define $D(x) = D(alph(t))$ and $Im(x) = Im(t)$ for some dependence graph t representing x. Note that if x is denoted by $(u,D(A))$, the sets $D(A) = Im(x)$ and $D(x) = D(u) \cup D(A)$ are well defined even if the alphabet A is not well defined.

2. Real traces

Throughout this section the multiplication on real traces is carried out in $G(X,D)$ and the symbol $\perp$ stands for the ideal of graphs with non-empty transfinite part.
In [Gas91] and [GPZ91] recognizable real trace languages are studied. There are several equivalent definitions for recognizability [Gas91], we will use the following one. A language $L \subseteq \mathbb{R}(X,D)$ is called recognizable if the word language $\varphi^{-1}(L) \subseteq X^\infty$ is recognizable in the usual sense [Eil74]. The set of recognizable real trace languages is denoted by $Rec(\mathbb{R}(X,D))$.
A result of [Gas91] states that if $K,L \subseteq \mathbb{R}(X,D)$ are recognizable then $KL \setminus \{\perp\} \subseteq \mathbb{R}(X,D)$ is a recognizable real trace language too. In order to extend this result to complex traces we need the following theorem. Although it concerns real traces only, it is the basic reason for which we can extend the results from real traces to complex traces.

Theorem 2.1. Let $K,L \subseteq \mathbb{R}(X,D)$ be recognizable. Then $KL^{-1} = \{u \in \mathbb{R}(X,D) \,/\, \exists\, v \in L$ with $uv \in K\}$ and $K^{-1}L = \{u \in \mathbb{R}(X,D) \,/\, \exists\, v \in K$ with $uv \in L\}$ are recognizable real trace languages.

The following result on real traces becomes crucial below. In the full paper we will give a proof which is simpler than the original one and uses intermediary tools of independent interest.

Theorem 2.2. [GPZ91] Let L be a language of $M(X,D)$ such that L^* is recognizable. Then L^ω is a recognizable real trace language too.

3. Recognizable complex trace languages

First we have to define recognizable languages in $\mathbb{C}(X,D)$. Our definition is motivated from the following considerations. With any reasonable definition of recognizability, the family of recognizable languages must be closed under union, intersection and concatenation by letters in X. Moreover any recognizable real trace language has of course to remain recognizable in $\mathbb{C}(X,D)$. In particular $\mathbb{R}(X,D)_A = \{u \in \mathbb{R}(X,D) \,/\, alphinf(u) = A\}$ has to be recognizable. It follows that $\mathbb{C}(X,D)_B = \{x \in \mathbb{C}(X,D) \,/\, Im(x) = B\}$ is recognizable, since it can be written as a finite union of sets of the form $\mathbb{R}(X,D)_A.v_{A,B}$ for some $A \subseteq X$ and $v_{A,B} \in M(X,D)$. Therefore, if a complex trace language L is recognizable, $L \cap \mathbb{C}(X,D)_B$ has to be recognizable too. Conversely, if $L \cap \mathbb{C}(X,D)_B$ is recognizable for every $B \subseteq X$ then L has to be recognizable too since L can be written as union of such languages. These facts lead to the following definitions.

Definition 3.1. Let L be a language of $\mathbb{C}(X,D)$. Then, define $L_B = \{x \in L \,/\, Im(x) = D(B)\}$. The language L is called recognizable if, for every $B \subseteq X$, the set of real traces $Re(L_B)$ is recognizable. The set of recognizable complex trace languages is denoted by $Rec(\mathbb{C}(X,D))$.

Our definition of recognizability satisfies the following basic requirements:
Remark 3.2. 1) Let $\psi: X^{\infty} \to \mathbb{C}(X,D)$ be the canonical mapping and $L \in Rec(\mathbb{C}(X,D))$. Then $\psi^{-1}(L) \subseteq X^{\infty}$ and $Re(L) \subseteq \mathbb{R}(X,D)$ are recognizable.
2) Let L be a language of $\mathbb{R}(X,D)$), then L is recognizable if and only if its image as complex trace language $\{(u,Im(u)) \,/\, u \in L\} \subseteq \mathbb{C}(X,D)$ is recognizable.
3) Let $L \in Rec(\mathbb{R}(X,D))$ such that $L = L_A$ for some $A \subseteq X$ and $v \in M(X,D)$ be a finite trace with $min(v) \subseteq D(A)$ and $D(A) \cup D(v) = D(B)$. Then the complex trace language $L.v = \{(u,D(B)) \,/\, u \in L\}$ is recognizable.

Before we continue, let us point out that it makes no sense to define recognizability via the inverse image with respect to $\psi : X^{\infty} \to \mathbb{C}(X,D)$ because this mapping covers only a minor part of $\mathbb{C}(X,D)$ or via the projection Re: $\mathbb{C}(X,D) \to \mathbb{R}(X,D)$. Again we would obtain a class which is much too large and which neither satisfies 1) above nor which is closed under intersection.

Proposition 3.3. i) $Rec(\mathbb{C}(X,D))$ is a boolean algebra, i.e., it is closed with respect to union, intersection and complementation.
ii) $Rec(\mathbb{C}(X,D))$ is the smallest family of complex trace languages which satisfies 2) and 3) of Remark 3.2 and which is closed under union.

Further evidence that our notion of recognizability is good will follow from the results below. For $L \subseteq \mathbb{C}(X,D)$ and $A \subseteq B \subseteq X$, define

$$\mu_{A,B}(L) = \{\mu_A(x) \,/\, x \in L \text{ and } \exists\, y \in \mathbb{C}(X,D) \text{ such that } Im(y) = D(A) \text{ and } Im(yx) = D(B)\}.$$

Recall that $\mu_A(x)$ is the maximal real prefix of x containing letters from I(A) only and note that "$\exists\, y \in \mathbb{C}(X,D)$" is equivalent to "$\forall\, y \in \mathbb{C}(X,D)$" in the definition above since Im(yx) depends on D(A) and x only. Moreover any set $\mu_A(L)$ can be written as the union of the sets $\mu_{A,B}(L)$ for $A \subseteq B$. It can be shown that if L is a recognizable language then $\mu_{A,B}(L)$ and $\mu_A(L)$ are recognizable real trace languages. Using this result, we obtain the following theorem.

Theorem 3.4. The class of recognizable complex trace languages is closed under concatenation, left-quotient and right-quotient.

4. Connected traces.

A basic property of the calculation with the error element $\perp$ in the monoid $\mathbb{R}(X,D) \cup \{\perp\}$ is that any product of infinite real traces becomes stationary. More precisely, let $(u_i)_{i \geq 1}$ be any sequence of infinite real traces, let $k = |X|$ and let $j \geq k$. Then $u_1 \ldots u_j \in G(X,D)$ has a non-empty transfinite part. Hence, in the monoid $\mathbb{R}(X,D) \cup \{\perp\}$ we have $u_1 \ldots u_j = \perp$ for all $j \geq k$. From the closure under union and concatenation of recognizable trace languages, it follows directly that if L is a recognizable real trace language which contains infinite traces only, then L^*

and L^ω are recognizable. This reduces the recognizability problem of the finite and infinite iterations in $\mathbb{R}(X,D) \cup \{\bot\}$ to languages of finite traces. Such arguments are not available for complex traces.
For instance, let $(X,D) = a — b — c \quad d$. Consider the real infinite trace $u = acd^\omega$. The calculation in $\mathbb{C}(X,D)$ yields $u^k = (ac)^k d^\omega \neq u^{k+1}$ for all $k \geq 1$ and $u^\omega = (acd)^\omega = a^\omega c^\omega d^\omega$. The set $\{u\}^* = \{ (ac)^k d^\omega, k \geq 0 \}$ is not recognizable. On the other hand, the set $\{u\}^\omega$ is recognizable, but this is a general fact (see the next section) which holds as soon as $\{u\}$ is recognizable. To see that the ω-iteration of a finite language of infinite traces is not recognizable in general, consider two infinite real traces $u = bd^\omega$ and $v = acd^\omega$. Then $\{u,v\}^\omega = \{b,ac\}^\omega d^\omega$ is not recognizable.
But of course, we can not expect that recognizable trace languages are closed under star or omega, in general. We are only interested in the closure under the concurrent star and omega iterations. The definition of these iterations requires a notion of connectivity on complex traces. Intuitively, a trace is connected if it cannot be splitted into two non-empty independent traces. Thus we have to introduce first the notion of independence for traces. This is clear in $\mathbb{R}(X,D)$ or $G(X,D)$: two traces r and s are independent, denoted by $(r,s) \in I$, if $\text{alph}(r) \times \text{alph}(s) \subseteq I$. Unfortunately, we cannot use this definition in $\mathbb{C}(X,D)$ since the alphabet of a complex trace is not well-defined. Nevertheless, the independence in $\mathbb{C}(X,D)$ admits a canonical definition as follows. Two complex traces $(u,D(A))$ and $(v,D(B))$ are called independent if $(\text{alph}(u) \cup A) \times (\text{alph}(v) \cup B) \subseteq I$. Note that this is well-defined. We define now connected traces and connected components of a trace.

Definition 4.1. A trace t in $\mathbb{R}(X,D)$, $\mathbb{C}(X,D)$ or $G(X,D)$ is said to be connected if $t = rs$ with $(r,s) \in I$ implies $r = 1$ or $s = 1$. A non-empty trace r is called a connected component of t if r is connected and there exists a trace s such that $t = rs$ and $(r,s) \in I$. The set of connected components of t is denoted by $C(t)$. The trace t is said to be finitary connected if it admits at most one finite connected component.

Remark 4.2. Every dependence graph is the disjoint union of its connected components. It follows that every trace t in $\mathbb{R}(X,D)$, $G(X,D)$ can uniquely be written as a product of independent connected components (called a decomposition of t). Moreover, if $t = t_1 \dots t_k$ is such a decomposition then $\{t_1, \dots, t_k\}$ is exactly the set $C(t)$ of connected components of t. This is not true for traces in $\mathbb{C}(X,D)$. A complex trace may have several decompositions and some of its connected components may not be independent. Nevertheless, every complex trace admits at least one decomposition and if a complex trace belongs to $\mathbb{R}(X,D) \subseteq \mathbb{C}(X,D)$ then it admits a unique decomposition into real connected components .

For example, let $(X,D) =$ 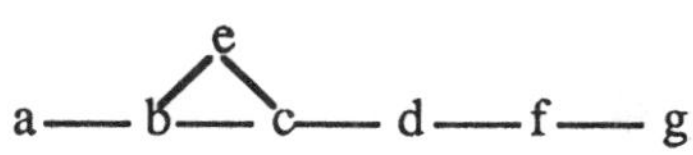 and let x be the finitary connected complex trace $(gg(ad)^\omega, X\setminus\{g\})$. Then $y = (gg,\emptyset)$, $y_1 = (a^\omega,D(a))$, $z_1 = (d^\omega,D(c,d)\})$, $y_2 = (a^\omega,D(a,b))$ and $z_2 = (d^\omega,D(d))$ are connected components of x. The trace x admits two decompositions $x = yy_1z_1 = yy_2z_2$ but neither (y_1,y_2) nor (z_1,z_2) nor (z_1,y_2) are independent. The components y,y_1,z_2 are pairwise independent but $yy_1z_2 = (gg(ad)^\omega, X\setminus\{e,g\}) \neq x$.

The notation C is obviously extended to trace languages and the following inclusions hold: $L^* \subseteq C(L)^*$, $L^\omega \subseteq C(L)^\omega$. We will see in next section that if L is recognizable then so are $C(L)^*$ and $C(L)^\omega$. For this, we need the following lemma.

Lemma 4.6. Let L be a recognizable language of complex traces. Then the language C(L) is recognizable too.

5. Closure properties

The aim of this section is to find sufficient conditions on a recognizable complex trace language L which ensure that L^* and L^ω are recognizable too. We prove that connectivity or finitary connectivity are such conditions. In fact, this follows directly from a stronger result that we are going to establish first.

For a language $L \subseteq \mathbb{C}(X,D)$, we define $\nu_A(L) = \mu_{A,A}(L)$. Note that each trace in $\nu_A(L)$ is finite since for a real trace u, we have that $alph(u) \subseteq I(A)$ and $Im(u) \subseteq D(A)$ imply $alphinf(u) = \emptyset$. For $A \subseteq B \subseteq X$, and $L \subseteq \mathbb{C}(X,D)$ define

$$L_{A,B} = \{x \in L \,/\, \exists\, y \in \mathbb{C}(X,D) \text{ such that } Im(y) = D(A) \text{ and } Im(yx) = D(B)\}.$$

Note that in this definition again, "$\exists\, y \in \mathbb{C}(X,D)$" is equivalent to "$\forall\, y \in \mathbb{C}(X,D)$" since $Im(yx)$ depends on $D(A)$ and x only. Clearly, we have $\mu_{A,B}(L) = \mu_A(L_{A,B})$. It is easy to verify that if L is a recognizable language then the languages $L_{A,B}$ are recognizable too.

Theorem 5.1. Let L be a recognizable language of complex traces such that $\nu_A(L)^* \subseteq M(X,D)$ is recognizable for all $A \subseteq X$. Then L^* and L^ω are both recognizable.

Proof. Let $x_1, x_2, \ldots$ be a finite or infinite sequence of members of L and let $x = x_1x_2\ldots$ Then there exist an integer k, a sequence of integers $0 = i_0 < i_1 < \ldots < i_k < i_{k+1} = \infty$ and a sequence of subalphabets $\emptyset = D(B_0) \subsetneq D(B_1) \subsetneq \ldots \subsetneq D(B_k) \subseteq X$ such that $Im(x_1 \ldots x_n) = D(B_j)$ for all $i_j \le n < i_{j+1}$ and $0 \le j \le k$.
Define the sequence (u_n) as follows: $u_n = \mu_{B_j}(x_n)$ if $i_j < n < i_{j+1}$ and $u_n = x_n$ if $n = i_j$. It follows $x_1 \ldots x_n = u_1 \ldots u_n$ for all n. By definition, $u_n \in \mu_{B_j,B_j}(L) = \nu_{B_j}(L)$ if $i_j < n < i_{j+1}$ and $u_n \in L_{B_{j-1},B_j}$ if $n = i_j$. Thus, we have:

$$x \in \nu_{B_0}(L)^* L_{B_0,B_1} \nu_{B_1}(L)^* \ldots L_{B_{k-1},B_k} \nu_{B_k}(L)^\dagger \text{ (where } \dagger = * \text{ or } \omega).$$

Therefore $L^\dagger$ is a finite union of languages $\nu_{B_0}(L)^* L_{B_0,B_1} \nu_{B_1}(L)^* \ldots L_{B_{k-1},B_k} \nu_{B_k}(L)^\dagger$ and the recognizability of $L^\dagger$ follows directly from Theorem 3.4 and Theorem 2.2. ◊

Many properties follow from the previous theorem .

Corollary 5.2 . Let L be a recognizable complex trace language. If L contains connected (resp. finitary connected) traces only then L^* and L^ω are recognizable too. In particular, $C(L)^*$ and $C(L)^\omega$ are recognizable for all L in $Rec(\mathbb{C}(X,D))$.

The following result (already announced in Section 4) can be proved using similar techniques.

Proposition 5.3. Let x be a complex trace such that the language $L = \{x\}$ is recognizable. Then the language $L^\omega = \{x^\omega\}$ is recognizable too.

6. Co-rational complex trace languages.

We turn now to the final result: the Kleene theorem for complex trace languages. As for finite traces [Och85] and for infinite real traces [GPZ91], define the concurrent iterations co-* and co-ω of a complex trace language L by $L^{co\text{-}*} = C(L)^*$ and $L^{co\text{-}\omega} = C(L)^\omega$.
Let the co-rational family Co-Rat($\mathbb{C}(X,D)$) be the smallest class of complex trace languages such that finite sets of finite traces are co-rational and such that co-rationals are closed under union, concatenation, co-star and co-omega. In [GPZ91] it is shown that in $\mathbb{R}(X,D) \cup \{\perp\}$, the family of recognizable languages is equal to the family of co-rational languages. We will extend this result to complex traces using the following lemma.

Lemma 6.1. If there exists a complex trace $(u,D(B))$ with $Im(u) = D(A)$ then there exists a trace $w_{A,B}$ of length at most $|X|$ such that in $\mathbb{C}(X,D)$ we have $v.w_{A,B} = (v,D(B))$ for all real traces v with $Im(v) = D(A)$, in particular, $w_{A,B} \in M(X,D)_{A,B}$.

Theorem 6.2. A complex trace language L is recognizable if and only if L is a finite union of languages $MN_1^\omega \ldots N_k^\omega.w$ where w is finite, M and N_i are recognizable languages of $M(X,D)$ such that all traces in N_i have the same connected alphabet A_i and $A_i \times A_j \subseteq I$ for all $i \neq j$.

Proof. Let L be recognizable. Since $L = \bigcup_{B \subseteq X} L_B$ we may assume that $L = L_B$ for some B.
By definition, $Re(L_B)$ is recognizable. From [GPZ91], in $\mathbb{R}(X,D) \cup \{\perp\}$ we have $Re(L_B) = \bigcup_{\text{finite}} MN_1^\omega \ldots N_k^\omega$ where M and N_i are recognizable languages of finite traces such that all traces in N_i have the same connected alphabet A_i and $A_i \times A_j \subseteq I$ for all $i \neq j$. In this particular case, calculations in $\mathbb{R}(X,D) \cup \{\perp\}$ and in $\mathbb{C}(X,D)$ are exactly the same. Thus, the equation $Re(L_B) = \bigcup_{\text{finite}} MN_1^\omega \ldots N_k^\omega$ holds also in $\mathbb{C}(X,D)$.
Note that all traces u in $MN_1^\omega \ldots N_k^\omega$ have the same alphabet at infinity $A = \bigcup_{1 \leq i \leq k} A_i$. Let $w = w_{A,B}$ be the finite trace defined in Lemma 6.1. Clearly, for all u in $MN_1^\omega \ldots N_k^\omega$, we have $uw = (u,D(B))$ in $\mathbb{C}(X,D)$. We obtain $L_B = \bigcup_{\text{finite}} MN_1^\omega \ldots N_k^\omega.w$ in $\mathbb{C}(X,D)$.
The converse follows from previous results. ◊

Corollary 6.3. The family Rec($\mathbb{C}(X,D)$) is the smallest family of complex trace languages which contains finite sets of finite traces and which is closed under finite union, concatenation and iterations * and ω restricted to languages of finite connected traces.

From Proposition 3.3, Theorem 3.4 and Corollary 5.2 it follows directly that any co-rational complex trace language is recognizable. Then using Ochmanski's results [Och85] and Theorem 6.2, we prove our main result.

Theorem 6.4. Rec($\mathbb{C}(X,D)$) = co-Rat($\mathbb{C}(X,D)$).

7. References

[AR88] I.J. AALBERSBERG and G. ROZENBERG, "Theory of traces", Theoretical Computer Science 60, p. 1-82, 1988.

[BMP89] P. BONIZZONI, G. MAURI and G. PIGHIZZINI, "About infinite traces", Proceedings of the ASMICS Workshop on Partially Commutative Monoids, Tech. Rep. TUM-I 9002, Technische Universität München, 1989.

[CP85] R. CORI and D. PERRIN, "Automates et commutations partielles", RAIRO Theoretical Informatics and Applications 19, p. 21-32, 1985.

[DGP91] V. DIEKERT, P. GASTIN, A. PETIT, "Recognizable complex trace languages", Tech. Rep. 640, LRI, Université Paris Sud, Orsay, France, 1991.

[Die90] V. DIEKERT, "Combinatorics on traces", Lecture Notes in Computer Science 454, 1990.

[Die91] V. DIEKERT, "On the concatenation of infinite traces", STACS'91, Lecture Notes in Computer Science 480, p. 105-117, 1991.

[Eil74] S. EILENBERG, "Automata, Languages and Machines", Academic Press, New York, 1974.

[Fli74] M. FLIESS, "Matrices de Hankel", J. Math. pures et appl. 53, p. 197-224, 1974.

[FR85] M.P. FLE and G. ROUCAIROL, "Maximal serializability of iterated transactions", Theoretical Computer Science 38, p. 1-16, 1985.

[Gas90] P. GASTIN, "Infinite traces", Proceedings of the Spring School of Theoretical Computer Science on "Semantics of concurrency", Lecture Notes in Computer Science 469, p. 277-308, 1990.

[Gas91] P. GASTIN, "Recognizable and rational languages of finite and infinite traces", STACS'91, Lecture Notes in Computer Science 480, p. 89-104, 1991.

[GPZ91] P. GASTIN, A. PETIT, W. ZIELONKA, "A Kleene theorem for infinite trace languages", ICALP'91, appear in Lecture Notes in Computer Science, Tech. Rep. 90.93, LITP, Université Paris 6, France, 1991.

[GR91] P. GASTIN and B.ROZOY, "The Poset of infinitary traces", to appear in Theoretical Computer Science, Tech. Rep. 91-07, LITP, Université Paris 6, France, 1991.

[Kwi90] M.Z. KWIATKOWSKA, "A metric for traces", Information and Processing Letters 35, p. 129-135, 1990.

[Maz86] A. MAZURKIEWICZ, "Trace theory", Advanced Course on Petri Nets, Lecture Notes in Computer Science 255, p. 279-324, 1986.

[Met86] Y. METIVIER, "On recognizable subsets in free partially commutative monoids", ICALP'86, Lecture Notes in Computer Science 226, p. 254-264, 1986.

[Och85] E. OCHMANSKI, "Regular behaviour of concurrent systems", Bulletin of EATCS 27, p. 56-67, October 1985.

[Per89] D. PERRIN, "Partial commutations", ICALP'89, Lecture Notes in Computer Science 372, p. 637-651, 1989.

[Roz90] B. ROZOY, "On Traces, Partial Order Sets and Recognizability", ISCIS V, Cappadocia, Turkey, proceedings to appear, 1990.

[Sak87] J. SAKAROVITCH, "On regular trace languages", Theoretical Computer Science 52, p. 59-75, 1987.

Solving Systems of Linear Diophantine Equations: An Algebraic Approach *

Eric Domenjoud
CRIN & INRIA Lorraine
BP 239 F-54506 Vandœuvre-Lès-Nancy Cedex
FRANCE
e-mail domen@loria.crin.fr

Abstract

We describe through an algebraic and geometrical study, a new method for solving systems of linear diophantine equations. This approach yields an algorithm which is intrinsically parallel. In addition to the algorithm, we give a geometrical interpretation of the satisfiability of an homogeneous system, as well as upper bounds on height and length of all minimal solutions of such a system. We also show how our results apply to inhomogeneous systems yielding necessary conditions for satisfiability and upper bounds on the minimal solutions.

Solving linear diophantine equations is a problem which appears in many fields, from linear programming to resolution of equations in semigroups or constrained logic programming. In the framework of semigroups, one needs an efficient test for the satisfiability of equations as well as a way to generate a basis for the set of all solutions. The efficiency of the satisfiability test becomes crucial in constrained equational logic because constraints are accumulated until their conjunction becomes unsatisfiable. The simplex method is quite convenient for checking the satisfiability of a set of homogeneous equations, but does not provide an algorithm for finding all solutions. Such algorithms were proposed by G. Huet [9], A. Fortenbacher [6], A. Herold & T. Guckenbiehl [8], M. Clausen & A. Fortenbacher [4] for one equation. In 1987, J.-L. Lambert [10] gave an upper bound on the components of all minimal solutions of a system, and a finer one on the length of minimal solutions of one equation, improving then greatly Huet's algorithm. In 1989, J.-F. Romeuf [14] described an algorithm for solving two equations, using a finite automaton recognizing the solutions. Definite improvements were then brought in 1989 by E. Contejean and H. Devie [5, 2] who found an extension of Fortenbacher's algorithm to systems of arbitrary size, and L. Pottier who described in [11] a similar algorithm and gave a new upper bound on the length of all minimal solutions of a system. Very recently, L. Pottier gave in [12] another algorithm using Gröbner bases and improved the upper bounds on the minimal solutions.

*This work was partly supported by the GRECO de programmation (CNRS)

Our approach and L. Pottier's one, seem to be the first attempts to find the solutions from algebraic considerations only. These two approaches are however completely different. All other algorithms have in common to implicitly or explicitly increase vectors starting from $\vec{0}$, until a solution or a failure is reached. On the contrary, our algorithm computes directly a finite set of solutions containing all minimal ones. This makes our algorithm very easy to parallelize because no comparison between solutions is needed until the very last step.

The paper is organized as follows. In section 1, we give our main definitions and notations, and two theorems used in all the rest of the paper. In particular, we show that the diophantine system $A^1x_1 + \cdots + A^nx_n = \vec{0}$ is satisfiable if and only if the convex hull of $\{A^1, \ldots, A^n\}$ contains $\vec{0}$. In sections 2 and 3, we describe how we get a basis of the set of solutions of this system by generating first an integer basis of its real positive solutions. In section 4, we give upper bounds on the length and the height of all minimal solutions and we show in section 5 how our results apply to inhomogeneous systems $A^1x_1 + \cdots + A^nx_n = C$. All these results have been first presented at the fourth international workshop on unification held in Leeds (UK) in July 1990.

1 Preliminaries

$\boldsymbol{N}$, $\boldsymbol{Z}$, $\boldsymbol{Q}$ and $\boldsymbol{R}$ denote respectively the set of natural, integer, rational and real numbers. If $X \in \boldsymbol{R}^n$, $\|X\|_1 = \sum_{i=1}^n |x_i|$ is its length, and $\|X\|_\infty = \max |x_i|$ is its height. The carrier of X is the set of indexes of its nonzero components. If V and W are vectors, $V \geq W$ iff $\forall i : v_i \geq w_i$, $V > W$ iff $V \geq W$ and $V \neq W$, $V \gg W$ iff $\forall i : v_i > w_i$. The null vector of $\boldsymbol{R}^n$ will be denoted by $\vec{0}_n$ or simply $\vec{0}$ if no ambiguity arises. For a matrix A, A^j is its jth column and if A is a $n \times n$ matrix, $|A|$ is its determinant and A is unimodular iff $|A| = \pm 1$. A and B are similar, written $A \sim B$, if there exists an integer unimodular matrix U such that $UA = B$. A linear diophantine system is an equation $AX = C, X \in \boldsymbol{N}^n$ where A is an $m \times n$ integer matrix and $C \in \boldsymbol{Z}^m$. If $C = \vec{0}$, the diophantine system is homogeneous, otherwise it is inhomogeneous. A solution S of this system is minimal if $S \neq \vec{0}$ and there does not exist a nonzero solution $S' < S$. If V is a set of vectors, card(V) is its cardinality and Conv(V) its convex hull. If $V = \{V_1, \ldots, V_k\}$, we write Conv$(V_1, \ldots, V_k)$ instead of Conv$(\{V_1, \ldots, V_k\})$.

$$\mathrm{Conv}(V_1, \ldots, V_k) = \{\lambda_1 V_1 + \cdots + \lambda_k V_k \mid \forall i : \lambda_i \geq 0 \text{ and } \sum_{j=1}^{k} \lambda_j = 1\}$$

At last, if A is an $(n-1) \times n$ matrix and e is the canonical basis of $\boldsymbol{R}^n$, $\begin{vmatrix} e \\ A \end{vmatrix}$ is the vector product of A's rows.

In the sequel, we shall use the following two theorems.

Theorem 1 (Carathéodory) *Let V be a subset of $\boldsymbol{R}^m$.*

$$\mathrm{Conv}(V) = \bigcup_{\substack{V' \subset V \\ \mathrm{card}(V') \leq m+1}} \mathrm{Conv}(V')$$

The proof of this theorem may be found in [3]. We shall actually use the following two corollaries of this theorem:

Corollary 1 *Let V be a subset of $\boldsymbol{R}^m$, the convex hull of which contains X. Then there exists $V' \subset V$ such that* $\mathrm{card}(V') \leq m+1$ *and* $X \in \mathrm{Conv}(V')$.

Corollary 2 *If $\vec{0} \in \mathrm{Conv}(V)$ and for all strict subsets V' of V, $\vec{0} \notin \mathrm{Conv}(V')$, then V has rank* $\mathrm{card}(V) - 1$.

Proof: There exists an isomorphism ϕ from the vector space spanned by V to $\boldsymbol{R}^k$, where k is the rank of V. Since $\mathrm{Conv}(\phi(V)) = \phi(\mathrm{Conv}(V))$, $\vec{0}_k \in \mathrm{Conv}(\phi(V))$. Now from corollary 1, there exist $V_1, \ldots, V_{k+1}$ in V such that $\vec{0}_k \in Conv(\phi(V_1), \ldots, \phi(V_{k+1}))$. Hence, $\vec{0}_m \in \mathrm{Conv}(V_1, \ldots, V_{k+1})$ and $k+1 = \mathrm{card}(V)$ otherwise, $\{V_1, \ldots, V_{k+1}\}$ is a strict subset of V, the convex hull of which contains $\vec{0}_m$. □

Theorem 2 *If A is an $m \times n$ integer matrix, the diophantine system $AX = \vec{0}, X \in \boldsymbol{N}^n$ has a nonzero solution iff $\vec{0} \in \mathrm{Conv}(A^1, \ldots, A^n)$.*

Proof:

[⇒] If $X \in \boldsymbol{N}^n - \{\vec{0}\}$ and $AX = \vec{0}$ then $\frac{x_i}{\|X\|_1} \geq 0$, $\left\|\frac{X}{\|X\|_1}\right\|_1 = 1$ and $A\frac{X}{\|X\|_1} = \vec{0}$. Hence $\vec{0} \in \mathrm{Conv}(A^1, \ldots, A^n)$

[⇐] From corollaries of theorem 1, there exists $\{A^{i_1}, \ldots, A^{i_k}\} \subset \{A^1, \ldots, A^n\}$, the convex hull of which contains $\vec{0}$ and rank is $k-1 \leq m$. There exists then an $m \times m$ integer unimodular matrix U such that $U[A^{i_1} \cdots A^{i_k}] = \begin{bmatrix} M \\ 0 \end{bmatrix}$ where M has $k-1$ rows and rank $k-1$. The kernel of M is thus spanned by one vector. Since $\vec{0} \in \mathrm{Conv}(A^{i_1}, \ldots, A^{i_k})$, there exists a nonzero positive solution of $MY = \vec{0}$ and thus all components of any solution of $MY = \vec{0}$ have the same sign. Now, $S = (|M^{i_2} \cdots M^{i_k}|, \ldots, (-1)^{k-1}|M^{i_1} \cdots M^{i_{k-1}}|)$ is an integer solution of $MY = \vec{0}$ which is not zero because the rank of M is $k-1$. Thus, S or $-S$ is a nonzero solution of $MY = \vec{0}, Y \in \boldsymbol{N}^k$ and $S' = s_1 e^{i_1} + \cdots + s_k e^{i_k}$ or $-S'$, where $e = (e^1, \ldots, e^m)$ is the canonical basis of $\boldsymbol{R}^m$, is a nonzero solution of $AX = \vec{0}, X \in \boldsymbol{N}^m$. □

Example 1 *The homogeneous diophantine system, the matrix of which is*

$$A = \begin{bmatrix} 4 & -1 & 5 & 2 & -2 \\ -3 & 5 & 2 & -1 & 3 \end{bmatrix}$$

has no nonzero solution because $\vec{0} \notin \mathrm{Conv}(A^1, \ldots, A^5)$.

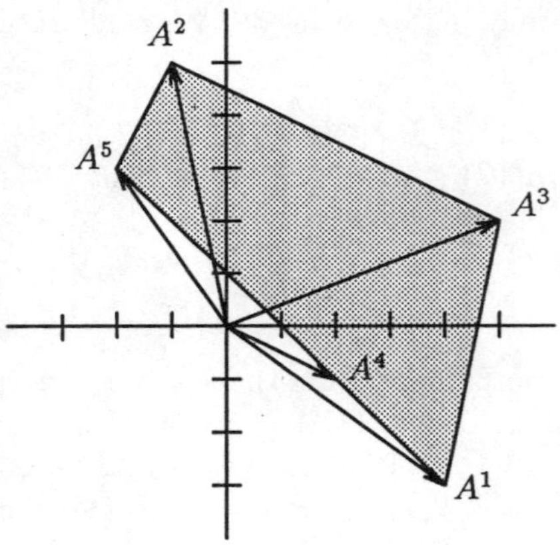

2 Basis of the set of positive solutions of $AX = \vec{0}$

From now on, A is an $m \times n$ integer matrix with rank m and $n \geq m+1$, and $\mathcal{A}$ denotes the set $\{A^1, \ldots, A^n\}$. Let $P_0(A)$ be the set of subsets of $\mathcal{A}$, the convex hull of which contains $\vec{0}$ and which are minimal with respect to the inclusion. For each $E = \{A^{i_1}, \ldots, A^{i_k}\} \in P_0(A)$, from the proof of theorem 2, $[A^{i_1} \cdots A^{i_k}]X = \vec{0}, X \in \boldsymbol{N}^k$ has a unique minimal solution $(x_1, \ldots, x_k)$. Let then $S^{(E)}$ be the vector $x_1 e^{i_1} + \cdots + x_k e^{i_k}$ where e is the canonical basis of $\boldsymbol{R}^n$, and $S_0(A)$ be $\{S^{(E)} \mid E \in P_0(A)\}$. The elements of $S_0(A)$ are known as the minimal solutions with minimal carrier. The following theorem is a refinement of a classical result which may be found in [10]:

Theorem 3 *Every positive solution of $AX = \vec{0}$ is a linear combination with positive coefficients of linearly independent vectors in $S_0(A)$.*

Proof: By induction on the carrier of a solution. The theorem obviously holds for every positive solution, the carrier of which is minimal. Now, let X be a positive solution of $AX = \vec{0}$, the carrier of which is not minimal, and let us assume that the theorem holds for any vector, the carrier of which is strictly contained in X's one. $S_0(A)$ contains a vector S, the carrier of which is contained in X's one, and $X' = X - \lambda S$ where $\lambda = \min\{\frac{x_i}{s_i} \mid s_i \neq 0\}$, is again a positive solution of $AX = \vec{0}$. Since the carrier of X' is strictly contained in X's one, by the induction hypothesis, $X' = \sum \lambda_i S_i$ where λ_i's are positive and S_i's are linearly independent vectors of $S_0(A)$. X is then equal to $\lambda S + \sum \lambda_i S_i$. S and S_i's are linearly independent because at least one component of S is zero in all S_i's. □

One may note that if X is rational, so is the coefficient λ constructed in the proof. Thus every positive solution of $AX = \vec{0}, X \in \boldsymbol{Q}^n$ is a linear combination with positive rational coefficients of linearly independent vectors in $S_0(A)$.

Example 2 *Consider the matrix*

$$A = \begin{bmatrix} 6 & 4 & -5 & -4 & -1 \\ 3 & 5 & 2 & -2 & -4 \end{bmatrix}$$

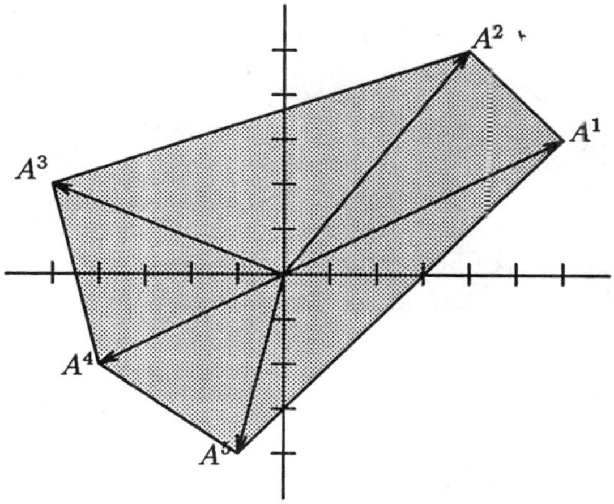

Since Conv($\mathcal{A}$) *contains* $\vec{0}$, *the system* $AX = \vec{0}, X \in \mathbf{N}^n$ *has a nonzero solution. In this case,*

$$P_0(A) = \{\{A^1, A^4\}, \{A^1, A^3, A^5\}, \{A^2, A^3, A^5\}, \{A^2, A^4, A^5\}\}$$

and

$$S_0(A) = \{(2,0,0,3,0), (22,0,21,0,27), (0,2,1,0,3), (0,14,0,11,12)\}$$

There is actually no need for looking for $P_0(A)$. The simplex algorithm gives a way to compute $S_0(A)$. We give here another way:

Theorem 4 *If* $S^{(E)} \in S_0(A)$, *there exists* $\{A^{i_1}, \ldots, A^{i_{m+1}}\} \subset \mathcal{A}$ *such that*

$$\begin{vmatrix} e^{i_1} \cdots e^{i_{m+1}} \\ A^{i_1} \cdots A^{i_{m+1}} \end{vmatrix} = K \cdot S^{(E)}$$

where K *is a nonzero integer.*

Proof: $E = \{A^{i_1}, \ldots, A^{i_k}\}$ and from corollary 2 of theorem 1, E has rank $k-1$. we get then

$$[A^{i_1} \cdots A^{i_k}] \sim \begin{bmatrix} B^{i_1} \cdots B^{i_k} \\ 0 \cdots 0 \end{bmatrix} \}k-1 \quad \text{and} \quad \begin{vmatrix} e^{i_1} \cdots e^{i_k} \\ B^{i_1} \cdots B^{i_k} \end{vmatrix} = K_1 \cdot S^{(E)}$$

Since A has rank m, we may find $A^{i_{k+1}}, \ldots, A^{i_{m+1}}$ such that $\{A^{i_1} \cdots A^{i_{m+1}}\}$ has rank m. We get then

$$\begin{vmatrix} e^{i_1} \cdots e^{i_{m+1}} \\ A^{i_1} \cdots A^{i_{m+1}} \end{vmatrix} = \begin{vmatrix} e^{i_1} \cdots e^{i_k} & e^{i_{k+1}} \cdots e^{i_{m+1}} \\ B^{i_1} \cdots B^{i_k} & R \\ 0 \cdots 0 & U \end{vmatrix} = |U| \cdot \begin{vmatrix} e^{i_1} \cdots e^{i_k} \\ B^{i_1} \cdots B^{i_k} \end{vmatrix} = |U| \cdot K_1 \cdot S^{(E)}$$

□

We compute then $S_0(A)$ as follows: for each subset $\{A^{i_1}, \ldots, A^{i_{m+1}}\}$ of $\mathcal{A}$, we compute the determinant $\begin{vmatrix} e^{i_1} \cdots e^{i_{m+1}} \\ A^{i_1} \cdots A^{i_{m+1}} \end{vmatrix}$. If it is a nonzero vector, all components of which have the same sign, we divide it by the greatest common divisor of its components and we get a vector in $S_0(A)$. Furthermore, from theorem 4, we get all vectors in $S_0(A)$ in this way.

3 Minimal solutions of $AX = \vec{0}, X \in N^n$

Theorem 5 *Vectors in $S_0(A)$ are minimal solutions of $AX = \vec{0}, X \in \boldsymbol{N}^n$.*

Proof: follows immediately from minimality of subsets in $P_0(A)$ and corollary 2. □

We saw in section 2 that every solution of $AX = \vec{0}, X \in \boldsymbol{N}^n$ is a combination with positive rational coefficients of linearly independent vectors in $S_0(A)$. Let M be a matrix the columns of which are k linearly independent vectors in $S_0(A)$. The problem is now to find all $\alpha \in (\boldsymbol{Q}^+)^k$ such that $M\alpha$ is an integer vector. There exists an integer unimodular matrix U such that $UM = \begin{bmatrix} T \\ 0 \end{bmatrix}$ where T is an upper triangular matrix with positive entries on the diagonal. Now, $M\alpha$ is integer iff $T\alpha$ is, that is to say, iff $\alpha = T^{-1}X$ where X is an integer vector. Since we are only interested in minimal solutions, we may consider only the case $0 \leq \alpha_i < 1$ for all i. Indeed we are only interested in the case $0 \leq \alpha_i$ for all i, and if $\alpha_i \geq 1$ for some i then $M\alpha \geq M^i$ which is a minimal solution we already have. For this purpose, we take $\alpha = T^{-1}X - \lfloor T^{-1}X \rfloor = \frac{1}{d}[dT^{-1}X]_d$ where d is the determinant of T, $\lfloor x \rfloor$ denotes the floor value of x and $[n]_d$ denotes the smallest positive integer congruent with n modulo d. The following theorem allows us to cut down the search space for convenient X.

Theorem 6 *If $D = (d_1, \ldots, d_k)$ is the diagonal of T and $d = d_1 \cdots d_k$ is its determinant, then:*

$$\forall X \in \boldsymbol{Z}^k : \exists X' \in \boldsymbol{Z}^k : \vec{0} \leq X' \ll D \text{ and } [dT^{-1}X']_d = [dT^{-1}X]_d$$

Proof: For all i, and all X,

$$\begin{aligned} dT^{-1}X &= dT^{-1}(X - Te^i + Te^i) \\ &= dT^{-1}(X - Te^i) + de^i \\ &\equiv dT^{-1}(X - Te^i) \pmod d \end{aligned}$$

$X - Te^i$ is an integer vector such that $\forall j > i : (X - Te^i)_j = X_j$ and $(X - Te^i)_i = X_i - d_i$. This proves the result. □

Example 3 (Example 2 continued)
$S_0(A) = \{(2,0,0,3,0), (22,0,21,0,27), (0,2,1,0,3), (0,14,0,11,12)\}$

$$M = \begin{bmatrix} 2 & 0 & 22 \\ 0 & 2 & 0 \\ 0 & 1 & 21 \\ 3 & 0 & 0 \\ 0 & 3 & 27 \end{bmatrix} \sim \begin{bmatrix} 1 & 0 & -4 \\ 0 & 1 & -3 \\ 0 & 0 & 6 \\ 0 & 0 & 0 \\ 0 & 0 & 0 \end{bmatrix} \quad \text{and } 6 \begin{bmatrix} 1 & 0 & -4 \\ 0 & 1 & -3 \\ 0 & 0 & 6 \end{bmatrix}^{-1} = \begin{bmatrix} 6 & 0 & 4 \\ 0 & 6 & 3 \\ 0 & 0 & 1 \end{bmatrix}$$

and we get $\alpha = \frac{1}{6}([6x_1 + 4x_3]_6, [6x_2 + 3k_3]_6, [x_3]_6) = \frac{1}{6}([4x_3]_6, [3k_3]_6, [x_3]_6)$ with $1 \leq x_3 \leq 6$. The table below contains for each value of k_3, the value of α and the corresponding solution *of* $AX = \vec{0}$.

k_3	6α	solution
1	$(4,3,1)$	$(5,1,4,2,6)$
2	$(2,0,2)$	$(8,0,7,1,9)$
3	$(0,3,3)$	$(11,1,11,0,15)$
4	$(4,0,4)$	$(16,0,14,2,18)$
5	$(2,3,5)$	$(19,1,18,1,24)$

4 Bounds on minimal solutions of $AX = \vec{0}, X \in N^n$

We prove now two bounds on minimal solutions of an homogeneous diophantine system. The former one bounds each component of a minimal solution while the latter one bounds its length.

Theorem 7 *Let* $\mathcal{M}$*, and* $\mathcal{M}'$ *be upper bounds on absolute values of* $m \times m$ *subdeterminants of* A*, and* $(m+1) \times (m+1)$ *subdeterminants of* $\begin{bmatrix} 1 \cdots 1 \\ A^1 \cdots A^n \end{bmatrix}$ *respectively. Then for any minimal solution* S *of* $AX = \vec{0}, X \in \boldsymbol{N}^n$:

1. $\|S\|_\infty \leq (n-m)\mathcal{M}$
2. $\|S\|_1 \leq (n-m)\mathcal{M}'$

Proof: We saw that every minimal solution is a combination with coefficients between 0 and 1 of vectors in $S_0(A)$. But $S_0(A)$ contains at most $n-m$ linearly independent vectors because it is the dimension of the kernel of A. Thus, each component of a minimal solution is bounded by $n-m$ times the maximum of a component of a vector in $S_0(A)$ and analogously, the length of a minimal solution is bounded by $n-m$ times the maximal length of a vector in $S_0(A)$. Now, we saw in section 2 that for any vector $S \in S_0(A)$, $S = \frac{1}{K}\begin{vmatrix} e^{i_1} \cdots e^{i_{m+1}} \\ A^{i_1} \cdots A^{i_{m+1}} \end{vmatrix}$ for some $A^{i_1}, \ldots, A^{i_{m+1}}$ and some nonzero integer K. Thus any component of S is bounded by a determinant $|A^{j_1} \cdots A^{j_m}|$ and the length of S is bounded by the absolute value of $\begin{vmatrix} 1 \cdots 1 \\ A^{i_1} \cdots A^{i_{m+1}} \end{vmatrix}$.
□

Bounds similar to 1 were already proved by I. Borosch & L.B. Treybig [1] and J. von zur Gathen and M. Sieveking [7] for the components of the smallest solution, and by J.-L. Lambert [10] for the components of all minimal solutions. Our bound is a slight improvement of Lambert's one, and was also discovered independently by L. Pottier [12]. On the other hand, our bound on the length of minimal solutions compares favorably with previous ones. L. Pottier gave in [11] a bound which is, in any case, much larger than ours and in [12] he proved that the length of minimal solutions is bounded by $(1 + \max \|A_i\|_1)^m$ which is also in general greater than our bound. At last, J.-F. Romeuf [13] gave $(2a)^{2^m - 1}$ as an upper bound, which is interesting since it does not depend on the number of variables but is doubly exponential and thus grows very fast. We conjecture that our bounds hold without the factor $n-m$, but we have no proof for this. I. Borosch & L.B. Treybig gave one, but only for the minimal solutions of an inhomogeneous system when the associated homogeneous system has no nonzero solution. From the proof of theorem 7, we see that these bounds would be sharp. Furthermore, the very sharp bound Lambert gave on the length of minimal solutions of one equation would become an instance of this one.

5 Inhomogeneous diophantine systems

The inhomogeneous case is a bit more complicated than the homogeneous one. Indeed, we are unable to find necessary and sufficient conditions for the satisfiability of such a system over natural numbers. However, we give here necessary conditions which are not sufficient and sufficient conditions which are not necessary.

Theorem 8 *The inhomogeneous diophantine system $AX = C, X \in \boldsymbol{N}^n$ has a solution only if $AX = C, X \in \boldsymbol{Z}^n$ has a solution and there exists a natural l such that $\frac{1}{l}C \in$* $\mathrm{Conv}(A^1, \ldots, A^n)$.

If $AX = C, X \in \boldsymbol{Z}^n$ has a solution and for all i, $AX = \vec{0}, X \in \boldsymbol{N}^n$ has a solution such that $x_i \neq 0$, then $AX = C, X \in \boldsymbol{N}^n$ has a solution.

Proof: It is obvious that $AX = C, X \in \boldsymbol{Z}^n$ has a solution if $AX = C, X \in \boldsymbol{N}^n$ has one. Now let S be a solution of $AX = C, X \in \boldsymbol{N}^n$. Then $A\frac{S}{\|S\|_1} = \frac{1}{\|S\|_1}C$ and $\frac{1}{\|S\|_1}C \in \mathrm{Conv}(A^1, \ldots, A^n)$ because $\left\| \frac{S}{\|S\|_1} \right\|_1 = 1$. This proves the first part of the theorem.

For the second part, let S be a solution of $AX = C, X \in \boldsymbol{Z}^n$. For each component s_i which is negative, we add to S a solution S' of $AX = 0, X \in \boldsymbol{N}^n$ such that $s'_i \neq 0$, until the component becomes positive. We get then a solution of $AX = C, X \in \boldsymbol{N}^n$.
□

One may note that the satisfiability over integers is very easy to check. Let k be the rank of A. There exist then two unimodular integer matrices L and R such that $LAR = \begin{bmatrix} D & 0 \\ 0 & 0 \end{bmatrix}$ where D is an $k \times k$ diagonal matrix and 0's stand for null matrices of suitable size (maybe 0). Any solution X is then of the form RY where Y is a solution of $\begin{bmatrix} D & 0 \\ 0 & 0 \end{bmatrix} Y = LC$. Checking the satisfiability of this last system is then straightforward.

To solve an inhomogeneous system, we add a variable x_{n+1}, solve the homogeneous system $[A\ -C] \begin{bmatrix} X \\ x_{n+1} \end{bmatrix} = \vec{0}$, and keep only solutions which satisfy $x_{n+1} \leq 1$. If $x_{n+1} = 0$ then X is a solution of the homogeneous system $AX = \vec{0}$, otherwise, X is a solution of the inhomogeneous system $AX = C$. Any solution of the inhomogeneous system is then of the form $X_0 + X_1$ where X_0 is a basic solution of $AX = C$ and X_1 is any linear combination with natural coefficients of solutions of $AX = \vec{0}$. We get then bounds on the solutions by applying theorem 7 to the homogeneous system $[A\ -C] \begin{bmatrix} X \\ x_{n+1} \end{bmatrix}$.

6 Conclusion

Our approach of linear diophantine equations is interesting for various reasons. First of all, it provides a geometrical interpretation of the satisfiability of a system and yields upper bounds on minimal solutions that are sharper than previous ones. From a more practical point of view, let us first notice that our algorithm may compare favorably with E. Contejean & H. Devie's one. The table below displays for some homogeneous systems, the matrix of the system, the number of minimal solutions, the computation time in seconds for our algorithm and for theirs. These comparisons are performed by a program written in the C programming language, running on a SUN 4/390 workstation.

1	$\begin{bmatrix} 1 & 2 & -3 & -2 & -4 \\ 2 & -1 & -3 & 2 & 5 \end{bmatrix}$	10	0.0035	0.0248
2	$\begin{bmatrix} 10 & -7 & -8 & 3 & -11 \\ 12 & -9 & -7 & 3 & 13 \end{bmatrix}$	240	0.7850	9.925
3	$\begin{bmatrix} 1 & 2 & -1 & 0 & -2 & -1 \\ 0 & -1 & -2 & 2 & 0 & 1 \\ 2 & 0 & 1 & -1 & -2 & 0 \end{bmatrix}$	13	0.0278	0.0740
4	$\begin{bmatrix} -10 & 0 & 20 & -1 & -21 \\ 9 & 1 & -17 & 2 & 19 \end{bmatrix}$	0	0.0013	15.94

The most interesting features of our algorithm are:

1. It is intrinsically parallel at various levels and could thus run much more quickly on a parallel machine. This did not hold for E. Contejean and H. Devie's algorithm. Indeed, the computation of $S_0(A)$ may be easily parallelized as well as the computation of all minimal solutions from $S_0(A)$ because our algorithm does not perform any comparison between the solutions until all have been computed.

2. Large solutions are found as easily as small ones. This fact is very interesting because algorithms which find solutions by increasing a vector one coordinate at a time are practically unable to find very large solutions.

3. It fails quickly when a system has no solution. This is demonstrated by example 4 above.

References

[1] I. Borosh and L. B. Treybig. Bounds on positive integral solutions of linear diophantine equations. *Proceedings of the American Mathematical Society*, 55:299–304, 1976.

[2] A. Boudet, E. Contejean, and H. Devie. A new AC unification algorithm with a new algorithm for solving diophantine equations. In *Proceedings 5th IEEE Symposium on Logic in Computer Science, Philadelphia (Pennsylvania, USA)*, pages 289–299, 1990.

[3] A. Brøndsted. *An Introduction to Convex Polytopes*, volume 90 of *Graduate Texts in Mathematics*. Springer-Verlag, 1983.

[4] M. Clausen and A. Fortenbacher. Efficient solution of linear diophantine equations. *Journal of Symbolic Computation*, 8:201–216, 1989. Special issue on unification. Part two.

[5] E. Contejean and H. Devie. Solving systems of linear diophantine equations. In H.-J. Bürckert and W. Nutt, editors, *Proceedings 3rd International Workshop on Unification, Lambrecht (Germany)*, 1989.

[6] A Fortenbacher. Algebraische unifikation. Diplomarbeit, Institut für Informatik, Universität Karlsruhe, 1983.

[7] J. von zur Gathen and M. Sieveking. A bound on solutions of linear integer equalities and inequalities. *Proceedings of the American Mathematical Society*, 72:155–158, 1978.

[8] T. Guckenbiehl and A. Herold. Solving linear diophantine equations. Technical Report SEKI-85-IV-KL, Universität Kaiserslautern, 1985.

[9] G. Huet. An algorithm to generate the basis of solutions to homogenous linear diophantine equations. *Information Processing Letters*, 7:144–147, 1978.

[10] J.-L. Lambert. *Le problème de l'accessibilité dans les réseaux de Petri*. PhD thesis, Université de Paris-sud, Centre d'Orsay, 1987.

[11] L. Pottier. Bornes et algorithme de calcul des générateurs des solutions de systèmes diophantiens linéaires. Technical report, INRIA Sophia Antipolis, 1990.

[12] L. Pottier. Minimal solutions of linear diophantine systems: Bounds and algorithms. In R.V. Book, editor, *Proceedings 4th Conference on Rewriting Techniques and Applications, Como, (Italy)*, volume 488 of *Lecture Notes in Computer Science*, pages 162–173. Springer-Verlag, 1991.

[13] J.-F. Romeuf. Solutions of a linear diophantine system, 1988. LIR & Université de Rouen.

[14] J.-F. Romeuf. A polynomial algorithm for solving systems of two linear diophantine equations. Technical report, Laboratoire d'Informatique de Rouen (France) and LITP, 1989.

A Second-Order Pattern Matching Algorithm for the Cube of Typed λ-Calculi

Gilles Dowek

INRIA*†

In [16] [17] Huet gives an algorithm for second-order pattern matching in the simply typed λ-calculus. We generalize this algorithm to the calculi of Barendregt's cube [1]. The same result is presented in a longer paper where proofs are given [8].

Introduction

The calculi of Barendregt's cube [1] are generalizations of the simply typed λ-calculus allowing functions from terms to types (dependent types), types to terms (polymorphism) and types to types (type constructors).

In the simply typed λ-calculus two classes of problems are known to be decidable: *first-order unification* (i.e. unification in a language with first-order existential variables and arbitrary order universal variables) (Robinson [21]) and *second-order pattern matching* (i.e. pattern matching in a language with at most second-order existential variables and at most third-order universal variables) (Huet [16] [17]). Recently Miller has proposed a generalization of first-order unification: *argument-restricted unification* [18].

First-order unification and argument-restricted unification have been generalized by Pfenning [19] to the calculi of the cube.

In this paper we generalize second-order matching in three ways: (1) we consider any calculus of the cube, (2) universal variables may have arbitrary order, (3) in some restricted cases, existential variables may have order greater than two. In the spirit of Miller and Pfenning we call these problems *second-order-argument-restricted* matching problems. If we want the first or the second generalization, the third is also required for technical reasons.

In this paper we assume some basic properties of calculi of the cube that are not yet fully demonstrated and which are out of the scope of this paper. We assume that in the eight calculi of the cube and in the Calculus of Constructions with universes the $\beta\eta$-reduction is Strongly Normalizable and Church-Rosser. Strong Normalization and the Church-Rosser property of β-reduction for all these calculi is proved in [2] [5]. Strong Normalization and the Church-Rosser property of $\beta\eta$-reduction for system F is proved in [11]. The Church-Rosser property of $\beta\eta$-reduction for $\lambda\Pi$ is proved in [22].

1 The Cube of Typed λ-Calculi

A *type system* [1] is a λ-calculus defined by a set S of sorts, a set Ax of pairs of sorts and a set R of triples of sorts. The *calculi of the cube* are the eight type systems such

*B.P. 105, 78153 Le Chesnay CEDEX, France. dowek@margaux.inria.fr

†This research was partly supported by ESPRIT Basic Research Action "Logical Frameworks".

that $S = \{Prop, Type\}$, $Ax = \{< Prop, Type >\}$, if $< s, s', s'' >\in R$ then $s' = s''$ and $< Prop, Prop, Prop >\in R$. Examples are the simply typed λ-calculus, $\lambda\Pi$ [14], system F, Fω [13] and the Calculus of Constructions without Universes [2] [4].

Syntax:

$$T \ ::= \ s \mid x \mid (T\ T) \mid [x:T]T \mid (x:T)T$$

In this paper we ignore variable renaming problems. A rigorous presentation would use de Bruijn indices. The terms s are sorts, the terms x are called variables, the terms $(T\ T')$ applications, the terms $[x:T]T'$ λ-abstractions and the terms $(x:T)T'$ products. The notation $T \to T'$ is used for $(x:T)T'$ when x has no free occurrence in T'.

Let t and t' be terms and x a variable. We write $t[x \leftarrow t']$ for the term obtained by substituting t' for x in t. We write $t \equiv t'$ when t and t' are $\beta\eta$-equivalent.

A *context* is a list of pairs $< x, T >$ (written $x : T$) where x is a variable and T a term.

Definition: We define inductively two judgements: Γ *is well-formed* and t *has type* T *in* Γ ($\Gamma \vdash t : T$) where Γ is a context and t and T are terms.

$$\frac{}{[]\ \text{well-formed}} \qquad \frac{\Gamma \vdash T : s}{\Gamma[x:T]\ \text{well-formed}}\, s \in S \qquad \frac{\Gamma\ \text{well-formed}}{\Gamma \vdash s : s'} < s, s' >\in Ax \qquad \frac{\Gamma\ \text{well-formed} \quad x : T \in \Gamma}{\Gamma \vdash x : T}$$

$$\frac{\Gamma \vdash T : s \quad \Gamma[x:T] \vdash T' : s'}{\Gamma \vdash (x:T)T' : s''} < s, s', s'' >\in R \qquad \frac{\Gamma \vdash (x:T)T' : s \quad \Gamma[x:T] \vdash t : T'}{\Gamma \vdash [x:T]t : (x:T)T'}\, s \in S$$

$$\frac{\Gamma \vdash t : (x:T)T' \quad \Gamma \vdash t' : T}{\Gamma \vdash (t\ t') : T'[x \leftarrow t']} \qquad \frac{\Gamma \vdash T : s \quad \Gamma \vdash T' : s \quad \Gamma \vdash t : T \quad T \equiv T'}{\Gamma \vdash t : T'}\, s \in S$$

In all these calculi, a term t well-typed in a context Γ has a unique type modulo $\beta\eta$-equivalence.

We assume the $\beta\eta$-reduction relation to be Strongly Normalizable and Church-Rosser. Thus each term has a unique $\beta\eta$-normal form. Two terms are equivalent if they have the same $\beta\eta$-normal form.

Definition: A term t is said to be atomic if it has the form $(u\ c_1\ ...\ c_n)$ where u is a variable or a sort. The symbol u is called the *head* of the term t.

Propositions: Let t be a normal well-typed term, t is either an abstraction, a product or an atomic term.

Let T be a well-typed normal term of type s for some sort s, T can be written in a unique way $T = (x_1 : P_1)...(x_n : P_n)P$ with P atomic.

Definition: Let Γ be a context and t be a $\beta\eta$-normal term well-typed in Γ and T the $\beta\eta$-normal form of its type. We define the *η-long form* of the term t:

• If t is an abstraction $[x:U]V$ (resp. a product $(x:U)V$) we let the η-long form of t be $t' = [x:U']V'$ (resp. $t' = (x:U')V'$) where U' and V' are the η-long form of U and V.

• If $t = (w\ c_1\ ...\ c_p)$ we let $T = (x_1 : P_1)...(x_n : P_n)P$ (P atomic). We let c'_i be the η-long form of c_i in Γ, P'_i be the η-long form of P_i in $\Gamma[x_1 : P_1; ...; x_{i-1} : P_{i-1}]$ and x'_i the η-long form of x_i in $\Gamma[x_1 : P_1; ...; x_i : P_i]$. We let the η-long form of t be $t' = [x_1 : P'_1]...[x_n : P'_n](w\ c'_1\ ...\ c'_p\ x'_1\ ...\ x'_n)$.

The well-foundedness of this definition is proved in the full paper.

The *β-normal η-long form* of a term t is the η-long form of its $\beta\eta$-normal form.

2 Quantified Contexts

Definition: A *quantified context* is a context in which a quantifier ($\forall$ or $\exists$) is associated to each variable[1].

Definition: A normal term t is said to be *flexible* if it is atomic and its head is an existential variable else it is said to be *rigid*.

Definition: A finite set σ of triples $< x, \gamma, t >$ where x is a variable, γ a context in which all the variables are existentially quantified and t a term is a *substitution* if for every variable x there is at most one triple $< x, \gamma, t >$ in σ. If there is one then x is said to be *bound* by σ.

Let x be a variable and σ a substitution. If there is a triple $< x, \gamma, t >$ in σ then we let $\sigma x = t$ else we let $\sigma x = x$. This definition extends straightforwardly to terms.

Definition: Let Γ be a quantified context and σ a substitution. We define by induction on the length of Γ a compatibility relation: *σ is well-typed in Γ*, and if σ is well-typed in Γ, a quantified context $\sigma\Gamma$.

• If $\Gamma = []$ then σ is well-typed in Γ and $\sigma\Gamma = []$.

• If $\Gamma = \Delta[\forall x : T]$ then if σ is well-typed in Δ we let $\Gamma' = \sigma\Delta$. If $\Gamma' \vdash \sigma T : s$ for some sort s, then σ is well-typed in Γ and $\sigma\Gamma = \Gamma'[\forall x : \sigma T]$. Otherwise σ is not well-typed in Γ.

• If $\Gamma = \Delta[\exists x : T]$ then let γ be the context associated to x by σ if x is bound by this substitution and $\gamma = [\exists x : \sigma T]$ if it is not. If σ is well-typed in Δ we let $\Gamma' = \sigma\Delta$. If $\Gamma'\gamma$ is well-formed and $\Gamma'\gamma \vdash \sigma x : \sigma T$ then σ is well-typed in Γ and $\sigma\Gamma = \Gamma'\gamma$. Otherwise σ is not well-typed in Γ.

Definition: Let σ and τ be two substitutions. We define the substitution $\tau \circ \sigma$ as:

$$\tau \circ \sigma = \{< x, \tau\gamma, \tau t > \mid < x, \gamma, t > \in \sigma\} \cup \{< x, \gamma, t > \mid < x, \gamma, t > \in \tau,\ x \text{ not bound by } \sigma\}$$

where $\tau\gamma$ is defined inductively by $\tau[] = []$ and $\tau(\gamma'[\exists y : U]) = (\tau\gamma')\gamma''$ where γ'' is the context associated to y by τ if y is bound by τ and $\gamma'' = [\exists y : \tau U]$ if it is not.

Definitions: Let Γ be a context and t a term well-typed in Γ. The term t is *ground* in Γ if for every symbol x of type T free in t, x is a universal variable in Γ.

The term t is *closed* in Γ if it is hereditarily ground i.e. if for every symbol x of type T free in t, x is a universal variable in Γ, and T is closed in the prefix of Γ defined before x. It is *open* otherwise.

Definition: Let Γ be a context and T be a well-typed normal term of type s for some sort s. The *order* of T in Γ is defined as an element of $N \cup \{\infty\}$ as follows:

• if $T = (y : U)V$ then $o(T) = max\{1 + u, v\}$ where u is the order of U in the context Γ and v is the order of V in the context $\Gamma[\exists y : U]$.

• if $T = (x\ t_1\ \ldots\ t_n)$ then if x is a universal variable of Γ then $o(T) = 1$, if x is an existential variable of Γ then $o(T) = \infty$ and if x is a sort then $o(T) = 2$.

We prove in the full paper that if T has order n in Γ and σ is a substitution well-typed in Γ then the order of σT is less or equal than n in $\sigma\Gamma$.

---

[1]These quantified contexts are very similar to Miller's *Mixed Prefixes* [18].

Definition: Let Γ be a context and t a term well-typed in Γ. The term t is *second-order* in Γ if for every symbol x of type T free in t, if x is existential in Γ then T is of order at most 2, and in all cases T is a second-order term in the prefix of Γ defined before x.

Definition: A *matching problem* is a triple $< \Gamma, a, b >$ such that Γ is a well-formed quantified context, the terms a and b are normal and well-typed in Γ and have the same type, and the term b is closed. A solution of this problem is a substitution θ well-typed in Γ such that $\theta a = b$.

3 Problems in Adapting Huet's Algorithm

In this section we discuss using examples the problems that arise when we generalize Huet's algorithm to the calculi of the cube.

Variables of Order Greater than Two

In Huet's algorithm, when we start with a problem that has a second-order open term, all subsequent problems also have a second-order open term. In the calculi of the cube, variables of order greater than two may appear. We give two examples.

In the first example, we unify a flexible term with a product. In the context:

$$[\forall T : Prop; \forall a : T; \forall Q : Prop; \exists f : T \to Prop]$$

we start with the problem:

$$(f\ a) = (P : T \to Prop)Q$$

which leads us to consider:

$$(P : (h\ a))(k\ a\ P) = (P : T \to Prop)Q$$

with $h : (x : T)s$ and $k : (x : T)((h\ x) \to Prop)$. The variable k is of order ∞. Even if we solve first the problem $(h\ a) = T \to Prop$, we get $k : (x : T)((T \to Prop) \to Prop)$ and k is still third-order.

In the second example we unify a flexible term with an atomic term whose head has a type of order greater than three[2]. In the context:

$$[\forall T : Prop; \forall F : ((T \to T) \to T) \to T; \forall a : T; \exists f : T \to T]$$

we start with the problem:

$$(f\ a) = (F\ ([x : T \to T](x\ a)))$$

which leads us to consider:

$$(F\ (h\ a)) = (F\ ([x : T \to T](x\ a)))$$

with a new existential variable $h : T \to (T \to T) \to T$ which is third-order.

To deal with these problems we generalize the class of matching problems we consider, in such a way that the open term is *second-order-argument-restricted* i.e. a term in which the arguments of existential variables are either terms of a first-order type, or η-equivalent

[2]In [16] [17] such universal variables are forbidden, so even in simply typed λ-calculus the algorithm presented here is a generalization of Huet's algorithm allowing arbitrary order universal variables. Actually, a much simpler algorithm is obtained in considering Huet's unification algorithm [15] [16] and in proving its termination when one of the terms is second-order-argument-restricted and the other is closed. Not all the extra complexity of the algorithm presented here is needed in simply typed λ-calculus.

to a universal variable or a variable bound higher in the term. These terms are related to the *argument-restricted* terms studied by Miller [18] and Pfenning [19]. Despite the variables of order greater than two, these problems can be solved. For instance in:

$$[\forall T : Prop; \forall a : T; \forall P : T \to Prop; \forall Q : T \to Prop; \exists k : T \to (T \to Prop) \to Prop]$$

we consider the problems:

$$(k\ a\ [x:T](P\ x)) = (P\ a) \quad \text{and} \quad (k\ a\ [x:T](P\ x)) = (Q\ a)$$

The projection on the higher-order argument of k leads us to the problems:

$$(P\ (k'\ a\ [x:T](P\ x))) = (P\ a) \quad \text{and} \quad (P\ (k'\ a\ [x:T](P\ x))) = (Q\ a)$$

The first simplifies to $(k'\ a\ [x:T](P\ x)) = a$ where the right term has been simplified, and the second is a failure problem. In both cases, infinite branches are avoided.

Definition: Let Γ be a context and t a term well-typed in Γ. The term t is said to be a *second-order-argument-restricted term* if either:

- $t = [x:T]u$ (resp. $t = (x:T)u$) with T second-order-argument-restricted in Γ and u second-order-argument-restricted in $\Gamma[\forall x : T]$,
- $t = (x\ c_1\ ...\ c_n)$ with x universal and $c_1, ..., c_n$ second-order-argument-restricted in Γ and the type of x is second-order-argument-restricted in the prefix of Γ defined before x,
- $t = (x\ c_1\ ...\ c_n)$ with x existential, for all i, c_i is either a term of a first-order type in Γ or η-equivalent to a universal variable of Γ, c_i is second-order-argument-restricted in Γ and the type of x is second-order-argument-restricted in the prefix of Γ defined before x,
- t is a sort.

A second-order term is second-order-argument-restricted and the η-long form of a second-order-argument-restricted term is second-order-argument-restricted.

If the term $t = (x\ c_1\ ...\ c_n)$ is an η-long flexible second-order-argument-restricted term and the type of x is $(x_1 : P_1)...(x_p : P_p)P$ (P atomic) then $p = n$.

The Meta Type System

Let $\mathcal{T}$ be any calculus of the cube. We consider in $[\exists x : Prop]$ the problem:

$$x = (P : Prop)P$$

which leads us to consider:

$$(P : h_1)(h_2\ P) = (P : Prop)P$$

The variable h_1 must be of type *Type*. Such a declaration is not allowed in the system $\mathcal{T}$, so we embed our type system in a larger one: the *meta type system*. In this meta type system we can declare a variable standing for every term well-typed in $\mathcal{T}$ and can also express a term $t : T'$ well-typed in $\Gamma[x : T]$ as $t = (f\ x)$ with f well-typed in Γ. This cannot be done in general in $\mathcal{T}$, because the term $f = [x : T]t$ may be ill-typed.

During the execution of the algorithm all variables are well-typed in this system. We prove in the full paper that upon termination, the substitution obtained is well-typed in the original system $\mathcal{T}$.

Definition: In this system we have three sorts: *Prop*, *Type* and *Extern*, two axioms *Prop* : *Type* and *Type* : *Extern* and the following rules:

$$< Prop, Prop, Prop >, < Prop, Type, Type >, < Type, Prop, Prop >,$$

$$< Type, Type, Type >, < Prop, Extern, Extern >, < Type, Extern, Extern >$$

All the terms typable in this system are typable in the Calculus of Constructions with universes [3], so we assume that in this system the reduction relation is Strongly Normalizable and Church-Rosser. Each term has a unique β-normal η-long form. Two terms are equivalent if they have the same β-normal η-long form.

If a term t is well-typed in a context Γ then it has a unique type modulo $\beta\eta$-equivalence.

Accounting Equations

In Huet's algorithm, when we perform an elementary substitution $\{< x, \gamma, t >\}$ we need to check that it is well-typed, i.e. that the variable x and the term t have the same type. As remarked by Elliott [9] [10] and Pym [20] in calculi of the cube these types may contain variables. Thus instead of checking that they are equal we have to unify them.

In general, this *accounting equation* is a unification problem i.e. may have variables on both sides[3]. But if the initial problem is second-order-argument-restricted then this accounting equation is also a second-order-argument-restricted *matching* problem and therefore can be solved. The key lemma is that if in a context Γ a second-order-argument-restricted term $(u\ c_1\ ...\ c_n)$ has a closed type and u has the type $(x_1 : P_1)...(x_n : P_n)P$ (P atomic) then P is closed in $\Gamma[\forall x_1 : P_1; ...; \forall x_n : P_n]$. The intuition is that the type of $(u\ c_1\ ...\ c_n)$ is $P[x_1 \leftarrow c_1, ..., x_n \leftarrow c_n]$ and P must be closed because the c_i are atomic terms and their substitution cannot cancel existential variables in P. The (rather long) proof is given in the full paper.

We also prove that if c_i is η-equivalent to a variable v which has a closed type in Γ then P_i is closed in the context $\Gamma[\forall x_1 : P_1; ...; \forall x_{i-1} : P_{i-1}]$.

Termination

We also have to prove that these accounting equations do not jeopardize termination. We prove in the full paper that a function that recurses both on a strict subterm and on the type of its argument is total, i.e. the well-foundedness of the smallest transitive relation $<$ such that if t is a strict subterm of t' then $t < t'$ and if T is the normal type of t in Γ then $T < t$.

Types of the new variables

In some calculi, a variable v whose type begins by p products may be applied more that p times. So when we perform an elementary substitution we do not have enough information to know the types of the new variables. For instance, in the context:

$$[\forall T : Prop; \forall v : (P : Prop)P; \forall a : T; \forall G : (Prop \rightarrow T) \rightarrow Prop; \exists f : Prop \rightarrow T]$$

we start with the problem:

$$(f\ (T \rightarrow T)) = (v\ (T \rightarrow T)\ a)$$

where we want to perform the substitution $f \leftarrow [p : Prop](v\ (h_1\ p)\ (h_2\ p))$.

The variable h_1 has type $Prop \rightarrow Prop$ but since the term $(v\ (h_1\ p))$ has type $(h_1\ p)$ which is not yet a product we cannot give a type to the variable h_2.

If we performed the substitution we would get the problem:

$$(v\ (h_1\ (T \rightarrow T))\ (h_2\ (T \rightarrow T))) = (v\ (T \rightarrow T)\ a)$$

[3]This remark is used in [6] to prove that third-order matching is undecidable in calculi with dependent types or type constructors.

which simplifies to the system:

$$(h_1\ (T \rightarrow T)) = (T \rightarrow T) \qquad (h_2\ (T \rightarrow T)) = a$$

The idea is to solve the first subsequent equation before we give a type to the variable h_2. This equation has two solutions $h_1 \leftarrow [p : Prop](T \rightarrow T)$ and $h_1 \leftarrow [p : Prop]p$.

In the first solution, $(v\ (T \rightarrow T))$ has type $T \rightarrow T$ so we give the type $Prop \rightarrow T$ to h_2 and we can solve the equation $(h_2\ (T \rightarrow T)) = a$. In the second solution, $(v\ p)$ has type p which is not a product, so $[p : Prop](v\ p\ (h_2\ p))$ is ill-typed whatever the type of h_2 may be, and thus this branch does not lead to a solution of the initial equation.

Non-linearity

In the last example the variable f has only one occurrence in the problem, thus h_2 did not appear in the first subsequent equation. Let us consider now:

$$(f\ (G\ f)) = (v\ (T \rightarrow T)\ a)$$

Considering again the substitution $f \leftarrow [p : Prop](v\ (h_1\ p)\ (h_2\ p))$, we may still give a type to h_1 but not to h_2. Also h_2 occurs in the first subsequent equation:

$$(h_1\ (G\ [p : Prop](v\ (h_1\ p)\ (h_2\ p)))) = (T \rightarrow T)$$

The idea is to keep the variable f in the equation:

$$(h_1\ (G\ f)) = (T \rightarrow T)$$

For each substitution solution of the subsequent equations σ, we want to instanciate f by the term $t_1 = [p : Prop](v\ (\sigma h_1\ p)\ (\sigma h_2\ p))$, but f may be already instanciated by $t_2 = \sigma f$. A priori we would have to unify t_1 and t_2.

Actually in a substitution solution of a matching problem, a variable is either invariant $(\sigma x = x)$ or instanciated by a ground term. Moreover if the variable is the head of the flexible term of the problem then it is instanciated by a ground term. So the term t_1 is ground and the term t_2 is either the variable f itself or a ground term. Three cases may occur: if t_2 is the variable f then $\sigma \cup \{< f, [], t_1 >\}$ is a solution of the initial problem, if t_1 is ground and different from t_2 then σ does not lead to a solution of the initial problem, if $t_1 = t_2$ then σ is a solution of the initial problem.

4 A Pattern Matching Algorithm

As motivated above we consider matching problems $< \Gamma, a, b >$ such that Γ is a well-formed context in the meta type system, a and b are normal terms well-typed in Γ, they have the same type and this type is also well-typed in Γ (i.e. is different from *Extern*). The term a is second-order-argument-restricted.

In comparison with Huet's algorithm we have to strengthen the control on the algorithm: we define recursively a function *Sol* that takes a problem and returns a set of substitutions solutions to the problem. We use the symbols $\Sigma, \Upsilon, \Phi, \Psi, \Omega$ for sets of substitutions obtained in the intermediate steps in the construction of $Sol < \Gamma, a, b >$.

If Σ is a set of substitutions and σ a substitution, we let $\Sigma \circ \sigma = \{\tau \circ \sigma \mid \tau \in \Sigma\}$.

In the full paper this definition is proved to be well-founded, sound and complete, i.e. every $\sigma \in Sol < \Gamma, a, b >$ is a solution of $< \Gamma, a, b >$ and if θ is a solution of this problem, there is $\sigma \in Sol < \Gamma, a, b >$ and ρ well-typed in $\sigma\Gamma$ such that for all x of Γ, $\theta x = (\rho \circ \sigma)x$.

Definition:

- If the terms a and b are both abstractions, modulo α-conversion we can consider that the variable bound in these abstractions is the same and is different from all the variables of Γ: $a = [x : U]c$ and $b = [x : U]d$. We take:
$$Sol < \Gamma, a, b >= Sol < \Gamma[\forall x : U], c, d >$$

- If the terms a and b are both products: modulo α-conversion we can consider that the variable bound in these products is the same and is different from all the variables of Γ, $a = (x : U)U'$ and $b = (x : V)V'$. If U and V have different types then $Sol < \Gamma, a, b >= \emptyset$, else we take:
$$\Sigma = Sol < \Gamma, U, V >$$
$$Sol < \Gamma, a, b >= \cup_{\sigma \in \Sigma}(Sol < \sigma\Gamma[\forall x : V], \sigma U', V' > \circ \sigma)$$

- If the terms a and b are both atomic and rigid with the same head and the same number of arguments: $a = (v\ c_1\ ...c_n)$ and $b = (v\ d_1\ ...d_n)$. We take:
$$\Sigma_0 = \{Id\}$$
$$\Sigma_{i+1} = \cup_{\sigma \in \Sigma_i}(Sol < \sigma\Gamma, \sigma c_{i+1}, d_{i+1} > \circ \sigma)$$
$$Sol < \Gamma, a, b >= \Sigma_n$$

- In all the other cases in which a and b are both rigid, we take $Sol < \Gamma, a, b >= \emptyset$.

- If a is atomic flexible $a = (u\ c_1\ ...\ c_n)$ and b is a product $b = (y : V)V'$, we have $u : (x_1 : P_1)...(x_n : P_n)P$ (P atomic). Let s be the type of V in Γ and s' be the type of V' in $\Gamma[\forall y : V]$, s and s' are sorts.

  We let Γ_1 be the insertion of the declarations: $\exists h : (x_1 : P_1)...(x_n : P_n)s$ and $\exists k : (x_1 : P_1)...(x_n : P_n)(y : (h\ x_1\ ...\ x_n))s'$ just before u in Γ. We take:
$$\Sigma = Sol < \Gamma_1, (h\ c_1\ ...\ c_n), V >$$
$$\Upsilon = \cup_{\sigma \in \Sigma}(Sol < \sigma(\Gamma_1[\forall y : V]), (k\ \sigma c_1\ ...\ \sigma c_n\ y), V' > \circ \sigma)$$
  For all $\sigma \in \Upsilon$ we let:
$$t_\sigma = [x_1 : \sigma P_1]...[x_n : \sigma P_n](y : ((\sigma h)\ x_1\ ...\ x_n))((\sigma k)\ x_1\ ...\ x_n\ y)$$
  We take:
$$Sol < \Gamma, a, b >= \{\sigma \cup \{< u, [], t_\sigma >\} \mid \sigma \in \Upsilon \text{ and } \sigma u = u\} \cup \{\sigma \mid \sigma \in \Upsilon \text{ and } \sigma u = t_\sigma\}$$

- If a is atomic flexible $a = (u\ c_1\ ...\ c_n)$ and b is a sort, we have $u : (x_1 : P_1)...(x_n : P_n)P$ (P atomic). We take:
$$Sol < \Gamma, a, b >= \{< u, [], [x_1 : P_1]...[x_n : P_n]b >\}$$

- If a is atomic flexible $a = (u\ c_1\ ...\ c_n)$ and b atomic $b = (v\ d_1\ ...d_p)$ with v universal, we have $u : (x_1 : P_1)...(x_n : P_n)P$ (P atomic). We let $\Gamma' = \Gamma[\forall x_1 : P_1; ...; \forall x_n : P_n]$. The term P is closed in Γ'.

  Let *Heads* be the set containing all the x_i such that c_i has a first-order type or is η-equivalent to v. Also, if the variable u is declared to the right of v, v is in *Heads*.

- For every variable x_i of *Heads* such that c_i has a first-order type, the terms P and P_i are well-typed in Γ' and their types are sorts. If they are different, then we take $\Omega_{x_i} = \emptyset$ else we take:
$$\Sigma_{x_i} = \{\{< u, [], [x_1 : \sigma P_1]...[x_n : \sigma P_n]x_i >\} \circ \sigma \mid \sigma \in Sol < \Gamma', P_i, P >\}$$
$$\Omega_{x_i} = \cup_{\sigma \in \Sigma_{x_i}} (Sol < \sigma\Gamma, \sigma a, b > \circ \sigma)$$
- For every w of *Heads* such that $w = v$ or $w = x_i$ with c_i η-equivalent to v, by induction on i we build $\Sigma_{w,i}$ a set of substitutions that may bind variables of Γ and also some extra variables $h_1, ..., h_i$.
$$\Sigma_{w,0} = \{Id\}$$
For each $\sigma \in \Sigma_{w,i}$, the term $(w\ ((\sigma h_1)\ x_1\ ...\ x_n)\ ...\ ((\sigma h_i)\ x_1\ ...\ x_n))$ is well-typed in $\sigma\Gamma'$. If its type is not a product then we take:
$$\Phi_{w,\sigma,i+1} = \emptyset$$
else let $(y : T)T'$ be its type. Let the context Γ_{i+1} be the insertion of the declaration $\exists h_{i+1} : (x_1 : P_1)...(x_n : P_n)T$ just before u in Γ'. We let also $H_{i+1} = T[x_1 \leftarrow \sigma c_1, ..., x_n \leftarrow \sigma c_n]$ which is the type of $(h_{i+1}\ \sigma c_1\ ...\ \sigma c_n)$ in $\sigma\Gamma_{i+1}$, and D_{i+1} be the type of d_{i+1}.
The terms H_{i+1} and D_{i+1} are well-typed in $\sigma\Gamma_{i+1}$ and their types are sorts. If they are different, then we take $\Phi_{w,\sigma,i+1} = \emptyset$ else we take:
$$\Upsilon_{w,\sigma,i+1} = Sol < \sigma\Gamma_{i+1}, H_{i+1}, D_{i+1} > \circ \sigma$$
$$\Phi_{w,\sigma,i+1} = \cup_{\tau \in \Upsilon_{w,\sigma,i+1}} Sol < \tau\Gamma_{i+1}, (h_{i+1}\ \tau c_1\ ...\ \tau c_n), d_{i+1} > \circ \tau$$
and:
$$\Sigma_{w,i+1} = \cup_{\sigma \in \Sigma_{w,i}} \Phi_{w,\sigma,i+1}$$
For each $\sigma \in \Sigma_{w,p}$, let $T_{w,\sigma}$ be the type in the context $\sigma\Gamma'$ of the term $(w\ ((\sigma h_1)\ x_1\ ...\ x_n)\ ...\ ((\sigma h_p)\ x_1\ ...\ x_n))$. We take:
$$\Psi_w = \{\sigma \in \Sigma_{w,p} \mid T_{w,\sigma} = P\}$$
For each σ of Ψ_w, we let:
$$t_{w,\sigma} = [x_1 : \sigma P_1]...[x_n : \sigma P_n](w\ ((\sigma h_1)\ x_1\ ...\ x_n)\ ...\ ((\sigma h_p)\ x_1\ ...\ x_n))$$
$$\Omega_w = \{\sigma \cup \{< u, [], t_{w,\sigma} >\} \mid \sigma \in \Psi_w \text{ and } \sigma u = u\} \cup \{\sigma \mid \sigma \in \Psi_w \text{ and } \sigma u = t_{w,\sigma}\}$$

Finally:
$$Sol < \Gamma, a, b > = \cup_{w \in Heads} \Omega_w$$

Conclusion

We have proved the decidability of second-order pattern matching in all the calculi of the cube under the assumption of Strong Normalisation and the Church-Rosser property of $\beta\eta$-reduction. In another paper [6], an analysis of this proof leads to the undecidability of third-order matching in the calculi allowing dependent types or type constructors. Higher-order matching is also undecidable in polymorphic λ-calculus [7]. The problem of the decidability of higher-order matching is still open for the simply typed λ-calculus.

References

[1] H. Barendregt, Introduction to Generalized Type Systems, To appear in *Journal of Functional Programming.*

[2] Th. Coquand, Une Théorie des Constructions, *Thèse de troisième cycle*, Université Paris VII, 1985.

[3] Th. Coquand, An analysis of Girard's paradox, *Proceedings of Logic in Computer Science*, 1986, pp. 227-236.

[4] Th. Coquand, G. Huet, The Calculus of Constructions, *Information and Computation*, 76, 1988, pp. 95-120.

[5] Th. Coquand. J. Gallier, A Proof of Strong Normalization For the Theory of Constructions Using a Kripke-Like Interpretation, Personal communication.

[6] G. Dowek, L'Indécidabilité du Filtrage du Troisième Ordre dans les Calculs avec Types Dépendants ou Constructeurs de Types (The Undecidability of Third Order Pattern Matching in Calculi with Dependent Types or Type Constructors), *Compte Rendu à l'Académie des Sciences*, 1991.

[7] G. Dowek, The Undecidability of Pattern Matching in Polymorphic λ-Calculus, In preparation.

[8] G. Dowek, A Second-Order Pattern Matching Algorithm for the Cube of typed λ-Calculi, *Rapport de Recherche INRIA*, 1991.

[9] C. M. Elliott, Higher-order Unification with Dependent Function Types, *Proceedings of the 3rd International Conference on Rewriting Techniques and Applications*, N. Dershowitz (Ed.), Lecture Notes in Computer Science, 355, Springer-Verlag, 1989, pp.121-136.

[10] C. M. Elliott, Extensions and Applications of Higher-order Unification, *PhD Thesis*, 1990, Carnegie Mellon University, Pittsburgh, Report CMU-CS-90-134.

[11] J. Gallier, On Girard's Candidats de Réductibilité, *Logic and Computer Science*, P. Odifreddi (Ed.), Academic Press, London, 1990, pp. 123-203.

[12] H. Geuvers, M.J. Nederhof, A Modular Proof of Strong Normalization for the Calculus of Constructions, Catholic University Nijmegen.

[13] J.Y. Girard, Interprétation fonctionnelle et élimination des coupures dans l'arithmétique d'ordre supérieur, *Thèse de Doctorat d'État*, Université de Paris VII, 1972.

[14] R. Harper, F. Honsell, G. Plotkin, A Framework for Defining Logics, *Proceedings of Logic in Computer Science*, 1987, pp. 194-204.

[15] G. Huet, A Unification Algorithm for Typed λ-calculus, *Theoretical Computer Science*, 1, 1975, pp. 27-57.

[16] G. Huet, Résolution d'Équations dans les Langages d'Ordre 1,2, ..., ω, *Thèse de Doctorat d'État*, Université de Paris VII, 1976.

[17] G. Huet, B. Lang, Proving and Applying Program Transformations Expressed with Second Order Patterns, *Acta Informatica*, 1978, 11, pp. 31-55.

[18] D.A. Miller, Unification Under a Mixed Prefix, To appear in *Journal of Symbolic Computation.*

[19] F. Pfenning, Unification and anti-Unification in the Calculus of Constructions, To appear in *Proceedings of Logic in Computer Science*, 1991.

[20] D. Pym, Proof, Search and Computation in General Logic, *PhD thesis*, University of Edinburgh. Report CST-69-90 also ECS-LFCS-90-125.

[21] J.A. Robinson, A Machine-Oriented Logic Based on the Resolution Principle, *Journal of the Association for Computing Machinery*, 12, 1, 1965, pp. 23-41.

[22] A. Salvesen, The Church-Rosser Theorem for LF with β/η-reduction, Manuscript, University of Edinburgh, 1989.

The lazy call-by-value λ-calculus[1]

Lavinia Egidi*, Furio Honsell**, Simona Ronchi della Rocca*

* Dipartimento di Informatica, Università degli Studi di Torino (Italy)
** Dipartimento di Matematica e Informatica, Università degli Studi di Udine (Italy)

Introduction.

In the implementation of many functional programming languages, parameters are evaluated before being passed (call-by-value evaluation) and function bodies are evaluated only when parameters are supplied (lazy evaluation). These two features first appeared in the implementation of the language ISWIM given by Landin using the SECD machine [5]. The ISWIM—SECD system can be seen as the paradigm implementation of the functional kernel of several programming languages such as ML.

The call-by-value λ-calculus ($\lambda\beta_v$) was introduced by Plotkin [8], in order to reason about equivalence and termination of ISWIM programs interpreted by the SECD machine. The $\lambda\beta_v$-calculus is obtained from the λ-calculus by restricting the β-rule to redexes whose operand is a value (i.e. a constant, a variable or a function). Plotkin showed that given a $\lambda\beta_v$-term M the SECD machine evaluates M to an output value v if and only if M β_v-reduces to v according to the lazy-leftmost reduction strategy. This strategy being the one which reduces the leftmost redex not inside a lambda abstraction.

Once we have a sequential interpreter I for a programming language, the most natural behavioural equivalence between programs is the one determined by termination observations of the computation processes. This relation, termed I-operational equivalence, is defined as:

$M \equiv_I N \Leftrightarrow$ for all closing context C[ ].(C[M] terminates $\Leftrightarrow$ C[N] terminates).

The $\lambda\beta_v$-calculus provides a sound (albeit incomplete) formal system for establishing SECD-operational equalities between $\lambda\beta_v$-terms ($\equiv_{SECD}$).

The model theory of the $\lambda\beta_v$-calculus is particularly interesting. In fact, contrary to ordinary pure λ-calculus, pure $\lambda\beta_v$-calculus has a canonical denotational model in any of the usual categories for denotational semantics. Namely the initial solution of the equation:

$$D \approx [D \to_\perp D]_\perp$$

where $[D \to_\perp D]_\perp$ denotes the lifted space of continuous strict functions from D to D.

In this paper we study some topics in the model theory of the $\lambda\beta_v$-calculus in view of a deeper understanding of the operational semantics of the SECD machine. First we deal with the general principles. Then we study the theory induced by the model D above. This theory however is too weak, since it is strictly included in $\equiv_{SECD}$. A phenomenon similar to those described in [3] and [9] arises in this context. The SECD machine is sequential and hence not all strict continuous functions are definable by an ISWIM term. In order to find a fully abstract model we have to take into account also models where not all the strict continuous functions are representable. We build then a fully abstract model by means of a notion of call-by-value applicative bisimulation. Similar constructions were carried out by Plotkin [10], Abramsky [1], Mulmuley [6] and Ong [7] for different calculi. This construction is interesting in itself since it provides an alternative observational characterization of the operational equivalence $\equiv_{SECD}$ in terms of a restricted class of "computational experiments", i.e. applicative contexts. Formally

$M \equiv^v N \Leftrightarrow$ for all term vectors $P_1 \ldots P_n$.($MP_1 \ldots P_n$ terminates $\Leftrightarrow$ $NP_1 \ldots P_n$ terminates).

The fully abstract model produced is very thin, it amounts to a term model equipped with a richer structure. However this result is remarkable since it is not immediate that $\equiv^v$ is a congruence relation.

---

1 Work partially supported by MURST 40% and 60% grants and EEC "Project Stimulation ST2J/0374/C(EDB): Lambda Calcul Typé".

Finally we give a logic for reasoning about lazy call-by-value programs, which takes the form of a type assignment system for λ-terms, and can be viewed as the call-by value analogue of the intersection type assignment system of Coppo-Dezani [2]. This is inspired by a "concrete" logical presentation of the canonical model.
Added in proof. Since this paper was submitted to MFCS 91, two papers related to this topic have appeared [12] and [13] .

1. The ISWIM-SECD system.

In this section we recall basic definitions and facts concerning the language ISWIM, the SECD machine [5] and the lazy-call-by-value λ-calculus $\lambda\beta_v$ [8]. We illustrate also the use of $\lambda\beta_v$ for describing the operational semantics of terms induced by the SECD machine.

1.1. Pure ISWIM.

As remarked in the introduction, ISWIM is the paradigm of many functional programming languages. For simplicity, here we deal with the pure version of it, i.e. ISWIM without constants. Thus ISWIM programs turn out to be terms of the pure λ-calculus.

The language of pure ISWIM is defined by the pair (Λ, Val), where Λ is the set of terms of pure λ-calculus over a set of variables *Var* , and $Val = Var \cup \{\lambda x.M \mid M \in \Lambda\}$ is the *set of values* . Free and bound variables are also defined as usual and, for any term M, FV(M) is the set of its free variables. $\Lambda^\circ \subset \Lambda$ is the set of closed terms. We also adopt usual conventions about parentheses, and usual notations for contexts.

A complete definition of a programming language is given once we specify an evaluation function describing the way programs are executed. Landin defined the evaluation function eval : $\Lambda \to \Lambda$ for ISWIM, by means of an abstract transition machine, the SECD machine[5]. Plotkin showed that the SECD machine can be described through a reduction strategy on Λ, using a suitable notion of reduction.

Definition 1. The rules for the *call-by-value reduction* are:
$(\alpha)\ (\lambda x.M) \to_v (\lambda y.M[y/x])$
$(\beta_v)\ (\lambda x.M)\,N \to_v M[x/N]$ if $N \in Val$.

The contextual, reflexive and transitive closure of $\to_v$ is denoted by $\twoheadrightarrow_v$; the equivalence induced by the reduction is denoted by $=_v$. We shall write $M \twoheadrightarrow_v Val$ to indicate that M $\twoheadrightarrow_v$ reduces to a value; we shall call such a term *valuable.*

Definition 2. The *SECD reduction strategy* $(\to_{SECD})$ is defined as follows:
let $M = M'[(\lambda x.P)Q]$, where $(\lambda x.P)Q$ is the leftmost β_v-redex which is not in the scope of a λ, then $M \to_{SECD} M'[P[x/Q]]$.

The reflexive and transitive closure of $\to_{SECD}$ is denoted by $\twoheadrightarrow_{SECD}$; the equivalence induced by $\twoheadrightarrow_{SECD}$ is denoted by $=_{SECD}$. We shall write $M \twoheadrightarrow_{SECD} Val$ to indicate that M SECD-reduces to a value.

The SECD reduction strategy is a faithful simulation of the behaviour of the SECD machine in the following sense (see [8]):
the SECD machine, given an input M, "halts on an output value V" $\Leftrightarrow M \twoheadrightarrow_{SECD} V$.

Therefore the reduction strategy SECD can safely be taken as defining the evaluation function of ISWIM.

In order to study the properties of pure ISWIM abstractly, we proceed as Plotkin [8] and introduce the *$\lambda\beta_v$-calculus* as ISWIM equipped with the $\twoheadrightarrow_v$ reduction defined above.

The calculus $\lambda\beta_v$ is *lazy* since all abstractions are considered to be values regardless of their body; it is *call-by-value* since a reduction is performed only if the argument to which the abstraction is applied is a value.

1.2. Equivalence between programs.

The $\lambda\beta_v$-calculus allows for the definition of a notion of equivalence between programs with respect to the termination of the SECD machine. Such a notion is called operational semantics and it amounts to Leibniz's principle for programs, i.e. a criterion for establishing equivalence on the basis of the behaviour of programs regarded as black boxes. Formally the operational semantics of the SECD machine is:

Definition 3. Let $M \in \Lambda$. The SECD *operational semantics* of M is defined as follows:
$Op_{SECD}(M) = \{N \mid \forall C[].C[M],C[N] \in \Lambda^\circ \Rightarrow (C[M] \twoheadrightarrow_{SECD} Val \Leftrightarrow C[N] \twoheadrightarrow_{SECD} Val)\}$.

In order to reason about the SECD operational semantics in the $\lambda\beta_v$-calculus we have to show that $\lambda\beta_v$ is "well behaved" with respect to the SECD reduction.

A calculus C is said to be *correct* with respect to a machine R if and only if $M =_C N \Rightarrow Op_R(M) = Op_R(N)$, where $=_C$ is the equivalence relation induced by the reduction rules of C; moreover it is said to be *complete* if and only if $Op_R(M) = Op_R(N) \Rightarrow M =_C N$.

Proposition 4. The $\lambda\beta_v$-calculus is correct with respect to the pure SECD machine. However it is not complete; consider for example the terms $(\lambda x.xx)(\lambda x.xx)$ and $(\lambda x.xxx)(\lambda x.xxx)$.

Since the calculus is correct with respect to the machine it makes sense to try to define a congruence in the calculus and to compare it with the equivalence between terms induced by the SECD operational semantics.

Definition 5. The congruence $\equiv_v$ is defined as follows:
$M \equiv_v N$ if and only if $\forall C[].C[M],C[N] \in \Lambda^\circ \Rightarrow (C[M] \twoheadrightarrow_v Val \Leftrightarrow C[N] \twoheadrightarrow_v Val)$.

Let $T_v = \{(M,N) \mid M,N \in \Lambda^\circ M \equiv_v N\}$ and $T_{SECD} = \{(M,N) \mid M,N \in \Lambda^\circ Op_{SECD}(M) = Op_{SECD}(N)\}$ be the λ-theories induced by the congruences above. We have the following:

Theorem 6. $T_v \equiv T_{SECD}$.
Proof . $T_v \supset T_{SECD}$ is trivial since $\twoheadrightarrow_{SECD}$ is just a particular β_v-reduction strategy. The other inclusion follows from the fact that $M \twoheadrightarrow_v Val \Rightarrow M \twoheadrightarrow_{SECD} Val$ which can be shown using the standardization theorem in [8]. QED

2. A model theoretic analysis of $\lambda\beta_v$.

The $\lambda\beta_v$-calculus doesn't provide a full account of the equivalence of programs. In order to gain a deeper understanding of the operational semantics we therefore introduce an appropriate model theoretic machinery.

A model for the call-by-value λ-calculus must provide a semantic account of valuable terms. This can be achieved by defining a subset of the domain of interpretation , the set of semantic values, where valuable terms are interpreted. Environments must map variables to semantic values since variables are values. Consequently, in order to model β_v-reduction, the interpretation of the abstraction needs to be defined only when applied to semantic values. A structure with these properties can be defined as follows:

Definition 7. A *syntactical model for $\lambda\beta_v$* is a structure $\langle V, U, \cdot, [\![\,]\!]\rangle$, where
$\cdot : (V \cup U) \times (V \cup U) \rightarrow (V \cup U)$ and $[\![\,]\!] : (Var \rightarrow V) \times \Lambda \rightarrow (V \cup U)$ satisfy
1) $[\![x]\!]_\rho = \rho(x)$
2) $[\![MN]\!]_\rho = [\![M]\!]_\rho \cdot [\![N]\!]_\rho$
3) $[\![\lambda x.M]\!]_\rho \cdot d = [\![M]\!]_{\rho(x/d)}$ if $d \in V$
4) $\rho(x) = \rho'(x)$ for all $x \in FV(M) \Rightarrow [\![M]\!]_\rho = [\![M]\!]_{\rho'}$
5) $y \notin FV(M) \Rightarrow [\![\lambda x.M]\!]_\rho = [\![\lambda y.M[x/y]]\!]_\rho$

6) $\forall d \in V.\ [\![M]\!]_{\rho(x/d)} = [\![N]\!]_{\rho(x/d)} \Rightarrow [\![\lambda x.M]\!]_\rho = [\![\lambda x.N]\!]_\rho$
7) $M \twoheadrightarrow_v Val \Rightarrow \forall \rho.\ [\![M]\!]_\rho \in V$
where $\rho(x/d)$ denotes the environment ρ' such that $\forall y \neq x. \rho'(y) = \rho(y)$ and $\rho'(x) = d$.

In section 2.1 we define the canonical syntactical domain model, *P*, of $\lambda\beta_v$. Sections 2.2 and 2.3 are devoted to the study of the denotational semantics of $\lambda\beta_v$ defined by *P*, i.e. the theory T_P induced by *P*. Our main result is:

Theorem 8. $T_P \subset T_v \equiv T_{SECD}$ but $T_P \neq T_v \equiv T_{SECD}$.

This result implies that the denotational semantics we have given is not fully abstract with respect to the operational one. We remind the reader that a model is said to be fully abstract with respect to a semantics S if the theory of the model coincides with S.

In order to prove the first part of Theorem 8 we need an Approximation Theorem for *P*. This theorem allows us to express the interpretation of a term as the l.u.b. of the interpretations of its approximants. It will be proved in section 2.2 using the notion of indexed reductions (see Wadsworth [11] and Hyland [4]).

The second part of Theorem 8 ($T_P \neq T_v$) will be proved in section 2.3 by directly exhibiting two terms which are equated in T_v but not in T_P, i.e. two terms which have distinct interpretations but have the same operational semantics.

2.1. Denotational semantics of the $\lambda\beta_v$-calculus.

Since $\lambda\beta_v$ is both lazy and call-by-value, we need a domain with appropriate features which models both properties. It is therefore expedient to recall a couple of useful definitions:
-A function from a domain D to a domain D' is *strict* if it is continuous and maps the bottom element of D into the bottom element of D'. The space of strict functions from D to D' is denoted $[D \rightarrow_\perp D']$.
- Given a domain D, $D_\perp$ (the *lifting* of D) denotes the domain obtained from D by adjoining a new bottom element.

The canonical model for $\lambda\beta_v$ is the initial solution of the equation $D \approx [D \rightarrow_\perp D]_\perp$. The reason for considering strict functions in the equation is to model call-by-value, by interpreting closed nonvaluable terms in $\perp$. We lift the domain $[D \rightarrow_\perp D]$ in order to keep interpretations of abstractions distinct from $\perp$ (i.e. the interpretation of closed non-valuable terms), thus modeling laziness.

Definition 9. The canonical model *P* is the structure $\langle P, \bullet, [\![\,]\!]^P \rangle$, where
- P is the initial solution of the domain equation $D \approx [D \rightarrow_\perp D]_\perp$, with isomorphisms $\Phi: P \rightarrow [P \rightarrow_\perp P]_\perp$ and $\Psi: [P \rightarrow_\perp P]_\perp \rightarrow P$, i.e., $P = \underleftarrow{\lim}\, P_n$, where $P_0 = \{\perp\}$, $P_{n+1} = [P_n \rightarrow_\perp P_n]_\perp$;
- $\bullet : P \times P \rightarrow P$ is defined as $e \bullet d =$ *if* $e = \perp$ *then* $\perp$ *else* $out(\Phi(e))(d)$
- $[\![\,]\!]^P: (Var \rightarrow (P \setminus \{\perp\})) \times \Lambda \rightarrow P$ is defined inductively as follows:
$[\![x]\!]_\rho^P = \rho(x)$
$[\![MN]\!]_\rho^P = ([\![M]\!]_\rho^P) \bullet ([\![N]\!]_\rho^P)$
$[\![\lambda x.M]\!]_\rho^P = \Psi(\, in(strict\,(\lambda d \in P.\ [\![M]\!]_{\rho(x/d)}^P)))$
where $\rho: Var \rightarrow (P \setminus \{\perp\})$, λ indicates the lambda abstraction in the meta language, *in* and *out* are the constructor and destructor of the lift operator and the function *strict* is defined as: *strict*(f)(x) = *if* $x = \perp$ *then* $\perp$ *else* f(x).

Proposition 10. *P* is a syntactical model for the $\lambda\beta_v$-calculus with set of semantic values $V = P \setminus \{\perp\}$, and set of semantic non values $U = \{\perp\}$.

In the following we will denote the interpretation function $[\![\,]\!]^P$ simply as $[\![\,]\!]$, since from now on we shall always refer to the model *P*. Moreover, we will write $[\![M]\!]$ instead of $[\![M]\!]_\rho$ when $M \in \Lambda^\circ$, since the interpretation of closed terms does not depend on the environment.

The theory induced by the model *P* is $T_P = \{(M,N) \mid M, N \in \Lambda^\circ,\ [\![M]\!] = [\![N]\!]\}$.

2.2. The Approximation Theorem.

We need to introduce an extension of the $\lambda\beta_v$-calculus, the $\lambda\beta_v{}^*$-calculus, in order to define the set of approximants of a $\lambda\beta_v$ term.

Definition 11. i) The set of terms Λ^* of $\lambda\beta_v{}^*$ is inductively defined as Λ, starting from $Var \cup \{\Omega\}$. The set of values is $Val^* = Var \cup \{\lambda x.M \mid M \in \Lambda^*\}$.
The reduction rules are (α) and
$(\beta_v{}^*)$ $(\lambda x.M)N \rightarrow^* M[x/N]$ if and only if $N \in Val^*$
(Ω) $\Omega M \rightarrow^* \Omega$ and $M\Omega \rightarrow^* \Omega$.
ii) The set of *approximants* of a term $M \in \Lambda$ is:
$A(M)=\{A \mid A\in\Lambda^*$, A is in normal form with respect to the $*$-reduction and A is obtained from some M', $M' =_v M$, by substituting some of its subterms with $\Omega\}$.

Definition 12. i) An *indexed term* (M,I) is a term $M \in \Lambda^*$, together with a map I from the subterms of M to the natural numbers. The number associated to each subterm is often denoted as a superscript to the subterm itself.
ii) The *indexed reduction* is given by the rules:
$(I\alpha)$ $(\lambda x.M^n)^m \rightarrow_I (\lambda y.(M[x/y])^n)^m$
$(I\beta_v)$ $(\lambda x.M^n)^{m+1}N^p \rightarrow_I (M[x/N^a])^b$ provided $N \in Val^*$, $p>0$, $b=\min(m,n)$, $a=\min(m,p)$;
$(I\Omega)$ $(\lambda x.M^n)^{m+1}N^0 \rightarrow_I \Omega$ and $(\lambda x.M^n)^0 N^p \rightarrow_I \Omega$;
$(I<)$ $(M^n)^m \rightarrow_I (M)^{\min(m,n)}$.
iii) As is standard for topological models obtained as inverse limits, we define for all n the functions $i_{n,\infty}$ and $j_{\infty,n}$ mapping respectively P_n into P and P onto P_n. Then it is possible to define the projections π_n:P—>P as $i_{\infty,n} \circ j_{n,\infty}$, yielding for each element $d \in P$ the image $d_n \in P$ of its n-th approximation. This allows us to define the value $[\![M^I]\!]_\rho$ of (M,I) as follows:
$[\![\Omega^I]\!]_\rho = \bot$; $[\![x^I]\!]_\rho = (\rho(x))_{I(x)}$;
$[\![(MN)^I]\!]_\rho = (([\![M^I]\!]_\rho) \cdot ([\![N^I]\!]_\rho))_{I(MN)}$;
$[\![(\lambda x.M)^I]\!]_\rho = (\Psi(in(strict(\lambda d. [\![M^I]\!]_{\rho(x/d)}))))_{I(\lambda x.M)}$
We denote in the same way the index map of a term and its restrictions to subterms.

Theorem 13. (Approximation Theorem) $[\![M]\!]_\rho = \sqcup\{ [\![A]\!]_\rho \mid A \in A(M)\}$.
Proof The proof is a reformulation for a lazy call-by-value calculus of the proof of the Approximation Theorem given by Hyland [4]. It follows from the fact that the interpretation of a term M is the l.u.b. of the interpretations of all the indexed terms of the shape (M,I), and that any indexed term (M,I) reduces to an (indexed) approximant of M.QED

2.3. Proof of Theorem 8.

We are now ready to prove that $T_P \subset T_v$. This amounts to showing that T_P does not equate valuable and non valuable terms, since any way of extending T_v amounts to equating valuable and non valuable terms. This is proved in the following:

Lemma 14. i) If M is not valuable, then there exists an environment ρ such that $[\![M]\!]_\rho = \bot$.
ii) If M is valuable, then $\forall\rho. [\![M]\!]_\rho \neq \bot$.
Proof (sketch) i) If M is not valuable, and $A \in A(M)$ then by definition of $*$-reduction, either $A=\Omega$ or $A=xA_1 \ldots A_m$ or $A= (\lambda x.P)B_1 \ldots B_m$ $(m\geq 1)$, $B_1 \notin Val^*$. Hence by complete induction, we have the result for the approximants and so for the terms. See [3] for similar arguments.
ii) If M is valuable the thesis follows from Proposition 10 and Definition 7.7.QED

The proof that $T_P \neq T_v$ is quite elaborate. We achieve it by introducing two terms equated in T_v but not in T_P, namely:
$M_1 = \lambda x.(\lambda xyz.\Delta\Delta)(x(\lambda x.\Delta\Delta)(\lambda xy.\Delta\Delta))(x(\lambda xy.\Delta\Delta)(\lambda x.\Delta\Delta))$
$M_2 = \lambda x.(\lambda xy.\Delta\Delta)(x(\lambda x.\Delta\Delta)(\lambda x.\Delta\Delta))$, where $\Delta = \lambda x.xx$.
First we need to give some technical definitions.

Definition 15. Let $M\in\Lambda$ and let $\lambda x.P$ be a subterm of M. Then $\lambda x.P$ is *relevant* in M if, for some Q, $M \twoheadrightarrow_{SECD} M'[(\lambda x.P)Q] \rightarrow_{SECD} M'[P[x/Q]]$, and the occurrence of $\lambda x.P$ in M' is a descendant of the occurrence of $\lambda x.P$ in M.

Informally a subterm of a term is relevant if it occurs as an operator of a redex which is reduced during the evaluation of the term by the SECD machine.

Definition 16. A *discriminator* of P and Q $(P,Q\in\Lambda)$, *of rank n*, is a context D[] such that $D[P],D[Q]\in\Lambda^\circ$ and $D[P]\twoheadrightarrow_v Val$ in n steps whereas D[Q] doesn't reduce to a value.

Lemma 17. $M_1=M_2$ holds in T_{SECD}.
Proof From the very structure of the terms, we have $[\![M_1]\!]\sqsupseteq[\![M_2]\!]$ and hence, if for any context C[] $C[M_2]\twoheadrightarrow_v Val$, then necessarily $C[M_1]\twoheadrightarrow_v Val$. Therefore it remains to show that there exists no discriminator of M_1 and M_2. Let us prove this by way of contradiction. Let D[] be a discriminator of minimum rank. Necessarily M_1 and M_2 must be relevant subterms of $D[M_1]$ and $D[M_2]$ respectively. It may be directly checked that whenever a relevant occurrence of M_1 in $D[M_1]$ and the corresponding relevant one of M_2 in $D[M_2]$ are reduced, then if both reductions yield a value, then both yield the same value (i.e. $\lambda x.\Delta\Delta$). Therefore D[] must be of the form [](T[]) for some context T[], and must be such that $M_1T[M_1]\twoheadrightarrow_v Val$ and $M_2T[M_2]$ is not valuable. But $M_1T[M_1]\twoheadrightarrow_v Val$ if and only if $(\lambda xyz.\Delta\Delta)(T[M_1]\ (\lambda x.\Delta\Delta)\ (\lambda xy.\Delta\Delta))\ (T[M_1]\ (\lambda xy.\Delta\Delta)\ (\lambda x.\Delta\Delta))\twoheadrightarrow_v Val$.
This holds if and only if both subterms $T[M_1]\ (\lambda x.\Delta\Delta)\ (\lambda xy.\Delta\Delta)$ and $T[M_1](\lambda xy.\Delta\Delta)\ (\lambda x.\Delta\Delta)$ reduce to values; but note that whenever $\lambda x.\Delta\Delta$ is relevant in a term, the term is non valuable. Therefore neither its first nor its second argument can be relevant in $T[M_1]\xi_1\xi_2$. i.e. $\forall\ \xi_1,\xi_2.\ T[M_1]\xi_1\xi_2\twoheadrightarrow_v Val$.
But $D[M_2]$ is non valuable if and only if $(\lambda xy.\Delta\Delta)(\ T[M_2]\ (\lambda x.\Delta\Delta)\ (\lambda x.\Delta\Delta))$ is non valuable, i.e. $T[M_2]\ (\lambda x.\Delta\Delta)\ (\lambda x.\Delta\Delta)$ is non valuable.
Using these last two facts, we have that D[] is a discriminator of M_1 and M_2 if and only if $D'[]=T[](\lambda x.\Delta\Delta)\ (\lambda x.\Delta\Delta)$ is also a discriminator of M_1 and M_2; clearly D'[] has smaller rank than D[], thus contradicting the hypothesis that D[] be of minimum rank. QED

Lemma 18. $M_1=M_2$ does not hold in T_P.
Proof Let $d\in P$ be such that: $d\bullet e =$ *if* $e=\bot$ *then* $\bot$ *else* (*if* $e=\lambda x.\bot$ *then* d' *else* $\lambda xy.\bot$) where $d'\bullet e =$ *if* $e\sqsubseteq\lambda x.\bot$ *then* $\bot$ *else* $\lambda x.\bot$. One can easily see that d and d' are continuous and $[\![M_1]\!]\bullet d = \lambda x.\bot$ whereas $[\![M_2]\!]\bullet d =\bot$. Clearly $[\![M_1]\!]\bullet d \neq [\![M_2]\!]\bullet d$. QED

3. An alternative description of the SECD operational semantics.

The argument given in the previous section showing that *P* is not fully abstract w.r.t. the SECD machine, applies to a wide class of domain models of $\lambda\beta_v$. For instance those which contain *all* strict functions and where all nonvaluable terms are equated to $\bot$. We are then led naturally to look for a fully abstract model in a wider class of models.

In this section we define a fully abstract structure utilizing a technique similar to those used by Plotkin [10], Mulmuley [6] and Ong[7] for different calculi. The model we obtain amounts to just the closed term model of the theory T_{SECD} equipped with a partial order relation. The interest of this construction however goes far beyond this as we will show.

The definition of the SECD-operational semantics is intuitively redundant if not overspecified. Moreover it is very complex to consider, as class of experiments on which termination of programs must be tested, the set of all contexts. It would be more pleasing to restrict our attention just to applicative contexts. We would rather like to use the following notion of equivalence:

Definition 19. The "applicative congruence" $\equiv^v$ on Λ° is defined as follows:
$M\equiv^v N$ if and only if $\forall n\ \forall Q_1,\dots,Q_n\in\Lambda^\circ .\ MQ_1\dots Q_n\twoheadrightarrow_v Val \Leftrightarrow NQ_1\dots Q_n\twoheadrightarrow_v Val$.

This operational equivalence corresponds more closely to our intuitive understanding of a $\lambda\beta_v$-term as a functiona' program. The interest of the fully abstract model, which we build, lies in the fact that it permits us to prove that these two notions of the operational semantics coincide. This illuminates the nature of call-by-value computations.

3.1. Call-by-value applicative bisimulation.

Our construction of the fully abstract model is based on the notion of call-by-value applicative bisimulation. This is a congruence relation on the subset P° of P consisting of the interpretations of closed terms, and can be viewed as a call-by-value version of Abramsky's notion of applicative bisimulation [1]. Now please recall the notions of projection π_n and n^{th} approximation of an element a_n introduced in definition 12.iii).

Proposition 20. P° is closed under approximations.
Proof(sketch) Define Π_n inductively as $\Pi_0=\lambda x.\Delta\Delta$; $\Pi_{n+1}=\lambda yx.\Pi_n(y(\Pi_n x))$.
It can be checked that $[\![\Pi_n]\!]=\pi_n$ for all $n\in\omega$. Therefore the projection functions are interpretations of closed terms and so are the approximations of elements of P°.QED

Definition 21. Let $(P^\circ)_n=\{d_n|d\in P^\circ\}$. The family of relations $\{\equiv_n|\equiv_n\subset P^\circ\times P^\circ\}_{n\in\omega}$ is defined inductively as follows:
- $\equiv_0=(P^\circ)_0\times(P^\circ)_0$
- $a\equiv_{n+1}b$ if and only if $a,b\in(P^\circ)_{n+1}$ and
  $(a=\bot\ \&\ b=\bot)\vee(a\neq\bot\ \&\ b\neq\bot\ \&\ (\forall d,e.\ d\equiv_n e\Rightarrow a\cdot d\equiv_n b\cdot e)$.

The *call-by-value applicative bisimulation* is the relation $\equiv_\omega\subset P^\circ\times P^\circ$ defined as follows:
- $\forall a,b\in P^\circ.(\ a\equiv_\omega b\Leftrightarrow\forall n.\ a_n\equiv_n b_n)$.

Proposition 22. The relation $\equiv_\omega$ is a congruence relation on P°.
Proof (sketch) The only difficult property to show is the reflexivity of $\equiv_\omega$ on P°. In order to prove it we need the following lemmata, all proved by appropriate structural inductions:
-$\forall n.\forall a,b\in P^\circ\ .\ a\equiv_n b\Rightarrow a\equiv_{n+1}b$
-$\forall n.\forall a,b\in P^\circ\ .\ a\equiv_{n+1}b\Rightarrow a_n\equiv_n b_n$
-$\forall n.\ \forall a,b\in P^\circ\backslash\{\bot\}\ .\ a\equiv_\omega b\Leftrightarrow\forall d,e\in P^\circ.\ d\equiv_\omega e\Rightarrow a\cdot d\equiv_\omega b\cdot e$
-$\forall\rho,\rho'.\ (\forall x.\rho(x)\equiv_\omega\rho'(x))\Rightarrow[\![M]\!]_\rho\equiv_\omega[\![M]\!]_{\rho'}$.

3.2. The model *H*.

We are now ready to define a fully abstract model for T_{SECD}.

Definition 23. $H=\langle H,\cdot_\omega,[\![\,]\!]^H\rangle$ where:
- H is $P^\circ/\equiv_\omega$,
- the application $\cdot_\omega$ between equivalence classes modulo $\equiv_\omega$ is defined as $[a]\cdot_\omega[b]=[a\cdot b]$
- the interpretation function $[\![\,]\!]^H:(Var\rightarrow(H\backslash\{[\bot]\}))\times\Lambda\rightarrow H$ is defined as $[\![M]\!]^H_{\rho^H}=[[\![M]\!]_\rho]$

where $[\![\,]\!]$ denotes the interpretation function in the model P and if ρ: $Var\rightarrow(P^\circ\backslash\{\bot\})$ then ρ^H: $Var\rightarrow(H\backslash\{[\bot]\})$ is such that $\forall x\in Var.\rho^H(x)=[\rho(x)]$.

Proposition 24. $\langle H,\ \cdot_\omega,[\![\,]\!]^H\rangle\models\lambda\beta_v$ where the set of semantic values is $V=H\backslash\{[\bot]\}$.
Proof. It is easy to check that Definition 23 is well posed and H satisfies the appropriate conditions given in Definition 7. QED

3.3. Comparing theories.

We now study the theory induced by the model H and we prove that it coincides with the equivalence relation induced by the applicative relation $\equiv^v$. Thus we have that H is a fully abstract model for the SECD operational semantics, that $\equiv^v$ defines a theory and that the black box congruence is the same as the applicative one.

Definition 25. $T_H=\{(M,N)|M,N\in\Lambda^\circ,\ [\![M]\!]^H=[\![N]\!]^H\}$; $T^v=\{(M,N)|\ M,N\in\Lambda^\circ,\ M\equiv^v N\}$.

One can see immediately that T_H is a theory but we do not know that T^v is a theory yet.

Theorem 26. $T^v \equiv T_v$.
Proof In order to show $T^v \equiv T_v$ we prove : i) $T_v \subset T^v$, ii) $T^v \subset T_H$ and iii) $T_H \subset T_v$.
i) Immediate.
ii) If the pair $(M,N) \in T^v$ doesn't belong to T_H, then by the definition of $\equiv_\omega$ we can find, say, an integer n and closed terms $Q_1,\ldots,Q_k$, $(k<n)$, such that $[\![Q_i]\!] \in (P^\circ)_{n-i}$ and $[\![(M)_nQ_1\ldots Q_k]\!]=\bot$ but $[\![(N)_nQ_1..Q_k]\!]\neq\bot$. But now one can easily see that $[\![(M)_nQ_1\ldots Q_k]\!]=[\![(MQ_1\ldots Q_k)_{n-k}]\!]$ and $[\![(N)_nQ_1\ldots Q_k]\!]=[\![(MQ_1\ldots Q_k)_{n-k}]\!]$.
Now, since only $\bot$ has approximations equal to $\bot$, as one can see from the limit construction of P, we have $[\![MQ_1\ldots Q_k]\!]=\bot$ and $[\![NQ_1\ldots Q_k]\!]\neq\bot$, which imply $NQ_1\ldots Q_k \twoheadrightarrow_v Val$ while $MQ_1\ldots Q_k$ does not reduce to a value, contradicting the hypothesis.
iii) Let $M,N \in \Lambda^\circ$ be such that $[\![M]\!]^H=[\![N]\!]^H$. Let us suppose $M \not\equiv_v N$ and take C[] such that $C[M],C[N] \in \Lambda^\circ$, but, say, $C[M] \twoheadrightarrow_v Val$ whereas C[N] is not valuable. Then $(C[M],C[N]) \notin T_H$. Using the facts that β_v-reduction preserves interpretation and bisimulation is reflexive we reach the contradiction $(C[M],C[N]) \in T_H$.QED

Corollary 27. T^v is a theory. H is a fully abstract model for T_{SECD}.

The theory T_v is maximal, as we can see from the following:

Proposition 28. T_v is a non trivial maximal theory.
Proof First note that all terms that are not valuable are equated in T_v. Suppose we add to T_v the equation V=N where V is a value and N is not valuable. Then for any term $M \in \Lambda$, $x \notin FV(M)$, $M = (\lambda x.M)V = (\lambda x.M)N = N$. The theory would then be trivial. Now suppose we add an equation V'=V'' where V' and V'' are both valuable but are not equated in the theory. If they are not equated in T_v, there must be a context C[] such that, say, C[V'] β_v-reduces to a value (call it V) and C[V''] is a non valuable term N. Thus from V'=V'' we have C[V']=C[V''] and V= N, and so by the above argument, the theory is trivial.QED

4. Call-by-value intersection types.

The $\lambda\beta_v$-calculus is similar to the standard λ-calculus in various respects. In this section we will briefly discuss the call-by-value analogue of the Intersection Type Assigment System of Coppo-Dezani [2]. Namely we define first a type language with a natural interpretation in call-by-value lambda models as defined in Section 2. Then we give a type assignment system for terms which is complete with respect to this interpretation. This system is suggested by a "concrete" logical account of the canonical model.

The language of types T is inductively generated from a set of basic constants including a distinguished one υ, intended to denote the predicate "being a value", using the constructors $\to$ and $\cap$. Types are interpreted in a model M of the λ-calculus. Given an arbitrary interpretation ρ of the constants as subsets of the set V of semantic values of M, the other types are inductively interpreted from ρ as follows:
$[\![\upsilon]\!]_\rho{}^M = V$, $\quad [\![\alpha\cap\beta]\!]_\rho{}^M = [\![\alpha]\!]_\rho{}^M \cap [\![\beta]\!]_\rho{}^M$,
$[\![\alpha\to\beta]\!]_\rho{}^M = \{d \in V \text{ s.t. } \forall e \in [\![\alpha]\!]_\rho{}^M. \text{ if } d \cdot e \in V \text{ then } d \cdot e \in [\![\beta]\!]_\rho{}^M\}$.
Let a *basis* be a partial function from variables to types. We denote a basis B by the set of pairs $\{x:\sigma \mid B(x)=\sigma\}$.
The *Intersection Type Assignment System by value*, IT_v, consists of the following rules:

$$\text{(var)}\ \frac{}{B\cup\{x:\sigma\} \vdash x:\sigma} \qquad\qquad (\upsilon)\ \frac{}{B \vdash \lambda x.N : \upsilon}$$

$$(\to E)\ \frac{B \vdash N:\tau\to\sigma \qquad B\vdash Q:\tau}{B\vdash NQ:\sigma} \qquad (\to I)\ \frac{B \cup \{x:\sigma\} \vdash N:\tau \quad x\notin dom(B)}{B \vdash \lambda x.N:\sigma\to\tau}$$

$$(\cap I)\ \frac{B \vdash N:\sigma \qquad B\vdash N:\tau}{B\vdash N:\sigma\cap\tau} \qquad (\cap E)\ \frac{B \vdash N:\sigma\cap\tau}{B \vdash N:\sigma} \qquad (\leq)\ \frac{B \vdash N:\sigma \qquad \sigma\leq\tau}{B \vdash N:\tau}$$

The judgement $\leq$ is regulated by the set of rules consisting of those given in [2] which do not mention ω, together with: $\tau\leq\upsilon$ for all τ. These rules formalize the intended interpretation of $\leq$ as inclusion between interpretations of types. We will omit them here.

The following completeness result holds (for simplicity we give it only for closed terms):

Theorem 29. Let $N\in\Lambda^{\circ}$, the judgement $\emptyset\vdash N:\sigma$ can be established in the system IT_{v} if and only if for all models M and constant type interpretations ρ, $[\![N]\!]^{M}\in[\![\sigma]\!]_{\rho}^{M}$.

We will not give the proof of this result since this goes along the same lines as that for the completeness of the Intersection Type Assignment System for standard λ-calculus.

It is worth noticing that the model P, introduced in Section 2.1, realizes exactly all the judgements true in all models which use only types built over υ. This allows us to understand the types built over υ only, as names for the finite elements of the model P different from $\bot$. Thus we can interpret the judgement $\emptyset\vdash M:\sigma$ as $\sigma\sqsubseteq[\![M]\!]^{P}$. We have then a finitary definition of the interpretation function. We can also view this as an instance of the duality paradigm of Abramsky [1] connecting domains and logical theories. The type assignment system is thus an endogenous logic for reasoning about ISWIM programs.

References.

1. Abramsky S. *The lazy lambda calculus*, in: D. Turner (ed.) *Declarative programming*, Addison-Wesley, 1988.
2. Barendregt H., Coppo M., Dezani M. *A filter lambda model and the completeness of type assignment*, The J. of Symbolic Logic, **48**, no.4, 1983, 931-940.
3. Honsell F.,Ronchi Della Rocca S. *An approximation theorem for topological lambda models and the topological incompleteness of lambda calculus*, J. of Computer and System Sciences, to appear.
4. Hyland M. *A syntactic characterization of the equality in some models for the lambda calculus*, J. of the London Mathematical Society (2), **12**, 1976, 361-370.
5. Landin P.J. *The mechanical evaluation of expressions*, Computer J. **6**, no.4, 1964, p. 308-320.
6. Mulmuley K. *Fully abstract submodels of typed lambda calculus*, J. of Computer and System Sciences, **33**, 1986, 2-46.
7. Ong C.-H. L. *The lazy lambda calculus: an investigation into the foundations of functional programming*, Ph.D. thesis, Imperial College of Science and Technology, University of London, 1988.
8. Plotkin G.D. *Call-by-name, call-by-value and the λ-calculus*. Theoretical Computer Science, **1**, 1975, 125-159.
9. Plotkin G.D. *LCF, considered as a Programming Language*, Theoretical Computer Science, **5**, 1977, 223—255.
10. Plotkin G.D. Personal communication.
11. Wadsworth C.P. *The relation between computational and denotational properties for Scott's D_∞-models of the λ-calculus*, SIAM J. of Computing, **5**, no.3, 1976, 488.
12. Boudol G. *Lambda-Calculi for (strict) Parallel Functions* , INRIA preprint 1991.
13. Pino-Perez R. *A Strict partial combinatory algebra which modelizes Partial Lambda Calculus* , preprint 1991.

THE LIMIT OF SPLIT$_n$-BISIMULATIONS FOR CCS AGENTS *

Roberto Gorrieri   Cosimo Laneve

Dipartimento di Informatica, Università di Pisa
Corso Italia 40, I-56125, PISA, ITALY
e-mail: {gorrieri, laneve}@dipisa.di.unipi.it

1 Introduction

The semantics usually described in the literature for process algebras are based on the strong assumption that the actions a system performs have to be atomic, i.e., they do not take time. Extending the well-established theories of concurrency to include also the time duration of actions is, at present, one of the main goals in concurrency research. This extension turns out to be crucial for refinement (see, e.g. [5]): an action, having duration in time, can be considered as an abstraction of a whole system, thus introducing the possibility of relating system descriptions at different levels of detail. Indeed, non atomic equivalences are possible candidates for congruences w.r.t. an operation of refinement.

Hennessy [8] was probably the first who relaxed the atomicity assumption by permitting actions to be observed in the middle of their evolution. In particular, he suggested atomic actions composed of two phases, their ***beginnings*** and their ***endings***. This proposal can be generalized to an arbitrary number of phases, yielding equivalences named split$_n$. The interesting point is to discover when the split$_n$- equivalences are more and more discriminating for increasing n and if there exists a limit to this chain.

In this paper we give an answer to this problem for CCS [9] and strong bisimulation equivalence. We prove that by increasing the number of phases, we get finer and finer equivalences, and that the limit of this chain is a non atomic equivalence already proposed in the literature under the name of ***ST-bisimulation equivalence***. ST equivalence was originally proposed in [3] on Petri Nets, by mimicking split equivalence.

We will define split$_n$-transition systems for CCS (for any n) by splitting each action of the language into n phases; hence transitions are labelled by action phases. Ordinary bisimulation equivalence on these detailed transition systems is by definition split$_n$-bisimulation equivalence. Similarly, also an ST transition system for CCS is introduced by slightly enriching the information in the phases of the split$_2$-semantics. To be more precise, to every ending phase we add a backward pointer towards its own beginning phase.

We introduce ST-transition systems to model for operational semantics; they are transition systems whith arcs labelled either by beginning phases or by ending phases enriched with a backward pointer. An operational ST-semantics in SOS style [11] for CCS is defined. ST-bisimulation equivalence is proved to be finer than all split$_n$-equivalences. When considering agents having ***autoconcurrency*** at most n, the identifications induced by ST-bisimulation are exactly the same as those induced by split$_{n+1}$-bisimulation.

The main result of the paper is that the ST-bisimulation is the intersection of all the split$_n$-bisimulations for guarded CCS terms (i.e., image finite agents). This result contrasts with the fact that ST-trace is not the limit of the split$_n$-trace for CCS, as shown in [4].

In this paper proofs are omitted, they can be found in [6].

*Work partially supported by ESPRIT Basic Research Action 3011, CEDISYS.

2 Background

We begin by defining Milner's Calculus of Communicating Systems (CCS for short) [9] with its standard operational semantics.

Let $\Delta = \{\alpha, \beta, \gamma, \ldots\}$, ranged over by α, be a (possibly infinite) set of *action names*, $\overline{\Delta} = \{\overline{\alpha}, \overline{\beta}, \overline{\gamma}, \ldots\}$ the set of *action conames* and τ a special *silent action*. We will call $\Lambda = \Delta \cup \overline{\Delta}$ the set of *visible actions* ranged over by λ, while $\mathcal{M} = \Lambda \cup \{\tau\}$ the set of *actions* ranged over by μ. Consider the set of *recursive terms* inductively defined by the following rules:

$$t ::= \; 0 \;\mid\; \mu.t \;\mid\; t + t' \;\mid\; t \,|\, t' \;\mid\; t \backslash \alpha \;\mid\; t[\Phi] \;\mid\; x \;\mid\; \mathrm{rec}\, x.\, t$$

where $\alpha \in \Delta$, Φ is a permutation of $\mathcal{M}$ preserving $^{-}$ and τ and x is any element in a (possibly infinite) set of variables $\mathcal{V}$. We suppose that the reader is familiar with the usual notions of *free* and *bound* variables and with *syntactic substitution*. The notation $t[^{t'}/_{x}]$ will mean that the term t' is syntactically substituted for the variable x in the term t. Let us call a recursive term $\mathrm{rec}\, x.\, t$ *guarded* whenever every occurence of the variable x in t is inside the scope of a prefixing μ-operator. We denote by $\mathrm{CCS}_{\mathcal{M}}$ the set of *closed guarded recursive terms*, also called *CCS agents*, which will be ranged over by the variable t.

The operational semantics of CCS is defined in terms of Labelled Transition Systems, *LTS* for short. Thus, first some general definitions on *LTS* are given, and then the CCS labelled transition system is introduced.

Definition 2.1 *A Labelled Transition System $\mathcal{H}$ is a triple $\langle \mathcal{S}, \mathcal{L}, \mathcal{T} \rangle$ where $\mathcal{S}$ is a set of states, $\mathcal{L}$ is a set of labels and $\mathcal{T} = \{\xrightarrow{a}, a \in \mathcal{L}\}$ is the transition relation, where each $\xrightarrow{a}$ is a binary transition relation on $\mathcal{S}$. We will write $s \xrightarrow{a} s'$ instead of $\langle s, s' \rangle \in \xrightarrow{a}$.* □

Definition 2.2 *Let $\mathcal{H} = \langle \mathcal{S}, \mathcal{L}, \{\xrightarrow{a}, a \in \mathcal{L}\} \rangle$ be an LTS, and let $s, s' \in \mathcal{S}$. Then, s' is reachable from s if $s \longrightarrow^{*} s'$, where $\longrightarrow^{*}$ is the reflexive and transitive closure of $\{\xrightarrow{a}, a \in \mathcal{L}\}$. Let $\mathcal{S}(s)$ denote the set of the states reachable from s; let $\mathcal{T}(s)$ be $\{\xrightarrow{a} \cap (\mathcal{S}(s) \times \mathcal{S}(s)), a \in \mathcal{L}\}$, the reachable transition relation. The LTS (with root s) $\mathcal{H}(s) = \langle \mathcal{S}(s), \mathcal{L}, \mathcal{T}(s) \rangle$ is called the (sub-)transition system reachable from s.* □

The *LTS* $\mathcal{G} = \langle \mathrm{CCS}_{\mathcal{M}}, \mathcal{M}, \{\xrightarrow{\mu}, \mu \in \mathcal{M}\} \rangle$ defining CCS operational semantics has agents as states and the transition relation is defined as follows.

Definition 2.3 *The transition relation $\{\xrightarrow{\mu}, \mu \in \mathcal{M}\}$ is defined as the least relation satisfying the following axiom and inference rules:*

$$Act)\;\; \mu.t \xrightarrow{\mu} t \qquad Res)\;\; \frac{t \xrightarrow{\mu} t'}{t\backslash\alpha \xrightarrow{\mu} t'\backslash\alpha}\;\; \mu \notin \{\alpha, \overline{\alpha}\} \qquad Rel)\;\; \frac{t \xrightarrow{\mu} t'}{t[\Phi] \xrightarrow{\Phi(\mu)} t'[\Phi]}$$

$$Sum_l)\;\; \frac{t_1 \xrightarrow{\mu} t_1'}{t_1 + t_2 \xrightarrow{\mu} t_1'} \qquad Sum_r)\;\; \frac{t_2 \xrightarrow{\mu} t_2'}{t_1 + t_2 \xrightarrow{\mu} t_2'} \qquad Par_l)\;\; \frac{t_1 \xrightarrow{\mu} t_1'}{t_1|t_2 \xrightarrow{\mu} t_1'|t_2}$$

$$Par_r)\;\; \frac{t_2 \xrightarrow{\mu} t_2'}{t_1|t_2 \xrightarrow{\mu} t_1|t_2'} \qquad Par_s)\;\; \frac{t_1 \xrightarrow{\lambda} t_1' \quad t_2 \xrightarrow{\overline{\lambda}} t_2'}{t_1|t_2 \xrightarrow{\tau} t_1'|t_2'} \qquad Rec)\;\; \frac{t[^{\mathrm{rec}\, x.\, t}/_{x}] \xrightarrow{\mu} t'}{\mathrm{rec}\, x.\, t \xrightarrow{\mu} t'}$$

□

The drawback of the above semantics is that it is too *intensional*: terms are differentiated even if no observer can distinguish them. The usual next step consists of introducing a (behavioural) equivalence among terms. A standard tool for defining behavioural equivalences relies upon the notion of bisimulation [10].

Definition 2.4 *Let* $\langle \mathcal{S}, \mathcal{L}, \{\xrightarrow{a}, a \in \mathcal{L}\rangle$ *be an LTS. A binary relation* $\Re \subseteq \mathcal{S} \times \mathcal{S}$ *is a (strong) bisimulation if* $\langle p, q\rangle \in \Re$ *implies,* $\forall a \in \mathcal{L}$*:*

- *whenever* $p \xrightarrow{a} p'$, $\exists q'$ *such that* $q \xrightarrow{a} q'$ *and* $\langle p', q'\rangle \in \Re$ *;*
- *whenever* $q \xrightarrow{a} q'$, $\exists p'$ *such that* $p \xrightarrow{a} p'$ *and* $\langle p', q'\rangle \in \Re$*.*

States p *and* q *are (strong) bisimilar, written* $p \sim q$*, if* $\langle p, q\rangle \in \Re$ *for some (strong) bisimulation* $\Re$*. That is,* $\sim = \bigcup\{\Re, \Re$ *is a bisimulation* $\}$*.* □

From now on, when we use the notation $\mathcal{H}(s) \sim \mathcal{H}(s')$, where $\mathcal{H}$ is a *LTS* and s and s' are two of its states, we mean that there exists a bisimulation between the two transition systems which contains the pair $\langle s, s'\rangle$.

3 Split semantics

Before introducing the split$_n$-operational semantics for CCS, let us show some problems we have tackled. First of all, we have to distinguish the phases which actions are splitted into; thus the resulting set of phases is the union $\mathcal{M}^n = \bigcup_{i \le n} \mathcal{M}^{(i)}$, where $\mathcal{M}^{(i)} = \{\mu_i, \mu \in \mathcal{M}\}$. We adopt the convention of distinguishing the various phases of an action by indexing the action itself with the corresponding natural number.

The first serious problem is concerned with the synchronization of two complementary actions: if a process starts a synchronization with the first phase of an action, then the interaction must be completed by the following phases of the same action. For example, consider the agent $\alpha \mid \overline{\alpha}$, which has the split$_2$-transition system depicted in Fig. 1. Note that we have to distinguish the

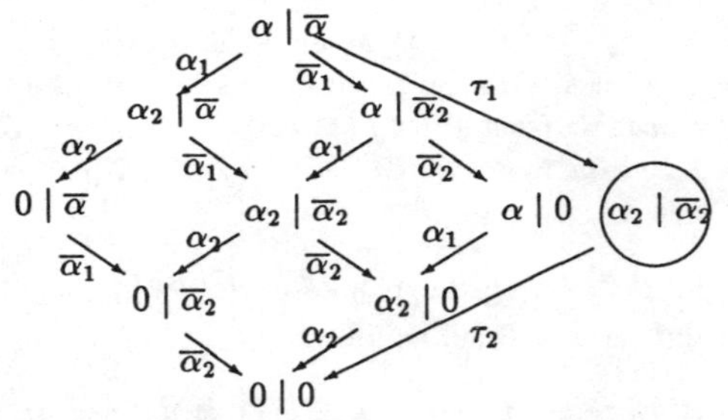

Figure 1: The expected split$_2$-transition system for $\alpha|\overline{\alpha}$.

two occurrences of the state $\alpha_2 \,|\, \overline{\alpha}_2$ because they have different behaviours. As a first approximation, we can imagine to superscript the phases resulting from an initial synchronization by an s, indicating that the phase must be performed synchronously. Therefore, the circled state $\alpha_2 \,|\, \overline{\alpha}_2$ becomes $\alpha_2^s \,|\overline{\alpha}_2^s$. The second problem, again concerning synchronization, is a consequence of autoconcurrency. Consider the agent $(\alpha|\overline{\alpha}) \,|\, (\alpha|\overline{\alpha})$; after the execution of the sequence $\tau_1 \cdot \tau_1$, which we can suppose to be due to the beginnings of the synchronizations of the first α with the first $\overline{\alpha}$ and of the second α with the second $\overline{\alpha}$, the state reached is $(\alpha_2^s \,|\, \overline{\alpha}_2^s) \mid (\alpha_2^s \,|\, \overline{\alpha}_2^s)$ from which only synchronization transitions can outgo. From this state, however, also the ending synchronizations of the first α_2 with the second $\overline{\alpha}_2$ and of the second α_2 with the first $\overline{\alpha}_2$ are possible. To avoid these (dangerous) behaviours we need to recover, just for synchronizing phases, the *causal dependencies* between the successive phases involved in a synchronization. A way to implement this requirement is by using, for the whole system, a global clock whose elapsing is set by the beginning of actions. Hence a state will be a pair whose first component is an agent and whose second component is the current value of the global clock. The phases which must synchronize are postfixed by a superscript showing the clock value of the time at which the corresponding action started. Therefore, only phases postfixed by the same clock value can be synchronized.

The presence of the clock value, which is never reset, avoids the possibility of cycles in the transition system. The (intensional) semantics of a CCS agent t will be the sub-transition system reachable from the state having t (restricted by some auxiliary actions) as first component and the clock value set to 1, as second component.

Summing up, the set $\mathcal{M}_n$ of labels of the transition system for the split$_n$-semantics will be $\mathcal{M}^n \cup \Lambda_n^s$, where $\mathcal{M}^n = \bigcup_{i \leq n} \mathcal{M}^{(i)}$, and $\Lambda_n^s = \{\lambda_i^k,\ \lambda \in \Lambda \wedge i \in [1,n] \wedge k \in \omega\}$, the set of *synchronous phases*, contains the phases of visible actions in $\mathcal{M}_n$ postfixed by a global clock value. This set will play a fundamental role in the synchronization mechanism. $\mathcal{M}_n$ and $\mathcal{M}^n$ will be ranged over by η and ν, respectively.

The set $\mathcal{S}_n$ of the states of the split$_n$-transition system consists of the terms s generated by the following grammar (t ranges over *CCS agents*, α ranges over Δ and z ranges over ω):

$$\begin{aligned} r &::= t \mid \eta.t \mid r\backslash\alpha \mid r[\Phi] \mid r|r \\ s &::= \langle r,\, z\rangle \mid \langle r\backslash\Lambda_n^s,\, z\rangle \end{aligned}$$

It is worth noting that we have augmented CCS syntax with a new unary operator $-\backslash\Lambda_n^s$ which will denote the restriction w.r.t. all the auxiliary synchronous phases. For example, the initial state associated to any *CCS agent* t is $\langle t\backslash\Lambda_n^s,\, 1\rangle$. As a consequence, all the states which are reachable from this are of the form $\langle r\backslash\Lambda_n^s,\, z\rangle$ and, thus, cannot exhibit a synchronous phase to an external observer. Rule Res_2) below takes care of this fact.

Definition 3.1 *The operational model for split$_n$-semantics is $\mathcal{G}_n = \langle \mathcal{S}_n, \mathcal{M}_n, \mathcal{T}_n\rangle$, where the set of transition relations $\mathcal{T}_n = \{\xrightarrow{\eta},\ \eta \in \mathcal{M}_n\}$ is the least set of relations satisfying the following axioms and inference rules:*

$Act_b)$ $\langle \mu.t,\, z\rangle \xrightarrow{\mu_1} \langle \mu_2.t,\, z+1\rangle$

$Act_m)$ $\langle \mu_i.t,\, z\rangle \xrightarrow{\mu_i} \langle \mu_{i+1}.t,\, z\rangle \quad i<n$

$Act_e)$ $\langle \mu_n.t,\, z\rangle \xrightarrow{\mu_n} \langle t,\, z\rangle$

$Act_b^s)$ $\langle \lambda.t,\, z\rangle \xrightarrow{\lambda_1^z} \langle \lambda_2^z.t,\, z+1\rangle$

$Act_m^s)$ $\langle \lambda_i^k.t,\, z\rangle \xrightarrow{\lambda_i^k} \langle \lambda_{i+1}^k.t,\, z\rangle \quad i<n$

$Act_e^s)$ $\langle \lambda_n^k.t,\, z\rangle \xrightarrow{\lambda_n^k} \langle t,\, z\rangle$

$$Res_1)\ \frac{\langle r,\, z\rangle \xrightarrow{\eta} \langle r',\, z'\rangle}{\langle r\backslash\alpha,\, z\rangle \xrightarrow{\eta} \langle r'\backslash\alpha,\, z'\rangle} \quad \eta \notin \{\alpha_i, \overline{\alpha}_i, \alpha_i^k, \overline{\alpha}_i^k\ ;\ 1 \leq i \leq n \wedge k \in \omega\}$$

$$Res_2)\ \frac{\langle r,\, z\rangle \xrightarrow{\eta} \langle r',\, z'\rangle}{\langle r\backslash\Lambda_n^s,\, z\rangle \xrightarrow{\eta} \langle r'\backslash\Lambda_n^s,\, z'\rangle} \quad \eta \notin \Lambda_n^s$$

$$Rel)\ \frac{\langle r,\, z\rangle \xrightarrow{\eta} \langle r',\, z'\rangle}{\langle r[\Phi],\, z\rangle \xrightarrow{\Phi(\eta)} \langle r'[\Phi],\, z'\rangle}$$

$$Sum_l)\ \frac{\langle t_1,\, z\rangle \xrightarrow{\eta} \langle r_1,\, z'\rangle}{\langle t_1+t_2,\, z\rangle \xrightarrow{\eta} \langle r_1,\, z'\rangle}$$

$$Sum_r)\ \frac{\langle t_2,\, z\rangle \xrightarrow{\eta} \langle r_2,\, z'\rangle}{\langle t_1+t_2,\, z\rangle \xrightarrow{\eta} \langle r_2,\, z'\rangle}$$

$$Par_l)\ \frac{\langle r_1,\, z\rangle \xrightarrow{\nu} \langle r_1',\, z'\rangle}{\langle r_1|r_2,\, z\rangle \xrightarrow{\nu} \langle r_1'|r_2,\, z'\rangle}$$

$$Par_l^a)\ \frac{\langle r_1,\, z\rangle \xrightarrow{\lambda_i^k} \langle r_1',\, z'\rangle \quad \langle r_2,\, z\rangle \not\xrightarrow{\overline{\lambda}_i^k}}{\langle r_1|r_2,\, z\rangle \xrightarrow{\lambda_i^k} \langle r_1'|r_2,\, z'\rangle}$$

$$Par_r)\ \frac{\langle r_2,\, z\rangle \xrightarrow{\nu} \langle r_2',\, z'\rangle}{\langle r_1|r_2,\, z\rangle \xrightarrow{\nu} \langle r_1|r_2',\, z'\rangle}$$

$$Par_r^a)\ \frac{\langle r_2,\, z\rangle \xrightarrow{\lambda_i^k} \langle r_2',\, z'\rangle \quad \langle r_1,\, z\rangle \not\xrightarrow{\overline{\lambda}_i^k}}{\langle r_1|r_2,\, z\rangle \xrightarrow{\lambda_i^k} \langle r_1|r_2',\, z'\rangle}$$

$$Par^s)\ \frac{\langle r_1,\, z\rangle \xrightarrow{\lambda_i^k} \langle r_1',\, z'\rangle \quad \langle r_2,\, z\rangle \xrightarrow{\overline{\lambda}_i^k} \langle r_2',\, z'\rangle}{\langle r_1|r_2,\, z\rangle \xrightarrow{\tau_i} \langle r_1'|r_2',\, z'\rangle}$$

$$Rec)\ \frac{\langle t[{}^{\mathrm{rec}\,x.\,t}/_x],\, z\rangle \xrightarrow{\eta} \langle r',\, z'\rangle}{\langle \mathrm{rec}\,x.\,t\,,\, z\rangle \xrightarrow{\eta} \langle r',\, z'\rangle}$$

where the function Φ satisfies also the constraints $\Phi(\mu_i) = \Phi(\mu)_i$ and $\Phi(\mu_i^k) = \Phi(\mu)_i^k$. □

Let us explain the intuition behind these inference rules. By performing action λ_1 or λ_1^k we actually are choosing whether the following phases have to be executed independently (Act_b)), or synchronously (Act_b^s)). Once it is decided to synchronize the first phase, also the remainder of the sequence has to be synchronized (Act_m^s)); this means that the split$_n$ view of a CCS one-step synchronization is a sequence of n-phase synchronizations. Indeed, the rule Par^s) accepts as premise only complementary synchronous phases, but gives rise to an asynchronous τ phase. On the other hand, rules Par) and Par^a) accept transitions with both asynchronous and synchronous phases; the latter can overcome the parallel construct only when the other component cannot offer the complementary phase.

The split$_n$-transition system $\mathcal{G}_n$ has several properties, listed in the proposition below.

Proposition 3.2 *i) If there are two outgoing transitions from a state $s \in \mathcal{S}_n$, labelled by phases η_i and θ_j reaching states s_1 and s_2, respectively, and $(i \neq 1 \vee j \neq 1)$, then there exists a state s_3 such that $s_1 \xrightarrow{\theta_j} s_3$ and $s_2 \xrightarrow{\eta_i} s_3$;*
ii) if $s \xrightarrow{\eta_i} s_1 \xrightarrow{\theta_j} s_2$ and $\theta_j, \eta_i \in \mathcal{M}^n$ and $(\eta \neq \theta \vee j \neq i+1) \wedge (i < n \vee j \neq 1)$ then there exist a states s_3, s_4, $s_3 \neq s_1$, such that $s \xrightarrow{\theta_j} s_3 \xrightarrow{\eta_i} s_4$. Furthermore, $s_2 = s_4$ except when $i = j = 1$ and $(\theta = \tau \vee \eta = \tau)$;
iii) if $s \xrightarrow{\theta_i} s_1 \in \mathcal{T}_n$ and $i < n$ then there exists $s_2 \in \mathcal{S}_n$ such that $s_1 \xrightarrow{\theta_{i+1}} s_2 \in \mathcal{T}_n$. □

Definition 3.3 *Two CCS agents t_1 and t_2 are split$_n$-bisimilar (denoted by $t_1 \overset{n}{\sim} t_2$) iff $\langle t_1 \backslash \Lambda_n^s, 1\rangle$ and $\langle t_2 \backslash \Lambda_n^s, 1\rangle$ in $\mathcal{G}_n$ are bisimilar.* □

We want to emphasize that, according to the above definition, two CCS agents are split$_n$-equivalent exactly when their split$_n$-transition systems are bisimilar. The first immediate result we prove is that $\overset{2}{\sim} \subseteq \sim$. The containment is strict as shown by a subsequent example. This result, which is not new (see, e.g. [8]), is given in order to explain in a simple case the proof technique we will use in the following.

Intuitively, a semantics A is *finer* than a semantics B if it is possible to derive the model of the B-semantics by transforming the model of the A-semantics. Transformations of models constitute the kernel of our proof technique. In the present case, the transformations are concerned with LTS's and consider just the scheleton of the graph and the labelling. These are the only two ingredients which are relevant for bisimulation. We obtain the inclusion between the two semantics by proving that the transformation preserves bisimulation equivalence classes. Our proof technique can be summarized by the following three steps:

1. define a labelled graph transformation which fits the above requirements;
2. prove that it works well, that is the transition system which is the output of the transformation is bisimilar to the one obtained directly by the other semantics;
3. verify that bisimilar transition systems, which are operational models of (CCS) agents, are transformed into bisimilar ones.

In particular, transformations are defined inductively, by means of inference rules which have neither state constants nor state constructors as premises (no information about the structure of the state is exploited). In the second step we have to show that the deductive system defining the input transition system of the transformation, enriched by the inference rules defining the transformation itself, is able to simulate the deductive system of the other semantics. In the third step, we check that there always exists at least a bisimulation (and, thus, always the maximal one) between the input transition systems which induces a bisimulation between the correspondent outputs.

Theorem 3.4 $\overset{2}{\sim} \subseteq \sim$.

Proof: The proposition of the theorem can be read as follows: $\mathcal{G}_2(\langle t\backslash\Lambda_2^s, 1\rangle) \sim \mathcal{G}_2(\langle t'\backslash\Lambda_2^s, 1\rangle)$ implies that $t \sim t'$. The proof consists of the steps 1-3 below:

1. Let $\mathcal{K}[\langle t\backslash\Lambda_2^s, 1\rangle] = \langle \mathcal{Q}, \mathcal{M}, \mathcal{R}\rangle$ be the atomic transition system where $\mathcal{Q}$ and $\mathcal{R}$ are the least fixpoints when the rules $a)$ and $b)$ below are applied to $\mathcal{G}_2(\langle t\backslash\Lambda_2^s, 1\rangle)$:

$$a)\ \frac{}{\langle t\backslash\Lambda_2^s, 1\rangle \in \mathcal{Q}} \qquad b)\ \frac{s_1 \in \mathcal{Q} \qquad s_1 \xrightarrow{\mu_1} s_2 \in \mathcal{T}_2(\langle t\backslash\Lambda_2^s, 1\rangle) \qquad s_2 \xrightarrow{\mu_2} s_3 \in \mathcal{T}_2(\langle t\backslash\Lambda_2^s, 1\rangle)}{s_3 \in \mathcal{Q} \qquad s_1 \xrightarrow{\mu} s_3 \in \mathcal{R}}$$

2. To show that $\mathcal{K}[\langle t\backslash\Lambda_2^s, 1\rangle] \sim \mathcal{G}(t)$, we prove that $\langle \zeta, \vartheta\rangle : \mathcal{K}[\langle t\backslash\Lambda_2^s, 1\rangle] \rightarrow \mathcal{G}(t)$ is well defined and transition preserving, where $\zeta(\langle r\backslash\Lambda_2^s, 1\rangle) = r$ and $\vartheta(s \xrightarrow{\mu} s') = \zeta(s) \xrightarrow{\mu} \zeta(s')$. For the well definition, it is enough to observe that, by construction, any state $\langle r\backslash\Lambda_2^s, 1\rangle \in \mathcal{K}[\langle t\backslash\Lambda_2^s, 1\rangle]$ is such that r is a CCS agent. We leave to the reader the proof that $\langle \zeta, \vartheta\rangle$ is a transition preserving morphism (as a matter of facts, $\langle \zeta, \vartheta\rangle$ defines an isomorphism).

3. To prove that $t \overset{2}{\sim} t'$ implies $\mathcal{K}[\langle t\backslash\Lambda_2^s, 1\rangle] \sim \mathcal{K}[\langle t'\backslash\Lambda_2^s, 1\rangle]$, we construct a bisimulation $\Re$ between $\mathcal{K}[\langle t\backslash\Lambda_2^s, 1\rangle]$ and $\mathcal{K}[\langle t'\backslash\Lambda_2^s, 1\rangle]$ from the bisimulation $\Re_2$ between $\mathcal{G}_2(\langle t\backslash\Lambda_2^s, 1\rangle)$ and $\mathcal{G}_2(\langle t'\backslash\Lambda_2^s, 1\rangle)$, by taking the following relation:

$$\Re = \{\langle\langle r_1\backslash\Lambda_2^s, z\rangle, \langle r_2\backslash\Lambda_2^s, z\rangle\rangle\ ,\quad \langle\langle r_1\backslash\Lambda_2^s, z\rangle, \langle r_2\backslash\Lambda_2^s, z\rangle\rangle \in \Re_2 \wedge r_1, r_2 \in CCS_{\mathcal{M}}\}.$$

$\Re$ is indeed a bisimulation. If $\langle p, q\rangle \in \Re$ and $p \xrightarrow{\mu} p'$ then $\exists p''$ such that $p \xrightarrow{\mu_1} p''$, $p'' \xrightarrow{\mu_2} p' \in \mathcal{T}_2(\langle t\backslash\Lambda_2^s, 1\rangle)$. Since $\Re_2$ is a bisimulation between $\mathcal{G}_2(\langle t\backslash\Lambda_2^s, 1\rangle)$ and $\mathcal{G}_2(\langle t'\backslash\Lambda_2^s, 1\rangle)$ then there exist states q' and q'' belonging to $\mathcal{S}_2(\langle t'\backslash\Lambda_2^s, 1\rangle)$ such that $\langle p', q'\rangle, \langle p'', q''\rangle \in \Re_2$. Therefore, by rule $c)$, $q \xrightarrow{\mu} q'$ is a transition in $\mathcal{K}_2[\langle t'\backslash\Lambda_2^s, 1\rangle]$ and $\langle p', q'\rangle \in \Re$. □

Example 3.5 The basic example discriminating atomic interleaving bisimulation equivalence and split$_2$-bisimulation equivalence is concerned with the two CCS agents $\alpha\,|\,\beta$ and $\alpha.\beta + \beta.\alpha$. They give rise to the same atomic unfolding trees but have two not bisimilar split$_2$ unfolding trees since the first agent can perform the sequence $\alpha_1 \cdot \beta_1 \cdot \alpha_2$ while the second cannot. □

The next theorem will prove that the split$_n$-equivalences are indeed a spectrum of finer and finer equivalences. The successive example will show that the inclusion is actually strict. In the following, given a relation R, we denote by R^* its *reflexive* and *transitive* closure.

Theorem 3.6 $\overset{n+1}{\sim} \subseteq \overset{n}{\sim}$.

Proof: The proof is similar to the one of Theorem 3.4, where the transformation is given below:

$$a)\ \frac{}{\langle t\backslash\Lambda_{n+1}^s, 1\rangle \in \mathcal{Q}} \qquad b)\ \frac{s \in \mathcal{Q} \qquad s \xrightarrow{\mu_i} s' \in \mathcal{T}_{n+1}(\langle t\backslash\Lambda_{n+1}^s, 1\rangle) \qquad i < n}{s' \in \mathcal{Q} \qquad s \xrightarrow{\mu_i} s' \in \mathcal{R}}$$

$$c)\ \frac{s_1 \in \mathcal{Q} \qquad s_1 \xrightarrow{\mu_n} s_2 \in \mathcal{T}_{n+1}(\langle t\backslash\Lambda_{n+1}^s, 1\rangle) \qquad s_2 \xrightarrow{\mu_{n+1}} s_3 \in \mathcal{T}_{n+1}(\langle t\backslash\Lambda_{n+1}^s, 1\rangle)}{s_3 \in \mathcal{Q} \qquad s_1 \xrightarrow{\mu_n} s_3 \in \mathcal{R}}$$
□

Example 3.7 This example shows that the inclusion of Theorem 3.6 is strict when $n = 2$, that is a CCS translation of the *owl example* originally introduced in [4] over prime event structures (see also [1]). Consider the process $p = (\eta.(\overline{\beta}\,|\,(\overline{\alpha} + p_1))\,|\,\theta.(\overline{\alpha}'\,|\,(\overline{\beta}' + p_2)))\backslash\alpha\backslash\alpha'\backslash\beta\backslash\beta'$ where $p_1 = \xi.(\alpha'.\varepsilon + \beta'.(\delta\,|\,\xi.\tau.\varepsilon))$ and $p_2 = \xi.(\beta.\delta + \alpha.(\varepsilon\,|\,\xi.\tau.\delta))$. Let q be identical to p but with the actions δ and ε interchanged, that is $q = (\eta.(\overline{\beta}\,|\,(\overline{\alpha} + q_1))\,|\,\theta.(\overline{\alpha}'\,|\,(\overline{\beta}' + q_2)))\backslash\alpha\backslash\alpha'\backslash\beta\backslash\beta'$ where $q_1 = \xi.(\alpha'.\delta + \beta'.(\varepsilon\,|\,\xi.\tau.\delta))$ and $q_2 = \xi.(\beta.\varepsilon + \alpha.(\delta\,|\,\xi.\tau.\varepsilon))$. It is possible to show that the split$_2$-transition system of p is bisimilar to that of q, but their split$_3$-transition systems are not bisimilar. Indeed the trace $\eta \cdot \xi_1 \cdot \xi_2 \cdot \theta \cdot \xi_1 \cdot \xi_3 \cdot \tau \cdot \varepsilon$ (each unsplitted action stands for a consecutive sequence of its phases) can be performed by p but not by q. □

The *owl example* can be generalized to any natural number (see [4]). Essentially, it shows, by exploiting the weakness of split semantics in recovering the individuality of actions, that every split$_n$-equivalence does not persist to splitting actions into more than n phases. As a consequence, split$_n$-equivalence is not a congruence w.r.t. $n+1$ *linear refinement*, i.e. a weak form of refinement according to which actions are detailed into a sequence of $n+1$ actions (see [7] for more details on this issue). However, when the language is CCS, split semantics is a congruence, as the following theorem states.

Theorem 3.8 *For every n, $\overset{n}{\sim}$ is a congruence w.r.t. the CCS syntax.*

Proof: The rules which specify the split$_n$-transition system fall within a slight generalization of the *GSOS format* [2], for which the same results hold. □

As a final remark, we want to emphasize that the interplay between communication, restriction and autoconcurrency is crucial to get the strict inclusion among the various split$_n$-semantics: indeed when one of the three features is absent in the language, the spectrum collapses to split$_2$-semantics.

4 ST semantics

ST-semantics is, essentially, split$_2$-semantics plus a mechanism to recover the individuality of actions. Indeed the drawback of split semantics, illustrated in Example 3.7, is the inability to properly match up endings with their own beginnings. This problem is overcome in ST-semantics by explicitly giving *causal links* connecting each ending phase to its beginning phase. If we observe the two agents p and q of Example 3.7, we can easily realize that they are not ST-bisimilar because the following ST-trace

$$\eta \cdot \xi_1 \cdot \theta \cdot \xi_1 \cdot \xi_2 \cdot \tau \cdot \varepsilon$$

(the arrow from the phase ξ_2 to ξ_1 represents the causal link) may be performed by p but not by q.

We recall that split semantics already exploit backward pointers, even if just to manage the causality needed for completing synchronizations correctly. The idea can be extended to all the actions: the beginning phase of an action a is denoted (with abuse of notation) by a, whilst the ending phase by a^k, where k is the clock value when a started. As a consequence there is no need to distinguish synchronous and asynchronous phases (rule Res_2 has no counterpart here). Therefore the definition of the ST-operational semantics will be simpler than the one for the split case.

Let $\mathcal{M}_{ST} = \mathcal{M} \cup \mathcal{M}^\omega$, where $\mathcal{M}^\omega = \{\mu^k \,,\, \mu \in \mathcal{M} \wedge k \in \omega\}$, be the set of *ST-labels*, ranged over, with abuse of notation, by η. The set $\mathcal{S}_{ST}$ of states of the transition system for the ST-semantics is the set of terms s generated by the following grammar (t ranges over *CCS agents*, α over Δ and z over ω):

$$\begin{aligned} r &::= \; t \;\mid\; \eta.t \;\mid\; r\backslash\alpha \;\mid\; r[\Phi] \;\mid\; r|r \\ s &::= \; \langle r,\, z\rangle. \end{aligned}$$

Definition 4.1 *The operational model for ST-semantics is $\mathcal{G}_{ST} = \langle \mathcal{S}_{ST}, \mathcal{M}_{ST}, \mathcal{T}_{ST}\rangle$, where the set of transition relations $\mathcal{T}_{ST} = \{\xrightarrow{\eta},\ \eta \in \mathcal{M}_{ST}\}$ is the least set of relations satisfying the following axioms and inference rules:*

$Act_b)$ $\langle \mu.t,\, z\rangle \xrightarrow{\mu} \langle \mu^z.t,\, z+1\rangle$ $\qquad$ $Act_e)$ $\langle \mu^k.t,\, z\rangle \xrightarrow{\mu^k} \langle t,\, z\rangle$

$Res)$ $\dfrac{\langle r,\, z\rangle \xrightarrow{\eta} \langle r',\, z'\rangle}{\langle r\backslash\alpha,\, z\rangle \xrightarrow{\eta} \langle r'\backslash\alpha,\, z'\rangle} \quad \eta \notin \{\alpha,\, \overline{\alpha},\, \alpha^k,\, \overline{\alpha}^k \;;\; k \in \omega\}$

$$Rel)\ \frac{\langle r,\, z\rangle \xrightarrow{\eta} \langle r',\, z'\rangle}{\langle r[\Phi],\, z\rangle \xrightarrow{\Phi(\eta)} \langle r'[\Phi],\, z'\rangle} \qquad Rec)\ \frac{\langle t[{}^{\mathrm{rec}\,x.\,t}/_x],\, z\rangle \xrightarrow{\eta} \langle r',\, z'\rangle}{\langle \mathrm{rec}\,x.\,t,\, z\rangle \xrightarrow{\eta} \langle r',\, z'\rangle}$$

$$Sum_l)\ \frac{\langle t_1,\, z\rangle \xrightarrow{\eta} \langle r_1,\, z'\rangle}{\langle t_1 + t_2,\, z\rangle \xrightarrow{\eta} \langle r_1,\, z'\rangle} \qquad Sum_r)\ \frac{\langle t_2,\, z\rangle \xrightarrow{\eta} \langle r_2,\, z'\rangle}{\langle t_1 + t_2,\, z\rangle \xrightarrow{\eta} \langle r_2,\, z'\rangle}$$

$$Par_b^l)\ \frac{\langle r_1,\, z\rangle \xrightarrow{\mu} \langle r_1',\, z'\rangle}{\langle r_1|r_2,\, z\rangle \xrightarrow{\mu} \langle r_1'|r_2,\, z'\rangle} \qquad Par_b^r)\ \frac{\langle r_2,\, z\rangle \xrightarrow{\mu} \langle r_2',\, z'\rangle}{\langle r_1|r_2,\, z\rangle \xrightarrow{\mu} \langle r_1|r_2',\, z'\rangle}$$

$$Par_e^l)\ \frac{\langle r_1,\, z\rangle \xrightarrow{\mu^k} \langle r_1',\, z'\rangle \quad \langle r_2,\, z\rangle \not\xrightarrow{\bar{\mu}^k}}{\langle r_1|r_2,\, z\rangle \xrightarrow{\mu^k} \langle r_1'|r_2,\, z'\rangle} \qquad Par_e^r)\ \frac{\langle r_2,\, z\rangle \xrightarrow{\mu^k} \langle r_2',\, z'\rangle \quad \langle r_1,\, z\rangle \not\xrightarrow{\bar{\mu}^k}}{\langle r_1|r_2,\, z\rangle \xrightarrow{\mu^k} \langle r_1|r_2',\, z'\rangle}$$

$$Par_b^s)\ \frac{\langle r_1,\, z\rangle \xrightarrow{\lambda} \langle r_1',\, z'\rangle \quad \langle r_2,\, z\rangle \xrightarrow{\bar{\lambda}} \langle r_2',\, z'\rangle}{\langle r_1|r_2,\, z\rangle \xrightarrow{\tau} \langle r_1'|r_2',\, z'\rangle} \qquad Par_e^s)\ \frac{\langle r_1,\, z\rangle \xrightarrow{\lambda^k} \langle r_1',\, z'\rangle \quad \langle r_2,\, z\rangle \xrightarrow{\bar{\lambda}^k} \langle r_2',\, z'\rangle}{\langle r_1|r_2,\, z\rangle \xrightarrow{\tau^k} \langle r_1'|r_2',\, z'\rangle}$$

where the relabelling function Φ satisfies also the constraint $\Phi(\mu^k) = \Phi(\mu)^k$. □

As an example we consider the agent $\alpha \mid \alpha$, whose ST-transition system $\mathcal{G}_{ST}(\langle \alpha \mid \alpha,\, 1\rangle)$ is depicted in Fig. 2.

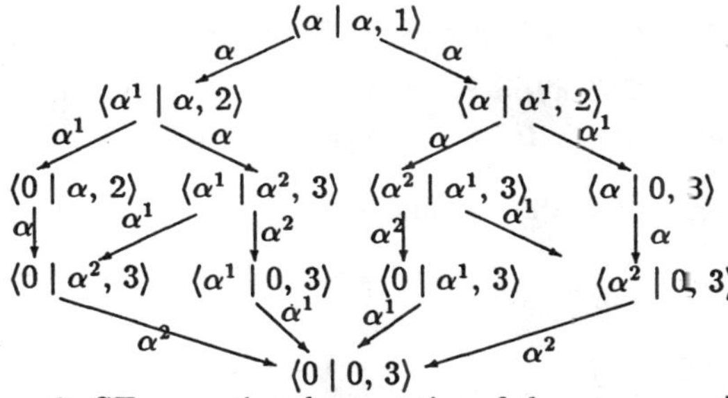

Figure 2: ST operational semantics of the process $\alpha \mid \alpha$.

By $\mathcal{G}_{ST}(\langle t,\, 1\rangle)$ we mean the (sub-)transition system in $\mathcal{G}_{ST}$ reachable from the state $\langle t,\, 1\rangle$. Similarly to Proposition 3.2, here we list some interesting properties of $\mathcal{G}_{ST}(\langle t,\, 1\rangle)$.

Proposition 4.2 *For every CCS agent t, the sub-transition system $\mathcal{G}_{ST}(\langle t,\, 1\rangle) = \langle \mathcal{S}_{ST}(\langle t,\, 1\rangle),\, \mathcal{M}_{ST},\, \mathcal{T}_{ST}(\langle t,\, 1\rangle)\rangle$ satisfies the following properties:*
i) for every state $\langle r,\, z\rangle$ and every path from $\langle t,\, 1\rangle$ to $\langle r,\, z\rangle$, the transitions with labels in $\mathcal{M}$ are $z-1$;
ii) in every path starting from the root, all the labels have different superscripts, if any; moreover an ending μ^k can occur only if there are at least k beginnings preceding it and the k-th beginning is μ;
iii) let s be a state with an outgoing μ-transition and let ϕ be a path from the root to s having $k-1$ beginning phases. Then every state reachable from s with a path not having a μ^k-transition has an outgoing transition labelled by μ^k;
iv) for every state s such that $s \xrightarrow{\mu} s_1 \xrightarrow{\nu} s_2$ and $\mu,\, \nu \in \mathcal{M}$ there exist states s_3 and s_4 such that $s \xrightarrow{\nu} s_3 \xrightarrow{\mu} s_4$ □.

Definition 4.3 *Two CCS agents t_1 and t_2 are ST-bisimilar (denoted $t_1 \overset{ST}{\sim} t_2$) iff $\mathcal{G}_{ST}(\langle t_1,\, 1\rangle) \sim \mathcal{G}_{ST}(\langle t_2,\, 1\rangle)$.*

Theorem 4.4 *ST-bisimulation is a congruence over CCS agents.* □

We are ready to prove the main result of the paper: the correspondence between ST-semantics and the spectrum of split-semantics. To this aim, we will argue by transforming transition systems according to the style developed in the previous Section.

Definition 4.5 *Let $\langle \mathcal{S}, \mathcal{L}_{ST}, \mathcal{T}\rangle$ be an ST transition system with root s, then $\nabla_n^{ST}(\langle \mathcal{S}, \mathcal{L}_{ST}, \mathcal{T}\rangle)$ is the split$_n$-transition system $\langle \mathcal{Q}, \mathcal{L}_n, \mathcal{R}\rangle$ where $\mathcal{L}_n = \{a_i\,,\, a \in \mathcal{L} \wedge 1 \leq i \leq n\}$, $\mathcal{Q}$ and $\mathcal{R}$ are the least fixpoints of the following rules:*

1. $$\frac{}{\langle \phi\,,\, s,\, 1\rangle \in \mathcal{Q}}$$

2. $$\frac{\langle p,\, q,\, z\rangle \in \mathcal{Q} \qquad q \xrightarrow{a} q' \in \mathcal{T}}{\langle p \cup \{a_1^z\},\, q',\, z+1\rangle \in \mathcal{Q} \qquad \langle p,\, q,\, z\rangle \xrightarrow{a_1} \langle p \cup \{a_1^z\},\, q',\, z+1\rangle \in \mathcal{R}}$$

3. $$\frac{\langle p,\, q,\, z\rangle \in \mathcal{Q} \qquad a_i^k \in p \qquad i < n-1}{\langle (p - \{a_i^k\}) \cup \{a_{i+1}^k\},\, q,\, z\rangle \in \mathcal{Q} \qquad \langle p,\, q,\, z\rangle \xrightarrow{a_{i+1}} \langle (p - \{a_i^k\}) \cup \{a_{i+1}^k\},\, q,\, z\rangle \in \mathcal{R}}$$

4. $$\frac{\langle p,\, q,\, z\rangle \in \mathcal{Q} \qquad a_{n-1}^k \in p \qquad q \xrightarrow{a^k} q' \in \mathcal{T}}{\langle (p - \{a_{n-1}^k\}),\, q',\, z\rangle \in \mathcal{Q} \qquad \langle p,\, q,\, z\rangle \xrightarrow{a_n} \langle (p - \{a_{n-1}^k\}),\, q',\, z\rangle \in \mathcal{R}}$$ □

Informally, rule 1. says that the root of the ST-transition system is turned into the root of the split$_n$-one. Rule 2. ensures that, whenever a beginning is performed by the ST-transition system, a begining phase is also possible by the corresponding split$_n$-state. Rule 3. accounts for the execution of intermediate phases. At last, rule 4 models the closure of an action.

Theorem 4.6 *i) For every CCS agent t, $\nabla_n^{ST}(\mathcal{G}_{ST}(\langle t,\, 1\rangle)) \sim \mathcal{G}_n(\langle t\backslash\Lambda_n^s,\, 1\rangle)$.*
ii) for every pair t_1, t_2 of CCS agents: $t_1 \overset{ST}{\sim} t_2$ implies $t_1 \overset{n}{\sim} t_2$. □

Definition 4.7 *A CCS agent t has autoconcurrency m if*

$$m \;=\; \max_{\substack{\mu \in \mathcal{M} \\ s \in \mathcal{S}_2(\langle t\backslash\Lambda_n^s,\, 1\rangle)}} \{\, |\, \{s' ,\quad s \xrightarrow{\mu_2} s' \wedge s' \in \mathcal{S}_2(\langle t\backslash\Lambda_n^s,\, 1\rangle)\}\, |\, \}$$

where $|-|$ is the set cardinality function. □

In order to provide the reverse of Theorem 4.6.*ii*), let us give the transformation from a split$_n$-transition system to an ST-one. Below, $\pi_i(-)$ will mean the *i*-th projection of its *t*-uple argument.

Definition 4.8 *Let $\langle \mathcal{Q}, \mathcal{L}_n, \mathcal{R}\rangle$ be a rooted split$_n$-transition system with root s, then $\Delta_n^{ST}(\langle \mathcal{Q}, \mathcal{L}_n, \mathcal{R}\rangle)$ is the ST-transition system $\langle \mathcal{S}, \mathcal{L}_{ST}, \mathcal{T}\rangle$ where $\mathcal{L}_{ST} = \mathcal{L} \cup \mathcal{L}^\omega$, $\mathcal{S}$ and $\mathcal{T}$ are the least fixpoints of the following:*

1. $$\frac{}{\langle \phi\,,\, s,\, 1\rangle \in \mathcal{S}}$$

2. $$\frac{\langle p,\, q,\, z\rangle \in \mathcal{S} \qquad q \xrightarrow{a_{j+1} \cdot a_j \cdot \ldots \cdot a_1} q' \in \mathcal{R}^* \qquad j = \min_i\{i\,,\; a_{i+1}^k \notin p\}}{\langle p',\, q',\, z+1\rangle \in \mathcal{S} \qquad \langle p,\, q,\, z\rangle \xrightarrow{a} \langle p',\, q',\, z+1\rangle \in \mathcal{T}}$$

$$\text{where} \quad p' = \; p[{}^{a_{x+1}^k}/_{a_x^k},\; 1 \leq x \leq j] \cup \{a_1^z\} \quad \text{and} \quad \min_i \phi \; = 0$$

3. $$\frac{\langle p,\, q,\, z\rangle \in \mathcal{S} \qquad a_i^k \in p \qquad \varphi \in \Im_{a_i^k}(q)}{\langle p - \{a_i^k\},\, \pi_{n-i+1}(\varphi),\, z\rangle \in \mathcal{S} \qquad \langle p,\, q,\, z\rangle \xrightarrow{a^k} \langle p - \{a_i^k\},\, \pi_{n-i+1}(\varphi),\, z\rangle \in \mathcal{T}}$$

$$\text{where } \Im_{a_i^k}(q) \;=\; \{\langle q_1, q_2, \ldots, q_{n-i+1}\rangle\,,\;\; q_1 = q \wedge q_j \xrightarrow{a_{i+j}} q_{j+1} \in \mathcal{R} \wedge$$
$$\wedge\, \forall y > i\,.\, \exists w\,.(a_y^w \in p \Rightarrow (\forall \psi \in \Im_{a_y^w}(q)\,.\, \pi_2(\psi) \xrightarrow{a_{i+1} \cdot a_{i+2} \cdot \ldots \cdot a_y} q_{y-i+2} \notin \mathcal{R}^*))\}$$ □

Definition 4.8 deserves some clearing up. We will apply Δ_n^{ST} to split$_n$-transition systems with autoconcurrency $n-1$. In this case, to obtain individuality of actions, the idea is to avoid those split$_n$-states for which at least two autoconcurrent actions have progressed to the same phase (*confusion states*). In the case of beginning moves, rule 2 ensures this fact by increasing the phases of the already started autoconcurrent actions. The upper bound on autoconcurrency guarantees the consistency of such increments.

The case of ending phases is more tricky. In a state where several actions a have already started we must be able to close each of them. Obviously, indiscriminate closures may make individuality lost: we have to perform a closure path where the same occurrence of the autoconcurrent action is actually progressed till to completition. More precisely, let a_i^s be the phase belonging to the first component p of a ST-state resulting from Δ_n^{ST}, and φ_{a_i} be the right closure path in the split$_n$-transition system. Since other autoconcurrent actions at a greater phase may belong to p, confusion states of the split$_n$-transition system may be passed through while reaching the final state of φ_{a_i}. As a consequence, in such states different choices for the next state of φ_{a_i} are possible. In order to choose the right transition, let us suppose that the paths φ_{a_j} for all the autoconcurrent actions belonging to p and at a phase greater than i have already been found. For semplicity, assume that $\varphi_{a_{i+1}}$ is one of these paths. Then, by performing the move a_{i+1} in the path φ_{a_i}, we can go either in a state which is reachable from the second state of the path $\varphi_{a_{i+1}}$ with an a_i-move (Proposition 3.2.*i*) or in a state which is not reachable by any move from the above state of $\varphi_{a_{i+1}}$. This latter state is the one we choose in the definition of $\Im_{a_i^s}(q)$, because the former corresponds to a move of the subprocess prefixed by the highest phase in q. We illustrate Δ_n^{ST} through an example.

Example 4.9 Let us run Δ_3^{ST} with our 'working' process $\alpha \mid \alpha$. For semplicity, in the figure below we drop the restriction $-\backslash\Lambda_3^s$ inside the split$_3$ states.

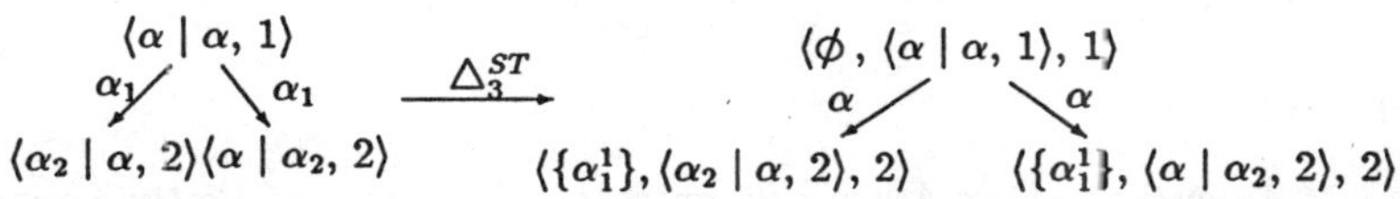

and, by applying Δ_3^{ST} to $\langle\alpha_2 \mid \alpha, 2\rangle$ we obtain:

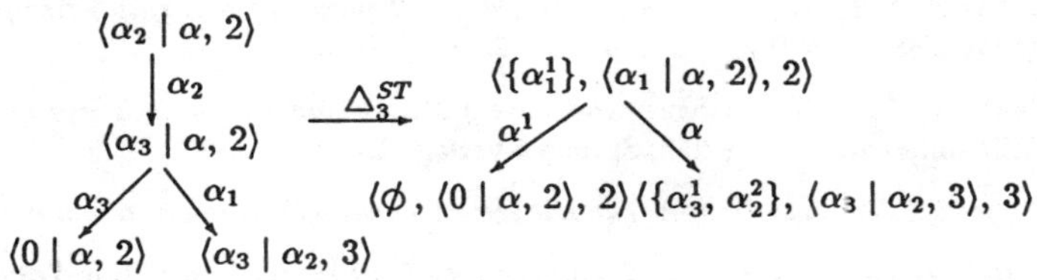

Note that, by performing the ST-transition labelled α, the already started autoconcurrent action is progressed to a greater phase.

In the state $\langle\alpha_3 \mid \alpha_2, 3\rangle$ we can complete the execution either of the left subprocess or of the right one. In the case we choose to end the action of the right subprocess, we have to select a proper path inside the split$_n$-transition system. The situation is depicted in the following picture. We leave out the path $\langle\alpha_3 \mid \alpha_2, 3\rangle \xrightarrow{\alpha_2} \langle\alpha_3 \mid \alpha_3, 3\rangle \xrightarrow{\alpha_3} \langle\alpha_3 \mid 0, 3\rangle$ because, by diamond property of Proposition 3.2.*i*, α_2 and α_3 are independent. □

Theorem 4.10 *i) Let t be a CCS agent such that $t \in \mathcal{A}_{n-1}$, then* $\Delta_n^{ST}(\mathcal{G}_n(\langle t\backslash\Lambda_n^s, 1\rangle)) \sim \mathcal{G}_{ST}(\langle t, 1\rangle)$.

ii) for every pair t_1, t_2 of CCS agents, such that $t_1, t_2 \in \mathcal{A}_{n-1}$, $t_1 \overset{n}{\sim} t_2$ implies $t_1 \overset{ST}{\sim} t_2$. □

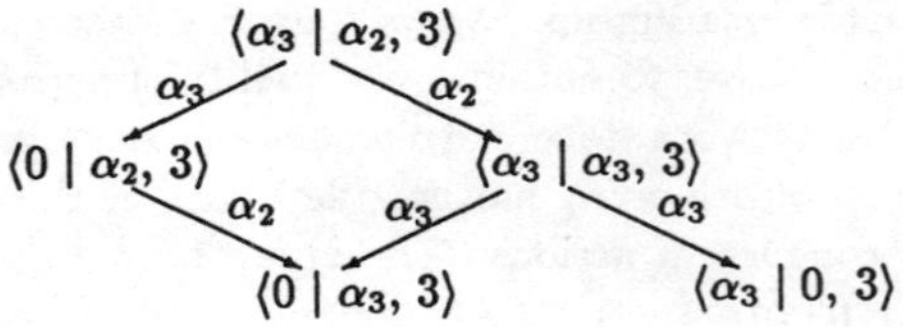

Theorem 4.11 *Let t_1 and t_2 be CCS agents. Then $t_1 \stackrel{ST}{\sim} t_2$ if and only if, for every n, $t_1 \stackrel{n}{\sim} t_2$.* □

5 Conclusions

Split$_n$- and ST-operational models for full CCS are defined here for the first time. ST-bisimulation semantics is proved to be the intersection of all the split$_n$-bisimulation semantics by means of graph transformations among the various transition systems. The crucial transformation is the one used in proving that, when autoconcurrent events are at most n, ST-bisimulation is as discriminating as split$_{n+1}$-bisimulation.

In a forthcoming paper [7], we address the problem of adding to CCS an operator $t[\lambda \rightsquigarrow t']$ for action refinement whose meaning is, intuitively, that actions λ within t should be seen more detailedly as t'. There we prove that ST-bisimulation is a congruence for refinement and all the CCS operations, and, moreover, by exploiting the results in this paper, that it is the coarsest bisimulation congruence for that language.

Acknowledgements

We are grateful to Luca Aceto, Philippe Darondeau, Pierpaolo Degano, Gian Luigi Ferrari, Frits Vaandrager and Walter Vogler for their comments.

References

[1] L. Aceto, M. Hennessy: *Adding Action Refinement to a Finite Process Algebra*, University of Sussex, Report No. 6/90, 1990.

[2] B. Bloom, S. Istrail, A.R. Mayer: *Bisimulation can't be Traced*, in Proc. 15th Principles of Programming Languages (POPL'88), San Diego, California, 229-239, 1988.

[3] R.J. van Glabbeek, F.W. Vaandrager: *Petri Nets Models for Algebraic Theories of Concurrency*, in Proc. PARLE Conference, LNCS 259, Springer-Verlag, 224-242, 1987.

[4] R.J. van Glabbeek, F.W. Vaandrager: *The Difference between Splitting in n and in n+1*, in preparation.

[5] R. Gorrieri: *Refinement, Atomicity and Transactions for Process Description Languages*, PhD Thesis, University of Pisa, TD-2/91, March 1991.

[6] R. Gorrieri, C. Laneve: *Split- and St-Semantics for CCS*, Internal Report, University of Pisa, 1991.

[7] R. Gorrieri, C. Laneve: *The Coarsest Bisimulation Congruence for CCS plus Refinement*, forthcoming.

[8] M. Hennessy: *Axiomatising Finite Concurrent Processes*, SIAM Journal on Computing, 17 (5), 997-1014, 1988.

[9] R. Milner: *Communication and Concurrency*, Prentice Hall, 1989.

[10] D. Park: *Concurrency and Automata on Infinite Sequences*, LNCS 104, Springer-Verlag, 1981.

[11] G. Plotkin: *A Structural Approach to Operational Semantics*, Report DAIMI-FN-19, Computer Science Dept, Arhus University, Denmark, 1981.

Stochastic automata and length distributions of rational languages [1]

Georges HANSEL [2] – Daniel KROB [3] – Christian MICHAUX [4]

0 Introduction

This paper is motivated by two questions of D. Niwinski (cf [6]) that we first recall. For any finite set S, let us denote by $|S|$ the number of elements of S. If w is a word, we also write $|w|$ the length of w. Let $p \in \mathbb{Z}[x]$ be a polynomial function which maps $\mathbb{N}$ into $\mathbb{N}$ and suppose that L is a rational subset of $\{a,b\}^*$. D. Niwinski raised the two following questions :

($Q1$) Does the set $\{n \in \mathbb{N} \mid |L \cap \{a,b\}^n| > 2^{n-1}\}$ belong to Rat($\mathbb{N}$) ?

($Q2$) Does the set $\{w \in \{a,b\}^* \mid |w^{-1}L \cap \{a,b\}^{p(|w|)}| > 2^{p(|w|)-1}\}$ belong to Rat($\{a,b\}^*$) ?

These questions are in fact motivated by complexity theory. Indeed the operator C which associates with every language L over $\{a,b\}$ and with every polynomial $p \in \mathbb{Z}[x]$ which maps $\mathbb{N}$ into $\mathbb{N}$, the language $C(L,p)$ given by :

$$C(L,p) = \{\, w \in \{a,b\}^* \mid |w^{-1}L \cap \{a,b\}^{\leq p(|w|)}| > 2^{p(|w|)-1} \,\} \qquad (C)$$

is used to define the counting hierarchy in complexity theory (see [1]). The "exact counting" operator $C_=(L,p)$ which is obtained by writing "$p(|w|)$" instead of "$\leq p(|w|)$" in (C), was introduced in [9] for studying several subclasses of the counting hierarchy. It appears now that question ($Q2$) asked exactly if the class of rational languages is stabilized by the operator $C_=$.

We will see here that the two questions ($Q1$) and ($Q2$) are related, that their answers are positive for "almost" every rational language, but that counterexamples to both assertions can be found. Moreover it appears that question ($Q1$) is related to the theory of one-letter stochastic automata that was developed by Turakainen (see [10]).

---

[1] This paper was supported by OMII "Algorithmique Stochastique" project, by Esprit BRA - ASMICS N° 3166 project and by cooperation project CGRI-CNRS 1990.

[2] Université de Rouen – Laboratoire d'Informatique de Rouen – Place Blondel – 76134 Mont Saint-Aignan Cedex – FRANCE

[3] Université Paris 6, Université de Rouen and CNRS(LITP) – Laboratoire d'Informatique de Rouen – Place Blondel – 76134 Mont Saint-Aignan Cedex – FRANCE

[4] Université de Mons-Hainaut – 15, Avenue Maistriau – B 7000 Mons – BELGIQUE

Let us now give the structure of our paper. After some preliminaries, the first section is devoted to the study of the rationality of one-letter stochastic languages. We give a new and more concise presentation of Turakainen's results concerning this class of stochastic languages. In the second section, we address the two questions of Niwinski. With any rational language L, we associate a stochastic automaton that computes the length frequencies of L. Then, applying Turakainen's results, we obtain general classes of rational languages for which question $(Q1)$ has a positive answer. Finally we show how to reduce question $(Q2)$ to the first one.

1 Preliminaries

Let A be a finite alphabet and K a semiring. A K-automaton of order n is a triple (I, μ, T) where I and T are respectively a row and a column vector of K^n and where μ is a morphism from the free monoid A^* into the monoid $\mathcal{M}_n(K)$ of square matrices of order n with entries in K. The element $I \cdot \mu(w) \cdot T$ is called the *multiplicity* of the word w. We refer the reader to [2] or [3] for more details.

In particular, an automaton in the usual meaning is a $\mathcal{B}$-automaton where $\mathcal{B} = \{0, 1\}$ is equipped with its boolean semiring structure. Since $\mathcal{B}$ can also be considered as a subset of the semiring $\mathbf{N}$, we can associate an $\mathbf{N}$-automaton with any $\mathcal{B}$-automaton. Moreover, a *deterministic* $\mathcal{B}$-automaton and the corresponding $\mathbf{N}$-automaton have the same multiplicities for every word $w \in A^*$ and consequently can be identified. Of course, we can replace $\mathbf{N}$ by $\mathbf{R}$ in these considerations.

We will also use the natural isomorphism $n \mapsto a^n$ between the monoid $\mathbf{N}$ and the free monoid over the one-letter alphabet $\{a\}$. With this identification, let us recall that the rational languages in a^* are exactly the ultimately periodic subsets of $\mathbf{N}$ (see [3]).

We refer the reader to [4] for the probabilistic vocabulary used in the sequel. In particular, let us recall that a *stochastic* matrix is a matrix with entries in $[0, 1]$ such that the sum of the entries of every row is equal to 1. We shall use the usual correspondence between stochastic matrices and finite homogeneous Markov chains (see [4]). If M is a stochastic matrix, the following facts are true : 1 is an eigenvalue of M; all eigenvalues of M have a modulus ≤ 1; all the eigenvalues of modulus 1 are roots of the unity.

Finally we say that a complex number z is *rational in argument* iff the argument of z belongs to $\pi\,\mathbf{Q}$ (i.e. is a rational multiple of π).

2 Stochastic automata

2.1 Stochastic languages

Stochastic automata were introduced by Rabin (see [7]). A stochastic automaton is just an $\mathbf{R}$-automaton with a "stochastic representation" as says the following definition.

DEFINITION 2.1 : A *stochastic automaton* of order n is an $\mathbf{R}$-automaton (π, μ, T) of order n where every matrix $\mu(a)$ is a stochastic matrix, where the initial vector π is a stochastic vector and where the terminal vector T has its values in the two-element set

$\{0,1\}$ and thus is naturally associated with a subset of $\{1,\ldots,n\}$. Its *behaviour* is the set $\{\,\pi\cdot\mu(w)\cdot T \mid w \in A^*\,\}$ of its multiplicities.

We now define the stochastic languages as follows (see also [8,10]).

DEFINITION 2.2 : Let A be a finite alphabet. Then a language S over A is said to be a *strict* (resp. *large*) *stochastic language* iff there exists a stochastic automaton $\mathcal{A} = (\pi,\mu,T)$ and a real number $\lambda \in [0,1]$ such that S is equal to the language :

$$S_{>}(\mathcal{A},\lambda) = \{\,w \in A^* \mid \pi\cdot\mu(w)\cdot T > \lambda\,\} \quad (\text{resp. } S_{\geq}(\mathcal{A},\lambda) = \{\,w \in A^* \mid \pi\cdot\mu(w)\cdot T \geq \lambda\,\})$$

λ is called the *cut-point* associated with S and S is said to be *recognized* by $\mathcal{A}$.

Notes : 1) Every rational language L is a strict and large stochastic language. Indeed, L can be recognized in the usual meaning by a complete deterministic automaton with a unique initial state. [5] This $\mathcal{B}$-automaton can be considered as a stochastic automaton (π,μ,T) and we have for every $w \in A^*$, $\pi\cdot\mu(w)\cdot T = 1$ (resp. $= 0$) if $w \in L$ (resp. $w \notin L$). Taking (for instance) the cut-point $\lambda = 1/2$, we get that L is both a strict and a large stochastic language.

2) The reader will find several interesting examples of stochastic languages in [7]. In particular, Rabin gives there non-recursively enumerable languages which are recognized by stochastic automata.

3) In [7], Rabin proved that every strict or large stochastic language is rational whenever the associated cut-point is isolated in the behaviour of the corresponding stochastic automaton.

4) Turakainen (see [8,10]) proved that a language is strictly (resp. largely) stochastic iff it is the strictly (resp. largely) positive support of some $\mathbb{R}$-rational series.

2.2 One-letter stochastic languages

We present here several results of Turakainen concerning one-letter stochastic automata. If (π,μ,T) is a stochastic automaton on the one letter alphabet $\{a\}$, the representation μ is completely defined as soon as the matrix $\Sigma = \mu(a)$ is given. Thus we can consider a one-letter stochastic automaton as a triple (π,Σ,T) where Σ denotes a stochastic matrix. Moreover we freely identify a language $L \subset a^*$ with the corresponding subset of $\mathbb{N}$.

LEMMA 2.1 : Let M be a strictly positive integer, let J be a subset of $\{0,\ldots,M-1\}$ and let L_j be a language in a^* for every $j \in J$. Then the two following assertions are equivalent :

1) $L = \bigcup_{j\in J} (M.L_j + j)$ is rational.

2) L_j is rational for every $j \in J$.

Proof : Notice first that a subset K of $\mathbb{N}$ is rational (i.e. ultimately periodic) iff $M.K+m$ is rational ($m \in \mathbb{N}$ being a fixed integer). Hence, if L is rational, we get for every $j \in J$

[5] For instance, the minimal automaton of L has these two properties.

that $M.L_j + j = L \cap (M.\mathbb{N} + j)$ is rational and consequently that L_j is rational. The other implication is obtained in the same way. ■

Let $\mathcal{A} = (\pi, \Sigma, T)$ be a one-letter stochastic automaton. Since every eigenvalue of modulus 1 of Σ is a root of the unity, there exists an integer M such that 1 is the unique eigenvalue of modulus 1 of the stochastic matrix Σ^M. Using the partition of $\mathbb{N}$ in classes modulo M, we have :

$$S_{\succ}(\mathcal{A}, \lambda) = \{n \in \mathbb{N} \mid \pi \cdot \Sigma^n \cdot T \succ \lambda\} = \bigcup_{i=0}^{M-1} (\{n \in \mathbb{N} \mid (\pi\Sigma^i) \cdot (\Sigma^M)^n \cdot T \succ \lambda\} \cdot M + i) \quad (\mathcal{DC})$$

where $\succ$ stands either for $>$ or for $\geq$ (we shall repeatedly use this notation in the sequel). Since a row vector of the form $\pi \cdot \Sigma^i$ is stochastic, the triple $\mathcal{A}_i = (\pi\Sigma^i, \Sigma^M, T)$ is a stochastic automata for every $i \in \{0, \ldots, M-1\}$. Then according to lemma 2.1 and to relation $(\mathcal{DC})$, the stochastic language $S_{\succ}(\mathcal{A}, \lambda)$ is rational iff the stochastic language $S_{\succ}(\mathcal{A}_i, \lambda)$ is rational for every $i \in \{0, \ldots, M-1\}$. Hence we reduced the study of the rationality of a one-letter stochastic language to the special case where 1 is the only eigenvalue of modulus 1 of the corresponding stochastic matrix.

Let us now consider a one-letter stochastic automaton $\mathcal{A} = (\pi, \Sigma, T)$ of order m where 1 is the only eigenvalue of modulus 1 of Σ. Let $(\lambda_i)_{i=0,\ldots,r}$ denote the sequence of the different eigenvalues of Σ sorted by order of decreasing modulus :

$$1 = \lambda_0 > |\lambda_1| \geq |\lambda_2| \geq \ldots \geq |\lambda_r| \qquad (EV)$$

For every eigenvalue λ_i of Σ, let m_i be its multiplicity and θ_i its argument. Then it is a classical consequence of the Jordan decomposition (see [5] for instance) that we have for every $n \in \mathbb{N}$:

$$\Sigma^n = \sum_{k=0}^{m_0-1} n^k \, \Sigma_{0,k} + \sum_{i=1}^{r} \lambda_i^n \, (\sum_{k=0}^{m_i-1} n^k \, \Sigma_{i,k})$$

where the matrices $\Sigma_{i,k}$ do not depend on n and are independent elements of the vector space $\mathcal{M}_m(K)$. Since the matrix Σ^n is stochastic for every $n \in \mathbb{N}$ and since $|\lambda_i| < 1$ for every $i \neq 0$, the first term of the second member in the previous equality is reduced to $\Sigma_{0,0}$. Hence we get that, for all $n \in \mathbb{N}$,

$$\pi \cdot \Sigma^n \cdot T = \lambda(\mathcal{A}) + \sum_{i=1}^{r} |\lambda_i|^n \, (\sum_{k=0}^{m_i-1} \alpha_k^{(i)} \, n^k \, \cos(n\theta_i + \varphi_k^{(i)})) \qquad (RS)$$

with $\lambda(\mathcal{A}) = \pi \cdot \Sigma_{0,0} \cdot T$ and where $\alpha_k^{(i)}$ and $\varphi_k^{(i)}$ are real numbers that depend only on Σ. Note that $\lambda(\mathcal{A})$ belongs to $[0, 1]$ and is intrinsically associated with $\mathcal{A}$. It follows from the previous equalities that the sequence $(\pi \cdot \Sigma^n \cdot T)_{n \in \mathbb{N}}$ converges to $\lambda(\mathcal{A})$. Hence we obtain the following result.

THEOREM 2.2 : (Turakainen [10]) Let $\mathcal{A} = (\pi, \Sigma, T)$ be a one-letter stochastic automaton such that 1 is the only eigenvalue of modulus 1 of Σ and let λ be a real number in $[0, 1]$. Then

$$\lambda \neq \lambda(\mathcal{A}) \implies S_{\succ}(\mathcal{A}, \lambda) \text{ is finite or cofinite}$$

Notes : 1) It follows from the previous theorem and from lemma 2.1 that a one-letter stochastic language is rational excepted possibly for a finite number of cut-points, namely the real numbers $\lambda(\mathcal{A}_i)$, $i \in \{0,\ldots,M-1\}$. Moreover there a unique possible exception $\lambda(\mathcal{A})$ when 1 is the only eigenvalue of modulus 1 of Σ.

2) Let us suppose that 1 is the unique eigenvalue of modulus 1 of Σ and that 1 is a simple eigenvalue of Σ. Let ν be the unique Σ-invariant probability row vector, i.e. the unique stochastic row vector such that $\nu\cdot\Sigma = \nu$. Then the sequence $(\Sigma^n)_{n\in\mathbb{N}}$ converges to a matrix whose rows are all equal to ν and consequently $\lambda(\mathcal{A}) = \nu\cdot T$.

Hence, according to theorem 2.2, it remains to study the rationality of the stochastic language $S_{\succ}\big(\mathcal{A},\lambda(\mathcal{A})\big)$ corresponding to the cut-point $\lambda(\mathcal{A})$ when $\mathcal{A}$ is a stochastic automaton whose stochastic matrix has 1 as unique eigenvalue of modulus 1. But, according to relation (RS), this problem is equivalent to study the rationality of the language

$$\{n \in \mathbb{N} \mid \sum_{i=1}^{r} |\lambda_i|^n \,(\sum_{k=0}^{m_i-1} \alpha_k^{(i)}\, n^k \cos(n\theta_i + \varphi_k^{(i)})\,) \succ 0\,\}.$$

Hence it appears that the rationality of $S_{\succ}\big(\mathcal{A},\lambda(\mathcal{A})\big)$ is related with the eigenvalues of Σ which are distinct from 1.

Let us now end this section by giving without proof a result of Turakainen that precises the nature of $S_{\succ}(\mathcal{A},\lambda(\mathcal{A}))$.

THEOREM 2.3 : (Turakainen; [10]) Let $\mathcal{A} = (\pi,\Sigma,T)$ be a one-letter stochastic automaton such that 1 is the unique eigenvalue of modulus 1 of Σ and such that every eigenvalue of Σ is rational in argument. Then $S_{\succ}\big(\mathcal{A},\lambda(\mathcal{A})\big)$ is a rational language.

Notes : 1) It follows from theorem 2.3 and from our study that $S_{\succ}(\mathcal{A},\lambda)$ is always rational if every eigenvalue of the stochastic matrix associated with $\mathcal{A}$ is rational in argument.

2) It should be noticed that $S_{\succ}(\mathcal{A},\lambda(\mathcal{A}))$ can be non-rational (see [10] for examples and sufficient conditions on the eigenvalues of Σ for $S_{\succ}(\mathcal{A},\lambda(\mathcal{A}))$ to be non-rational).

3 Length distributions of a rational language

This section is centered around the two questions of D. Niwinski that we presented in the introduction of this paper. We denote by A a finite alphabet.

3.1 Length distributions

Let us first notice that every transition matrix of a complete deterministic automaton is stochastic and consequently every convex combination of these matrices is stochastic. This gives sense to the following definition.

DEFINITION 3.1 : Let $\mathcal{A} = (I,\mu,T)$ be a complete deterministic automaton over A. We associate with $\mathcal{A}$ the following stochastic matrix :

$$\Sigma(\mathcal{A}) = \frac{1}{|A|} \sum_{a\in A} \mu(a)$$

Let L be a language over A. For every $n \in \mathbb{N}$, let $F(n,L) = |L \cap A^n|/|A|^n$ be the frequency of the words of L of length n among all words of length n.

PROPOSITION 3.1 : Let $\mathcal{A} = (I, \mu, T)$ be a complete deterministic automaton over A, let L be the language recognized by $\mathcal{A}$ and let $\Sigma = \Sigma(\mathcal{A})$ be its associated stochastic matrix. Then we have for all $n \in N$:

$$I \cdot \Sigma^n \cdot T = \frac{|A^n \cap L|}{|A|^n} = F(n, L).$$

Proof : Since μ is a representation, we get that for all $n \in \mathbb{N}$,

$$\Sigma^n = \frac{1}{|A|^n} \sum_{w \in A^n} \mu(w).$$

But, for every word $w \in A^*$, the integer $I \cdot \mu(w) \cdot T$ is equal to 1 or to 0 according as w belongs or not to L. Hence we have

$$I \cdot \Sigma^n \cdot T = \frac{1}{|A|^n} \sum_{w \in A^n} I \cdot \mu(w) \cdot T = \frac{|L \cap A^n|}{|A|^n} = F(n, L)$$

which proves the proposition. ∎

Proposition 3.1 allows to link the study of the length distribution of a rational language with stochastic automata theory. In particular, we can give the following corollary of proposition 3.1 and of the results of section 2.

COROLLARY 3.2 : Let L be a rational language, let $\mathcal{A}$ be any complete deterministic automaton recognizing L which has a unique initial state and let Σ be its associated stochastic matrix. Then with the possible exception of a finite family of real numbers λ in $[0, 1]$, the following subset of $\mathbb{N}$ is always rational :

$$S_{\succ}(L, \lambda) = \{\, n \in \mathbb{N} \mid |A^n \cap L| \succ \lambda |A|^n \,\}$$

Moreover there are no exceptions when every eigenvalue of Σ is rational in argument.

Remark : As we will see in the sequel, there are examples of rational languages such that $S_{\succ}(L, \lambda)$ is non-rational for some λ. However there are also classes of rational languages for which $S_{\succ}(L, \lambda)$ is always rational. For instance, the class of factorial languages is such a class (a language is called *factorial* iff all factors of elements of L belong to L). Indeed, for such a language L, the inequality $|L \cap A^{n+m}| \leq |L \cap A^n| \, |L \cap A^m|$ holds for every $n, m \in \mathbb{N}$ and thus we get for every $n, m \in \mathbb{N}$:

$$F(n + m, L) \leq F(n, L)\, F(m, L) \qquad (1)$$

Let us now suppose that $L \neq A^*$. Then there is some integer $N \geq 1$ such that $F(N, L) < 1$. Therefore it follows from relation (1) that we have for every $n \in \mathbb{N}$

$$F(n, L) \leq F(N, L)^{[n/N]}$$

and thus $\lim_{n \to +\infty} F(n, L) = 0$. From this, we can conclude that $S_{\succ}(L, \lambda)$ is either finite or equal to $\mathbb{N}$ and hence is rational.

Note : Niwinski's question ($Q1$) is equivalent to asking if the one-letter language $S_{>}(L, 1/2)$ is rational. Therefore question ($Q1$) has a positive answer whenever 1/2 is not one the possible finite exceptions mentioned in corollary 3.2. Hence question ($Q1$) has in a certain sense almost always a positive answer.

3.2 A counterexample to Niwinski's question $(Q1)$

Here is an explicit example of a three-state automaton $\mathcal{A}$ over a two-letter alphabet such that $S_>(\mathcal{A}, 1/2)$ is not rational (and hence if L is the language recognized by this automaton, the one-letter language $S_>(L, 1/2)$ is not rational). Note that 1 is the unique initial and final state in this automaton.

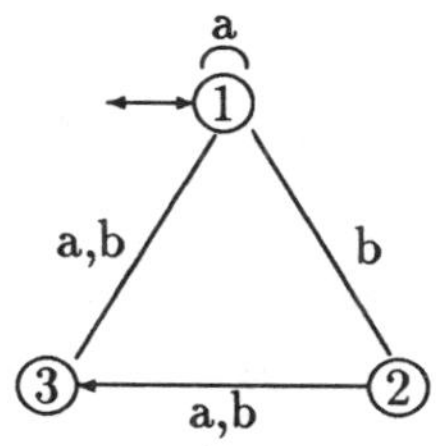

Its associated stochastic matrix Σ is equal to

$$\Sigma = \begin{pmatrix} 1/2 & 1/2 & 0 \\ 0 & 0 & 1 \\ 1 & 0 & 0 \end{pmatrix}$$

The eigenvalues of Σ are 1, $\lambda = (-1+\sqrt{7}i)/4$ and $\overline{\lambda} = (-1-\sqrt{7}i)/4$ or equivalently 1, $\lambda = 1/\sqrt{2}\, e^{i\theta}$ and $\overline{\lambda} = 1/\sqrt{2}\, e^{-i\theta}$ where θ is defined by :

$$\cos(\theta) = -\frac{1}{2\sqrt{2}} \qquad \text{with } \theta \in \,]\pi/2, \pi[$$

Note first that θ is not rational in argument since if it were the case, 2θ would also be rational in argument. But this is not possible since $\cos(2\theta) = -3/4$ and since the only rational values of the cosinus of a rational in argument angle are 0, ± 1, $\pm 1/2$ (see [10]). By formula (RS), there exist α, β and γ such that, for all $n \in \mathbb{N}$

$$I \cdot \Sigma^n \cdot T = \alpha + \beta\lambda^n + \gamma\overline{\lambda}^n$$

An easy computation shows that $\alpha = 1/2$, $\beta = \overline{\gamma} = (\sqrt{7} - i)/4\sqrt{7}$ and after some computation we get

$$I \cdot \Sigma^n \cdot T = 1/2 - \frac{1}{\sqrt{7}(\sqrt{2})^{n-1}} \sin((n-1)\theta)$$

Hence we have :

$$S_>(\mathcal{A}, 1/2) = \{\, n \in \mathbb{N} \mid \sin((n-1)\theta) > 0 \,\}$$

Since θ is not rational in argument, the set $\{\sin((n-1)\theta),\ n \in \mathbb{N}\}$ is dense in $[-1, 1]$ and therefore $S_>(\mathcal{A}, 1/2)$ is infinite. Hence, if $S_>(\mathcal{A}, 1/2)$ was rational, there would exist a periodic sequence of the form $(a + nk\theta)_{n \in \mathbb{N}}$ such that $\sin(a + nk\theta) > 0$ for all $n \in \mathbb{N}$. But, using again the fact that θ is not rational in argument, it follows that the set $\{\sin(a + nk\theta),\ n \in \mathbb{N}\}$ is dense in $[-1, 1]$ and we get therefore a contradiction.

Remark : An exhaustive enumeration allows to show that the preceding automaton is the unique three-state automaton (up to an isomorphism) whose associated stochastic matrix has a non-rational in argument eigenvalue. Hence it is also the only possible counterexample to Niwinski's question with three states.

3.3 Word distributions

Let f be a function from $\mathbb{N}$ into $\mathbb{N}$, let L be a language and let $\lambda \in [0,1]$. Then let us define the language

$$WS_{\succ}(L,f,\lambda) = \{\, w \in A^* \mid F(f(|w|), w^{-1}L) \succ \lambda \,\}.$$

The original question of D. Niwinski concerned the rationality of $WS_{>}(L,f,1/2)$ when f is a polynomial of $\mathbb{Z}[x]$ which maps $\mathbb{N}$ into $\mathbb{N}$ and L is a rational language over $\{a,b\}^*$. We will show here how to answer to a generalization of this problem. For this purpose, let us introduce the following definition.

DEFINITION 3.2 : A function f from $\mathbb{N}$ into $\mathbb{N}$ is said to be a *good* function iff the inverse image of every ultimately periodic subset of $\mathbb{N}$ is ultimately periodic.

Note : Since an ultimately subset of $\mathbb{N}$ is a finite union of sets of the form $k + K.\mathbb{N}$, a function f from $\mathbb{N}$ into $\mathbb{N}$ is good iff $f^{-1}(k+K.\mathbb{N})$ is ultimately periodic for every $k, K \in \mathbb{N}$.

In particular, we can state :

PROPOSITION 3.3 : Every polynomial function $p \in \mathbb{Z}[x]$ which maps $\mathbb{N}$ into $\mathbb{N}$ is a good function.

Proof : Let $p(x)$ be a polynomial in $\mathbb{Z}[x]$ which maps $\mathbb{N}$ into $\mathbb{N}$ and let $k, K \in \mathbb{N}$. Let us now consider $n \in p^{-1}(k+K.\mathbb{N})$, i.e. $n \in \mathbb{N}$ such that $p(n) \in k+K\mathbb{N}$. But a straightforward computation shows that $p(n+K) = p(n)+K.m$ for some integer $m \in \mathbb{N}$. Hence we proved that $n \in p^{-1}(k+K.\mathbb{N}) \Longrightarrow n+K \in p^{-1}(k+K.\mathbb{N})$. It follows from this last property that $p^{-1}(k+K.\mathbb{N})$ is ultimately periodic and therefore that p is a good function. ■

Notes : 1) It can also be shown that every exponential function $x \longrightarrow a^x$ with $a \in \mathbb{N}$ is a good function.

2) The image by a good function of an ultimately periodic subset of $\mathbb{N}$ is not ultimately periodic in general. For instance the set $\{\, n^2,\ n \in \mathbb{N} \,\}$ is not ultimately periodic (cf [3]).

Let L be a language over A, let λ be a real number in $[0,1]$ and let $\succ$ denote either $>$ or $\geq$. Then we can introduce the two following sets :

$$S_{\succ}(L,f,\lambda) = \{\, n \in \mathbb{N} \mid F(f(n), L) \succ \lambda \,\}$$
$$VS_{\succ}(L,f,\lambda) = \{\, w \in A^* \mid F(f(|w|), L) \succ \lambda \,\}$$

Let us denote by l the monoid morphism from A^* into $\mathbb{N}$ which associates with every word $w \in A^*$ its length $l(w) = |w|$. Then we have the following proposition which gives us the first link between length and word distributions :

PROPOSITION 3.4 : Let T be a rational language over A, let L be a language over A, let f be a good function, let λ be in $[0,1]$. Then the rationality of the two following languages is equivalent :

$$T \cap VS_{\succ}(L,f,\lambda) \in \mathrm{Rat}(A^*) \quad \Longleftrightarrow \quad l(T) \cap S_{\succ}(L,f,\lambda) \in \mathrm{Rat}(\mathbb{N})$$

Proof : Since $l^{-1}(S_{\succ}(L,f,\lambda)) = VS_{\succ}(L,f,\lambda)$ and $l(VS_{\succ}(L,f,\lambda)) = S_{\succ}(L,f,\lambda)$, we have the two following relations :

$$l\big(T \cap VS_{\succ}(L,f,\lambda)\big) = l(T) \cap l(VS_{\succ}(L,f,\lambda)) = l(T) \cap S_{\succ}(L,f,\lambda)$$

$$T \cap l^{-1}(l(T) \cap S_{\succ}(L,f,\lambda)) = T \cap VS_{\succ}(L,f,\lambda)$$

The proposition follows now from the fact that rational languages are closed under morphisms, inverse morphisms and intersection. ■

LEMMA 3.5 : Let L be a rational language over A and let $w_0 \in A^*$. Then the following language over A is rational :

$$T(L,w_0) = \{\, w \in A^*,\ w^{-1}L = w_0^{-1}L \,\}$$

Proof : $T(L,w_0)$ is rational since it is exactly the language that consists of the words going from the state L to the state $w_0^{-1}L$ in the minimal automaton of L. ■

Let us finally give the following result that gives a necessary and sufficient condition for a language $WS_{\succ}(L,f,L)$ to be rational :

PROPOSITION 3.6 : Let L be a rational language over A, let $(w_k^{-1}L)_{k=1,r}$ be the distinct left quotients of L, let λ be in $[0,1]$, let f be a good function and let $\succ$ stands either for $>$ or for $\geq$. Then the two following assertions are equivalent :

1) $WS_{\succ}(L,f,\lambda) \in \mathrm{Rat}(A^*)$
2) $\forall\, k \in \{1,\ldots,r\},\ S_{\succ}(w_k^{-1}L,f,\lambda) \cap l(T(L,w_k)) \in \mathrm{Rat}(\mathbf{N})$

Proof : Since $(T(w_k,L))_{k=1,r}$ is a partition of A^*, we get the relation :

$$WS_{\succ}(L,f,\lambda) = \bigcup_{k=1}^{r} \big(T(L,w_k) \cap VS_{\succ}(w_k^{-1}L,f,\lambda)\big)$$

But since $(T(w_k,L))_{k=1,r}$ is a partition of A^* that consists of rational languages according to lemma 3.5, it follows that $WS_{\succ}(L,f,\lambda)$ is rational iff $T(w_k,L) \cap VS_{\succ}(w_k^{-1}L,f,\lambda)$ is a rational language for every $k \in \{1,\ldots,r\}$. Our proposition follows now from proposition 3.4 and lemma 3.5. ■

Note : Since $l(L)$ is a rational subset of $\mathbf{N}$ for every rational language L, it follows from lemma 3.5, proposition 3.6 and from the results of section 3.1 that $WS_{\succ}(L,f,\lambda)$ is always rational at the possible exception of a finite number of λ when f is a good function. In particular, this shows that question $(Q2)$ has almost always a positive answer.

Example : Let us consider again the rational language L recognized by the automaton given in the example of section 3.1. Then we have :

$$S_{>}(L,x,1/2) \cap l(T(L,1)) = S_{>}(L,x,1/2) \notin \mathrm{Rat}(\mathbf{N})$$

Using the above proposition, we get that $WS_{>}(L,x,1/2)$ is not a rational language. Hence we obtain a counterexample to question $(Q2)$. Moreover it can be proved that $WS_{>}(L,p,1/2)$ is here never rational for every polynomial $p \in \mathbf{Z}[x]$ which maps $\mathbf{N}$ into $\mathbf{N}$.

References

[1] Allender E.W., Wagner K.W., *Counting hierarchies : polynomial time and constant depth circuits,* Bull. of the EATCS, **40**, pp. 182-194, 1990

[2] Berstel J., Reutenauer C., *Rational series and their languages,* Springer Verlag, 1988

[3] Eilenberg S., *Automata, Languages and Machines,* Vol. 1, Academic Press, 1974

[4] Feller W., *An introduction to probability theory and its applications,* Vol. 1, Wiley, 1968

[5] Lang S., *Algebra,* Addison-Wesley, 1965

[6] Niwinski D., [in "Logics and recognizable sets", Report on the Workshop ESPRIT - ASMICS N° 3166, Dersau, Germany, October 8-10, 1990], p. 24, 1990

[7] Rabin M.O., *Probabilistic automata,* Inf. and Control, **6**, pp. 230-245, 1963

[8] Salomaa A., Soittola M., *Automata-theoretic aspects of formal power series,* Springer Verlag, 1978

[9] Torán J., *An oracle characterization of the counting hierarchy,* [in "Proc. 3rd Annual Conference on Structure in Complexity Theory"], pp. 213-223, IEEE Computer Soc., Washington, 1988

[10] Turakainen P., *On probabilistic automata and their generalizations,* Ann. Acad. Scient. Fennicae, Series A, **429**, pp. 7-53, Helsinki, 1969

Towards a Categorical Semantics of Type Classes

Barney P. Hilken and David E. Rydeheard *

Abstract. This is an exercise in the description of programming languages as indexed categories. Type classes have been introduced into functional programming languages to provide a uniform framework for 'overloading'. We establish a correspondence between *type classes* and *comprehension schemata* in categories. Coherence results allow us to describe subclasses and implicit conversions between types.

In programming, there is a temptation to classify types by their attributes. Some versions of ML distinguish equality types (types with a given equality on their values) from other types. Such classifications prevent the proliferation of arguments which should properly be derived from types. The mechanism for this is 'overloading', in which a variety of functions is given a single name and types resolve the ambiguity. A systematic treatment which allows the definition of type classes and function overloading has become available in languages such as Haskell [Hudak et. al. 88]. Type classes should be distinguished from program 'modules' as present in, for example, ML.

The aim of this paper is to give a category-theoretic account of languages with type classes. We start with the observation (see, for instance, [Pitts 87], [Seely 87] and [Hyland, Pitts 89]) that languages with terms, types and kinds correspond to indexed categories. Kinds are objects in the base category, types are objects in the fibres and terms are arrows in the fibres. Arrows in the base are built from type constructors. The relationship between types and type classes we describe as a comprehension schema in the indexed category. Comprehension schemata, introduced by Lawvere [1970], capture the universality of set comprehension, whereby subsets are defined through predicates as $x \in \{a : A|\varphi\}$ iff $x \in A \wedge \varphi(x)$. In type theory, type classes are Σ-types of the form of a kind-sum of types whose result is a kind (using Barendregt's [1989] notation the sorting is $(\Box, *, \Box)$). The universality of this 'sum' defines the comprehension schema.

In addition to comprehension, an account of type classes should admit a notion of 'inclusion' or 'subsumption'. Languages with type inclusion (or subtyping) are well established. Here we elevate the inclusion to the level of kinds so as to allow subclasses. This is implicit in the following 'class elimination' rule:

$$\frac{\Gamma; \Theta \vdash S : \{a : K | o_1 : T_1, \ldots, o_n : T_n\}}{\Gamma; \Theta \vdash S : K}$$

Such rules separate *form* and *formation* so that the structure of a sequent no longer determines its derivation. Coherence is the means by which we restore the link between sequents and their derivations [Mac Lane 82]. In this paper, we consider an equational theory of indexed

*This is an extended abstract. A full paper is available from the authors at the address: Department of Computer Science, University of Manchester, Oxford Road, Manchester M13 9PL, U.K.

categories with comprehension schemata and present a decision procedure based on conversion to canonical forms. An interesting aspect of the development is that the language of type classes we present corresponds to the canonical forms of this equational theory. The correspondence arises through *type inference* in that a language with subsumption is converted into a language with explicit conversions between types. We define a semantics of type classes, using the coherence to ensure that the semantics of derivable sequents is defined and independent of their derivation.

The language defined in this paper admits type classes, including classes of multiple types (n-ary type classes), and an implicit conversion between types, from subclasses to superclasses. The language is extensible in the sense that it is defined in terms of a family of function constants and type constructors. The constants may take type arguments so that there is a form of explicit (parametric) polymorphism. The implicit polymorphism of languages like ML and Haskell introduces further coherence requirements which interact with those of type classes. The requirement for coherence means that modularity becomes compromised: subtypes and subkinds cannot be treated simply as additional features of a language and incorporated in a strictly modular fashion. The approach here admits extensibility; for instance we could include function types via cartesian closure and this would not interfere with coherence. However, additional features which themselves incorporate subsumption lead us, in general, into the difficult problem of combining coherence results. The Haskell group have addressed this problem with syntactic characterisations of well-formed contexts to prevent ambiguity arising. A further restriction in the language is that variables occur at most once in contexts, so that natural contexts such as $a : Eq, a : Print$ (declaring a type a to support an equality function and a print function) are not allowed. This somewhat simplifies the treatment of the language but, we believe, is not essential to our treatment.

Categorical semantics enables us to isolate several layers of structure in languages (see for example [Jacobs, Moggi, Streicher 91] and [Moggi 91]). The category of basic structures (e.g. categories, indexed categories, fibrations etc) required to model the language captures the context structure and general interdependency of the levels of the language (called the 'setting' by Jacobs). Within the language, constructs such as type and term constructors ('features') determine a monad on this category. Finally, a specific language is an algebra of this monad. As an illustration, consider finite products. These have two roles in modelling languages; firstly, as part of the setting, they model contexts of more than one variable; secondly, where present, they model product types (where they are part of the features and therefore described as a monad). If we consider evaluation as part of the structure of a language then, because we evaluate terms but not contexts, these two products are different; for instance those modelling product types may be lax. A similar distinction can be drawn between two roles for comprehension schemata in describing languages. Ehrhard [1988] and Jacobs [1990], use comprehension schemata (as 'settings' to capture the structure of languages with dependent types. Here, we consider comprehension schemata as modelling type classes (a 'feature'). Other work related to that of this paper is that of Curien and Ghelli [Curien 90], [Curien, Ghelli 90], especially that on the coherence of subsumption. Nipkow and Snelting use order-sorted unification for type inference in languages with type classes [Nipkow, Snelting 90].

Comprehension

For indexed category $p : \mathcal{C}^{op} \to \mathbf{Cat}$, $\mathcal{C}$ is the *base category* and $p(K)$ the *fibre* over K. For arrow s in the base, denote the functor $p(s)$ by p_s, or, where there is only one indexed category under consideration, by s^*. Composition of arrows in categories is written as ';' in diagrammatic order.

Definition 1 Let $p : \mathcal{C}^{op} \to \mathbf{Cat}$ be an indexed category such that each fibre has a terminal object $\top$ preserved under the functors s^*. A **comprehension schema** in p consists of a functor, for each object L of $\mathcal{C}$, $\{L|\_\} : p(L) \to \mathcal{C}/L$ and a family of bijections

$$\begin{aligned} \theta_{s,\varphi} : p(K)(\top, s^*(\varphi)) &\simeq (\mathcal{C}/L)(s, \{L|\varphi\}) \\ \text{natural in } s &: K \to L \\ \text{and in } \varphi &\in p(L) \end{aligned}$$

Terminology and notation: $\mathcal{C}/L$ is the slice category of arrows into L. Write $\hat{\varphi} : \{L|\varphi\} \to L$ for the image of φ under the functor $\{L|\_\}$ and call such arrows *conversions*. The inverse of θ is denoted $\bar{\theta}$. Lawvere [1970] defines $\{L|\_\} : p(L) \to \mathcal{C}/L$ as right adjoint to the functor $\Sigma\_(1) : \mathcal{C}/L \to p(L)$. The above definition is equivalent, but does not depend on the existence of Σ_s, the left adjoint of s^*. Comprehension schemata in an indexed category are unique to within an isomorphism. For examples of comprehension schema, see [Lawvere 70].

The naturality conditions in the definition are given explicitly as follows. Naturality in s is, for $h : s' \to s$ in $\mathcal{C}/L$:

$$h; \theta_{s,\varphi}(\alpha) = \theta_{s',\varphi}(h^* \alpha) \tag{1}$$

Naturality in φ is the following, for any $\alpha : \varphi \to \psi$ in $p(L)$:

$$\theta_{s,\varphi}(\beta); \{L|\alpha\} = \theta_{s,\psi}(\beta; s^*(\alpha)) \tag{2}$$

Naturality of $\bar{\theta}$ follows from that of θ.

The treatment of languages as indexed categories depends upon a connection between certain arrows in the base and objects in the fibres. We set out the required connection in the following definition.

Definition 2 Let $p : \mathcal{C}^{op} \to \mathbf{Cat}$ be an indexed category. A **generic object** is an object Ω of $\mathcal{C}$ with functions $\mu_K : |\mathcal{C}(K, \Omega)| \to |p(K)|$ natural in K. Let ω be $\mu_\Omega(\mathrm{id}_\Omega)$. Then, by naturality, $\mu_K(t : K \to \Omega) = t^*(\omega)$. This provides a definition of μ in terms of Ω and ω.

We introduce the following notation $\delta_\varphi = \bar{\theta}_{\hat{\varphi},\varphi}(\mathrm{id}) : \top \to \hat{\varphi}^*(\varphi)$. Naturality allows the following definition of $\bar{\theta}$: $\bar{\theta}_{s,\varphi}(h) = h^* \delta_\varphi$. Moreover, from (2) we may define $\{L|\alpha\}$ as: $\{L|\alpha : \varphi \to \psi\} = \theta_{\hat{\varphi},\psi}(\delta_\varphi; \hat{\varphi}^*(\alpha))$. Using these definitions, the following set of three equations is equivalent to the above definition (Definition 1) of comprehension schemata.

$$\begin{aligned} \theta_{s,\varphi}(\alpha); \hat{\varphi} &= s \\ (\theta_{s,\varphi}(\alpha))^* \delta_\varphi &= \alpha \\ \theta_{s;\hat{\varphi},\varphi}(s^* \delta_\varphi) &= s \end{aligned} \tag{3}$$

The first two equations are β-rules; the third is an η-rule.

Let Σ be a family of constants of the following forms: Constant arrows in the base of the form $k_L : L \to \Omega$ and constant arrows in the fibres of the form $f_{K,\varphi} : \varphi \to s^* \omega$ in the fibre over K. Let $\mathbf{Comp}_\Sigma$ (or **Comp** where Σ is understood) be the language of (strict) indexed categories with:

- A comprehension schema defined by (3).

- Finite products in the base category and fibres, with those in the fibres strictly preserved by the functors s^*. The terminal object in the base is denoted 1 with terminal arrows $1 : K \to 1$ and the binary product in the base is $\times$. In each fibre, the terminal object is denoted $\top$ with terminal arrows $\top : \varphi \to \top$ and the binary product is $\wedge$. Product pairing is $\langle\, ,\, \rangle$ and projections are π_i, $(i = 1, 2)$.
- A generic object Ω (with object ω in the fibre over Ω).
- The constants of Σ.

For this treatment of languages, *strict* structures (including indexed categories) suffice. More generally, we may consider fibrations and related coherence problems (see [Mac Lane, Paré 85]).

We now introduce a decision procedure for **Comp**. A simple application of rewriting techniques to provide normal or canonical forms will not suffice because of the interplay of β-rules and η-rules. If the η-rules are oriented to be expansive then the system is not strongly normalising; if they are oriented to be reductions then confluence is difficult to ensure, even with critical-pair completion. Huet [1976] recognised this problem for λ-calculi and introduced so-called 'long $\beta\eta$' normal forms (see also [Jay 91]). We adapt this technique to define canonical forms for the language **Comp**.

We begin with a result corresponding to a form of 'cut-elimination'. There are three relevant 'cut' rules in indexed categories: Composition of arrows in the base, composition of arrows in the fibres and applications $s^*(\alpha)$ of base arrows to fibre arrows. The results of this conversion we call 'intermediate forms'.

Lemma (Intermediate Forms)

Every object and arrow expression e in **Comp** is equal (in **Comp**) to a unique expression $\mathsf{E}(e)$ of the following grammar (of 'intermediate' forms):

$$\begin{array}{rcl} K & ::= & 1 \mid K_1 \times K_2 \mid \{K|\varphi\} \mid \Omega \\ \varphi & ::= & \top \mid \varphi_1 \wedge \varphi_2 \mid s^*\omega \\ s & ::= & \mathrm{id} \mid 1 \mid \langle s_1, s_2\rangle \mid s;\pi_i \mid \theta_{s,\varphi}(\alpha) \mid s;\hat{\varphi} \mid s;k_K \\ \alpha & ::= & \mathrm{id} \mid \top \mid \langle \alpha_1, \alpha_2\rangle \mid \alpha;\pi_i \mid \top; s^*\delta_\varphi \mid \alpha; s^* f_{K,\varphi} \end{array}$$

Indeed, there is an effective procedure for converting expressions into intermediate form (see the full paper for details).

We now introduce a conversion of expressions e in intermediate form to $\mathsf{C}(e)$ also in intermediate form. We call $\mathsf{C}(e)$ the canonical form of e. The conversion consists of repeated applications of β-rules and single applications of η-rules depending upon the codomain type of arrows. This conversion is defined on objects and arrows for the β conversions, and on pairs $s : L$ ('typed arrows') of arrows with their codomain for η-conversions. The result of the η-conversion is that arrows with product codomain become pairs and arrows with comprehension codomain become θ-forms.

Lemma (Canonical Forms)

The following defines a unique ('canonical') form $\mathsf{C}(e)$ for each object and arrow expression e in intermediate form such that $e = \mathsf{C}(e)$ in **Comp**. Moreover, $\mathsf{C}(e)$ is also in intermediate form.

$$\begin{aligned}
&\mathsf{C}(s:1) = 1\\
&\mathsf{C}(s:K\times L) = \langle \mathsf{C}(s;\pi_1:K), \mathsf{C}(s;\pi_2:L)\rangle\\
&\mathsf{C}(s:\{K|\varphi\}) = \theta_{t,\mathsf{B}(\varphi)}(\mathsf{C}(\top;s^*\delta_\varphi:\mathsf{E}(t^*(\varphi)))) \text{ where } t = \mathsf{C}(s;\hat{\varphi}:K)\\
&\mathsf{C}(s:\Omega) = \mathsf{B}(s)\\
&\mathsf{C}(\alpha:\top) = \top\\
&\mathsf{C}(\alpha:\varphi\wedge\psi) = \langle \mathsf{C}(\alpha;\pi_1:\varphi), \mathsf{C}(\alpha;\pi_2:\psi)\rangle\\
&\mathsf{C}(\alpha:s^*\omega) = \mathsf{B}(\alpha)
\end{aligned}$$

The function B is repeated β-conversion on arrows in the base:

$$\begin{aligned}
&\mathsf{B}(\mathrm{id}) = \mathrm{id}\\
&\mathsf{B}(1) = 1\\
&\mathsf{B}(\langle s,t\rangle) = \langle \mathsf{B}(s), \mathsf{B}(t)\rangle\\
&\mathsf{B}(s;\pi_i) = \text{ if } \mathsf{B}(s) = \langle s_1, s_2\rangle \text{ then } s_i \text{ else } \mathsf{B}(s);\pi_i\\
&\mathsf{B}(\theta_{s,\varphi}(\alpha)) = \theta_{\mathsf{B}(s),\mathsf{B}(\varphi)}(\mathsf{B}(\alpha))\\
&\mathsf{B}(s;\hat{\varphi}) = \text{ if } \mathsf{B}(s) = \theta_{t,\psi}(\alpha) \text{ then } t \text{ else } \mathsf{B}(s);\mathsf{B}(\hat{\varphi})\\
&\mathsf{B}(s;k_K) = \mathsf{C}(s:K);k_{\mathsf{B}(K)}
\end{aligned}$$

Similarly for arrows in the fibres, together with the clauses:

$$\begin{aligned}
&\mathsf{B}(\top;s^*\delta_\varphi) = \text{ if } \mathsf{B}(s) = \theta_{t,\psi}(\alpha) \text{ then } \mathsf{E}(\top;\alpha) \text{ else } \top;\mathsf{B}(s)^*\delta_{\mathsf{B}(\varphi)}\\
&\mathsf{B}(\alpha;s^*f_{K,\varphi}) = \mathsf{C}(\alpha:\mathsf{E}(s^*\varphi));(\mathsf{C}(s:K))^*f_{\mathsf{B}(K),\mathsf{B}(\varphi)}
\end{aligned}$$

Finally, B is defined on objects as follows:

$$\begin{aligned}
&\mathsf{B}(1) = 1 && \mathsf{B}(\top) = \top\\
&\mathsf{B}(K\times L) = \mathsf{B}(K)\times\mathsf{B}(L) && \mathsf{B}(\varphi\wedge\psi) = \mathsf{B}(\varphi)\wedge\mathsf{B}(\psi)\\
&\mathsf{B}(\{K|\varphi\}) = \{\mathsf{B}(K)|\mathsf{B}(\varphi)\} && \mathsf{B}(s^*(\omega)) = \mathsf{B}(s)^*(\omega)\\
&\mathsf{B}(\Omega) = \Omega
\end{aligned}$$

The presence of E in the above definition ensures that constructed expressions are in intermediate form. The non-trivial part of the proof lies in showing that C is well-defined, i.e. that the above definition terminates as a rewrite system.

These conversions provide a decision procedure for **Comp** in the following sense.

Proposition (Decidability)

Arrow expressions (in either the base category or in the fibres) $h_1 : A_1 \to B_1$ and $h_2 : A_2 \to B_2$ are equal in **Comp** iff $\mathsf{B}(\mathsf{E}(A_1)) \equiv \mathsf{B}(\mathsf{E}(A_2))$ and $\mathsf{C}(\mathsf{E}(h_1) : \mathsf{E}(B_1)) \equiv \mathsf{C}(\mathsf{E}(h_2) : \mathsf{E}(B_2))$, where $\equiv$ is syntactic equality.

A language of type classes

The language **Comp** is a categorical language of combinators satisfying equational constraints. We now introduce a language, which we call **Class**, with variables and contexts to describe type classes. The relationship between **Class** and **Comp**, which we set out in the next section, involves the canonical forms and decision procedure above. The language **Class** is carefully constructed: As usual, the products and projections of **Comp** are replaced by contexts and their structural rules, so that if Γ is a context and x a variable, then a well-formed sequent $\Gamma \vdash x : A$ corresponds to a unique projection. A similar treatment of type classes and conversions requires that the conversions are implicit (just as projections are implicit) and that 'witnesses' to class membership are accumulated in a context, which we call an *equation context*. Moreover, the language admits n-ary type classes, extensibility with constants (both type constructors and functions for term formation) and a form of explicit (parametric) polymorphism (with type arguments to terms).

Judgements

$\vdash K\,\mathrm{kind}(a)$ means that K is a well-formed kind which matches pattern a (patterns are described below).

$\Gamma;\Theta \vdash T : K$ means that T is a well-formed type expression of kind K, with free type variables declared in context Γ. It implies that Γ, Θ and K are well-formed.

$\Gamma;\Theta;\Delta \vdash E : T$ means that E is a well-formed term of type T, with free type variables declared in Γ and free value variables declared in Δ. It implies that Γ, Θ, Δ and T are well-formed, and that T has kind Ω.

Note that kind judgements have no context and type judgements have no value context. This ensures that kinds cannot have free variables, and types cannot have free value variables.

The following grammar defines the forms of the syntactic classes. The metavariables are x for variables and o for labels; ε is the empty sequence (empty contexts in sequents may be omitted).

$$\begin{array}{rcl}
\text{patterns } a & ::= & x \mid (a_1, \ldots, a_n) \\
\text{kind contexts } \Gamma & ::= & \Gamma, a : K \mid \varepsilon \\
\text{equation contexts } \Theta & ::= & \Theta, o[T] = E \mid \varepsilon \\
\text{type contexts } \Delta & ::= & \Delta, x : T \mid \varepsilon \\
\text{kinds } K & ::= & \Omega \mid K_1 \times \cdots \times K_n \mid \{a : K | o_1 : T_1, \ldots, o_n : T_n\} \\
\text{type expressions } T & ::= & x \mid kT \mid (T_1, \ldots, T_n) \\
\text{terms } E & ::= & x \mid o[T] \mid f[T](E_1, \ldots, E_n)
\end{array}$$

for $n \geq 0$. We denote the empty product by 1 and the empty tuple by (). Note that patterns are a subclass of type expressions.

The type and kind contexts are as expected. The equation contexts need some explanation. These contexts hold expressions for the components of types. The language incorporates implicit conversion between types, so that, to belong to a class, a type does not come equipped with suitable functions, but the functions are retrievable from the context. This is the equation context and holds 'witnesses' to kinding judgements[1]. Patterns are trees of variables and are used, *inter alia*, to describe the structure of type expressions in classes (see the kind formation rules). Strictly speaking, this use of patterns requires them to be identified under α-convertibility (renaming of variables) and could be avoided by introducing unlabelled trees as an additional syntax class.

The families of type constructors k and functions f have kinds and types, respectively, given by: $k : K \to \Omega$ and $a : K \vdash f : S_1 \times \cdots \times S_n \to T$ for well-formed kind K and well-formed types

[1] Equational contexts correspond loosely to what are called (multi)dictionaries in the draft operational semantics of Haskell [Peyton Jones, Wadler 90].

$S_1,\dots,S_n,T$ of kind Ω in context $a : K$. Notice that the type of f is dependent on type a and so f is polymorphic. The restricted form of these types and kinds does not reduce generality, since a type constructor yielding a pair can be considered as a pair of type constructors, and a constructor yielding a type class may be replaced by a constructor of the above form using equation contexts.

We need the concept of '$o[T]$ occurring in $\Gamma;\Theta$', defined by

$o[T]$ occurs in $o[T] = E$

$o_j[a]$ occurs in $a : \{b : K | o_1 : T_1,\dots,o_n : T_n\}$ for $j \leq n$

$o[a]$ occurs in $a : \{b : K | o_1 : T_1,\dots,o_n : T_n\}$ otherwise, iff $o[a]$ occurs in $a : K$

$o[a]$ occurs in $(a_1,\dots,a_n) : K_1 \times \cdots \times K_n$ iff $o[a]$ occurs in $a_j : K_j$ for some $j \leq n$.

Structural rules

The structural rules are weakening, exchange and assumption (projection) for contexts Γ, Θ and Δ. Only weakening and assumption for Θ need display:

$$\frac{\Gamma;\Theta;\Delta \vdash \varphi \quad \Gamma;\Theta \vdash S : K \quad \Gamma;\Theta \vdash E : T}{\Gamma;\Theta, o[S] = E;\Delta \vdash \varphi}$$

provided $o[S]$ does not occur in $\Gamma;\Theta$,

$$\frac{\Gamma;\Theta \vdash S : K \quad \Gamma;\Theta \vdash E : T}{\Gamma;\Theta, o[S] = E \vdash o[S] : T}$$

provided $o[S]$ does not occur in $\Gamma;\Theta$.

Kinds

$$\frac{}{\vdash \Omega\,\mathrm{kind}(x)} \quad \text{Omega}$$

$$\frac{\vdash K_1\,\mathrm{kind}(a_1) \quad \cdots \quad \vdash K_n\,\mathrm{kind}(a_n)}{\vdash K_1 \times \cdots \times K_n\,\mathrm{kind}((a_1,\dots,a_n))} \quad \text{Product formation}$$

where the variables in the a_i's are pairwise disjoint.

$$\frac{a : K \vdash T_1 : \Omega \quad \cdots \quad a : K \vdash T_n : \Omega}{\vdash \{a : K | o_1 : T_1,\dots,o_n : T_n\}\,\mathrm{kind}(a')} \quad \text{Class formation}$$

where a' is α-equivalent to a.

Types

The product rules are standard. We give the introduction and elimination rules for type classes.

$$\frac{\begin{array}{c} a : K \vdash T_1 : \Omega \;\cdots\; a : K \vdash T_n : \Omega \qquad \Gamma;\Theta \vdash S : K \\ \Gamma;\Theta \vdash o_1[S] : T_1[S/a] \;\cdots\; \Gamma;\Theta \vdash o_n[S] : T_n[S/a] \end{array}}{\Gamma;\Theta \vdash S : \{a : K | o_1 : T_1,\dots,o_n : T_n\}} \quad \text{Class intro}$$

$$\frac{\Gamma;\Theta \vdash S : \{a : K | o_1 : T_1,\dots,o_n : T_n\}}{\Gamma;\Theta \vdash S : K} \quad \text{Class elim}$$

$$\frac{\Gamma;\Theta \vdash T : K}{\Gamma;\Theta \vdash kT : \Omega} \quad \text{Type constructor}$$

for $k : K \to \Omega$ a type constructor.

Terms

$$\frac{\Gamma;\Theta\vdash S:\{a:K|o_1:T_1,\ldots,o_n:T_n\}}{\Gamma;\Theta\vdash o_j[S]:T_j[S/a]}\quad\text{Overloaded operator}$$

$$\frac{\Gamma;\Theta;\Delta\vdash E_1:T_1[S/a]\ \cdots\ \Gamma;\Theta;\Delta\vdash E_n:T_n[S/a]\qquad \Gamma;\Theta\vdash S:K}{\Gamma;\Theta;\Delta\vdash f[S](E_1,\ldots,E_n):T[S/a]}\quad\text{Function appln}$$

for $a:K\vdash f:T_1\times\cdots\times T_n\to T$ a constant.

Semantics

We present a semantics of the language **Class**, interpreting sequents as arrows in indexed categories. The definition proceeds by induction on the structure of sequents rather than on the structure of their derivations. The semantics is *partial* – not all sequents correspond to arrows. A sequent is said to be *well-formed* when it has a semantics. Later we show that all derivable sequents have a semantics and that structural induction on derivations is admissible.

Let $p:\mathcal{C}^{\mathrm{op}}\to\mathbf{Cat}$ be an indexed category with a comprehension schema, finite products in the base and fibres (preserved by functors s^*) and with a generic object Ω (and ω in $p(\Omega)$). Constants $k:K\to\Omega$ are interpreted as arrows $k_{[\![K]\!]}:[\![K]\!]\to\Omega$ in the base and constants $a:K\vdash f:S_1\times\cdots\times S_n\to T$ are interpreted as arrows $f_{[\![K]\!],\varphi}:\varphi\to t^*\omega$ where $t=[\![a:K\vdash T]\!]$, $\varphi=s_1{}^*\omega\wedge\cdots\wedge s_n{}^*\omega$ and $s_i=[\![a:K\vdash S_i]\!]$. We extend this to the language **Class** as follows. For brevity, we omit the standard semantic treatment of products to concentrate on type classes.

Kinds

The semantics of kind K for which $\vdash K\,\mathrm{kind}(a)$ is derivable for some pattern a is an object $[\![K]\!]$ of $\mathcal{C}$, defined by induction:

$$\begin{aligned}[\![\Omega]\!] &= \Omega\\ [\![\{a:K|o_1:T_1,\ldots,o_n:T_n\}]\!] &= \{[\![K]\!]|t_1^*\omega\wedge\cdots\wedge t_n^*\omega\}\\ \text{where}\quad t_j &= [\![a:K\vdash T_j:\Omega]\!]:[\![K]\!]\to\Omega\end{aligned}$$

Type expressions

The semantics of kind contexts is defined by:

$$[\![a_1:K_1,\ldots,a_n:K_n]\!]=[\![K_1]\!]\times\cdots\times[\![K_n]\!]$$

and the semantics of type variables is defined by induction on the structure of the context:

$$\begin{aligned}[\![\Gamma,a:K,\Gamma'\vdash x]\!] &= \pi_a;c\text{ if }x\text{ occurs in }a\\ \text{where}\quad\pi_a &: [\![\Gamma,a:K,\Gamma']\!]\to[\![K]\!]\\ &\quad\text{is the projection determined by }a\\ c &= [\![a:K\vdash x]\!]\\ [\![a:\{b:K|\Delta\}\vdash x]\!] &= \hat{\varphi};c\\ \text{where}\quad\hat{\varphi} &: [\![\{b:K|\Delta\}]\!]\to[\![K]\!]\text{ is the relevant conversion}\\ c &= [\![a:K\vdash x]\!]\\ [\![x:\Omega\vdash x]\!] &= \mathrm{id}\end{aligned}$$

The semantics of the derivable sequent $\Gamma;\Theta \vdash T : K$ is an arrow

$$[\![\Gamma;\Theta \vdash T : K]\!] : [\![\Gamma]\!] \to [\![K]\!]$$

in $\mathcal{C}$, defined by induction on the structure of T and K:

$$\begin{array}{rcl}
[\![\Gamma;\Theta \vdash S : \{a : K | o_1 : T_1, \ldots, o_n : T_n\}]\!] & & \\
 & = & \theta_{s,\varphi}(\langle e_1, \ldots, e_n \rangle) : [\![\Gamma]\!] \to \{[\![K]\!] | \varphi\} \\
\text{where} & s & = [\![\Gamma;\Theta \vdash S : K]\!] : [\![\Gamma]\!] \to [\![K]\!] \\
 & \varphi & = t_1^* \omega \wedge \cdots \wedge t_n^* \omega \\
 & t_j & = [\![a : K \vdash T_j : \Omega]\!] : [\![K]\!] \to \Omega \\
 & e_j & = [\![\Gamma;\Theta \vdash o_j(S) : T_j[S/a]]\!] : \top \to s^* \varphi_j \\
[\![\Gamma;\Theta \vdash kT : \Omega]\!] & = & t; [\![k]\!] : [\![\Gamma]\!] \to \Omega \\
\text{where} & k & : K \to \Omega \\
 & t & = [\![\Gamma;\Theta \vdash T : K]\!] : [\![\Gamma]\!] \to [\![K]\!] \\
[\![\Gamma;\Theta \vdash x : \Omega]\!] & = & [\![\Gamma \vdash x]\!]
\end{array}$$

Terms

The semantics of type contexts is defined by

$$\begin{array}{rcl}
[\![\Gamma;\Theta;x_1 : T_1, \ldots, x_n : T_n]\!] & = & t_1^* \omega \wedge \cdots \wedge t_n^* \omega \in p([\![\Gamma]\!]) \\
\text{where} & t_j & = [\![\Gamma;\Theta \vdash T_j : \Omega]\!] : [\![\Gamma]\!] \to \Omega
\end{array}$$

and the semantics of the derivable sequent $\Gamma;\Theta;\Delta \vdash E : T$ is an arrow

$$[\![\Gamma;\Theta;\Delta \vdash E : T]\!] : [\![\Gamma;\Theta;\Delta]\!] \to [\![\Gamma;\Theta \vdash T : \Omega]\!]^* \omega$$

in $p([\![\Gamma]\!])$, defined by induction:

$$\begin{array}{rcl}
[\![\Gamma;\Theta, o[S] = E, \Theta';\Delta \vdash o[S] : T]\!] & = & \top; [\![\Gamma;\Theta,\Theta' \vdash E : T]\!] : \varphi \to t^* \omega \\
\text{where} & t & = [\![\Gamma;\Theta,\Theta' \vdash T : \Omega]\!] \\
 & \varphi & = [\![\Gamma;\Theta, o[S] = E, \Theta';\Delta]\!] \\
[\![\Gamma, b : K, \Gamma';\Theta;\Delta \vdash o[a] : T]\!] & = & \top; \pi_b^* e : \varphi \to t^* \omega \text{ if } o[a] \text{ occurs in } b : K \\
\text{where} & \pi_b & : [\![\Gamma, b : K, \Gamma']\!] \to [\![K]\!] \text{ the unique projection} \\
 & e & = [\![b : K \vdash o[a]]\!] \\
 & t & = [\![\Gamma, b : K, \Gamma';\Theta \vdash T : \Omega]\!] \\
 & \varphi & = [\![\Gamma, b : K, \Gamma';\Theta;\Delta]\!] \\
[\![\Gamma;\Theta;\Delta \vdash f[S]\,(E_1, \ldots, E_n) : T]\!] & = & \langle e_1, \ldots, e_n \rangle; s^* [\![f]\!] : [\![\Gamma;\Delta]\!] \to s^* t^* \omega \text{ in } p([\![\Gamma]\!]) \\
\text{where} & & a : K \vdash f : S_1 \times \cdots \times S_n \to T' \\
 & e_j & = [\![\Gamma;\Theta;\Delta \vdash E_j : S_j[S/a]]\!] \\
 & s & = [\![\Gamma;\Theta \vdash S : K]\!] \\
 & t & = [\![a : K \vdash T' : \Omega]\!] \\
[\![\Gamma;\Theta;x_1 : T_1, \ldots, x_n : T_n \vdash x_j : T_j]\!] & = & \pi_j
\end{array}$$

where the semantics of overloaded operators is defined by induction on the structure of the context:

$$\begin{array}{rcl}
[\![a : \{b : K | o_1 : T_1, \ldots, o_n : T_n\} \vdash o_j[a]]\!] & = & \delta_\varphi; \pi_j : \top \to \hat{\varphi}^* t_j^* \omega \text{ if } j \leq n \\
\text{where} & \varphi & = t_1^* \omega \wedge \cdots \wedge t_n^* \omega \\
 & t_i & = [\![b : K \vdash T_i : \Omega]\!] \\
[\![a : \{b : K | o_1 : T_1, \ldots, o_n : T_n\} \vdash o[a]]\!] & = & \hat{\varphi}^* e \text{ if } o[a] \text{ occurs in } a : K \\
\text{where} & \varphi & = t_1^* \omega \wedge \cdots \wedge t_n^* \omega \\
 & e & = [\![a : K \vdash o[a]]\!] \\
 & t_j & = [\![b : K \vdash T_j : \Omega]\!]
\end{array}$$

Coherence

We now describe the relationship between the language **Class** and its derivations; the semantics; and the categorical language **Comp**.

Proposition

All derivable sequents have a semantics and, conversely, all well-formed sequents (those given a semantics) are derivable.

Proof. The proof uses the intermediate and canonical forms introduced for **Comp**. By inspection of the semantic definition, we have (1) a derivation of a sequent corresponds to an intermediate form for which there is an equal (in **Comp**) derivation in canonical form, and which yields a semantics for the sequent; (2) a well-formed sequent corresponds to a canonical form, which is an intermediate form and corresponds to a derivation of the sequent.

Conclusions

Usually a stronger correspondence between categories and languages is established, for instance the exact correspondence (equivalence of categories) between cartesian closed categories and *λ-theories*. Such a correspondence should be available here through a free construction and extensions of the equality derived from equation contexts. The admissibility of cut rules in **Class** needs examination. As the Haskell designers have pointed out only restricted forms of cut will be admissible. Recall that **Class** has explicit polymorphism with type arguments. For a more realistic programming language, we must examine the interplay of *implicit* ML-style polymorphism and the coherence result of this paper.

Acknowledgements

Fu Yuxi, here at Manchester, pointed out a mistake in an earlier version of the paper which caused us to revise the notion of 'canonical form'. Phillip Wadler has provided useful comments on the relationship between the language of this paper and the programming language Haskell. The research was undertaken under S.E.R.C. funding. Thanks also to the M.F.C.S. reviewers.

References

H. Barendregt (1989) Introduction to Generalised Type Systems. *Proc. Third Italian Conference on Theoretical Computer Science*, Ed. U. Moscati. Montova, November, 1989. World Scientific Publishing, Singapore.

P.-L. Curien (1990) Substitution up to Isomorphism. Research Report LIENS-90-9, Laboratoire d'Informatique de l'Ecole Normale Supérieure, 45 Rue d'Ulm, 75230 Paris.

P.-L. Curien and G. Ghelli (1990) Coherence of Subsumption. Research Report LIENS-90-10, Laboratoire d'Informatique de l'Ecole Normale Supérieure, 45 Rue d'Ulm, 75230 Paris.

Th. Ehrhard (1988) A categorical semantics of constructions. In Proc. *Logic in Computer Science* I.E.E.E. publ. Computer Society Press, Washington.

J.M.E. Hyland and A.M. Pitts (1989) Theory of constructions: categorical semantics and topos-theoretic models. In J.W. Gray and A. Scedrov (editors), *Proc. A.M.S. Conference on Categories in Computer Science and Logic, Boulder, Colorado (1987).* American Mathematical Society.

P. Hudak, P. Wadler, et al. (1988) Report on the Functional Programming Language, Haskell. Draft proposed standard. Preprint, Dept. Computer Science, University of Glasgow, U.K.

G. Huet (1976) Résolutions d'équations dans des langages d'ordre $1,2,\ldots,\omega$. *Thèse d'Etat, Université de Paris VII.*

B. Jacobs (1990) Comprehension Categories and The Semantics of Type Dependency. Preprint, Dept. Comp. Sci. Toernooiveld, 6525 ED Nijmegen, The Netherlands.

B. Jacobs, E. Moggi and T. Streicher (1991) Relating Models of Impredicative Type Theories. Preprint, Dept. Comp. Sci. Toernooiveld, 6525 ED Nijmegen, The Netherlands.

C.B. Jay (1991) Long $\beta\eta$ Normal Forms in Confluent Categories. Preprint: LFCS, Department of Computer Science, University of Edinburgh, The King's Buildings, Mayfield Road, Edinburgh.

J. Lambek and P. Scott (1986) *Introduction to higher order categorical logic.* Cambridge University Press.

F.W. Lawvere (1970) Equality in Hyperdoctrines and the Comprehension Schema as an Adjoint Functor. In *Proc. Symp. in Pure Math., XVII: Applications of Categorical Algebra*, Am. Math. Soc. pp. 1–14.

S. Mac Lane (1982) Why commutative diagrams coincide with equivalent proofs (presented in honour of Nathan Jacobson). In *Contemporary Mathematics 13*, American Mathematical Society. pp 387–401.

S. Mac Lane and R. Paré (1985) Coherence for Bicategories and Indexed Categories. *J. Pure and Applied Algebra*, 37. pp 59–80.

E. Moggi (1991) A category-theoretic account of program modules. Math. Structures in Comp. Sci. 1.1. pp 103–139.

T. Nipkow and G. Snelting (1990) Type Classes and Overloading Resolution via Order-Sorted Unification. Technical Report No. 200, Computer Laboratory, University of Cambridge, Cambridge, U.K.

S.L. Peyton Jones and P. Wadler (1990) A static semantics for Haskell. Preprint, Dept Computer Science, University of Glasgow, U.K.

A.M. Pitts (1987) Polymorphism is Set Theoretic, Constructively. *Proc. Summer Conference on Category Theory and Computer Science*, Edinburgh 1987, LNCS 283.

R.A.G. Seely (1987) Categorical semantics for higher order polymorphic lambda calculus. *J. Symbolic Logic*, 52, pp 969–989

P. Wadler and S. Blott (1989) How to make *ad hoc* polymorphism less *ad hoc*. In *Proceedings of 16th ACM Symposium on Principles of Programming Languages*, A.C.M.

Single-Path Petri Nets

Rodney R. Howell
Dept. of Computing and Information Sciences
Kansas State University
Manhattan, KS 66506
U.S.A.

Petr Jančar
Dept. of Computer Science
Dvořákova 7
University of Ostrava
701 00 Ostrava 1
Czechoslovakia

Louis E. Rosier
Dept. of Computer Sciences
The University of Texas at Austin
Austin, TX 78712
U.S.A.

Abstract

We examine a subclass of persistent Petri nets called single-path Petri nets. Our intention is to consider a class of Petri nets whose study might yield some insight into the mathematical properties of persistent Petri nets or even general Petri nets. We conjecture that the Karp-Miller coverability tree for a persistent net is small enough to be searched in polynomial space. Although we are unable to prove this conjecture, we do show that single-path Petri nets have this property. We then use this fact to show that the canonical analysis problems (i.e., boundedness, reachability, containment, and equivalence) for single-path Petri nets are PSPACE-complete in the strong sense. Furthermore, we show that the problem of recognizing a single-path Petri net is also PSPACE-complete.

1 Introduction

A Petri net is a formalism that has been used extensively to model parallel computations (see, e.g., [Pet81, Rei85]). As is typical regarding automata-theoretic models, the decidability and computational complexity of a number of decision problems involving Petri nets have been studied in order to gain a better understanding of the mathematical properties of the model. These problems include boundedness, reachability, containment, and equivalence. These four problems may be considered to be the canonical problems regarding Petri nets because most other Petri net problems have been shown to be polynomial-time many-one equivalent to one of these (see, e.g., [Pet81]). Lipton [Lip76] and Rackoff [Rac78] have shown exponential space lower and upper bounds, respectively, for the boundedness problem, whereas Rabin [Bak73] and Hack [Hac76] have shown the containment and equivalence problems, respectively, to be undecidable (in particular, both problems are Π_1-complete). On the other hand, the complexity of the reachability problem has remained open for many years. Lipton's lower bound for the boundedness problem [Lip76] also yields an exponential space lower bound for the reachability problem, and Mayr [May84] has shown the problem to be decidable, though his algorithm is not primitive recursive (see also [Kos82, Lam87]). No one has yet succeeded in tightening these bounds. Such a large disparity between the known lower and upper bounds for this problem suggests that the fundamental mathematical properties of Petri nets still are not well understood.

Early efforts to show the reachability problem to be decidable included the study of various restricted subclasses of Petri nets. For many of these subclasses, tight complexity bounds for all four problems listed above have been shown. A notable exception is the class of persistent Petri nets. All four problems regarding persistent Petri nets are PSPACE-hard [JLL77]. As far as known upper bounds are concerned, the boundedness problem can be solved in exponential space [Rac78], and the other three problems are known to be decidable [Gra80, May81, Mul81], though none are known to be primitive recursive. Thus, the disparities in the known upper and lower bounds for all four problems regarding persistent Petri nets are even larger than the corresponding disparities for general Petri nets. Furthermore, even the problem of recognizing a persistent Petri net — a problem that is also PSPACE-hard [JLL77] — is not known to be primitive recursive (cf. [Gra80, May81, Mul81]).

It is instructive to compare the strategies given by E. Mayr in showing the decidability of the reachability problems for both persistent and general Petri nets [May81, May84]. Both algorithms involved the construction of a tree similar to the coverability tree given by Karp and Miller [KM69]. Unfortunately, for general Petri nets, the size of this tree is potentially nonprimitive recursive [MM81, Mul85]. It is not currently known whether a primitive recursive upper bound can be given for the size of the coverability tree for an arbitrary persistent Petri net. If such a bound could be shown, it would be a significant step toward giving a primitive recursive upper bound for the reachability problem for persistent Petri nets. On the other hand, if a primitive recursive algorithm could be given for this problem without showing a primitive recursive upper bound on the size of the coverability tree, this could be a significant step toward finding a primitive recursive algorithm for the general reachability problem. The understanding of persistent Petri nets therefore seems to be crucial to the understanding of general Petri nets. Our conjecture is that persistent Petri nets have relatively small coverability trees, perhaps no more than exponential depth; however, we have not yet been able to prove this conjecture.

Since the problems surrounding persistent Petri nets seem to be almost as difficult to analyze as for general Petri nets, a logical strategy appears to be to narrow the scope of the investigation still further to subclasses of persistent Petri nets. One subclass of persistent Petri nets that is now well-understood is the class of conflict-free Petri nets [CRM75, HR88, HRY87, LR78]. However, the study of conflict-free nets does not seem to yield much insight into the properties of persistent Petri nets. The main reason seems to be that persistent Petri nets are defined solely in terms of their behavior, whereas conflict-free Petri nets may be defined in terms of their structure (cf. [HRY89]). For example, the problem of recognizing a conflict-free Petri net may be solved in polynomial time by a straightforward examination of the structure of the net, whereas recognizing a persistent Petri net seems to require a costly reachability analysis (see [Gra80, May81, Mul81]). Therefore, we consider in this paper a subclass of persistent Petri nets called *single-path Petri nets*, which are defined solely in terms of their behavior. Persistent Petri nets are characterized by the fact that at any reachable marking, the only way to disable an enabled transition is to fire it; thus, conflicts are avoided at all reachable markings. In a conflict-free Petri net, conflicts are avoided at all markings, whether reachable or not. On the other hand, a single-path Petri net avoids conflicts due to the fact that it has only one firing sequence. Single-path Petri nets are therefore a proper subset of persistent Petri nets, but neither contain nor are contained in the class of conflict-free Petri nets. Furthermore, we are able to show that the coverability tree for a single-path Petri net — a tree with exactly one leaf — has at most exponential depth.

In showing our exponential bound on the depth of the coverability tree for single-path Petri nets, we must overcome two hurdles. The first hurdle concerns paths in the coverability tree that terminate upon iterating a loop (i.e., a firing sequence causing a nonnegative gain in all places). Lipton [Lip76] has shown that for general Petri nets, the shortest path containing a loop can be doubly exponential in the size of the Petri net. The second hurdle is potentially more serious: there can exist, in general Petri nets, paths with a nonprimitive recursive length that contain no loops [MM81, Mul85]. In order to overcome these hurdles, we show that if the path in a single-path Petri net exceeds a certain exponential length, it must contain a loop-like structure that we call an r-semi-loop. This structure is such that it must be iterated (possibly forever) until the path terminates; furthermore, if this iteration does not continue forever, it continues for at most an exponential length. Combining this result with a PSPACE-hardness proof given by Jones, Landweber, and Lien [JLL77], we show that the four canonical problems regarding single-path Petri nets and the problem of recognizing a single-path Petri net are all PSPACE-complete in the strong sense; that is, the problems are PSPACE-complete even if all integers in the input are expressed in unary. It is hoped that the techniques given here might be generalized to apply to persistent Petri nets.

The remainder of this paper is organized as follows. In Section 2, we give some preliminary definitions. In Section 3, we derive a bound on the depth of the coverability tree for a single-path Petri net and show the canonical problems to be PSPACE-complete. We then give some concluding remarks in Section 4.

2 Preliminaries

In this section, we introduce some basic terminology and define single-path Petri nets. Let $A \backslash B$ denote the set difference of sets A and B, and let $A \times B$ denote their cartesian product. $|A|$ is the cardinality of a set A, and $f \downarrow A$ is the restriction of a function f to a domain A. N is the set of nonnegative integers, Z the set of all integers. A^* (A^ω, respectively) denotes the set of finite (infinite, respectively) sequences of elements of A; ε is the empty sequence. For $u \in A^*, k \in N, (u)^k$ stands for $uu \dots u$, u being written k times, and $(u)^\omega$ stands for the infinite sequence $uuu \dots$.

A *Petri net* (PN, for short) is a tuple (P, T, φ, μ_0), where P is a finite set of *places*, T is a finite set of *transitions*, $P \cap T = \emptyset$, φ is a *flow function* $\varphi : (P \times T) \cup (T \times P) \longrightarrow N$ and μ_0 is the *initial marking*; a *marking* is a function $\mu : P \longrightarrow N$. $\mathbf{0}$ denotes the zero marking $\mathbf{0} : P \longrightarrow \{0\}$. A transition $t \in T$ is *enabled* at a marking μ if $\mu(p) \geq \varphi(p, t)$ for every $p \in P$; we then write $\mu \xrightarrow{t}$. A transition t may *fire* at a marking μ if t is enabled at μ; we then write $\mu \xrightarrow{t} \mu'$, where $\mu'(p) = \mu(p) - \varphi(p, t) + \varphi(t, p)$ for all $p \in P$. In the obvious way, the definitions can be extended for the case $\mu \xrightarrow{u}$, $\mu \xrightarrow{u} \mu'$, where $u \in T^*$.

A sequence $\sigma = \mu_0 \xrightarrow{t_1} \mu_1 \xrightarrow{t_2} \mu_2 \dots \xrightarrow{t_n} \mu_n$ is a (finite) *path*; $\sigma_T = t_1 t_2 \dots t_n$ is then a *firing sequence*. We extend these notions to infinite paths and firing sequences in the obvious way. A path or a firing sequence is *complete* if it can not be extended; all infinite paths and firing sequences will be considered to be complete. By the *length* of a path we mean the length of the corresponding firing sequence. In describing (a part of) a path, we often write only some "passed through" markings explicitly; e.g. we write $\mu_1 \xrightarrow{u} \mu_2 \xrightarrow{v} \mu_3$ for $u, v \in T^*$. A *segment* of a path may be formed by removing a prefix and/or a suffix from the path. The *effect* of $u \in T^*$ (on markings), denoted by $\Delta(u)$, is defined as follows:

$$\Delta : T^* \longrightarrow Z^P, \text{ where } \begin{array}{l} \Delta(\varepsilon) = \mathbf{0}, \\ \Delta(t)(p) = \varphi(t, p) - \varphi(p, t), \text{ and} \\ \Delta(tu) = \Delta(t) + \Delta(u). \end{array}$$

For a PN $\mathcal{P} = (P, T, \varphi, \mu_0)$, the *reachability set* of $\mathcal{P}$ is the set $R(\mathcal{P}) = \{\mu \mid \mu_0 \xrightarrow{u} \mu$ for some

$u \in T^*\}$. Given a PN $\mathcal{P}$ and a marking μ of $\mathcal{P}$, the *reachability problem* (RP) is to determine whether $\mu \in R(\mathcal{P})$. Given a PN $\mathcal{P}$, the *boundedness problem* (BP) is to determine whether $R(\mathcal{P})$ is finite. Given PNs $\mathcal{P}_1, \mathcal{P}_2$, the *containment problem* (CP) is to determine whether $R(\mathcal{P}_1) \subseteq R(\mathcal{P}_2)$; the *equivalence problem* (EP) is to determine whether $R(\mathcal{P}_1) = R(\mathcal{P}_2)$.

For our purposes, it suffices to define the *size* of a PN $\mathcal{P} = (P, T, \varphi, \mu_0)$ as the maximum of $|P|, |T|$ and $\log_2 m$, where m is the maximum integer in the description of φ and μ_0 ($\log_2 m$ approximates the number of bits needed to write m).

A set M of markings is *linear* if there is a finite set of markings $\mu_i : P \longrightarrow N, 0 \leq i \leq n$, such that

$$M = \{\mu_0 + \sum_{i=1}^{n} c_i \mu_i \mid c_i \in N, 1 \leq i \leq n\}.$$

M is a *semilinear set* (SLS) if it is a finite union of linear sets. A linear set M is *r-bounded* if μ_i can be chosen in such a way that $range(\mu_i) \subseteq \{0, 1, \ldots, r\}$ for every $i, 0 \leq i \leq n$. A semilinear set M is *r-bounded* if it is a finite union of r-bounded linear sets.

Definition 2.1. A *single-path Petri net* (SPPN) $\mathcal{P}$ is a PN with only one complete path (i.e., at most one transition is enabled at every reachable marking). This path is denoted by $\sigma_{\mathcal{P}}$.

3 PSPACE-completeness of the canonical problems

In this section, we show the four canonical problems for SPPNs and the problem of recognizing a SPPN to be PSPACE-complete in the strong sense. The difficult part of the proof is showing membership in PSPACE. We will discuss the solution of this problem after we show the lower bound, which follows from results given by Jones, Landweber, and Lien [JLL77].

Theorem 3.1.

a) For SPPNs, BP, RP, CP and EP are PSPACE-hard in the strong sense.

b) The problem of recognizing a SPPN is PSPACE-hard in the strong sense.

Proof. Jones, Landweber, and Lien [JLL77] have given a simulation of a nondeterministic linear bounded automaton (LBA) by a 1-conservative PN. For the purposes of this proof, it is not necessary to understand what a 1-conservative PN is. What is important is to note that in the given simulation, if the LBA is deterministic, the PN is, in fact, a SPPN. Since the LBA acceptance problem is PSPACE-complete even for deterministic LBAs [Kar72], it is a straightforward matter to use this simulation to show each of the five problems to be PSPACE-hard. Since the construction gives a PN whose size is polynomial in the size of the LBA such that no integer in the description of φ and μ_0 exceeds 1, the problems are PSPACE-hard in the strong sense. □

We will spend the remainder of this section showing the four canonical problems to be in PSPACE. Recall the well-known fact that every infinite sequence of nonnegative vectors $v_1, v_2, \ldots$ has an infinite subsequence $v_{i_1} \leq v_{i_2} \leq \cdots$, where $i_1 < i_2 < \cdots$ (see, e.g., [KM69]), and notice that

$$(\mu \leq \mu' \ \& \ \mu \xrightarrow{t}) \text{ implies } \mu' \xrightarrow{t} .$$

Then it is easy to see that the complete firing sequence for a given SPPN is either finite or of the form $u(v)^\omega$ for the first (i.e., leftmost) nonempty v such that $\Delta(v) \geq 0$. Furthermore, the coverability tree contains exactly one path; this path corresponds to some prefix of the firing sequence uv^2 (see [KM69] for a formal definition of the coverability tree). Roughly said, we will show that the loop v must occur within the first exponentially many moves (firings). This result is significant because there exist infinitely many general PNs in which every path containing a loop is at least doubly exponential in length [Lip76]. Our strategy will be to first generalize the problem of finding a loop; i.e., we will show that if the complete path exceeds a certain exponential length, it must contain a more general type of segment, which we will call an r-semi-loop. If the r-semi-loop is, in fact, a nonnegative loop, we will be done; otherwise, we will show that the complete path must be of a "short" finite length. This result is quite strong, because even in general PNs that are bounded, there can exist finite complete paths of nonprimitive recursive length [MM81, Mul85]. We will then be able to use techniques from [HR88] to show the four canonical problems to be in PSPACE. We now define an r-semi-loop.

Definition 3.2. For a PN (P, T, φ, μ_0), a segment $\rho = \mu' \xrightarrow{v} \mu'', v \neq \varepsilon$, of a path is an *$r$-semi-loop* ($r \in N$) if every $p \in P$ meets at least one of the following conditions:

(1) $\mu'(p) \geq \mu''(p)$ (i.e., $\Delta(v)(p) \leq 0$), or

(2) throughout ρ, the marking on p is never less than r.

The following lemma shows the role of r-semi-loops in SPPNs.

Lemma 3.3. Let $\mathcal{P} = (P, T, \varphi, \mu_0)$ be a SPPN where $range(\varphi) \subseteq \{0, 1, 2, .., r\}$. If $\sigma_{\mathcal{P}}$ is in the form $\sigma_{\mathcal{P}} = \mu_0 \xrightarrow{u} \mu' \xrightarrow{v} \mu'' \xrightarrow{\alpha}$, $\alpha \in T^* \cup T^\omega$, where $\mu' \xrightarrow{v} \mu''$ is an r-semi-loop, then α is a prefix of $(v)^\omega$.

Proof. If α is not a prefix of $(v)^\omega$, then we can write

$$v = v_1 t_1 v_2,\ \alpha = (v)^k v_1 t_2 \alpha' \text{ where } t_1 \neq t_2 \text{ (and } k \geq 0).$$

Thus, $\sigma_{\mathcal{P}} = \mu_0 \xrightarrow{u} \mu' \xrightarrow{v_1} \mu_1 \xrightarrow{t_1 v_2} \mu'' \xrightarrow{(v)^k v_1} \mu_2 \xrightarrow{t_2 \alpha'}$. Because $\mu_1 \xrightarrow{t_1}$, we have $\neg(\mu_1 \xrightarrow{t_2})$, since $\mathcal{P}$ is a SPPN. Hence there is a $p \in P$ such that $\mu_1(p) < \varphi(p, t_2) \leq r$; i.e., $\Delta(v)(p) \leq 0$, due to the definition of an r-semi-loop. Thus we have $\mu_2(p) = \mu_1(p) + (k+1)(\Delta(v)(p)) \leq \mu_1(p) < \varphi(p, t_2)$, which is a contradiction with $\mu_2 \xrightarrow{t_2}$. □

Thus, the complete firing sequence of the SPPN $\mathcal{P}$ in Lemma 3.3 can be constructed by first firing u, then iterating v until the next transition cannot be fired. (Note that this procedure might not terminate.) In particular, if $\Delta(v) \not\geq \mathbf{0}$, the complete path must be finite.

We now show a simple sufficient condition for a segment of a path to be an r-semi-loop. Note that Lemmas 3.4 and 3.7 below apply to general PNs.

Lemma 3.4. If $\mu \xrightarrow{(v)^r} \mu' \xrightarrow{v} \mu'', v \neq \varepsilon, r \in N$, is a segment of a path of a PN then $\mu' \xrightarrow{v} \mu''$ is an r-semi-loop.

Proof. It suffices to consider the places p for which $\Delta(v)(p) > 0$. Suppose $\mu' \xrightarrow{v_1} \mu_1 \xrightarrow{v_2} \mu''$ such that $v = v_1 v_2$ and $\mu_1(p) < r$. Then $\mu \xrightarrow{v_1} \mu_2$, where $\mu_2(p) + r \cdot \Delta(v)(p) = \mu_1(p) < r$, and $\mu_2(p) < r(1 - \Delta(v)(p)) \leq 0$ — a contradiction. □

We now introduce another technical notion.

Definition 3.5. Let a PN $\mathcal{P} = (P, T, \varphi, \mu_0)$ and one of its (finite or infinite) paths $\sigma = \mu_0 \xrightarrow{t_1} \mu_1 \xrightarrow{t_2} \mu_2 \xrightarrow{t_3} \ldots$ be fixed. A subsequence $S = \mu_{i_1}, \mu_{i_2}, \ldots, \mu_{i_m}$ ($m \geq 1$) of the sequence $\mu_0, \mu_1, \mu_2, \ldots$ will be called *P'-constant*, for $P' \subseteq P$, if $\mu_{i_1} \downarrow P' = \mu_{i_2} \downarrow P' = \ldots = \mu_{i_m} \downarrow P'$; S will be called *in-maximal P'-constant* if S is P'-constant and, for every j, $1 \leq j \leq m-1$, there is no k, $i_j < k < i_{j+1}$, such that $\mu_k \downarrow P' = \mu_{i_j} \downarrow P'$ (S can not be extended with an "inner member").

Remark 3.6. It is clear that every P'-constant sequence can be extended with "inner members" yielding an in-maximal P'-constant sequence.

Notice that if we have a segment of a path $\mu_1 \xrightarrow{u} \mu_2$ such that μ_1, μ_2 is a P-constant sequence, where P is the set of places in the PN, u is an r-semi-loop for any r ($\Delta(u) = \mathbf{0}$). In what follows, we will show how to find, in a "long enough" segment of the complete path of a SPPN, P'-constant sequences for successively larger P' until either $P' = P$ or we have the situation described in Lemma

3.4. In either case, we have an r-semi-loop, where r is the largest integer in the description of the flow function φ. We first need the following lemma.

Lemma 3.7. Let $\mathcal{P} = (P, T, \varphi, \mu_0)$ be a PN and σ one of its paths. Let

$$\rho = \mu_1 \xrightarrow{u_1} \mu_2 \xrightarrow{u_2} \mu_3 \ldots \xrightarrow{u_{m-1}} \mu_m$$

be a segment of σ, where $S = \mu_1, \mu_2, \ldots \mu_m$ is in-maximal P'-constant for some $P' \subseteq P$. Then u_j is not a proper prefix of u_i for any $i, j, 1 \leq i, j \leq m-1$.

Proof. If $u_i = u_j v$, the segment $\mu_i \xrightarrow{u_i} \mu_{i+1}$ can be written $\mu_i \xrightarrow{u_j} \mu \xrightarrow{v} \mu_{i+1}$, where $\mu \downarrow P' = \mu_i \downarrow P' + (\mu_{j+1} \downarrow P' - \mu_j \downarrow P') = \mu_i \downarrow P'$. Hence $v = \varepsilon$ due to the in-maximality of S. □

The following lemma is the crucial one.

Lemma 3.8. Let $\mathcal{P} = (P, T, \varphi, \mu_0)$ be a SPPN, where $range(\varphi) \subseteq \{0, 1, 2, .., r\}$. Let us have $P' \subseteq P$, where $|P \backslash P'| = k > 0$, and a segment

$$\rho = \mu_1 \xrightarrow{u_1} \mu_2 \xrightarrow{u_2} \mu_3 \ldots \xrightarrow{u_{m-1}} \mu_m$$

of $\sigma_{\mathcal{P}}$, where $S = \mu_1, \mu_2, \ldots \mu_m$ is in-maximal P'-constant and m is even.

Then one of the following holds:

(1) $u_1 = u_2 = \ldots = u_{m-1}$; or

(2) there is a subsequence of the sequence S which is P''-constant for some $P'' \supset P'$; furthermore, this subsequence has at least $m/(2kr)$ members.

Proof. Suppose that (1) does not hold, and let u denote the maximal common prefix of $u_1, u_2, \ldots, u_{m-1}$ (u may be empty). We can write $u_i = uu'_i$, where $u'_i \neq \varepsilon$ $(i = 1, 2, \ldots, m-1)$ due to Lemma 3.7; thus ρ can be written in the form

$$\rho = \mu_1 \xrightarrow{u} \mu'_1 \xrightarrow{u'_1} \mu_2 \xrightarrow{u} \mu'_2 \xrightarrow{u'_2} \mu_3 \ldots \xrightarrow{u} \mu'_{m-1} \xrightarrow{u'_{m-1}} \mu_m.$$

It now suffices to show that (2) holds for the sequence $\mu'_1, \mu'_2, \ldots \mu'_{m-1}$, since $\mu'_i = \mu_i + \Delta(u)$ for all i. Due to the pigeonhole principle and the fact that $\mathcal{P}$ is a SPPN, there is a $t \in T$ such that $\mu'_i \xrightarrow{t}$ for at least 1 index i and $\neg(\mu'_i \xrightarrow{t})$ for at least $(m-1)/2$ indices i; i.e., for at least $m/2$ indices i (because m is even). It follows that there is a $p \in P \backslash P'$ such that $\mu'_i(p) < \varphi(p, t) \leq r$ for at least $m/(2k)$ indices i. Hence $\mu'_i(p)$ is the same number for at least $m/(2kr)$ indices i. □

We can now establish a bound on the length of the segment sufficient to guarantee the existence of an r-semi-loop.

Lemma 3.9. Let $\mathcal{P} = (P, T, \varphi, \mu_0)$ be a SPPN, where $|P| \leq n$ and $range(\varphi) \subseteq \{0, 1, 2, .., r\}$, $r \geq 1$. Any segment of $\sigma_{\mathcal{P}}$ of length at least $2(2r)^n n!$ contains an r-semi-loop.

Proof. We can use Lemma 3.8 with Remark 3.6 several times (at most $|P|$ times) until one of the following two cases occurs:

(1) $\rho = \mu_1 \xrightarrow{u_1} \mu_2 \xrightarrow{u_2} \cdots \xrightarrow{u_{m-1}} \mu_m$ such that $u_1 = u_2 = \cdots = u_{m-1}$ and $m \geq 4r$. To see why $m \geq 4r$, note that when $P = \emptyset$, $m \geq 2(2r)^n n!$. If we apply Lemma 3.8 n times, we get case (2) below; on the other hand, if we apply Lemma 3.8 no more than $n-1$ times, $m \geq 4r$. Now from Lemma 3.4, $\mu_{m-1} \xrightarrow{u_{m-1}} \mu_m$ is an r-semi-loop.

(2) ρ contains a P-constant sequence with at least two members. Thus, ρ contains a segment $\mu_1 \xrightarrow{u} \mu_2$ such that $\mu_1 = \mu_2$. Clearly, this segment is an r-semi-loop. □

Let $\mathcal{P} = (P, T, \varphi, \mu_0)$ be a SPPN of size n. Then $|P| \le n$ and $range(\varphi) \subseteq \{0, 1, 2, ..., 2^n\}$. Suppose $\sigma_{\mathcal{P}}$ is of length at least $k = 2(2 \cdot 2^n)^n n! \le 2^{dn^2}$ for some constant d. Then the initial segment of $\sigma_{\mathcal{P}}$ of length k has a 2^n-semi-loop. Let $\mu_0 \xrightarrow{u} \mu_1 \xrightarrow{v} \mu_2$ such that $|uv| \le k$ and $\mu_1 \xrightarrow{v} \mu_2$ is a 2^n-semi-loop. Suppose $\Delta(v) \not\geq \mathbf{0}$; i.e., there is some place p such that $\Delta(v)(p) < 0$. Since $|u| \le k$, $\mu_1(p) \le \mu_0(p) + k2^n \le (k+1)2^n$. Thus, v cannot be iterated from μ_1 more than $(k+1)2^n$ times. Since $|v| \le k$, from Lemma 3.3, $\sigma_{\mathcal{P}}$ is no longer than $k + k(k+1)2^n \le 2^{cn^2}$ for some constant c. We therefore have the following theorem.

Theorem 3.10. There is a constant c such that, for any SPPN $\mathcal{P}$ of size n, any segment of $\sigma_{\mathcal{P}}$ of length at least 2^{cn^2} contains a nonempty nonnegative loop $v, \Delta(v) \geq \mathbf{0}$. Thus, the coverability tree of $\mathcal{P}$ has depth no more than 2^{cn^2+1}.

Remark 3.11. The bound from the theorem can not be improved: it is not difficult, for any n, to construct a SPPN of size $2n$ which generates all 2^n-adic numbers with at most n digits; in addition, the complete path has length 2^{n^2} and does not contain a nonnegative loop. Note also that if $range(\varphi) \subseteq \{0, 1, ..., n\}$, the bound given in Theorem 3.10 becomes $2^{cn \log n}$. If we modify the SPPN described above to generate all n-adic numbers with at most n digits, we show this bound to be tight as well.

We can now make use of a result from [HR88] to show all four of the canonical problems to be PSPACE-complete.

Theorem 3.12. For SPPNs, BP, RP, CP and EP are PSPACE-complete.

Proof. BP and RP follow immediately from Theorems 3.1 and 3.10. It is not quite as obvious that CP and EP are in PSPACE. Clearly, if we can decide CP in PSPACE, we can also decide EP in PSPACE. Therefore, let us consider the problem of deciding whether $R(\mathcal{P}_1) \subseteq R(\mathcal{P}_2)$ for arbitrary SPPNs $\mathcal{P}_1$ and $\mathcal{P}_2$ with size at most n. If $\mathcal{P}_1$ is bounded, then from Theorem 3.10, the problem is in PSPACE. Furthermore, if $\mathcal{P}_1$ is unbounded and $\mathcal{P}_2$ is bounded, the containment cannot hold. Hence, we only need to consider the case in which both $\mathcal{P}_1$ and $\mathcal{P}_2$ are unbounded. Then the complete firing sequence of each $\mathcal{P}_i$ is of the form $u_i v_i^{\omega}$, where each marking reached during the process of firing $u_i v_i$ is bounded in each place by 2^{cn^2} for some constant c. Then $R(\mathcal{P}_i)$ is a 2^{cn^2}-bounded semilinear set in which each linear set either contains exactly one marking or is of the form $\{\mu + c\Delta(v_i) \mid c \in N\}$. We can clearly determine in PSPACE whether each singleton linear set in $R(\mathcal{P}_i)$ is contained in $R(\mathcal{P}_2)$. Let ν be the least upper bound of the singleton linear sets from $R(\mathcal{P}_2)$ Clearly, ν is bounded by 2^{cn^2} in each place, and we can determine in PSPACE whether there is a $\mu \le \nu$ such that $\mu \in R(\mathcal{P}_1) \backslash R(\mathcal{P}_2)$.

Let $\mathcal{S}_1$ be the set obtained by removing the singleton linear sets and all markings not greater than ν from $R(\mathcal{P}_1)$. Clearly, $\mathcal{S}_1$ is a 2^{cn^2+1}-bounded semilinear set in which each linear set is of the form $\{\mu + c\Delta(v_1) \mid c \in N\}$. Also, let $\mathcal{S}_2$ be the set obtained by removing all markings not greater than ν from $R(\mathcal{P}_2)$. Again, $\mathcal{S}_2$ is a 2^{cn^2+1}-bounded SLS in which each linear set is of the form $\{\mu + c\Delta(v_2) \mid c \in N\}$. Thus, we have reduced our problem to deciding whether $\mathcal{S}_1 \subseteq \mathcal{S}_2$. In [HR88], it was shown that

1. if $\Delta(v_1)$ is not an integer multiple of $\Delta(v_2)$, then $\mathcal{S}_1 \not\subseteq \mathcal{S}_2$; and

2. if $\Delta(v_1)$ is an integer multiple of $\Delta(v_2)$ and $\mathcal{S}_1 \not\subseteq \mathcal{S}_2$, then there must be a marking μ bounded in each place by $n(2^{cn^2+1})^{2n+1} + 2^{cn^2+1}$ such that $\mu \in \mathcal{S}_1 \backslash \mathcal{S}_2$.

Thus, CP and EP can be decided in PSPACE, and by Theorem 3.1 are PSPACE-complete. □

We conclude this section by showing the problem of recognizing a SPPN to be PSPACE-complete.

Theorem 3.13. The problem of recognizing a SPPN is PSPACE-complete.

Proof. From Theorem 3.1, we only need to show the problem to be in PSPACE. Let $\mathcal{P} = (P, T, \varphi, \mu_0)$ be a PN of size n, and let c be the constant given by Theorem 3.10. We first try to traverse a path σ of length 2^{cn^2} in $\mathcal{P}$, verifying at each step that at most one transition is enabled. During this traversal, we only store the current marking, the next marking being constructed, and a counter of the path length; thus, we only need a polynomial amount of space. If we reach a marking at which two or more transitions are enabled, we immediately reject $\mathcal{P}$. If we reach a marking at which no transitions are enabled, we immediately accept $\mathcal{P}$. Otherwise, when we have traversed σ to a length of 2^{cn^2}, we decide in PSPACE whether σ has a nonempty nonnegative loop; if not, by Theorem 3.10, we reject $\mathcal{P}$. Otherwise, we store μ_1 and μ_2 such that $\mu_0 \xrightarrow{u} \mu_1 \xrightarrow{v} \mu_2$, $\mu_1 \leq \mu_2$, and $|uv| \leq 2^{cn^2}$. Clearly, $\mathcal{P}$ is a SPPN iff the path $\sigma' = \mu_0 \xrightarrow{u} \mu_1 \xrightarrow{v} \mu_2 \xrightarrow{v} \mu_3 \xrightarrow{v} \cdots$ never reaches a marking at which two transitions are enabled. For each transition t, let μ_t be the minimum marking at which t is enabled. For each pair of transitions t and t', we can now determine in PSPACE whether σ' reaches a marking $\mu \geq \max(\mu_t, \mu_{t'})$, where max is defined componentwise: simulate the unique firing sequence from μ_1 to μ_2, and at each marking μ, check to see whether $\mu + a(\mu_2 - \mu_1) \geq \max(\mu_t, \mu_{t'})$ for some $a \in N$. We accept $\mathcal{P}$ iff no such μ is found for any pair of transitions. □

4 Conclusions

We have shown the boundedness, reachability, containment, equivalence, and recognition problems for single-path Petri nets to be PSPACE-complete. In so doing, we showed that the depth of the coverability tree for single-path Petri nets is at most exponential in the size of the net. Crucial to our proof was the notion of an r-semi-loop. We believe that Lemma 3.8 can be extended to persistent Petri nets, so that we can be guaranteed to have r-semi-loops given that "long enough" paths (i.e., paths of a certain exponential length) exist in the Petri net. However, it is not clear how useful such an extension would be, since Lemma 3.3 clearly does not hold for persistent Petri nets. Thus, it is apparent that an analysis of the problems for persistent Petri nets must depend upon more than simply finding an r-semi-loop. Still, our conjecture is that the depth of the coverability tree for persistent Petri nets is at most exponential in the size of the Petri net.

References

[Bak73] H. Baker. Rabin's proof of the undecidability of the reachability set inclusion problem of vector addition systems. Memo 79, MIT Project MAC, Computer Structure Group, 1973.

[CRM75] S. Crespi-Reghizzi and D. Mandrioli. A decidability theorem for a class of vector addition systems. *Information Processing Letters*, 3:78–80, 1975.

[Gra80] J. Grabowski. The decidability of persistence for vector addition systems. *Information Processing Letters*, 11:20–23, 1980.

[Hac76] M. Hack. The equality problem for vector addition systems is undecidable. *Theoret. Comp. Sci.*, 2:77–95, 1976.

[HR88] R. Howell and L. Rosier. Completeness results for conflict-free vector replacement systems. *J. of Computer and System Sciences*, 37:349–366, 1988.

[HRY87] R. Howell, L. Rosier, and H. Yen. An $O(n^{1.5})$ algorithm to decide boundedness for conflict-free vector replacement systems. *Information Processing Letters*, 25:27–33, 1987.

[HRY89] R. Howell, L. Rosier, and H. Yen. Normal and sinkless Petri nets. In *Proceedings of the 7th International Conference on Fundamentals of Computation Theory*, pages 234–243, 1989. LNCS 380.

[JLL77] N. Jones, L. Landweber, and Y. Lien. Complexity of some problems in Petri nets. *Theoret. Comp. Sci.*, 4:277–299, 1977.

[Kar72] R. Karp. Reducibility among combinatorial problems. In R. Miller and J. Thatcher, editors, *Complexity of Computer Computations*, pages 85–103. Plenum Press, 1972.

[KM69] R. Karp and R. Miller. Parallel program schemata. *J. of Computer and System Sciences*, 3:147–195, 1969.

[Kos82] R. Kosaraju. Decidability of reachability in vector addition systems. In *Proceedings of the 14th Annual ACM Symposium on Theory of Computing*, pages 267–280, 1982.

[Lam87] J. Lambert. Consequences of the decidability of the reachability problem for Petri nets. In *Proceedings of the Eighth European Workshop on Application and Theory of Petri Nets*, pages 451–470, 1987. To appear in *Theoret. Comp. Sci.*

[Lip76] R. Lipton. The reachability problem requires exponential space. Technical Report 62, Yale University, Dept. of CS., Jan. 1976.

[LR78] L. Landweber and E. Robertson. Properties of conflict-free and persistent Petri nets. *JACM*, 25:352–364, 1978.

[Mul81] H. Müller. On the reachability problem for persistent vector replacement systems. *Computing, Suppl.*, 3:89–104, 1981.

[Mul85] H. Müller. Weak Petri net computers for Ackermann functions. *Elektronische Informationsverarbeitung und Kybernetik*, 21:236–244, 1985.

[May81] E. Mayr. Persistence of vector replacement systems is decidable. *Acta Informatica*, 15:309–318, 1981.

[May84] E. Mayr. An algorithm for the general Petri net reachability problem. *SIAM J. Comput.*, 13:441–460, 1984. A preliminary version of this paper was presented at the *13th Annual Symposium on Theory of Computing*, 1981.

[MM81] E. Mayr and A. Meyer. The complexity of the finite containment problem for Petri nets. *JACM*, 28:561–576, 1981.

[Pet81] J. Peterson. *Petri Net Theory and the Modeling of Systems.* Prentice Hall, Englewood Cliffs, NJ, 1981.

[Rac78] C. Rackoff. The covering and boundedness problems for vector addition systems. *Theoret. Comp. Sci.*, 6:223–231, 1978.

[Rei85] W. Reisig. *Petri Nets: An Introduction.* Springer-Verlag, Heidelberg, 1985.

The Bisection Problem for Graphs of Degree 4 (Configuring Transputer Systems)

extended abstract

Juraj Hromkovič[1] Burkhard Monien[2]

University of Paderborn
Paderborn, West Germany

Abstract

It is well known that for each $k \geq 3$ there exists such a constant c_k and such an infinite sequence $\{G_n\}_{n=8}^{\infty}$ of k-degree graphs (each G_n has exactly n vertices) that the bisection width of G_n is at least $c_k \cdot n$. It this paper some upper bounds on c_k's are found. Let $\sigma_k(n)$ be the maximum of bisection widths of all k-bounded graphs of n vertices. We prove that

$$\sigma_k(n) \leq \frac{(k-2)}{4} \cdot n + o(n)$$

for all $k = 2^r, r \geq 2$. This result is improved for $k = 4$ by constructing two algorithms A and B, where for a given 4-degree-bounded graph G_n of n vertices

(i) A constructs a bisection of G_n involving at most $n/2+4$ edges for even $n \leq 76$ (i.e., $\sigma_4(n) \leq n/2+4$ for even $n \leq 76$)

(ii) B constructs a bisection of G_n involving at most $n/2+2$ edges for $n \geq 256$ (i.e. $\sigma_4(n) \leq n/2+2$ for $n \geq 256$).

The algorithms A and B run in $O(n^2)$ time on graphs of n vertices, and they are used to optimize hardware by building large transputer systems.

1 Introduction

The problem investigated here is related to the bisection problem, where for a given graph G a balanced partition with a minimal number of crossing edges has to be found. This problem is well known and well studied. It has many applications, especially in the field of VLSI layout. The problem is NP-complete [GJ76, BCLS87], and many heuristic algorithms have been proposed, [KL70, JAGS85, BCLS87]. We don't deal directly with the classical problem of finding a minimal bisection of a given graph but with the question how large the minimal bisections of k-degree graphs (graphs whose degree is bounded by k) may be. Let $\sigma_k(n)$ be the maximum of bisection widths of all k-degree graphs of n vertices. Using expander graphs (an (n,b,d)-expander is a b-regular graph of n nodes which has for any natural number $m \leq n/2$ at least dm edges between any two components of the sizes m and $n-m$) [A186, GG86, BS87] we know that $\sigma_k(n) \in \Omega(n)$ for each $k \geq 3$. Namely, $\sigma_4(n) \geq n/7$ and $\sigma_3(n) \geq n/16$ follows from [A186, BS87]. Generally, the following lower bounds follows from [A186, AM85, LPS88]

$$\sigma(n) \geq (k - 2\sqrt{k-1})n/4 \qquad (*)$$

[1]On the leave of Comenius University, Bratislava

[2]The work of this author has been supported by the grant Mo 285/4-1 from the German Research Association (DFG)

(Note that we conjecture that these lower bounds are not very closed to the real values $\sigma_k(n)$ because they are proved by probabilistic methods for almost all random k-regular graphs). The aim of this paper is to give some upper bounds on $\sigma_k(n)$ for $k \geq 3$.

In Section 2 we shall use a new, combinatorial optimization technique to show that

$$\sigma_k(n) \leq (k-2)n/4 + o(n) \tag{1}$$

for each $k \in \{2 \cdot 2^{r+1} | r \in N\}$, where N denotes the set of all nonnegative integers. Using this technique we also get a more general result: Each k-regular graph of n vertices can be partitioned into d equal-sized components by removing at most

$$\frac{(k-2)}{2} \cdot n \cdot \frac{d-1}{d} + o(n)$$

edges for $k \in \{2 \cdot 2^{r+1} | r \in N\}$. The proof of the above stated facts is constructive because we give an algorithm C which finds for any given k-degree graph a bisection whose width is bounded by (1). Unfortunately, this algorithm C has an exponential complexity.

This unpleasant property of C will be overviewed in Section 3 by constructing two algorithms A and B, where for a given 4-degree graph G_n of n vertices

(i) A constructs a bisection of G_n involving at most $n/2+4$ edges for even $n \leq 76$, i.e.

$$\sigma_4(n) \leq n/2 + 4 \quad \text{for} \quad n \leq 76$$

(ii) B constructs a bisection of G_n involving at most $n/2+2$ edges for $n \geq 256$, i.e.

$$\sigma_4(n) \leq n/2 + 2 \quad \text{for} \quad n \geq 256$$

Both algorithms A and B run in $O(n^2)$ time on graphs of n vertices.

In Section 4 we use the results of Section 3 to give an effective algorithm for configuring transputer systems. This new algorithm has led to an optimization of the use of hardware in the process of modular configuring of transputer systems. This optimization question was the original question which has started the research presented in this paper.

2 Asymptotic Estimates

In this section we shall give an asymptotic estimate for the general task of bounded graph partitioning into $d \geq 2$ equal-sized components. This estimate implies that the bisection of 4-bounded graphs can be done by removing $n/2 + o(n)$ edges, and that the 4-partition of 4-bounded-degree graphs can be done by removing $3n/4 + o(n)$ edges (note that these cases are the most important from the configuring transputer system point of view). Now, let us start with partitioning of regular graphs because they are the hardest ones among bounded graphs from the bisection width point of view.

Theorem 2.1: Let G be a 4-regular graph of n vertices, where $n = d \cdot h$ for some $d, h \in N$. Let $\mathcal{M} = \{m \in N \mid m \geq 2,\ h = mb, \text{ for some } b \in N\} \neq \emptyset$, and let $e(n,d,m) = (n + 2(md-1)) \cdot ((d-1)/d) \cdot (m/(m-1/d))$. Then one can partition G into d components, each of h vertices, by removing at most

$$min\{e(n,d,m)|m \in \mathcal{M}\} \leq n \cdot \frac{d-1}{d} + o(n)$$

edges.

Proof. First, for any $m \in \mathcal{M}$ we give an algorithm $d-PART(m,G)$ that finds a d-partitioning of G, and then we prove that this d-partitioning is achieved by removing at most $e(n,d,m)$ edges.

Algorithm $d-PART(m,G)$

Input: A 4-regular graph $G=(V,E)$ of n vertices, positive integers d,m such that $n=dmb$ for some positive integer b

Output: d components of $G: D_1=(V_1,E_1),\dots,D_d=(V_d,E_d)$ and the set of removed edges $H_m = E - \bigcup_{i=1}^{d} E_d$ with the property $|H_m| \leq e(n,d,m)$.

1. Remove at most n edges from G in order to obtain a graph $G'=(V,E')$ with the degree bounded by 2.
{ This can be achieved by constructing the Euler Cycle in G and then removing each odd edge in the cycle}

2. Remove at most $2(md-1)$ edges from G' in such a way that we obtain $m \cdot d$ components $C_i = (V_i', E_i'), i \in \{1,\dots,md\}$, each exactly of n/md vertices. { So, we have md-partitioning by removing at most $n+2(md-1)$ edges }

3. Set $H := E - \bigcup_{i=1}^{md} E_i'$.
For each $i,j \in \{1,\dots,md\}$ compute c_{i_j} - the number of edges from H leading between the component C_i, and the component C_j.
Divide the set $I=\{1,2,\dots,md\}$ into d disjoint sets $I_1,\dots,I_d$, each of m elements, in such a way that the sum $\sum_{k=1}^{d}\sum_{i,j\in I_k} c_{ij}$ is maximal for all possible partitions of I.

4. For each $k \in \{1,\dots,d\}$ set $V_k = \bigcup_{i\in I_k} V_i'$ and set $D_k=(V_k,E_k)$ to be the induced graph of G by the set of vertices $V_k \subseteq V$.
{ For each $k: |V_k| = m\cdot(n/md) = n/d$}.
Set $H_m := H - \bigcup_{k=1}^{d}(\bigcup_{i,j\in I_k} V_i' \times V_j')\{= E - \bigcup_{k=1}^{d} E_k\}$

Now, let us prove that $|H_m| \leq e(n,d,m)$ for any $m \in \mathcal{M}$. Let, for any $i,j \in \{1,\dots,md\}$, E_{ij} be the set of edges connecting the components $C_i=(V_i',E_i')$ and $C_j=(V_j',E_j')$, i.e., $E_{ij} = E\cap(V_i'\times V_j')$. Clearly, there are exactly $\binom{md}{2}$ distinct sets E_{ij}. Let $a=\sum_{i<j}|E_{ij}| \leq |H| \leq n+2(dm-1)$.

By constructing the components $D_1,\dots D_d$ (each D_k as union of m components $\bigcup_{i\in I_k} C_i$) we choose $d\cdot\binom{m}{2}$ sets E_{ij} from the set of edges H partitioning G into md components. Now, one can prove that

$$|H_m| \leq a\cdot(1-d\binom{m}{2}/\binom{md}{2}), \tag{1}$$

i.e., that there exist such $I_1,\dots,I_d$ that the number of edges removed from H is proportional to the number of sets E_{ij} removed from H.

Thus the inequality (1) implies

$$
\begin{aligned}
|H_m| &\leq a(1 - \tfrac{d \cdot m(m-1)}{2} / \tfrac{md(md-1)}{2}) \\
&\leq (n + 2dm - 2)^{\frac{md(md-1)-md(m-1)}{md(md-1)}} = \\
&\leq (n + 2dm - 2)^{\frac{md-m}{md-1}} = (n + 2dm - 2)^{\frac{d-1}{d} \cdot \frac{m}{m-1/d}} = \\
&= e(n, d, m)
\end{aligned}
$$

Clearly

$min\{e(n,d,m)|m \in \mathcal{M}\} \leq n \cdot \frac{d-1}{d} + o(n)$ □

Concluding this section we shall still generalize our result for k-regular graphs with $k \in \{2 \cdot 2^{r+1}, |r \in N\}$.

Theorem 2.2 Let k be an integer from $\{2 \cdot 2^{r+1}|r \geq 0\}$. Let G be a k-regular graph of n vertices, where $n = d \cdot h$ for some $d, h \in N$. Then one can partition G into d components, each of h vertices, by removing at most

$$\frac{(k-2)}{2} n \cdot \frac{d-1}{d} + o(n)$$

edges. □

Proof. Let $k = 2 \cdot 2^{r+1}$. First we shall $r + 1$ times perform the step 1 of the algorithm $d - PART(m, G)$ (i.e. we decrease $r + 1$ times the degree of the graph to one half of the original degree by removing all odd edges of an Euler Cycle). Thus we remove $(k - 2) \cdot n/2$ edges in order to obtain a graph of the degree two. Then the algorithm $d - PART(m, G)$ runs exactly in the way described in the previous proof. □

Concluding Section 2 we note that both Theorem 2.1 and Theorem 2.2 can be formulated also for 4-degree and k-degree (instead of k-regular) graphs respectively. To see this it is sufficient to realize that one can construct for each k-degree bounded graph G of n vertices a k-regular graph G' with the following properties:

(i) G is a subgraph of G' (G can be obtained from G' by removing some node and some edges)

(ii) G' has at most $n + k$ nodes.

3 Improved Estimates

In the previous section we have presented some results regarding the partitioning k-regular graphs into d equal-sized components for $d \geq 2$. In this section we improve these results for $k = 4$ and $d = 2$. We note that we were not able to improve the asymptotic estimates established in Theorem 2.1 for $d \geq 3$, and we conjecture that the technique used there provides better estimates for larger d's than for smaller d's.

We shall present two algorithms for partitioning 4-regular graphs into two equal-sized components. The first one works very well for small numbers n, and the second one shows that $\sigma(n) \leq n/2 + 2$ for $n \geq 256$. Before giving our algorithms we claim that small graphs can require the removal of more than $n/2$ edges in order to be 2-partitioned. To illustrate this fact we show the hardest graphs for $n = 16$ in Figures 2. Besides these examples we have shown by using a computer program that $\sigma_4(8) = 8, \sigma_4(12) = 10, \sigma_4(16) = 12, \sigma_4(20) = 14$ and $\sigma_4(26) = 16$.

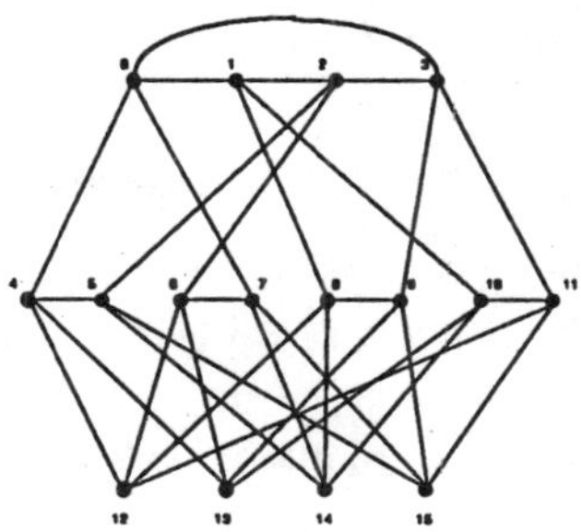

Figure 2, A graph with 16 nodes, $\sigma_4 = 12$

Section 3 contains our main results improving the asymptotic estimations for $\sigma_4(n)$ established in the previous section. To achieve this improvements we develop two techniques based on graph-theoretical approaches.

Both techniques use the so-called "Balancing Lemma" established here, which shows that we are able to construct a balanced partition from an "almost balanced" (specified later) partition by increasing the number of edges between the two components at most by 2.

The first technique works very well for small $n \leq 76$, and it is used to prove that

$$\sigma_4(n) \leq n/2 + 4 \qquad \text{for all } n \equiv 0 \; mod 4$$
$$\sigma_4(n) \leq n/2 + 3 \qquad \text{for all } n \equiv 2 \; mod 4.$$

This technique is based on an iterative approach. Starting with one small component (the smallest cycle in the graph) on one side and one large component on the other side we iteratively increase the amount of vertices in the small component by removing some small subgraphs from the large component to the small one. The small subgraphs removed are chosen in such a way that the number of its edges is large as possible in the comparison with the number of its vertices. This iterative process stops after reaching an "almost balanced" partition. The Balancing Lemma is used to obtain the balanced partition of the given graph. Considering the above stated examples we see that some small graphs also require $n/2 + 4$ removed edges to be divided into two equal-sized components.

The second technique is used to show that

$$\sigma_4(n) \leq n/2 + 2 \quad \text{for} \quad n \geq 256.$$

This technique is based on local optimization, i.e., it uses local replacements between two given components. The algorithm based on this technique starts with a balanced partition of the given graph and decreases the number of edges leading between the two components by repeatedly replacing some subgraphs of the two components. The algorithm halts either in a balanced partition with at most $n/2$ edges or in an "almost balanced" partition with at most $n/2 - 2$ edges. Obviously, the Balancing Lemma can be used in the second case in order to obtain the required partition.

Both presented algorithms are simple to use and they work in $O(n^2)$ time. The main difficulty is of the mathematical nature: to give the proofs showing that the resulting partitions have a small number of edges between the two components.

Now, we introduce our "Balancing Lemma". Before the formulation we need the following definitions.

Def.: Let $G = (V, E)$ be an undirected regular graph of degree 4. Let $n = |V|$, n even. A partition $\pi = \pi(G)$ is a mapping $\pi : V \to \{1,2\}$. Set $V_i = V_i(\pi) = \{u \in V; \pi(u) = i\}, i = 1,2$, and $r(\pi) = |V_1(\pi)|$. For our convenience and w.l.o.g. we will always assume that $r(\pi) \leq n/2$ holds. We call $bal(\pi) = n/2 - r(\pi)$ the balance of π and we call the partition π balanced if $bal(\pi) = 0$ holds.

Of course we are interested only in finding good balanced partitions but in constructing them we have to consider also partitions which are not balanced. Let $Ext(\pi) = \{e \in E; e = \{u,v\}$ with $\pi(u) \neq \pi(v)\}, ext(\pi) = |Ext(\pi)|$ be the set of crossing edges and number of crossing edges, respectively, and let $Int(\pi) = \{e \in E; e = \{u,v\}$ with $\pi(u) = \pi(v) = 1\}, int(\pi) = |Int(\pi)|$ be the set (respectively number) of internal edges. Note that $ext(\pi) = 4 \cdot r(\pi) - 2 \cdot int(\pi)$ holds.

Now, we are giving a definition of the almost-balanced partition. Note, that the "almost-balanced" property is relative here because it is related also the the number $ext(\pi)$.

A partition π is called almost balanced if there exists such number z that

(i) $ext(\pi) \geq z \cdot bal(\pi) = z(n/2 - r(\pi))$, and

(ii) $4 \cdot (n/2 + bal(\pi)) \leq 5 \cdot ext(\pi) + 2 \cdot z$.

We characterize the nodes according to the number of crossing edges incident to them. For $u \in V$ let $ext_\pi(u) = |\{v \in V; \{u,v\} \in E, \pi(u) \neq \pi(v)\}|$ be the number of crossing edges incident to u. We call a node u an A node, B node, C node or D node, respectively, if $ext(u) \geq 3$ or $ext(u) = 2, 1$ or 0, respectively.

We can now formulate and prove our main lemma which will be very useful for both following bisection algorithm.

Lemma 3.1 (Balancing Lemma)
Let π be an almost balanced partition with $ext(\pi) \geq 4$. Then we can construct a balanced partition $\hat{\pi}$ with $ext(\hat{\pi}) \leq ext(\pi) + 2$.

Proof: Consider the partition π. While $r(\pi) < \frac{n}{2}$ and there exists an A or B node u in V_2 shift this node u to V_1, i.e. redefine π by setting $\pi(u) = 1$ and leave $\pi(v)$ unchanged for $v \neq u$. Note that the number of crossing edges decreases by shifting an A node and remains unchanged by shifting a B node. If the number of crossing edges has decreased then we can still shift C nodes or D nodes to V_1. In this way we finally reach a balanced partition π' with $ext(\pi') \leq ext(\pi)$ or we reach a partition π' with $bal(\pi') \leq bal(\pi)$, $r(\pi) \leq r(\pi') < n/2, ext(\pi') = ext(\pi)$ and $V_2(\pi')$ contains no A nodes and no B nodes. In the first case we have proved the lemma, in the second case the new partition π' (which of course may be the old partition π) still fulfills the conditions of the lemma and additionally we know that $V_2(\pi')$ contains no A nodes and no B nodes.

For the sake of convenience we rename our partition and write π instead of π'.
Consider $V_2 = V_2(\pi)$. Let F_C be the set of C nodes and F_D be the set of D nodes in V_2 and let $c = |F_C|$ and $d = |F_D|$. Then $c + d = |V_2| = n - r(\pi)$ and $c = ext(\pi)$.

Now let $U \subset F_C$ be a set of C nodes and let $Int(U) = \{e = \{u,v\}; e \in E, u,v \in U\}$ be its set of internal edges. Every node from U is connected via exactly one of its edges to a node from V_1 and therefore there are $3 \cdot |U| - 2 \cdot |Int(U)|$ edges connecting U with nodes in $V_2 - U$. If we shift all nodes from U to V_1 then we have defined a new partition π' with $ext(\pi') = ext(\pi) + 2 \cdot (|U| - |Int(U)|)$.

Note that every connected graph of m nodes has at least $m-1$ edges. We proceed as follows. We compute the connected components of the graph $H = (F_C, Int(F_C))$. If there exists a connected component of at least $bal(\pi) = n/2 - r(\pi)$ nodes then there exists also a connected graph $(U, Int(U)), U \subset F_C$, with $|U| = bal(\pi)$. Since U is connected, we have $|Int(U)| \geq |U| - 1$, and therefore the balanced partition $\hat{\pi}$ obtained by shifting the nodes from U to V_1 fulfills $ext(\hat{\pi}) \leq ext(\pi) + 2$.

From now on we assume that every connected component of H has less than $bal(\pi) = n/2 - r(\pi)$ nodes. We will show that in this case there exists a cycle consisting only of C nodes. Let x be the number of connected components of H. Then $ext(\pi) = c < x \cdot (\frac{n}{2} - r(\pi))$ holds, i.e. $x > z$, for any z satisfying conditions (i) of π. Now let us assume that H contains no cycles, i.e. H is a forest.

$\Longrightarrow$ $|Int(F_C)| = c - x$

$\Longrightarrow$ There are $3c - 2(c - x) = c + 2x$ edges connecting C nodes with D nodes within V_2
(if this is not clear then see the fact above that $3 \cdot |U| - 2|Int(U)|$ edges connect U with $V_2 - U$)

$\Longrightarrow$ $4d \geq c + 2x$, since every D node is incident to at most 4 of these edges

$\Longrightarrow$ $4 \cdot (n - r(\pi)) - 4 \cdot ext(\pi) = 4d \geq c + 2x = ext(\pi) + 2x$ since $d = n - r(\pi) - ext(\pi)$

This is a contradiction because $x > z$ for any z and each z satisfies (ii) $4(n - r(\pi)) \leq 5ext(\pi) + 2z$.

Therefore we can assume now that there exists a cycle within F_C, i.e. there exist $U \subset F_C$ with $|Int(U)| = |U|$. Since U induces a connected graph we can also assume that $|U| < bal(\pi)$. We can shift the nodes from U to V_1 and get this way a new partition π' with $r(\pi) < r(\pi') < n/2$ and (i.e., $bal(\pi') < bal(\pi)$) $ext(\pi') = ext(\pi)$. With this partition fulfilling (i) and (ii) we can iteratively start the whole construction described in this proof. Finally we will reach in this way the partition $\hat{\pi}$ we are aiming for. □

Since we do not have enough space to present the algorithms A and B completely in this extended abstract we shall only outline them in what follows.

Theorem 3.2: Let n be an even number, $n \leq 76$. Then

$$\sigma_4(n) \leq n/2 + 4 \quad \text{, if } n \equiv 0 \ mod 4\text{, and} \qquad \sigma_4(n) \leq n/2 + 3 \quad \text{, if } n \equiv 2 \ mod 4.$$

Sketch of the proof: We will prove the result here only for $16 \leq n \leq 44$, $n \equiv 0 \ mod \ 4$ in order to outline this idea of algorithm A. For all the other cases the result can be proved in a similar way. Let $G = (V, E), |V| = n, 16 \leq n \leq 44, n \equiv 0 \ mod \ 4$.
Since $n \geq 16$, G contains a cycle of at most $n/4$ nodes. Using this cycle we can construct a subgraph $(\hat{U}, Int(\hat{U}))$ of G with $|\hat{U}| = n/4$ and $|Int(\hat{U})| \geq \frac{n}{4}$.

It is not difficult to find some U with $|U| \geq n/4 + 2$ and $4 \cdot |U| - 2 \cdot |Int(U)| = n/2 + 2$. Define π by $\pi(u) = 1 \Leftrightarrow u \in U$, i.e. $r(\pi) = |U|, ext(\pi) = n/2 + 2$.
Then $ext(\pi) \geq 2 \cdot (n/2 - r(\pi))$ and $4 \cdot (3n/4 - 2) = 4 \cdot (n - r(\pi)) \geq 5 \cdot ext(\pi) + 4 = 5n/2 + 14$ holds if and only if $n \leq 44$.
Thus by applying Lemma 3.1 we find a balanced partition π' with $ext(\pi') \leq n/2 + 4$.
□

Theorem 3.3 $\quad \sigma_4(n) \leq n/2 + 2 \quad$ for $n \geq 254$.

Idea of the proof.

Our second technique uses local transformations. We start with an arbitrary partition. In each step we are looking for some set of nodes lying on one side of the partition whose shift to the other side reduces the number of crossing edges. We call a subset of nodes t-helpfull, $t \in N$, if the number of crossing edges is decreased by $2t$ this way. Figure 3 shows two 1-helpful graphs.

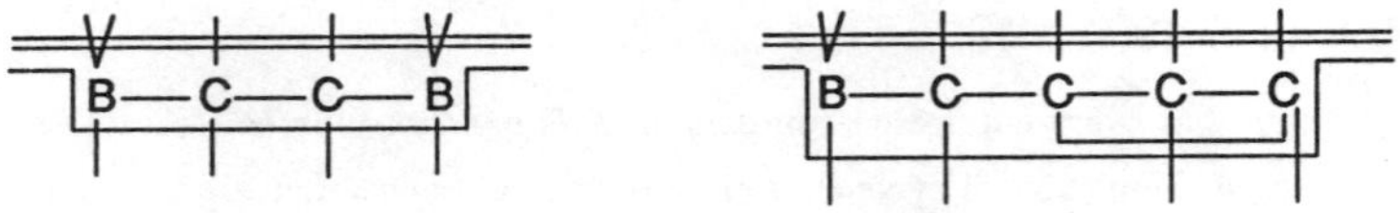

Figure 3: Two 1-helpful sets

Our method now is the following. Let π be a partition. Determine a 2-helpful set of small cardinality and define a new partition π' by shifting the nodes of this set to the other side. Then π' is not balanced, but $ext(\pi') \leq ext(\pi) - 4$. Now compute a balanced partition $\hat{\pi}$ with $ext(\hat{\pi}) \leq ext(\hat{\pi}) + 2 \leq ext(\pi) - 2$ by using the Balancing Lemma. The most difficult part of this proof consists in the constructive proof of the existence of t-helpful sets of small cardinality. □

4 Optimization of Hardware by Building Transputer Systems

The research presented in this paper is motivated by the problem of configuring transputer systems.

A transputer system can be realized as a system which consists of m processors ($m = 32$, for instance) and two switches (the upper switch and the lower switch) containing some physical links (see Fig. 5). Each processor has 4 fast communication links, two of them may be connected to two links in the upper switch and two of them may be connected to two links in the lower switch. Each link in a switch may be connected with at most two processors P_1 and P_2, which means that P_1 and P_2 communicate via this link. Thus, we can embed any 4-degree graph G of at most m vertices (any 4-degree interconnection network of at most m processors) to the transputer system (TS) by assigning processors of TS to the vertices of G, and by connecting any pair of processors of TS corresponding to a pair of vertices connected by an edge in G. The proof of this fact is based on the classical result of Petersen [Pet91]. For instance, the lower switch at Figure 4 is used to connect the following pairs of processors: $(P_1, P_m), (P_1, P_3), (P_2, P_3), (P_2, P_m)$.

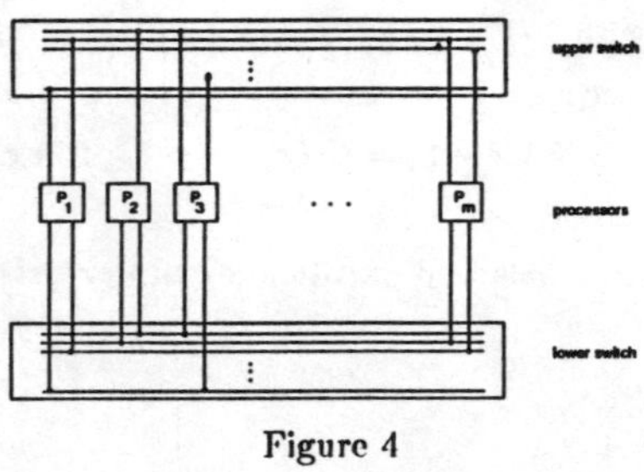

Figure 4

Modern systems allow to configure the network automatically, i.e. given a logical description of the network (parallel architecture) the system computes a mapping and some setting of switches so that every logical link in the network corresponds to a physical link of the transputer system. Thus the main computational advantage of TS is that it can be reconfigured to realize any parallel architecture (of the degree bounded by 4) given by users, and the users don't need to deal with the problem how to embed their networks into TS. Since current TS allows to work several users simultaneously (obviously, only in the case that the sum of processors of all networks of users does not exceed the number of processors of TS), and there are users interested in building large networks the computer companies are required to build larger and larger transputer systems.

A natural way how to do it is the modular approach consisting in connecting a few small transputer systems into a larger one. The task with which we are dealing here is to build a new TS from two basic TSs described above. This task is depicted at Figure 5, where the new TS is obtained by connecting some links of the switches of basic TSs. The links laying between the two upper (lower) switches are called upper external links (lower external links). The links leading between the upper switches on one side and the lower switches on the other side are called crossing external links.

Since we are interested to optimize hardware (to minimize the number of external links) in such a way that the new modular TS of $2m$ processors will be still able to realize any 4-degree-bounded network of at most $2m$ processors, our optimization task can be formulated as follows.

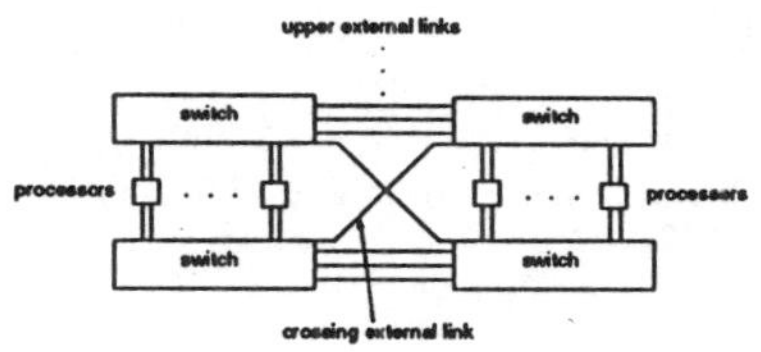

Figure 5: A modular parallel architecture

Find such number $b(2m)$, that each graph of $2m$ vertices can be embedded into the modular TS of $2m$ processors in such a way that

(i) the number of external links used is at most $b(2m)$

(ii) the number of upper external links used differs from the number of lower external links used at most by 1, and

(iii) the number of the crossing external links used has to be as small as possible.

Now, we are able to prove the following theorem which really reduces this complex problem of configuring transputers to the problem of finding a reasonable upper bound on $\sigma_4(n)$.

Theorem 4.1 For each partition of a logical network into two-equal sized parts there is an assignment of logical processors to the physical transputers with the properties:

(1) the number of upper links and the number of lower links differs at most by 1, and

(2) there are at most two crossing links laid out as depicted at Fig. 6. □

This Theorem 4.1 provides the optimal solution for the tasks (ii) and (iii) and we see that the only one problem to be solved now is the determination of $\sigma_4(n)$. (Note, that this means that $\sigma_4(n)/2$ (physical) upper links, $\sigma_4(n)/2$ (physical) lower links and two (physical) crossing links are sufficient for the realization of any logical network of the size smaller than two times the size of the basic transputer system module by two-connected modules).

References

Al86 N. Alon: Eigenvalues and expanders. Combinatorica 6 (1986), 85-95.

AM85 N. Alon — V.D. Milman: λ_1, isometric inequalities for graphs, and superconcentrators. J. Combinatorial Theory B 38 (1985), 73-88.

BCLS87 T.N. Bui — S. Chanduri — F.T. Leighton — M. Sipser: Graph bisection algorithms with good average case behavior. Combinatorica 7 (1987), 171-191.

BS87 A. Broder — E. Shamir: On the second eigenvalue of random regular graphs. In: Proc. 28th Annual Symp. on FOCS, IEEE 1987, 286-294.

GG81 Gabber — Z. Galil: Explicit constructions of linear-sized superconcentrators. J. Comput. Syst. Sci. 22 (1981), 407-420.

GJ76 M.R. Garey — D.S. Johnson: Some simplified NP-complete graph problems, Theor. Comp. Science 1 (1976), 237-267.

JAGS85 D.S. Johnson — C.R. Aragon — L.A. Mc Geoch — C. Schevon: Optimization by simulated annealing: An experimental evaluation (Part I), Preprint, AT + T Bell Labs, Murray Hill, NY (1985).

KL70 B.W. Kernighan — S. Lin: An efficient heuristic procedure for partitioning graphs, Bell Systems Techn. J. 49 (1970), 291-307.

LPS88 A. Lubotzky — R. Phillips — P. Sarnak: Ramanujan graphs. Combinatorica 8 (1988), No. 3, 261-277.

MKPR89 H. Mühlenbein — O. Krämer — G. Peise — R. Rinn: The Megaframe Hypercluster - A reconfigurable architecture for massively parallel computers, IEEE Conference on Computer Architecture, Jerusalem 1989.

Nic88 D.A. Nicole, Esprit Project 1085, Reconfigurable Transputer Processor Architecture, Proc. CONPAR 88, 12-39.

Pet91 J. Petersen: Die Theorie der regulären Graphs, Acta Mathematica 15 (1891), 193-220.

Some Results Concerning 2-D On-line Tessellation Acceptors and 2-D Alternating Finite Automata

Oscar H. Ibarra*, Tao Jiang† and Hui Wang*

Abstract

A two-dimensional nondeterministic on-line tessellation acceptor (2-NOTA) is a special type of real-time two-dimensional nondeterministic cellular automaton in which data flows from the upper-left corner to the lower-right corner. A two-dimensional alternating finite automaton (2-AFA) is an alternating finite automaton with a two-dimensional rectangular input whose input head can move in all four directions on the input. In this paper, we show that 2-NOTA's and 2-AFA's are incomparable. This answers in the negative an open question in [IT89a]. Closure properties of the classes of languages (i.e., sets of two-dimensional patterns) accepted by two-way, three-way, and four-way two-dimensional alternating finite automata and two-dimensional alternating finite automata with only universal states are also obtained which answer several open questions in [IN88].

1. Introduction

Blum and Hewitt were the first to study finite automata and marker automata operating on a two-dimensional input tape [BL67]. Since then, many new types of automata with two-dimensional input tape have been introduced. Two models that have attracted some attention in recent literature are the two-dimensional alternating finite automaton (2-AFA) [IN83] and the two-dimensional tessellation acceptor (2-OTA) [IN77a] which is a special type of real-time rectangular array bounded cellular automaton. There are also restricted versions of these automata, such as the two-way 2-AFA (TW2-AFA) which is a 2-AFA whose input head can only move in two directions (right and down, in addition to no move), the three-way 2-AFA (TR2-AFA) which is a 2-AFA whose input head can only move in three directions (left, right and down) and the 2-AFA with only universal states (2-UFA). Many researchers have investigated the properties and relationship between these automata ([HR89], [IB74], [IN77a], [IN85], [IT82], [IT88a], [IT88b] and [IT89b]). Recently, Ito, Inoue and Takanami [IT89a] showed that the TW2-AFA's are equivalent to deterministic 2-OTA's (2-DOTA's) through 180°-rotation. They conjectured that the class of languages accepted by 2-AFA's is included in the class of languages accepted by nondeterministic 2-OTA's (2-NOTA's).

In section 3, we show that 2-AFA's and 2-NOTA's are incomparable, disproving the above-mentioned conjecture. The proof is rather interesting in that we use the planar embedding of acyclic directed bipartite graphs to show the incomparability. This is the first time that this type of language has been used in showing a negative result. All previous negative proofs are essentially based on pattern matching. In section 4, we show that the class of languages accepted by 2-AFA's is not closed under complementation and that the class of languages accepted by TR2-AFA's is closed under complementation. These results answer two open questions in [IN88]. Some interesting closure properties of these automata under rotations are shown in section 5. We make some concluding remarks in section 6.

* Current address: Department of Computer Science, University of California, Santa Barbara, CA 93106, USA. Research supported in part by NSF Grants CCR-8918409 and CCR90-96221.
† Current address: Department of Computer Science and Systems, McMaster University, Hamilton, Ontario, Canada L8S 4K1. Research supported in part by NSERC Operating Grant OGP 0046613 and a grant from SERB, McMaster University, Canada.

2. Preliminaries

We adopt most of the definitions and notation in [IN77a], [IN83], and [IT89a]. Below we give brief descriptions of the devices. For more details, the reader is referred to [IN77a], [IN83] and [IT89a].

Definition 2.1. Let Σ be a finite set of input symbols. A pattern over Σ is a rectangular two-dimensional array of symbols from Σ surrounded by boundary symbols, #. The set of all input patterns over Σ is denoted by $\Sigma^{(2)}$. For a pattern x, we denote row(x) and col(x) to be the number of rows and columns of x. A pattern x is square if row(x) = col(x). A pattern can be rotated clockwise 90°, 180° and 270° to obtain new patterns.

A two-way nondeterministic on-line tessellation acceptor (2-NOTA) M is an infinite mesh-connected array of cells. Each cell of the array consists of a nondeterministic finite-state machine. The cells can be identified by a pair of integers denoting its coordinate. A cell is called the (i,j)-cell if its coordinate is (i,j). An input to M is a pattern x with symbol x[i,j] placed at the (i,j)-cell of the array, where $1 \leq i \leq$ row(x) and $1 \leq j \leq$ col(x) and subject to the condition that the boundary symbols are '#', i.e., x[1,j] = x[row(x),j] = # for $1 \leq j \leq$ col(x) and x[i,1] = x[i,col(x)] = # for $1 \leq i \leq$ row(x). The interior symbols are from Σ. See Figure 1. The 2-NOTA M on input pattern x works as follows. At time t = 0, all cells in M except the (1,1)-cell are in the "quiescent state" and the (1,1)-cell is in the "motive state". At time t = 1, the (1,1)-cell enters an active state which depends on symbol x[1,1]. At time t = k (k > 1), each (i,j)-cell such that (i-1) + (j-1) = k-1 enters an active state which depends on the active states of (i-1,j)-cell and (i,j-1)-cell at time t = k-1, and on the symbol x[i,j]. We assume that if a cell's neighbor does not exist, the state of that neighbor is quiescent. M accepts input pattern x if and only if the active state of the (m,n)-cell at time t = m+n-1 is one of the specified final states, where m = row(x) and n = col(x).

Definition 2.2. A two-dimensional nondeterministic on-line tessellation acceptor (2-NOTA) is a 7-tuple M = (Q, E^2, $\Sigma\cup\{\#\}$, δ, q_e, q_0, F), where

Q is the finite set of states,
E^2 is the set of all pairs of integers,
Σ is the finite set of input symbols,
$\# \notin \Sigma$ is the boundary symbol,
$\delta: Q^3 \times (\Sigma\cup\{\#\}) \to 2^{Q'} \cup \{\{q_0\}\}$ is the state transition function, where $Q' = Q - \{q_e, q_0\}$,
$q_e \in Q$ is the motive state,
$q_0 \in Q$ is the quiescent state,
$F \subseteq Q - \{q_e, q_0\}$ is the set of final (accepting) states.

The cell state transition function δ prescribes state transitions of cells with coordinates in E_2. The state transition function is defined as follows. Let $q_{(i,j)}(t) \in Q$ denote the state of the (i,j)-cell at time t. Then

$$q_{(i,j)}(t+1) \in \delta(q_{(i,j)}(t), q_{(i-1,j)}(t), q_{(i,j-1)}(t), a)$$

where a is the input symbol on the (i,j)-cell. In addition, δ has the property that for any $a \in \Sigma\cup\{\#\}$ and any $p_i \in Q$ $(1 \leq i \leq 3)$

$$\delta(p_1, p_2, p_3, a) = q_0 \text{ iff } p_1 = q_0 \text{ and } p_2, p_3 \in \{q_e, q_0\}$$

This property allows each cell to remain in the quiescent-state q_0 before being activated.

A two-way on-line tessellation acceptor is called deterministic (2-DOTA) if the finite-state machine in each cell of the array is deterministic.

Definition 2.3. Let $M = (Q, E^2, \Sigma\cup\{\#\}, \delta, q_e, q_0, F)$ be a 2-NOTA and $x \in \Sigma^{(2)}$ be an input pattern to M. Then, a run of M on x is a two-dimensional pattern z over $Q - \{q_e, q_0\}$ which satisfy the following conditions:

(1) row(z) = row(x) and col(z) = col(x);

(2) $z(1,1) \in \delta(q_e, q_0, q_0, x(1,1))$;

(3) $z(i,j) \in \delta(q_0, z(i-1,j), z(i,j-1), x(i,j))$ for $(i+j \geq 2)$ and $(1 \leq i \leq \text{row}(z), 1 \leq j \leq \text{col}(z))$,
where $z(i,0) = z(0,j) = q_0$.

A run of M on x is a state configuration of M's computation on x.

Definition 2.4. Let $M = (Q, E^2, \Sigma\cup\{\#\}, \delta, q_e, q_0, F)$ be a 2-NOTA. The language accepted by M, L(M), is the set of all patterns over $\Sigma^{(2)}$ that are accepted by M, i.e.,

$L(M) = \{x \in \Sigma^{(2)} \mid$ there is a run z of M on x such that $z(row(z),col(z)) \in F\}$.

A two-dimensional alternating finite automaton (2-AFA) is an alternating finite automaton with a two-dimensional pattern (as defined in Definition 2.1) as its input. A 2-AFA has a read-only input head attached to a finite control as shown in Figure 2.

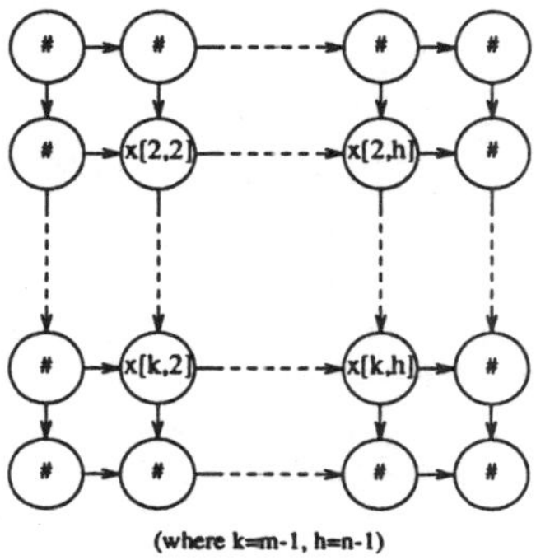

Figure 1. A 2-D nondeterministic on-line tessellation acceptor

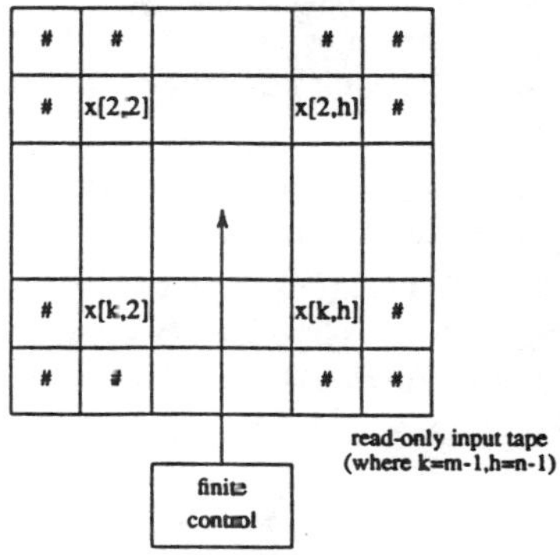

Figure 2. A 2-D alternating finite automaton.

Definition 2.5. A two-dimensional alternating finite-automaton (2-AFA) is a six-tuple $M = (Q, \Sigma \cup \{\#\}, \delta, q_0, U, F)$, where

- Q is the finite set of states,
- Σ is the finite input alphabet ($\# \notin \Sigma$ is the boundary symbol),
- $\delta : (Q \times (\Sigma \cup \{\#\})) \rightarrow (Q \times \{\text{left,right,up,down,no move}\})$ is the transition function,
- $q_0 \in Q$ is the initial state,
- $U \subseteq Q$ is the set of universal states,
- $Q - U$ is the set of existential states,
- $F \subseteq Q$ is the set of final (accepting) states.

A 2-AFA is called 2-UFA if it does not have existential states.

A configuration of M is a triple $c = (x, (i,j), q)$, where x is the input pattern to M, (i,j) is the input head position and q is the state of the finite control when input head is at position (i,j). The initial configuration of M on x is $I_M(x) = (x, (1,1), q_0)$. A configuration is called universal or existential if the state associated with it is universal or existential, respectively. A configuration is called accepting if the state associated with it is both final and halting. We assume, without loss of generality, that the input head never falls off the input tape. Let c and c′ be two configurations of M. We write $c \vdash_M c'$ if configuration c′ follows from configuration c in one step of M.

Definition 2.6. Let $M = (Q, \Sigma \cup \{\#\}, \delta, q_0, U, F)$ be a 2-AFA and x be an input pattern to M. A computation tree of M on x is a (possibly infinite) nonempty labeled tree which satisfies the following conditions:

(1) each node u is labeled with a configuration c(u);

(2) if u is an internal node of the tree, c(u) an universal configuration and $\{c \mid c(u) \vdash_M c\} = \{c_1, \cdots, c_k\}$, then u has exactly k children $v_1, \cdots, v_k$ such that $c(v_i) = c_i$ for $1 \le i \le k$;

(3) if u is an internal node of the tree and c(u) is an existential configuration, then u has exactly one child v such that $c(u) \vdash_M c(v)$.

For any configuration c, a c-computation tree of M is a computation tree of M whose root is labeled with configuration c. A c-accepting computation tree of M is a finite c-computation tree whose leaves are all labeled with accepting configurations.

A three-way two-dimensional alternating finite automaton (TR2-AFA) is a 2-AFA whose input head movement is restricted to left, right, down or no move. Similarly, a two-way two-dimensional alternating finite automaton (TW2-AFA) is a 2-AFA with its input head movement restricted to left, down or no move. The definitions of a two-dimensional nondeterministic or deterministic finite automaton (2-NFA or 2-DFA) and a two-dimensional alternating, nondeterministic or deterministic Turing machine (2-ATM, 2-NTM or 2-DTM) and their restricted versions are straightforward. The language accepted by a two-dimensional machine M, denoted by L(M), is the set of the patterns accepted by M. The class of languages accepted by 2-NOTA's, 2-AFA's, TR2-AFA's, TW2-AFA's, 2-NFA's, or 2-DFA's is denoted by £(2-NOTA), £(2-AFA), £(TR2-AFA), £(TW2-AFA), £(2-NFA), or £(2-DFA), respectively.

In the following definition, we are concerned with two-dimensional TM's with square inputs. We use n to denote the height (or width) of a pattern.

Definition 2.7. Let S(n) be a function and M be a 2-ATM (or 2-DTM, or 2-NTM). A computation tree of M (on some input pattern) is S(n) space-bounded if all nodes of the tree are labeled with configurations using at most S(n) worktape space. (Note that 2-DTM's and 2-NTM's are special cases of 2-ATM's.)

The complement of a language L is denoted by $\bar{L}$. For any $m > 0$, $\lceil \log_2 m \rceil$ is simply written as log m.

3. £(2-NOTA) and £(2-AFA) are incomparable

Lemma 3.1. £(2-AFA) notinclude £(2-NOTA).

Proof. Consider the planar embedding of directed bipartite graphs with equal number of vertices on both sides. Let $\Sigma = \{v, l, r, u, d, +, x, 0\}$ be the alphabet used for the embedding. We use the following embedding rule. The symbol v represents a vertex, symbol + means an intersection of two edges (i.e. where they join or split), symbol x is for a cross-over of two edges, symbol 0 represents a blank space, and symbols u, d, l and r are the symbols needed to form upward, downward, leftward and rightward edges, respectively. Let pattern P be a planar embedding of a directed bipartite graph with n vertices on both sides. The size of P will be (4n+3)×(4n+2) (including the boundary markers). Row 2n+2 of P defines 2n vertices of a bipartite graph, where the left n vertices form one group and the right n vertices form the other. The 2n v's are placed such that there are two blanks separating the first n vertices from the second n vertices, and there is a blank between consecutive v's in both the left and right groups. The upper 2n+1 rows of P specify the set of (directed) edges from the left group of vertices to the right group of vertices and the lower 2n+1 rows specify the set of edges from the right group of vertices to the left group of vertices. An example of such embedding is given in Figure 3. It is easy to see that every directed bipartite graph with equal number of vertices on both sides can be embedded in the plane following the above rule.

Consider language $L_1 = \{ P \mid P \in \Sigma^{(2)}$ and P is a planar embedding of some acyclic directed bipartite graph with equal number of vertices on both sides$\}$. We show that L_1 can be accepted by a 2-AFA but not by any 2-NOTA.

We describe a 2-AFA A which accepts L_1. For a given pattern P, A first checks the correctness of embedding so that the nonboundary symbols of P are all from Σ and that the vertices are placed according to the embedding rule. A also verifies that every edge above the 2n+2-nd row connects one vertex in the left group with one vertex in the right group and that every edge below the 2n+2-nd row connects one vertex in the right group with one vertex in the left group. A then systematically scans the 2n+2-nd row of the pattern P. For every vertex v encountered, A checks that each directed path leading from vertex v does not enter any loop. If P is not an planar embedding of any acyclic directed bipartite graph, A will eventually enter an infinite loop and thus will reject P, otherwise A will accept P. Hence, L_1 can be accepted by a 2-AFA.

Suppose that L_1 is accepted by a 2-NOTA M. For each pattern P that is the embedding of an acyclic directed bipartite graph with equal number of vertices on both sides, fix an accepting computation of M on P and let c(P) denote the set of active states of M at row 2n+2. Clearly, there are k^{4n+2} possible c(P)'s, where k is the number of states of M.

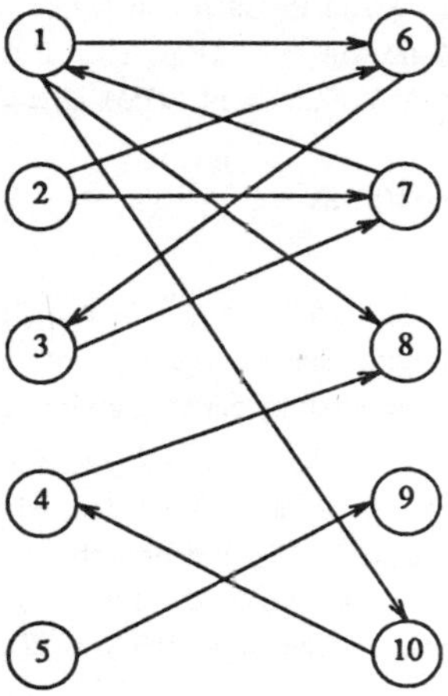

(a) A directed bipartite graph G.

```
######################
#+rrrrrrrrrr+rrr+rrr+#
#u          d   d   d#
#u +rrrrrrrr+r+ d   d#
#u u        d d d   d#
#u u +rrrrrrxr+ d   d#
#u u u      d d d   d#
#u u u +rrrrxrxr+   d#
#u u u u    d d d   d#
#u u u u +rrxrxrxr+ d#
#u u u u u  d d d d d#
#v0v0v0v0v00v0v0v0v0v#
#u   u u    d d     d#
#u   +lxllll+ d     d#
#u     u      d     d#
#+lllllxllllll+     d#
#      u            d#
#      u            d#
#      u            d#
#      u            d#
#      u            d#
#      +llllllllllll+#
######################
```

(b) The planar embedding of G.

Figure 3. An example of embedding.

Let P_U (P_L) be the upper (lower) half (i.e., 2n+1 rows) of a pattern as described above. Note that the 2n+2-nd row is for vertices. We fix an ordering on the edges from the right group of vertices to the left group of vertices as follows: for edges (a,b) and (c,d) in a graph, (a,b) < (c,d) if a is to the left of c or a = c and b is to the left of d in the embedding. We also fix an ordering on the P_L's of the patterns of the same size as follows. Let $P_L^1 = \{ e_1^1, e_2^1, \cdots, e_s^1 \}$ and $P_L^2 = \{ e_1^2, e_2^2, \cdots, e_t^2 \}$, where $e_1^1 > e_2^1 > \cdots > e_s^1$ and $e_1^2 > e_2^2 > \cdots > e_t^2$. Then $P_L^1 < P_L^2$ if there exists some m such that $m < t$, $e_1^1 = e_1^2, \ldots, e_m^1 = e_m^2$, and either $m = s$ or $m < s$ and $e_{m+1}^1 < e_{m+1}^2$.

For each P_U, let $L(P_U) = \max \{ P_L \mid$ pattern P formed by P_U, 2n+2-nd row and P_L is acyclic$\}$. $L(P_U)$ will be called the maximum match of P_U. Let $MAXLP(n) = \{ L(P_U) \mid$ for some P_U of size $(2n+1)\times(4n+2)$ $\}$. The following proposition can be shown.

Proposition 3.1. $|MAXLP(n)| \geq n!$.

Proof. We will construct n! P_U's whose maximum matches are pairwise different. Let $n \geq 1$ be any integer and $VL = \{1, 2, \ldots, n\}$ and $VR = \{n+1, n+2, \ldots, 2n\}$ denote the two groups of vertices, where vertex i is placed to the left of vertex i+1. Let $\pi = v_1, v_2, \cdots, v_n$ be any permutation of VL. Construct a P_U, denoted by $P_U(\pi)$, as follows:

$$\begin{aligned} P_U(\pi) = \{ & (v_n, n+1), (v_n, n+2), (v_n, n+3), \ldots, (v_n, 2n-1), \\ & (v_{n-1}, n+1), (v_{n-1}, n+2), \ldots, (v_{n-1}, 2n-2), \\ & (v_{n-2}, n+1), \ldots, (v_{n-2}, 2n-3), \\ & \cdots \\ & (v_2, n+1) \}. \end{aligned}$$

Then one can easily verify that

$$\begin{aligned} L(P_U(\pi)) = \{ & (2n, v_1), (2n, v_2), (2n, v_3), \ldots, (2n, v_n), \\ & (2n-1, v_1), (2n-1, v_2), \ldots, (2n-1, v_{n-1}), \\ & (2n-2, v_1), \ldots, (2n-2, v_{n-2}), \\ & \cdots \\ & (n+1, v_1) \}. \end{aligned}$$

Clearly, for any two different permutations π_1 and π_2 of VL, $L(P_U(\pi_1)) \neq L(P_U(\pi_2))$. Since there are n! different permutations of VL, $|MAXLP(n)| \geq n!$.

(Proof of Lemma 3.1 continued) Let n be sufficiently large such that $n! > k^{4n+2}$. For each P_U, define $P(P_U)$ as the pattern formed by P_U, the 2n+2-nd row and $L(P_U)$. Then, there exist P_U^1 and P_U^2 such that $L(P_U^1) < L(P_U^2)$ and $c(P(P_U^1)) = c(P(P_U^2))$. Now let P be the pattern formed by P_U^1, the 2n+2-nd row and $L(P_U^2)$. Then P must also be accepted by M. Since M accepts L_1, P is acyclic. But this contradicts the definition of $L(P_U^1)$. Hence, L_1 is not accepted by any 2-NOTA. □

Lemma 3.1 answers in the negative the open question in [IT89a] and [IN88].

Lemma 3.2. For every 2-AFA A, $\overline{L}(A)$ is accepted by a 2-NOTA.

Proof. Let A = (Q, Σ∪{#}, δ, q_0, U, F) be a 2-AFA. Define the complement 2-AFA of A to be $\overline{A}$ = (Q, Σ∪{#}, δ, q_0, Q–U, Q–F). That is, $\overline{A}$ is obtained by swapping the universal and existential states and the accepting and nonaccepting states of A. Note that in general, $\overline{L}(A) \neq L(\overline{A})$, since A may reject an input by entering an infinite loop.

We construct a 2-NOTA M to accept $\overline{L}(A)$. Let x be an input pattern. Given x, M tries to guess and verify the existence of a (possibly infinite) computation tree of $\overline{A}$ on x whose leaves are all labeled with accepting configurations. Let π denote the computation tree of $\overline{A}$ on x that M will guess. Let R(i,j) denote the set of all states of $\overline{A}$ when its input head is at the (i,j)-cell, $1 \le i \le row(x)$, $1 \le j \le col(x)$, in the guessed computation tree π. For each q ∈ R(i,j), call (x,i,j,q) a configuration (of $\overline{A}$) represented by q. For convenience, let R(i,0) = R(i,col(x)+1) = ∅, $1 \le i \le row(x)$, and R(0,j) = R(row(x)+1,j) = ∅, $1 \le j \le col(x)$.

Generally, the (i,j)-cell of M operates as follows. It receives the sets R(i,j–1) and R(i,j) from the (i,j–1)-cell and the sets R(i–1,j) and R(i,j) from the (i–1,j)-cell. It guesses the sets R(i,j+1) and R(i+1,j) and verifies that R(i,j) is consistent with the neighboring sets R(i–1,j), R(i,j–1), R(i+1,j), R(i,j+1). That is, the following conditions must hold: (a) None of the members of R(i,j) represents a terminating nonaccepting configuration; (b) If q ∈ R(i,j) and q is universal, then all immediate successors of the configuration (x,1,1,q) are represented by the states contained in R(i–1,j)∪R(i,j–1)∪R(i+1,j)∪R(i,j+1)∪R(i,i); and (c) If q ∈ R(1,1) and q is existential, then at least one of the immediate successors of the configuration (x,1,1,q) are represented by the states contained in R(i–1,j)∪R(i,j–1)∪R(i+1,j)∪R(i,j+1)∪R(i,i). Also, the (i,j)-cell passes the sets R(i,j) and R(i,j+1) to the (i,j+1)-cell and the sets R(i,j) and R(i+1,j) to the (i+1,j)-cell. In addition, the (1,1)-cell makes sure that R(1,1) contains q_0.

The 2-NOTA M constructed above verifies that for every configuration in the guessed tree π, either it is a terminating accepting configuration or it is nonterminating and all (or at least one, depending on whether it is universal or existential) of its immediate successor configurations exist. In other words, M verifies that π is a (possibly infinite) computation tree of $\overline{A}$ on x whose leaves are all labeled with accepting configurations. It is easy to see that, if x is rejected by A, then there exists a (possibly infinite) computation tree of $\overline{A}$ on x whose leaves are all labeled with accepting configurations, and vice versa. Hence, M accepts $\overline{L}(A)$. □

Lemma 3.3. £(2-NOTA) notinclude £(2-AFA).

Proof. Suppose that £(2-NOTA) ⊆ £(2-AFA). Let L_1 be the same language that we considered in the proof of Lemma 3.1. Since L_1 is accepted by a 2-AFA, $\overline{L}_1$ is also accepted by some 2-NOTA, by Lemma 3.2. Thus, by hypothesis, $\overline{L}_1$ is accepted by a 2-AFA. By Lemma 3.2, L_1 is accepted by a 2-NOTA. But we have already shown in the proof of Lemma 3.1 that L_1 is not accepted by any 2-NOTA. Hence, £(2-NOTA) notinclude £(2-AFA). □

Combining Lemmas 3.1 and 3.3, we have the following result, which answers in the negative the open question in [IT89a].

Theorem 3.1. £(2-NOTA) and £(2-AFA) are incomparable.

Theorem 3.1 would still hold if a 2-NOTA is replaced by a TR2-NTM (i.e., three-way two-dimensional nondeterministic TM) operating in a suitable space. Let TR2-NSPACE(S(n)) stand for the class of languages accepted by S(n) space-bounded TR2-NTM's. Clearly, £(2-NOTA) ⊆ TR2-NSPACE(n). It was an open question in [IN88] whether £(2-AFA) ⊆ TR2-NSPACE(n). By Lemma 3.3, TR2-NSPACE(n) notinclude £(2-AFA). It is easy to see that the proof of Lemma 3.1 still works if £(2-NOTA) is replaced by TR2-NSPACE(S(n)) for any S(n) = o(nlogn). Thus, L_1 cannot be accepted by any S(n) space-bounded TR2-NTM and £(2-AFA) notinclude TR2-NSPACE(S(n)) for any S(n) = o(nlogn). Hence, we have the following theorem, which answers in the negative the open question in [IN88].

Theorem 3.2. For any function S(n) such that $S(n) = \Omega(n)$ and $S(n) = o(n\log n)$, TR2-NSPACE(S(n)) and £(2-AFA) are incomparable.

Also, it is easy to show the following theorem.

Theorem 3.3. £(TR2-AFA) $\subsetneq$ £(2-NOTA).

Proof. We shall show in Theorem 4.4 that £(TR2-AFA) is closed under complementation. The inclusion follows from Lemma 3.2. That it is proper follows since £(TR2-AFA) $\subseteq$ £(2-AFA) and £(2-AFA) is incomparable with £(2-NOTA). □

4. Closure properties under complementation

Now we consider closure properties under complementation for 2-NOTA's, 2-AFA's, TR2-AFA's and 2-UFA's.

The following result has already been shown in [IN77b]. Using Lemmas 3.1 and 3.2, we can give a very simple proof.

Theorem 4.1 [IN77b]. £(2-NOTA) is not closed under complementation.

Proof. Suppose that £(2-NOTA) is closed under complementation. Let L_1 be the language used in the proof of Lemma 3.1. Since $L_1 \in$ £(2-AFA), $\overline{L}_1 \in$ £(2-NOTA) by Lemma 3.2. Thus, by the hypothesis, $L_1 \in$ £(2-NOTA). This contradicts Lemma 3.1. □

The next result answers an open question in [IN88].

Theorem 4.2. £(2-AFA) is not closed under complementation.

Proof. Suppose that £(2-AFA) is closed under complementation. Let L_1 be the language used in the proof of Lemma 3.1. Then $L_1 \in$ £(2-AFA). By the assumption, $\overline{L}_1 \in$ £(2-AFA). Thus, $L_1 \in$ £(2-NOTA) by Lemma 3.2. This contradicts Lemma 3.1. □

The results in [IT82] imply that £(TR2-UFA) is not closed under complementation. Using the above results, we can easily show that £(2-UFA) is not closed under complementation.

Theorem 4.3. £(2-UFA) is not closed under complementation.

Proof. Consider the language L_1 used in the proof of Lemma 3.1. One can easily see that the 2-AFA A that accepts L_1 has only universal states. Thus L_1 is accepted by a 2-UFA. If £(2-UFA) is closed under complementation then $\overline{L}_1$ is also in £(2-UFA). By Lemma 3.2, L_1 will be in £(2-NOTA). But we know that $L_1 \notin$ £(2-NOTA). □

It is obvious that the class of languages accepted by TW2-AFA's is closed under complementation. We can show that the same result holds even for TR2-AFA's. This answers another open question in [IN88].

Theorem 4.4. £(TR2-AFA) is closed under complementation.

Proof. Let A be a TR2-AFA and $\overline{A}$ be the complement of A as described in the proof of Lemma 3.2. Note that, $L(\overline{A}) \neq \overline{L}(A)$ generally. The problem again is that $\overline{A}$ may enter an infinite loop when it should accept. But we can construct a TR2-AFA A′ from $\overline{A}$ such that $L(A') = \overline{L}(A)$. The idea is for A′ to simulate $\overline{A}$ and, meanwhile, detect if it enters an infinite loop. Observe that $\overline{A}$ can enter an infinite loop only when its head moves back and forth along a same row of the input pattern.

Let x be a given input pattern. A′ operates as follows. A′ simulates $\overline{A}$ on x. When $\overline{A}$ arrives at a new row, A′ existentially performs the following two tasks:

(1) simulates $\overline{A}$, or

(2) verifies that $\overline{A}$ will enter an infinite loop on this row.

To do (2), we need to define two relations L and R, over Q, where Q is the state set of $\overline{A}$. For row i of the input pattern,

$L(j) = \{ (q_1, q_2) \mid$ from configuration (x,i,j, q_1), $\bar{A}$ can enter configuration (x,i,j, q_2) after one or more steps without shifting its head to the right of cell $x[i,j]$ $\}$;

$R(j) = \{ (q_1, q_2) \mid$ from configuration (x,i,j, q_1), $\bar{A}$ can enter configuration (x,i,j, q_2) after one or more steps without shifting its head to the left of cell $x[i,j]$ $\}$.

Suppose that the input head reaches row i at cell x[i,j]. Then A′ guesses the relations L(j) and R(j), and enters a universal state with three branches. Two of the branches will have A′ shift its input head towards the left and right ends of row i, respectively. In the left-to-right sweep (the right-to-left sweep is similar), A′ calculates L(j′) from L(j′−1), guesses R(j′) and verifies that it is consistent with R(j′−1) for each j′ = j+1, j+2, ..., col(x). The purpose of these two branches is to verify that L(j) and R(j) are correctly guessed. It is easy to see that $\bar{A}$ enters an infinite loop iff there exists a j_0, $1 \le j_0 \le col(x)$, such that either $R(j_0)$ or $L(j_0)$ contains (q_1, q_1) for some q_1. The existence of such a j_0 can easily be detected by A′ using the third branch which moves existentially towards the left or the right end of row i, computes L(j′)'s and R(j′)'s, and searches for the j_0. Hence A′ accepts $\bar{L}(A)$. □

Corollary 4.1.
(1) Every 2-DFA can be converted to a halting one;
(2) Every TR2-AFA can be converted to a halting one;
(3) There are 2-AFA's that cannot be converted to halting ones;
(4) There are 2-UFA's that cannot be converted to halting ones.

Proof. (1) A 2-DFA can be converted to a halting one by using Sipser's technique [SI78].
(2) This is implied by the proof of Theorem 4.4.
(3) It is easy to see that halting 2-AFA's can be simulated by 2-NOTA's (by using the technique in the proof of Lemma 3.2). By Lemma 3.1, we know that there are 2-AFA's that cannot be simulated by 2-NOTA's.
(4) It is similar to the proof of (3). □

5. Closure properties under rotation

As mentioned in Definition 2.1 that input patterns can be rotated clockwise 90°, 180° and 270°. In this section, we consider some closure properties under rotation of input patterns. Note that all computations start from the upper-left corner of the input pattern. The following theorem is easy to prove.

Theorem 5.1. £(2-DFA), £(2-NFA), £(2-UFA), £(2-AFA) and £(2-NOTA) are closed under 90°, 180° and 270°-rotations.

It is obvious that the classes of languages accepted by restricted 2-DFA's and 2-NFA's (i.e., TR2-DFA's, TW2-DFA's, TR2-NFA's, and TW2-NFA's) are not closed under rotations. The next two theorems show that the same thing is true for restricted 2-AFA's. The results are shown by considering different candidate languages.

Theorem 5.2. £(TR2-AFA) is not closed under 90°, 180° and 270°-rotations.

Proof. The proof that £(TR2-AFA) is not closed under 90°-rotation was given in [IT88a]. For the completeness of the paper, here we include the proof.

For 90°-rotation, let

$$L_2 = \{ x[1..n,1..n] \mid x \in \{0,1,\#\}^{(2)}, x[1..n,2] = x[n..1,2], x[i,j] = 0, \text{ for } 2 \le i \le n\text{-}1 \text{ and } 2 < j \le n\text{-}1 \}.$$

Since £(TR2-AFA) is closed under complementation, it is sufficient to show that $\bar{L}_2$ is accepted by a TR2-AFA A. Let x be an input pattern. The TR2-AFA A scans along the second column of the input pattern x. It guesses the cell x[i,2] which will be the discrepancy (i.e., x[i] ≠ x[n-i+1,2]). The symbol x[i,2] is remembered. A then moves the input head along the diagonal until the middle row is reached. A enters a universal state with two branches. One verifies that it is indeed the middle row. Another moves the input head along the opposite diagonal until the second column is reached, and when this happens, A's input head will be under the cell x[n-i+1]. The discrepancy can then be verified. Hence, L_2 is accepted by a TR2-AFA.

Let L'_2 be the language obtained through 90°-rotation of L_2. It was shown in [IT88a] that L'_2 is not accepted by any TR2-AFA's. (The proof is based on the well-known crossing sequence technique.)

For 180°-rotation, let

$$L_3 = \{ x[1..n,1..n] \mid x \in \{0,1,\#\}^{(2)}, x[n-1,1..n] = x[n-1,n..1], x[i,j] = 0, \text{ for } 2 \le i < n-1 \text{ and } 2 \le j \le n-1 \}.$$

Similarly, L_3 is accepted by by a TR2-AFA, but the 180°-rotation of L_3 is not accepted by any TR2-AFA's.

The candidate language for 270°-rotation is

$$L_4 = \{ x[1..n,1..n] \mid x \in \{0,1,\#\}^{(2)}, x[1..n,n-1] = x[n..1,n-1], x[i,j] = 0, \text{ for } 2 \le i \le n-1 \text{ and } 2 \le j < n-1 \}.$$

One can easily see that L_4 is accepted by a TR2-AFA, but the 270°-rotation of L_4 is not accepted by any TR2-AFA's. □

Theorem 5.3. £(TW2-AFA) is not closed under 90°, 180° and 270°-rotations.

Proof. For simplicity, we only consider 180°-rotation. The 90° and 270°-rotations are similar and we leave the proofs to the reader.

For 180°-rotation, consider language

$$L_5 = \{ x \mid x \in \{0,1,\#\}^{(2)}, \text{ there is some } i \ge 2, x[2,i] = x[i,2] = 1, \text{ all other entries of } x \text{ are } 0, \text{ except for the boundaries } \}.$$

It can easily be shown that the 180° rotation of L_5 is accepted by a TW2-AFA.

Suppose that L_5 is accepted by TW2-AFA M. Let x be a pattern in L_5 with col(x) = m and row(x) = n. Without loss of generality assume that M moves on every step. Let T be an accepting computation tree of M on x. Consider M's operations on the first and the second column of x and let T' be the part of T corresponding to the second column of x. Note, since M is two-way (input head can only move right or down), once the input head leaves the second column, it cannot come back. Thus, T' is a subtree of T obtained by purging nodes involving columns 3, ..., m. Let S(h) denote the set of states at the h+1-st level of T', i.e., S(h) is the set of all states that M will enter after h steps when the input head is shifted to entry x[h,2]. Let k be the size of M's state set. It is easy to see that, if n is sufficiently large (the value of n will depend on k) and we choose i = n/2, then there exist

$$2 \le c_1 < c_2 < \dots < c_f < i < p_1 < p_2 < \dots < p_g \le n-1$$

such that

$$S(c_1) = S(c_2) = S(c_3) = \dots = S(c_{f-1}) = S(c_f),$$
$$S(p_1) = S(p_2) = S(p_3) = \dots = S(p_{f-1}) = S(p_g),$$
$$c_{j+1} - c_j \le 2^k, \text{ for } 1 \le j < f,$$
$$p_{j+1} - p_j \le 2^k, \text{ for } 1 \le j < g,$$

where f and g are numbers much larger than 2^k. Let

$$d_j = c_{2j} - c_{2j-1}, \; j = 1, 2, \dots, f/2,$$
$$q_j = p_{2j} - p_{2j-1}, \; j = 1, 2, \dots, g/2.$$

Then, clearly there exist subset C of {1, 2, ..., f/2} and subset P of {1, 2, ..., g/2} such that

$$\sum_{j \in C} d_j = \sum_{j \in P} q_j = t, \text{ for some } t$$

Thus, if we cut off levels c_{2j-1} to c_{2j} for all j in C, and duplicate levels p_{2j-1} to p_{2j}, for all j in P, in the subtree T', we still get a valid subtree. Let T'' denote the new subtree. Clearly if we replace T' by T'' in tree T, we get an accepting computation tree for the following pattern x'

$$x'[2,i] = x'[i\text{-}t,2], \text{ and all other interior entries of } x' \text{ are } 0.$$

But x' is not in L_5. A contradiction! □

It was shown in [IT89a] that £(2-DOTA) is equivalent to £(TW2-AFA) through 180°-rotation. Combining this result with Theorem 5.3, we have the following Corollary.

Corollary 5.1. £(2-DOTA) is not closed under 90°, 180° and 270° rotations.

6. Conclusion

It is straightforward to see that all results in sections 3, 4, and 5 hold for square input patterns. It is still unknown whether £(2-NFA) and £(TR2-DFA) are closed under complementation. It also remains an open question whether connected pictures (the definition of connected pictures is given in, e.g., [SE72]) can be accepted by 2-DFA's, 2-NFA's or 2-NOTA's.

References

[BL67] Blum, M. and C. Hewitt, Automata on a 2-dimensional tape, *IEEE Symp. on Switching and Automata Theory*, (1967), pp. 155-160.

[HO67] Hopcroft, J.E. and J.D. Ullman, Some results on tape-bounded Turing machines, *J. ACM* 16-1, 1967, pp. 168-177.

[HR89] Hromkovic, J., K. Inoue and I. Takanami, Lower bounds for language recognition on two-dimensional alternating multihead machines, *J. of Computer and System Sciences* 38, (1989), pp. 431-451.

[IB74] Ibarra, O. and R. Melson, Some results concerning automata on two-dimensional tapes, *Intern. J. Comp. Math.* 4-A, (1974), pp. 269-279.

[IN77a] Inoue, K., and A. Nakamura, Some properties of two-dimensional on-line tessellation acceptors, *Info. Sci.*, (1977), pp. 95-121.

[IN77b] Inoue, K., and A. Nakamura, Nonclosure properties of two-dimensional on-line tessellation acceptors and one-way parallel sequential array acceptors, *IECE of Japan Trans. (E)*, Sept., 1977, pp. 475-476.

[IN83] Inoue, K., I. Takanami and H. Taniguchi, Two-dimensional alternating Turing machines, *Theoret. Comp. Sci.* 27, (1983), pp. 61-83.

[IN85] Inoue, K., A. Ito, I. Takanami and H. Taniguchi, A space hierarchy result on two-dimensional alternating Turing machines with only universal states, *Info. Sci.* 35, (1985), pp. 79-90.

[IN88] Inoue, K. and I. Takanami, A survey of two-dimensional automata theory, *IMYCS*, (1988), pp. 21-35.

[IT82] Ito, A., K. Inoue, I. Takanami and H. Taniguchi, Two-dimensional alternating Turing machines with only universal states, *Info. and Control* 55, (1982), pp. 193-221.

[IT83] Ito, A., K. Inoue, I. Takanami and H. Taniguchi, A note on space complexity of nondeterministic two-dimensional Turing machines, *IECE of Japan Trans. (E)* (1983), pp. 508-509.

[IT88a] Ito, A., K. Inoue and I. Takanami, A note on three-way two-dimensional alternating Turing machines, *Info. Sci.* 45, (1988), pp. 1-22.

[IT88b] Ito, A., K. Inoue and I. Takanami, A relationship between one-dimensional bounded cellular acceptors and two-dimensional alternating finite automata, *Informatik-Skripten* 21, T. U. Braunschweig (1988), pp. 60-76.

[IT89a] Ito, A., K. Inoue and I. Takanami, Deterministic two-dimensional on-line tessellation acceptors are equivalent to two-way two-dimensional alternating finite automata through 180°-rotation, *Theoret. Comp. Sci.* 66, (1989), pp. 273-287.

[IT89b] Ito, A., K. Inoue and I. Takanami, Some closure properties of the class of sets accepted by three-way two-dimensional alternating finite automata, *The Trans. of the IEICE* E 72, No. 4, (1989), pp. 348-350.

[SE72] Selkow, S., One-pass complexity of digital picture properties, *J. ACM* 19 (2), (1972), pp. 283-295.

[SI78] Sipser M., Halting space-bounded computations, *Symp. on Foundations of Comp. Sci.*, (1978), pp. 73-74.

[SZ89] Szepietowski, A., On three-way two-dimensional Turing machines, *Info. Sci.* 47, (1989), pp. 135-147.

Infinite Normal Forms for Non-linear Term Rewriting Systems

P. Inverardi ♥, *M. Nesi* ♥♦

♥ IEI - CNR, via S. Maria 46, I-56126 Pisa, ITALY

♦ Computer Laboratory, University of Cambridge, Cambridge, UK

1. Introduction

Recently, a great amount of work has been dedicated to the study of non-terminating rewriting relations [DK89, DKP89, DKP91, KKSV90, KKSV91, Giv90]. These attempts have the merit of sheding light on the nature of non-terminating relations thus permitting to extend the rewriting setting to a number of interesting equational theories.

In this paper we discuss the applicability of the framework defined in [DK89, DKP89, DKP91] about the existence of infinite normal forms on a particular class of non-left-linear term rewriting systems.

The approach has been used in [IN90b] to prove the existence of infinite normal forms for recursive (finite state) CCS expressions [Mil80] with respect to a correct and complete axiomatization of the observational congruence given by Milner [Mil89]. In fact, our interest in non-terminating non-linear rewriting systems comes from the experience we have made by developing verification techniques for the CCS language based on term rewriting [DIN90, IN90a]. In that framework it results that all the axiomatic characterizations of the various behavioural equivalences contain non-left-linear rules. On the other hand, non-termination arises as soon as one wants to consider recursive processes.

In particular, we deal with term rewriting systems in which the non-terminating rules are *unfolding* rules that model the operational semantics of a recursive operator. The left-linearity requirement is replaced by a *retraction* property of the supporting term algebra that allows the definition of a rewriting relation modulo an equivalence relation induced on the set of terms by the unfolding rules. With these two assumptions we can still restrict to consider, as in [DK89, DKP89, DKP91], only a subset of infinite derivations, i.e. fair derivations. Actually, we go further on by focussing on those rewriting systems which admit a peculiar kind of fair derivations, i.e. *uniform* systems and *structured* fair derivations. The ω-confluence of these rewriting systems can then be proved by properly constraining the possible interaction between the non-terminating rules and the remaining rules. In this respect the notions of *independence* and *preservation* on the rules of the rewriting system are introduced.

2. Basic Definitions

We assume that the reader is familiar with the basic concepts of term rewriting systems. We summarize the most relevant definitions below, while we refer to [DK89, DKP89, DKP91] for more details.

Let Σ be a set of operators, V be a set of variables and $T_\Sigma(V)$ denotes the set of terms over Σ and V. A *term rewriting system* (TRS) R is any set $\{(l_i, r_i) \mid l_i, r_i \in T_\Sigma(V), \mathcal{V}(r_i) \subseteq \mathcal{V}(l_i)\}$. The pairs (l_i, r_i) are

Work partially supported by Progetto Finalizzato Sistemi Informatici e Calcolo Parallelo, Obiettivo LAMBRUSCO.

called *rewriting rules* and written $l_i \rightarrow r_i$. The *rewriting relation* $\rightarrow_R$ on $T_\Sigma(V)$ is defined as the smallest relation containing R that is closed under monotonicity and substitution. A term t *rewrites* to a term s, written $t \rightarrow_R s$, if there exist a rule $l_k \rightarrow r_k$ in R, a substitution σ and a subterm t/u at the occurrence u, called *redex*, such that $t/u = l_k\sigma$ and $s = t[u \leftarrow r_k\sigma]$. A term t is an *instance* of a term s if there exists a substitution σ such that $t = s\sigma$. A TRS R is *left-linear* if the left-hand side l of each rule $l \rightarrow r$ in R has at most one occurrence of any variable. A term t is said to *overlap* a term t' if t unifies with a non-variable subterm of t' (after renaming the variables in t so as not to conflict with those in t'). A TRS R is *non-overlapping* if no left-hand side overlaps another.
Let $\xrightarrow{+}$ and $\xrightarrow{*}$ denote the transitive and transitive-reflexive closure of $\rightarrow$, respectively.
A TRS R is *finitely terminating* if there is no infinite sequence $t_1 \rightarrow_R t_2 \rightarrow_R \ldots$ of terms. A TRS R is *confluent* if whenever $s \;{}_R\!\xleftarrow{*}\; t \xrightarrow{*}_R q$, then there exists a term t' such that $s \xrightarrow{*}_R t' \;{}_R\!\xleftarrow{*}\; q$, while R is *locally confluent* if whenever $s \;{}_R\!\leftarrow t \rightarrow_R q$, then there exists a term t' such that $s \xrightarrow{*}_R t' \;{}_R\!\xleftarrow{*}\; q$.
A term t is *in R-normal form* if there exists no term s such that $t \rightarrow_R s$. A term s is an *R-normal form* of t if $t \xrightarrow{*}_R s$ and s is in R-normal form. A TRS R is *canonical* if it is finitely terminating and confluent.

Let $T_\Sigma{}^\infty(V)$ denote the set of finite and infinite terms over Σ and V. It is possible,[DKP91], to form a complete ultra-metric space on $T_\Sigma{}^\infty(V)$, by defining a notion of *distance d* between two terms *s*, *t* such that $d(s,t) = 1/2^{v(s,t)}$ where $v(s,t)$ is the smallest depth of a symbol occurrence at which terms *s* and *t* differ, with the convention that $d(t,t) = 0$. Given a TRS R, it is straightforward to define $\rightarrow_R$ over $T_\Sigma{}^\infty(V)$ [DKP91]. Let $\rightarrow$ be a (possibly non-terminating) rewriting relation. A term t (ω-)*rewrites* to t', written $t \rightarrow^\omega t'$, if $t \xrightarrow{*} t'$ or if there exists an infinite *derivation* $t = t_0 \rightarrow t_1 \rightarrow \ldots \rightarrow t_n \rightarrow \ldots$ such that $\lim_{n\rightarrow\infty} t_n = t'$. The relation $\rightarrow$ is ω-*terminating* if for any infinite derivation $t = t_0 \rightarrow t_1 \rightarrow \ldots \rightarrow t_n \rightarrow \ldots$ of terms, the limit $\lim_{n\rightarrow\infty} t_n$ exists. The relation $\rightarrow$ is *top-terminating* if there are no infinite derivations $t = t_0 \rightarrow t_1 \rightarrow \ldots \rightarrow t_n \rightarrow \ldots$ of terms with infinitely many rewrites at the topmost occurrence. The relation $\rightarrow$ is ω-*confluent* if ${}^\omega\!\leftarrow \bullet \rightarrow^\omega$ implies $\rightarrow^\omega \bullet {}^\omega\!\leftarrow$. In other words, for any t, s, q such that $s \;{}^\omega\!\leftarrow t \rightarrow^\omega q$, there exists t' such that $s \rightarrow^\omega t' \;{}^\omega\!\leftarrow q$. The relation $\rightarrow$ is ω-*canonical* if it is ω-terminating and ω-confluent. A term t' is an ω-*normal form* of t if $t \rightarrow^\omega t'$ and t' is minimal for $\rightarrow$, i.e. if $t' \rightarrow t''$, then $t'' = t'$. Thus, an ω-normal form need not be irreducible.

Among all the infinite derivations from any term, it is possible to single out "interesting" derivations. A derivation $t_0 \rightarrow t_1 \rightarrow \ldots \rightarrow t_n \rightarrow \ldots$ is *fair* if whenever there is a rule $l \rightarrow r$ and an occurrence u such that, for all n past some N, the subterm t_n/u is a redex for $l \rightarrow r$, then (at least) one of the rule applications $t_n \rightarrow t_{n+1}$ ($n \geq N$) is an application of $l \rightarrow r$ at u.
Thus, a fair derivation guarantees that a redex does not persist forever. Note that this definition does not prevent the fact that the same rewriting rule is applicable infinitely many times at different occurrences.

3. Dealing with Non-left-linear Term Rewriting Systems

In this section, we introduce a particular class of non-left-linear term rewriting systems and discuss which hypotheses have to be assumed in order to guarantee the existence of ω-normal forms. The first result we have to show is that fair derivations are still the only derivations we have to look at. This means that the limit of a fair derivation is an ω-normal form and, viceversa, the ω-normal form of any term can be computed as the limit of a fair derivation. This result has been shown in [DK89, DKP89, DKP91] by using the hypothesis of left-linearity of the term rewriting system.

Definition 1 (unfolding) An *unfolding* is an equation $G = H[G]$, where G is a non-variable subterm of t_i, $1 \le i \le k$, in the context $H[\ ] = op(t_1, \dots, t_k)$ for some $op \in \Sigma$ of arity k.
Definition 2 (unfolding rule) Given an unfolding $G = H[G]$, the non-terminating rule $G \to H[G]$ is an *unfolding rule.*
From now on, let us call *recursive* the terms that can be rewritten by an unfolding rule.
Definition 3 (retraction) Let S be a set of unfolding rules over $T_\Sigma(V)$. $T_\Sigma(V)$ is *retractile* if it is possible to define on it a minimal equivalence relation, $=_C$, such that for any two terms $t_1, t_2 \in T_\Sigma(V)$, $t_1 =_C t_2$ iff:
i. it exists a term $t' \in T_\Sigma^\infty(V)$ such that $t_1 \to_S^\omega t' \ {}^\omega{}_S\!\leftarrow t_2$ and t' is an ω-normal form with respect to $\to_S$;
ii. for each equivalence class of $=_C$ it is possible to determine a unique (finite) canonical representative.

The retraction property means that, given any class of finite terms which rewrite via $\to_S$ into the same unfolded term $t \in T_\Sigma^\infty(V)$, it is possible to single out a unique (finite) canonical representative of the class, $C(t) \in T_\Sigma(V)$. Thus, an equivalence, $=_C$, on $T_\Sigma(V)$, can be found such that for any two terms t_1, t_2, the equivalence $t_1 =_C t_2$ holds if and only if $C(t_1) = C(t_2)$.

Definition 4 ($\to_T$) Let $T = R \cup S$ be a term rewriting system, defined over a retractile $T_\Sigma(V)$ equipped with $=_C$, such that R is a finitely terminating term rewriting system and S only contains unfolding rules. The rewriting relation $\to_T$ is defined as follows: $t \to_T s$ if there exist a rule $l \to r$ in T, a substitution σ and a subterm t/u at the occurrence u, such that $t/u =_C l\sigma$ and $s = t[u \leftarrow r\sigma]$.
Let us denote the application of only rules in R, S with $\to_{T\setminus R}$ and $\to_{T\setminus S}$, respectively.
The rewriting relation $\to_T$ could have been defined relying only on the equivalence relation $=_C$, i.e. $\to_T =_{def} =_C \bullet \to_R \bullet =_C$, that is avoiding the use of $\to_S$. But this is stronger than requiring the decidability of the equivalence $=_C$, since it implies the generation of the related equivalence classes.
Note that the application of $\to_{T\setminus R}$ might depend on rewriting by $\to_{T\setminus S}$. Roughly, it might be necessary to rewrite a term t by $G \to H[G]$ in order to obtain redexes for $\to_{T\setminus R}$ that arise in t by replacing G with H[G], see Section 4 for an example.

Example 1
The term rewriting system $T = R \cup S$ where

R	$h(x, e) \to x$	S	$f(x) \to g(f(x))$
	$h(x, x) \to x$		

is not finitely terminating and non-left-linear. R is a finitely terminating term rewriting system and S contains only one non-terminating rule. In this particular case, the equivalence $=_C$ is trivially defined by the unfolding $f(x) = g(f(x))$, which collapses the terms $g^n(f(t))$, for all n and t, to a canonical representative f(t') for some fixed t'. Consequently the canonical representative of all the other terms can be obtained. Note that, as soon as S contains more than one rule, it may easily happen that the equivalence $t_1 =_C t_2$ cannot be proved by applying a finite number of unfolding steps. Let us, for example, consider $T = R \cup S'$ where S' is as follows:

S' $\quad f_1(c, g_1(a, h(c,b))) \to g_1(a, h(f_1(c, g_1(a, h(c,b))), b))$

$\quad\ \ f_1(c, h(g_1(a, c), b)) \to h(g_1(a, f_1(c, h(g_1(a, c), b))), b)$

and the two terms $t_1 = f_1(c, g_1(a, h(c,b)))$ and $t_2 = g_1(a, f_1(c, h(g_1(a, c), b)))$. In this case, $t_1 =_C t_2$, but it does not exist a term $t' \in T_\Sigma(V)$ such that $t_1 \overset{*}{\to}_{T\setminus S} t' \ {}_{T\setminus S}\!\overset{*}{\leftarrow} t_2$. ♦

Note that, as regards the equivalence $=_C$, when dealing with recursive terms that are regular set of equations, results exist that allow to compute the unique canonical representative in the class of the terms with the same (tree) semantics, e.g. [CKV74]. This means that our notion of retraction actually permits coping with a reasonably interesting class of infinite rewriting systems.

In all the following definitions and propositions, when not explicitely stated, it is assumed to deal with the above characterized term rewriting systems T, with the further hypothesis of top-termination. Note that, the ω-termination of $\rightarrow_T$ follows from top-termination by Theorem 11 in [DKP89].

We are now in the position to derive the result about fair derivations and ω-normal forms similar to the result in [DK89, DKP89, DKP91] by replacing the left-linearity hypothesis with the retraction condition on the supporting algebra $T_\Sigma(V)$. Then, we introduce and discuss some requirements on the rewriting relation $\rightarrow_T$, which allow us to guarantee its ω-confluence.

Proposition 1 Given the rewriting relation $\rightarrow_T$ and any term $t \in T_\Sigma(V)$, then:

i. if t admits an ω-normal form t', then it exists a fair derivation $t_0 \rightarrow_T t_1 \rightarrow_T \ldots \rightarrow_T t_n \rightarrow_T \ldots$ with $\lim_{n\rightarrow\infty} t_n = t'$;

ii. for any fair derivation $t_0 \rightarrow_T t_1 \rightarrow_T \ldots \rightarrow_T t_n \rightarrow_T \ldots$ with $\lim_{n\rightarrow\infty} t_n = t'$, t' is an ω-normal form of t.

Sketch of the proof

i. This part of the proof carries on similarly to [DKP91]. Given a non-fair derivation with an ω-normal form as limit, by definition of $\rightarrow_T$ it is possible to build a corresponding fair derivation.

ii. Let D be a fair derivation $t = t_0 \rightarrow_T t_1 \rightarrow_T \ldots \rightarrow_T t_n \rightarrow_T \ldots$ with $\lim_{n\rightarrow\infty} t_n = t'$. By contradiction, let us suppose that t' is not an ω-normal form of t. Due to the fairness hypothesis we have only to consider the case in which t' can be reduced on an infinite redex by a non-left-linear rule in T, whose application was never possible on any of the finite terms t_n along D. In fact, in order to be applied, such rules may require the equivalence of syntactically different subexpressions which represent the same infinite term. Thus, it could happen that a reduction by $\rightarrow_T$ is never detected on the finite terms in D, because it involves subexpressions which are semantically equivalent, but syntactically different. Such subexpressions become syntactically equivalent at the limit and the reduction can then be applied. Since the applicability of the rules in T is checked modulo $=_C$, this situation can never occur. ♦

Proposition 1 allows us to restrict our attention to fair derivations, as they compute ω-normal forms at the limit. Actually, there are many cases in which it is possible to identify a subclass of fair derivations which have a peculiar structure.

Definition 5 (structured derivation) A derivation $t_0 \rightarrow_T t_1 \rightarrow_T \ldots \rightarrow_T t_n \rightarrow_T \ldots$ over $T_\Sigma(V)$ is *structured* if there exists an index N such that, for all $n \geq N$, $t_n \rightarrow_{T\setminus S} t_{n+1}$ and it never happens that $t_n \rightarrow_{T\setminus R} t_{n+1}$ can be applied.

Thus, for any structured derivation it is possible to single out an index N which splits the infinite derivation into a finite subderivation of terms t_n ($n < N$), in which $\rightarrow_T$ is applied, and an infinite subderivation of terms t_n ($n \geq N$), in which only $\rightarrow_{T\setminus S}$ can be applied.

Note that, in general, there is no guarantee that even a fair derivation is structured.

Example 2

The term rewriting system $T = R \cup S$ where

R $\quad g(a, g(c,x)) \to g(a,x)$

S $\quad f(g(c, g(a,x))) \to g(c, g(a, f(g(c, g(a,x)))))$

allows the following fair derivation that is not structured since every rewriting step by $\to_{T\setminus S}$ generates a reduction for $\to_{T\setminus R}$:

D: $f(g(c, g(a,b))) \to_{T\setminus S} g(c, g(a, f(g(c, g(a,b))))) \to_{T\setminus S} g(c, g(a, g(c, g(a, f(g(c, g(a,b))))))) \to_{T\setminus R} g(c, g(a, g(a, f(g(c, g(a,b)))))) \to_{T\setminus S} \ldots$ ♦

Definition 6 (uniformity) A term rewriting system T which admits fair derivations is *uniform* if for any fair derivation D: $t \to_T t_1 \to_T \ldots \to_T t_n \to_T \ldots$ with $\lim_{n\to\infty} t_n = t'$, there exists a structured fair derivation D': $t \to_T t'_1 \to_T \ldots \to_T t'_N \to_{T\setminus S} \ldots$, for some N, with $\lim_{n\to\infty} t'_n = t'$.

Our interest in uniform term rewriting systems is twofold. First, in order to show the ω-confluence of an infinite rewriting relation, it is possible to factorize the proof in two steps: i) to prove the confluence of $\to_T$ restricted to the finite part of the fair derivations, thus retrieving all the results valid for finitely terminating rewriting relations, e.g. local confluence; ii) to prove ω-confluence only for $\to_{T\setminus S}$. Second, given a uniform term rewriting system, it might be possible to determine a bound N on the number of the rewriting steps of a fair derivation, which guarantees that a finite representation of the ω-normal form has been reached. In case of confluent uniform term rewriting systems, this provides a procedure for deciding the equivalence of two terms by computing a finite representation of their ω-normal forms.

In order to show the ω-confluence of $\to_T$, some additional requirements on the nature of R and S have to be stated.

Definition 7 (independence) Given $R = \{l_i \to r_i \mid 1 \le i \le n\}$ and $S = \{G_j \to H_j[G_j] \mid 1 \le j \le m\}$, then R and S are *independent* if l_i and G_j do not overlap, $1 \le i \le n$ and $1 \le j \le m$.

Example 3

In the following term rewriting system

R $\quad g(a) \to b \qquad$ S $\quad f(g(x)) \to f(f(g(x)))$

R and S are not independent. Moreover, it is easy to verify that the term f(g(a)), for example, admits an infinite number of fair derivations leading to different ω-normal forms. ♦

Proposition 2 Let $R = \{l_i \to r_i \mid 1 \le i \le n\}$ and $S = \{G \to H[G]\}$ such that R is canonical and S consists of a single non-overlapping unfolding rule. If R and S are independent, then $\to_T$ is locally confluent.

Sketch of the proof

We have to show that whenever $t' \ {}_T\!\leftarrow t \to_T t''$ at the occurrences u and u' respectively, then there exists a term q such that $t' \overset{*}{\to}_T q \ {}_T\!\overset{*}{\leftarrow} t''$. Since R is canonical and S is a non-overlapping unfolding rule, we have only to consider the cases in which t can be rewritten with both $\to_{T\setminus R}$ and $\to_{T\setminus S}$. In fact, in this case it is easy to show that also $\to_{T\setminus R}$ is confluent. Let us consider the two cases:

a) t/u and t/u' are disjoint redexes. Straightforward.

b) the redex t/u contains the redex t/u'. Note that, since R and S are independent, the redex t/u' can only be an instance of a variable x of the left-hand side of the rule that rewrites t/u.

Let us first consider the situation in which a redex for $\rightarrow_{T\setminus R}$ contains a redex for $\rightarrow_{T\setminus S}$.

$$\begin{array}{ccc} & t\,[l_i\sigma'[x \leftarrow G\sigma]] & \\ & \downarrow_{T\setminus S} \qquad \downarrow_{T\setminus R} & \\ t\,[l_i\sigma'[x \leftarrow H[G]\sigma]] & & t\,[r_i\sigma'[x \leftarrow G\sigma]] \\ & \downarrow_{T\setminus R} \qquad *\!\downarrow_{T\setminus S} & \\ & t\,[r_i\sigma'[x \leftarrow H[G]\sigma]] & \end{array}$$

Note that the right-hand side of the rule $l_i \rightarrow r_i$ that rewrites t/u, may or may not contain the variable x, this explains why the last rewriting on the right can consist of zero rewritings by S.

On the other hand, if a redex t/u for $\rightarrow_{T\setminus S}$ contains a redex t/u' for $\rightarrow_{T\setminus R}$, t/u is an instance $G\sigma$ of G for some substitution σ and, since R and S are independent, t/u' can only occur if σ substitutes a variable x of G with an instance $l_i\sigma'$ for some $l_i \rightarrow r_i$ in R and substitution σ'. The following diagram shows how the local confluence can be obtained:

$$\begin{array}{ccc} & t\,[G\sigma[x \leftarrow l_i\sigma']] & \\ & \downarrow_{T\setminus S} \qquad \downarrow_{T\setminus R} & \\ t\,[H[G]\sigma[x \leftarrow l_i\sigma']] & & t\,[G\sigma[x \leftarrow r_i\sigma']] \\ & *\!\downarrow_{T\setminus R} \qquad \downarrow_{T\setminus S} & \\ & t\,[H[G]\sigma[x \leftarrow r_i\sigma']] & \end{array}$$

The last rewriting on the left can consist of more than one application of $\rightarrow_{T\setminus R}$ because H[G] can contain more than one occurrence of the variable x. ♦

Corollary 1 Let $R = \{l_i \rightarrow r_i \mid 1 \leq i \leq n\}$ and $S = \{G_j \rightarrow H_j[G_j] \mid 1 \leq j \leq m\}$ such that R is canonical and S consists of non-overlapping unfolding rules. If R and S are independent and $T = R \cup S$ is uniform, then $\rightarrow_T$ is locally confluent.

Proposition 3 (ω-confluence) Let $T = R \cup S$ where $R = \{l_i \rightarrow r_i \mid 1 \leq i \leq n\}$ and $S = \{G_j \rightarrow H_j[G_j] \mid 1 \leq j \leq m\}$ such that $\rightarrow_{T\setminus S}$ is ω-confluent. If $\rightarrow_T$ is locally confluent and T is uniform, then $\rightarrow_T$ is ω-confluent.

Sketch of the proof

Under the uniformity hypothesis we can restrict to fair structured derivations. Thus, in order to prove that $\rightarrow_T$ is ω-confluent, we note that the local confluence hypothesis implies the confluence of $\rightarrow_T$ on the finite subderivations of fair structured derivations. Then, the ω-confluence of $\rightarrow_T$ follows from ω-confluence of $\rightarrow_{T\setminus S}$. ♦

Independence is quite a strong condition on the syntactic nature of the rewriting rules. It is, anyhow, weaker than the "non-overlapping" condition on the whole term rewriting system, which is required to guarantee ω-confluence in case of non-terminating left-linear system [DK89, DKP89, DKP91]. Now, we introduce a notion of *preservation* between the components R and S of a term rewriting system, which guarantees that a reduction by R on a term denoting an infinite data structure cannot destroy its infinite nature. Thus, if a reduction applies on a term denoting an infinite data structure, the term can only be rewritten into a term denoting another infinite data structure. The intuition behind the notion of preservation can be made clearer by considering the term rewriting system in Example 3. There, the fact that the rule in R can destroy the redexes for $\rightarrow_{T\setminus S}$ implies that even confluence is necessarily compromised. Some syntactic conditions on the term rewriting systems can be determined in order to preserve S, that weaken the constraint on the left-hand sides of the rules by permitting overlapping, but imply some constraints on the right-hand sides of the rules.

Definition 8 (preservation) Let $R = \{l_i \to r_i \mid 1 \le i \le n\}$, $S = \{G_j \to H_j[G_j] \mid 1 \le j \le m\}$ and $t \in T_\Sigma(V)$ be reducible by $G_k \to H_k[G_k]$, $1 \le k \le m$. R *preserves* S if whenever $t \to_{T \setminus R} t'$, then t' still contains a redex for $G_k \to H_k[G_k]$.

Corollary 2 Let $R = \{l_i \to r_i \mid 1 \le i \le n\}$ and $S = \{G_j \to H_j[G_j] \mid 1 \le j \le m\}$. If R preserves S then

i) at least one occurrence of every variable of l_i still occurs in r_i;

ii) for each rule $l_i \to r_i$ such that l_i overlaps the left-hand side G_k of an unfolding rule at an occurrence u via a substitution σ, the term $(G_k[u \leftarrow r_i])\sigma$ unifies with G_k, $1 \le k \le m$.

Example 4

In the following term rewriting system

R $\quad g(x, g(y,x)) \to g(x,y) \qquad$ S $\quad f(g(a,x)) \to g(a, f(g(a,x)))$

R and S are not independent (the left-hand sides overlap), but R preserves S: R maintains an occurrence for both x and y, and the term f(g(a,y)) obtained by overlapping R and S is an instance of the left-hand side of the unfolding rule. ♦

Proposition 4 Let $R = \{l_i \to r_i \mid 1 \le i \le n\}$ and $S = \{G \to H[G]\}$ such that R is canonical and S consists of a non-overlapping unfolding rule. If R preserves S and G does not overlap l_i, for each rule $l_i \to r_i$, then $\to_T$ is locally confluent.

Sketch of the proof

We have to show that whenever $t' \;{}_T\!\!\leftarrow t \to_T t''$ at the occurrences u and u' respectively, then there exists a term q such that $t' \overset{*}{\to}_T q \;{}_T\!\overset{*}{\leftarrow} t''$. Since R is canonical and S is a non-overlapping unfolding rule, we have only to consider the cases in which t can be rewritten with both $\to_{T \setminus R}$ and $\to_{T \setminus S}$. In fact, in this case it is easy to show that also $\to_{T \setminus R}$ is confluent. Let us consider the two cases:

a) t/u and t/u' are disjoint redexes. Straightforward.

b) the redex t/u contains the redex t/u'.

From the hypotheses we have only to consider the situation in which a redex for a rule $l_i \to r_i$ is contained in a redex for $G \to H[G]$. The other case reduces to independence for which Proposition 2 holds.

A redex t/u for $\to_{T \setminus S}$ contains a redex t/u' for $\to_{T \setminus R}$. The following diagram shows how the local confluence can be obtained:

$$
\begin{array}{ccc}
 & t\,[G\sigma[l_i\sigma']] & \\
\swarrow_{T \setminus S} & & \searrow_{T \setminus R} \\
t\,[(H[G]\sigma)[l_i\sigma']] & & t\,[G\sigma[r_i\sigma']] \\
{}^*\!\downarrow_{T \setminus R} & & \downarrow_{T \setminus S} \text{ by the preservation hypothesis} \\
 & t\,[(H[G]\sigma)[r_i\sigma']] &
\end{array}
$$

The last rewriting on the left can consist of more than one application of $\to_{T \setminus R}$ because H[G] can contain more than one subterm $l_i\sigma'$. ♦

Corollary 3 Let $R = \{l_i \to r_i \mid 1 \le i \le n\}$ and $S = \{G_j \to H_j[G_j] \mid 1 \le j \le m\}$ such that R is canonical and S consists of non-overlapping unfolding rules. Let R preserve S and for each rule $l_i \to r_i$ and $G_j \to H_j[G_j]$, G_j does not overlap l_i. If $T = R \cup S$ is uniform, then $\to_T$ is ω-confluent.

4. An Application Example

In this section we show a term rewriting system that can be considered representative of the class we intend to deal with. The signature that we consider defines the syntax of a language of regular expressions. This is a subset of the language defined by a process algebra like CCS [Mil80]. In particular, we consider a restricted set of actions Act = L $\cup$ {τ}, where L is the set of labels {a,b}.

It is possible to equip this language with several different semantic equivalences that define which terms can be considered as equivalent with respect to a certain *behaviour*. Axiomatic presentations of several behavioural equivalences for CCS do exist in the literature, e.g. trace equivalence, branching bisimulation, observational congruence and testing equivalence. All these presentations differ only for the axioms for the unobservable action τ.

The rewriting rules define the semantics of the operators. In particular, we consider as R the rules expressing the behavioural semantics and as S the operational semantics for the recursion operator. The rewriting system R we present is a simplified version of the canonical term rewriting system which characterizes branching bisimulation congruence [GW89]. Note that, in this example, we are also considering $\rightarrow_T$ modulo AC, since the + operator is associative and commutative. As regards S, we consider the semantics for the recursion operator only for two specific patterns. Note that the general axiom for the recursion operator, recX.E = E{recX.E/X}, is actually an axiom schema.

E ::= nil | a•E | b•E | τ•E | E + E | rec(id, E) | id

R	x + x $\rightarrow$ x	a• (τ• (x+y) + x) $\rightarrow$ a• (x+y)	
	x + nil $\rightarrow$ x	b• (τ• (x+y) + x) $\rightarrow$ b• (x+y)	
	a•τ•x $\rightarrow$ a•x	τ• (τ• (x+y) + x) $\rightarrow$ τ• (x+y)	
	b•τ•x $\rightarrow$ b•x	a• (τ• x + x) $\rightarrow$ a• x	
	τ•τ•x $\rightarrow$ τ•x	b• (τ• x + x) $\rightarrow$ b• x	τ• (τ• x + x) $\rightarrow$ τ• x
S	rec(id, a•(id + b•nil)) $\rightarrow$ a•(rec(id, a•(id + b•nil)) + b•nil)		
	rec(id, (a•id) + b•nil) $\rightarrow$ (a•rec(id, (a•id) + b•nil)) + b•nil		

The above system T = R $\cup$ S is not finitely terminating and non-left-linear. R is canonical modulo AC, S is top-terminating and non-overlapping. T is top-terminating and ω-terminating. $T_\Sigma(V)$ is retractile since it is possible to define a relation $=_C$ according to Definition 3: the recursive terms are regular set of equations, thus it is possible to compute the unique canonical representative in the class of the terms with the same (tree) semantics [CKV74]. R and S are independent. Let us now see if T is uniform as regards its fair derivations. We have to show that each fair derivation can be splitted in two parts the latter being only made of reductions with $\rightarrow_{T\backslash S}$. It is easy to verify this by showing that rewritings by $\rightarrow_{T\backslash S}$ cannot produce infinitely many redexes for $\rightarrow_{T\backslash R}$. Thus, the hypotheses of Corollary 1 are satisfied and $\rightarrow_T$ is locally confluent, furthermore since the rules in S are non-overlapping, $\rightarrow_{T\backslash S}$ is ω-confluent and for Proposition 3, also $\rightarrow_T$ is ω-confluent.

Note that there exist terms for which rewritings by $\rightarrow_{T\backslash S}$ are necessary to make a reduction to normal form possible. For example, let us consider the following derivation: rec(id, (a•id) + b•nil) + b•nil $\rightarrow_{T\backslash S}$ (a•rec(id, (a•id) + b•nil)) + b•nil + b•nil $\rightarrow_{T\backslash R}$ (a•rec(id, (a•id) + b•nil)) + b•nil.

Let us now briefly discuss what happens if we deal with the recursion operator in its full extent. In this case, R and S are not independent any more, but R preserves S. Moreover, uniformity is not guaranteed any more, since it is possible to find recursive terms on which rewritings by $\rightarrow_{T\setminus S}$ produce infinitely many redexes for the rule $\mu \cdot \tau \cdot x \rightarrow \mu \cdot x$, where $\mu \in$ Act. However, new rewriting rules can be added to prevent this situation and assure uniformity [IN90b].

5. Final Remarks

In this paper we have presented an extension to the framework defined in [DK89, DKP89, DKP91] for a specific class of non-left-linear term rewriting systems. In doing this, we have been driven by the experience made dealing with non-terminating rewriting relations related to the axiomatic presentation of behavioural semantics for process algebras.

The notions we have introduced, namely the retraction property of the supporting algebra, the uniformity of term rewriting systems and the independence and preservation requirements on the rewriting rules, appeared to be much more *natural* in our theories than the left-linearity requirement. On the other hand, the general framework defined in [DK89, DKP89, DKP91], namely the notions of ω-termination, ω-normal form, fair derivations as the only interesting derivations one has to look at, is very suitable to study for a notion of normal forms of recursive process algebras terms. Thus, the contribution of this paper has been to formalize and properly generalize the features of our theories in order to release the left-linearity condition while retaining the general framework.

References

[CKV74] Courcelle, B., Kahn, G., Vuillemin, J. Algorithmes d'Equivalence et de Reduction a des Expressions Minimales dans une Classe d'Equations Recursives Simples, Proc. Automata, Languages and Programming, 2nd Colloquium, Saarbrucken, LNCS **14**, (1974), 200-213.
[DIN90] De Nicola, R., Inverardi, P., Nesi, M. Using the Axiomatic Presentation of Behavioural Equivalences for Manipulating CCS Specifications, in Proc. Workshop on Automatic Verification Methods for Finite State Systems, LNCS **407** (1990), 54-67.
[DK89] Dershowitz, N., Kaplan, S. Rewrite, Rewrite, Rewrite, Rewrite, Rewrite..., Proc. 16th Annual ACM Symposium on POPL, Austin, Texas, (1989), 250-259.
[DKP89] Dershowitz, N., Kaplan, S., Plaisted, D.A. Infinite Normal Forms, Proc. 16th ICALP, Stresa, Italy, LNCS **372**, (1989), 249-262.
[DKP91] Dershowitz, N., Kaplan, S., Plaisted, D.A. Rewrite, Rewrite, Rewrite, Rewrite, Rewrite..., Theoretical Computer Science, in press.
[Giv90] Givler, J.S. Continuous Rewriting Systems, IBM T.J. Watson Res. Center Report, (1990).
[GW89] van Glabbeek, R.J., Weijland, W.P. Branching Time and Abstraction in Bisimulation Semantics, Proc. IFIP 11th World Computer Congress, San Francisco, (1989).
[IN90a] Inverardi, P., Nesi, M. A Rewriting Strategy to Verify Observational Congruence, Information Processing Letters Vol. **35**, No. **4**, (August 1990), 191-199.
[IN90b] Inverardi, P., Nesi, M. Deciding Observational Congruence of Finite-State CCS Expressions by Rewriting, I.E.I. Internal Report nº B4-10, (March 1990).
[KKSV90] Kennaway, J.R., Klop, J.W., Sleep, M.R., de Vries, F.J. An Infinitary Church-Rosser Property for Non-collapsing Orthogonal Term Rewriting Systems. CWI Report CS-R9043, (1990).
[KKSV91] Kennaway, J.R., Klop, J.W., Sleep, M.R., de Vries, F.J. Transfinite Reductions in Orthogonal Term Rewriting Systems, CWI Report CS-R9042, Amsterdam, (1990), Proceedings of the RTA'91 Conference, LNCS **488**, (1991), 1-12.
[Mil80] Milner, R. A Calculus of Communicating Systems, LNCS **92**, (1980).
[Mil89] Milner, R. A Complete Axiomatization for Observational Congruence of Finite-State Behaviours, Information and Computation **81**, (1989), 227-247.

Two algorithms for approximate string matching in static texts *

(Extended Abstract)

Petteri Jokinen Esko Ukkonen

Department of Computer Science, University of Helsinki
Teollisuuskatu 23, SF-00510 Helsinki, Finland

Abstract. The problem of finding all approximate occurrences P' of a pattern string P in a text string T such that the edit distance between P and P' is $\leq k$ is considered. We concentrate on a scheme in which T is first preprocessed to make the subsequent searches with different P fast. Two preprocessing methods and the corresponding search algorithms are described. The first is based suffix automata and is applicable for edit distances with general edit operation costs. The second is a special design for unit cost edit distance and is based on q-gram lists. The preprocessing needs in both cases time and space $O(|T|)$. The search algorithms run in the worst case in time $O(|P||T|)$ or $O(k|T|)$, and in the best case in time $O(|P|)$.

Introduction

The *approximate string matching* problem is to find, given a pattern string P and a text string T, the approximate occurrences of P in T. Typically one wants to find all occurrences that are good enough in some measure of the approximation quality.

There are several situations where it is necessary to allow for approximate matches instead of exact ones. Some natural variation in the occurrences of P (e.g. due to morphological variation of the same base word in natural languages) sometimes takes place. In other cases, P or T or both may have been slightly distorted through noisy communication channels or through different types of errors (measurement error, typing error).

We concentrate on the important special case where T stays unchanged for searches with numerous different P, and we have the whole T available before the searches. Such a static T can first be preprocessed into a suitable form (an *index* for approximate searches) that makes the subsequent searches faster. Hence we want to find a preprocessing of T and the associated algorithm to search for approximate occurrences of P using the preprocessed T.

The edit distance will be used as the measure for the approximation quality.

Definition. Let P and P' be strings in alphabet Σ. The *edit distance* form P to P' is the minimum possible total cost of a sequence of editing steps that convert P' to P. Each

*Research supported by the Academy of Finland and by the Alexander von Humboldt Foundation (Germany). The work of the second author was in part carried out when visiting Institut fuer Informatik, University of Freiburg, Germany.

editing step is a rewriting step of the form $a \rightarrow \epsilon$ (a deletion), $\epsilon \rightarrow a$ (an insertion), or $a \rightarrow b$ (a change), where a, b in Σ are any symbols, $a \neq b$, and ϵ is the empty string. Each editing operation $x \rightarrow y$ has a cost $c(x \rightarrow y) > 0$. In the conversion from P to P' rewriting of each symbol is allowed only at most once; this makes it possible to use dynamic programming algorithms for edit distances. The special case where $c(x \rightarrow y) = 1$ for all edit operations $x \rightarrow y$ is called the *unit cost model* of the edit distance.

Definition (approximate string matching problem). Given two strings, *text* $T = t_1 t_2 \ldots t_n$ and *pattern* $P = p_1 p_2 \ldots p_m$ in alphabet Σ, and a threshold value $k \geq 0$, find the end locations j of all substrings P' of T such that the edit distance from P to P' is at most k. If the unit cost model is used, the problem is called the *k differences problem.*

The on-line version of the problem in which no preprocessing of T is allowed has recently received lot of attention [5, 6]. Standard solution is by dynamic programming in time $O(mn)$. For the k differences problem fast special methods are possible, including $O(kn)$ time algorithms [10, 7, 16, 14, 2].

In the case of *exact string matching* ($k = 0$) preprocessing of T leads to optimal time searches. If T is preprocessed into a suffix tree [17, 12] or into a suffix automaton [1, 3], the queries of P can be accomplished in time $O(m + \text{size of output})$. If the suffix array [11] is used, the search time becomes $O(m + \log n + \text{size of output})$.

In the case of the approximate matching ($k \geq 0$) we develop in this paper two data structures for representing a static T and give the corresponding search algorithms. The first solution combines suffix automata and dynamic programming, and is applicable for general edit operation costs c. Text T is represented as annotated suffix automaton. The search is performed by dynamic programming over P and the transition graph of the automaton. Based on certain properties of suffix automata we develop a search strategy that avoids entering the same state of the automaton repeatedly. This gives a time bound that is in the worst case the same as for the standard on-line dynamic programming but in the best case is essentially better.

The second data structure is a simple special design for the k differences problem. The structure is based on the so-called q-grams that are simply any strings of q symbols. The preprocessing phase creates for each q-gram of T a chain that links together all occurrences of the q-gram in T. This structure can be understood as an abridged version of the suffix array or—when the headers of the link chains are organized as a trie—as a suffix tree which has been cut to the depth q.

The search phase marks the areas of T that have a sufficient number of q-grams in common with P. The marked areas are then checked by dynamic programming for occurrences of P with at most k differences. The method has a predecessor in the work of Owolabi & McGregor [13], and related 'signature' methods have been used e.g. in spelling correction; see e.g. [9]. We show how the different parameters of the method should be chosen to solve the given k differences problem.

Annotated suffix automaton $SA(T)$

The *suffix automaton* [3, 4] (also known as *DAWG*, *directed acyclic word graph*, [1]) for a string $T = t_1 t_2 \cdots t_n$ is the smallest DFA recognizing all the suffixes $T_i = t_i \cdots t_n$, $1 \leq i \leq n+1$, of T. We let *root* denote its initial state and *goto* its transition function;

there is a transition from state s to state r on input symbol a if $r = goto(root, a)$. The suffix automaton can be constructed in time $O(n)$ by the methods given in [3, 4, 1]. The suffix automaton for T has at most $3n - 4$ *goto* transitions and at most $2n - 1$ states. It can be viewed as the *suffix tree* for T, with the identical subtrees merged. As a graph, it is a dag.

The *depth* of a state s of the automaton, denoted $depth(s)$, is the length of the (unique) longest string x such that there is a *goto* path from *root* to s, $goto(root, x) = s$; here we have extended the *goto* function for strings in the obvious way. Similarly, $mindepth(s)$ denotes the length of the shortest string y such that $goto(root, y) = s$.

The following property of a suffix automaton is an immediate consequence of the fact that the automaton accepts all the suffixes of a string and nothing more.

Lemma 1 *For a state s, let x be the longest string such that $goto(root, x) = s$. Then the set of strings y such that $goto(root, y) = s$ consists of all suffixes of x of length at least $mindepth(s)$.*

The important *fail* function on the states of the automaton has the following characterization.

Lemma 2 ([4]) *Let $s = goto(root, x)$ for some string x, and let w be the longest suffix of x such that $s \neq goto(root, w)$. Then, $goto(root, w) = fail(s)$ and $|w| = depth(fail(s))$.*

Corollary 1 $mindepth(s) = depth(fail(s)) + 1$.

The suffix automaton serves an an index giving the locations of different substrings of T. There are different ways to attach the location information to the automaton. For our purposes the following is suitable.

The states of the automaton are divided into two classes: the *primary states* and the *secondary states.* The primary states are the states $s_i = goto(root, t_1 \cdots t_i)$ for $0 \leq i \leq n$. These states are disjoint, and $depth(s_i) = i$. The other states are secondary.

A string x is said to *occur at location j* in T if $x = t_{j-|x|+1}t_{j-|x|+2} \cdots t_j$.

Lemma 3 *Let $goto(root, x) = s$, and let $L = \{depth(r) \mid r$ is primary and $s = fail^i(r)$ for some $i \geq 0\}$. Then L is the set of all locations at which x occurs in T.*

Hence the occurrences of a string leading to s can be found by finding the primary states from which there is a $fail$ transition path to s. Therefore we also need the inverse of $fail$: with each state r we attach a list of links, the $cofail$ links, pointing to states r' such that $fail(r') = r$.

The *annotated suffix automaton* for T, denoted $SA(T)$, is the suffix automaton of T (i.e., the states and the *goto* function) with the states marked primary or secondary and with the $fail$ and *cofail* links and the *depth* value for each state. The annotations do not increase the construction time; in fact, both $fail$ and *depth* are needed in the construction, so the only extra work is to reverse $fail$ and mark states primary or secondary which clearly does not increase the asymptotic time requirement.

Proposition 1 *The annotated suffix automaton $SA(T)$ can be constructed in time and in space $O(n)$.*

Approximate string matching with $SA(T)$

The approximate string matching problem for text $T = t_1t_2\cdots t_n$ and pattern $P = p_1p_2\cdots p_m$ can be solved on-line, without preprocessing T, with the following well-known dynamic programming method.

Let D be a $m+1$ by $n+1$ table such that for $0 \le i \le m$, $0 \le j \le n$, $D(i,j)$ is the minimum edit distance from $p_1\cdots p_i$ to the substrings of T ending at t_j. Clearly, there is an approximate occurrence of P in T, ending at t_j, with edit distance $\le k$ from P, if and only if $D(m,j) \le k$. Such indexes j can be found by evaluating D from

$$D(0,j) = 0, \quad 0 \le j \le n; \tag{1}$$

$$D(i,j) = \min \begin{cases} D(i-1,j) + c(p_i \to \epsilon) \\ D(i-1,j-1) + \text{ if } p_i = t_j \text{ then } 0 \text{ else } c(p_i \to t_j) \\ D(i,j-1) + c(\epsilon \to t_j) \end{cases} \tag{2}$$

for $1 \le i \le m$, $0 \le j \le n$.

As $D(i,j)$ depends only on entries $D(i-1,j)$, $D(i-1,j-1)$, and $D(i,j-1)$ of D, the evaluation conveniently proceeds column-by-column: Column $D(*,j)$ can be evaluated from column $D(*,j-1)$, proceeding in the order $D(0,j),\ldots,D(m,j)$. The total time is $O(mn)$.

The length $L(i,j)$ of the *shortest* suffix of $t_1\cdots t_j$, whose edit distance from P is $D(i,j)$, can be computed together with $D(i,j)$ itself. Clearly, for $0 \le j \le n$ we have $L(0,j) = 0$, and for $1 \le i \le m$, $0 \le j \le n$:

$$\begin{aligned} L(i,j) \;=\; & \text{if } D(i,j) = D(i-1,j) + c(p_i \to \epsilon) \text{ then } L(i-1,j) \\ & \text{elsif } D(i,j) = D(i-1,j-1) + (\text{if } p_i = t_j \text{ then } 0 \text{ else } c(p_i \to t_j)) \\ & \qquad \text{then } L(i-1,j-1)+1 \\ & \text{else } L(i,j-1)+1. \end{aligned}$$

Then $D(i,j)$ equals the edit distance from $p_1\cdots p_i$ to $t_{j'}\cdots t_j$ where $j' = j - L(i,j) + 1$.

Next we develop a method that performs a similar dynamic programming over P and $SA(T)$ to find the approximate occurrences of P in T. The method will attach with the states of $SA(T)$ similar columns of $m+1$ entries as are the columns of matrices D and L. The column representing edit distances at state r is denoted as $dcol(r)$, and the column representing the corresponding lengths is denoted as $lcol(r)$.

The method will work, roughly formulated, in the following steps.

1. Traverse the *useful subtree* $U(P,k)$ of $SA(T)$ starting from *root* and using a modified Dijkstra's shortest path algorithm to control the traversing order;

2. When the traversal enters state r along a transition $goto(s,a) = r$, evaluate $dcol(r)$ and $lcol(r)$ by dynamic programming from a, $dcol(s)$, and $lcol(s)$;

3. If $dcol(r)(m) \le k$, mark all states that can be reached from r along *cofail* links and are not already marked. Output $depth(q)$ for each primary state q that gets a mark.

Next we refine the above description of the algorithm, starting from step 2.

To understand the use of *dcol* and *lcol* some further notation is necessary. For any string x, we let $d(i,x)$ denote the minimum edit distance between $p_1\cdots p_i$ and any suffix of x, and $l(i, x)$ denote the length of the shortest such a suffix. Then, for example, $D(i, j) = d(i, t_1 \cdots t_j)$ and $L(i, j) = l(i, t_1 \cdots t_j)$. It should be clear that d anf l can be evaluated in the same way as D and L from a recursion similar to (1) and (2); string x now takes the role of $t_1 \cdots t_j$.

The traversal over $SA(T)$ starts from *root*. Initially, $dcol(root) = d(*, \epsilon)$ and $lcol(root) = l(*, \epsilon)$, where $d(i, \epsilon) = \sum_{h=1}^{i} c(p_h \to \epsilon)$ and $l(i, \epsilon) = 0$, for $0 \le i \le m$. For other states s the columns $dcol(s)$ and $lcol(s)$ will be such that $dcol(s) = d(*, x)$ and $lcol(s) = l(*, x)$, where x is the string spelled out by the path from *root* to s in the traversed subtree $U(P, k)$. This property is preserved if, when the traversal takes transition $goto(s, a) = r$, the new columns $dcol(r) = d(*, xa)$ and $lcol(r) = l(*, xa)$ are evaluated by dynamic programming from $dcol(s)$, $lcol(s)$, and a. For example, for $dcol(r)$ this evaluation gets the form

$$dcol(r)(0) \;=\; 0 \tag{3}$$

$$dcol(r)(i) \;=\; \min \begin{cases} dcol(r)(i-1) + c(p_i \to \epsilon) \\ dcol(s)(i-1) + \;\text{ if } p_i = a \text{ then } 0 \text{ else } c(p_i \to a) \\ dcol(s)(i) + c(\epsilon \to a) \end{cases} \tag{4}$$

for $i = 1, 2, \ldots, m$.

Next consider step 1. Our goal is to develop a traversing order that guarantees that all approximate occurrences of P will be found but extra traversing is avoided. This should be done in such a way that each *goto* transition of $SA(T)$ is traversed at most once. As there are $O(n)$ transitions and taking a transition needs time $O(m)$ (for evaluating *dcol* and *lcol*), this would give an $O(mn)$ time bound for the whole method. It turns out that it suffices to traverse over subtree $U(P, k)$ which we shall define next.

We denote as $\lambda(x)$ the length of the longest suffix y of a string x such that the edit distance from some prefix of P to y is $\le k$. Obviously, $\lambda(x) = l(i, x)$ where $i \le m$ is the largest index such that $d(i, x) \le k$. We say that an entry $d(i, x)$ is *essential*, if $d(i, x) \le k$. Hence $\lambda(x)$ expresses the length of the part of x on which the essential part of $d(*, x)$ depends.

Let $goto(root, a_1 a_2 \cdots a_h) = s$ for some $a_i \in \Sigma$. The *goto* path $a_1 \cdots a_h$ is called *useful*, if $\lambda(a_1 \cdots a_i) \ge mindepth(goto(root, a_1 \cdots a_i))$ for all $1 \le i \le h$. State s is useful, if all *goto* paths from *root* to s are useful. In particular, *root* is useful.

Definition. The *useful subtree* $U(P, k)$ of $SA(T)$ is the subgraph of $SA(T)$ that contains all the useful states and for each such state s, it also contains the *goto* transitions on the longest useful *goto* path from *root* to s.

Useful subtree $U(P, k)$ is really a tree because every initial segment of a useful path is useful.

It is sufficient to restrict the traversal on $U(P, k)$. To prove this, we need first a lemma.

Lemma 4 *Let x be a string and y its suffix such that $|y| \ge \lambda(x)$. Then $d(*, x)$ and $d(*, y)$ are identical when restricted to the essential entries, and the correspondingly restricted $l(*, x)$ and $l(*, y)$ are identical.*

Let J be the set of all locations j that our algorithm will output (step 3) when performing dynamic programming over $U(P, k)$, and let J' be the correct set of locations we want to find, that is, $J' = \{j \mid D(m, j) \le k\}$.

Theorem 1 $J = J'$.

Finally we need an efficient way to isolate and traverse the useful subtree $U(P,k)$. This will be done by finding a slightly larger tree that consists of $U(P,k)$ and of some additional leaves.

A *goto* path $a_1 \cdots a_h$ is called a *bounding path*, if path $a_1 \cdots a_{h-1}$ is useful but path $a_1 \cdots a_h$ is not useful (that is, $\lambda(a_1 \cdots a_h) < mindepth(goto(root, a_1 \cdots a_h))$. A state s of $SA(T)$ is a *boundary state* if there is to s at least one bounding path but no useful path.

Definition. The *extended useful subtree* $U^+(P,k)$ of $SA(T)$ consists of $U(P,k)$ and of all boundary states of $SA(T)$ and of longest possible bounding paths to them.

Again, subgraph $U^+(P,k)$ is really a tree because the longest bounding path to each boundary state is unique, and its each initial segment is useful and longest possible and hence belongs to $U(P,k)$.

Assume for a moment that we know a priori the nodes of $U^+(P,k)$. Then its arcs can be found by Dijkstra's shortest path algorithm. We define the cost $w(s,r)$ of an arc (s,r) (i.e., $goto(s,a) = r$ for some a) as $w(s,r) = depth(r) - depth(s) - 1$, if s is a useful state. If s is a boundary state, then we set $w(s,r) = \infty$; hence, in effect, such arcs are removed from $SA(T)$.

Then find with Dijkstra's algorithm the minimum cost paths with respect to cost function w from *root* to all states in $U^+(P,k)$. Consider the path $s_0 = root, s_1, \ldots, s_{h-1}, s_h = s$ found in this way to some s.

Lemma 5 *The length h of the path to s is largest possible.*

The useful states and the boundary states are not known a priori, but we can recognize them easily during the execution of the Dijkstra's algorithm. The dynamic programming is performed at each state in the traversal order determined by the algorithm: When the algorithm reaches a new state r along transition $goto(s,a) = r$, columns $dcol(r)$ and $lcol(r)$ are computed from $dcol(s)$, $lcol(s)$, and a, as already explained.

Let x be the path from *root* along which r is found. Then $\lambda(x) = lcol(r)(i)$ where i is the largest index such that $dcol(r)(i)$ is essential. Hence $\lambda(x)$ can be evaluated locally at r, and we may write $\lambda(r) = \lambda(x)$.

Now the status of r can be decided.

Lemma 6 *If $\lambda(r) < mindepth(r)$, then r is a boundary state, otherwise r is a useful state.*

By Lemmas 5 and 6, our algorithm finds the boundary states and the useful states correctly along longest possible paths. Therefore $U^+(P,k)$ is found correctly which means, by Theorem 1, that the approximate occurrences of P are found correctly.

Theorem 2 *The described algorithm can be implemented such that it works in time $O(mn)$ in the worst case and, for the unit cost model of the edit distance, in time $O(m)$ in the best case.*

Proof. As there are $O(n)$ states in $U^+(P,k)$ and a dynamic programming step of time $O(m)$ is performed once at each, the time for dynamic programming is $O(mn)$. In Dijkstra's algorithm we use bucket sort instead of heap to get time $O(n)$, and hence total time $O(mn)$. For the best case bound consider P and T such that they do not have common symbols.

Remark 1. The algorithm of Theorem 2 satisfies the natural requirement that the worst case time $O(nm)$ is not larger than the time of the on-line solution, without preprocessing T. The best case time is $O(m)$ showing that we have achieved some progress with the preprocessing of T. Without it also the best case has to grow proportional to n.

When $k = 0$, the algoritm requires time $O(m^2)$. The time seems to grow very fast with k but we leave open a more complete analysis of this dependency.

Remark 2. The simplest way to find approximate P's from automaton $SA(T)$ would be to follow each *goto* path from the *root* until the corresponding string has an edit distance $> k$ from all prefixes of P. Such paths can have total length $O(mn)$. It can be shown that this leads to $O(mnk)$ time search.

The q-gram method

This section considers the k differences problem, that is, $c(x \to y) = 1$ for all editing operations $x \to y$.

A *q-gram* in Σ is any string in Σ^q. The usefulness of the q-grams is based on the following lemma.

Lemma 7 *Let an occurrence of P with at most k differences end at t_j in T. Then at least $m+1-(k+1)q$ of the $m-q+1$ q-grams of P occur in $t_{j-m+1}\ldots t_j$.*

Proof. Let P' be the approximate version of P that ends at t_j. Hence P' is a suffix of $t_{j-m+1}\cdots t_j$ or $t_{j-m+1}\cdots t_j$ is a suffix of P'. String P' is obtained from P with at most k insertions, deletions or changes. A deletion or a change at character p_i of P destroys at most q q-grams of P, namely those that contain p_i. An insertion between p_i and p_{i+1} destroys at most $q-1$ q-grams of P, namely those that contain both p_i and p_{i+1}. Hence at most $k_1q+k_2(q-1)$ q-grams of P are missing in P', where k_1 is the total number of deletions and changes and k_2 is the total number of insertions. As $|P'| \le m+k_2$, string $t_{j-m+1}\cdots t_j$ contains all q-grams of P' except for at most k_2. Hence at most $k_1q + k_2(q-1) + k_2 = kq$ q-grams of P are not present in $t_{j-m+1}\cdots t_j$, which proves the lemma.

Using the lemma the areas of T that may contain a good enough approximate occurrence of P can be found fast. These are separately checked with dynamic programming.

Text T is preprocessed as follows: For each q-gram G in Σ^q we construct a list $L(G)$ consisting of all j such that T has an (exact) occurrence of G starting at t_j. The lists for all G can be created in one scan over T either by using a natural encoding of q-grams into integers to the base $|\Sigma|$ (c.f. [8]) or by using a modified suffix automaton with *fail*-transitions representing different q-grams of T [15]. We also create a search structure for finding fast the list $L(G)$, given G. A suitable stucture is an array indexed by the integer code of G, or a trie representing the different q-grams of T. The preprocessing time is $O(n + |\Sigma|^q)$ for the method based on integer codings; the size of the resulting structure is also $O(n + |\Sigma|^q)$ where n represents the total length of the lists and $|\Sigma|^q$ the search structure.

Assume then that we have to find the occurrences of P in T with $\leq k$ differences. The first phase of the search traverses all lists $L(G)$ where G occurs in P. The occurrences listed in $L(G)$'s are counted into initially zero buckets B_i, $0 \leq i \leq \lceil n/(m-1) \rceil + 1$. Bucket B_i is increased by 1 when the next element j of $L(G)$ satisfies $(i-1)(m-1)+1 \leq j \leq (i+1)(m-1)$. Hence the width of each bucket is $2(m-1)$ and two successive buckets have an overlap of length $m-1$; the overlap ensures that no occurrences of P are lost. (For simplicity, we assume that $m \geq 2$.) The rule for updating the buckets can be stated simply $B_{\lfloor \frac{j-1}{m-1} \rfloor} \leftarrow * + 1$; $B_{\lfloor \frac{j-1}{m-1} \rfloor + 1} \leftarrow * + 1$.

When B_i achieves value $m + 1 - (k+1)q$ we know by Lemma 7 that an approximate occurrence of P can end somewhere in $t_{i(m-1)} \cdots t_{(i+1)(m-1)}$. As an occurrence is of length $\leq m+k$, its leftmost possible starting character is t_j where $j = i(m-1)-m-k+1$. Hence we check by dynamic programming whether or not there is an approximate occurrence in $t_j \cdots t_{(i+1)(m-1)}$.

Because the total length of the q-gram lists $L(G)$ for G in P is $\leq n$, they can be traversed as described above in time $O(m+n)$. Under the random string assumption (each symbol in T is chosen uniformly and independently from Σ) the expected length of each list is $n/|\Sigma|^q$, hence the expected traversal time is $O(m + (m-q+1)n/|\Sigma|^q)$. In the best case each list is empty, hence time $O(m)$ suffices. Let r be the number of the buckets checked by dynamic programming. Using the $O(kn)$ version of dynamic programming [7, 16], the total time for the checking phase is $O(rkm)$ which in the worst case is $O(kn)$.

Theorem 3 *The q-gram lists for T can be constructed in time and in space $O(n + |\Sigma|^q)$. The search for occurrences of P with at most k differences can be done in time $O(m + n + rkm)$ where r is the number of buckets checked with dynamic programming. In the best case time $O(m)$ suffices for the search.*

The bound $m + 1 - (k+1)q$ for the number of q grams in Lemma 7 is non-trivial only if $q < (m+1)/(k+1)$. Hence it is possible that the q used in preprocessing T is too large for the present m and k. Fortunately, we can in this case use a smaller q' that is $< (m+1)/(k+1)$. The list $L(G)$ for a q'-gram G is the catenation of the q-gram lists $L(GX)$ where X is in $\Sigma^{q-q'}$.

Annotated suffix automaton $SA(T)$ is also a complete '$*$-gram' index for T containing q-gram lists for all $0 \leq q \leq n$. For a gram G of any length, $L(G)$ consists of all values $depth(s)$ such that s is primary and reachable from $goto(root, G)$ along *cofail*-links (Lemma 3). The q-gram method could be based on $SA(T)$ as well.

References

[1] Blumer,A., Blumer,J., Haussler, D., Ehrenfeucht, A., Chen, M.T. and Seiferas, J. (1985): The smallest automaton recognizing the subwords of a text. *Theor. Comp. Sci. 40,* 31-55.

[2] Chang,W. and Lawler,E (1990): Approximate string matching in sublinear expected time. *FOCS'90*, pp. 116-124.

[3] Crochemore, M. (1986): Transducers and repetitions. *Theor. Comp. Sci. 45*, 63-86.

[4] Crochemore, M. (1988): String matching with constraints. *Proc. MFCS'88. SLNCS 324*, pp. 44-58.

[5] Dowling, G. R. & Hall, P. (1980): Approximate string matching. *ACM Comput. Surv. 12*, 381-402.

[6] Galil, Z. & Giancarlo, R. (1988): Data structures and algorithms for approximate string matching. *J. Complexity 4*, 33-72.

[7] Galil, Z. & Park, K. (1989): An improved algorithm for approximate string matching. *ICALP'89. SLNCS 372*, pp. 394-404.

[8] Karp,R.M. and Rabin,M.O. (1987): Efficient randomized pattern matching. *IBM J. Res. Dev. 31*, 249-260.

[9] Kohonen,T. & Reuhkala,E. (1978): A very fast associative method for the recognition and correction of misspellt words, based on redundant hash-addressing. *Proc. 4th Int. Joint Conf. on Pattern Recognition*, 1978, Kyoto, Japan, pp. 807-809.

[10] Landau, G. & Vishkin, U. (1988): Fast string matching with k differences. *JCSS 37*, 63-78. (Also 26th *FOCS*, pp. 126-136).

[11] Manber, U. & Myers, G. (1990): Suffix arrays: a new method for on-line string searches. *SODA'90*, pp. 319-327.

[12] McCreight, E. M. (1976): A space economical suffix tree construction algorithm. *J. ACM* 23, 262-272.

[13] Owolabi, O. & McGregor, D. R.(1988): Fast approximate string matching. *Software - Practice and Experience* 18(4), 387-393.

[14] Tarhio, J. & Ukkonen, E. (1990): Boyer-Moore approach to approximate string matching. *2nd Scand. Workshop on Algorithm Theory (SWAT90), SLNCS 447*, pp. 348-359.

[15] Ukkonen, E. (1991): Approximate string matching with q-grams and maximal matches. *Theor. Comp. Sci.*, to appear.

[16] Ukkonen, E. & Wood, D. (1990): Approximate string matching with suffix automata. Report A-1990-4. Department of Computer Science, University of Helsinki.

[17] Weiner, P. (1973): Linear pattern matching algorithms. *Proc. 14th IEEE Symp. Switching and Automata Theory*, pp. 1-11.

EFFICIENT CONSTRUCTIONS OF TEST SETS FOR REGULAR AND CONTEXT-FREE LANGUAGES

Juhani KARHUMAKI
Department of Mathematics, University of Turku
Wojciech RYTTER and Stefan JAROMINEK
Institute of Informatics, Warsaw University

ABSTRACT: We present a simple construction of linear size test sets for regular languages and of single exponential test sets for context free languages. In the case of regular sets the size of our test set is exactly the number of transitions of the automaton. This improves the best known upper bounds: exponential for regular and doubly exponential for context-free languages. We give also an $O(n \log n)$ time algorithm for the morphism equivalence and an $O(n^3 \log n)$ time algorithm to test the gsm equivalence on a regular language. An $O(n^2 \log n)$ time algorithm is given to test the equivalence of two deterministic gsm's as well as that of two deterministic finite transducers.

One of the fundamental areas of research in formal language theory is that related to the so called Ehrenfeucht conjecture, see [Ka1], [Ka2]: each language $L \subseteq \Sigma^*$ has a finite subset F being a *test set*.
A language F is a *test set* of a given language $L \subseteq \Sigma^*$ iff $F \subseteq L$ and for any pair of morphisms $h,g:\Sigma^* \longrightarrow \Delta^*$ we have

$$\forall_{x \in L}\ h(x)=g(x) \Leftrightarrow \forall_{x \in F}\ h(x)=g(x).$$

In other words: for each pair of morphisms if they "agree" on a finite subset F then they agree on the whole set L. It was shown that the Ehrenfeucht conjecture is true, see [Sa], [AL], however in general the construction of a finite test set is noneffective. For regular and context free languages test sets can be constructed effectively, but so far constructed test sets were very large.
In our paper we deal with the size of the set F in the two cases:

- L is a regular language accepted by a given nondeterministic finite automaton with n transitions. The number n is the size of the problem.
- L is a context-free language generated by context-free grammar with n symbols (nonterminal and terminal together) . Assume for simplicity that the grammar is in Chomsky normal form. Then the size of the whole grammar is polynomially related to the number n of nonterminals (the number of productions is $O(n^3)$). Assume that n is the size of the problem in this case.

The bounds previously given for regular sets were exponential, and doubly exponential for context-free sets, see [ACK] and [CuS1]. The sets F consisted of all words of linear size and exponential size, respectively. If the alphabet has at least two symbols then there are exponentially and, respectively, doubly exponentially many words of such lengths. The basic argument in the previously known constructions of test sets for regular languages was the pumping lemma. Our construction is different, it gives another proof of the existence of finite test sets for regular languages.

Let Σ, Δ be arbitrary alphabets and h, g arbitrary morphisms $\Sigma^* \longrightarrow \Delta^*$. Being a morphism means that $h(\varepsilon) = \varepsilon$, and $h(uw) = h(u)h(w)$. Denote by Eq(h,g) the equality set of morphisms h, g.

$$Eq(h,g) = \{ x \in \Sigma^* : h(x) = g(x) \}.$$

A Test set F can be defined as a subset of L satisfying

$$F \subseteq Eq(h,g) \Rightarrow L \subseteq Eq(h,g)$$

for any pair of morphisms $h,g : \Sigma^* \longrightarrow \Delta^*$.

Denote by $F(\Delta)$ the free group generated by Δ. Then Δ^* as well as $(\Delta^*)^{-1}$,i.e. the set of inverses of words in Δ^*, are free submonoids of $F(\Delta)$. Let us define the *overflow* $Ovf_{h,g}(u) : \Sigma^* \longrightarrow \Delta^* \cup (\Delta^*)^{-1}$ as follows :

$$Ovf_{h,g}(u) = g(u)^{-1}h(u).$$

Since the range is restricted, $Ovh_{h,g}$ is a partial mapping, and $Ovh_{h,g}(u)$ is defined iff one of the words h(u) and g(u) is a prefix of another. Moreover, it is, if defined, an element of Δ^* iff $h(u) = g(u)z$ for some word z, i.e. h is "ahead" g.

The overflows have the following property

(1) $$Ovf_{h,g}(uv) = g(v)^{-1}\, Ovf_{h,g}(u)\, h(v)$$

Lemma 1. (key lemma)

a) If $\{xy, zy\} \subseteq Eq(h,g)$ then $Ovf_{h,g}(x) = Ovf_{h,g}(z)$;

b) If $Ovf_{h,g}(x) = Ovf_{h,g}(z)$ then $Ovf_{h,g}(xa) = Ovf_{h,g}(za)$.

Proof. If $\{xy, zy\} \subseteq Eq(h,g)$ then $Ovf_{h,g}(xy) = Ovf_{h,g}(zy) = \varepsilon$, where ε is the empty word (unit element of $F(\Delta)$). Consequently by (1) we have:

$$g(y)^{-1}\, Ovf_{h,g}(x)\, h(y) = g(y)^{-1}\, Ovf_{h,g}(z)\, h(y) .$$

This implies that $Ovf_{h,g}(x) = Ovf_{h,g}(z)$ and completes the proof of point (a). The point (b) follows trivially from (1) . ■

Let $A=(\Sigma,Q,q_i,\delta,q_f)$ be an automaton accepting the language L, where Q is a (finite) set of states, δ is the transition function, q_i is the initial state and q_f is the (only) accepting (final) state. We assume that there is exactly one accepting state with no transition outgoing that state (a special sink node can be added). We also assume that there are no useless states (each state is on a path from the initial to the terminal state).

Let G be the graph of the automaton A, its nodes are states of A, edges correspond to transitions and are labelled by input symbols. Formally we denote the transition $p \xrightarrow{a} q$ from the state p to the state q on an input symbol a by the triple (p,a,q). There can be several edges (labelled by distinct symbols) between any two given nodes. On the other hand each edge or path determine unique input symbol or word, respectively. By a path we mean a sequence of edges.

For a path π denote by $[\pi]$ the word "spelled" by this path (concatenation of labels of consecutive edges).

For a set Π of paths denote $[\Pi] = \{[\pi] : \pi \in \Pi\}$.

Let Π be a subset of paths from the initial node to the terminal node. We define two conditions:

(*) {the unique continuation property}

if $\{\pi_1\pi_1', \pi_2\pi_2'\} \subseteq \Pi$ and π_1, π_2 end at the same node then
$\{\pi_1\pi', \pi_2\pi'\} \subseteq \Pi$ for some path π';

(**) {the edge covering property}

for each edge in G there is a path $\pi\in\Pi$ traversing this edge.

Lemma 2.
Let Π be a subset of paths from the initial to the terminal node. If Π satisfies conditions (*) and (**) then $F = [\Pi]$ is a test set for the language L accepted by the automaton A.

Proof. Let G be the graph of the automaton A. Assume that we are given two morphisms h, g which agree on the set $[\Pi]$. This means that

$$[\Pi] \subseteq Eq(h,g).$$

Our aim is to prove that for each initial-terminal path π we have

$$[\pi] \in Eq(h,g).$$

For a path $\pi \in \Pi$ and morphisms h,g define

$$Ovf_{h,g}(\pi) = Ovf_{h,g}([\pi]).$$

Claim 1.

If prefixes $\pi 1'$, $\pi 2'$ of two paths $\pi 1$, $\pi 2 \in \Pi$ end at the same node then

$$Ovf_{h,g}(\pi 1') = Ovf_{h,g}(\pi 2').$$

Observe that for all paths $\pi 1$, $\pi 2 \in \Pi$ we have

$$[\pi 1],\ [\pi 2] \in Eq(h,g).$$

Our claim follows directly from Lemma 1.a and the property (*). The claim guarantees that the following definition makes sense:

for each node v denote by $OVF_{h,g}(v)$ the value of $Ovf_{h,g}(\pi')$, where π' is a prefix of a path in Π and π' ends at v.

Claim 2.

Let π be any path starting at the initial node and ending at a node v (not necessarily the terminal node). Then

$$Ovf_{h,g}(\pi) = OVF_{h,g}(v).$$

The proof of the claim is by induction on the length of the path π. If π is of zero length then its overflow is the empty word and it equals the overflow of the initial node. Assume that the path π ends with the edge (w,a,v). We have $\pi = \pi'(w,a,v)$. Using the inductive assumption (π' is shorter) we have $Ovf_{h,g}(\pi') = OVF(w)$.We use the fact that there is a path $\phi \in \Pi$ which passes through the edge (w,a,v) due to the property (**). Take a prefix ϕ' of this path which ends at the node w. Then we have $Ovf_{h,g}(\phi') = OVF_{h,g}(w)$ because of Claim 1. Hence $Ovf_{h,g}(\pi') = Ovf_{h,g}(\phi')$. Now the paths spelled by paths $\phi'(w,a,v)$ and $\pi = \pi'(w,a,v)$ are of the form xa, za, where x, z are words and a is a single letter. We know that $Ovf_{h,g}(x) = Ovf_{h,g}(z)$. We use the second point of Lemma 1, to obtain $Ovf_{h,g}(xa) = Ovf_{h,g}(za)$. But $\phi'(w,a,v)$ is a prefix of the path $\phi \in \Pi$ so that $Ovf_{h,g}(\phi'(w,a,v)) = OVF_{h,g}(v)$ by Claim 1. Hence

$$Ovf_{h,g}(\pi) = Ovf_{h,g}(za) = Ovf_{h,g}(xa) = Ovf_{h,g}\phi'(w,a,v) = OVF_{h,g}(v).$$

This completes the proof of claim.

Now the statement of the lemma follows from the last claim,

because $OVF_{h,g}(q_f) = \varepsilon$. Indeed, the overflow of every path in the graph from the initial node q_i to the terminal node q_f equals the empty word. This completes the proof. ■

Lemma 3

Assume that the graph of the automaton has n edges and that it has exactly one accepting state and the outdegree of this state is zero. Then there is a set Π of paths satisfying conditions (*) and (**) whose size is n. The set Π consists of paths not longer than the number of states and it can be constructed in time proportional to its total size.

Proof. Let G be the graph of the automaton. Define two trees - the in-tree T1 and the out-tree T2. T2 is the BFS tree of G with the root in the initial node q_i and T1 is the tree of shortest paths from each node to the terminal node q_f. T1 can be constructed by reversing the directions of all edges and then computing the BFS tree rooted in q_f. Afterwards the directions are reversed to have their original direction in G. It has to be emphasized that in the construction of above trees if there are transitions $p \xrightarrow{a} q$ and $p \xrightarrow{b} q$ in A then at most one of them is taken to the tree T1 or T2. Consequently in the trees there is at most one edge between any given pair of nodes (as opposed to the graph of A). Now we take

$$\Pi = \{ \text{path}(v,a,w) : (v,a,w) \text{ is an edge of } G \},$$

where path(v,a,w) is the path consisting of three parts:

- the path from q_i to v in T2 (first part),
- the edge (v,a,w) (special edge of path(v,a,w)) and
- the pathfrom w to q_f in T1 (last part).

Obviously the set Π satisfies the condition (**). It is very easy to prove that (*) also holds. ■

Theorem 1

For each nondeterministic finite automaton with n transitions we can construct a test set F of size n in time proportional to the total length of words in F. The length of each word in F does not exceed the number of states of the automaton.

Proof. Assume that the graph of the automaton has n edges. We assumed in Lemma 2 that it has exactly one accepting state and the outdegree of this state is zero. Now we explain how to drop this assumption.

Assume that we added a new sink state v to the original automaton A. Next in the graph of A for each edge $v1 \xrightarrow{a} v2$ with v2 accepting we can add the edge $v1 \xrightarrow{a} v$. A new equivalent automaton A' is obtained

whose only accepting state is v. The set of paths chosen in Lemma 1 was Π = { path(v,a,w) : (v,a,w) is an edge of G }.
Now it is easy to see that the path of the form path(v1,a,v2), can be deleted. The role of each such path is played by path(v1,v). Hence the number of paths is now exactly n, where n is the number of transitions of the original automaton A. This completes the proof. ■

Theorem 2
For each context-free grammar G of size n we can construct in $2^{O(n\log(n))}$ time a test set of size $2^{O(n\log(n))}$.
Proof. It was shown in [ACK] that it is enough to take as a test set F the set of all words with a derivation tree of height O(n). However it is easy to see that F can be accepted by a nondeterministic pushdown automaton with pushdown store of size O(n). This automaton can be treated as a nondeterministic finite automaton, whose state is the contents of the pushdown store. There are $n^{O(n)} = 2^{O(n\log(n))}$ possible contents of the pushdown store, hence the finite automaton has $2^{O(n\log(n))}$ states. Now as the final test set take the test set of the test set F given by Theorem 1. ■
Let h, g be two morphism. The morphism equality problem for the family of regular languages consists in verifying for a regular language L whether h(x) = g(x) for all words in L, see [CuS1]. Assume that the sizes of h(a) for single letters are of constant size.

Theorem 3
The morphism equality problem for regular languages is solvable in O(nlogn) time, where n is the number of transitions of the finite automaton A accepting the regular language.
Proof. Take the out-tree T of a given nondeterministic automaton with one accepting state (whose outdegree is zero), the same as in the proof of Lemma 3. We are given homomorphisms h, g: $\Sigma^* \to \Delta^*$.
Construct two trees T_h and T_g whose edges are labelled by symbols of Δ, and whose paths represent all prefixes of words h(x), g(x), respectively, for all paths x in T.
Then for each node $v \in T$ there is a corresponding node $h(v) \in T_h$ and the corresponding node $g(v) \in T_g$. If x is the path from the root to v in T then h(x) is the path from the root to h(v) in T_h. Similarly for g(v). The key observation is the following:

$OVF_{h,g}(v)$ is a suffix of a path from the root of T_h to h(v) or it is a inversed suffix of a path from the root of T_g to g(v).

This enables to process overflows eficiently. Suppose that the

morphism h is "ahead" of g on a path x of T which ends at the node v. Then $OVF_{h,g}(v)$ is a suffix π' of the path π from the root of T_h to h(v) in T_h and it can be represented by a constant size information consisting of three objects: the length k of π', the starting node v1 and the end node v2 = h(v) of π'.
One can identify $OVF_{h,g}(v)$ with the triple (v1,v2,k). All these triples can be computed easily in linear time traversing the tree T in, for example, BFS order.

Assume that overflows of two paths x, y ending at node u, v, respectively, in the tree T are represented by triples (u1,u2,k2), (v1,v2,k1) in T_h and there is a transition (u,a,v).
We have to check the equality

$$(*) \qquad Ovf_{h,g}(xa) = Ovf_{h,g}(y)$$

Claim
Assume that overflows on branches of the tree T are well defined and overflow on the branch ending at the accepting state is ε.Then $L(A) \subseteq EQ(h,g)$ iff for each transition (u,a,v) of A the equality (*) holds, where x,y are paths in T from the root to u,v.
Proof (of the claim). Assume that for each transition (u,a,v) of A the equality (*) holds. Let γ be a path starting at the root (not necessarily a branch of the tree T). Let v be the end-node of γ. It can be easily proved by induction on $|\gamma|$ that there is a branch γ' of T starting at the root, ending at v and satisfying $Ovf_{h,g}(\gamma) = Ovf_{h,g}(\gamma')$.
Each word of L(A) corresponds to a path γ from the root to the terminal node. However it is the same as overflow of the path γ' of T ending at the same node. However we assumed that the overflow of the branch of T ending at the accepting state is ε. This completes the proof of the claim.

Equalities (*) can be tested effciently by using the information about (u1,u2,k2), (v1,v2,k1) and a suitable data structure [CR].

We have O(n) questions of the form "$Ovf_{h,g}(xa)=Ovf_{h,g}(y)$?", where (u,a,v) is a transition of A and x, y are paths in T from the root to, respectively, nodes u,v, see Fig.5. Each such question can be easily reduced to the equality between certain subpaths. The cost of one question is O(logn), hence the total complexity of the algorithm is O(nlogn). This completes the proof. ∎

The gsm equivalence problem for regular languages is to verify whether M1(x) = M2(x) for all $x \in L$, where M1, M2 are given gsm's and L is a given regular language. We refer the reader to [HU] for the definition of the gsm. The size of the problem is in this case the total size of nondeterministic finite automaton accepting L plus the total sizes of the gsm's M1, M2.

Theorem 4

The gsm equality problem for regular languages can be solved in $O(n^3 \log n)$ time.

Proof. Assume that the automata A, M1, M2 have state sets Q, Q1, Q2,
respectively and the input alphabet is Σ. A new automaton M is constructed with the state set $Q \times Q1 \times Q2$ which is a parallel composition of machines A, M1 and M2. The input alphabet of M is $\Sigma' = Q1 \times Q2 \times \Sigma$.

Let δ, $\delta 1$, $\delta 2$ be transition functions of A, M1, M2, respectively. Then from state (q,q1,q2) the automaton M can go to state $(\delta(q,a),\delta_1(q1,a),\delta_2(q2,a))$ reading the input symbol (q1,q2,a). The accepting states are those triples whose first component is an accepting state of A.

For i=1,2 we define morphisms $h_i(q_1,q_2,a) = \lambda_i(q_i,a)$, where λ_1, λ_2 is the output function of the gsm M1, M2, respectively.

It is easy to see that the gsm mappings agree on the language L accepted by the automaton A iff the morphisms h_1, h_2 agree on the language accepted by M. The total number of transitions of M is cubic with respect to the total size of A, M1 and M2. Hence the statement of the theorem follows from Theorem 3. This completes the proof. ∎

We consider now deterministic finite transducers (dft's), see [AU1] and [BH]. We recall that the dft is a generalization of the deterministic general sequential machine (gsm's). The basic difference is the dft capability to make an output without reading the input, these are so called ε-moves. However if a machine makes an ε-move in a given state then no other moves are allowed from this state. Hence the states are divided into two disjoints classes:

- the first one consists of states in which symbols are read and output are produced deterministically ;
- the second class consists of states in which outputs are deterministically produced without reading any input symbol.

Lemma 4

For two given dft's M1, M2 we can construct in linear time deterministic gsm's A1, A2 such that M1 and M2 are equivalent iff A1 and A2 are so.

Proof. We can assume w.l.o.g that on ε-moves the output is nonempty, otherwise all such transitions can be compressed.
We call the transducer compressed iff all ε-transitions are between accepting states or from the initial state to an accepting state (the output is always nonempty). It is straightforward to transform a given dft to an equivalent compressed dft (some ε-transitions are compressed). For a given compressed transducer M we create the gsm GSM(M) by replacing each ε-transition with output α by a non ε-transition with the same output. A new input symbol '$' is placed at each such transition. Then it is easy to see that two compressed transducers M1, M2 are equivalent iff the gsm's GSM(M1) and GSM(M2) are equivalent. This completes the proof. ■
As a consequence of our proof of Theorem 4 and Lemma 3 an alternative, see [BH] and [GI], and more efficient algorithm for the dft equivalence problem is given. The automaton A can be assumed to have a single state.

Theorem 5
The equivalence of two deterministic dft's can be tested in $O(n^2 \log n)$ time. ■

It is worth noting here that for single-valued finite transducers the equivalence problem is P-space complete, see [GI]. Such transducers are in between deterministic and nondeterministic ones.

The test sets constructed in Theorem 1 are of size n, where n is the total size (number of transition) of the automaton. For unbounded alphabets this matches the lower bound. We can take two-state automaton, with one initial and one accepting state and such that each transition requires a distinct symbol of the alphabet. In the case when alphabet is bounded the problem of the size of the test set is open.

REFERENCES

[ACK] Albert,J.,Culik II,K. and Karhumaki J., Test sets for context free languages and algebraic systems of equations, Inform. Control 52 (1982) 172-186.

[AL] Albert,M.H. and Lawrence J., A proof of Ehrenfeucht conjecture, Theoret. Comp. Science 41 (1985) 121-123

[AU] Aho A., Ullman J., The theory of parsing translation and compiling, vol.I, Prentice-Hall (1972)

[BH] Blattner M., Head T., The decidability of equivalence for deterministic finite transducers, Journal of Comp. and System Sciences 19 (1979) 45-49.

[CR] Crochemore M., Rytter w., Parallel computations on strings and arrays, STACS (1990)

[Cu] Culik II,K., Homomorphisms: Decidability, Equality and Test Sets, in R. Book (ed.), Formal Language Theory, Perspectives and Open Problems (Academic Press, New York, 1980).

[CuS1] Culik II,K. and Salomaa,A., On the decidability of homomorphism equivalence for languages. J. Comput. System Sci. 17 (1978)163-175.

[CuS2] Culik II,K. and Salomaa,A., Test sets and checking words for homomorphism equivalence, J. Comput. System Sci. 20 (1980) 375-395.

[GI] Gurari, E M., Ibarra, O H., A note on finite valued and finitely ambiguous transducers, Mathematical Systems Theory 16 (1983) 61-66

[Ka1] Karhumaki,J., The Ehrenfeucht conjecture: a compactness claim for finitely generated free monoids, Theoret.Comp.Science 29 (1984) 285-308

[Ka2] Karhumaki,J., On recent trends in formal language theory, Lect. Notes in Comp. Science 267 (1987) 136-162

[EKR] Ehrenfeucht,A., Karhumaki,J. and Rozenberg,G., On binary equality sets and a solution to the test set conjecture in the binary case, J. Alg. 85 (1983) 76-85

[HU] Hopcroft J., Ullman J., Introduction to the theory of automata, langauges and computation, Addison Wesley (1979)

[KMR] Karp R., Miller R., Rosenberg A., Rapid identification of repeated patterns in strings, arrays and trees, 4th STOC (1972) 125-136

[Ko] Kosaraju R., Eficient tree pattern matching, 30th FOCS (1989) 178-183

[Sa] Salomaa, A., Jewels of formal language theory, Russian edition Mir (1986), page 150.

The Complexity of The Reliable Connectivity Problem[1]

Dimitris Kavadias[3], Lefteris M. Kirousis[2,3] and Paul Spirakis[2,3,4]
E-addresses: ⟨lastname⟩@grpatvx1.bitnet

Abstract

Let $G = (V, E)$ be a graph together with two distinguished nodes s and t, and suppose that to every node $v \in V$, a nonnegative integer $f(v) \leq \text{degree}(v)$ is assigned. Suppose, moreover, that each node v can cause at most $f(v)$ of its incident edges to "fail" (these $f(v)$ edges can be arbitrarily chosen). The Reliable Connectivity Problem is to test whether node s remains connected with t with a path of non-failed edges for all possible choices of the failed edges. We first show that the complement of the Reliable Connectivity Problem is NP-complete and that this remains true even if G is restricted to the class of directed and acyclic graphs. However, we show that the problem is in P for directed and acyclic graphs if we assume that the edges caused to fail by each node v are chosen only from the edges *incoming* to v. Concerning the parallel complexity of this version of the problem, it turns out that it is P-complete. Moreover, approximating the maximum d such that for any choice of failed edges there is a directed path of non-failed edges that starts from s and has length d turns out to be P-complete as well, for any given degree of relative accuracy of the approximation. On the contrary, given that every node v will cause *at least* $f(v)$ incoming edges to fail, the question whether there is a choice of failed edges such that s remains connected with t via non-failed edges turns out to be in NC, even for general graphs.

1 Introduction

In this paper we introduce the problem of **Reliable Connectivity** (RCP, in short): given a graph $G = (V, E)$ with two distinguished nodes s and t and an assignment of a nonnegative integer $f(v) \leq \text{degree}(v)$ for each $v \in V$, assume that each node v can cause at most $f(v)$ of its incident edges to "fail" and inquire whether s and t remain connected via non-failed edges under all possible choices of failed edges. The motivation for the problem originates from distributed systems where the sites are represented by the nodes of a graph whose edges represent communication lines; a communication line is connected with each site through a *port* and these ports are supposed to be prone to failure, but in a way that the number of failures at each site cannot exceed a given safety threshold. Formally, if F is a function that to each node v assigns a set $F(v)$ of cardinality $\leq f(v)$ and with elements edges incident with the node v, determine whether for all choices of F, s remains connected with t via a path of edges not in $\bigcup_{v \in V} F(v)$ (i.e. not in the set of failed edges).

Notice that according to our definition, the total number of failed edges incident with v *can* exceed $f(v)$. This may happen when an edge $e = \{v, w\}$ is an element of $F(w) - F(v)$. Intuitively, this is due to the assumption that the number of failed ports, and not the number of failed connections, cannot exceed a given threshold. Therefore, a site v may be disconnected from w either by a failure of the port at v or by a failure of the port at w. Another approach would have been to assume that for every node v, the total number of failed edges incident with v cannot exceed $f(v)$. Formally, that would account to assuming that the cardinality of $\{e \in \bigcup_v F(v) \text{ and } e \text{ incident with } v\} \leq f(v)$. Although these two models are, as we will show,

[1]This research was partially supported by the ESPRIT II Basic Research Actions Program of the EC under contract no. 3075 (project ALCOM).

[2]Department of Computer Science and Engineering, University of Patras, Patras 26110, Greece.
[3]Computer Technology Institute, P.O. Box 1122, Patras 26110, Greece.
[4]Courant Institute of Mathematical Sciences, NYU, U.S.A.

equivalent as far as their complexities are concerned, we prefer the first approach since it is closer to our motivation.

Reliability problems like this one have been previously investigated, but usually under the assumption that each edge (or node) of the graph has a given probability to fail [Rosenthal, 1974] [Ball, 1977]. Under that assumption the question asked was whether any two nodes remain connected with at least a given probability. Notice that in that model it is possible to have a failure of *all* edges of the graph (with a very small probability, in general), while in our model, the number of edges that a node can cause to fail is bounded by $f(v)$.

Other problems related to the reliable connectivity problem have been introduced in [Yannakakis, 1981]. The approach in that paper was to determine the least number of edges that should be deleted from G in order to have the remaining graph satisfy a given property, like planarity, which is not satisfied by G itself. In our model, the question is whether a property satisfied by G is not destroyed by all deletion patterns.

In Section 2 of this paper we prove that the complement of the Reliable Connectivity Problem is NP-complete. By a simple modification of the reduction, it turns out that we can assume that $f(v) \leq 1$, for all $v \in V$ and that degree(G) is bounded. Moreover, it turns out that RCP is co-NP-complete if, in addition to these constraints, we assume that G is a directed acyclic graph (for this case a path is defined to mean either a path directed from s towards t or, alternatively, a path of the underlying undirected graph; the edges caused to fail by a node v may be both incoming and outgoing).

In Section 3 we introduce the Reliable Connectivity Problem for failures of incoming edges (RCP-IN, in short): given a directed and acyclic graph $G = (V, E)$ with two distinguished nodes s and t and an assignment of a nonnegative integer $f(v) \leq$ in-degree(v) for every node $v \in V$, assume that each node v can cause at most $f(v)$ of its *incoming* edges to fail and inquire whether, under all failure patterns, s and t remain connected via a *directed* from s towards t path of non-failed edges. We prove that this version of the Reliable Connectivity Problem is polynomially solvable, and moreover that it is P-complete. It is worth pointing out that it is not the restriction to the class of directed and acyclic graphs (dags, in short) that lowers the complexity of the problem, since we show that RCP is co-NP-complete even when restricted to dags, but rather the restriction of failure only to *incoming* edges. This restriction imposes a direction to the failure patterns which is globally compatible with the direction of the graph and and this is, intuitively, the reason of the lower complexity.

The way to handle P-complete problems is, in analogy with NP-complete problems, to try to find approximation algorithms that can be efficiently parallelized. There exist P-complete problems that are amenable to approximations, i.e., they turn out to be efficiently parallelizable for any given degree of permissible relative error [Serna and Spirakis, 1991]; other of P-complete problems exhibit a *threshold* behavior, i.e. they remain P-complete for approximations up to a certain degree of relative accuracy, but they turn out to be in NC for approximations with relative accuracy worse than a certain value [Anderson and Mayr, 1984], [Kirousis, Serna and Spirakis, 1989]. Unfortunately, the RCP-IN does *not* yield to approximation techniques: we show that for any given $\epsilon > 0$ the problem of approximating the largest d for which, under all failure patterns of incoming edges, there is a directed path starting from s and of length at least d cannot be approximated by an algorithm in NC with a relative error $\leq \epsilon$, unless P=NC.

In Section 4 we introduce what we call the Feasible Connectivity Problem: determine whether there exists at least one failure pattern that leaves s connected with t. Of course, for the problem to make sense, we assume that in this case every node causes *at least* $f(v)$ of its incident edges to fail. We show that the Feasible Connectivity Problem is in NC. It is interesting to contrast this result with a result in [Yannakakis, 1981] where it is proved that for any fixed $r \geq 2$, to find the minimum number of edges that must be deleted from a graph in order to get a connected graph

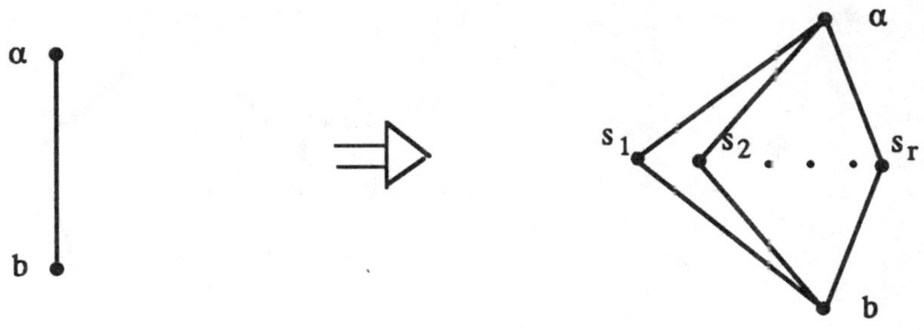

Figure 1

with maximum degree r is an NP-complete problem (the catch is that Yannakakis requires a deletion that would leave connected the whole graph, while in our case, given a pair of nodes, we look for a deletion that would leave the pair connected).

2 NP-completeness results

Since detecting whether in a given graph two given nodes can be connected by a path is in P, it immediately follows that the complement of the RCP is in NP: simply guess a failure pattern and show that under this pattern s is disconnected from t.

To show the completeness of the complement of RCP, we use 3SAT for the reduction. Let us first describe a simple gadget to be used in the reduction. The utility of this gadget will be to exclude the possibility of failure of a given edge of a graph. This gadget is depicted in Figure 1. It consists of two nodes a and b plus a number of nodes $s_1, \ldots, s_r$, connected as shown. If we want to maintain a connection between two given nodes a and b of a graph for all choices of failed edges, it suffices to connect a and b with such a gadget with $r > \max(f(a), f(b))$ and let $f(s_i) = 0$, for $i = 1, \ldots, r$. Such a connection between nodes a and b will be called, by a slight abuse of terminology, a **safe** edge (it is not actually a single edge, but rather a collection of edges and nodes). In graph diagrams, a safe edge will depicted by a jagged line. Let us now proceed with the reduction. Given a Boolean formula ϕ in conjunctive normal form with clauses $C_1, \ldots, C_k$ and Boolean variables $X_1, \ldots, X_n$, such that each C_j contains at most three literals, define a graph G_ϕ as follows:

- The nodes of G_ϕ are: two distinguished nodes s and t; one node c_j for each clause C_j (we call these nodes clause nodes) and three nodes x_i, $\overline{x_i}$ and a_i, respectively, for each Boolean variable X_i (we call the nodes x_i and $\overline{x_i}$ literal nodes, the former corresponding to a positive literal of the variable X_i and the latter to a negative one).

- The edges of G_ϕ are defined as follows: s is connected with all c_j's, via *safe* edges (literally, this means that G_ϕ has more nodes than the ones already mentioned, but it is more illuminating to ignore the structure of the gadget and think of it as an edge). The value of the function f for the nodes s and c_j, which must be known in order to determine the structure of the safe edges, is given below. Each c_j is connected with an x_i or an $\overline{x_i}$ according to whether C_j contains the literal X_i or $\neg X_i$ (these are not assumed to be safe connections). Each a_i is connected with both x_i and $\overline{x_i}$ (again by ordinary, i.e. non-safe, edges), and finally t is connected with all a_i's with safe edges (see Figure 2).

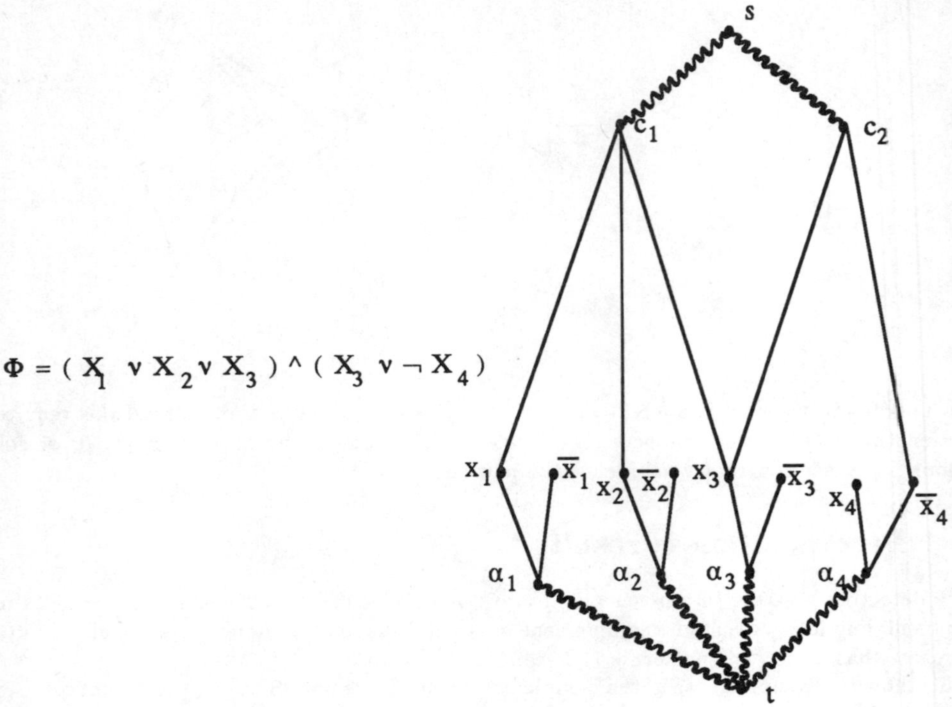

Figure 2

- $f(s) = f(t) = f(x_i) = f(\overline{x_i}) = 0$ and $f(a_i) = 1$, for all $i = 1, \ldots, n$, while $f(c_j)$ = number of literal nodes that are connected with c_j minus 1. Obviously then, $f(c_j) \leq 2$.

Now it is not hard to see that the formula ϕ is satisfiable if and only if in G_ϕ there is a failure pattern that disconnects s from t. Indeed, if ϕ is satisfied, then each clause C_j contains at least one true literal. Define then a failure pattern as follows: each node a_i causes the unique one of its incident edges that connects it to a node corresponding to a true literal to fail; on the other hand, c_j causes all edges connecting it to a node corresponding to a literal to fail, except a single one (arbitrarily chosen, if there are more than one) that connects it to a node corresponding to a true literal. It can be easily checked that this failure pattern disconnects s from t. Conversely, suppose that there is a failure pattern that disconnects s from t. Define a variable X_i of ϕ to be true if the edge connecting x_i with a_i fails, and false otherwise. Since s is connected with all clause nodes with safe edges and also t is connected with all nodes a_i with safe edges, we conclude that for s and t to be disconnected, each c_j must be disconnected from all the nodes a_i. But since c_j can cause all but one of its edges leading to a literal node to fail, we conclude that c_j must be connected to a literal node which is not connected with its corresponding a_i. Therefore, each clause must contain a true literal, and so we have proved that:

Theorem 1-a *The complement of the Reliable Connectivity Problem is NP-complete.*

Notice that because of the specific reduction used, it turns out that RCP is co-NP-complete

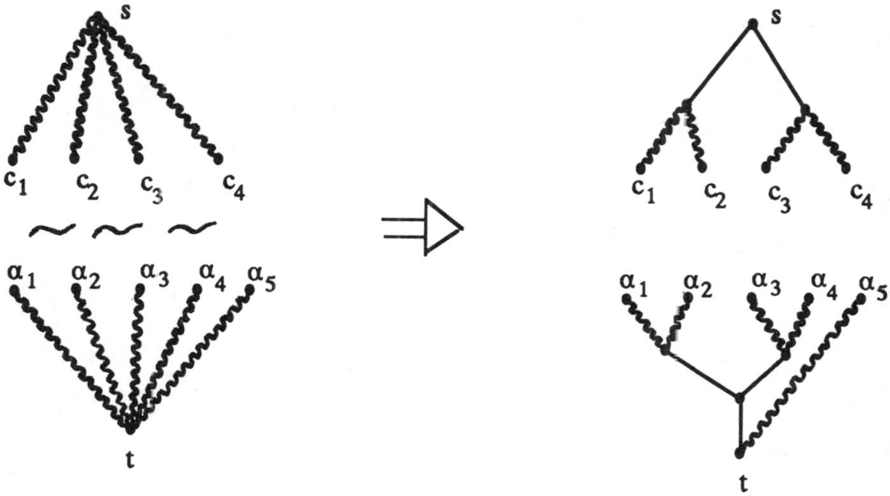

Figure 3

even under the more restrictive model where we require that

$$|\{e \in \bigcup_v F(v) \text{ and } e \text{ incident with } v\}| \leq f(v).$$

Also, if we give to all edges a direction from s towards t, then it turns out that RCP is co-NP-complete even for directed and acyclic graphs where paths are assumed either to be directed or, alternatively, to exist in the underlying undirected graph. Moreover, it can be proved that the degree of the graph can be assumed to be bounded. This is not immediate from the reduction described, but it can be easily proved if in the reduction, instead of connecting s with each c_j and t with each a_i with a single safe edge, we make these connections via fanout trees (see Figure 3). The value of the function f on the internal nodes of these trees is defined to be 0. Also, for the reduction we use not just 3SAT, but rather the satisfiablity problem where each clause contains at most three literals and each variable appears in at most three clauses. To show that we may restrict the problem to graphs where $f(v) \leq 1$ without destroying the NP-completeness, we substitute, in the reduction, each connection of a clause node c_j with three literal nodes l_1, l_2, l_3 with a connection like the one depicted in Figure 4, where a new node b_j is introduced, and where we set $f(c_j) = f(b_j) = 1$. Observe that when $f(v) \leq 1$, the gadget implementing a safe edge need only contain two intermediate nodes, therefore it follows that G has degree at most 5. Therefore, we have proved that:

Theorem 1-b *The complement of the Reliable Connectivity Problem is NP-complete, even when $f(v) \leq 1$ and degree(G) ≤ 5. Moreover, the problem remains NP-complete if in addition to these constraints, G is assumed to be a directed acyclic graph (a path, in this case, is assumed to be either a path directed from s towards t, or, alternatively, to exist in the underlying undirected graph). Finally, the problem remains NP-complete even under the assumption that*

$$|\{e \in \bigcup_v F(v) \text{ and } e \text{ incident with } v\}| \leq f(v).$$

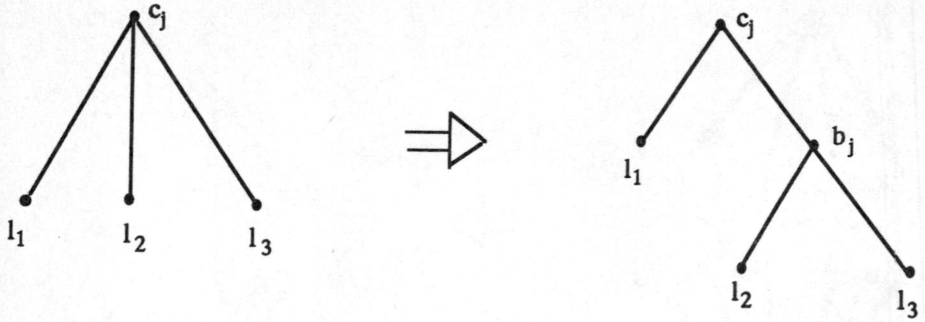

Figure 4

3 P-completeness results

In contrast with the results of the preceding section, the Reliable Connectivity Problem turns out to be polynomially solvable for directed acyclic graphs, *when the edges that a node v causes to fail are chosen among its incoming edges.* Moreover, we prove that this version of the problem is actually P-complete.

Formally, let $G = (V, E)$ be a dag together with two distinguished nodes s and t, and an assignment of an integer $f(v) \leq$ in-degree(v), for all nodes v. A choice of failed edges for a node v is a set $F(v)$ of edges incoming to v and of cardinality $\leq f(v)$. The Reliable Connectivity Problem for failures of incoming edges is the question whether s and t remain connected by a *directed s-t* path consisting of edges not in $\bigcup_v F(v)$, for all choices of the function F. Let us point out that for the rest of the paper when we say that a node s (the source) is connected with a node t (the sink) we mean that there exists a *directed* path starting from s and ending in t.

Theorem 2 *The Reliable Connectivity Problem for failures of incoming edges is polynomially solvable.*

Proof The algorithm we give is a greedy one: simply delete all edges (incoming and outgoing) of a node $v \neq s$ if $f(v) \geq$ the current in-degree(v) and then repeat the same until no node $v \neq s$ with $f(v) \geq$ the current in-degree(v) remains. Formally, first run the procedure delete_edges given in Figure 5 and then check whether in the resulting graph, call it G', s remains connected with t (with a path directed from s towards t). We show that there exists a failure pattern under which s is disconnected from t if and only if s is disconnected from t in G'. Indeed, if s is disconnected from t in G', then, obviously, there is a failure pattern that disconnects s from t. It is the converse that is more interesting.

We first show that for any edge $e = (t_2, t_1)$ in G', if $t_1, t_2 \neq s$ then t_2 can be connected with s in G' (by a path directed from s towards t_2). Indeed, assume, looking for a contradiction, that there is no such path in G' from s towards t_2. Since the procedure delete_edges has not eliminated (t_2, t_1), there must exist an edge (t_3, t_2) in G' with $t_3 \neq s$ and no path from s towards t_3. If we continue this way defining a sequence $t_1, t_2, \ldots, t_k, \ldots$, we must reach a point (since the graph is finite) where $t_n = t_k$ with $n > k$. But that will give a circle, a contradiction.

We now return to the proof of the converse of the theorem. Suppose that s and t are connected in G'. Let d be the length of the longest directed path from s to t. We will show by induction on d that under any failure pattern of G, s and t remain connected. Let F be

```
procedure delete_edges (E)
begin
for all nodes v ∈ V do in-degree(v) := number of edges in E incoming into v od;
while ∃v ∈ V : v ≠ s and f(v) ≥ in-degree(v) do
      E := E − {e : e is incident (incoming or outgoing) with a v such that
                              f(v) ≥ in-degree(v) and v ≠ s};
      for all nodes v ∈ V do in-degree(v) := number of edges in E incoming into v od od
end
```

Figure 5

an arbitrary failure pattern. By the definition of G', $f(t) <$ in-degree(t). Therefore there is an edge (t_1, t) in G' which does not fail under F. Then, by what has been shown in the preceding paragraph, t_1 is connected in G' with s and therefore, by the inductive hypothesis, it is connected with s under any failure pattern of G. Since the choice of F was arbitrary, the same is true for t. □

Theorem 3 *The RCP-IN is P-complete.*

Proof We exhibit a reduction from the Monotone Circuit Value Problem. Let C be a monotone circuit (i.e. one with only OR and AND gates) with inputs $s_1, \ldots, s_k$ and output node t. Consider C as a dag with the direction of its edges being from the inputs towards the output. Add to this graph a new node s and connect it with those of the input nodes of C that have value 'true'. The direction of these new edges are from s towards the s_i's. Call the dag thus obtained G_C. Let the function f take the value 0 for all nodes of this dag, except the AND gates, where f is defined to take the value 1. We claim that the output of C is true iff for any failure pattern of incoming edges, s remains connected with t. To show this, we prove that for any node $t' \neq s$, s remains connected with t' under any failure pattern iff the value of t' in C is true. This is proved by induction on the length of the longest path from an input node towards t'. The claim is trivially true if t' is an input node. Suppose therefore that t' is a gate. If it is an OR gate, then since no edge incoming to t' is caused to fail by t', the result is proved using the induction hypothesis. If t' is an AND gate, again the result follows by the induction hypothesis and by the fact that at one of the edges incoming into t' can be caused to fail by t'. □

Next we consider approximations to RCP-IN. RCP-IN is a decision problem, therefore we have first to formulate a corresponding optimization problem. Thinking of the node s as the source of a message in a distributed system, a natural approach is to try to maximize the largest d such that under all failure patterns of incoming edges, there is a directed path of length at least d starting from s. Computing this d has polynomial time-complexity. Indeed, as in proof of Theorem 3, apply the procedure 'delete_edges' to G to obtain G' and show that the length of the longest path in G' that starts from s is equal to d.

Now, approximating this d with relative error $\leq \epsilon$, where ϵ is a positive real, means finding an algorithm that outputs a d' satisfying: $d \geq d' \geq (1-\epsilon)d$. We show that for any given $\epsilon > 0$, no such algorithm exists in NC, unless P=NC. The method we use is reducing the decision RCP-IN to the approximation of the optimization RCP-IN. The reduction employs the padding technique: given a dag G with distinguished nodes s and t and a failure function f, append to G a new path starting from t and of length $\lceil (1/1-\epsilon)n \rceil$ (n is the number of nodes of G) and let $f(v) = 0$ for all new nodes of this path. Call this padded graph G^p. Now, it is easy to see that an approximation algorithm for the optimization RCP-IN otputs a number $\geq n$ when applied to G^p iff in G there is a path from s to t under all failure patterns. So we have proved:

Theorem 4 *Approximating the largest d for which there is, under all failure patterns, a path starting from s and having length at least d is P-complete.*

4 The Feasible Connectivity Problem is in NC

Let again $G = (V, E)$ be an undirected graph together with two distinguished nodes s and t and a function f that assigns a nonegative integer $f(v) \leq$degree(v) to every node v and assume that each node v causes *at least* $f(v)$ of its incident edges to fail. The **Feasible Connectivity Problem** is defined to be to test whether there is at least one failure pattern such that s remains connected with t via non-failed edges.

Theorem 5 *The Feasible Connectivity Problem is in NC.*

Proof Without loss of generality, we may assume that $f(s) \leq$degree$(s)-1$ and $f(t) \leq$degree$(t)-1$ (otherwise, the problem is trivial). Now, delete from G all nodes other than s and t for which $f(v) \leq$ degree$(v) - 2$ and let G' be the subgraph of G *induced* by the remaining nodes. Since testing connectivity between two nodes in a given graph is known to be in NC, it suffices to prove that there is a failure pattern under which s and t remain connected iff s and t are connected in G'. Suppose, in order to prove the only if part, that s and t remain connected under a failure pattern F. Let P be a path connecting s and t under F. Then for all nodes of P other than s and t it must be true that $f(v) \leq$ degree$(v)-2$, because for each such node there are at least two incident edges that do not belong to the set of failed edges. Therefore, all nodes of P belong to G', and so s and t are connected in G'. For the converse, suppose that s and t are connected in G' and let P be a path in G' connecting them. Because for every node v in P other than s and t we have that $f(v) \leq$ degree$(v) - 2$ and because $f(s) \leq$degree$(s) - 1$ and $f(t) \leq$degree$(t) - 1$, it is obvious that there is a failure pattern that leaves all edges of P intact. □

References

R. Anderson and E.W. Mayr [1984] *A P-Complete Problem and Approximations to It*, Technical Report, Computer Science Dept., Stanford University, California.

M.O. Ball [1977], *Network Reliability and Analysis: Algorithms and Complexity*, Doctoral Thesis, Operations Research Dept., Cornell University, Ithaca, New York.

L.M. Kirousis, M. Serna and P. Spirakis [1989], The parallel complexity of the subgraph connectivity problem, *Proceedings of the 30th Annual Symposium on Foundations of Computer Science*, (IEEE Computer Society Press), 294-299.

A. Rosenthal [1974], *Computing Reliability of Complex Systems*, Doctoral Thesis, Dept. of Electrical Engineering and Computer Science, University of California, Berkeley, California.

M. Serna and P. Spirakis [1991], "Tight RNC approximations to max flow," to appear in: *Proceedings of the 8th Symposium on Theoretical Aspects of Computer Science* (Lecture Notes in Computer Science, Springer-Verlag).

M. Yannakakis [1981], Edge-deletion problems, *SIAM J. Computing* **10**, 297-309.

Pattern Matching in Order-Sorted Languages*

Delia Kesner

INRIA Rocquencourt
Domaine de Voluceau, BP 105
78153 Le Chesnay Cedex, France

LRI and CNRS UA 410
Bât 490, Université de Paris-Sud
91405 ORSAY Cedex, France

Abstract

We study the problem of pattern matching in languages whose type system is hierarchical and whose evaluation strategy is lazy. We propose an extension of the Puel-Suárez compilation scheme to function definitions via order-sorted patterns. Pattern matching trees (PMT's) are defined to have edges labelled not only with structure, but also with subsort constraints. Due to this latter kind of edges, terms are reduced only as far as required to make either a structure or a subsort verification decidable. We show that the PMT is optimal if a decidable property of sequentiality holds for the sets generated during the compilation process. Our method turns out to be applicable for strict languages as well.

1 Introduction

Many programming languages use pattern matching in a many-sorted (such as those in the ML family [HMM86]) or order-sorted (such as those in the OBJ family [GW88]) term algebra for function argument-passing. Function definitions consist of an ordered set of rewrite rules. Usually, this order corresponds to the appearance of the rules in the text but any other priority will suffice [Lav88]. The rules are often **ambiguous** as some left-hand sides (LHSs) of the same function definition may overlap. Thus direct access to the relevant rule based on the LHS's structure is not possible in general. The naïve operational semantics amounting to sequential lookup until a calling pattern matches a rule LHS obviously leads to poor performance. In addition, a lazy evaluation strategy allows the manipulation of infinite objects, but it is a difficult question to define what it means pattern matching in lazy languages: the reduction scheme should be such that the only part of a term which is evaluated is the one required for the pattern matching. Recently, Puel and Suárez [PS90] devised a clever compilation scheme to generate statically a PMT in lazy languages. Such a tree is then used at run-time for fast rule-indexing and takes full advantage of the nature of the LHSs in a definition. Their work simplified and generalized seminal ideas by Huet and Lévy [HL79] that were in turn sharpened by Laville [Lav88]. The gist of the Puel-Suárez method rests on generalized notions of constructor terms and sequentiality. They called the new terms **constrained terms**.

Although partially ordered sorts provide a substantially improved expressiveness over many-sorted languages, in an order-sorted system with a lazy reduction strategy, pattern matching is more complex than with non-ordered sorts in that it consists now of two kinds of verifications: the first one, as in the conventional case, is **structure** matching; the other one, is to ascertain that the argument's sort is a **subsort** of that of the formal parameter. Moreover, as functions can have a non-strict semantics, they can yield a result even for some arguments whose evaluation is non-terminating. Therefore, the arguments need only be evaluated enough to make either a structure or subsort verification decidable;

*This work was partly done while the author was a research intern at the *Paris Research Laboratory* of *Digital Equipment Corporation*.

but it is not clear how many steps of reduction must be performed on a given term in order to precise its sort. Thus, lazy order-sorted pattern matching becomes a non-trivial problem deserving careful attention.

We propose here an extension of the Puel-Suárez compilation scheme that accommodates order-sorted constructor-based functions definitions. Ambiguous patterns are eliminated by introducing order-sorted constrained terms. Moreover, as order-sorted pattern matching consists of two kind of verifications, discrimination trees are now quite complex since some edges are labelled with sort restrictions. We restrict our interest to syntactic pattern matching[1]. Sorts are partially ordered. Minimal sorts are assumed to be pairwise disjoint and non-minimal sorts are assumed to be the union of their subsorts. Functions can have more than one declaration [GM89]. We will also restrict our attention to linear LHSs, *i.e.*, without repeated variables. We lose no expressive power, though we lose some notational convenience.

Given a pattern matching problem S, the **strict set** of S is the set of terms for which every PMT associated to S will fail to terminate and an **optimal** PMT is a PMT that will only fail to terminate on the strict set of S. We show that optimality of an order-sorted PMT is a decidable property equivalent to a generalization of the notions of strong **sequentiality** presented in [HL79] and [PS90]. Sequentiality of a pattern matching problem S is the possibility of systematically expanding any term step by step until either it matches a pattern of S or it is clear that a positive matching is impossible. Our notion of sequentiality takes not only the structure of terms into account, but also the sort system.

The paper is organized as follows. Section 2 presents unitary signatures and function definitions by order-sorted equations. Section 3 defines the syntax and semantics of patterns and order-sorted constrained terms. Section 4 describes the more interesting ideas of the compilation method. Finally, in section 5, the new notion of sequentiality for order-sorted constrained terms is presented. We show that sequentiality and optimality of pattern-matching problems are equivalent.

2 Functions Defined by Order-Sorted Pattern Matching

All the conventional notions regarding substitutions, instantiation, and unification of unsorted terms are readily extended to order-sorted terms [SNGM89].

A **signature** $\Sigma = \langle \mathcal{S}, \leq, \mathcal{F}, \mathcal{C}, \mathcal{V}, \mathcal{D} \rangle$ consists of a set of sort symbols $\mathcal{S} = \{\sigma, \eta, \delta, \ldots\}$, a partial order $\leq$ on $\mathcal{S}$, a set of function symbols $\mathcal{F} = \{F, G, H, \ldots\}$, a set of constructor symbols $\mathcal{C} = \{f, g, h, \ldots\}$, a set of $\mathcal{S}$-indexed variables $\mathcal{V} = \{x^\sigma, y^\sigma, z^\sigma, \ldots\}$ with countably infinite variables for each sort symbol σ, and a set of declarations $\mathcal{D}$ of the form $q : \sigma_1 \ldots \sigma_n \to \sigma$ where $q \in \mathcal{F} \cup \mathcal{C}$. The sets $\mathcal{S}$, $\mathcal{F}$, $\mathcal{C}$ and $\mathcal{V}$ are mutually disjoint. We write $s \in \Sigma$ for any symbol s in $\mathcal{S}$, $\mathcal{F}$, $\mathcal{C}$, $\mathcal{V}$ or $\mathcal{D}$ and we will use $\vec{\sigma}, \vec{\eta}, \ldots$ to denote possibly empty strings of sorts.

Σ-terms are constructed in the usual manner with the additional constraint that they be **well-sorted**. Formally, a variable $x^\sigma \in \Sigma$ is a well-sorted Σ-term of sort η if $\sigma \leq \eta$, and $q(t_1 \ldots t_n)$ is a well-sorted Σ-term of sort η if and only if there exists a declaration $q : \sigma_1 \ldots \sigma_n \to \sigma \in \Sigma$ such that $\sigma \leq \eta$ and for all $1 \leq i \leq n$, t_i is a well-sorted Σ-term of sort σ_i. Where Σ is understood, we will refer simply to terms instead of Σ-terms.

A signature Σ is **regular** if all terms have a least sort. We may emphasize the fact that a term t has least sort σ by writing it as t^σ. A signature Σ is **unitary** if it is regular and:

- $(\mathcal{S}, \leq)$ is a boolean[2]lattice with least upper bound operation $\sqcup$, greatest lower bound operation $\sqcap$, and greatest element $\top$ and least element $\bot$.
- No function or constructor declaration contains the sort symbol $\bot$.

[1]See [JK90] for a discussion of unification and matching in equational theories.

[2]A lattice $(\mathcal{S}, \leq)$ is said to be boolean iff $\forall \sigma \in \mathcal{S}, \exists ! \, \sigma^c \in \mathcal{S}$ such that $\sigma \sqcap \sigma^c = \bot$ and $\sigma \sqcup \sigma^c = \top$.

- **(Minimal codomain sort)** If $f \in \mathcal{C}$, then there exists a declaration $f : \vec{\sigma} \to \sigma \in \Sigma$ and σ is a minimal sort (*i.e.*, if $\delta \leq \sigma$ then $\delta = \bot$ or $\delta = \sigma$).
- **(Disjoint domain sort)** If $f \in \mathcal{C}$ and $f : \sigma_1 \ldots \sigma_n \to \sigma \in \Sigma$ and $f : \eta_1 \ldots \eta_m \to \eta \in \Sigma$ are two different declarations of f in Σ, then $n \neq m$ or $n = m \geq 1$ and $\vec{\sigma}$ and $\vec{\eta}$ are disjoint (*i.e.*, $\exists i, 1 \leq i \leq n, \sigma_i \sqcap \eta_i = \bot$).

Any signature Σ satisfying the first two conditions can be transformed into a unitary one such that, when restricting its set of sort symbols to that of Σ, the same set of well-sorted terms can be constructed. This allows us, w.l.o.g, to consider only unitary signatures in order to deal with some quite complex concepts in a simple and more readable way. Full details can be found in [Kes91].

Constructor terms do not contain function symbols. A term is called **linear** if its variables occur at most once and **ground** if it contains no variables. We say that t' **matches** t, denoted as $(t \sqsubseteq t')$, iff there exists a substitution θ such that $t' = \theta(t)$. When only linear terms are on the left-hand side, then $x^\sigma \sqsubseteq t^\eta$ if $\eta \leq \sigma$, and $f(t_1 \ldots t_n) \sqsubseteq f(h_1 \ldots h_n)$ if and only if for all $(1 \leq i \leq n)$ $t_i \sqsubseteq h_i$. Unification is a least upper bound operation for $\sqsubseteq$. We will note the least upper bound of two Σ-terms t and t' as $t \sqcup t'$. Two terms are said to **overlap**, or to be **ambiguous**, if they are unifiable. Regular signatures are finitary unifying and they make order-sorted term unification well-behaved (see [SNGM89] for a discussion). If in addition it is unitary, then a unique unifier is produced.

A **function definition** is specified by a set of **rewrite rules** $\{F(t_i) = p_i\}_{i=1}^m$, where $F \in \mathcal{F}$, the t_i's are (possibly mutually ambiguous) linear constructor terms (the **patterns** of F) and each p_i is a term containing no variables not in t_i. A **program** $\mathcal{P}$ is a set of function definitions.

3 Order-Sorted Constrained Pattern Matching

3.1 Syntax

For a signature Σ, we define the syntax and semantics of constructor Σ-terms, Σ-constraints, constrained Σ-terms and Σ-patterns. We will drop the prefix Σ where it is understood.

Let t be a term, l a linear term, σ a sort and $\mathcal{T}$ and $\mathcal{F}$ the two logical constants denoting truth and falsehood, respectively. Then $\mathcal{T}, \mathcal{F}, t : \sigma$ and $t \diamond l$ are **atoms**. A **constraint** is recursively defined as an atom or as $C_1 \vee C_2$ or as $C_1 \wedge C_2$ where C_1,C_2 are constraints. We will write $t \diamond S$ for a constraint $t \diamond t_1 \wedge \ldots \wedge t \diamond t_n$ where $S = \{t_1, \ldots, t_n\}$ and $A \in C$ for an atom A of the constraint C. When $f : \sigma_1 \ldots \sigma_n \to \sigma \ \in \ \Sigma$ and $x_1 \ldots x_n$ are pairwise distinct variables, we may write an atom $t \diamond f(x_1^{\sigma_1} \ldots x_n^{\sigma_n})$ as $t \diamond f^\sigma$ or $t \diamond f$ if the sorts are clear from the context. The intended interpretations of a **sort constraint** $t : \sigma$ and a **structure constraint** $t \diamond l$ are, respectively, it is decidable that term t has sort σ, and it is decidable that t is structurally different from l.

If t is a constructor term and C a constraint, then $t \,|\, C$ is a **constrained term**. If t is linear (resp. ground), then $t \,|\, C$ is a linear (resp. ground) constrained term. A **pattern** is a non-empty list of linear constructor terms $p_1 \ldots p_n$. A **constrained pattern** is a non-empty list of linear constrained terms $P_1 \ldots P_n$.

For brevity, we will refer to either a term or a constraint as an object. The set of **free variables** of an object h, denoted $\mathcal{V}(h)$, is defined as expected except that $\mathcal{V}(t \diamond l) = \mathcal{V}(t : \sigma) = \mathcal{V}(t)$. Substitution, denoted $\theta(h)$, is also defined as expected except that $\theta(t \diamond l) = \theta(t) \diamond l$ and $\theta(t : \sigma) = \theta(t) : \sigma$. We will also refer to the **restriction** of a constraint C to a set of variables V, denoted $C|_V$. For atomic constraints C, $C|_V = C$ if $\mathcal{V}(C) \subseteq V$, otherwise $C|_V = \mathcal{T}$. For constraints formed from $\vee$ and $\wedge$, the restriction distributes through to their arguments.

3.2 Semantics

With a lazy reduction scheme, functions can yield a result even when applied to arguments whose evaluation is non-terminating; and thus a new element is necessary in order to give semantics to such functions. We will introduce a new symbol $\bullet^\sigma$ for each sort σ in the signature. Formally, an **augmented signature** $\Sigma^\bullet$ is a unitary signature Σ without variables and without function symbols $\{F, G, H, \ldots\}$ but with a 0-ary constructor $\bullet^\sigma$ for each sort symbol $\sigma \in \Sigma$ different from $\bot$, that denotes those terms of sort σ that cannot be reduced to a term having a constructor symbol at the root (the so-called **head-normal form**). Note that all $\Sigma^\bullet$-terms are ground.

The free order-sorted term algebra on the signature Σ is denoted by $\mathcal{T}_\Sigma$. An interpretation of a signature Σ over its $\Sigma^\bullet$-term algebra $\mathcal{T}_{\Sigma^\bullet}$ satisfies:

- $\sigma^{\mathcal{T}_{\Sigma^\bullet}} := \{s \mid s \text{ is a } \Sigma^\bullet\text{-term of sort } \sigma\}$, $\bot^{\mathcal{T}_{\Sigma^\bullet}} = \emptyset$ and $\top^{\mathcal{T}_{\Sigma^\bullet}}$ is the universe of $\mathcal{T}_{\Sigma^\bullet}$.
- $\sigma \leq \eta$ implies $\sigma^{\mathcal{T}_{\Sigma^\bullet}} \subseteq \eta^{\mathcal{T}_{\Sigma^\bullet}}$.
- If $f \in \mathcal{C}$ and $f : \vec{\sigma} \to \sigma \in \Sigma$, then $f^{\mathcal{T}_{\Sigma^\bullet}}$ is a function $\vec{\sigma}^{\mathcal{T}_{\Sigma^\bullet}} \to \sigma^{\mathcal{T}_{\Sigma^\bullet}}$ such that $f^{\mathcal{T}_{\Sigma^\bullet}}(\vec{s}) = f(\vec{s})$

Sorts and subsorts sometimes allow us to decide if a term matches a pattern even if its evaluation is non-terminating. We show how to characterize such nice terms.

To each constructor term t, constrained term $t|C$ and constrained pattern $P_1 \ldots P_n$ we associate three disjoint sets of $\Sigma^\bullet$-terms such that the union of these three sets is the $\mathcal{T}_{\Sigma^\bullet}$-algebra. The first one is the **denotation** set, denoted $[\![\ ]\!]_\mathcal{T}$ or simply $[\![\ ]\!]$, the second one is the **uncertain or strict** set, denoted by $[\![\ ]\!]_\mathcal{U}$, and the last one is the **complement** set, denoted $[\![\ ]\!]_\mathcal{F}$.

If t is a constructor term, $[\![t]\!]_\mathcal{T}$ is the set of $\Sigma^\bullet$-terms that are instances of t, $[\![t]\!]_\mathcal{U}$ is the set of all $\Sigma^\bullet$-terms for which we cannot decide if they are instances of t and $[\![t]\!]_\mathcal{F}$ is the set of $\Sigma^\bullet$-terms that are not instances of t. Formally,

$[\![t]\!]_\mathcal{T} = \{\theta(t) | \theta \text{ is a } (\mathcal{V}, \Sigma^\bullet)\text{-assignment}\}$ and $[\![t]\!]_\mathcal{F} = \mathcal{T}_{\Sigma^\bullet} - [\![t]\!]_\mathcal{T} - [\![t]\!]_\mathcal{U}$, where
$[\![x^\sigma]\!]_\mathcal{U} = \{\bullet^\eta | \eta \sqcap \sigma \neq \bot \text{ and } \eta \not\leq \sigma\}$ and
$[\![f(t_1 \ldots t_n)^\sigma]\!]_\mathcal{U} = \{\bullet^\eta | \eta \geq \sigma\} \cup \{f(a_1 \ldots a_n) \mid \exists i\ a_i \in [\![t_i]\!]_\mathcal{U} \text{ and } \forall j\ a_j \notin [\![t_j]\!]_\mathcal{F}\}$

To each constraint C, we associate a three-valued **truth value** — *true*, *false* or *uncertain* — under a $(\mathcal{V}(C), \Sigma^\bullet)$-assignment θ, denoted $[\![C]\!]_\theta$. The important cases are defined as follows. The other cases follow standard three-valued logic with $\wedge$ being the minimum of its arguments and $\vee$ the maximum (*false* < *uncertain* < *true*).

$$[\![t \diamond l]\!]_\theta = \begin{cases} true & \text{if } \theta(t) \in [\![l]\!]_\mathcal{F} \\ false & \text{if } \theta(t) \in [\![l]\!]_\mathcal{T} \\ uncertain & \text{if } \theta(t) \in [\![l]\!]_\mathcal{U} \end{cases} \qquad [\![t : \sigma]\!]_\theta = \begin{cases} true & \text{if } \theta(t) \in \sigma^{\mathcal{T}_{\Sigma^\bullet}} \\ false & \text{if } \theta(t) \text{ is of sort } \eta \\ & \text{and } \eta \sqcap \sigma = \bot \\ uncertain & \text{otherwise} \end{cases}$$

If $t|C$ is a constrained term, $[\![t|C]\!]_\mathcal{T}$ contains the instances of t that satisfy C, $[\![t|C]\!]_\mathcal{U}$ is the set of $\Sigma^\bullet$-terms for which we cannot decide if they are instances of t or if they satisfy C and $[\![t|C]\!]_\mathcal{F}$ is the set of $\Sigma^\bullet$-terms that are not instances of t or do not satisfy C. Formally,

- $[\![t|C]\!]_\mathcal{T} = \{\theta(t) \mid \theta \text{ is a } (\mathcal{V}(t), \Sigma^\bullet)\text{-assignment and } [\![C|_{\mathcal{V}(t)}]\!]_\theta = \text{true}\}$.
- $[\![t|C]\!]_\mathcal{U} = \{\theta(t) \mid \theta \text{ is a } (\mathcal{V}(t), \Sigma^\bullet)\text{-assignment and } [\![C|_{\mathcal{V}(t)}]\!]_\theta = \text{uncertain}\} \cup [\![t]\!]_\mathcal{U}$
- $[\![t|C]\!]_\mathcal{F} = \mathcal{T}_{\Sigma^\bullet} - [\![t|C]\!]_\mathcal{U} - [\![t|C]\!]_\mathcal{T}$

A constrained term is **consistent** if its denotation is a nonempty set. If $C_1 \vee \ldots \vee C_n$ is the disjunctive normal form of C, the denotation of $t|C$ is the union of the denotations of $t \mid C_1 \ldots t \mid C_n$.

Proposition 1 *The following equivalences will be used where required:*

- $[\![x^\eta]\!] = [\![x^\sigma | x^\sigma : \eta]\!]$ *if* $\sigma \geq \eta$ *and* $[\![t]\!]_\mathcal{F} = [\![x^\top | x^\top \diamond t]\!]_\mathcal{T}$
- $[\![t \mid C_1 \vee C_2]\!] = [\![t \mid C_1]\!] \cup [\![t \mid C_2]\!]$ *and* $[\![t \mid C_1 \wedge C_2]\!] = [\![t \mid C_1]\!] \cap [\![t \mid C_2]\!]$

- $[\![t \mid \mathcal{F}]\!] = \{\}$, $[\![t \mid \mathcal{T}]\!] = [\![t]\!]$ *and, if t is a ground constructor term, then* $[\![t]\!] = \{t\}$.

If $P_1 \ldots P_n$ is a constrained pattern (and thus in particular a pattern), $[\![P_1 \ldots P_n]\!]_{\mathcal{T}}$ is the set of $\Sigma^\bullet$-terms t for which there exists a constrained pattern P_i such that t matches P_i and it is decidable that t does not match any $P_k, k < i$; $[\![P_1 \ldots P_n]\!]_{\mathcal{U}}$ is the set of $\Sigma^\bullet$-terms t such that there exists a P_i for which we cannot decide whether t matches P_i and t is not in the denotation of any preceding prefix $P_1 \ldots P_k$, $k < i$ of this pattern and finally, $[\![P_1 \ldots P_n]\!]_{\mathcal{F}}$ is the set of $\Sigma^\bullet$-terms that, decidably, are not solutions of $P_1 \ldots P_n$. Formally,

- $[\![P_1 \ldots P_n]\!]_{\mathcal{T}} = \{t \mid \exists i\ (1 \leq i \leq n)\ t \in [\![P_i]\!]_{\mathcal{T}} \text{ and } \forall k\ (k < i)\ t \in [\![P_k]\!]_{\mathcal{F}}\}$
- $[\![P_1 \ldots P_n]\!]_{\mathcal{U}} = \{t \mid \exists i\ (1 \leq i \leq n)\ t \in [\![P_i]\!]_{\mathcal{U}} \text{ and } \forall k\ (k < i)\ t \notin [\![P_1 \ldots P_k]\!]_{\mathcal{T}}\}$
- $[\![P_1 \ldots P_n]\!]_{\mathcal{F}} = \Sigma^\bullet - [\![P_1 \ldots P_n]\!]_{\mathcal{T}} - [\![P_1 \ldots P_n]\!]_{\mathcal{U}}$.

Example 1 *Consider the following subsort order:*

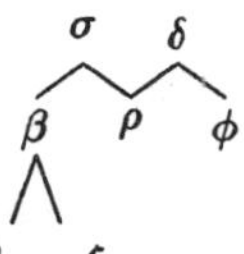

$b :\rightarrow \eta$
$a :\rightarrow \xi$
$p : \eta \times \rho \rightarrow \rho$
$q : \xi \times \rho \rightarrow \rho$
$f : \sigma \times \sigma \rightarrow \phi$

Let h be a constructor whose domain sort is $\top \times \top$. Let t be the term $h(x^\sigma, y^\delta)$ and C be the constraint $x^\sigma : \eta \wedge y^\delta \Diamond q(a, z^\rho)$. We have for example $h(b, \bullet^\phi) \in [\![t]\!]_{\mathcal{T}}$,$h(b, \bullet^\sigma) \in [\![t]\!]_{\mathcal{U}}$ and $h(\bullet^\phi, b) \in [\![t]\!]_{\mathcal{F}}$. On the other hand, $[\![C]\!]_{[x^\sigma/b, y^\delta/f(a,a)]} = true$,$[\![C]\!]_{[x^\sigma/\bullet^\sigma, y^\delta/f(a,a)]} = uncertain$ and $[\![C]\!]_{[x^\sigma/\bullet^\xi, y^\delta/q(a,\bullet^\rho)]} = false$. Now, $h(b, f(a,a)) \in [\![t|C]\!]_{\mathcal{T}}$, $h(\bullet^\sigma, f(a,a)) \in [\![t|C]\!]_{\mathcal{U}}$ and $h(\bullet^\xi, q(a, \bullet^\rho)) \in [\![t|C]\!]_{\mathcal{F}}$. Consider the constrained pattern $P_1, P_2 = f(x^\beta, b) \mid x^\beta : \xi\ ,\ f(y^\sigma, z^\sigma) \mid z^\sigma : \beta \ \wedge\ z^\sigma \Diamond a$. $f(\bullet^\sigma, b) \in [\![P_2]\!]_{\mathcal{T}}$ and $f(\bullet^\sigma, b) \notin [\![P_1]\!]_{\mathcal{F}}$, since $f(\bullet^\sigma, b) \in [\![P_1]\!]_{\mathcal{U}}$. Hence, $f(\bullet^\sigma, b) \notin [\![P_1, P_2]\!]$. We will use this subsort order along all the examples of the paper.

4 Compilation Rules

Our compilation method consists of three kinds of rules acting on constrained terms. The **simplification** rules transform restrictions on terms into restrictions on variables. **Partitioning** transforms an ambiguous order-sorted pattern into an equivalent set (modulo simplification) of disjoint order-sorted constrained terms. The **normalization** rules transform a set of disjoint order-sorted constrained terms into a set of simpler ones that facilitates the construction of the pattern discrimination tree. We will present here the central ideas, referring the interested reader to [Kes91] for a full exposition.

The simplification rules define a reduction relation $\longrightarrow_s$ on constraints that transforms a structure or sort constraint on terms into an equivalent constraint on variables, that is, either $\mathcal{T}$, $\mathcal{F}$, or of the form $x \Diamond t$ or $x : \sigma$.

If $x^\sigma : \eta$ appears in a constraint C, then the variable x^σ is said to be **restricted** by η in C, otherwise it is restricted by σ. We shall say that a constraint C is in **simplified form** (irreducible by $\longrightarrow_s$), denoted $C \downarrow_s$, if and only if it is either $\mathcal{T}$ or $\mathcal{F}$ or

- If $x^\sigma : \eta$ is in C then $\sigma > \eta$ and $\eta \neq \bot$
- If $x^\sigma \Diamond t^\eta$ is in C then t is not a variable and $\sigma \geq \eta$.
- If $x \Diamond \{f_1, \ldots, f_n\} \in C$ and $f_i : \vec{\rho_i} \rightarrow \delta_i \in \Sigma$ $(i = 1 \ldots n)$, then $f_1, \ldots, f_n$ are not all the constructors of $\{\delta_1 \ldots \delta_n\}$.

Example 2 *The constraint $x^\sigma : \eta \wedge y^\delta \Diamond f(a,a)$ is in simplified form while $x^\sigma \Diamond z^\xi$ and $y^\delta \Diamond \{p, q\}$ are not.*

The definition of a pattern's denotation (section 3.2) suggests splitting a pattern into an equivalent set of constrained terms, whose set denotations are disjoint, and whose union is the set denotation of the pattern. Partitioning a constrained term $x^\sigma | x^\sigma \Diamond S$ according to a pattern $p_1 \dots p_n$, consists of transforming an ambiguous list of order-sorted terms (a pattern) into an unambiguous set of order-sorted constrained terms.

To illustrate, suppose we wish to partition $x^\top | \mathcal{T}$ according to $p_1 \dots p_n$. The first set generated is $\{p_1 | \mathcal{T}\}$ and we go on to recursively partition the decidable $x^\top$-complement of p_1, that is, $x^\top | x^\top \Diamond p_1$, according to the rest of the pattern $p_2 \dots p_n$. Note that the order of the pattern is respected.

Example 3 *Let α be the sort $\sigma \sqcup \delta$ and g a new constructor of ϕ whose domain sort is $\alpha \times nat$, with nat and α disjoint sorts. Partitioning $x^\top | \mathcal{T}$ according to the pattern $g(x^\sigma, 0), g(y^\delta, z^{nat}), x^\top$ yields* $\{g(x^\sigma, 0),\ g(y^\delta, z^{nat}) | y^\delta : \phi,\ g(y^\delta, z^{nat}) | z^{nat} \Diamond 0,\ x^\top | x^\top \Diamond g(x^\sigma, 0) \wedge x^\top \Diamond g(y^\delta, z^{nat})\}$.

A set S of constrained terms is a **complete decomposition** if and only if there exists a pattern $p_1 \dots p_n$ such that partitioning $x^\top | \mathcal{T}$ according to the list $[p_1 \dots p_n\ x^\top]$ yields S. When $x^\top$ is appended to the list of patterns, both the success and the failure of the matching are considered. We don't change the original problem because discrimination trees covers anyway all the cases that may appear during the pattern matching process.

Finally, in order to simplify the construction of the PMT, we use the normalization rules that operate over sets of disjoint constrained terms. We shall say that $t | C$ is in **normalized form**, if and only if C is in simplified form but is different from $\mathcal{F}$ and whenever $x \Diamond f(t_1 \dots t_n)^\sigma$ appears in C and $f : \sigma_1 \dots \sigma_n \rightarrow \sigma \in \Sigma$, then x is restricted by σ in C and $t_1 \dots t_n$ are mutually distinct variables of sort $\sigma_1 \dots \sigma_n$ respectively. We shall say that a set of constrained terms $\{t_1 | C_1, \dots, t_n | C_n\}$ is in normalized form (irreducible by $\longrightarrow_n$), if and only if every $t_i | C_i$ is in normalized form and, whenever there exist two terms t_i and t_j and there exists a position u such that for all $v <_{pos} u$, t_i and t_j have the same structure symbol at position v, t_i / u and t_j / u are variables restricted by the same sort and by nonempty sets s_i and s_j of structure symbols in C_i and C_j respectively, then $s_i = s_j$.

Example 4 *$h(x^\sigma, y^\phi) | y^\phi \Diamond f(v^\sigma, w^\sigma)$ is in normalized form while $f(x^\sigma, y^\sigma) | x^\sigma \Diamond a$ and $h(x^\sigma, y^\phi) | y^\phi \Diamond f(v^\eta, w^\xi)$ are not.*

5 Sequentiality and Optimality

Compiling pattern matching consists of transforming a function defined by order-sorted patterns into a case-expression presented as a discrimination tree. In a lazy evaluation framework, some trees may fail to terminate due to unnecessary verifications that try to reduce subterms that do not have a head-normal form. The tree obtained is not always optimal, that is, it could fail to terminate on some terms that are not in the strict set of the pattern. As the evaluation mechanism is sequential, we must choose an order of verification running the risk of losing some solutions. In a many-sorted framework, consider Berry's example [Ber78] formed by the patterns $f(true, true, z)$, $f(false, y, true)$, $f(x, false, false)$. Given a term $f(\_, \_, \_)$, we must choose an argument position in order to start the matching. If we start at position three, the term $f(true, true, \bullet^{Bool})$ will not be matched, even though it belongs to the denotation of the first pattern. The same happens with the terms $f(false, \bullet^{Bool}, true)$, $f(\bullet^{Bool}, false, false)$ if we start at the second or first position, respectively.

In our framework, the sort of each term is examined before its structure, because the sort can be refined after usually only a few reduction steps whereas to examine the structure, more reduction steps are required to obtain head-normal form. The construction method for PMT's that we present here chooses a direction (intuitively, a position in a term at which to start reduction) and thus decides whether a subsort or structure verification is required. At each level of the tree, the structures and sorts are more precise than those of preceding levels.

We propose a notion of **sequentiality** of the pattern matching predicate that takes not only the structure of terms into account, but also the sort system. Intuitively, a disjoint set S of patterns is sequential in the sense of [HL79, PS90] if it is possible to decide the matching property without doing some look-ahead. That is, for any constrained term T not matching any pattern of S, there exists a position (so-called **direction**) which we necessary need to reduce in order decide the matching property; and furthermore this place can be determined without looking at the subterms of T which are not computed yet. Our definition of sequentiality requires the set S of patterns to have also the **sort property**: intuitively, S has the sort property if whenever u is a position of a constrained term T to be evaluated and two different patterns of S that are compatible with T have variables x^σ, y^ρ at position u respectively, then σ and ρ are either disjoint or comparable sorts. This property enriches the known notions of sequentiality by taking into account the sort system. In fact, if σ and ρ have a common subsort δ different from $\perp, \sigma$ and ρ, the position u cannot be taken as a direction because unnecessary reductions of T may be performed in order to distinguish between σ and ρ and some solutions will be lost.

For example, consider the patterns $f(true, y^\sigma, z^{bool})$, $f(false, y^\delta, true)$, $f(x^{bool}, y^\phi, false)$, where the subsort order is that of example 1. If pattern matching starts at the first position (resp. at the third), it will fail to terminate on the term $f(\bullet^{bool}, f(a,a), false)$ (resp. on $f(true, a, \bullet^{bool})$). On the other hand, if the chosen direction is the second one, there are two cases to be considered: if matching ask a term to be or not of sort σ, it will fail to terminate on $f(false, \bullet^\delta, true)$ whereas if it ask a term to be or not of sort δ it will fail to terminate on $f(true, \bullet^\sigma, true)$. Even though the first one (resp. the second one) is in the denotation of the second (resp. first) pattern.

Optimal PMT's will only fail to terminate on the strict set of the problem. It turns out that sequentiality of a pattern matching problem is equivalent to optimality of its tree. Thus, sequentiality becomes a necessary and sufficient condition for the construction of an optimal tree. We show an effective decision procedure for sequentiality on disjoint sets of patterns.

5.1 Sequentiality

The set of **positions** of a constrained term $t|C$, denoted $\mathcal{O}(t|C)$, is the set of positions of t, which is defined as usual. We use $<_{pos}$ to denote the lexical ordering between positions. We write $(t|C)/u$ to denote the constrained term $(t/u)|D$, where D is the constraint of all the atoms in C restricting variables of t/u. For example, $(h(x^\sigma, y^\rho)|x^\sigma : \eta)/1 = x^\sigma|x^\sigma : \eta$. We use $T[u \leftarrow t]$ to denote the constrained term obtained from T by replacing the variable at position u with t.

Let $T_1 = t_1|C_1$ and $T_2 = t_2|C_2$ be two constrained terms. We say T_2 **matches** T_1, denoted $T_1 \sqsubseteq T_2$, iff there exists a substitution θ such that $t_1 = \theta(t_2)$ and $C_2 \Rightarrow \theta(C_1)$, where $\Rightarrow$ is logical implication. T_1 and T_2 are **compatible** if $\exists M$ such that $T_1 \sqsubseteq M$ and $T_2 \sqsubseteq M$ and T_1 is **compatible** with a set of disjoint constrained patterns S if and only if there exists $P \in S$ such that T_1 and P are compatible.

We can think a set of disjoint patterns S as a predicate on constrained terms such that $S(M)$ is true if and only if $\exists P \in S$ such that $P \sqsubseteq M$. If truth values are ordered by $false \sqsubseteq true$, S is a monotonic predicate on constrained terms. Increasing information about the given term (in the sense of $\sqsubseteq$) can only change the value of the predicate to a favorable one.

A position $u \in \mathcal{O}(t|C)$ is said to be a **direction** of $T = t|C$ in a set of disjoint constrained patterns $S = \{t_1|C_1, \ldots, t_n|C_n\}$ if and only if

- T/u has the form $x^\sigma|P$.
- For every constrained term M such that $T \sqsubseteq M$ and $S(M)$ is true $M/u \not\sqsubseteq T/u$.
- **(Sort property)** If $\exists i, j$ such that $\forall v, v <_{pos} u$, t_i and t_j have the same constructor symbol at position v; t_i/u is a variable restricted by η_i in C_i and t_j/u is a variable restricted by η_j in C_j, then $\eta_i \sqcap \eta_j = \perp$ or $\eta_i \leq \eta_j$ or $\eta_j \leq \eta_i$.

The following lemma gives a characterization of directions which in turn is used in order to construct the PMT. See [Kes91] for a proof.

Lemma 1 *Let T be a constrained term $t|C$. A position u is a direction of T in $S = \{S_1, \ldots, S_n\}$ if and only if T/u has the form $x^\sigma|P$ and for every constrained pattern $S_i \in S$ compatible with T, we have that u is an occurrence of S_i, $S_i/u \not\sqsubseteq T/u$ and the sort property holds.*

By normalization, a complete decomposition S reduces to another complete decomposition S' and the set of directions of any term T in S is the set of directions of T in S'.

A set of disjoint constrained patterns S is **sequential in a constrained term** T if and only if, whenever $S(T)$ is false but it is compatible with S, then there exists a direction of T in S. We say that S is **sequential** if and only if it is sequential in all constrained terms in normalized form. Sequentiality of a predicate S is the possibility of systematically expanding any constrained term step by step until either the predicate is true or it is clear that a positive answer is impossible.

Example 5 *The set* $\{\ h(p, y^\top, z^\rho)|y^\top : \sigma,\ h(x^\rho, y^\top, p)|x^\rho \Diamond p \wedge y^\top : \delta,\ h(x^\rho, y^\top, q)|y^\top : \phi\ \}$ *is not sequential.*

5.2 Construction of pattern matching trees

A pattern matching tree (PMT) for a constrained term T and a complete decomposition $S = \{S_1, \ldots, S_n\}$ is defined as:

- T is the root and each node is a constrained linear pattern in simplified form.
- If u is a direction of P in S and $T_1, \ldots, T_k$ are the children of P, then
  $T_1 \ldots T_k$ are pairwise incompatible constrained terms,
  P/u has the form $x^\sigma|\mathcal{T}$, T_i and P only differs on u and $P/u \sqsubset T_i/u$,
  for all T_i, there exists a pattern S_j such that $T_i \sqsubseteq S_j$.
- If $H_1 \ldots H_m$ are the leaves of the tree, $\{S_1, \ldots, S_n\} \longrightarrow_n^* \{H_1, \ldots, H_m\}$.

A PMT of a complete decomposition S is a pattern matching tree for the constrained term $x^\top|\mathcal{T}$ and S. A PMT of $\{S_1 \ldots S_n\}$ is **optimal** if and only if it fails to terminate only for the strict set of $S_1 \ldots S_n$.

The algorithm $\mathcal{TREE}$ constructs a PMT for a constrained term $T = t|C$ and a complete decomposition $S = \{S_1 \ldots S_n\}$ If C is non-consistent, return the empty tree. Otherwise, if T is a S_i, return the single-node tree T. Otherwise, normalize S into the set H and search a direction u of T in H. If such a direction cannot be found, H is not sequential in T and then fail. Otherwise, proceed with $\mathcal{DIR}(T, H, u)$ where $T = t|C$; $T/u = x^\sigma|\mathcal{T}$; $H = \{H_1 \ldots H_m\}$; $H_i = h_i|C_i$ and $\mathcal{DIR}$ is defined as follows:

$\mathcal{DIR}\ (T, \{H_1, \ldots, H_m\}, u) =$
  Let *Sorts* be the maximal sorts of $\{\ \eta \mid T \sqsubseteq H_i,\ h_i/u$ is a variable restricted by $\eta\ \}$ and
  Let *Forms* be $\{\ f_i \mid T \sqsubseteq H_i, h_i/u = f_i(\ldots)\ \}$ in
  if *Sorts* $= \emptyset$
  then A structure step: build a tree rooted at T, with children:
    $\mathcal{TREE}(T[u \leftarrow f(\ldots)] \downarrow_s, \{H_1, \ldots, H_m\})$ for each f in *Forms* and
    $\mathcal{TREE}(t \mid (C \wedge x^\sigma \Diamond \mathit{Forms}) \downarrow_s, \{H_1, \ldots, H_m\}))$
  else A sort step: build a tree rooted at T, with children:
    $\mathcal{TREE}(t \mid (C \wedge x^\sigma : \eta) \downarrow_s, \{H_1, \ldots, H_m\})$ for each $\eta \in$ *Sorts* and
    $\mathcal{TREE}(t \mid (C \wedge x^\sigma : \sigma - \mathit{Sorts}) \downarrow_s, \{H_1, \ldots, H_m\}))$

where $\sigma - \eta = \sigma \sqcap \eta^c$ and $\sigma - \{\eta_1, \ldots, \eta_k\} = (\ldots(\sigma - \eta_1) - \ldots) - \eta_k$.

Example 6 *With the same hypothesis of example 3, normalization of the pattern* $g(x^\sigma, y^{nat}), g(x^\alpha, 0), g(x^\delta, 1)$ *yields:*

- $g(x^\sigma, y^{nat})$
- $y^\top | y^\top : nat \sqcup \sigma$
- $g(x^\alpha, 0) | x^\alpha : \phi$
- $g(x^\alpha, y^{nat}) | x^\alpha : \phi \wedge y^{nat} \diamond 0 \wedge y^{nat} \diamond 1$
- $g(x^\delta, 1) | x^\delta : \phi$

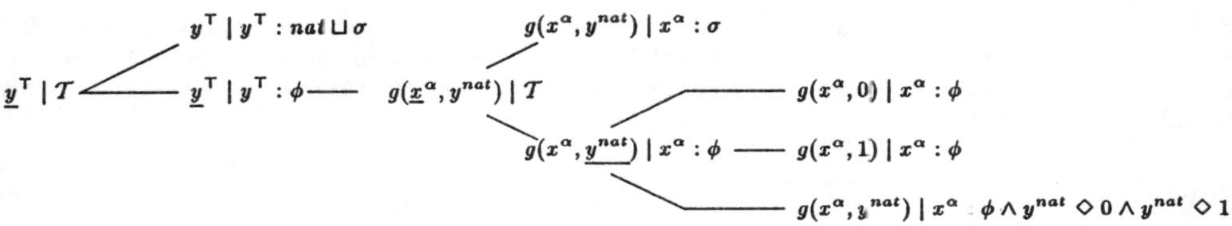

Figure 1: Pattern Matching Tree

Theorem 1 *A finite complete decomposition in normalized form S is sequential in every normalized term if and only if it is sequential in every node of its associated PMT.*

Proof Sketch. Since nodes of the PMT are in normalized form the left to right implication is evident. Conversely, let M be a constrained term in normalized form such $S(M)$ is false and M is compatible with S. Then, there exists a node $T = t|C$ whose children are $T_1 \ldots T_m$ and whose direction in S is u such that $T \sqsubseteq M$ and for all $1 \leq i \leq m$, $T_i \not\sqsubseteq M$. We can prove that u is a direction of M in S by considering the structure and the subsort step cases separately. Full details of the proof are given in [Kes91].

Theorem 2 *A PMT of a complete decomposition S in normalized form is optimal iff S is sequential.*

Proof Sketch. By theorem 1, S is sequential if and only if there exists a discrimination tree in which each node $t|C$ has a direction in the set S and the sort property holds. The set of terms for which the algorithm does not terminate is generated at each node $t|C$ of the PMT by some terms of the form $(t|C)[u \leftarrow \bullet^\rho]$, where u is the chosen direction of $t|C$ in S. By definition S is optimal if and only if it fails to terminate only for the strict set of S. We can see that the algorithm fails to terminate in $(t|C)[u \leftarrow \bullet^\rho]$ if and only if it is in the strict set of S. The right to left implication is evident and for the converse, both the sort and the structure level cases must be considered. We omit here the details that can be found in [Kes91].

Conclusions

The method of treating ambiguous linear order-sorted pattern matching presented in this paper generalizes previous work on non-ambiguous linear patterns [HL79], ambiguous linear patterns [Lav88] and ambiguous constrained linear patterns [PS90]. We extend several notions in [PS90], such as constrained terms, non-reducible terms $\bullet$, strict sets of patterns, sequentiality and pattern-matching trees, to the order-sorted case. We define discrimination trees to have edges labelled not only with structure constraints, but also with subsort restrictions. This feature allows to decide pattern matching without reducing terms to normal forms, taking advantage in this way of the lazy evaluation strategy. It turns out that our method constructs optimal order-sorted PMT's for sequential order-sorted pattern matching problems and can be used either with a lazy or strict evaluation strategy. As in [PS90], our method can also be used for non-sequential problems.

Unsorted and many-sorted signatures are particular cases of unitary ones, and therefore our work remains also applicable to them. Our general order-sorted framework allows also to accommodate pattern matching in languages with order-sorted type systems like **features types** [SAK89] and polymorphic order-sorted types restricted to free constructors [Smo88].

Acknowledgements

I wish to thank Hassan Aït-Kaci, Laurence Puel and Ascánder Suárez for proposing this topic for my *DEA* and for many fruitful discussions. I am also very grateful to Kathleen Milsted for her patient help in the whole preparation of the first version of this paper and to Roberto Di Cosmo for helpful suggestions on the presentation of the final version.

References

[Ber78] Gérard Berry. Séquentialité de l'évaluation formelle des lambdas-expressions. Proc. 3rd International Colloquium on Programming, Paris, France, 1978.

[GM89] Joseph Goguen and José Meseguer. Order-Sorted Algebra I: Equational Deduction for Multiple Inheritance, Overloading, Exceptions and Partial Operations. Technical report, SRI International-CSL, 1989.

[GW88] Joseph Goguen and Timothy Winkler. Introducing OBJ3. Technical Report SRI-CSL-88-9, SRI International, Computer Science Lab, 1988.

[HL79] Gérard Huet and Jean-Jacques Lévy. Call by need computations in non ambiguous linear term rewriting systems. Rapport IRIA Laboria 359, INRIA, Le Chesnay, France, 1979.

[HMM86] Robert Harper, David MacQueen, and Robin Milner. Standard ML. Technical Report ECS-LFCS-86-2, Departament of Computer Science, University of Edinburgh, 1986.

[JK90] Jean-Pierre Jouannaud and Claude Kirchner. Solving Equations in Abstract Algebras: A Rule-Based Survey of Unification. Technical Report 561, Laboratoire de Recherche en Informatique. Université Paris-Sud, Centre d'Orsay, 1990.

[Kes91] Delia Kesner. Pattern Matching in Order-Sorted Languages. Technical report, Laboratoire de Recherche en Informatique. Université Paris-Sud, Centre d'Orsay, 1991.

[Lav88] Alain Laville. Evaluation des Filtrages avec Priorité. Application au Langage ML. Thèse de Doctorat. Université Paris VII, 1988.

[PS90] Laurence Puel and Ascánder Suárez. Compiling Pattern Matching by Term Decomposition. In *1990 ACM Conference on Lisp and Functional Programming.* ACM Press. Also in PRL Research Report Number 4, Digital Equipment Corporation, Paris Research Laboratory, 1990.

[SAK89] Gert Smolka and Hassan Aït-Kaci. Inheritance Hierarchies: Semantics and Unification. Volume 7. In *Journal of Symbolic Computation.* Academic Press, 1989.

[Smo88] Gert Smolka. Logic Programming over Polymorphically Order-Sorted Types. PhD thesis. Universität Kaiserslautern, Germany, 1988.

[SNGM89] Gert Smolka, Werner Nutt, Joseph Goguen, and José Meseguer. Order Sorted Equational Computation. In Hassan Aït-Kaci and Maurice Nivat, editors, *Resolution of Equations in Algebraic Structures. Volume 2: Rewriting Techniques*, pages 297–367. Academic Press, 1989.

Two over three: a two-valued logic for software specification and validation over a three-valued predicate calculus

Beata Konikowska

Institute of Computer Science, Polish Academy of Sciences
00-901 P.O. Box 22, Warsaw, Poland
Project MetaSoft

1 Introduction

When we evaluate dynamically predicates with free variables in the process of running a program, we obviously may ecounter some cases when the value of a formula is undefined. Thus we have to assume that the formulas (Boolean expressions) appearing in programs represent either partial two-valued functions, or total functions with values in the set {**tt**, **ff**, **ee**}, where **ee** represents "undefinedness". We choose the latter solution, since it allows us to define a calculus of the three logical values (together with a corresponding predicate calculus) in which undefinedness is treated in a non-strict, lazy way (see [Blikle 88]).

In order to define the semantics of three-valued formulas (Boolean expressions, assertions etc.), we have to construct a three-valued predicate calculus. Obviously, that calculus should agree with the classical one on the classical logical values. Hence in fact we have to choose one of the three-valued extensions of the two-valued predicate calculus proposed in the literature (see e.g. [Kleene 51], [McCarthy 61]). Following the arguments given in [Blikle 88], our choice here is the hybrid MK-calculus, combining McCarthy's connectives with Kleene's quantifiers, proposed and examined in [Blikle 81a,81b,87,88].

The next decision to make is how to employ this calculus in software specification and validation. There are two major strategies: one is to develop a special three-valued logic, and the other - to couple the chosen calculus with a two-valued logic. The first approach has beeen studied e.g. in [Hoogewijs 79,83,87], [Barringer, Chang, Jones 84], [Owe 85], [Jones 86,87]. In particular, two three-valued logical systems corresponding to the MK-calculus have been developed in [Konikowska, Tarlecki, Blikle 88].

However, such strategy has obvious drawbacks: there are very few people which can feel at ease in a three-valued logical "environment". Moreover, the existing proof support systems are also tailored to classical logic.

For the above reasons, in the present paper we apply the second strategy, pursuing an idea which stems from [Blikle 88]. Before we outline our approach, we have first to sketch the relevant context.

When dealing with software specification and validation, we encounter four types of logical expressions:

- *Boolean expressions*, which appear in program statements and are evaluated during their execution. They are three-valued, and represent computable functions; hence they cannot contain quantifiers;
- *Assertions*, which appear as preconditions, postconditions, invariants etc. in dynamic formulas (see below). They are also three-valued, but need not be computable. In particular, they can contain quantifiers;
- *Static formulas*, which express facts about data and appear in proofs. They are quite similar to assertions, but form a distinct syntactic category for the reasons explained below;
- *Dynamic formulas*, which express facts about programs (typical examples of dynamic formulas are Hoare's statements). They differ from static formulas in that they contain some expressions representing programs.

In order to prove some facts about programs, i.e. to prove the validity of some dynamic formulas, one usually first reduces them to static formulas (e.g. by means of Hoare's rules), and then proves the validity of the resulting static formulas. The language developed here is meant to be a language of static formulas, which - unlike that of dynamic formulas - is independent of the choice of a concrete programming language.

Now let us give some particulars about our approach. In [Blikle 88] it has been shown that the static formulas, to which Hoare's dynamic formulas representing basic properties of programs are reduced with help of Hoare's rules, can be expressed as logical combinations of static formulas of the form $p \Rightarrow q$, where p, q are program assertions and $\Rightarrow$ is a two-valued "predicate implication" such that $p \Rightarrow q$ holds iff q is true whenever p is true. Hence there is no need to introduce three-valued logic on the level of static formulas: the "undefined" value is necessary only at the computation level, and on the assertion level, but not in reasoning about program properties. To obtain an adequate logical system for static formulas, one can just develop a special two-valued logic featuring $\Rightarrow$, in which three-valued assertions will be only *components* of formulas - but not formulas in their own right. This is just the approach we adopt in our paper.

As $\Rightarrow$ is non-monotonic, then in order to express it within our language we must augment the monotonic MK-calculus we have chosen by some non-monotonic operator. Our choice is the operator *is_true* such that *is_true*(**tt**)=**tt** and *is_true*(**ff**)=*is_true*(**ee**)=**ff**, denoted in the language by **T**. The above operator is used for constructing atomic formulas out of assertions. All other formulas are built of the atomic ones with help of the classical, two-valued connectives and quantifiers. The considered language will be denoted by $L_{2/3}$ in order to stress the fact that it is a language of a two-valued logic defined over three-valued assertions.

Due to a low page limit imposed on the final version of the paper, a lot of material - including all proofs - had to be deleted. For this reason, some aspects will perhaps be not as clear as they could (and should) be.

2 Syntax of the language $L_{2/3}$

For any sets A, B, by $A \longrightarrow B$ we denote the set of all total functions from A to B. The expression $A \longrightarrow B \longrightarrow C$ is understood as $A \longrightarrow (B \longrightarrow C)$. We write $f.a$ instead of $f(a)$, and $g.a.b$ instead of $(g.a).b$. For any $a \in A$, any $b \in B$ and any $f : A \longrightarrow B$, by $f[b/a]$ we denote a modification of f such that $f[b/a].a = b$ and $f[b/a].x = f.x$ for $x \neq a$.

The three logical values considered in this paper are **tt** (truth), **ff** (falsity) and **ee** (undefinedness). We denote

$$Boolerr = \{\mathbf{tt}, \mathbf{ff}, \mathbf{ee}\}, \qquad Bool = \{\mathbf{tt}, \mathbf{ff}\}.$$

The expresions of the language $L_{2/3}$ we are going to define belong to three categories:

- *Fterm* - functional terms,
- *Assr* - program assertions,
- *Form* - static formulas.

Note that Boolean expressions form a subset of assertions, but do not appear as a separate syntactical category, since this is not necessary for our purposes

Semantically, assertions are interpreted as three-valued predicates (of the MK-calculus) mapping object environments into *Boolerr.* On the other hand, static formulas are interpreted as two-valued predicates, taking values in *Bool.* Naturally enough, functional terms are interpreted as functions assigning values to environments.

The elements of *Fterm* are denoted by $t, u, \ldots$, the elements of *Assr* - by $\alpha, \beta, \ldots$, and the elements of *Form* - by $\varphi, \psi, \ldots$ (possibly with indices). Identifiers are denoted by $x, y, \ldots$.

In what follows, *Ide* is a countable set of identifiers, Fun_m and $Pred_m$ - a countable set of m-ary function symbols and a countable set of m-ary predicate symbols, respectively, m=0,1,2,.... **T** is the symbol denoting the *is_true* operator; NOT , OR , AND , IMP , A , E are symbols denoting McCarthy's three-valued connectives and Kleene's three-valued quantifiers, respectively, and $\neg, \vee, \wedge, \rightarrow, \forall, \exists$ are symbols for classical connectives and quantifiers.

It should be noted that though (from a strictly semantic viewpoint) MK-connectives and quantifiers agree with the classical connectives and quantifiers on classical logical values, yet we make a clear syntactical distinction between them, since they play quite different roles here. Namely, in $L_{2/3}$ the only bona fide connectives and quantifiers are in fact the classical ones - the MK-operators are just constructors of special "logical" terms (assertions).

The *set Fterm of functional terms* is defined in the usual way over *Ide* and Fun_m, $m = 0, 1, \ldots$.

By an *atomic assertion* we mean any expression of the form $P(t_1, \ldots, t_m)$, where all the t_i's are in *Fterm* and $P \in Pred_m$ $(m = 0, 1, \ldots)$. The *set Assr of assertions* is defined as the least set containing all atomic assertions and closed under the three-valued connectives and quantifiers of the MK-calculus.

By an *atomic formula* we mean any expression of the form $\mathbf{T}(\alpha)$, where $\alpha \in Assr$. The *set Form of (static) formulas* is the least set containing all atomic formulas and closed under classical two-valued connectives and quantifiers.

An occurrence of an identifier x in an expression $\xi \in Assr \cup Form$ is said to be *free* if it is not bound by any of the quantifiers A , E , $\forall, \exists$. The set of all identifiers which have a free occurrence in ξ is denoted by $Free(\xi)$.

3 Semantics of the language $L_{2/3}$

The semantics of our language is defined in terms of the interpretation of its expressions in a logical structure, analogous to the notion of a *model* commonly used in logics.

By a *logical structure* we mean any triple

$$S = (\mathit{Value}, (F_m)_{m=0}^{\infty}, (P_m)_{m=0}^{\infty}),$$

where *Value* is a nonempty set of values, and for any m

$$F_m : \mathit{Fun}_m \longrightarrow \mathit{Value}^m \longrightarrow \mathit{Value},$$

$$P_m : \mathit{Pred}_m \longrightarrow \mathit{Value}^m \longrightarrow \mathit{Boolerr}.$$

By an *object environment* (shortly: *environment*) over S we mean any element of the set

$$\mathit{Env} = \mathit{Ide} \longrightarrow \mathit{Value}.$$

By *Classpred* and *Predicate* we denote the set of all two-valued predicates over S and the set of all three-valued predicates over S, respectively, i.e.

$$\mathit{Classpred} = \mathit{Env} \longrightarrow \mathit{Bool}, \qquad \mathit{Predicate} = \mathit{Env} \longrightarrow \mathit{Boolerr}.$$

The *interpretation of functional terms in a structure* S is a function

$$T_S : \mathit{Fterm} \longrightarrow \mathit{Env} \longrightarrow \mathit{Value}$$

defined in the standard way with help of the F_m functions.

The *interpretation of assertions in a structure* S is a function

$$A_S : \mathit{Assr} \longrightarrow \mathit{Predicate}$$

defined inductively as follows:

- $A_S.p(t_1, \dots, t_m).e = P_m.p.(T_S.t_1.e, \dots, T_S.t_m.e)$ for any $p \in \mathit{Pred}_m$, any $t_1, \dots, t_m \in \mathit{Fterm}$ and any $e \in \mathit{Env}$;
- for any $\alpha, \beta \in \mathit{Assr}$, any $x \in \mathit{Ide}$ and any $e \in \mathit{Env}$:

1. $A_S.\,\mathrm{NOT}\,\alpha.e = \begin{cases} \mathbf{tt} & \text{if } A_S.\alpha.e = \mathbf{ff}, \\ \mathbf{ff} & \text{if } A_S.\alpha.e = \mathbf{tt}, \\ \mathbf{ee} & \text{otherwise;} \end{cases}$

2. $A_S.\alpha\,\mathrm{OR}\,\beta.e = \begin{cases} \mathbf{tt} & \text{if either } A_S.\alpha.e = \mathbf{tt}, \\ & \text{or } A_S\alpha.e = \mathbf{ff} \text{ and } A_S.\beta.e = \mathbf{tt}, \\ \mathbf{ff} & \text{if } A_S.\alpha.e = A_S.\beta.e = \mathbf{ff}, \\ \mathbf{ee} & \text{otherwise;} \end{cases}$

3. $A_S.\alpha\,\mathrm{AND}\,\beta.e = A_S.\,\mathrm{NOT}\,(\,\mathrm{NOT}\,\alpha\,\mathrm{OR}\,\mathrm{NOT}\,\beta).e,$

4. $A_S.\alpha\,\mathrm{IMP}\,\beta.e = A_S.\,\mathrm{NOT}\,\alpha\,\mathrm{OR}\,\beta.e,$

5. $A_S.(\,\mathrm{E}\,x)\alpha.e = \begin{cases} \mathbf{tt} & \text{if } A_S.\alpha.e[a/x] = \mathbf{tt} \text{ for some } a \in \mathit{Value}, \\ \mathbf{ff} & \text{if } A_S.\alpha.e[a/x] = \mathbf{ff} \text{ for every } a \in \mathit{Value}, \\ \mathbf{ee} & \text{otherwise;} \end{cases}$

6. $A_S.(\,\mathrm{A}\,x)\alpha.e = A_S.\,\mathrm{NOT}\,(\,\mathrm{E}\,x)\,\mathrm{NOT}\,\alpha.e.$

As we see, NOT , OR , AND , IMP are interpreted as McCarthy's connectives (note that neither OR nor AND are commutative!), whereas A and E are interpreted as Kleene's qunatifiers.

The *interpretation of formulas in a structure S* is a function

$$F_S : Form \longrightarrow Classpred.$$

The interpretation of atomic formulas is defined by

$$F_S.\mathbf{T}(\alpha).e = \begin{cases} \mathbf{tt} & \text{if } A_S.\alpha.e = \mathbf{tt}, \\ \mathbf{ff} & \text{otherwise} \end{cases}$$

for any $\alpha \in Assr$ and any $e \in Env$, and F_S is extended to composite formulas in the standard way by interpreting $\neg, \vee, \wedge, \rightarrow, \forall, \exists$ as classical connectives and quantifiers, respectively.

We say that a formula $\varphi \in Form$ is *true in a logical structure S* iff $F_S.\varphi.e = \mathbf{tt}$ for every environment e over S. We say that φ is *valid* iff it is true in every structure S.

4 The expressive power of $L_{2/3}$

Consider any logical structure S. Then the predicate implication (over S) is a binary function $\Rightarrow$ such that $\Rightarrow : Predicate \times Predicate \longrightarrow Bool$ and

$$p \Rightarrow q \text{ iff, for each } e \in Env, \; p.e = \mathbf{tt} \text{ implies } q.e = \mathbf{tt},$$

where *Env* and *Predicate* are the set of all environments over S and the set of all three-valued predicates over S, respectively. The two lemmas given below show that within $L_{2/3}$ we can express both predicate implication itself and semantic entailment involving predicate implication (the latter is a "must", if we want to derive correctness of some programs from the correctness of other programs).

Lemma 1 *For any $\gamma, \delta \in Assr$ and any logical structure S, $A_S.\gamma \Rightarrow A_S.\delta$ iff the formula*

$$\gamma \Rrightarrow \delta \stackrel{\mathrm{df}}{=} (\forall x_1)(\forall x_2)\ldots(\forall x_n)(\mathbf{T}(\gamma) \longrightarrow \mathbf{T}(\delta)), \tag{1}$$

where $\{x_1, \ldots, x_n\} = Free(\gamma) \cup Free(\delta)$, is true in the structure S.

Lemma 2 *For any $\alpha_i, \beta_i, \alpha, \beta \in Assr$ $(i = 1, \ldots, n)$, and any logical structure S, $A_S.\alpha_i \Rightarrow A_S.\beta_i$ $(i = 1, \ldots, n)$ implies $A_S.\alpha \Rightarrow A_S.\beta$ iff the formula*

$$\bigwedge_{i=1}^{n} (\alpha_i \Rrightarrow \beta_i) \longrightarrow (\alpha \Rrightarrow \beta)$$

(where $\bigwedge$ denotes multiple conjunction and $\Rrightarrow$ is defined as in (1)) is true in S.

Thus the language $L_{2/3}$ meets the basic requirements for a language of static formulas supporting software validation. The reader is referred to [Blikle 88] for comments on employing $\Rightarrow$ in reasoning about programs.

It should be noted that within $L_{2/3}$ we can express the famous "definiteness" operator Δ on assertions by defining

$$\Delta(\alpha) \stackrel{\mathrm{df}}{=} \mathbf{T}(\alpha) \vee \mathbf{T}(\text{NOT}\,\alpha).$$

Moreover, one can also express a strong equality $\equiv$ between assertions, taking

$$\alpha \equiv \beta \stackrel{\mathrm{df}}{=} (\mathbf{T}(\alpha) \longleftrightarrow \mathbf{T}(\beta)) \wedge (\mathbf{T}(\text{NOT}\,\alpha) \longleftrightarrow \mathbf{T}(\text{NOT}\,\beta)).$$

5 The proof theory of $L_{2/3}$

Though our language features classical connectives and quantifiers, its atomic formulas are in fact composed constructs involving the non-monotonic operator **T** and a composite expression representing a three-valued predicate of the MK-calculus. Hence in order to provide a complete deduction mechanism for $L_{2/3}$, we must augment some proof system for the classical first-order logic by special rules expressing the properties of **T** and laws of the MK-calculus to be obeyed by the operators NOT, OR, AND, IMP, E, A.

As the proof system for classical logic to be extended in this way we choose the set of decomposition rules for sequences of formulas proposed in [Rasiowa, Sikorski 63]. This choice is motivated by two facts: first, the above system is almost a ready-made validation algorithm; second, the task of extending it by the aforesaid special rules is particularly easy. It should be noted that this deduction system belongs to the wide class of the so called natural deduction systems, and is dual to the tableaux method (see [Beth 59]).

Let us begin with introducing some necessary notions and notation. For any $\alpha \in$ *Assr*, we define

$$\mathbf{NT}(\alpha) \stackrel{\text{df}}{\equiv} \neg(\mathbf{T}(\alpha)).$$

One can easily note that the $\mathbf{T}(\alpha)$ and $\mathbf{NT}(\alpha)$ formulas have the following important properties:

$$F_S.\mathbf{T}(\alpha).e = \mathbf{tt} \text{ iff } A_S.\alpha.e = \mathbf{tt},$$

$$F_S.\mathbf{NT}(\alpha).e = \mathbf{tt} \text{ iff } A_S.\alpha.e \in \{\mathbf{ff}, \mathbf{ee}\},$$

$$A_S.\alpha.e = \begin{cases} \mathbf{tt} & \text{iff} \quad F_S.\mathbf{T}(\alpha).e = \mathbf{tt}, \\ \mathbf{ff} & \text{iff} \quad F_S.\mathbf{T}(\text{NOT}\,\alpha).e = \mathbf{tt}, \\ \mathbf{ee} & \text{iff} \quad F_S.\mathbf{NT}(\alpha).e = \mathbf{tt} \\ & \text{and} \quad F_S.\mathbf{NT}(\text{NOT}\,\alpha).e = \mathbf{tt}. \end{cases}$$

Thus formulas of the form $\mathbf{T}(\alpha), \mathbf{NT}(\alpha)$ are analogous to the signed formulas used in [Konikowska, Tarlecki, Blikle 88] to define an axiom system for a three-valued logic based on the MK-calculus via the semantic tableaux method. In our case they provide a very convenient tool for expressing the decomposition rules for formulas of $L_{2/3}$ induced by the properties of the underlying MK-calculus.

The letter Γ (possibly with indices) will always denote a finite (or empty) sequence of formulas.

A formula $\varphi \in$ *Form* is said to be *indecomposable* iff it is in one of the following forms:

$$\mathbf{T}(p(\mathbf{u})),\ \mathbf{NT}(p(\mathbf{u})),\ \mathbf{T}(\text{NOT}\,p(\mathbf{u})),\ \mathbf{NT}(\text{NOT}\,p(\mathbf{u})),$$

where $p \in \mathit{Pred}_m$ for some $m, \mathbf{u} = (u_1, \ldots, u_m), u_i \in \mathit{Fterm},\ i = 1, \ldots, m$. A sequence Γ of formulas is said to be *indecomposable* if all its elements are indecomposable. Otherwise it is said to be *decomposable.*

A sequence $\Gamma = \varphi_1, \ldots, \varphi_n$ is said to be *valid* if the disjunction $\varphi_1 \vee \varphi_2 \vee \ldots \vee \varphi_n$ is valid.

A sequence Γ is said to be *fundamental* if it contains one of the following pairs of formulas:

1. φ and $\neg\varphi$ (in particular, $\mathbf{T}(\alpha)$ and $\mathbf{NT}(\alpha)$), where $\varphi \in$ *Form*,
2. $\mathbf{NT}(\alpha)$ and $\mathbf{NT}(\text{NOT}\,\alpha)$, where $\alpha \in$ *Assr*.

It can be easily seen that the notion of a "fundamental" sequence is dual to that of a "closure rule" in the tableaux method. Obviously, we have:

Lemma 3 *Every fundamental sequence is valid.*

By a *decomposition rule* we understand either a pair Γ, Γ_1 or a triple $\Gamma, \Gamma_1, \Gamma_2$ of non-empty sequences of formulas, written usually in the form

$$\frac{\Gamma}{\Gamma_1} \quad (2) \qquad \text{and} \qquad \frac{\Gamma}{\Gamma_1 \ ; \ \Gamma_2} \quad (3),$$

respectively. Γ is called the *conclusion* of the rule, and Γ_1, Γ_2 - its *premises*. A rule is said to be *sound* provided that its conclusion is valid iff its premises are valid (i.e. a single premise in case of (2), and two premises in case of (3)).

In the rules given below Γ' always denotes an indecomposable sequence of formulas (which may also be empty). For the sake of brevity, we omit the decomposition rules corresponding to the definable connectives and quantifiers $\wedge$, $\longrightarrow$, $\forall$ and AND , IMP , A . By $Free(t, \chi)$, where $\chi \in Form \cup Assr$ and $t \in Fterm$, we denote the set of all identifers in *Ide* which are free for the term t in χ; thus $x \in Free(t, \chi)$ iff after replacing each occurrence of x in χ by t no variable which possessed a free occurrence in t becomes bound by any of the quantifiers $\forall, \exists$, E , A .

THE DECOMPOSITION RULES FOR $L_{2/3}$ (DR$_{2/3}$)
I. Rules for the classical predicate calculus:

$(\neg\neg)$
$$\frac{\Gamma', \neg\neg\varphi, \Gamma''}{\Gamma', \varphi, \Gamma''}$$

$(\vee)$
$$\frac{\Gamma', \varphi \vee \psi, \Gamma''}{\Gamma', \varphi, \psi, \Gamma''}$$

$(\neg\vee)$
$$\frac{\Gamma', \neg(\varphi \vee \psi), \Gamma''}{\Gamma', \neg\varphi, \Gamma'' \ ; \ \Gamma', \neg\psi, \Gamma''}$$

$(\exists)$
$$\frac{\Gamma', (\exists x)\varphi(x), \Gamma''}{\Gamma', \varphi(t), \Gamma'', (\exists x)\varphi(x)},$$

where t is an arbitrary term in *Fterm* with $x \in Free(t, \varphi)$,

$(\neg\exists)$
$$\frac{\Gamma', \neg(\exists x)\varphi(x), \Gamma''}{\Gamma', (\forall x)\neg\varphi(x), \Gamma''}$$

II. Special rules for the interaction of T and NT with MK-connectives and quantifiers:

$(\neg$T)
$$\frac{\Gamma', \neg\mathbf{T}(\alpha), \Gamma''}{\Gamma', \mathbf{NT}(\alpha), \Gamma''}$$

(T NOT NOT)
$$\frac{\Gamma', \mathbf{T}(\text{NOT NOT}\,\alpha), \Gamma''}{\Gamma', \mathbf{T}(\alpha), \Gamma''}$$

(NT NOT NOT)
$$\frac{\Gamma', \mathbf{NT}(\text{NOT NOT}\,\alpha), \Gamma''}{\Gamma', \mathbf{NT}(\alpha), \Gamma''}$$

(T OR) $$\frac{\Gamma', \mathbf{T}(\alpha \text{ OR } \beta), \Gamma''}{\Gamma', \mathbf{T}(\alpha), \mathbf{T}(\text{NOT } \alpha), \Gamma'' \; ; \; \Gamma', \mathbf{T}(\alpha), \mathbf{T}(\beta), \Gamma''}$$

(NT OR) $$\frac{\Gamma', \mathbf{NT}(\alpha \text{ OR } \beta), \Gamma''}{\Gamma', \mathbf{NT}(\alpha), \Gamma'' \; ; \; \Gamma', \mathbf{NT}(\text{NOT } \alpha), \mathbf{NT}(\beta), \Gamma''}$$

(T NOT OR) $$\frac{\Gamma', \mathbf{T}(\text{NOT } (\alpha \text{ OR } \beta)), \Gamma''}{\Gamma', \mathbf{T}(\text{NOT } \alpha), \Gamma'' \; ; \; \Gamma', \mathbf{T}(\text{NOT } \beta), \Gamma''}$$

(NT NOT OR) $$\frac{\Gamma', \mathbf{NT}(\text{NOT } (\alpha \text{ OR } \beta)), \Gamma''}{\Gamma', \mathbf{NT}(\text{NOT } \alpha), \mathbf{NT}(\text{NOT } \beta), \Gamma''}$$

(T E) $$\frac{\Gamma', \mathbf{T}((\text{E } x)\alpha(x)), \Gamma''}{\Gamma', \mathbf{T}(\alpha(t)), \Gamma'', \mathbf{T}((\text{E } x)\alpha(x))},$$

where t is an arbitrary term in *Fterm* with $x \in Free(t, \alpha)$,

(NT E) $$\frac{\Gamma', \mathbf{NT}((\text{E } x)\alpha(x)), \Gamma''}{\Gamma', \mathbf{NT}(\alpha(y)), \Gamma''},$$

where y is an arbitrary identifier in *Ide* which does not occur above the double line,

(T NOT E) $$\frac{\Gamma', \mathbf{T}(\text{NOT } ((\text{E } x)\alpha(x)), \Gamma''}{\Gamma', \mathbf{T}((\text{A } x) \text{ NOT } \alpha(x)), \Gamma''}$$

(NT NOT E) $$\frac{\Gamma', \mathbf{NT}(\text{NOT } (\text{E } x)\alpha(x)), \Gamma''}{\Gamma', \mathbf{NT}((\text{A } x) \text{ NOT } \alpha(x)), \Gamma''}$$

Lemma 4 *All the decomposition rules in $DR_{2/3}$ are sound.*

Using the above rules, we form a *decomposition tree of a formula* $\varphi \in$ *Form*, denoted by $DT(\varphi)$, which is a binary tree with vertices labeled by sequences of formulas. The root of $DT(\varphi)$ is labeled by φ. Let w be an end vertex (a leaf) of the already constructed part of $DT(\varphi)$, labeled by a sequence Γ. If Γ is either fundamental or indecomposable, then we terminate the branch of the tree ending with w. If Γ is non-fundamental and decomposable, then it is a conclusion of exactly one rule R in $\text{DR}_{2/3}$. We attach to w a single son labeled by Γ_1, if R is of the form (2), and two sons labeled by Γ_1 and Γ_2, respectively, if R is of the form (3). $DT(\varphi)$ is meant to be a maximal tree constructed according to the above principles.

Under certain assumptions (given in [Rasiowa, Sikorski 63]) about the choice of terms t and identifiers y appearing in the quantifier rules, $DT(\varphi)$ is uniquely determined by φ.

Obviously, $DT(\varphi)$ may be infinite in view of the rules for existential quantifiers. It is easy to see that a vertex w of $DT(\varphi)$ is an end vertex of this tree iff the sequence Γ labeling w is either fundamental or indecomposable.

We have the following completeness result for the considered deduction mechanism:

Theorem 1 *A formula $\varphi \in$ Form is valid iff $DT(\varphi)$ is finite and all its end sequences are fundamental.*

(By the end sequences of $DT(\varphi)$ we mean here the sequences labeling the end vertices of this tree)

In the proof of the above result we make use of the following lemma:

Lemma 5 *An end sequence* Γ *of* $DT(\varphi)$ *is valid iff it is fundamental.*

6 Final remarks

The system developed here forms a logical "shell" which - used with any specific data theory - allows us to have at our disposal the MK-calculus at assertion level, together with the classical predicate calculus at the static formulas level. For example, an axiom of the form "for every $x \neq 0$, $x \cdot x^{-1} = 1$" can be rewritten as

$$(\forall x)(\mathbf{T}(x \neq 0) \longrightarrow \mathbf{T}(x \cdot x^{-1} = 1)).$$

Once we have adopted a finite set Ax of such axioms, we can employ it in the normal way to determine which formulas are valid in the respective theory. Namely, from the deduction theorem of classical logic (which obviously holds in $L_{2/3}$) and from the completenes of the decomposition rules $\mathrm{DR}_{2/3}$ we conclude that a formula φ holds in every logical structure which satisfies Ax iff the formula $\bigwedge_{\psi \in \mathrm{Ax}} \psi \longrightarrow \varphi$ has a finite decomposition tree with only fundamental sequences of formulas at its leaves (see Section 5). It should be noted that speaking about the "deduction theorem" we of course refer to formula level. In fact, in our approach there is no sense in talking about a "deduction theorem" on the assertion level, since assertions in $L_{2/3}$ are just terms, and there can be no "deduction" on the term level; all the rules governing the MK-operations are taken care of on the formula level.

As we see, the properties of the MK-calculus are "sewn up" in the proof theory of $L_{2/3}$ in the guise of the special decomposition rules of group II. This is of course reasonable once we have decided that the MK-calculus is the only predicate calculus we want to use on the assertion level.

However, another approach is also posssible. Instead of expressing the properties of the MK-calculus in the logical system itself, we can represent this calculus by means of some specific axioms $\mathrm{A_{MK}}$ as a theory in normal classical logic, to be used in the same way as any other theory over this logic. Of course, in this case $\mathrm{A_{MK}}$ should be added to the axioms of any specific data theory in all cases when we decide to use the MK-calculus. It should be noted that such a solution allows us to employ in a straightforward way any existing proof support system for classical logic to facilitate reasoning about program correctness in terms of the static formulas of our logic. The discussed approach will be considered in another paper.

7 References

Barringer H., Cheng J.H., Jones C.B.,

[84] *A logic covering undefinedness in program proofs*, Acta Informatica **21** (1984), 251-269,

Beth E.W.,

[59] *The foundations of mathematics*, North Holland 1959,

Blikle A.,

[81 a] *On the devlopment of correctly specified programs*, IEEE Trans. on Soft. Eng. **SE-7** (1981), 519-527,

[81 b] *The clean termination of iterative programs*, Acta Informatica **16** (1981), 199 -217,

[87] MetaSoft Primer, *Towards a Metalanguage for Applied Denotational Semantics*, LNCS **288**, Springer-Verlag 1985,

[88] *Three-valued predicates for software specification and validation*, Proceedings of the VDM 88 Symposium, LNCS **328**, 243-266, Springer-Verlag 1988,

Hogevijs A.,

[79] *On a formalization of the non-definedness notion*, Zeitscrift f. Math. Logik u. Grundlagen d. Math., Vol. **25** (1979), 213-221,

[83] *A partial predicate calculus in two-valued logic*, ibid., Vol. **29** (1983), 239-243,

[87] *Partial predicate logic in computer science*, Acta Informatica **24**, (1987), 381-393,

Jones C.B.,

[86] *Systematic software development using VDM*, Prentice-Hall International Series in Computer Science, 1986,

[87] *VDM proof obligations and their justifications*, Proc. VDM Symposium 1987, 260-286, LNCS **252**, Springer-Verlag 1987,

Kleene S.C.,

[52] *Introduction to Metamathematics*, North Holland 1952,

Konikowska B., Tarlecki A., Blikle A.,

[88] *A three-valued logic for software specification and validation*, Proceedings of the VDM 88 Symposium, LNCS **328**, 218-242, Springer-Verlag 1988; a revised and extended version is to appear in Fundamenta Informatica,

McCarthy J.,

[61] *A basis for a mathematical theory of computation*, Western Joint Conference, 1961, later published in Computer Programming and Formal Systems, North Holland 1967, 33-70,

Owe O.,

[85] *An approach to program reasoning based on first order logic for partial functions*, Res. Rep. Institute of Informatics, University of Oslo, no. **89**, 1985,

Rasiowa H., Sikorski R.,

[63] *The mathematics of metamathematics*, Polish Scientific Publishers, Warsaw 1963.

A SOLUTION OF THE COMPLEMENT PROBLEM IN ASSOCIATIVE-COMMUTATIVE THEORIES

KOUNALIS Emmanuel* CRIN-INRIA **LUGIEZ Denis*** CRIN-INRIA **POTTIER Loic**** INRIA

ABSTRACT: We show in this paper that the question of checking whether there are ground instances of a term t which are not instances of the terms **t1**, ..., **tn** modulo sets of associativity and commutativity axioms is decidable. This question belongs to the the well-known class of *complement problems*.

Its solution provides a formal basis for automating the process of learning from examples, verifying the sufficient-completeness property of algebraic specifications, designing associative-commutative compilation algorithms, finding solutions of systems of equations and disequations in associative and commutative theories, etc.

KEYWORDS: The Subsumption Lattice of First-Order terms, Associative and Commutative Reasoning, AC-Disunification, Inductive Learning, Pattern-Matching, Sufficient-Completeness.

1. INTRODUCTION

The need to be able to reason about (first-order) *terms* with associative and commutative functions is fundamental in many Computer Science applications including logic and functional programming, automated reasoning, specification analysis, machine learning etc... Associativity and commutativity are typical axioms that are more naturally viewed as *structural* axioms (defining a congruence relation on terms) rather than *simplifiers* (defining a reduction relation).

We denote **variables** by x, y, z... and denote **function symbols** by f, g, h,... Each function symbol has an associated arity. Function symbols with arity 0 (zero) are called **constant symbols** and denoted by a,b,c,c_1, c_2,.... A **(first-order) term** is a variable, a constant or of the form $f(t_1, ..., t_n)$, where f is a function symbol with arity n and t_1, ..., t_n are terms. The symbols t, s, r, t_1, ... denote terms. Let **T(F,X)** denote a set of terms built out of function symbols taken from a finite vocabulary (signature) **F** and a denumerable set **X** of variables. We assume that **F** contains at least one constant symbol. Thus, the set of **ground terms** (the **Herbrand Universe**), i.e., terms containing no variable, is non-empty. An equation over T(F,X) is an element of **T(F,X)** X **T(F,X)**. The equation (t,s) will be written as t = s. Given a n set **E** of equations, the **equational theory of E** is the set of equations derivable from **E** by a finite proof, using

*) Campus scientifique - BP. 239, 54506 Vandoeuvre-lès-Nancy cedex, FRANCE, {kounalis, lugiez}@loria.crin.fr

**) 2004 route des Lucioles, Sophia Antipolis, 06565 ValbonneCedex, FRANCE, pottier@mirsa.inria.fr

reflexivity, symmetry, transitivity, replacement of equals, and instantiation as inference rules over equations. A function symbol f in **F** is **commutative** if and only if it satisfies the equation of the form **f(x,y) = f(y,x)**, where x, y are distinct variables. A function symbol f in **F** is **associative,** if and only if, it satisfies the equation of the form **f(f(x,y),z) = f(x,f(y,z))**, where x, y, z are distinct variables. We often refer to a function symbol that is both **associative** and **commutative** as an **AC-function.** Two terms t and s are said to be **associative-commutative equivalent**, written as $t =_{AC} s$, if and only if they are equivalent under the equational theory of the associative and commutative axioms (**AC-theory**).

A **substitution**, $\sigma = \{x_1 \leftarrow t_1, ..., x_n \leftarrow t_n\}$ where $x_i \neq t_i$, assigns terms to variables. The term, $t\sigma$, which is the result of applying the substitution σ to a term t, is said to be an **instance** of t. $t\sigma$ is obtained by simultaneously replacing all occurrences of each variable x_i in t by t_i. If each term t_i in σ is ground, then $t\sigma$ is said to be a **ground instance** of t. If s and t are terms and AC is an associative-commutative theory, then s is an **AC-instance** of t, if there is an substitution σ such that $s =_{AC} t\sigma$.

In this paper we deal with the following Complement Problem in Associative-Commutative theories (for a short **CPAC-problem**) *:*

INSTANCE: A finite set t, t_1, ..., t_n of first-order terms over T(F,X) with AC-functions.
QUESTION: Does there exist a ground instance of t which is not an AC-instance of any t_1, ..., t_n ?

Motivation:

Generalization is an important operation for programs that **learn from examples (inductive learning)**. A formal model for inductive inference has been proposed by Plotkin [PL,71], Angluin and Smith [AS, 83]. A formula of the form **G**:= t/ t_1,..., t_n is a **generalization** iff the set of ground instances of t which are not instances of t_1,..., t_n is non-empty. Recently Lassez and Marriot [LM,87] solved the CPAC-problem in the case where no AC-functions are involved in the terms t, t_1,..., t_n. However, in the learning from examples paradigm most of the symbols appearing in the set of examples and counter examples are AC-functions (notarious examples are the connectives *and, or* etc. (see Michalski [MICH, 83]). For instance, if a set of concept examples leads to the term **or(x,y)** and a set of counter examples is **or(x,x)**, **or(f(z),c)**, **or(f(y1), f(y2))**, then from immediate inspection it is not obvious that the formula **or(x,y) / or(x,x)**, **or(f(z),c)**, **or(f(y1),f(y2))**, provides no concept to be learned. Now, if in the **CPAC**-problem the term t is the generalization of a set of examples and the terms t1, t2,..., tn are counter examples, then its solution allows to check whether the formula **G**:= t/ t1, t2,..., tn is a generalization.

Functional languages allow to define functions by a set of rules such as: **0 + y → y, x + succ(y) → s(x+y)** together with a priority on this set of rules to deal with ambiguous patterns (i.e., a set of left-hand sides). However, for efficiency reasons the above definition should be compiled in a piece of code which is more suited to machine: (x+y) = **if** x = 0 **then** y **elsif** y=s(y') **then** let y=s(y') in s(x+y') **else no_match.** Nowdays, many algorithms for **compiling pattern-matching** definitions exist (see Peyton-Jones [P-J,87], Schnoebelen [SCH,88], Kounalis-Lugiez [KL,91]). Unfortunately, these algorithms apply either to a set of patterns without AC-functions, or to linear patterns. However, if in the **CPAC**-problem the term t is f(x1,x2,...,xk) and the terms t1, t2,..., tn are the patterns of the k-ary function f, then its solution will allow us to design effective algorithms to compile pattern-matching definitions.

The sufficient-completeness property of (equational) specifications of **Abstract Data Types** allows to ensure that every data item can be constructed using only constants and operations of a signature (*no junk*). This property has been investigated for specifications under certain conditions (Huet and Hullot[HH,82], Kounalis [KOU, 85],etc). Unfortunately, these algorithms cannot deal with specifications which include AC-axioms. However, if in the **CPAC**-problem the term t is f(x1,x2,...,xk), where f is the function to be checked for sufficient-completeness, and the terms t1, t2,..., tn are the left-hand sides of an equational specification, then its solution will allow us to design algorithms to verify the sufficient completeness property of convergent specifications with AC-functions.

Systems of equations and disequations $\{l_i = r_i: 1 \le i \le n, d_j = g_j: 1 \le j \le n\}_{AC}$ have many applications to Artificial Intelligence. The solution of such a system is a set of substitutions of the variables occurring in l_i, r_i, d_j, g_j, such that the instantiated terms of equations become equal, and the instantiated terms of disequations become inequal with respect to a given AC-theory. Colmerauer [COL,84] has discussed this problem in the framework of **Logic Programming** and given an algorithm to solve this problem in the case of an empty theory (see also Lassez and Marriott [LM,87] , Maher [MAH,88]). However, if we assume that some of the functions appearing in the system of equations and disequations are AC-functions then their results do not hold anymore.Consider, for instance, the system **S = {x = f(a,y)), y ≠ x}$_{AC}$**.The problem here consists of finding substitutions of the variables x and y such that the system S has a solution. However, if in the **CPAC-** problem the term t is P(f(a,y), y)) and the set of terms t1, t2,..., tn consists only of the term P(z,z), then its solution will allow us to find a complete set of solutions of a system S of equations and disequations. Note that there is an easy algorithm to transform S into the set t, t1, t2,..., tn.

How to solve the Complement problem (CPAC) ?

The main tool to solve the **CPAC**-problem is the concept of *pattern trees* which are trees whose nodes are labeled by flattened terms. However, when the signature contains several constant symbols and only one function symbol of arity greater than zero, then the solution of the **CPAC**-problem is reduced to verify the emptyness of differences of particular subsets of N^m using properties of additive semi-groups of N. In both cases, the correctness proof is non-trivial and requires the use of sequences of terms with complex combinatorial properties. At first glance this result may appear surprising since closely related problems are undecidable (see Kapur, Narendran, and Zhang [KNZ, 86], Treinen [TRE,90]).

Layout of the paper:

The structure of this paper is as follows: In Section 2 we present an overview of the proposed methods on two examples. In Section 3 we summarize the additional basic material which is relevant to this work. In Section 4 we provide the decision procedures for the **CPAC**-problem. In particular, we first show how to solve the **CPAC**-problem in the case where the signature contains several constant symbols and at least two function symbols of arity greater than zero, and next consider signatures that contain several constant symbols and only one function symbol of arity greater than zero.

2.OVERVIEW OF THE SOLUTION: TWO EXAMPLES

Before discussing the technical details of how to check whether there exist ground instances of a term t which are not AC-instances of a set S = { t_1, ..., t_n }, we first describe how these methods work on two simple examples. Each of the method solely depends on the structure of the signatures to be used : *Basic* signatures are those containing several constant symbols and only one function symbol of arity greater than zero. *Extended* signatures are those containing several constant symbols and at least two function symbols of arity greater than zero

EXAMPLE A: Consider, for instance, the term t = f(x,y) and the set S of terms { t_1= f(x1,x1), t_2=f(x1, f(x1,y)), t_3 =f(x1, g(y1)) }, over T({f,g,0}, X} where f is an AC-function symbol, g is a function of arity 1, and 0 is a constant symbol. Here the signature is extended.

The first step of our method consists of considering terms in a *flattened* version (see preliminaries definitions below) i.e., in a form in which all nested occurrences of associative-commutative function symbols are stripped, and where the order of arguments of such operators is not significant. For example, the flattened version of the term t2 is the term f(x1, x1, y).

Having computed the flattened version of the terms t, t_1, ..., t_n our second step consists of constructing a *complete pattern tree* for the flattened term t (see definition 4). By a *pattern tree* of t (see definition1) we mean a finitely branching tree whose root is labeled by the term t and such that an internal node labeled by s has successors all possible different flattened terms (modulo variable renaming) derived from s, i.e., obtained by replacing some variable in s by some term of the form g(x1, ..., xn), where g is a function symbol of arity n (in the signature F) and x1, ..., xn are distinct variables not appearing in s. For instance the following pattern tree is suitable for checking the **(CPAC)**-problem:

f(x,y)

f(0,y) f(g(x1),y) f(x1,x2,y)

f(0,0) f(0, g(y1)) f(0,y1, y2) f(0,x2,y) f(g(x3),x2,y) f(x3,x4,x2,y)

f(0, 0, y2) f(0, g(y3), y2) f(0,y3,y4, y2)

The key idea undrlying the construction of such a pattern tree **T** consists of extending variables in node labels whenever these nodes are AC-unifiable with some term t_j in S. In general this process is infinite and therefore we need some way to stop it: a node label s in a pattern tree is a leaf if

1) s is a term which is an AC-instance of some t_j in S (e.g., the terms f(g(x1),y) , f(0,0), f(0, g(y1)), f(0, 0, y2), f(0, g(y3), y2), f(g(x3),x2,y).

2) s is a term which is not AC-unifiable with some term in S (In the example here each leaf label is AC-unifiable with some term in S).

3) s is a term whose any non-variable subterm s/u has either no variable which corresponds to a function symbol or to a non linear variable in some t_j in S or a number of arguments greater than the maximal number of arguments of an AC-function in S (e.g., the terms f(0,y3,y4, y2), f(x3,x4,x2,y)).

The terms as partitioned above are said to be of *type* 1, 2, and 3 (see definition 4). A pattern tree whose all leaf labels are of type 1, 2, 3 is said to be *complete* (see definition 4).

Having computed a complete pattern tree our next step consists of testing whether there exist leaf labels in **T** of type 3. The reason for that is the following: if all leaf labels in T is of type 1 or 2, then the

(**CPAC**)-problem is resolved since terms of type 1 have all ground instances AC-equivalent to the ground instances of terms in S, whereas terms of type 2 have no ground instance AC-equivalent to the ground instances of terms in S. Unfortunately, the situation with terms of type 3 is more complicated.
Terms of type 3 are further partitioned into two sets according to the number of arguments of AC-function symbols occurring in them (see definition 6). In the present example all terms of type 3 are of type 3b (see definition 6) since the number of arguments of the function symbol f in the terms f(0,y3,y4, y2), f(x3,x4,x2,y)) is greater than 3: the maximal number of arguments of an AC-function in S. We must note here that the existence of terms of type 3a ensures that (**CPAC**)-problem is resolved since terms of type 3a have ground instances which are not AC-equivalent to any ground instance of terms in S (see theorem 4). However, in the absence of terms of type 3a we have to check whether the terms of type 3b contain ground instances which are not AC-equivalent to the ground instances of terms in S.

This consists of the last step of the method. To check terms of type 3b we compute an *answer set for* S (see definition 7) which in our case is the set A(S) = { 0, f(0,0), f(0,g(x)), g(0), g(f(x,y)), f(0,0,0), f(0,0,g(x)), f(g(z),0,g(x)), f(g(z),g(w),g(x)), f(0,0,0,y), f(0,0,g(x),y), f(g(z),0,g(x),y), f(g(z),g(w),g(x), y). We then verify whether all instances of a term of type 3b which are obtained by using terms in A(S) are AC-instances of S. Here, we can easily verify that all instances of the terms f(0,y3,y4, y2), f(x3,x4,x2,y)) by using elements of A(S) are AC-instances of the terms in S and this resolves the (**CPAC**)-problem (see theorem 5): *each ground instance of the term* t = f(x,y) *is an* AC-*ground instance of the terms* t_1= f(x1,x1), t_2=f(x1, f(x1,y)), t_3 =f(x1, g(y1)) }.

EXAMPLE B: Consider now the following example of the **CPAC**-problem. t = +(x,x,y,y) and t_1= +(x1,x1,x1, y)) over T({+,a}, X}, where + is an AC-function symbol, and a is a constant symbol. Therefore the signature here is basic.

In general, to tackle the **CPAC**-problem in the case where the signature is basic, we first consider terms as sums of variables, where + is the usual AC-function symbol. To denote such terms we shall use the notation $\Sigma_i a_i x_i + \Sigma_j a_j C_j$ which means that the variable x_i appears a_i times and the constant C_j appears a_j times in a term. In other words to any term of the **CPAC**-problem we associate vectors $(a_1, a_2, \dots a_k)$ of N^k. We then reduce the search of solutions of the **CPAC**-problem in a finite subset of N^k (**a test domain**).

For instance, in our case , let t be the term $2z_1+2z_2$ and t_1 be the term $3x_1+x_2$. Since the number of constant symbols in the signature is one , by applying theorem 6, we get that the test domain is the segment [0;8]. This means that we must verify whether there exists an instance of t obtained by replacing the variables in t by elements from the finite domain n*a, where n varies from 1 to 8 and a = (2,2) , which is not an AC-instance of t_1. However, all these instances of t are instances of t_1, we may conclude by using theorem 6 *that every ground instance of* t *is an AC-instance of* t_1.

3. PRELIMINARIES

In order that this paper be self-contained, this section contains an outline of further definitions (besides the ones presented in the introduction of this paper) and results related to AC-terms.

Let AC be an associative-commutative theory. By $[t]_{AC}$ (or simply [t]) we denote the AC-equivalence class containing a term t in T(F,X). If s and t are terms and AC is an associative-commutative theory, then: s is an AC-instance of t iff there exists s' in [s] such that s' is an instance of t. Further, suppose s and t are terms in an AC-theory. Then s and t are said to be **AC-unifiable** if there is a substitution σ such that $s\sigma =_{AC} t\sigma$. In this case we say that σ is an **AC-unifier** of s and t.

A term that involves associative -commutative function symbols is represented in **flattened** form, that is, no argument to an AC-function f is a term whose outermost symbol is f itself. For example, if f is AC, then f(a,f(b,c)) is represented as f(a,b,c). (In other words f is treated as a *varyadic* symbol). Flattening a term with respect to a function f can be done as follows: *firstwe* represent the term in the right-associative form. Such a term will be of the form $f(t_1, f(t_2, ..., f(t_{n-1}, t_n)...)$ where $t_1, ..., t_n$ do not start with f. *Then* we simply represent the term as $f(t_1, ..., t_n)$. If t is a term, then **Flat(t)** denotes the flattened form of t. In the following we shall make free use of the following lemma:

Lemma: *Suppose s and t are terms. Then $s =_{AC} t$ iff Flat(s) $=_P$ Flat(t), and $s =_{AC} t\sigma$ iff Flat(s) $=_P$ Flat(Flat(t)Flat(σ)), where $=_P$ is the permutative congruence on subterms of AC-functions.*

Given a term t, **arg(t)** denotes the set of arguments of t, i.e., if $t = f(t_1, ..., t_n)$, then **arg(t)**= $\{t_1, ..., t_n\}$. Further , **|arg(t)|** denotes the cardinality of arg(t). Also the **depth(t)** of a term t is defined to be the depth of the maximal path in the tree representation of t.

If t is a term, then **dom(t)** denotes the set of positions (occurrences) in a term t and the expresion t/u denotes the **subterm of t at position u.** Also **t(u)** denotes the symbol of t at position u. A position u in a term t is a **variable position** if $t(u) \in X$. Further, a position u in t is said to be a **non-linear variable position** if t(u) is a variable position and the variable x=t(u) appears more than once in t. Let **sdom (t)** denote the set of function symbols or non-linear variable positions in t and **V(t)** to denote the set of all variables that occur in t. A term t is **linear** iff no variable in **V(t)** appears more than once in t.

In the second part of the next section we need another representation of flattened terms. Let + be an AC-function. A flattened term $t = +(s_1,...,s_k)$ where the s_i are either variables or constants will be written as $\Sigma_i \alpha_i s_i$, where α_i is the number of occurrences of s_i in t. Some α_i can be zero, but the sum of α_i cannot be zero. For instance the term t = +(x,a,b,x,a) is written as 2x+2a+ b. Let N be the the set of non-negative integers. By [a; b] we denote a **closed segment** of N. Also by aN we denote the set of non-negative multiples of a. Let $e_1,...,e_n$ be a canonical basis of N^n. The ith **coordinate** of a vector x in N^n is denoted by x_i. Further, let $\alpha_1,...,\alpha_n$ be positive integers. Then we denote by (α) the **additive semigroup** of N generated by the α_i's. Also gcd(α) denotes their **greatest common divisor**, and lcm(α) their **least common multiple**. The following result is well-known from number theory (**Frobenious problem**):

Lemma: *There exists a least integer c(α) such that every integer c(α) + k*gcd(α) is in N, where $k \in N$. c(α) is said to be the* ***conductor*** *of (α).*

4. DECISION PROCEDURES

In this section we set the machinery to resolve the **CPAC**-problem stated in the introductory remarks of this paper, i.e., the problem of whether, in a sequence t, t1,...,tn of terms, there exist ground instances of t which are not instances of any t1,...,tn modulo the associativity and commutativity axioms.

Clearly, the major problem in solving the **CPAC**-problem is the unbounded number of ground substitutions of t one has to verify for instances of the terms t1,...,tn. To motivate our efforts in removing this problem we first consider the case of signatures with several constants and at least two function symbols of arity greater than zero **(extended signatures)**, and next consider the case of signatures with several constants and only one function symbol of arity greater than zero **(basic signatures)**. The reason of such a partitition is that the extended signatures allows the construction of flattened terms of any depth. Note that this is not the case of basic signatures.

4.1 (CPAC)-PROBLEM FOR EXTENDED SIGNATURES

Let us first consider the case of signatures with several constants and at least two function symbols of arity greater than zero and several constant symbols(extended signatures). The key to our approach in this case is the construction of *Pattern trees* :

Definition 1: Suppose t is a flattened term. Then the **sons** of t at a fixed variable x are all possible different flattened terms (modulo variable renaming) *derived* from t i.e., obtained by replacing the variable x in t by some term of the form g(x1, ..., xn), where g is a function symbol of arity n (in a signature F) and x1, ..., xn are distinct variables not appearing in s.

Example 1: Suppose that F={a, f, h} with f to be an AC-function symbol. Then the sons of f(x,y,z,y) are the terms: {f(x,a,z,a), f(x, x_1,x_2, z, x_1,x_2) f(x, h(x_1),z, h(x_1)) }. These terms are obtained by replacing the variable y in f(x,y,z,y) by the terms a, h(x_1), and f(x_1,x_2).

Definition 2: A **Pattern tree** T of a flattened term t is a labeled tree T whose root is labeled by t, and such that an internal node labeled by a flattened term s has successors the sons of s.

Example 2: Let F={a, f, h} and let f be an AC-function symbol, then a pattern tree of f(x,y,z,y) is:

f(x,y,z,y)

f(x,a,z,a) f(x, x_1,x_2, z, x_1,x_2) f(x, h(x_1),z, h(x_1))

f(x,a,a,a) f(x,a,h(z_1),a) f(x,a,z_1,z_2,a)

It follows from the above definition that in a pattern tree of a flattened term t (i) the label of any node is an instance of t,(ii) the set of ground instances of t is equal to the set of ground instances of all leaf labels, and (iii) for any ground substitution η there exists a leaf label r and a ground substitution σ, such that $t\eta = r\sigma$.

As we already have stated in the introductory example of this paper, the method of solving the **CPAC-problem** for extended signatures is based on the construction of a suitable finite pattern tree **T** of a flattened term t. In trying to compute such a tree there are several questions that come naturally in mind:

Q1) *Which node labels must be expanded* ?

Q2) *Which variables in them must be instantiated*?

Q3) *When the construction halts*?

Let us now set the machinery to resolve these queries. The following definition gives the answer of the first two queries:

Definition 3: Suppose t is a flattened term and S= {t_1, ..., t_n} is a set of flattened terms. Let k be the maximal number of arguments of an AC-function in S= $\{t_1, ..., t_n\}$. t is said to be **extensible** with respect to S=$\{t_1, ..., t_n\}$ if there exists a term s in $[t]_p$, a variable position u.i in s, an integer $j \leq n$ such that either u.i is in **sdom**(t_j) [i.e, u.i is a function symbol or a non-linear variable position in t_j], or $|arg(s/u))| \leq k$. Otherwise, is said to **cover** S = $\{t_1, ..., t_n\}$.

Example 3: Let F={a, f, h} and let f be an AC-function. The term t= $f(a, x_1, z_1, h(x_2))$ is extensible with respect to the flattened term t_j = f(x,y,z,y) since the variable x_1 is in the position 1 in the term s $(x_1, a, z_1, h(x_2)) =_P t$, and 1 is a non-linear variable position in t_j. However, the term t= $f(h(x_1), a, h(z_1), h(x_2))$ is not extensible with respect to t_j since no term in $[t]_p$ contains a variable position which belongs to sdom(t_j). The term t= $f(x, x_1, x_2, z, x_3, x_4)$ covers t_j since $|arg(f(x, x_1, x_2, z, x_3, x_4))| > |arg(f(x,y,z,y))|$.

It follows from the above definition that if a non-ground term t covers a set S = $\{ t_1,...,t_n\}$ of flattened terms, then for all $j \leq n$, and for every variable x in $[t]_p$ ***either*** x is in position that corresponds to a function symbol or to a non-linear variable in t_j ***or*** x is an argument of an AC-function subterm in $[t]_p$ the number of arguments of which is greater than k.

Morever, definition 3 provides a way to compute the pattern trees we are interested in: these pattern trees are obtained by deriving node labels which are extensible with respect to given set $\{ t_1,...,t_n\}$ of flattened terms. Let us now resolve the last query stated above. To do it we need the following definition :

Definition 4: Given a set $\{t_1, ..., t_n\}$ of flattened terms and a flattened term t, let **T** be a pattern tree of t. A node label s in **T** is said to be of

type 1, if s is an AC-instance of some t_j,

type 2, if s is not AC-unifiable with t_j, for any $j \leq n$,

type 3, if s is neither of type 1 nor of type 2 and covers $\{t_{i1}, ..., t_{ik}\}$, where $\{ i1, ..., ik \} \subset \{1,...,n\}$ and s is AC-unifiable with t_{ij}, for any $j \leq k$.

A pattern tree **T** of t is said to be **complete** if each node label of type 1, 2, 3 is a leaf.

Example 4: Let F={ a, b, f, h} and let f be an AC-function. Let t = f(x,y), t_1 = f(x,y,y) , t_2 = f(x,a) , t_3 = f(b,b). Term s=f(x,y,z) is not of type 1,2,3 since s is AC-unifiable with the term t_1 and does not cover it. The term f(a,x) is of type1 since it is an AC-instance of t_2. Term f(h(x),b) is of type 2 since is not AC-unifiable with S= $\{t_1,t_2,t_3\}$. Term s=f(h(x),h(y), h(z) is of type 3 since s is AC-unifiable with the term t_1 and covers it.

Given a flattened term t and a set S= $\{t_1, ..., t_n\}$ of flattened terms, there <u>always</u> exists a complete pattern tree T for t which is finite (see theorem 1). The reason for that is that the number of terms s whose depth is bounded by d (the depth of S) and such that no subterm of s has more than k +1 arguments is finite.

Theorem 1: *If S= $\{t_1, ..., t_n\}$ is a set of terms, a complete pattern tree T of a term t is finite*

Let us now show how complete pattern trees allow to solve the **CPAC**-problem. This will be done by studing the fundamental properties of leaf labels of a complete pattern tree.

Definition 5: Let $S=\{t_1, ..., t_n\}$ be a set of flattened terms, and let **T** be a complete pattern tree of a flattenned term t. A leaf label s in T is a **quasi-AC-instance** of S if for every ground instance $s\eta$ of s there exists a term t_j in S such that $s\eta$ is an AC-instance of t_j.

The previous definition allows us to solve the **CPAC**-problem by checking whether a leaf label of a complete pattern tree of a flattened term t is a quasi-AC-instance of a set $S = \{t_1, ..., t_n\}$ of flattened terms: **each ground instance of t is AC-equivalent to ground instances of S iff each leaf label of a complete pattern tree of t is a quasi-AC-instance of a set $S = \{t_1, ..., t_n\}$.**
The following theorem deals with leaf labels of type 1:

Theorem 2: *Let $S = \{ t_1, ..., t_n\}$ be a set of terms, and let T be a complete pattern tree of a flattened term t. Leaf labels in T of type 1 are quasi-AC-instances of $S= \{t_1, ..., t_n\}$.*

Leaf labels of type 2 enjoy the following property:

Theorem 3: *Let $S = \{ t_1, ..., t_n\}$ be a set of terms, and let T be a complete pattern tree of a flattened term t . Leaf labels in T of type 2 are not quasi-AC-instances of $S= \{ t_1, ..., t_n\}$.*

The situation with terms of type 3 is more complicated.To proceed with, we first need the following:

Definition 6: Let $\{t_1, ..., t_n\}$ be a set of flattened terms, and let **T** be a complete pattern tree of a flattenned term t. Assume that k is the maximal number of arguments of an AC-function in $\{t_1, ..., t_n\}$. A node label s in **T** of type 3 is said to be of **type 3a** if no subterm of s has more than k arguments. Otherwise s is said to be of type **3b**.

Example 5: Let F={ a, b, f, h} with f to be an AC-function symbol. Assume t_1 = f(x,y,y) , t_2 = f(x,a), and t_3 = f(b,b). The term s= f(h(x),h(y), h(z)) is of type 3 since t is AC-unifiable with t_1 and covers it. Further, s is of type 3a since it has no more than 3 arguments. On the other hand, the term s= f(x,y,z,w) is of type 3b.

The following theorem shows that terms of type 3a are not quasi-AC-instances of a set $\{t_1, ..., t_n\}$. The correcteness proof of this theorem is based on the following arguments:

1) every variable position in the P-equivalence class of a term s of type 3a corresponds either to a linear variable in $\{t_1, ..., t_n\}$ or it is not in the domain of $\{t_1, ..., t_n\}$, and

2) There exist ground terms of any depth since the signature is extended (i.e., terms of the form f(...g(...,f(...))),...). By the first argument a ground instance of s cannot be an AC-instance of a linear-term t_j in $\{t_1, ..., t_n\}$, otherwise s would have been an AC-instance of t_j. The second argument allows to build ground instances of s of any depth and width that prevent non-linear terms in $\{t_1, ..., t_n\}$ to match them.

Theorem 4: *Let $S = \{t_1, ..., t_n\}$ be a set of terms, and let T be a complete pattern tree of a flattened term t . Leaf labels in T of type 3a are not quasi-AC-instances of $\{ t_1, ..., t_n \}$.*

Let us now deal leaf labels of type 3b. The main problem with those leaf labels that there is no direct way to reason about the behavior of their ground instances. So some indirect way is needed. As we already have pointed out in the introductory example of this paper the check of whether a leaf label of type 3b is a quasi-AC-instance of $S=\{t_1, ..., t_n\}$ requires an *answer set* for S to be computed:

Definition 7: Let $S = \{ t_1, ..., t_n \}$ be a set of flattened terms. Assume that d is the depth of S and k is the maximal number of arguments of an AC-function in $\{t_1, ..., t_n\}$. The set **A(S)** = {r | r is a linear term in T(F,X) of depth bounded by d and no subterms with more than k+1 arguments and such that variables can occur only at depth d, and at the $k+1^{th}$-argument of some subterm of r } is said to be an **answer set for S.**

Example 6: Let F = { f,g,a} with f to be AC. Assume that **S** is a set of terms such that depth(S) = 3 and k= 3. The set {a, g(a), g(g(x)), g(f(x,y)), g(f(x,y,z)), g(f(x,y,z,w)), f(a,a), f(a,g(x)), f(g(x),g(y)), f(a,a,a), f(a,a,g(x)), f(a,g(y),g(x)), f(g(z),g(y),g(x)), f(a,a,a,w), f(a,a,g(x),w), f(a,g(y),g(x),w), f(g(z),g(y),g(x),w) is an answer set for S. Note that the term s=f(x,y,z,w) is not in A(S) since the variables x, y, and z occur at depth 2. However, in the term s= f(g(z),g(y),g(x),w) the variable w may occur in s (depth(s)< 3) since w is the fourth argument of s. Of course any term in the answer set for S that is an instance of another answer term can be omitted (since any ground instance of the former is also a ground instance of the latter). Further it is clear that such a set is finite and any ground term of any depth and of any width is an AC-instance of a term in A(S).

Definition 8: Let $S = \{ t_1, ..., t_n \}$ be a set of flattened terms, and let **A(R)** be an answer set for S. If t is a term of type 3b, then the set $\mathbf{I(t)_{A(S)}}$ of the instances of t which are obtained by using terms in A(S) (up to variable renaming) is said to be the **answer substitution set of t.**

When an answer set for $S = \{ t_1, ..., t_n \}$ is computed, the problem of determining whether a term of type 3b is a quasi-AC-instance of S can be decided as the following theorem shows :

Theorem 5: *Let $S = \{t_1, ..., t_n\}$ be a set of terms, and let T be a complete pattern tree of a flattened term l. Leaf labels s in T of type 3b are quasi-AC-instances of $\{ t_1, ..., t_n \}$ iff any term in $I(s)_{A(S)}$ is an AC-instance of $t_1, ..., t_n$. .*

The correcteness proof of this theorem is based on the following arguments:

1) The answer-substitutions of a term s of type 3b cover the set of its ground instances. This proves the "if-part" of the theorem.

2) For the "only if- part " the existence of ground terms of any depth allows to build instances $r\eta$ of r in $I(s)_{A(S)}$. The terms $r\eta$ have subterms that below to a constant d are all *very* different. Because of this construction $r\eta$ cannot be AC-instances of any non-linear term t_j in $\{t_1, ..., t_n\}$ if r is not itself an AC-instance of a term t_j. On the other hand side, because of the structure of A(S), $r\eta$ is an AC-instance of a linear term t_j in $\{t_1, ..., t_n\}$ iff r is an AC-instance of the term t_j.

4.2 THE (CPAC)-PROBLEM FOR BASIC SIGNATURES

Let us now consider the case of signatures with several constants and only one function of arity greater than zero **(basic signatures)**. Let F be the signature $\{+, C_1,...,C_m\}$ where C_i are all distinct constants, and + is an AC symbol. Let $t = \Sigma_i a_i x_i + \Sigma_k \alpha_k C_k$, $t_1 = \Sigma_i b^1{}_i x_i + \Sigma_k \beta^1{}_k C_k$, ... , $t_p = \Sigma_i b^p{}_i x_i + \Sigma_k \beta^p{}_k C_k$, be p+1 arbritary terms, where x_i are dinstinct variables (the upper symbols 1,..,p are not power exponents but indices). Let G be the set of ground instances of t that are not instances of one of $t_1,...,t_p$. Elements of G are of the form $\Sigma_k c_k C_k$. We identify G with the set of vectors $(c_1,...,c_m)$ associated to its elements. The following theorem shows that the CPAC-problem for basic signatures is decidable:

Theorem 6: *Suppose B is the maximun of the p+1 integers $\Sigma_i b^j_i + c(b^j) + sup(\beta^j)$ and $\Sigma_i a_i + c(a) + sup(\alpha)$. Let D be the lcm of the p+1 integers $gcd(b^j)$ and $gcd(a)$. Then G is empty, if and only if, $G \cap [0; B+ D]^m$ is empty.*

The proof of this theorem is based on the following lemma:

lemma: Let $E_i = E_i' + (a_i N)^n$, i =0,....,p be p+1 subsets of N^n with $a_i \geq 0$ such that E_i is a subset of $[0;b]^n$ for all $i \leq p$. Let $E = E_0 - \cup_{i>0} E_i$, and $a = lmc(a_i)$. Then E is empty, iff, $E \cap [0; a + b]^n$ is empty.

Examples 7: a) Let F = { +, c}, and t = 2x1 + 2x2, and t1 = x1 +3x2. We then have that p = 1, a = (2,2), b1= (1,3), n=2, m =1, d(a) = 2, $d(b^1)$ = 1, D=2, c(a) = 0, $c(b^1)$ = 3, and B= 7. We can easily verify that $G \cap [0;8] = \emptyset$, so G is empty. Therefore each ground instance of t is an AC- ground instance of t1.

b) Let F = { +, c1, c2 }, t = c1 +c2 + x +y, t1 = c1+2c2+x, t2=3c1+c2, t3 = 4c1 +c2. Then B = 4 and D = 1. Therefore G is non-empty since $G \cap [0;5]^2 \neq \emptyset$ and the terms (5+m)c1 +c2 are the ground instances of t which are not AC-instances of t1,t2,t3,t4.

5. BIBLIOGRAPHY

[AS, 83] ANGLUIN,D., and SMITH, C.H: Inductive Inference: Theory and Methods *Computing Surveys, Vol. 15, No. 3, Sept. 1983.*

[COL, 84] COLMERAUER, A.: Equations and Inequations on finite and infinite trees. *Proceeding of the FGCS conference pp. 85-99 ,1984 .*

[HH,82] HUET, G. and HULLOT J.M. : Proofs by induction in equational theories with constructors. *J. Comput. System Sci. No 25-2 , 1982.*

[KNZ,86]KAPUR, D.and NARENDRAN, P. and ZHANG, H. : Complexity of Sufficient Completeness and Quasi-reducibility *Proceeding of the Conference of in Foundations of Software Technology ,1986.*

[KOU,90] KOUNALIS, E.: Pumping lemmas for tree languages generated by rewrite systems. *Proc. EUROCAL 85 , LNCS No 204, Springer-Verlag (1985) .*

[KR,91] KOUNALIS, E. and LUGIEZ, D., Compiling pattern matching with AC-functions. *Proc. 16th Conf. of CAAP, 1991 LNCS Springer-Verlag (1990) , to appear.*

[LM,87] LASSEZ,J.L. and MARRIOTT, K. :Explicit representation of term defined by counter examples. *Journal of Automated reasoning 3 (1987) , pp. 301-317.*

[MAH,88] MAHER, M., Complete axiomatization of the algebra of finite, rational and infinite trees. *In Proc. of 3rd LICS, 1988.*

[MIC,83] MICHALSKI, R.S.:A theory and Methodology of Inductive Learning. *Artificial intelligence 20 (1983) , pp. 111-161*

[PJ, 87]: PEYTON-JONES: *The Implementation of Functional Programming Languages.* Prentice-Hall, 1987.

[PLO,71] PLOTKIN, G.:A Further note on Inductive Generalization . In *Machine Intelligence 6(1971)*

[SCH, 88] SCHNOEBELEN, P.: "Refined compilation of pattern-matching for functional languages". *Science of Computer Programming 11, 1988, 133-159.*

[TRE,90] TREINEN. R., A new method for undecidability proofs of first-order theories, In Proc. of FCT-TCS conference (INDIA), 1990 to appear.

A Model for Real-Time Systems

Padmanabhan Krishnan
Department of Computer Science
Ny Munkegade 540
Aarhus University
DK 8000, Aarhus C, Denmark
E-mail: paddy@daimi.aau.dk

Abstract

In this paper we define an equivalence and a modal logic for real-time systems. The equivalence is based on timed processes and the timing specifications they have to satisfy. While the equivalence we define is not a congruence, it does satisfy many laws.

1 Introduction

While process formalisms such as CCS [4], cannot express time delays between events (actions), there are extensions [6, 9] which do. Most of these extensions are called real-time extension. We feel that while the addition of time is an important step towards characterizing real-time, it is by no means sufficient. As [2] explain, a real-time system is one which has to satisfy certain properties given certain resource restrictions. The properties that a system is to satisfy play an important role in building architectures/schedulers for it.

Various notions of equivalence in a concurrent setting have been well established. Consider for example the notion of bisimilarity [8]. If two concurrent terms are bisimilar it means that any behavior exhibited by one can be exhibited by the other. If the bisimulation is a congruence, one item can be substituted for the other in 'arbitrary' contexts without affecting 'behavior'. Real-time systems have timing requirements and if context includes timing requirements, two terms to be equivalent in all contexts will necessarily have identical timing. For example, let timing constraints impose an upper bound on the time by which an execution must terminate. If two terms are equivalent under all timing constraints, they will have to terminate at the same time, as one can specify the upper bound for all times.

While the definitions which are a straight generalization of those used in concurrency results in a general theory which can be used in arbitrary contexts, it is not useful for specific real-time systems. Systems rarely impose precise times when actions are to be taken. Usually, they permit a range of times in which the actions can be taken. So one could define equivalence under a given set of timing constraints rather than arbitrary timing constraints. The notion of predictability is important in real-time systems. Thus, simulation must be defined in a more deterministic context. As a scheduler is an important implementation feature that introduces determinacy, simulation should consider schedulers. In the definition of simulation for concurrent systems, the definition of equivalence etc. is quite general. For example, ((a | b) | c) is trace equivalent to (a | (b | c)) meaning that any trace exhibited by the first can also be exhibited by the second and vice versa. However in the presence of timing constraints and a scheduler the effect of replacing one by the other could have disastrous effects on the safety of the system. Continuing with the above example, let (a after 10) be the timing specification t1 and (a before 30) be the timing specification t2. Let a scheduler for a uniprocessor environment for the above processes translate them as follows: sched((a | b) | c) = c·a·b, sched(a | (b | c)) = a·b·c,

If each · takes 20 units of time (starting at 0), then p1 is equivalent to p2 given the scheduler and timing constraint t1 but not under timing constraint t2. While this example may be rather contrived, the scheduler may behave in this particular fashion due to the presence of certain other timing constraints that it is trying to satisfy. In this paper we introduce definitions which are relevant in a real-time setting. As we are interested in implementation of real-time processes, we concentrate on simulations with bisimulation (or replaceability) defined as a symmetric simulation.

The aim of this paper is to propose a calculus for real-time systems which expresses delays between events and also properties that the system has to satisfy. The calculus assumes a time domain well ordered by $\leq$. We like [6, 9] use CCS as the basic untimed language. We also outline how the effect of a scheduler and architecture on a real-time program can be studied.

In section 2 we describe the syntactic elements of our calculus viz., a language to express behavior and a language to express timing constraints. In section 3 we describe an operational semantics for the behavioral aspect of the calculus. In section 4 we develop a notion of equivalence induced by the timing constraints, while in section 5 we present a logical characterization of the equivalence. In section 6 we outline how these ideas can be used to study system issues such as schedulers.

2 Syntax

We define our language RTCCS (for real-time CCS). We as in [4, 5, 6, 9], assume a set of atomic actions Λ on which a bijection $\bar{\cdot}$ can be defined such that for all $a \in \Lambda \bar{\bar{a}} = a$. The time domain we assume is integers (actually any discrete well ordered set would suffice).

The basic syntax of our language is as follows.

$$\mathrm{P} = \mathrm{Nil} \mid a;\mathrm{P} \mid (\mathrm{t})\mathrm{P} \mid (\mathrm{P} \mid \mathrm{P}) \mid \mathrm{P} + \mathrm{P} \mid \mathrm{P} \setminus \mathrm{H} \mid \mathrm{X} \mid \mathrm{rec}\widetilde{X}\colon \widetilde{E}$$

Nil is a process which can exhibit no further action. a;P is a process which performs action a and then evolves to process P. The action a is atomic and takes unit time. (It can also be defined to take time greater than 1.) (t)P defines a process which can behave as P after time t. P | Q denotes parallel composition. P+Q denotes non-deterministic choice. P \H restricts the behavior of P to those actions *not* in H. rec$\widetilde{X}$:$\widetilde{E}$ represents guarded recursion and we only consider closed terms.

The language defined above is similar to that in [6]. Unlike [9] we do not consider passing of time as values to processes. Rather we specify timing constraints which processes have to satisfy. Generally speaking, there are two types of constraints in real-time systems 1) absolute and 2) relative [1]. Absolute constraints are expressed in terms of what actions must or must not happen in time intervals. For example, A(a,t1,t2)=T requires an 'a' action to occur in the time interval [t1,t2] while A(a,t1,t2)=F requires that an 'a' action should not occur in the interval. An absolute constraint (represented by A) is an element of $\Lambda \times$ Time $\times$ Time $\rightarrow$ {T,F}. Relative constraints are expressed in terms of what actions must or must not happen in time intervals after a particular action has occurred. For example, R(a,b,t1,t2) = T requires a 'b' action to occur after t1 time units and within t2 time units after 'a' has occurred viz., t1 and t2 are to be measured after the occurrence of 'a'. Similarly, R(a,b,t1,t2)=F requires that a 'b' action should not occur in the duration interval [t1,t2] after 'a' has occurred. A relative constraint (represented by R) is an element of $\Lambda \times \Lambda \times$ Time $\times$ Time $\rightarrow$ {T,F}.

Example 1 *Consider the following process definitions: $A = (1)a + (2)a + (3)a + \ldots + (10)a$. $B = (2)b + (4)b + (6)b + (8)b + (10)b$. Let the constraint on the process $(A \mid B)$ be: R(a,b,0,5)=T. If an execution of $(A \mid B)$ is to satisfy the temporal constraint given one processing element, it cannot select (2)a and (10)b or (10)a and (1)b. The first selection violates the quantitative requirement, while the second violates ordering. However it can choose, (2)a and (4)b or (4)a and (8)b etc. So the available non-determinism is restricted by the temporal constraints.*

Example 2 *Periodic tasks which are common in real-time systems can be specified as: PT = a; B;b; PT , where a and b indicate the start and finish of the periodic task of periodicity P and B a purely sequential process (sequence of atomic actions). The constraints on it are $\forall$ i in 0 .. : A(a,i*P,(i+1)*P)=T which requires the task to start in the assigned period and $\forall$ i in 0 .. : A(b,i*P,(i+1)*P)=T which requires the task to finish appropriately.*

Example 3 *While a periodic task might be interrupted by a scheduler, (i.e., interleaved with other processes) there may be certain parts of the task which once started should finish quickly (such as reading a sensor which has continuous input). This can be specified as PT = a; B1; s; B2; f; B3; b; PT with a timing constraint R(s,f,0,ϵ)=T, where ϵ is the maximum permissible delay.*

Example 4 *Jitter control is also an important aspect in real-time systems and can be specified as: PT = a; B1; s; B2; b; PT with the constraint R(s,s,(P-ϵ_1), (P+ϵ_2))= T, where ϵ_1 and ϵ_2 represent the permissible jitter.*

3 Semantics

Given the set of basic actions (Λ), define a set of actions $Act = \Lambda \cup \{\tau, \delta\}$, where τ represents synchronization and δ idling (or stepping of time without executing an action). Let $\mathcal{P}$ be the set of all processes

Definition: 1 *Define a transition relation $\longrightarrow$ as the smallest sub-set of ($\mathcal{P} \times Act \times \mathcal{P}$), which satisfies the rules in figure 1.*

We do not define an operational semantics for the timing constraints; rather we define relations induced by them. The semantics of processes is almost identical to the usual

$$a;P \xrightarrow{a} P \qquad\qquad (t)P \xrightarrow{\delta} (t\text{-}1)P$$

$$\frac{P \xrightarrow{a} P'}{(0)P \xrightarrow{a} P'} \qquad\qquad \frac{P \xrightarrow{a} P' \quad a \neq \delta}{P+Q \xrightarrow{a} P',\ Q+P \xrightarrow{a} P'}$$

$$\frac{P \xrightarrow{\delta} P',\ Q \xrightarrow{\delta} Q'}{P + Q \xrightarrow{\delta} P' + Q'} \qquad\qquad \frac{P \xrightarrow{a} P',\ Q \xrightarrow{\bar{a}} Q',}{P \mid Q \xrightarrow{\tau} P' \mid Q'}$$

$$\frac{P \xrightarrow{a} P',\ Q \xrightarrow{\delta} Q',}{P \mid Q \xrightarrow{a} P' \mid Q' \quad Q \mid P \xrightarrow{a} Q' \mid P'} \qquad\qquad \frac{P \xrightarrow{a} P'}{P|Q \xrightarrow{a} P' \mid Translate(Q) \quad Q|P \xrightarrow{a} Translate(Q) \mid P'}$$

$$\frac{P \xrightarrow{a} P',\ a\ \bar{a} \notin H}{P \backslash H \xrightarrow{a} P' \backslash H} \qquad\qquad \frac{E_i(rec\ \widetilde{X}:\ \widetilde{E}/X) \xrightarrow{a} P'}{rec_i X{:}E \xrightarrow{a} P'}$$

Figure 1: Operational Semantics

semantics of timed CCS. The only difference is that we take a different view regarding the passing of time than [6]. As a waiting process does not need the processor, one can deduct time from it when executing other processes. Towards that we define a function called *Translate* as follows:

$$Translate(P) = \begin{cases} (t-1)Q & if\ P\ is\ (t)Q \\ Translate(P1) + Translate(P2) & if\ P\ is\ P1 + P2 \\ Translate(P1) \mid Translate(P2) & if\ P\ is\ P1 \mid P2 \\ Translate(Q) \setminus H & if\ P\ is\ Q \setminus H \\ rec\ \widetilde{X}: Translate(\widetilde{E}) & if\ P\ is\ rec\ \widetilde{X}{:}\widetilde{E} \\ P & otherwise \end{cases}$$

Example 5 *Notice that Translate forces the elapsing of time across choice. Our intuition behind the semantics can be interpreted as that the execution of an action can wait as long as the resource is not allocated. But delaying does not require the processor. Thus, a | ((1)b + c) after exhibiting a can exhibit b as the next action.*

An informal explanation of the transition rules follows. a;P performs 'a' and then behaves as P. Though not shown in the operational rules, this takes unit time. An execution of a step is treated as one clock tick. Hence if a process has to wait for time t, after a step it has to wait for only (t-1); but the processor was unused during that step. The rule for non-determinism exhibiting an action is as usual. However, if both branches of the choice can delay, the choice can be delayed. The rules that determine the behavior under parallel composition are as follows. The first two require the behaviors of processes P and Q to either be synchronizable or at least one of them is idling for the parallel composition to be successful. The third interleaves the execution. The rules for hiding and recursion are as usual.

4 Simulations

In this section we define simulations induced by temporal constraints. [6] show that strong bisimulation of a timed process (defined as: P $\sim$ Q iff P $\xrightarrow{a}$ then $\exists$ Q' such that Q $\xrightarrow{a}$ and P' $\sim$ Q' and if P $\overset{t}{\leadsto}$ P' then $\exists$ Q' such that Q $\overset{t}{\leadsto}$ Q' and P' $\sim$ Q' and vice versa) is a congruence. Thus, the definition requires identical timing of actions for two terms to be bisimilar.

Our concern here is to define a simulation which is more flexible hence which is substitutive in restricted contexts viz., the contexts defined by timing constraints. For example, if a takes 10 units of time in P and a takes 12 units of time in Q, and the constraint is that a occurs no later than 15 units, P can be replaced by Q and vice-versa. However this is not the case if the requirement is that a occurs no later than 11 units of time.

Clearly such a definition cannot be a congruence. What is the use of such a definition? We believe that our work will be relevant in two areas: 1) Reasoning about implementations satisfying specifications and 2) Fault-tolerant real-time systems.

The application of approximations to discuss implementations satisfying specifications is well understood. For example, it is required that any behavior exhibited by the implementation is permitted by the specification, but every behavior permitted by the specification need not be exhibited by the implementation. Real-time systems which have to be fault-tolerant clearly have to satisfy temporal constraints. Also, as they have to be fault tolerant, the occurrence of a fault requires one to 'replace' all the affected processes by 'equivalent' ones. Thus the equivalence need only be defined within the given system and the equivalence need not hold in general.

What are the possible scenarios? Consider two processes P and Q which are to satisfy the timing constraint A(a,0,n)=T. Let P perform an action different from 'a', and evolve to P', while Q be forced to delay for time t and become Q'. Q' should have the option of performing the same action as P and evolve to Q". How should P' and Q" be related? Firstly they must be observationally related. As P and Q should be related if and only if both satisfy the given constraint the same should hold for P' and Q" However as Q consumed time t units of time while P consumed only 1 unit of time, Q" should be able to produce an 'a' action in time (n-t) while P' should produce an 'a' action within time (n-1). So P' and Q" are no longer related by a single timing constraint. They are related by two constraints which differ only in the timing aspect of the constraint and not the observation part.

To accommodate the fact that one needs (bi)similarity under a pair of timing constraints, we define (bi)similarity as induced by a pair of timing constraints. In the next few paragraphs we introduce a definition of simulation indexed by a pair of constraints (indicated by $\precsim_{C1,C2}$). If $\precsim_{C1,C2}$ is a symmetric relation we write $\sim_{C1,C2}$. For notational convenience we write P $\lesssim_C$ Q, iff P $\lesssim_{C,C}$Q and P $\sim_C$ Q iff P $\sim_{C,C}$ Q. These definitions use observational simulation which is defined as follows.

Definition: 2 *Define* $P \overset{t}{\leadsto} Q$ *if there is a sequence of transitions rules such that* $P \overset{\delta}{\longrightarrow} P_1 \ldots P_{t-1} \overset{\delta}{\longrightarrow} Q$. *If t is 0, then P is identical to Q.*

Definition: 3 *Define* $P \lesssim Q$ *(i.e., P observationally simulates Q) iff* $P \overset{t_1}{\leadsto} P' \overset{a}{\longrightarrow} P'' \overset{t_2}{\leadsto} P_1$ *then* $\exists\, Q_1$ *such that* $Q \overset{t_3}{\leadsto} Q' \overset{a}{\longrightarrow} Q'' \overset{t_4}{\leadsto} Q_1$.

Definitions 4, 5, 6 and 7 define the four types of equivalences induced by the four types of timing constraints.

Definition: 4 $P \lesssim_{A(a,t1,t2)=T,A(a,t3,t4)=T} Q$ *iff*

1. *If* $P \overset{tp}{\leadsto} P' \overset{b}{\longrightarrow} P''$ *and* $tp \leq (t1-1)\ \exists\ Q \overset{tq}{\leadsto} Q' \overset{b}{\longrightarrow} Q''$ *and* $tq \leq (t3-1)$ *and* $P'' \lesssim_{A(a,x1,x2)=T,A(a,y1,y2)=T} Q''$ *where* $x1 = t1 - tp - 1, x2 = t2 - tp - 1, y1 = t3 - tq - 1, y2 = t4 - tq - 1$

2. *If* $P \overset{tp}{\leadsto} P' \overset{a}{\longrightarrow} P''$ *and* $(t1-1) \leq tp \leq (t2-1)\ \exists\ Q \overset{tq}{\leadsto} Q' \overset{a}{\longrightarrow} Q''$ *and* $(t3-1) \leq tq \leq (t4-1)$ *and* $P'' \lesssim Q''$

3. *If* $P \overset{tp}{\leadsto} P' \overset{b}{\longrightarrow} P''$ *and* $a \neq b$ *and* $(t1-1) \leq tp \leq (t2-1)\ \exists\ Q \overset{tq}{\leadsto} Q' \overset{b}{\longrightarrow} Q''$ *and* $(t3-1) \leq tq \leq (t4-1)$ *and* $P'' \lesssim_{A(a,x1,x2)=T,A(a,y1,y2)=T} Q''$ *where* $x1 = 0, x2 = t2 - tp - 1, y1 = 0, y2 = t4 - tq - 1$

The first rule relates the behavior before the interval while the second and third relates relevant behavior within the interval. Once the required action is observed then processes are only required to be 'observationally' related. Note that only processes which 'satisfy' the condition are related and thus P $\not\lesssim_{C(0,0),C(0,0)}$ Q as each action takes at least unit time.

Example 6 *Consider the two processes (10)a and (1)a + (3)a. They are bisimilar under the constraint A(a,0,11)=T; as both of them will always satisfy the timing constraint. Thus* $(10)a \sim_{A(a,0,11)=T} (1)a + (3)a$. *Similarly,* $(4)a \lesssim_{A(a,3,9)=T} (1)a + (6)a$

Proposition 1 $P \sim_{A(a,t1,t2)=T} Q$ *implies 1)* $P \sim_{A(a,t1-1,t2)=T} Q$ *and 2)* $P \sim_{A(a,t1,t2+1)=T} Q$

Example 7 *The above propositions are intuitive but would not be valid if P and Q were related if both violated the constraint as shown by the following example.* $(10)a \sim_{A(a,4,6)=T} (2)a$ *would be true if processes which violated the requirement were related. However, A(a,3,6) would be satisfied by (2)a but not (10)a and hence no longer bisimilar. Similarly (1)a would be bisimilar to (2)a under the above constraint but not under A(a,3,6). If processes which violate the condition have to be related one has to define similarity between the type and magnitude of error magnitude of error and is not considered here.*

Definition: 5 $P \lesssim_{A(a,t1,t2)=F,A(a,t3,t4)=F} Q$ *iff*

1. $P \lesssim Q$ *implies* $P \lesssim_{A(a,0,0)=F,A(a,0,0)=F} Q$

2. *If* $P \overset{tp}{\rightsquigarrow} P' \xrightarrow{b} P''$ $tp \leq (t1-1)$ $\exists\, Q \overset{tq}{\rightsquigarrow} Q' \xrightarrow{b} Q''$ *and* $tq \leq (t3-1)$ *and* $P'' \sqsubseteq\!\!\approx_{A(a,x1,x2)=F,A(a,y1,y2)=F} Q''$ *where* $x1 = t1 - tp - 1, x2 = t2 - tp - 1, y1 = t3 - tq - 1, y2 = t4 - tq - 1$

3. $P \overset{tp}{\rightsquigarrow} P' \xrightarrow{b} P''$ *and* $tp \geq t2$ *then* $\exists\, Q \overset{tq}{\rightsquigarrow} Q' \xrightarrow{b} P''$ *and* $tq \geq t4$ *and* $P'' \sqsubseteq\!\!\approx Q''$

4. $P \overset{tp}{\rightsquigarrow} P' \xrightarrow{b} P''$ *and* $a \neq b$ *and* $(t1-1) \leq tp \leq (t2-1)$ *then* $\exists\, Q \overset{tq}{\rightsquigarrow} Q' \xrightarrow{b} Q''$ *and* $(t3-1) \leq tq \leq (t4-1)$ *and* $P'' \sqsubseteq\!\!\approx_{A(a,x1,x2)=F,A(a,y2,y2)=F} Q''$ *where* $x1 = 0, x2 = t2 - tp - 1, y1 = 0, y2 = t4 - tq - 1$

The first rule relates processes which are observationally related. As all actions take at least unit time, no action can occur in the interval [0,0]. The second and third rules relate processes outside the interval, while the final rule requires that the desired action not occur in the interval.

Proposition 2 *$P \sim_{A(a,t1,t2)=F} Q$ implies 1) $P \sim_{A(a,t1+1,t2)=F} Q$ and 2) $P \sim_{A(a,t1,t2-1)=F} Q$.*

Example 8 *It is easy to check that $(1)a + (2)a + (10)a \sim_{A(a,5,9)=F} (3)a + (12)a$. Note that $(2)a$ is not bisimilar under the above constraint to $(12)a$, as the definition requires that behavior before and after the range be similar.*

This finishes the definitions for the absolute case. Now we turn our attention to the relative case.

Definition: 6 $P \sqsubseteq\!\!\approx_{R(a,b,t1,t2)=T,R(a,b,t3,t4)=T} Q$ *iff*

1. $P \overset{tp}{\rightsquigarrow} P' \xrightarrow{a} P''$ $\exists\, Q \overset{tq}{\rightsquigarrow} Q' \xrightarrow{a} Q''$ *and* $P'' \sqsubseteq\!\!\approx_{NC} Q''$ *and* $P'' \sqsubseteq\!\!\approx_{R(a,b,t1,t2)=T,R(a,b,t3,t4)=T} Q''$ *where NC is* $A(b,t1,t2)=T, A(b,t3,t4)=T$

2. $P \overset{tp}{\rightsquigarrow} P' \xrightarrow{c} P''$ *and* $a \neq c$ $\exists\, Q \overset{tq}{\rightsquigarrow} Q' \xrightarrow{c} Q''$ *and* $P'' \sqsubseteq\!\!\approx_{R(a,b,t1,t2)=T,R(a,b,t3,t4)=T} Q''$

The first condition starts the clock when an a occurs. The processes after the result of an 'a' action are required to be similar under two conditions. The first condition (NC) is that b occurs within the specified time while the second requires that the relative constraint be satisfied in future. The second condition in the definition ensures conditional similarity in the future if an 'a' action was not performed.

Proposition 3 *$P \sim_{R(a,b,t1,t2)=T} Q$ implies 1) $P \sim_{R(a,b,t1-1,t2)=T} Q$ and 2) $P \sim_{R(a,b,t1,t2+1)=T} Q$*

Definition: 7 $P \sqsubseteq\!\!\approx_{R(a,b,t1,t2)=F,R(a,b,t3,t4)=F} Q$ *iff*

1. $P \overset{tp}{\rightsquigarrow} P' \xrightarrow{a} P''$ $\exists\, Q \overset{tq}{\rightsquigarrow} Q' \xrightarrow{a} Q''$ *and* $P'' \sqsubseteq\!\!\approx_{NC} Q''$ *and* $P'' \sqsubseteq\!\!\approx_{R(a,b,t1,t2)=F,R(a,b,t3,t4)=F} Q''$ *where NC is* $A(b,t1,t2)=F, A(b,t3,t4)=F$

2. $P \overset{tp}{\rightsquigarrow} P' \xrightarrow{c} P''$ *and* $a \neq c$ *then* $\exists\, Q \overset{tq}{\rightsquigarrow} Q' \xrightarrow{c} Q''$ *and* $P'' \sqsubseteq\!\!\approx_{R(a,b,t1,t2)=F,R(a,b,t3,t4)=F} Q''$

As in the previous definition, the first condition starts the clock when an 'a' occurs (but now disallowing 'b'), while the second ensures conditional similarity in the future.

Proposition 4 *$P \sim_{R(a,b,t1,t2)=F} Q$ implies 1) $P \sim_{R(a,b,t1+1,t2)=F} Q$ and 2) $P \sim_{R(a,b,t1,t2-1)=F} Q$*

Proposition 5 *Though $\sim_C$ is not a congruence, the following hold.*

- *if $P \stackrel{\sqsubset}{\sim}_{C(t1,t2)} Q$, where $C(t1,t2)$ is $A(a,t1,t2)=T$ then*
  1) $(t)P \stackrel{\sqsubset}{\sim}_{C(t1+t,t2+t)} (t)Q$ 2) $b;P \stackrel{\sqsubset}{\sim}_{C(t1+1,t2+1)} b;Q$
  3) $P+R \stackrel{\sqsubset}{\sim}_{C(t1,t2)} Q+R$ 4) $P \mid R \stackrel{\sqsubset}{\sim}_{C(t1,t2)} Q \mid R$

- *if $P \stackrel{\sqsubset}{\sim}_{C(t1,t2)} Q$, where $C(t1,t2)$ is $A(a,t1,t2)=F$ then*
  1) $(t)P \stackrel{\sqsubset}{\sim}_{C(t1+t,t2+t)} (t)Q$ 2) $b;P \stackrel{\sqsubset}{\sim}_{C(t1+1,t2+1)} b;Q$
  3) $P+R \stackrel{\sqsubset}{\sim}_{C(t1,t2)} Q+R$ 4) $P \mid R \stackrel{\sqsubset}{\sim}_{C(t1,t2)} Q \mid R$

- *if $P \stackrel{\sqsubset}{\sim}_{C(t1,t2)} Q$, where $C(t1,t2)$ can either be $R(a,b,t1,t2)=T$ or $R(a,b,t1,t2)=F$ then*
  1) $(t)P \stackrel{\sqsubset}{\sim}_{C(t1,t2)} (t)Q$ 2) $c;P \stackrel{\sqsubset}{\sim}_{C(t1,t2)} c;Q$ and $c \neq a$
  3) $P+R \stackrel{\sqsubset}{\sim}_{C(t1,t2)} Q+R$ 4) $P \mid R \stackrel{\sqsubset}{\sim}_{C(t1,t2)} Q \mid R$

5 Logic

In this section, we introduce an extension of the Hennessy-Milner logic with recursion [3] to handle time. The formulae in our logic are defined as follows

$$\mathrm{L} = \, < a > \mathrm{L} \mid \bigwedge_{i \in I} \mathrm{L}_i \mid \neg \mathrm{L} \mid [\mathrm{t}]\mathrm{L} \mid \{\mathrm{t}\}\mathrm{L} \mid \mathrm{X} \mid \nu\mathrm{X{:}L}$$

True is defined to be the conjunction over an empty index set, while Terminated is defined to be the conjunction over the set of actions of ($\neg < a >$ True).

The formulae are interpreted on processes and is defined as follows. Let $\mathcal{P}$ represent the set of all processes.

$$[\![< a > \mathrm{L}\,]\!] = \{\, \mathrm{P} \in \mathcal{P} \;\; \exists\, \mathrm{P'},\, \mathrm{P} \xrightarrow{a} \mathrm{P'} \text{ and } \mathrm{P'} \in [\![\mathrm{L}\,]\!] \}$$

$$[\![\bigwedge_{i \in I} \mathrm{L}_i \,]\!] = \bigcap_{i \in I} [\![\mathrm{L}_i\,]\!]$$

$$[\![\neg \mathrm{L}\,]\!] = \mathcal{P} \text{ - } [\![\mathrm{L}\,]\!]$$

$$[\![[\mathrm{t}]\,\mathrm{L}\,]\!] = \{\, \mathrm{P} \in \mathcal{P} \;\; \exists\, \mathrm{Q}{:}\ \mathrm{P} \xrightarrow{a_1} \mathrm{P}_1 \ldots \mathrm{P}_{t-1} \xrightarrow{a_t} \mathrm{Q} \text{ and } \mathrm{Q} \in [\![\mathrm{L}\,]\!] \}$$

$$[\![\{\mathrm{t}\}\mathrm{L}\,]\!] = \{\, \mathrm{P} \in \mathcal{P} \;\; \exists\, \mathrm{Q}{:} \text{ with } \mathrm{n} \leq \mathrm{t},\ \mathrm{P} \xrightarrow{a_1} \mathrm{P}_1 \ldots \mathrm{P}_{n-1} \xrightarrow{a_n} \mathrm{Q} \text{ and } \mathrm{Q} \in [\![\mathrm{L}\,]\!] \}$$

$$[\![\nu\mathrm{X{:}L}\,]\!] = \cup\{\, \mathrm{Pr} \subseteq \mathcal{P} \text{ such that } [\![\mathrm{L}\,]\!] \supseteq \mathrm{Pr} \,\}$$

The intuition in using [t] and {t} is that the first characterizes the passing of exactly t units of time while {t} characterizes behavior with in the interval [0,t]. The other components have their usual meaning.

The temporal constraints that we have used can be translated into the above logic as follows.

A(a,t1+1,t2+1)=T = $[t1]\{t2\}< a >$ True

A(a,t1+1,t2+1)=F = $\neg([t1]\{t2\}< a >$ True)

R(a,b,t1+1,t2+1)=T = νC: ((< a > True $\rightarrow$ [t1]{t2}< b > True)) $\wedge$ (Terminated $\vee$ [1] C)

R(a,b,t1+1,t2+1)=F = νC: ($\neg$(< a > True $\wedge$ [t1]{t2}< b > True) $\wedge$ (Terminated $\vee$ [1] C)

As expected, R(a,b,t1,t2)=T/F represent safety properties and map to a maximal fixed point formula.

Proposition 6 *$P \sim_C Q$ implies that $P \in [\![L]\!]$ and $Q \in [\![L]\!]$ where L is the translation of C.*

Proposition 7 *($\forall$ L, ($P \models L$) iff ($Q \models L$)) implies $P \sim_{st} Q$. where $\sim_{st}$ represents strong timed bisimulation.*

6 System Issues

The above definitions can be used to develop a system for reasoning about timed processes. We outline how the concepts of schedulers can be formalized. Note that we do not present a language in which schedulers can be defined.

Definition: 8 *A scheduler, Sch, is a function which given a process P, yield a process such that Sch(P) $\precsim$ P.*

Schedulers as defined above, can be considered to be implementors of a specification [7]. The general problem of optimal scheduling is usually NP-complete and hence very rarely used. Usually, one either uses a scheduler which is 'satisfactory' or requires more information from the process. Thus, we do *not require* an implementation to satisfy the timing constraints specified for a system. However, the notion of equivalence is considered only for schedulers which can satisfy a given constraint, i.e., as there are many ways of implementing a specification, it is natural to identify similar satisfactory ones.

Definition: 9 *Two schedulers S1 and S2 are defined to be similar with respect to a process P and timing constraint C written as ($S1 \sim_{P,C} S2$), iff $S1(P) \sim_C S2(P)$.*

Definition: 10 *Process P and Q are similar under scheduler Sch and a given constraint C written as ($P \sim_{Sch,C} Q$), iff $Sch(P) \sim_C Sch(Q)$.*

Example 9 *$a|b$ is not similar to $b|a$ under A(a,0,1)=T and a 'FCFS' scheduler defined as follows scheduler($p|q$) = schedule(p);schedule(q) and schedule(a) = a*

Proposition 8 *Given a timing constraint C. If $S1 \sim_{P,C} S2$ and $P \sim_{S1,C} Q$ then $S1 \sim_{Q,C} S2$ iff $P \sim_{S2,C} Q$.*

We extend the above definitions to capture the process of translating a program in a high level language into a more low level language in the context of developing a real-time system. We consider a compilation as converting a program in a some language into a process in RTCCS. We assume that this process of compilation has knowledge of the architecture and hence assigns times to each action, i.e., prefixes each action by (n) for some n. It could assign a range by using the non-deterministic operator. Thus, a compiler might assign times ranging from 5 to 10 for an action a, while another might assign times ranging from 20 to 30.

Given a program (with timing constraints), and a scheduler one can define 'equivalence' of compilations as follows.

Definition: 11 *Given a program Pgm, a timing constraint C and a scheduler S. Compiler1* $\sim_{C,Pgm,S}$ *Compiler2 iff S(Compiler1(Pgm))* $\sim_{C}$ *S(Compiler2(Pgm))*

Proposition 9 *Given a program Pgm and timing constraint C and schedulers S1 and S2. If two compilers Cmp1* $\sim_{C,Pgm,S1}$ *Cmp2 and S1* $\sim_{Cmp1(Pgm),C}$ *S2 then Cmp1* $\sim_{C,Pgm,S2}$ *Cmp2 iff S1* $\sim_{Cmp2(P),C}$ *S2.*

Acknowledgment

The author is grateful to Uffe Engberg and Peter Mosses for their comments and encouragement.

References

[1] F. Jahanian and A. K. Mok. A graph-theoretic approach for timing analysis and its implementation. *IEEE Transactions on Computers*, pages 961–975, August 1987.

[2] M. Joseph and A. Goswami. What's 'Real' about real-time systems? In *IEEE Real-Time Systems Symposium*, pages 78–85, 1988.

[3] K. G. Larsen. Proof systems for Hennessy-Milner logic with recursion. In *13th Colloquim on Trees in Algebra and Programming*. Springer Verlag, 1988.

[4] R. Milner. *A Calculus of Communicating Systems*. Lecture Notes on Computer Science Vol. 92. Springer Verlag, 1980.

[5] R. Milner. *Communication and Concurrency*. Prentice Hall International, 1989.

[6] F. Moller and C. Tofts. A temporal calculus of communicating systems. In J. C. M. Baeten and J. W. Klop, editors, *CONCUR 90, LNCS-458*. Springer Verlag, 1990.

[7] E. R. Olderog and C. A. R. Hoare. Specification-oriented semantics for communicating processes. In *ICALP -83, LNCS 154*. Springer Verlag, 1983.

[8] D. Park. Concurrency and automata on infinite sequences. In *Proceedings of the 5th GI Conference, LNCS-104*. Springer Verlag, 1981.

[9] W. Yi. Real-time behavior of asynchronous agents. In J. C. M. Baeten and J. W. Klop, editors, *CONCUR 90, LNCS-458*. Springer Verlag, 1990.

ON STRICT CODES

Nguyen Huong Lam and Do Long Van
Institute of Mathematics *)

Abstract. This paper continues an earlier paper of the authors. The maximality, the decomposability etc. for infinitary strict codes are considered. Every infinitary strict code is shown to be included in an infinitary strict-maximal one. The so-called Theorem of Defect for infinitary strict codes is proved. Some conditions for an infinitary strict code to be written by an indecomposable one are stated.

1. PRELIMINARIES

The concept of strict codes has been first mentioned in [3]. The classical definition of a code says that the (finite) identity relations are the only (finite) relations satisfied by the code. For a strict code the demand is stronger: any relation, finite or infinite, which is satisfied by the code is an identity one. So, strict codes form a subclass of codes. In [7] a particular case of strict codes, namely, that of finitary strict codes was considered. In our paper [6] we studied infinitary strict codes; we proposed some procedures to verify whether a given language is a strict code. Also, we characterized strict codes by ∞-submonoids generated by them. In the present article, which is a sequel to [6], we mostly adapt some well-known notions and properties of codes for strict ones such as maximality, decomposability, Theorem of Defect, etc.

In what follows we mostly use standard terminology and notation (see, e.g. [5], [1]). Let A be an alphabet, finite or countable, A^* the free monoid generated by A whose elements are called finite words. We denote A^N the set of infinite words over A and $A^\infty = A^* \cup A^N$ whose elements we call simply words. We make A^∞ a monoid equipping it with the product defined as:

For any words α, β of A^∞,

$$\alpha.\beta = \begin{cases} \alpha & \text{if} \quad \alpha \in A^N, \beta \in A^\infty \\ \alpha\beta & \text{if} \quad \alpha \in A^*, \beta \in A^\infty \end{cases}$$

where $\alpha\beta$ means the catenation of α and β (see [6]). Clearly, the empty word, denoted by ϵ, is the unit of A^∞.

We call a subset X of A^∞ (respectively, of A^*) an infinitary language (respectively, finitary language). For a finite subset X, $\mathrm{Card}\, X$ denotes its cardinality; also, to simplify the notation we often identify a singleton set with its element. For a word $x \in A^*, |x|$ denotes the length of x and we say by convention that $|\epsilon| = 0$ and $|x| = \omega$ if $x \in A^N$.

For any infinitary language X we denote:

$X_{\mathrm{fin}} = X \cap A^*, X_{\mathrm{inf}} = X \cap A^N$

$XY = \{\alpha\beta : \alpha \in X, \beta \in Y\}, \quad Y \subseteq A^\infty$

- the product extended to languages.

$X^2 = XX$

$X^{n+1} = XX^n, \quad n = 1, 2, \ldots$

$X^* = \bigcup_{n\geq 0} X^n$
- the smallest submonoid of A^∞ containing X.

$X^+ = X^* - \epsilon$

$X^\omega = \{x_1x_2\cdots : x_i \in X_{\text{fin}}, i = 1, 2, \ldots\}$
- the set of all infinite products of elements in X_{fin}.

$X^\infty = X^* \cup X^\omega$

$X^{+\infty} = X^\infty - \epsilon.$

For every $n \geq 1$ we introduce the set $X_{(n)}$ of n-tuples defined as:

$X_{(n)} = \{(x_1, x_2, \ldots, x_n) : n \geq 1, x_1, \ldots, x_{n-1} \in X_{\text{fin}}, x_n \in X\}$

and the set $X_{(\omega)}$ of ω-tuples defined as:

$X_{(\omega)} = \{(x_1, x_2, \ldots) : x_i \in X_{\text{fin}}, i = 1, 2, \ldots\}$

and we put

$X_{(*)} = \bigcup_{n\geq 1} X_{(n)}$

$X_{(\infty)} = X_{(*)} \cup X_{(\omega)}$

We say that a word x of A^∞ admits a $*$-*factorization* (resp. an ω-*factorization*) $(x_1, x_2, \ldots)$ over X provided $x = x_1x_2\ldots$ with $(x_1, x_2, \ldots) \in X_{(*)}$ (resp. $(x_1, x_2, \ldots) \in X_{(\omega)}$); we say that X admits an ∞-*factorization* over X if it admits either a $*$-factorization or an ω-factorization over X. A given subset X is said to be an *infinitary code* (resp. an *infinitary strict code*) if every word of A^∞ admits at most one $*$-factorization (resp. one ∞-factorization) [6].

Finally, for any two subsets X, Y of A^∞, we define :

$XY^{-1} = \{\alpha \in A^\infty : \exists\beta \in Y : (\alpha\beta \in X) \wedge ((|\alpha| = \omega) \Rightarrow \beta = \epsilon)\}$

$Y^{-1}X = \{\alpha \in A^\infty : \exists\beta \in Y : (\beta\alpha \in X) \wedge ((|\beta| = \omega) \Rightarrow \alpha = \epsilon)\}.$

2. MAXIMALITY

In this section we consider maximality properties of strict codes and ∞-submonoids generated by them. First, we show that each strict code is included in a strict-maximal one. A strict code X is called *strict-maximal* if it is not contained properly in any other strict code. X is called *relatively strict-maximal* if for every finite word $w \in A^*, X \cup w$ is no more a strict code.

THEOREM 2.1. *Every strict code is contained in a strict-maximal one over A.*

PROOF: First, we prove that every strict code is included in a relatively strict-maximal one. To do this we enumerate all finite nonempty words in some order

$$A^+ = \{w_1, w_2, \ldots\}$$

and define an increasing sequence of strict codes $X_0 \subseteq X_1 \subseteq X_2 \subseteq \ldots$ as follows :

Put $X_0 = X$ and suppose for some $n \geq 0, X_n$ has been defined. Let $i(n)$ be the smallest integer such that $X_n \cup w_{i(n)}$ is a strict code (if no such $i(n)$ exists, we put $X_{n+1} = X_n$). Put $X_{n+1} = X_n \cup u_n$, where u_n is any word in $\{X_n \cup w_{i(n)}\}^* w_{i(n)}^\omega$. Since $X_n \cup w_{i(n)}$ is a strict code, so is X_{n+1}. Thus X_{n+1} is defined and by induction our sequence is built.

Consider the set $Y = \bigcup_{n\geq 1} X_n$. Since $Y = X \cup \{u_n : n \geq 0\}$ and all u_n's are infinite words it follows that every ∞-factorization over Y is also one over X_n for some n. This means that Y is a strict code because each X_n is a strict code. Also, it is easy to see that Y is relatively strict-maximal by construction of the sequence X_n. Thus every strict code is contained in a relatively strict-maximal one.

Next, we prove that the class of relatively strict-maximal codes is inductive, i.e. every chain (by inclusion) has an upper bound. Indeed, let

$$X_\alpha \subseteq X_\beta \subseteq X_\gamma \subseteq \ldots$$

be a chain in this class, indexed by a set I. Since each member of this chain is relatively strict-maximal, we have

$$X_{\alpha\,\mathrm{fin}} = X_{\beta\,\mathrm{fin}} = X_{\alpha\,\mathrm{fin}} = \ldots.$$

Putting

$$Z = \bigcup_{\gamma \in I} X_\gamma$$

we have $Z_{\mathrm{fin}} = X_{\gamma\,\mathrm{fin}}$ for every $\gamma \in I$ that means Z is relatively strict-maximal and thus Z is an upper bound of the chain. So, by Zorn's Lemma, every relatively strict-maximal code is included in a maximal one, but it is easy to see that every maximal element of this class is a strict-maximal code. Theorem is proved.

It has been known that there exists an algorithm to decide whether a given finitary language is a maximal code (in the class of finitary codes) [1]. Below we state a similar result for finitary strict codes. Recall that a finitary language X is said to be *complete* if for a.e. w of A^* : $A^*wA^* \cap X^* \neq \emptyset$. A word w of A^* is called *overlapping* if $w = ux = yu$ for some $u, x, y \in A^+$. It is easy to see that for every w of A^* (Card $A \geq 2$) there exist x and y of A^* such that xwy is not overlapping. Consider now in the class of finitary strict codes a maximal one by inclusion. We call such a code *finitary-strict-maximal*, or, following just defined terminology, X is finitary relatively strict-maximal code. We have then :

PROPOSITION 2.2. *Every finitary-strict-maximal code X is complete.*

PROOF: Let $w \in A^* - X$. As noted $xwy = w'$ is not overlapping for some x, y in A^*. If $w' \in X^*$, nothing is to be proved. Suppose $w' \notin X^*$, then $X \cup w'$ is not a strict code, it has to exist an infinite equality

$$x_1 x_2 \cdots = y_1 y_2 \ldots$$

over $X \cup w'$ with $x_1 \neq y_1$. Let $x_k = w', k \geq 1$ and m and n be respectively the largest and the smallest integers such that $x_k = w'$ is a subword of $y_m y_{m+1} \cdots y_n$. Since w' is not overlapping, there is no w' among $y_m, \ldots, y_n$ meaning that they are all in X. Consequently, w' is a subword of $y_m y_{m+1} \cdots y_n \in X^*$ and so is w. This completes the proof.

As a consequence of Proposition 2.2., we have

THEOREM 2.3. *There exists an algorithm to decide whether a given finitary finite subset X is a finitary-strict-maximal code*

PROOF: First, by Theorem 2.6. of [6], we can verify whether X is a strict code. Second, if X is finitary-strict-maximal code it must be complete. But it has been known that a finite (strict) code is complete if and only if it is maximal as a code (and therefore finitary-strict-maxial as a strict code). Thus it suffices to test the maximality of X as a code and this could be done, for example, by means of a Bernouille distribution. Because X is finite the test is always effective. Theorem is proved.

We now return to the general framework of infinitary words and languages. We state some properties of strict-maximal codes analogous to the case of ordinary ones, but let us first define some notions.

A language $X \subseteq A^\infty$ is said to be *dense* if for every α of $A^\infty, A^\infty \alpha A^\infty \cap X \neq \emptyset$, or, which amounts to the same, for every w of A^N : $A^* w \cap X_{\text{inf}} \neq \emptyset$; X is said to be *complete* (resp. *weakly complete*) if X^* (resp. X^∞) is dense.

THEOREM 2.4. *Every strict-maximal code is weakly complete.*

PROOF: If the alphabet A is a singleton $A = \{a\}$ then a language X is a strict code if and only if $X = \{a^n\}$ for some positive interger n or $X = \{a^\omega\}$. So every strict code is strict-maximal and weakly complete in this case.

Suppose now $\text{Card}\, A \geq 2$ and X is a strict-maximal code. We prove that $A^* \alpha \cap X_{\text{inf}} \neq \emptyset$ for every α in A^N. If $\alpha \in X_{\text{inf}}$, we are done, otherwise $X \cup \alpha$ is not a strict code, so that we have an equality :

$$(1) \qquad x_1 x_2 \cdots = y_1 y_2 \cdots$$

with two possibilities :

(i) $(x_1, x_2, \ldots) \in (X \cup \alpha)_{(n)}$ and $(y_1, y_2, \ldots) \in (X \cup \alpha)_{(m)}$ for $m, n \geq 1$.

(ii) $(x_1, x_2, \ldots) \in X_{(\omega)}$ and $(y_1, y_2, \ldots) \in (X \cup \alpha)_{(m)}$.

If (ii) holds, we are done. Now suppose that we have (i). If $x_n \neq \alpha$, we are through again, otherwise, from (1) and from the fact that α must occur among y_i's, we have $y_m = \alpha$. Hence $\alpha = p^\omega$ for some primitive word $p \in A^*$. We choose a letter b different from the last letter of p. Consider the word $bp^\omega = b\alpha$, which we can suppose not to belong to X (in the contrary, we are done), therefore the set $X \cup bp^\omega$ is not a strict code. But now with bp^ω playing the role of α, we have the equality (1) for $X \cup bp^\omega$. The case (i) with $x_n = bp^\omega = b\alpha = y_m$ is already impossible, otherwise $bp^\omega = q^\omega$ for some another primitive word q. Certainly $q = bq', q' \in A*$, hence $p^\omega = (q'b)^\omega$. Since $q'b$ is also primitive, we get $p = q'b$ which contradicts the fact that the last letter of p is not b. So we have now either (i) with $x_n \neq b\alpha, y_m = b\alpha$ (or $x_n = b\alpha, y_m \neq b\alpha$) or (ii). Consequently, $A^* b\alpha \cup X \neq \emptyset$ and the theorem follows.

It is well-known that every recognizable complete finitary code is a maximal one, but we cannot state such an analog for strict codes as shown in the next example.

Example 2.5. Consider the language $X = \{a^2, bA^\omega\}$ over the binary alphabet $A = \{a, b\}$. It is easy to see that X is a recognizable weakly complete strict code (even more so, complete one), but not a strict-maximal code, since $X \cup abA^\omega$ is still a strict code.

Given two languages X, Y of A^∞, X is said to be *written* by Y, in notation $X \prec Y$, if $X \subseteq Y^\infty$ and no proper subset Z of Y has this properties i.e. $\forall Z \subset Y, X \not\subseteq Z^\infty$. It is easy to verify that in the class of strict codes $\prec$ is a partial order. A strict code X is said to be *indecomposable* over A if for any strict code Y, $X \prec Y$ implies $Y = X$ or $Y \subseteq A$.

Example 2.6. (i) Consider the strict code $X = \{a^2, b, (ba)^\omega\}$ over the alphabet $A = \{a, b\}$; X is not indecomposable because $X \prec \{a^2, b, (ab)^\omega\}$.
(ii) Consider now the strict code $X = \{b^3a, b^2, a, abab^\omega\}$. We show that X is indecomposable. Indeed, if X is written by a strict code $Y \not\subseteq A$ then Y_{fin} equals either $\{b^3a, b^2, a\}$ or $\{ba, b^2, a\}$, and $abab^\omega = y_1y_2$ for some $y_1 \in Y_{\text{fin}}^*, y_2 \in Y_{\text{inf}}$ (the case $abab^\omega \in Y_{\text{fin}}^\omega$ is impossible). If $Y_{\text{fin}} = \{b^3a, b^2, a\}$ then y_1 must be ϵ, otherwise $y_2 = bab^\omega$, which is impossible because we would have then $b^3a.(b^2)^\omega = b^2y_2$. Hence $abab^\omega = y_1y_2 = y_2 \in Y$ meaning that $Y = X$.

If now $Y_{\text{fin}} = \{ba, b^2, a\}$, then $y_1 = \epsilon, y_1 = a$ or $y_1 = aba$, correspondingly $y_2 = abab^\omega, y_2 = bab^\omega$, or $y_2 = b^\omega$. In all cases, it is easy to see that Y is not a code, which is a contradiction.

We recall some notations introduced in [6]. A subset M of A^∞ is called ∞-*submonoid* if $M^\infty = M$ and it is called freeable if $M^{-1}M \cap MM^{-1} = M$. Every subset X of an ∞-submonoid M such that $X^\infty = M$ is called a *generator set* of M. It was proved in [6] that every ∞-submonoid possesses a smallest generator set in the sense that it is contained in every generator set of M and freeable ∞-submonoids are always generated by strict codes which are the smallest generator sets of them. An ∞-submonoid of A^∞ is said to be *freeable-maximal* provided it is not contained in any freeable ∞-submonoid other than itself and A^∞.

We state now a result relating indecomposability of a strict code and freeable-maximality of the ∞-submonoid generated by it.

THEOREM 2.7. *An ∞-submonoid M is freeable-maximal if and only if it is generated by an indecomposable strict-maximal code.*

PROOF: Let M be freeable-maximal and X its smallest generator set which is a strict code. If X is not strict-maximal, then there exists a word x of $A^\infty - X$ such that $X \cup x$ is a strict code. Hence $M = X^\infty \subset (X \cup x)^\infty$ which implies $(X \cup x)^\infty = A^\infty$. Since $X \cup x$ is a strict code, we have $X \cup x = A$, therefore x belongs to A. On the other hand $X \cup x^\omega$ is also a strict code and we have

$$M \subset (X \cup x^\omega)^\infty \subset A^\infty$$

i.e. M is not freeable-maximal: a contradiction. That X is not indecomposable also leads to a contradiction. Indeed, if $X \prec Y$ and $X \neq Y, Y \not\subseteq A$ then $M \subset Y^\infty \subset A^\infty$ which is a contradiction.

To prove the converse, we suppose that X is an indecomposable strict-maximal code and M' is a freeable ∞-submonoid such that $M \subseteq M' \subseteq A^\infty$. This yields $X \subseteq X^\infty = M \subseteq M' = X'^\infty$ with X' being a strict code generating M'. Thus $X \prec X$"

for some subset $X" \subseteq X'$. We show that $X"$ is also a strict-maximal code. If it is not so, then $X" \cup x$ is a strict code for some x in $A^\infty - X"$. Since X is strict-maximal, $X \cup x$ ($x \notin X$, because $x \notin (X")^\infty \supseteq X$) is not a strict code, we have then two different ∞-factorizations $(x_1, x_2, \dots)$ and $(y_1, y_2, \dots)$ over $X \cup x$ of some word of A^∞. Since every x_i, y_i that differ from x are in $X \subseteq X"^\infty$ they admit then ∞-factorizations over $X"$. Now, we replace every entry $x_i \neq x$ and $y_i \neq x$ in $(x_1, x_2, \dots)$ and $(y_1, y_2, \dots)$ with their ∞-factorizations over $X"$ and as a result we obtain two ∞-factorizations over $X" \cup x$ of the same word. Since $X" \cup x$ is a strict code, they must be identical, from which it follows that either $x \in X"$ or X is not a strict code. This contradiction shows that $X"$ must be strict-maximal, therefore $X" = X'$ and by indecomposability of X we have either $X' = X$ or $X' \subseteq A$. Hence $X' = X$ or $X' = A$, in other words, $M' = M$ or $M' = A^\infty$. The proof is completed.

Example 2.8. Consider the subset $M = \{w \in A^\infty : |w| \geq p\}, p$ is a prime number$\}$. It is clear that M is a freeable ∞-submonoid generated by the uniform (strict) code A^p which is strict-maximal and indecomposable. Therefore M is freeable-maximal by Theorem 2.7.

3. DECOMPOSITION

In this section we study the relation $\prec$, namely, we are concerned with the question, does there exist for an arbitrary infinitary strict code X an indecomposable one by which X is written? Such a problem is not to be posed for finitary codes simply because the Zorn's Lemma guarantees this for them. In general, we do not know the answer to this question, but below we state some conditions under which a strict code can be written by an indecomposable one.

THEOREM 3.1. *If X is a strict code with X_{fin} a finite finitary maximal code then X can be written by an indecomposable strict code.*

PROOF: First, we observe that for each strict code Y such that $X \prec Y$, we have $X_{\text{fin}} \subseteq Y^*_{\text{fin}}$ and it is not difficult to see that if X_{fin} is finite maximal code so is Y_{fin} and $X_{\text{fin}} \prec Y_{\text{fin}}$ (see [5], for example). Next, we denote $S(X) = \{Y \subseteq A^\infty : Y \text{is a strict code} : X \prec Y\}$. Also, we denote $\| Y \| = \sum_{y \in Y_{\text{fin}}} | y |$, so that for each $Y \in S(X)$ we have $\| Y \| < \infty$. Let N be the smallest value of $\| Y \|$ as Y runs through $S(X) : N = \min\{\| Y \| : Y \in S(X)\}$. We can see that if $Y_1 \prec Y_2$ with Y_1, Y_2 in $S(X)$ then $\| Y_1 \| \geq \| Y_2 \|$. Let Y be a strict code of $S(X)$ with $\| Y \| = N$. We show that Y is written by an indecomposable strict code and since $\prec$ is an order relation, so does X.

Consider $S(Y)$. Certainly $S(Y) \subseteq S(X)$. We use Zorn's Lemma to show that $S(Y)$ contains a maximal element which, therefore, is an indecomposable strict code and by which X can be written. For each $Z \in S(Y) \subseteq S(X)$, we have $X \prec Y \prec Z$ and $\| Z \| \leq \| Y \|$. As noted above $Y_{\text{fin}} \prec Z_{\text{fin}}$, and thus both Z_{fin}, Y_{fin} are maximal codes and since $\| Y \|$ is of minimum value, it follows that $\| Z \| = \| Y \|$ and $Z_{\text{fin}} = Y_{\text{fin}}$.

Consider first an arbitrary countable chain in $S(Y) : Y_1 \prec Y_2 \prec \dots$. We have $Y_{1\,\text{fin}} = Y_{2\,\text{fin}} = \dots$. For each $s \geq 1$ and $x \in Y_{s\,\text{inf}}, x$ does not belong to $Y^\omega_{s+1\,\text{fin}}$, since $Y_{s+1\,\text{fin}} = Y_{s\,\text{fin}}$ and Y_s is a strict code. From $Y_s \prec Y_{s+1}$ it follows

$$x = x^{(s+1)}_{\text{fin}} x^{(s+1)}_{\text{inf}},$$

where $x_{\text{fin}}^{(s+1)} \in Y_{s+1\,\text{fin}}^*, x_{\text{inf}}^{(s+1)} \in Y_{s+1\,\text{inf}}$. By the same argument, we have

$$x_{\text{inf}}^{(s+1)} = x_{\text{fin}}^{(s+2)} x_{\text{inf}}^{(s+2)},$$

where $x_{\text{fin}}^{(s+2)} \in Y_{s+2\,\text{fin}}^*, x_{\text{inf}}^{(s+2)} \in Y_{s+2\,\text{inf}}$, and so on. Clearly, there exist only finitely many i such that $x_{\text{fin}}^{(s+i)} \neq \epsilon$, otherwise $x \in Y_{\text{fin}}^{\omega}$. Thus there must be an integer $n(x,s)$ such that for a.e. $m \geq n(x,s)$

$$x_{\text{inf}}^{(m)} = x_{\text{inf}}^{(m+1)} = x(s).$$

Therefore $x(s) \in Y_{m\,\text{inf}}$ for a.e. $m \geq n(x,s)$.

Now, consider the set P of A^∞ with $P_{\text{fin}} = Y_{1\,\text{fin}}, P_{\text{inf}} = \{x(s) : x \in Y_{s\,\text{inf}}, s = 1,2,\ldots\}$. We verify that
(i) P is a strict code. In fact, if we have a relation, for example

$$x_1 \ldots x_m \alpha = y_1 \ldots y_n \beta$$

where $(x_1, \ldots, x_m, \alpha) \in P_{(m+1)}, (y_1, \ldots, y_n, \beta) \in P_{(n+1)}$, then there exist $s, t \geq 1, x \in Y_{s\,\text{inf}}, y \in Y_{t\,\text{inf}}$ such that $\alpha = x(s), \beta = y(t)$. Let $l = \max\{n(x,s), n(y,t)\}$, then the words $x_1, \ldots, x_m, \alpha, \beta, y_1, \ldots, y_n$ all are in Y_l and thus the above relation must be an identity. The other cases are treated similarly.
(ii) For all $i : Y_i \prec P$. As shown above, each $x \in Y_i$ is written in the form $x = x'x(i)$, where $x' \in Y_{1\,\text{fin}}^* = Y_{\text{fin}}^*$, so $x \in P_{\text{fin}}^* P_{\text{inf}}$. Thus $Y_i \subseteq P^\infty$ (moreover, $Y_i \subseteq P^*$). Further, for every $\alpha \in P_{\text{inf}}$, there is a positive integer s such that $x \in Y_{s\,\text{fin}}$ and $\alpha = x(s)$. If $n(x,s) \leq i$ then

$$\alpha = x(s) = x_{\text{inf}}^{(n(x,s))} = x_{\text{inf}}^{(i)} \in Y_{i\,\text{inf}}.$$

If $n(x,s) > i$, since $Y_i \prec Y_{n(x,s)}$ it follows $\alpha = x(s) \in Y_{n(x,s)}$ is present in the expression of some element of Y_i as a product of elements of $Y_{n(x,s)}$. Hence $Y_i \prec P$.

Thus we have proved that there exists an upper bound for any countable chain. Let now

$$Y_\alpha \prec Y_\beta \prec \ldots$$

be an uncountable chain (with cardinality at most continuum). We obviously can derive from this chain a countable subchain

$$Y_1 \prec Y_2 \prec \ldots$$

such that for each Y_γ from the uncountable chain there exists Y_n from the countable subchain satisfying $Y_\gamma \prec Y_n$. For the latter one there exists a maximal element Y and it is easy to see that Y is also a maximal element for the uncountable chain. Now, in virtue of Zorn's Lemma Y is followed by a maximl element, i.e. Y is written by an indecomposable strict code. Proof is completed

In the next proposition we try to weaken the heavy demand of maximality of X_{fin}, but in compensation to this, the finiteness of X is required. We call X an *alphabetical code* if X is a subset of the alphabet A, otherwise we call it *nonalphabetical* one.

THEOREM 3.2. *Each finite finitary nonalphabetical strict code is written by an indecomposable nonalphabetical strict code.*

PROOF: We note that for any finite finitary strict codes $X, Y : X \prec Y$ implies $\| X \| \geq \| Y \|$ as mentioned in the proof of the preceding theorem. The equality holds if and only if every word of Y occurs just once in the $*$-factorization of just one word of X, or equivalently, there exists a partition $Y = Y_1 \cup \cdots \cup Y_n$ $(n \geq 1)$ such that for any $i : 1 \leq i \leq n$ there exists a word x (thus uniquely) such that x is a product of the words in Y_i (in some order). Hence if $\| X \| = \| Y \|$, we have $n = \operatorname{Card} X \leq \operatorname{Card} Y$ and in addition to this, if $\operatorname{Card} X = \operatorname{Card} Y$ then each Y_i is a singleton, which means $X = Y$.

We turn now to the proof. Let X be a finite finitary strict code. If X is indecomposable, we are done. Otherwise, we assume the contrary that X cannot be written by any indecomposable nonalphabetical strict code and as a consequence of this, we have an infinite chain of finite finitary codes: $X = X_0 \prec X_1 \prec X_2 \prec \ldots$, where $X_i \neq X_{i+1}$ and X_i is nonalphabetical for all $i = 1, 2, \ldots$. Certainly, we have: $\| X_0 \| \geq \| X_1 \| \geq \| X_2 \| \geq \ldots$. Since $\| X_0 \| < \infty$ it follows that for some integer $N : \| X_N \| = \| X_{N+1} \| = \ldots$. On the other hand, as we noted above: $\operatorname{Card} X_N \leq \operatorname{Card} X_{N+1} \leq \ldots$. But for every $i \geq 1 : \| X_{N+1} \| \geq \operatorname{Card} X_{N+i} \cdot \min\{|x| : x \in X_{N+i}\} \geq \operatorname{Card} X_{N+i}$. Hence, there must be an integer M such that $\operatorname{Card} X_{M+N} = \operatorname{Card} X_{M+N+1}$, therefore $X_{N+M} = X_{N+M+1}$ which is a contradiction with the assumpion that $X_i \neq X_{i+1}$ for all i. The proof is completed.

As a consequence of the preceding theorem,we shall have a decomposition theorem for finite finitary strict codes, but we recall some notations first. For more details one can consult [5] or [1]. Let X, Y be finitary codes over A and $X \prec Y$. Consider an alphabet B of the same cardinality as Y and a bijection $f : B \longrightarrow Y$. Because Y is a code, we can extend f to an isomorphism of B^* and Y^*, which we denote by the same $f, f : B^* \longrightarrow Y^*$. Let $Z = \{f^{-1}(x) : x \in X \subseteq Y^*\}$ and it is not difficult to see that Z is a code over B and it is a strict code if X and Y are strict codes. In this case we write $X = Y \otimes_B Z$. Conversely, if $Y \subseteq A^*, Z \subseteq B^*$ are codes (resp. strict codes) then the expression $X = Y \otimes_B Z$ stands for the following: there exists an isomorphism $f : B^* \longrightarrow Y^*$ such that $X = f(Z) \subseteq Y^*$. In this way X becomes a code (resp. strict code) and we have $X \prec Y$ if and only if B is the least alphabet such that $Z \subseteq B^*$. It is noteworthy that $\otimes$ is associative. Now we state our theorem.

THEOREM 3.3. *Every finite finitary strict code X of A^* admits a finite decomposition:*

$$X = X_1 \otimes_B X_2 \otimes_C \cdots \otimes_D X_n,$$

where $X_1, X_2, \ldots, X_n$ are indecomposable strict codes over the corresponding alphabets $A, B, \ldots, D$.

PROOF: The proof is proceeded by induction on $\| X \|$. If $\| X \| = 1$ then X is a letter, so it is indecomposable by definition. Suppose now that for every strict code X with $\| X \| < k$ the assertion is valid. Let $\| X \| = k$. If X is indecomposable, we are done; if not, by the preceding theorem, X is written by a nonalphabetical indecomposable strict code Y over $A^* : X \prec Y$. Thus, using the notations mentioned above, we have

$$X = Y \otimes_B Z.$$

Certainly, $|\, f^{-1}(x)\,| \leq |\, x\,|$ for each $x \in X$, therefore $\|\, Z \,\| \leq \|\, X \,\| = k$ and the equality holds if and only if $|\, f^{-1}(x)\,| = |\, x\,|$ for every x, i.e. when each word of Y is a letter meaning that $Y \subseteq A$ which is a contradiction. Thus we must have $\|\, Z \,\| < \|\, X \,\| = k$. By induction hypothesis Y admits a finite decomposition and so does X. Theorem is proved.

4. THEOREM OF DEFECT

In this concluding section we establish for strict codes a result, which is an analog of Theorem of Defect in the theory of finitary codes [2]. Note that Theorem of Defect was also proved for infinitary codes [4].

THEOREM 4.1. *For any language X of A^∞, if X is not a strict code then X is written by a strict code of cardinality at most* $\operatorname{Card} X - 1$.

PROOF: If $\operatorname{Card} X = \infty$ nothing is to be done because X is always written by A or a subset of A and $\operatorname{Card} A - 1 \leq \infty$. So we can assume $\operatorname{Card} X < \infty$ and the proof is done by induction on $\operatorname{Card} X$.

If $\operatorname{Card} X = 1$ then X is a singleton strict code. Now suppose that for every X of cardinality not exceeding n the assertion is true. Let now X be a language of cardinality $n+1$ and X not be a strict code. We have then two different ∞-factorizations $(u_1, u_2, \dots)$ and $(v_1, v_2, \dots)$ over X with $u_1 \neq v_1$ of a word $\alpha \in A^\infty$. Further, we can suppose that $|\, v_1 \,| > |\, u_1 \,|$. Hence $v_1 = u_1\beta$ for some $\beta \in A^{+\infty}$. Consider two cases :

(i) If $v_1 \in A^N$, then we have $v_1 = u_1u_2\dots$, therefore $\beta = u_2u_3\dots$. Consider the language $X_1 = X - v_1$. If v_1 occurs among $u_2, u_3, \dots$, say, $v_1 = u_k$ with k the smallest possible, $k > 1$; then we have

$$v_1 = u_1u_2\dots u_{k-1}v_1$$

hence $v_1 = (u_1\dots u_{k-1})^\omega \in X_1^\infty$.

If $v_1 \neq u_i$ for $i = 2, 3, \dots$ then v_1 is obviuosly in X_1^∞. So we have $X \subseteq X_1^\infty$ and $\operatorname{Card} X_1 < \operatorname{Card} X$. If X_1 is not a strict code, then by the induction hypothesis, X_1 is written by a strict code of cardinality $< \operatorname{Card} X_1 < \operatorname{Card} X$.

(ii) If $v_1 \in A^*$, then $\beta \in A^+$ and we put $X_1 = X - v_1 \cup \beta$. Clearly, $X \subseteq X_1^\infty$ and $\|\, X_1 \,\| < \|\, X \,\|$. There are two possibilities

(ii.1) Replacing all the occurrences of v_1 by $u_1\beta$ in the equation:

$$u_1u_2\dots = v_1v_2\dots, \tag{1}$$

it becomes an identity. This means that $\beta = u_2 \in X - v_1$ and therefore $\operatorname{Card} X_1 \leq \operatorname{Card} X - 1 = n$. The assertion follows by induction hypothesis applied to X_1.

(ii.2) If (1) does not become an identity after the replacement (ii.1), then we repeat the argument with X_1 until (i) or (ii.1) occurs. The process cannot go into infinity avoiding (i) or (ii.1) since $\|\, X_i \,\|, i = 1, 2, \dots$ decreases strictly each time the argument is repeated. Thus we should obtain a finite sequence $X = X_0, X_1, \dots, X_s$ with $X_i \subset X_{i+1}^\infty, \operatorname{Card} X_{i+1} \leq \operatorname{Card} X_i$ for $i = 1, 2, \dots, s-1$ and $X_s \subset C$, where C is some finite strict code with $\operatorname{Card} C < \operatorname{Card} X_s$. So $X \subseteq C^\infty$ and $\operatorname{Card} C \leq \operatorname{Card} X - 1$. The proof is complete.

COROLLARY 4.2. *Every two-element language is a strict code if and only if it is a code.*

PROOF: If $X = \{\alpha, \beta\}$ is not a code then X is not a strict code. Conversely, if X is not a strict code then by Theorem 4.1 $X \subseteq \{\gamma\}^\infty$ for some $\gamma \in A^\infty$. Since $\alpha \neq \beta, \gamma$ must belong to A^*. Hence X is not a code.

REFERENCES

[1] Berstel J., Perrin D., "Theory of Codes," Academic Press, New York, 1985.

[2] Berstel J., Perrin D., Perrot J.F., Restivo A., *Sur la théorème du défaut*, Journal of Algebra **60** (1979), 169–180.

[3] Do Long Van, "Contribution to Combinatorics on Words," Thesis, Humboldt University, Berlin, 1985.

[4] Do Long Van, *Langages écrits par un code infinitaire. Théorème du défaut*, Acta Cybernetica **7** (1986), 247–257.

[5] Lallement G., "Semigroups and Combinatorial Applications," John Wiley, New York, 1979.

[6] Nguyen Huong Lam, Do Long Van, *On a Class of Infinitary Codes*, Theoretical Informatics and Applications 24 (1990), 441-458

[7] Staiger L., *On Infinitary Finite Length Codes*, Theoretical Informatics and Applications **20** (1986), 483–494.

*) P.O. Box 631, Bo Ho , Hanoi, Vietnam.

A Decidable Case of the Semi-Unification Problem

Hans Leiß
Centrum für Informations- und Sprachverarbeitung
Universität München
Leopoldstraße 139
D-8000 München 40

Fritz Henglein
Dept. of Computer Science,
University of Copenhagen
Universitetsparken 1
DK-2100 Copenhagen Ø
Internet: henglein@diku.dk

Abstract

The semi-unification problem has recently been shown to be undecidable [8]. We present a new subclass of decidable semi-unification problems, properly containing those over monadic languages. In our 'quasi-monadic' problems, function symbols may be of arity > 1, but only terms with at most one free variable are admitted.

1 Introduction

The semi-unification problem, in its simplest form, is the problem to find out whether, given two first-order terms s and t, there is a substitution T_0 such that the refined term $T_0(s)$ *matches* the refined term $T_0(t)$. That is, an *in*equation $s \leq t$ is to be solved by finding two substitutions T_0 and T such that $T(T_0(s)) = T_0(t)$, or detecting that no such solution exists.

In this simple form, the problem was introduced by Kapur e.a.[5] and shown to be decidable in polynomial time. Kapur e.a. were interested in a criterion for nontermination of the Knuth-Bendix completion algorithm.

In a more general form, the same problem arose in efforts to generalize the type inference algorithm of the programming language ML, independently in work of Henglein[2], Leiß[11], and Kfoury, Tiuryn, and Urcyczyn[7]. In this case, several inequations $s_1 \leq_1 t_1, \ldots, s_k \leq_k t_k$ have to be solved simultaneously, with possibly different matching substitutions $T_1, \ldots, T_k$, one for each inequation.

A third source of the semi-unification problem was the work of Baaz (see [14]) on Kreisel's problem on the length of proofs in Peano Arithmetic (see Footnote 3 in the Appendix to Takeuti[15]). In this context, Pudlák[14] showed that the semi-unification problem for more than two inequality relations can be reduced to the problem with two inequality relations.

In the second and third application, the point is to find a most general type scheme for a recursive function, and a most general proof, respectively.

Solvable instances of the problem always do have most general solutions [3, 14, 8], but only recently Kfoury e.a.[8] could prove that the solvability question itself is undecidable. With quite different methods, Dörre and Rounds[1] showed that the semi-unification problem for feature structures, where infinite rational terms are admitted in the solutions, is also undecidable.

For the application to type inference, the undecidability of the semi-unification problem implies the undecidability of the typability of recursive function definitions with respect to a 'polymorphic recursion' typing scheme [13, 2, 12, 7]. From the decidability result of Kapur e.a. it follows that typability for polymorphic *linear* recursive functions - which occur only once in their defining term - is decidable. We do not claim that our result on semi-unification below has a similarly nice application to type inference. On the other hand, we do not have *any* examples from programming for which the type inference algorithms based on semi-unification (see [2,11]) fail to terminate, in contrast to Mycroft's[13] semi-algorithm.

For the applications in proof theory, the solvability of the relevant instances is given in advance, but information on the form of the most general solution is very much needed [6].

A few decidable cases of the semi-unification problem have been presented so far: *uniform* semi-unification[5, 14], where only one inequality relation occurs, *left-linear* semi-unification[8, 4], *acyclic* semi-unification[9], and semi-unification in at most two variables[12].

In the present paper we give a new class of decidable semi-unification problems[1], including the semi-unification problem for monadic languages. In our *quasi-monadic* problems, function symbols may be of arity > 1, but terms must have at most one free variable, though they may be non-linear. Left-linear semi-unification also extends the semi-unification problem for monadic languages, but in a different dimension. Due to space limitations, we cannot compare these two cases.

The semi-unification problem appears to be an important foundational problem, and isolating decidable subproblems is a way to improve our understanding of it. We think the method of proof below can be adapted to show decidability of further subproblems.

2 The Semi-Unification Problem

Let L be a first-order functional language, with a countably infinite set Var of variables. Let $R = \{\leq_i |\ i \in I\}$ be a finite set of binary relation symbols, fixing $I = \{1, \ldots, k\}$ for the rest of the paper. A finite set S of equations and inequations in the language $L \cup R$ is an *instance of the semi-unification problem over L and R*. A *solution* for S is a sequence $T = (T_0, \ldots, T_k)$ of substitutions such that $T_0(s) \equiv T_0(t)$ for each equation $s = t$ in S, and $T_i(T_0(s)) \equiv T_0(t)$ for each inequation $s \leq_i t$ in S.

If $T = (T_0, \ldots, T_k)$ is a solution of S, we call its *main* substitution T_0 a *semi-unifier of S*, and $T_1, \ldots, T_k$ its *residual* or *matching* substitutions.

Another solution $\tilde{T} = (\tilde{T_0}, \ldots, \tilde{T_k})$ is *at least as general as T*, if there is a substitution U such that $T_0(x) = U(\tilde{T_0}(x))$ for each variable x occcurring in S. A *most general* solution is a solution that is at least as general as any other solution. A *most general semi-unifier* for S is the main substitution of a most general solution for S.

Lemma 1 *(Pudlák[14], Henglein[2], Kfoury e.a.[8]) A solvable instance of the semi-unification problem has a most general solution.*

Example 1 [11] Let S be $\{x \leq_i y,\ f(z) \leq_i x\}$. Then $T = (T_0, T_i) = ([f(z)/x, f(z)/y], Id)$ is a solution of S, but it is not the most general one. A more general solution, and in fact the most general one, is $\tilde{T} = (\tilde{T_0}, \tilde{T_i}) = ([f(z_1)/x, f(z_2)/y],\ [z_2/z_1, z_1/z])$:

$$\tilde{T_i}(\tilde{T_0}(x)) = \tilde{T_i}(f(z_1)) = f(z_2) = \tilde{T_0}(y), \quad \text{and} \quad \tilde{T_i}(\tilde{T_0}(f(z))) = \tilde{T_i}(f(z)) = f(z_1) = \tilde{T_0}(x).$$

Indeed, T_0 is $[z/z_1, z/z_2] \circ \tilde{T_0}$ on the free variables of S. Note that, in contrast to unifiers, the semi-unifier $\tilde{T_0}$ here refers to variables that do not occur in the problem instance.

2.1 An Algebraic Form of the Semi-Unification Problem

The semi-unification problem can also be stated in an equational language with unary function variables, rather than using inequality relations. In this formulation, a simple transformation

[1] extending a partial result announced in Theorem 2, (19), of [12]

calculus for semi-unification problems can be given that is very close to, and a generalization of, the transformations often used to present unification algorithms.

Instead of extending L by the relation symbols of R, we augment L by the set I of unary function symbols, which act as L-homomorphisms, and are written as right exponents. That is, for each i and j in I we consider additional compound terms

$$f(t_1, \dots, t_m)^i := f(t_1^i, \dots, t_m^i)$$

and additional compound 'variables' x^i, $(x^i)^j$ etc. We have to consider x^i to be free in $(x^i)^j$, since we want to substitute terms for x^i in $(x^i)^j$. We write x^{ij} instead of $(x^i)^j$ etc. Let Var*be $\{x^v \mid x \in \text{Var}, v \text{ a word over } I\}$, and L^* the language with function symbols of L and Var* as its variables. If v is the empty word ϵ, x^v will be identified with x. The set of free variables of a term t is defined via free$(x^\epsilon) = \{x\}$ and free$(x^{vi}) = \{x^{vi}\} \cup \text{free}(x^v)$.

Definition 1 *For a sequence $T = (T_0, \dots, T_k)$ of substitutions T_j : Var $\to$ L-terms and an L^*-term t, define $T(t)$ inductively by*

$$T(x) = T_0(x), \quad T(x^{vi}) = T_i(T(x^v)), \quad T(f(t_1, \dots, t_m)) = f(T(t_1), \dots, T(t_m)).$$

Proposition 1 *If $T = (T_0, \dots, T_k)$ is a sequence of substitutions, then for each variable x^v and L^*-terms s and t, we have*

i) $T(t^i) = T_i(T(t))$, for each $i \in I$ and ii) $T(x^v) = T(t)$ implies $T(s) = T(s[t/x^v])$.

Definition 2 A *solution* of a set S of L^*-equations is a sequence $T = (T_0, \dots, T_k)$ of substitutions T_j : Var $\to$ L-terms such that $T(s) \equiv T(t)$ for each equation $s = t$ in S. The notions of *most general solution* and *semi-unifier* extend naturally to sets of L^*-equations.

Proposition 2 *1. For any instance S of the semi-unification problem there is a set S^* of L^*-equations such that S has a (most general) solution if and only if S^* has.*

2. For any set S^* of L^*-equations there is an instance S of the semi-unification problem such that S^* has a (most general) solution if and only if S has.

Proof: 1. Let S^* be $\{s^i = t \mid s \leq_i t \in S\}$. Obviously, T solves S if and only if it solves S^*.
2. First, replace $S^*(x_1^{v_1}, \dots, x_n^{v_n})$ by $S^*[y_1/x_1^{v_1}, \dots, y_n/x_n^{v_n}] \cup \{x_1^{v_1} = y_1, \dots, x_n^{v_n} = y_n\}$, with fresh variables $y_1, \dots, y_n$.[2] Then, repeatedly replace an equation $x^{vi} = y$, with $i \in I$ and $v \in I^*$, $v \neq \epsilon$, by the equations $x^v = z$, $z^i = y$, with fresh variable z. Finally, replace $z^i = y$ by an inequation $z \leq_i y$. To show that the resulting instance S of the semi-unification problem is solvable if and only if S^* is, appropriately extend a solution T of S to the new variables and use Proposition 1, ii). □

The unification problem asks for systems $S(x_1, \dots, x_n)$ of L-equations whether the *first-order* sentence $\exists x_1 \dots \exists x_n. \bigwedge S(x_1, \dots, x_n)$ holds in the term algebra for L. In contrast, the semi-unification problem asks for systems $S^*(x_1, \dots, x_n, i_1, \dots, i_k)$ of $L \cup \{i_1, \dots, i_k\}$-equations, whether the *second-order* condition

$$\exists i_1 \dots i_k.(\bigwedge{}^j\ i_j \text{ is an } L\text{-endomorphism } \wedge \exists x_1 \dots x_n. \bigwedge S^*(x_1, \dots, x_n, i_1, \dots, i_k))$$

holds in the term algebra for L. It seems natural to study the semi-unification problem over algebras other than the term algebra, but we are not aware of such work. More generally, we are not aware of any work in monadic second-order logic "L_{hom}", i.e. first-order logic over the language L extended by quantified unary function variables that range over L-endomorphisms only, or L-homomorphisms, in the many-sorted case.

[2] If, as in Kfoury e.a.[7], an instance of the semi-unification problem must not contain equations between terms, we replace $s = t$ by $z \leq_i s$, $z \leq_i t$, with some fresh variable z (obtaining an instance that is not left-linear).

2.2 A Transformation Calculus for Semi-Unification Problems

To avoid discussing symmetry of equality, in the following we consider an *equation* between terms s and t of L^*, written as $s = t$, to be a multiset of two L^*-terms. An equation in a system S is *solved in* S, if it has the form $x^v = t$, where x^v occurs only once in S and, moreover, if t is a variable y^w, then $| v | \geq | w |$. In this case, we call x^v the *solved variable* of the equation, choosing one if both are solved. An occurrence of a variable x^v in S is *maximal*, if it is not inside some x^{vw}, $1 \leq | w |$. The *maximal variables of* S are those having a maximal occurrence in S.

Proposition 3 *The number of solved equations of a system S is bounded by the number of its maximal variables,* $| \operatorname{maxvar}(S) |$.

We use the following rules to transform systems, where $s, s_1, \ldots, s_m, t, t_1, \ldots, t_m$ are L^*-terms and x^v, y^w are L^*-variables:

Transformation Rules

Decomposition of terms

$$S \,\dot{\cup}\, \{f(s_1, \ldots, s_m) = f(t_1, \ldots, t_m)\} \longrightarrow S \cup \{s_1 = t_1, \ldots, s_m = t_m\}$$

Substitution of variables

$$S \,\dot{\cup}\, \{x^v = t\} \longrightarrow S[t/x^v] \cup \{x^v = t\},$$
$$\text{if } x^v \notin \operatorname{free}(t),\ x^v = t \text{ is not solved in } S \cup \{x^v = t\}, \text{ and } | w | \leq | v | \text{ if } t \equiv y^w.$$

By induction on the length of the transformation sequence, using Proposition 1, we obtain:

Lemma 2 *If S_1 can be transformed into S_2, then the sets of solutions of S_1 and S_2 are the same.*

It is useful to admit further transformation rules, in order to remove trivial equations and to stop the transformation process in case an unsolvable equation is found:

Simplification Rules

Elimination of trivial equations

$$S \cup \{x^v = x^v\} \longrightarrow S$$

Function clash

$$S \cup \{f(s_1, \ldots, s_m) = g(t_1, \ldots, t_n)\} \longrightarrow \textit{fail}, \qquad \text{if } f \not\equiv g$$

Extended 'occurs check'

$$S \cup \{x^v = t\} \longrightarrow \textit{fail}, \qquad \text{if } x^v \in \operatorname{free}(t) \text{ and } t \text{ is not a variable.}$$

The occurs check is called *extended* because a variable x^v also "occurs" in the variable x^{vw}. Note that $x^v = f(x^{vw})$ cannot be solved, since $| T(x^v) | < | T(f(x^{vw})) |$ for any T.

Lemma 3 *[12] If S is irreducible under the Transformation and Simplification rules, then it has a most general solution. If S can be reduced to fail, it has no solution.*

In order to decide solvability of a system S, we want to transform it into an irreducible one. The above two lemmata establish partial correctness of this method. However, by the undecidability result of Kfoury e.a., there can be no reduction strategy which terminates on all instances S.

Let us recall why the transformations ensure termination in the case of unification, when all variables have exponent ϵ. In this case, termination obviously follows from three facts:

- Term decompositions decrease the sum of the sizes of the equations in the system, while they do not decrease the number of solved equations.
- Substitutions of variables increase the number of solved equations, which is bounded by the number of variables occurring in the system.
- The number of variables is not increased by the transformations.

In the case of semi-unification, however, the second and third properties no longer hold:

- Substitutions of variables need not increase the number of <u>solved</u> equations; even for monadic languages, the number of solved equations may go down:

$$\{\underline{y^j = g(z)}, \underline{x^{ijk} = z}, x^i = f(y)\} \overset{[f(y)/x^i]}{\longrightarrow} \{y^j = g(z), f(y^{jk}) = z, \underline{x^i = f(y)}\}$$

- The number of <u>maximal</u> L^*-variables may be increased by substitutions of variables:

$$\{\underline{x}^i = c,\ \underline{x} = f(\underline{y}, \underline{z})\} \quad\longrightarrow\quad \{f(\underline{y}^i, \underline{z}^i) = c,\ \underline{x} = f(\underline{y}, \underline{z})\}.$$

Thus, substitution transformations do not make a system simpler in an obvious sense.

3 Systems With Incomparable Dependent Variables

In this section, we will show that any system can be transformed into one with incomparable dependent variables, where x^v and y^w are *incomparable* when $x \not\equiv y$ or none of v and w is a prefix of the other. Let us first look at substitutions with equations between variables only.

Lemma 4 *Each system can be transformed into one where all equations between variables are trivial or solved, without increasing the size of the system.*

Proof: By induction on the size of the system. Let $x^v = y^w \in S = S_0 \dot{\cup} \{x^v = y^w\}$ be an unsolved and nontrivial equation with $x^v \notin \text{free}(y^w)$. By $S_0 \cup \{x^v = y^w\} \longrightarrow S_0[y^w/x^v] \cup \{x^v = y^w\}$ we obtain a system S' which is strictly smaller than S, when $|\, v \,| > |\, w \,|$. If $|\, v \,| = |\, w \,| =: n$, use induction on the number of solved variables of size $\leq n$. □

In the following, the dot-notation $\dot{s}$ is used to indicate that s is a *compound term*, that is, not in Var*. By $s(\dot{x}^v)$ we indicate that x^v occurs at least once in s.

Definition 3 The *dependent variable of an equation* $x^v = \dot{s}$ in S is x^v, if $x^v \notin \text{free}(\dot{s})$. The dependent variable of a nontrivial equation $x^v = y^w$ in S is the variable with larger size, if there is one; if both variables are of the same size, it is the solved variable (resp. one of the two solved variables) in case the equation is solved in S, and taken from $\{x^v, y^w\}$ by some choice function, otherwise. No further equation has a dependent variable.

Lemma 5 *Each system can be transformed, by the substitution and the occurs check rules only, to fail or into a system S satisfying the following property*

$\Delta(S)$: *each dependent variable has only one occurrence that is not inside a compound term.*

In particular, the dependent variables of any two equations in S are incomparable.

Proof: By induction on the number $n(S)$ of equations between variables. Suppose the claim is true for all systems S' with $n(S') < n(S)$. We may assume that all equations in S between variables are solved or trivial, which, by the proof of Lemma 4, can be achieved without increasing n. In particular, we can assume:

$\Phi(S)$: In S, each dependent variable of an equation between variables has only one occurrence that is not in a compound term.

$\Psi(S)$: S does not contain an equation $x^v = \dot{s}(\dot{x}^{vw})$.

If $\Psi(S)$ does not hold, S can be reduced to *fail* by the occurs check rule. Let $m(S)$ be the number of equations $x^v = \dot{s}$, such that $x^v \notin \text{free}(\dot{s})$ and x^v has another occurrence in S that is not inside a compound term.

Case 1: $m(S) = 0$. The claim is true by assumption $\Phi(S)$ on equations between variables.

Case 2: $m(S) > 0$. Let S be $S_0 \cup \{x^v = \dot{s}\}$, such that $x^v \notin \text{free}(\dot{s})$ and x^v has another occurrence in some left or right variable side of an equation in S_0. Apply the substitution

$$S = \quad S_0 \cup \{x^v = \dot{s}\} \longrightarrow S_0[\dot{s}/x^v] \cup \{x^v = \dot{s}\} \quad =: S'.$$

Subcase 2.1: $n(S') < n(S)$. The claim holds by induction on n.

Subcase 2.2: $n(S') = n(S)$. In S there must be an equation of the form $x^{vw} = \dot{t}$ different from $x^v = \dot{s}$. We will show that $m(S') < m(S)$ and S' satisfies $\Phi(S')$, whence the claim follows by induction on m.

Each occurrence of some x^{vu} in S_0 has turned into a compound term $\dot{s}^u$ in S'. Hence, a variable side of an equation in S either has not changed or has become a compound term. This implies, first, that x^v has only one occurrence in S' that is not in a compound term, and, second, that every dependent and every solved variable in S of an equation between variables has at most one occurrence in S' that is not inside a compound term.

By $n(S') = n(S)$, $\Phi(S)$ and $\Psi(S)$, each dependent variable in S' of an equation of the form $z^u = \dot{r}'$ is the dependent variable of an equation $z^u = \dot{r}$ in S. Hence we obtain $m(S') < m(S)$. If $\Psi(S')$ does not hold, we can reduce S' to *fail*. To show $\Phi(S')$, one has to use a suitable choice of dependent variables for equations between variables in S'. □

Example 2 Let S_0 be $\{f(x, g(x, y)) \leq_1 f(h(y), z),\ g(z, h(x)) \leq_2 g(g(z, x), y)\}$. Transforming this according to Proposition 2 gives the following set of L^*-equations:

$$S_1 = \{f(x^1, g(x^1, y^1)) = f(h(y), z),\ g(z^2, h(x^2)) = g(g(z, x), y)\}.$$

While $\Delta(S_1)$ trivially holds, we lose property Δ when applying decompositions, obtaining

$$S_2 = \{x^1 = h(y),\ z = g(x^1, y^1),\ z^2 = g(z, x),\ y = h(x^2)\},$$

with $n(S_2) = 0$, but $m(S_2) = 1$, as z and z^2 are comparable dependent variables. By Case 2 of the lemma, using $[g(x^1, y^1)/z]$, this reduces to

$$S_3 = \{x^1 = h(y),\ z = g(x^1, y^1),\ g(x^{12}, y^{12}) = g(g(x^1, y^1), x),\ y = h(x^2)\},$$

with $n(S_3) = n(S_2) = 0$ and $m(S_3) = 0$. By $\Delta(S_3)$, all dependent variables are incomparable.

To each system S we associate a function f_S that gives the number $f_S(n)$ of variables x^v solved in S such that $|v| = n$. The number $f_S(n)$ for small n is more important than that for large n, because a solved equation $x^{vw} = s$ can get transformed by a substitution with some equation $x^v = t$. This motivates the following relation $\prec$ between functions, which is well-founded on a restricted class of functions only (see Lemma 9 in Section 4).

Definition 4 For $f, g : \omega \to \omega$, we define $\quad f \prec g \ :\Leftrightarrow\ \exists n.(f(n) > g(n) \wedge \forall k < n. f(k) = g(k))$.

If a system S is transformed to S' by term decompositions, then clearly $f_{S'} \preceq f_S$. For substitutions of variables, this need not be true: the two equations for y^i and y^j in

$$\{x^v = f(y),\ y^i = t,\ y^j = s,\ u_1 = r_1(x^{vi}),\ u_2 = r_2(x^{vj})\}$$

become unsolved when we solve the equation for x^v by substitution, as new occurrences $r_1(f(y^i))$, $r_2(f(y^j))$ arise. But *if* we then can turn y^i and y^j into solved variables (again), without destroying the solvedness of x^v or other variables, we have gained according to $\prec$.

Lemma 6 *Let S be a system that is irreducible under substitution of variables, and let S' be a system obtained from S by decompositions of terms. Suppose S'' is a system irreducible under substitution transformations, obtained from S' using substitutions only, and does not allow an applciation of the extended occur's check rule. Then $f_{S''} \prec f_S$.*

Proof: Let $x^v = s$ be an equation of S' that arose from the decomposion of an equation between compound terms of S. Suppose x^w is the dependent variable of an equation $x^w = t$ of S. Since S was substitution-irreducible, w is not a prefix of v. Hence either x^v is incomparable with x^w, or v is a *strict* prefix of w. Thus, for some n, there are more pairwise incomparable dependent variables of size $\leq n$ in S' than in S. Let m be the least such n, and S'' a substitution-irreducible system obtained from S' by substitution transformations only. We may assume that no substitution into the dependent variable of an equation between variables occurs; this can happen finitely often only and would increase the set of dependent variables of small size even further. We find that $f_{S''}(m) > f_S(m)$ and $f_{S''}(n) = f_S(n)$ for $n < m$, so $f_{S''} \prec f_S$: because of incomparability, no substitution *into* these dependent variables can occur, and the set of dependent variables of exponents with size $\leq m$ is not changed by substitutions. Hence, in S'' exactly these variables will be the solved variables of size $\leq m$. □

Example 3 The system $S_0 = \{x^i = f(y),\ y^j = g(z),\ h(x) = h(f(g(z^k)))\}$ is substitution-irreducible and satisfies Δ. Decompose the third equation and substitute into the first to get

$$S_1 = \{f(g(z^{ki})) = f(y),\ y^j = g(z),\ x = f(g(z^k))\}.$$

Again, S_1 is irreducible under substitutions and satisfies Δ. Note that $f_{S_0}(0) = 0,\ f_{S_1}(0) = 1$ and hence $f_{S_1} \prec f_{S_0}$. Decomposing the first equation and substituting into the second gives

$$S_2 = \{y = g(z^{ki}),\ g(z^{kij}) = g(z),\ x = f(g(z^k))\},$$

with $f_{S_2}(0) = 2,\ f_{S_2}(1) = f_{S_2}(2) = f_{S_2}(3) = 0$; in particular, $f_{S_2} \prec f_{S_1}$. By decomposition, we arrive at an irreducible, hence solvable, system

$$S_3 = \{y = g(z^{ki}),\ z^{kij} = z,\ x = f(g(z^k))\}.$$

Note that $f_{S_3}(0) = 2,\ f_{S_3}(1) = f_{S_3}(2) = 0$, but $f_{S_3}(3) = 1$, thus $f_{S_3} \prec f_{S_2}$.

4 Quasi-Monadic Semi-Unification

The termination of transformation sequences is not obvious for two reasons: substitution of variables may both increase the number of L^*-variables and decrease the number of solved equations in a system.

It is clear that in a monadic language, no substitution can increase the number of (maximal) variables in a system. But for possible applications to type inference, at least one binary function symbol is needed. The next best restriction is that each term must have at most one (maximal) free variable. [3]

[3] Due to this severe restriction, we do not expect our decidability result by itself to have applications in deciding typability for polymorphic recursive functions. However, note that in what follows, we use the restriction only to guarantee that 'each term being substituted' has at most one maximal variable.

Definition 5 The *quasi-monadic* semi-unification problem is the semi-unification problem restricted to instances in which each term has at most one free variable (one free maximal variable in the L^*-formulation).

For example, $\{f(x, g(x, 0, x)) \leq_1 f(g(0, y, y), y)\}$ is quasi-monadic, but $\{x^1 = g(0, y^1, y)\}$ is not. Note that decompositions can make a system quasi-monadic, as we go down to small subterms.

Lemma 7 *If S_1 can be transformed to S_2 and S_1 is quasi-monadic, then S_2 is also quasi-monadic, and $|\ maxvar(S_2)\ | \leq |\ maxvar(S_1)\ |$.*

Proof: By induction on the transformation. In case $S_1 \longrightarrow S_2$ is a substitution $S \cup \{x^v = t\} \longrightarrow S[t/x^v] \cup \{x^v = t\}$. then, since S_1 is quasi-monadic, we have $|\ \text{maxvar}(t)\ | \leq 1$, hence for any w, $|\ \text{maxvar}(x^{vw}[t/x^v])\ | = |\ \text{maxvar}(t^w)\ | \leq 1 = |\ \text{maxvar}(x^{vw})\ |$. □

With Proposition 3, it follows that the number of solved equations in a transformation of a system S is bounded by $|\ \text{maxvar}(S)\ |$.

Next we will strengthen Lemma 5 for quasi-monadic systems. Recall that a substitution with $y^w = f(x)$, say, may turn a solved equation $x^v = s$ into an unsolved one, since new occurrences of variables may arise, as in $y^{wv}[f(x)/y^w] = f(x^v)$. We now show that when starting with a quasi-monadic system, we do not run into an infinite sequence of substitutions:

Lemma 8 *Let S be a quasi-monadic system satisfying property $\Delta(S)$ of Lemma 5. Every sequence of substitution transformations beginning in S finally produces a system S' that either can be turned to fail by the occurs check rule or is irreducible under substitutions.*

Proof: For each variable x^v, we define the set of exponents of x^v in S by

$$exp(x^v, S) := \{w \mid x^{vw} \text{ occurs in some compound term of } S\}.$$

Note that if $x^v = t$ is a solved equation of S, with x^v as solved variable, then $exp(x^v, S)$ is empty. We perform an induction on $e(S) := \sum_{x^v \in dep(S)} |\ exp(x^v, S)\ |$, where $dep(S)$ is the set of dependent variables of S.[4]

Case 1: $e(S) = 0$. Let $x^v = s$ be an equation of S, with x^v as dependent variable. By $\Delta(S)$ and $e(S) = 0$, there is only a single occurrence of x^v in S, whence $x^v = s$ is solved. Thus S is irreducible under substitution transformations.

Case 2: $e(S) > 0$. We may assume that $\Psi(S)$ from the proof of Lemma 5 holds, since otherwise we can reduce S to *fail*. Let S' be obtained from S by performing the substitution

$$S := S_0 \cup \{x^v = s\} \longrightarrow S_0[s/x^v] \cup \{x^v = s\} =: S'.$$

Then $x^v \notin \text{free}(s)$, and $exp(x^v, S) \neq \emptyset$. If s is a constant term, then clearly $exp(x^v, S') = \emptyset$ and $e(S') < e(S)$, using $\Psi(S)$ to ensure that no new dependent variables arise. By induction, every sequence of substitution transformations beginning in S' is finite. If s is not a constant term, s contains one maximal variable, y^w say, since S is quasi-monadic. We distinguish two cases.

Case 2 a: There is an equation $y^u = t \in S$, with dependent variable y^u, such that $u \leq w$, i.e. u is a prefix of w. Let $(u\backslash w)$ stand for the suffix of w obtained by removing the prefix u, and $(u\backslash w)V = \{(u\backslash w)v \mid v \in V\}$ for any set V of words over I. By $\Delta(S)$, x^v and y^u are incomparable, and we obtain:

$$\begin{aligned} exp(x^v, S') &= \emptyset, \\ exp(y^u, S') &= exp(y^u, S) \cup (u\backslash w)exp(x^v, S), \\ exp(z^r, S') &= exp(z^r, S), \quad \text{for all other dependent variables } z^r. \end{aligned}$$

(We may assume that S and S' have the same dependent variables) It follows that

[4]To follow the argument, it will be helpful to draw a picture of the 'trees of variable exponents' $exp(x^v, S)$.

i) $e(S') \leq e(S)$, and $x^v = s$ is solved in S', but unsolved in S.

Any variable occurrence in S' that was not in S is obtained from a substitution $x^{vr}[s(y^w)/x^v] = s(y^{wr})$. Since $y^u = t$ is an unsolved equation of S and there are no incomparable dependent variables (in S), no solved equation of S has become an unsolved equation of S'. This shows:

ii) the number of solved equations in S' is greater than the number of solved equations in S.

As the number of solved equations is bounded, we can stay in Case 2 a) finitely often only.

Case 2 b: There is no equation $y^u = t \in S$, with dependent variable y^u, such that $u \leq w$. It is sufficient to show $e(S') < e(S)$.

Suppose $y^w \in \mathrm{free}(x^v)$. Then $y \equiv x$ and $w < v$. For a dependent variable x^r with $w \leq r$,

$$exp(y^r, S') = \begin{cases} exp(y^r, S) \cup r\backslash(w \cdot exp(x^v, S)), & r \neq v \\ r\backslash(w \cdot exp(x^v, S)), & r = v. \end{cases}$$

For different such r, the $r\backslash(w \cdot exp(x^v, S)) := \{r\backslash(wu) \mid u \in exp(x^v, S)\}$ are pairwise disjoint strict subsets of $exp(x^v, S)$, because the x^r are incomparable, by $\Delta(S)$. Hence

$$\textstyle\sum_{x^r \in dep(S'),\, w \leq r} \mid exp(x^r, S') \mid \quad < \quad \sum_{x^r \in dep(S),\, w \leq r} \mid exp(x^r, S) \mid .$$

As $exp(z^r, S') = exp(z^r, S)$ for all remaining dependent variables, we obtain $e(S') < e(S)$.

Suppose $y^w \notin \mathrm{free}(x^v)$. Then, using $0 = \mid exp(x^v, S') \mid$ and the same disjointness argument as above, we obtain

$$\begin{aligned} &\mid exp(x^v, S') \mid + \textstyle\sum_{y^r \in dep(S'),\, w \leq r} \mid exp(y^r, S') \mid \\ &< \ \mid exp(x^v, S) \mid + \textstyle\sum_{y^r \in dep(S),\, w \leq r} \mid exp(y^r, S) \mid . \end{aligned}$$

Again, $exp(z^r, S') = exp(z^r, S)$ for all remaining dependent variables, and $e(S') < e(S)$. □

The essential fact that substitution sequences terminate is guaranteed by other conditions than S being quasi-monadic. For example, it holds obviously when the dependent variables of S depend on variables of smaller (or equal) size only; this seems to correspond to generalized matching (or unification) problems.

Lemma 9 *For each $s : \omega \rightarrow \omega$ and $M \in \omega$, the relation $\prec$ is well-founded on*

$$F(s, M) \ := \ \{f : \omega \rightarrow \omega \mid \forall n. f(n) \leq s(n) \ \mathit{and} \mid \{n \in \omega \mid f(n) \neq 0\} \mid < M\}.$$

This is clear by the bounds M and s. We are now ready for the main result:

Theorem 1 *The quasi-monadic semi-unification problem is decidable.*

Proof: Suppose that S is quasi-monadic. For $M = \mid \mathrm{maxvar}(S) \mid$ and $s(n) = M$ for all n, by Lemma 9 the relation $\prec$ is well-founded on $F(s, M)$. If S' is obtained from S by a sequence of transformations, note that $f_{S'}(n) \leq \mid \mathrm{maxvar}(S') \mid \leq M$, by Lemma 7, and thus $f_{S'} \in F(s, M)$. The claim now follows by Lemma 5, 6 and 8, using the following reduction strategy: repeatedly apply substitutions as long as possible, followed by some term decompositions. As soon as the occurs check is applicable, stop. □

The outline of our proof can be used to prove decidability of classes of semi-unification problems that are (i) closed under transformations, (ii) admid a recursive bound on the number of maximal variables of transformed systems, in terms of an initial system, and (iii) guarantee that substitution sequences from instances of the class must terminate or lead to failure due to the occurs check.

References

[1] J. Dörre and B. Rounds: On Subsumption and Semiunification in Feature Algebras. Proc. 5th Annual IEEE Symposium on Logic in Computer Science, Pittsburgh 1990

[2] F.Henglein: Type Inference and Semi-Unification. Proc. ACM Conference on Lisp and Functional Programming, Snowbird, Utah, July 25-27, 1988, pp. 184–197.

[3] F.Henglein: Polymorphic Type Inference and Semi-Unification. PhD Thesis, New York University, Dept. of Computer Science, Tech. Report 443, May 1989.

[4] F.Henglein: Fast Left-Linear Semi-Unification. Proc. of the International Conference on Computing and Information, Niagara Falls, Canada, May 1990. Springer LNCS 468, pp 82–91.

[5] D. Kapur and D. Musser and P. Narendran and J. Stillman: Semi-Unification. Proc. of the 8th Conference on Foundations of Software Technology and Theoretical Computer Science, Pune, India, December 21 - 23, 1988. Springer LNCS 338, pp. 435-454.

[6] G.Kreisel, personal communication, Nov. 1989

[7] A.J.Kfoury, J.Tiuryn, and P.Urcyczyn: Computational Consequences and Partial Solutions of a Generalized Unification Problem. 4th IEEE Symposium on Logic in Computer Science, Asilomar, California, June 5-8, 1989.

[8] A.J.Kfoury, J.Tiuryn, and P.Urcyczyn: The Undecidablity of the Semi-Unification Problem. Proceedings of the 22nd Annual ACM Symposium on Theory of Computing, Baltimore, Maryland, May 14–16, 1990, pp. 468–476.

[9] A.J.Kfoury, J.Tiuryn, and P.Urcyczyn: ML-Typability is DEXPTIME-complete. 15th Colloquium on Trees in Algebra and Programming, CAAP'90, Arnold (ed.). Springer LNCS 431, pp. 206 - 220, 1990.

[10] H.Leiß, On Type Inference for Object-Oriented Programming Languages. E. Börger e.a. (eds.) CSL'87. Proc. of the 1st Workshop on Computer Science Logic, Universität Karlsruhe, 1987. Springer LNCS 329, pp. 151–172

[11] H.Leiß, Semi-Unification and Type Inference for Polymorphic Recursion. Technical Report INF2-ASE-5-89, Siemens AG, München, May 1989.

[12] H.Leiß: Polymorphic Recursion and Semi-Unification. E. Börger e.a. (eds.) CSL '89. Proc. of the 3rd Workshop on Computer Science Logic, Universität Kaiserslautern, Oct 2 - 6, 1989. Springer LNCS 440, pp. 211–224.

[13] A. Mycroft: Polymorphic Type Schemes and Recursive Definitions. In: International Symposium on Programming, 6th Colloquium. Toulouse, April 17 - 19, 1984. Springer LNCS 187, pp. 217 - 228.

[14] P. Pudlák: On a Unification Problem related to Kreisel's Conjecture. Commentationes Mathematicae Universitatis Carolinae, no. 29 (3), 1988, pp. 551-556.

[15] G. Takeuti: Proof Theory. Second Edition. North-Holland, Amsterdam 1975.

Maintaining Dictionaries in a Hierarchical Memory

Claudia Leopold
Fachbereich Informatik, Humboldt-Universität zu Berlin
O-1086 Berlin, PO Box 1297, Germany

Abstract

Aggarwal, Chandra and Snir ([ACS87]) developed a model of hierarchical memory with block transfer. It is an extension of the common RAM-model, where access to location x takes time f(x) and copying of memory blocks is supported. We will specify this hierarchical memory model by assuming f to be a nondecreasing step function, which comes closer to real architectures of hierarchical memories. We propose a data structure for the dictionary problem and prove its optimality according to a modified measure of asymptotic complexity introduced in this paper.

1. Introduction

Although in real life computers the range of speeds from the fastest memory to the slowest can be very large, algorithms are usually designed to run on random access machines (RAM), where access to all locations is equally fast. As computing grows increasingly more data oriented, the speed of storage functions becomes one of the limiting factors in the overall performance of modern computers. In particular, Aggarwal and Chandra ([AC88]) have shown that algorithms being optimal in a RAM can be necessarily nonoptimal in a hierarchical memory. This background in mind, the RAM model seems to be insufficient.

Aggarwal et al. have developed other models for studying algorithms and data structures - the Hierarchical Memory Model HMM_f ([AACS87]), the Hierarchical Memory Model with Block Transfer BT_f ([ACS87]) and the Virtual Memory Model VM_f ([AC88]). They assumed that the memory access time increases continuously, but suggested to study discrete cases, too, since they may better reflect the physical situation in real machines. Two-level-hierarchies have already been investigated extensively (e.g. [F72], [AV87]). We will devote to the general case. Similar to BT_f, our $BT_{f^{(s)}}$ model is like a RAM, except that the memory locations are numbered: 1,2,3.., and access to location x takes time

$$f^{(s)}(x) = \begin{cases} a_0 & x=1 \\ a_i & \sum_{j=0}^{i-1} b_j < x \le \sum_{j=0}^{i} b_j \end{cases}$$

where $a_i \in \mathbb{R}$ $(1 \le i \le m)$, $a_0=0$, $b_i \in \mathbb{N}$ $(1 \le i \le m-1)$, $b_0=1$, $b_m=\infty$ $(m \in \mathbb{N})$ and $a_i >> a_{i-1}$ as well as $b_i >> b_{i-1}$.

We denote the interval $(\sum_{j=0}^{i-1} b_j , \sum_{j=0}^{i} b_j]$ of length b_i by i-th level.

Furthermore, a block copy operation $[x-l,x] \longrightarrow [y-l,y]$ is defined for any disjoint intervals $[x-l,x]$ and $[y-l,y]$. It moves the contents of location x-i into location y-i for $0 \le i \le l$ and has a running time of respectively $f(x) + l$ if $x>y$ or $f(y) + l$ if $x<y$.

All standard RAM operations also exist. We assume that the program itself is stored outside the numbered memory.

So far, $BT_{f^{(s)}}$ is similar to the Memory Hierarchy Model MH of Alpern, Carter and Feig ([ACF90]). However we use a different approach to overcome the problem that the common measure of asymptotic complexity is not applicable to the discrete model, but exact calculations create a not justifiable technical expense already for simple problems ([L90], [L91]). Instead of simplifying the hierarchy to permit exact calculations, we introduce the following modified measure of asymptotic complexity.

For a fixed m, we define a function $t(x,a_i,b_i)$ as belonging to a class $O^{BT_{f^{(s)}}}(g(x,a_i,b_i))$ iff there is a constant $c \in \mathbb{N}^+$ and $x^*, b_i^* \in \mathbb{N}^+$ such that $t(x,a_i,b_i) \le c\, g(x,a_i,b_i)$ for all $x \ge x^*, a_i, b_i \ge b_i^* \in \mathbb{N}$ which obey the above conditions (hence, c does not depend on a_i and b_i but may depend on m)

$\Omega^{BT_{f^{(s)}}}$ and $\Theta^{BT_{f^{(s)}}}$ are defined analogously.

The $BT_{f^{(s)}}$ model is pretty robust. For example, changing the block copy time to $f(x)+f(y)+1$ or to $\sum_{\text{level } s \text{ intersects } [x-l,x]} a_s + \sum_{\text{level } s \text{ intersects } [y-l,y]} a_s + l$ increases the running time only within the constant factor.

In this paper we investigate how to implement a dictionary in $BT_{f^{(s)}}$ such that the operations search(y), insert(y,content) and delete(y) can be done efficiently. We take the view that update operations include a search to avoid double entries and deletes to not existing key values.

The paper is organized as follows. In Section 2, an efficient data structure and the corresponding algorithm are explained. They realize the tight bound of $\Theta^{BT_{f^{(s)}}}\left[\log n \max_{i=1..m}\left\{\frac{a_i}{\log(b_{i-1}+1)}\right\}\right]$, which is proved in Section 3. Section 4 is devoted to conclusions and possibilities for future research.

2. Data Structure

To simplify the demonstration, we will for the present ignore the contents connected with the keys.

Our data structure which is called rectree5 is a generalization of [a,b]-trees (see e.g. [M84]). We build [a,b]-trees recursively, such that the buckets of a large [a,b]-tree are again [a,b]-trees and so on. The arising problem is that a split (similarly, a fuse) is now a split of an [a,b]-tree into two physically separated trees. Since this can normally not be done efficiently (i.e., in O(height) worst case running time), we resort to online transformations. The cost of a bucket split is distributed over a period of time (split state). In this period the original bucket (A-block) is split into two trees stored in blocks N_1 and N_2. Updates are not immediately executed but only stored in I- and D-blocks to be executed in a period after the split state.

More precisely, a rectree5 is defined by the following recursion.

Let u_i (i=1..m-1) be natural numbers with $u_i \gg u_{i-1}$.

Definition: A rectree5 is a rectree5(m).

A rectree5(i) for i=3..m is a leaf-oriented $[u_{i-1},5u_{i-1}]$-tree that buckets (called order(i-1)-buckets) are connected by pointers. In contrast to an ordinary [a,b]-tree, a bucket is not simply organized as a linear list but has a more complicated structure consisting of

- an Actual-block (A-block) which is organized as a rectree5(i-1) and contains between $2u_{i-1}$ and $4u_{i-1}$ entries (the term entry here denotes the real elements of the order(i-1)-bucket, i.e., the leaf elements of the corresponding rectree5(i-1), in contrast to the non-leaf elements added to make the tree leaf-oriented)
- two New-blocks (N-blocks 1 and 2) which are organized as rectree5(i-1) and contain up to $2u_{i-1}$ entries each
- two Insert-blocks (I-blocks 1 and 2) which are organized as rectree5(i-1) and contain up to $2u_{i-1}$ entries each
- two Delete-blocks (D-blocks 1 and 2) which are organized as rectree5(i-1) and contain up to u_{i-1} entries each and
- a State block (S-block) of length up to u_{i-1}

A rectree5(2) is an $[u_1, 2u_1]$-tree.

The dictionary operations are done in a rectree5 as follows. (Our description reduces them to the corresponding operations in $[u_1, 2u_1]$-trees).

1) Search: Search recursively rectree5(m). A rectree5(i) is searched similar to an [a,b]-tree, i.e., search first the root bucket and determine the next bucket to be searched, search this bucket then, and so on. For searching a bucket, proceed as follows. Search the A-block, but ignore the entries which are marked as deleted. Search the I-block, then, and put the informations you got together in order to determine the next bucket to be searched.

2) Insert: Each insert operation is preceded by a search. After it we know into which bucket (leaf) of the rectree5(m) the new element y must be inserted.
How to insert y into a bucket Z of a rectree5(i) depends on the state (basic, split, normalization state) Z is in.
If Z is in its basic state, i.e., the N-, I- and D-blocks are empty, y is inserted into the A-block. For it, first determine by a search into which leaf of the corresponding rectree5(i-1) y must be inserted and then proceed recursively.
As soon as Z contains $4u_{i-1}$ entries, it changes over into its split state. This fact as well as the middle element and a pointer to the smallest element of the A-block are entered into the S-block. From now, the A-block remains untouched for the next u_{i-1} update operations. Insertions are directed to I-block 1 in case y is smaller or to I-block 2 in case y is greater than the middle element. Each update operation is followed by 4 transmit operations explained now. Although the N-blocks have been empty so far, they are 'prepared', i.e., empty buckets and pointers of a rectree5(i-1) structure for $2u_{i-1}$ entries have already been established. A transmit operation copies the next smallest entry of the A-block (addressed by the pointer in S-block) to N-block 1 or 2 depending on the middle element. Non-leaf elements are also inserted into the N-blocks. The pointer is increased.
After the u_{i-1} updates, all elements of the A-block have been transmitted to the N-blocks. The I- and D-blocks contain between 0 and u_{i-1}, Z between $3u_{i-1}$ and $5u_{i-1}$ entries. Now, we split Z physically into two new buckets and the middle element. N-block 1 becomes the A-block of the first, N-block 2 that of the second bucket. The I- and D-blocks change over respectively. New N- and S-blocks are also

established and 'prepared'. The new buckets become sons of the middle element and the middle element is inserted into the father bucket of the rectree5(i) as described. Pointers are treated appropriately.
The new rectree5(i) buckets are in their normalization state. This state is similar to the basic state, but here each update operation is followed by two others carrying out the inserts and, if there are not any more, the deletes stored in the I- and D-blocks over the A-block.

3) Delete: Each delete operation is preceded by a search which determines the bucket of the rectree5(m) containing the element.
An element y can be deleted from a bucket Z of a rectree5(i) as follows.
If Z is in its basic state and contains more than u_{k-1} entries, the element is deleted from the A-block with the help of the same, recursively applied procedure, which results in a delete from an $[u_1,2u_1]$-tree.
If Z is in its split state, y is only marked as deleted, but not physically deleted in the A- or I-block. Furthermore, y is inserted into D-block 1 or 2 depending on the middle element.
If Z is in its normalization state, but contains more than u_{i-1} entries, y is either deleted from the A-block or marked as deleted in the I-block.
If Z contains only u_{i-1} entries, we must consider its brother buckets in the rectree5(i). If one of them, for instance the left brother W, contains more than u_{i-1} entries, an overflow is possible. We delete the greatest element from W, change it with the middle element between W and Z and insert this into Z. Now the delete can be carried out as described above.
If both brothers of Z contain only u_{i-1} entries, too, a fuse becomes necessary. We notice that Z and its brothers, for instance the left brother W, have empty I-blocks. After carrying out the delete in Z, we copy the A-block of Z, which is in this state identical with the whole Z, to W and remove Z. Furthermore we delete the middle element between W and Z from their father bucket and insert it into W.

In the following, we describe how to store a rectree5 in $BT_{f^{(s)}}$. On principle, the whole rectree5 is stored in the m-th memory level. The other memory levels are used for carrying out the dictionary operations. The recursion levels of the rectree5 correspond to the memory levels of $BT_{f^{(s)}}$, i.e., an order(i)-bucket (i=1..m-1) is intended to be moved to the i-th memory level in case of an operation with it. All buckets occupy a fixed number of locations independent on how full they are. An appropriate relation between b_i and u_i is $u_i = \left\lfloor \frac{b_i}{32^i} \right\rfloor$. It will be established in Lemma 1.

Each rectree5(i) is stored within a separate memory interval. Its buckets, however, are stored in arbitrary order within this interval. They are connected by pointers containing local addresses.

The constituents of the rectree5 structure are the order(1)-buckets which contain the following informations:

- parent address (address of an order(1)-bucket)
- $u_1..2u_1$ entries and corresponding pointers (the smallest entry is connected with a left and a right, the others only with right pointers.)
- informations about the order(i)-buckets this order(1)-bucket belongs to, for i=1..m-1:
  - total number of elements in that subtree of the rectree5(i) rooted in the order(1)-bucket
  - if the order(1)-bucket lies along the path from the root to the middle element of the order(i)-bucket, additionally
    - → the greatest element f_1 of the order(1)-bucket smaller and the smallest element f_2 of it greater than the middle elment of the order(i)-bucket
    - → number of entries in the order(i)-bucket smaller f_1
    - → number of them greater f_2

  (These informations are necessary for determining the middle element)

LEMMA 1: Let $b_1 \geq 5m$. Then the algorithm works correctly, for $u_i = \left\lfloor \frac{b_i}{32^i} \right\rfloor$

Proof (sketch): The procedures search, insert and delete described above work correctly in rectree5, since the number of entries in the blocks does not accumulate.

As we will now prove inductively, the assumption $u_i = \left\lfloor \frac{b_i}{32^i} \right\rfloor$ ensures that the i-th memory level is capable to hold an order(i)-bucket.

An order(1)-bucket occupies at most $4u_1+4m-2 < \frac{1}{8}b_1+\frac{4}{5}b_1 < b_1$ locations.

An order(i)-bucket needs place for altogether $15u_i$ entries in its blocks. On the example of the A-block, $4u_i$ entries must fit into $\frac{4}{15}b_i$ locations. An order(i-1)-bucket contains at least u_{i-1} entries. The at most $\left\lceil \frac{8u_i}{u_{i-1}} \right\rceil$ order(i-1)-buckets fit into $\frac{4}{15}b_i$ locations, from the assumption. ■

So far, we have ignored the contents connected with the keys. How to store them effectively, depends on their size. If they are large, we can store each content in a separate normal-sized leaf of the rectree5(m), if necessary even in a list of memory fields handled as rectree5(m) buckets.

3. Running Time is Optimal

In the following we assume $b_{i-1} \leq a_i$. This assumption does not really restrict the generality of our considerations. If b_{i-1} increased beyond a_i (taken as constant), algorithms were faster if they would not exhaust the whole possible block length. Hence, such a memory organization was inexpedient. Furthermore we only regard comparison-based algorithms.

Theorem 1: The complexity of the dictionary operations search, insert and delete is $\Theta^{BT_{f^{(s)}}}\left(\log n \cdot \underset{i=1..m}{\text{MAX}}\left\{\frac{a_i}{\log(b_{i-1}+1)}\right\}\right)$.

<u>Proof</u> (sketch): a)upper bound: As easily to be seen, the time to deal with and the number of accesses to any bucket are for insert and delete by at most a constant factor greater than for search.

From the inequality about the height of [a,b]-trees given in [M84], the height h_i of an order(i)-bucket ($i \leq 2$), which is defined as the height of the corresponding rectree5(i), can be estimated by $h_i \leq c_i \frac{\log u_i}{\log u_{i-1}}$ for some constant c_i.
Let $T_{trans}(5u_i)$ denote the sum of the time to copy an order(i)-bucket from the i+1-th to the i-th memory level and that to search it. Then the running time of the search, insert and delete algorithms obeys the recurrence formula

$$T(n) \leq c_m \cdot \frac{\log n}{\log u_{m-1}} \cdot T_{trans}(5u_{m-1})$$

$$T_{trans}(5u_i) \leq a_{i+1} + b_i + c_i \cdot \frac{\log u_i}{\log u_{i-1}} \cdot T_{trans}(5u_{i-1}) \qquad (i=3..m-1)$$

$$T_{trans}(5u_2) \leq a_3 + b_2 + c_2 \cdot \frac{\log u_2}{\log u_1} \cdot T_{trans}(2u_1)$$

$$T_{trans}(2u_1) \leq a_2 + b_1 + c_1 \cdot a_1 \cdot \log u_1$$

It implies the explicit formula $T(n) \le c \cdot \left[\sum_{i=0}^{m-1} \frac{\log n}{\log u_i} (a_{i+1}+b_i)\right]$

with c being some constant and $u_0=2$.

$$\begin{aligned}\text{Hence, } T(n) &= O^{BT_{f^{(s)}}}\left[\log n \cdot \operatorname*{MAX}_{i=0..m-1}\left\{\frac{a_{i+1}}{\log u_i}\right\}\right] \\ &= O^{BT_{f^{(s)}}}\left[\log n \cdot \operatorname{MAX}\left\{\operatorname*{MAX}_{i=2..m}\left\{\frac{a_i}{\log(b_{i-1}/32^{i-1})}\right\}, a_1\right\}\right] \\ &= O^{BT_{f^{(s)}}}\left[\log n \cdot \operatorname{MAX}\left\{\operatorname*{MAX}_{i=2..m}\left\{\frac{a_i}{\log b_{i-1}-5i}\right\}, a_1\right\}\right] \\ &= O^{BT_{f^{(s)}}}\left[\log n \cdot \operatorname*{MAX}_{i=1..m}\left\{\frac{a_i}{\log(b_{i-1}+1)}\right\}\right] \qquad (b_i^*=32^{i+1})\end{aligned}$$

b)lower bound: since we assumed that inserts and deletes include a search, it suffices to regard searches.
For each i (i=1..m), b_{i-1} divides the memory into two parts forming a two-level-hierarchy. We ignore all cost except of the access cost a_i for a transfer across the b_{i-1} border. The algorithm can be represented by a multi-ary decision tree. Each node is marked by at most b_{i-1} elements moved across the border in one transfer. Since b_{i-1} elements form $b_{i-1}+1$ intervals according to the order of the universum, the outdegree of each node is at most $b_{i-1}+1$. Consequently, the height of the tree is at least $\frac{\log(n+1)}{\log(b_{i-1}+1)}$. Because of the node cost a_i, $T(n) \ge a_i \cdot \frac{\log(n+1)}{\log(b_{i-1}+1)}$. The claim follows.

4. Conclusions

The aim of this paper was to investigate the relationship between memory organization and efficiency of algorithms on the example of the dictionary problem. We refered to a discretization of the BT_f model introduced in [ACS87]. We suggested an adapted measure of asymptotic complexity and proved a tight bound for the operations search, insert and delete with it.
Although our algorithms are asymptotically optimal, they base on online transformations which waste space and cause very large constants. One aim of further research could be to improve the algorithms.
First of all, we tried to design algorithms increasing the efficiency of dictionary implementations in a given memory hierarchy. But on the other hand, the formula of the tight bound suggests that the quotient between a_i and $\log(b_{i-1}+1)$ should have about the same size for all

i=1..m. Although such a result must be qualified according to the regarded operations and to practical influences (e.g. price of production of memory devices), it could be interesting for hardware designers.

The memory access complexity of other basic computation problems should be investigated in a similar way as it was already done for the problem of reading each input ([L91]) and for matrix operations ([ACF90]). One could also modify the memory model, for instance in direction to the VM_f model or by incorporating parallelism as in [VS90].

Acknowledgement

Thanks to Herrmann Jung for reading and discussing earlier versions of this paper.

References

[AACS87] Aggarwal, A.; Alpern, B.; Chandra, A.K.; Snir, M.: A Model for Hierarchical Memory. -Proc. 19th Annual ACM Symp. on Theory of Comp.; 305-314

[AC88] Aggarwal, A.; Chandra, A.K.: Virtual Memory Algorithms. -Proc. 20th Annual ACM Symp. on Theory of Comp.; 173-185

[ACF90] Alpern, A.; Carter, L.; Feig, E.: Uniform Memory Hierarchies. Proc. 31th IEEE Symp. on Foundations of Comp. Science; 600-608

[ACS87] Aggarwal, A.; Chandra, A.K.; Snir, M.: Hierarchical Memory with Block Transfer. -Proc. 28th IEEE Symp. on Foundations of Comp. Science; 204-216

[AV87] Aggarwal, A.; Vitter, J.: The I/O Complexity of Sorting and Related Problems. -Proc. 14th Int. Coll. on Automata, Languages and Programming.; 467-478

[F72] Floyd, R.W.: Permuting Information in Idealized Two-Level Storage. -In: Miller, R.E.; Thatcher, J.W. (editors): Complexity of Computer Computations. Plenum Press. New York 1972; 105-109

[L90] Leopold, C.: Schnelles Suchen in Speicherhierarchien. -Diplomarbeit Humboldt-Universität zu Berlin

[L91] Leopold, C.: Reading Data in a Hierarchical Memory. EIK 4/91

[M84] Mehlhorn, K.: Data Structures and Algorithms. Vol. 1: Sorting and Searching. Berlin 1984

[VS90] Vitter, J.S.; Shriver, E.A.M.: Optimal Disk I/O with Parallel Block Transfer. -Proc. 22th Symp. on Theory of Comp.; 159-169

Upper and lower bounds for certain GRAPH-ACCESSIBILITY-PROBLEMs on bounded alternating ω-BRANCHING PROGRAMs

Christoph Meinel
Fachbereich Informatik
Humboldt-Universität zu Berlin
D-1086 Berlin, PF 1297

Stephan Waack
Karl-Weierstraß-Institut für Mathematik
D-1086 Berlin, PF 1304

Abstract

In the following we investigate the computational complexity of various ω-*GRAPH ACCESSIBILITY PROBLEMs* on the most general restricted type of ω-branching programs for which, up to now, exponential lower bounds on the size can be proved. By means of exponential lower bounds on various ranks of certain communication matrices we prove that ω*GRAPH ACCESSIBILITY PROBLEMs* can not be computed by bounded alternating ω-branching programs within polynomial size. In contrast, ω-*GRAPH ACCESSIBILITY PROBLEMs* restricted to monotone graphs can by computed by such devices.

1. Introduction

The interest in the various *GRAPH ACCESSIBILITY PROBLEMs* is mainly based on the property that *GRAPH ACCESSIBILITY PROBLEMs* are natural complete problems in logarithmic space-bounded complexity classes [e. g. Mei89]. Hence it is a favorite aim to characterize the complexity of such problems in more detail.

Relating the different solutions of the path problem in the various *GRAPH ACCESSIBILITY PROBLEMs* to the algebraic structure of certain semirings we are able to unify investigations substantially by considering ω-*GRAPH ACCESSIBILITY PROBLEMs*, $\omega : \mathbb{N} \twoheadrightarrow R$ semiring homomorphism.

In the following we investigate the computational complexity of such ω-*GRAPH ACCESSIBILITY PROBLEMs* on the most general restricted type of ω-branching programs for which, up to now, exponential lower bounds on the size can be proved [MW91]. By means of communication complexity theoretic considerations we prove that the ω-*GRAPH ACCESSIBILITY PROBLEM* can not be computed by such so-called bounded alternating ω-branching programs within polynomial size

(Theorem 7). This result is nicely contrasted by the fact, that the ω-*GRAPH ACCESSIBILITY PROBLEM* restricted to monotone graphs is computable by polynomial size bounded alternating ω-branching programs (Proposition 3).

In particular, we prove the following lower bounds

- The usual *GRAPH ACCESSIBILITY PROBLEM GAP* is not computable by polynomial size bounded alternating nondeterministic branching programs.
- *NO-PATH-GAP* is not computable by polynomial size bounded alternating co-nondeterministic branching programs.
- *ODD-PATH-GAP* is not computable by polynomial size bounded alternating parity branching programs.
- *MOD-p-GAP* is not computable by polynomial size bounded alternating MOD_p-branching programs.

By the way, these bounds are the first exponential lower bounds on the size of bounded alternating ω-branching programs for natural problems. So far lower bounds could be proved merely for some artificial problems [MW91] or for more restricted types of branching programs [AM86, KW89, KMW89, DKMW90]. Due to the theory of bounded alternating ω-branching programs developed in [MW91] we derive the mentioned lower bounds from some exponential lower bounds on certain R-ranks of communication matrices of the ω-*GRAPH ACCESSIBILITY PROBLEM* that contribute to the investigation of the communication complexity of graphs problems [e. g. HMT88, Lov89].
The paper is structured as follows. In Section 1 we define ω-*GRAPH ACCESSIBILITY PROBLEMs*, $\omega : \mathbb{N} \twoheadrightarrow R$, semiring homomorphism onto the finite semiring R. In Section 2 we introduce the concept of ω-branching programs and present some previous results concerning ω-branching programs, and the completeness of ω-*GRAPH ACCESSIBILITY PROBLEMs*. In Section 3 we consider bounded alternating ω-branching programs and prove that ω-*GRAPH ACCESSIBILITY PROBLEMs* for monotone graphs can be computed by such bounded alternating ω-branching programs.
Then, in Section 4 we derive some exponential lower bounds on the R-rank of communication matrices of the ω-*GRAPH ACCESSIBILITY PROBLEM* which are of great interest also in the context of communication complexity theory. These bounds provide, due to the lower bound criterion of Lemma 4, Section 3, lower bounds on the size of bounded alternating ω-branching programs which are presented beside of some separation results in the final section.

2. ω-Graph accessibility problems

The classical *GRAPH ACCESSIBILITY PROBLEM GAP* $= \{GAP_N\}_{N \in \mathbb{N}}$ for digraphs consists in the decision whether there is a path in a given directed n-node graph $G = (V, E), V = \{1, \ldots, n\}$, that leads from node 1 to node n. As usual, let G be given by its adjacency matrix $G = (a_{ij})_{1\le i,j\le n, i\ne j}$ with

$$a_{ij} = a(i,j) = \begin{cases} 1 & \text{if } (i,j) \in E, \\ 0 & \text{otherwise.} \end{cases}$$

For $N = n^2 - n, GAP_N : \{0,1\}^N \longrightarrow \{0,1\}$ is defined by

$$(a_{ij}) \longrightarrow \begin{cases} 1 & \text{if there is a path in the graph } G = (a_{ij}) \text{ from 1 to } n, \\ 0 & \text{otherwise.} \end{cases}$$

The problem *GAP* has been extensively studied in the past. For example, it has been shown to be complete in the complexity class *NL* of nondeterministic logarithmic space-bounded computations (complete via logspace reductions [Sav70], projection translations [Imm87], and via p-projection reductions for nonuniform *NL* [Mei86]). Soon it could be realized that certain modified *GRAPH ACCESSIBILITY PROBLEMs* have similar properties for complexity classes like *co-NL*, $\oplus L$, MOD_p-L, p prime, defined by logarithmic space-bounded computations under modified accepting modes. Interestingly, the structural backround of all such results is of algebraic nature. Relating the different solutions of the path problem in the modified *GRAPH ACCESSIBILITY PROBLEMs* under consideration to the algebraic structure of certain semirings we are able to unify considerations substantially. Let $R = [R; +_R, O_R, \cdot_R, 1_R]$ be a semiring with a *null*, O_R, and a *one*, 1_r, and let ω,

$$\omega : I\!N \twoheadrightarrow R,$$

be a semiring homomorphism from $I\!N = [I\!N; +, 0, \cdot, 1]$ onto R.
We define the ω-*GRAPH ACCESSIBILITY PROBLEM* ω-$GAP = \{\omega\text{-}GAP_N\}$ by

$$\omega\text{-}GAP_N(G) = 1 \text{ iff } \omega(\#[1 \xrightarrow{G} n]) = 1_R,$$

where $\#[i \xrightarrow{G} j]$ denotes the number of (directed) paths leading, in G, from node i to node j. Now, if $\omega_\vee, \omega_\oplus$, and ω_p, p prime, denote the semiring homomorphisms

$$\begin{aligned} \omega_\vee \; : \; I\!N & \twoheadrightarrow I\!B = [\{0,1\}; \vee, 0, \wedge, 1], \\ n & \mapsto \begin{cases} 1 & \text{iff } n > 0 \\ 0 & \text{otherwise,} \end{cases} \\ \omega_\oplus \; : \; I\!N & \twoheadrightarrow I\!F_2 = [\{0,1\}; \oplus, 0, \wedge, 1], \\ n & \mapsto \begin{cases} 1 & \text{iff } n \text{ is odd} \\ 0 & \text{otherwise,} \end{cases} \\ \text{and } \omega_p \; : \; I\!N & \twoheadrightarrow I\!F_p = [\{0,\ldots,p-1\}; +(\text{mod } p), 0, \cdot(\text{mod } p), 1], \\ n & \mapsto n \text{ mod } p, \end{aligned}$$

for each prime $p, p \in I\!N$, then, in the usual denotations of literature [e. g. Mei89], we have

$$GAP = \omega_\vee\text{-}GAP, \; ODD\text{-}PATH\text{-}GAP = \omega_\oplus\text{-}GAP, \; MOD\text{-}p\text{-}GAP = \omega_p\text{-}GAP.$$

Hence, ω-*GRAPH ACCESSIBILITY PROBLEMs* provide a comfortable frame for investigating modified *GRAPH ACCESSIBILITY PROBLEMs*.

Beside of the ω-*GRAPH ACCESSIBILITY PROBLEMs* $\omega\text{-}GAP = \{\omega\text{-}GAP_N\}$ the complementary problems $\neg(\omega\text{-}GAP) := \neg\{\omega\text{-}GAP_N\}$, are of interest. While $\neg(\omega_\oplus\text{-}GAP)$ and $\omega_\oplus\text{-}GAP$ as well as $\neg(\omega_p\text{-}GAP)$ and $\omega_p\text{-}GAP$ are computationally equivalent with respect to a variety of computation devices [Mei89, DKMW90] this is not true for $\neg(\omega_\vee\text{-}GAP)$, well-known as *NO-PATH-GAP*,

$$NO\text{-}PATH\text{-}GAP = \neg(\omega_\vee\text{-}GAP).$$

In the following w.l.o.g. we generally assume the acyclic digraphs to be of outdegree ≤ 2 [Mei89]. If we additionally restrict our considerations to monotone digraphs G (i. e. for each edge (i,j) of G it holds $i < j$) we write $\omega\text{-}GAP\text{mon}$. In particular, we have

$$\omega_\vee\text{-}GAP\text{mon} = GAP\text{mon}2, \neg(\omega_\vee\text{-}GAP\text{mon}) = NO\text{-}PATH\text{-}GAP\text{mon}2,$$

$$\omega_\oplus\text{-}GAP\text{mon} = ODD\text{-}PATH\text{-}GAP\text{mon}2, \omega_p\text{-}GAP\text{mon} = MOD\text{-}p\text{-}GAP\text{mon}2,$$

p prime, and

$$id_{\mathbb{N}}\text{-}GAP\text{mon} = GAP\text{mon}1$$

if, in the case $\omega = id_{\mathbb{N}}$, we restrict ourself to the consideration of digraphs of outdegree 1.

3. ω-Branching programs and previous results

In order to characterize and investigate nonuniform logarithmic space-bounded complexity classes the combinatorial computation model of branching programs has proved to be of great importance [see e. g. Mei89].

An ω-branching program $P, \omega : \mathbb{N} \twoheadrightarrow R$ a semiring homomorphism, is a directed acyclic graph where each node has outdegree $\leq \#R$. There is a distinguished node v_s, the source, which has indegree 0. Exactly two nodes, v_0 and v_1, are of outdegree 0. They are called 0-sink and 1-sink of P, respectively. Nodes of outdegree $\#R$ are labelled by variables x over $R, x \in X = \{x_1, \ldots, x_n\}$, or remain unlabelled as the remaining nodes of P. Each edge starting in a labelled node is labelled by an element $r, r \in R$, such that no two edges have the same label.

Each input $a = a_1 \ldots a_n \in R^X$ defines some computation paths in P : Starting at the source v_s, a node v of P is connected with one of its successor nodes. If v is labelled by the variable x_i then v is connected with the successor node of v that is reached from v via the edge labelled by a_i. A computation path is said to be accepting if it ends up in the 1-sink v_1. The number of accepting computation paths is denoted by $\text{acc}_P(a)$, $\text{acc}_P(a) \in \mathbb{N}$. Now, P is said to accept $a \in R^X$ iff $\omega(\text{acc}_P(a)) = 1_R$.

Obviously, ω-branching programs generalize ordinary branching programs where each non-sink node is assumed to be labelled by a Boolean variable. Hence, for each input, there is exactly one computation path which is accepting if it ends up in the 1-sink.

Moreover, ω-branching programs generalize Ω-branching programs introduced in [Mei88], $\Omega \subseteq \mathbb{B}_2$, and MOD_p-branching programs introduced in [DKMW90], $p \in \mathbb{N}$. For example, using the semiring homomorphisms $\omega_\vee, \omega_\oplus, \omega_p$ defined in Section 1 it can easily be seen that

- disjunctive $\{\vee\}$-branching programs are $\omega_\vee$-branching programs (they accept if at least one computation path is accepting),
- parity $\{\oplus\}$-branching programs are $\omega_\oplus$-branching programs (they accept if the number of accepting computation paths is odd),

  and
- MOD_p-branching programs, p prime, are ω_p-branching programs (they accept if the number of accepting computation paths equals 1 modulo p).

Due to this correspondence and the results of [PŽ83, Mei88, DKMW90] we immediately obtain the following characterization of the well-known nonuniform logarithmic space-bounded complexity classes $\mathcal{L} = L/poly, \mathcal{NL} = NL/poly, co\text{-}\mathcal{NL} = co\text{-}NL/poly, \oplus\mathcal{L} = \oplus L/poly, MOD_p\text{-}\mathcal{L} = (MOD_p\text{-}L)/poly$ by means of (sequence of) polynomial size ω-branching programs. Recall, the size of an ω-branching program P, Size(P), is the number of non-sink nodes of P. By $\mathcal{P}_{\omega\text{-}BP}$ we denote the complexity class of all languages $A, A \subseteq R^*$ that can be accepted by sequence of polynomial size ω-branching programs.

Fact 1.

(1)	$\mathcal{P}_{BP}$	$=$	$\mathcal{L}$	[PŽ83],
(2)	$\mathcal{P}_{\omega_\vee-BP}$	$=$	$\mathcal{NL}$	[Mei88],
(3)	$\neg(\mathcal{P}_{\omega_\vee-BP})$	$=$	$co\text{-}\mathcal{NL}$	[Mei88],
(4)	$\mathcal{P}_{\omega_\oplus-BP}$	$=$	$\mathcal{P}_{\omega_2-PB} = \oplus\mathcal{L}$	[Mei88], and
	$\mathcal{P}_{\omega_p-PB}$	$=$	$MOD_p\text{-}\mathcal{L}, p$ prime ,	[DKMW90]. □

Due to these branching program descriptions of nonuniform logarithmic space-bounded complexity classes it could be proved that each of these classes can be characterized by the completeness of a certain *GRAPH ACCESSIBILITY PROBLEM* (for details we refer to [Mei89]).

Fact 2.
Let $\omega : \mathbb{N} \twoheadrightarrow R$ be a semiring homomorphism of $\mathbb{N}$ onto R. Then ω-GAP and ω-GAPmon are (p-projection) complete in $\mathcal{P}_{\omega-BP}$. In detail,

- $GAP1$ and GAPmon1 are (p-projection) complete in $\mathcal{L} = \mathcal{P}_{BP}$.
- $GAP1$ and GAPmon2 are (p-projection) complete in $\mathcal{NL} = \mathcal{P}_{\omega_\vee-BP}$.
- $NO\text{-}PATH\text{-}GAP$ and $NO\text{-}PATH\text{-}GAPmon2$ are (p-projection) complete in $co\text{-}\mathcal{NL} = \neg(\mathcal{P}_{\omega_\vee-BP})$.
- $ODD\text{-}PATH\text{-}GAP$ and $ODD\text{-}PATH\text{-}GAPmon2$ are (p-projection) complete n $\oplus\mathcal{L} = \mathcal{P}_{\omega_\oplus-BP}$.
- $MOD\text{-}p\text{-}GAP$ and $MOD\text{-}p\text{-}GAP$mon2 are (p-projection) complete in $MOD_p\text{-}\mathcal{L}$ for each p, p prime. □

4. Bounded alternating ω-Branching programs

In order to establish differences in the computational power of polynomial size ω-branching programs we consider bounded alternating ω-branching programs.
Let X be a set of variables over a finite semiring R, and let $Y, Z \subseteq X, \#Y = \#Z = \varepsilon n, \varepsilon > 0$, be two disjoint subsets of X. An ω-branching program $P, \omega : I\!N \twoheadrightarrow R$ semiring homomorphism onto R, that tests variables of X is said to be of alternation length α with respect to Y and Z if each path of P can be devided into α segments in which, alternating, variables of Y or variables of Z are not tested. A sequence of ω-branching programs P_n is called bounded alternating if there exist two disjoint subsets $X, Z \subseteq X, \#Y = \#Z = \varepsilon n, \varepsilon > 0$, such that P_n is of alternation length $n^{o(1)}$ with respect to Y and Z.

By $\mathcal{P}_{ba\ \omega-BP}$ we denote the set of languages $A \subseteq R^*$ that can be accepted by sequences of polynomial size, bounded alternating ω-branching programs. For the particular semiring homomorphisms $\omega = \omega_\vee, \omega_\oplus, \omega_p$ (p prime), we write, due to Fact 1, $\mathcal{NL}_{ba}, \oplus\mathcal{L}_{ba}$, and MOD_p-$\mathcal{L}_{ba}$ instead of $\mathcal{P}_{ba\ \omega_\vee-BP}$, $\mathcal{P}_{ba\ \omega_\oplus-BP}$, and $\mathcal{P}_{ba\ \omega-BP}$, respectively.

Although polynomial size, bounded alternating ω-branching are definitively less powerful than unbounded alternating ones [MW91] they are quite powerful computation devices. This can be demonstrated considering the monotone ω-*GRAPH ACCESSIBILITY PROBLEMs* ω-GAPmon, that, due to Fact 2, are known to be complete in $\mathcal{L}, \mathcal{NL}$, *co*-$\mathcal{NL}, \oplus\mathcal{L}$, and MOD_p-$\mathcal{L}$, respectively.

Proposition 3.
Let $\omega : I\!N \twoheadrightarrow R$ be a semiring homomorphism. Then $\omega - GAP\text{mon} \in \mathcal{P}_{ba\ \omega-BP}$.

In detail,
- $GAP\text{mon}1 \in \mathcal{P}_{ba\ BP} = \mathcal{L}_{ba}$.
- $GAP\text{mon}2 \in \mathcal{P}_{ba\ \omega_\vee-BP} = \mathcal{NL}_{ba}$.
- $NO\text{-}PATH\text{-}GAP\text{mon}2 \in \neg(\mathcal{P}_{ba\ \omega_\vee-BP}) = co\text{-}\mathcal{NL}_{ba}$.
- $ODD\text{-}PATH\text{-}GAP\text{mon}2 \in (\mathcal{P}_{ba\ \omega_\oplus-BP}) = \oplus\mathcal{L}_{ba}$.
- $MOD\text{-}p\text{-}GAP\text{mon}2 \in (\mathcal{P}_{ba\ \omega_p-BP}) = MOD_p\text{-}\mathcal{L}_{ba}$ for each prime p. □

The lower bound technique for proving exponential lower bounds on the size of bounded alternating ω-branching programs is mainly based on considerations of invariants of the communications matrix $M_{Y,Z}(f)$ of a function f. If

$$f : R^Y \times R^Z \longrightarrow \{0,1\}$$

then $M_{Y,Z}(f)$ is a $(\#R)^{\#Y} \times (\#R)^{\#Z}$ matrix which is defined, for $a_y \in \{0,1\}^Y, a_Z \in \{0,1\}^Z$, by

$$(M_{Y,Z}(f))_{a_Y,a_Z} = f(a_Y, a_Z).$$

The matrix-invariants we work with are certain matrix-ranks. Recall, if R is a semiring and if $M \in I\!M_R(n,m)$ is a $n \times m$ matrix over R then the rank of M, $\text{rank}_R(M)$, is the minimum k such that M can be written as $M = A \cdot B$ with $A \in I\!M_R(n,k)$ and $B \in I\!M_R(k,m)$.

In order to derive exponential lower bounds on the size of bounded alternating ω-branching programs, $\omega : \mathbb{N} \twoheadrightarrow R$, that compute a function f, due to the following lemma, it suffices to derive exponential lower bounds on the R-ranks of the communication matrices of certain subfunctions of f. If $Y, Z \subseteq X$ are disjoint subsets of X, and if $c \in \{0,1\}^{X-(Y \cup Z)}$ then we denote by

$$f^c : R^{Y \cup Z} \longrightarrow \{0,1\}$$

the subfunction

$$f^c(y,z) := f(c,y,z) \text{ for all } y \in R^Y, z \in R^Z.$$

Lemma 4 (The lower bound criterion).
Let $f : R^X \longrightarrow \{0,1\}$ be a discrete function, let $Y, Z \subseteq X$ be two disjoint subsets of X, and let P be an ω-branching program of alternation length α with respect to Y and Z that computes f. Then

$$\text{Size}(P) \geq (\text{rank}_R\, M_{Y,Z}(f^c))^{1/3\alpha}$$

for each $c \in R^{X-(Y \cup Z)}$. □

5. Some lower bounds for the rank of communication matrices

In the following we derive exponential lower bounds on the R-rank of certain communication matrices of the ω-*GRAPH ACCESSIBILITY PROBLEM*, $\omega : \mathbb{N} \twoheadrightarrow R$. In order to do this we start with an easy observation concerning the *SEQUENCE EQUALITY TEST* $SEQ_{2n} = \{SEQ_{2n}\}$,

$$SEQ_{2n} : R^Y \times R^Z \longrightarrow \{0,1\},$$

$\#Y = \#Z = n$, defined by

$$SEQ_{2n}(y,z) = 1 \text{ iff } y = z.$$

Proposition 5.
Let R be a finite semiring. Then it holds

$$\text{rank}_R\, M_{Y,Z}(SEQ_{2n}) = (\#R)^n. \quad \square$$

Using *SEQUENCE EQUALITY TESTs* we derive exponential lower bounds on the rank of communication matrices $M_{Y,Z}$ for the ω-*GRAPH ACCESSIBILITY PROBLEMs*, $Y, Z \subseteq X$ disjoint subsets. The conceptual idea is to identify, by means of p- projections respecting Y, Z submatrices M' of $M_{Y,Z}$ corresponding to an *SEQUENCE EQUALITY TEST*. Since, due to Proposition 5, the rank of M' is exponential the same must be true for the rank of $M_{Y,Z}$. Recall, a mapping π_m

$$\pi_m : \{x_1, \ldots, x_n\} \longrightarrow \{u_1, \bar{u}_1, \ldots, u_m, \bar{u}_m, 0, 1\}$$

is called a *p-projection* from $M \subseteq \{0,1\}^m$ to $N \subseteq \{0,1\}^n$ if

$$M(u_1,\ldots,u_m) = N(\pi_m(x_1),\ldots,\pi_m(x_n))$$

and if m and n are polynomially related. π_m is said to respect Y, Z if $(\pi_m(Y) \cup \overline{\pi_m(Y)}) \cap (\pi_m(Z) \cup \overline{\pi_m(Z)}) = \emptyset$.

Proposition 6.
Let $X = \{x_{ij}, 1 \leq i,j \leq n, i \neq j\}$, $N := \#X = n^2 - n$, and $Y, Z \subseteq X$ be disjoint subsets with $\#Y = \#Z = \theta(N)$. Then there is a partial assignment $c \in \{0,1\}^{X-(Y\cup Z)}$ such that

$$\text{rank}_R\ M_{Y,Z}(\omega\text{-}GAP_N^c) = \exp(n),$$

for $\omega = \omega_\vee, \omega_\oplus$ or ω_p (p prime). □

6. Conclusion

The results of the last section imply, due to the lower bound criterion of Lemma 4, exponential lower bounds on the size of bounded alternating ω-branching programs that solve the ω-*GRAPH ACCESSIBILITY PROBLEM.*

Theorem 7.
Let $\omega : I\!N \twoheadrightarrow R$ be a semiring homorphism onto finite semiring R, and let P be a bounded alternating ω-branching program, that computes the ω-*GRAPH ACCESSIBILITY PROBLEM* ω-GAP_N. Then

$$\text{Size}(P) = \exp(\Omega(\sqrt{N})). \quad \square$$

Theorem 7 implies the following corollar, that contrasts Proposition 3.

Corollary 8.
(1) $GAP1 \notin \mathcal{L}_{ba}$,
(2) $GAP2 \notin \mathcal{NL}_{ba}$,
(3) $NO\text{-}PATH\text{-}GAP2 \notin co\text{-}\mathcal{NL}_{ba}$,
(4) $ODD\text{-}PATH\text{-}GAP2 \notin \oplus\mathcal{L}_{ba}$, and
(5) $MOD_p\text{-}GAP2 \notin MOD_p\text{-}\mathcal{L}_{ba}$.

Due to Fact 2, the logarithmic space-bounded complexity classes related to bounded alternating ω-branching programs and those related to unbounded ones can be separated from each other by means of ω-*GRAPH ACCESSIBILITY PROBLEMs*.

Corollary 9.
Let $\omega = \omega_\vee, \omega_\oplus$, or ω_p (p prime). Then

$$\mathcal{P}_{ba\ \omega-BP} \subsetneq \mathcal{P}_{\omega-BP}.$$

In detail,

$$\mathcal{L}_{ba} \subsetneqq \mathcal{L},\ \mathcal{NL}_{ba} \subsetneqq \mathcal{NL},\ co\text{-}\mathcal{NL}_{ba} \subsetneqq co\text{-}\mathcal{NL},\ \oplus\mathcal{L}_{ba} \subsetneqq \oplus\mathcal{L},\ \text{and } MOD_p\text{-}\mathcal{L}_{ba} \subsetneqq MOD_p\text{-}\mathcal{L}, p \text{ prime.} \quad \square$$

Finally, we remark only that the complexity classes $\mathcal{L}_{ba}, \mathcal{NL}_{ba}, co\text{-}\mathcal{NL}_{ba}, \oplus\mathcal{L}_{ba}, MOD_p\text{-}\mathcal{L}_{ba}$ (p prime) related to bounded alternating ω-branching programs are separated from each other. The proof of the separation results is based on considerations of the ω-*ORTHOGONALITY TEST* that are similar to Prop. 3 (upper bounds) and Prop. 6 (lower bounds). For details we refer to [MW91].

References

[AM86] N. Alon, W. Maass: Meanders, Ramsey theory and lower bounds, Proc. 27th ACM STOC, 1986, 30–39.

[DKMW90] C. Damm, M. Krause, Ch. Meinel, S. Waack: Separating Restricted MOD_p-Branching Program Classes, Informatik-Preprint 3; Humboldt-Universität Berlin, 1990.

[Imm87] N. Immerman: Languages that Capture Complexity Classes, SIAM J. Comput., Vol. 16, No. 4, 1987, 760–778.

[HMT88] A. Hajnal, W. Maass, G. Turan: On the Communication Complexity of Graph Problems. Proc. 20th STOC (1988) 186–191.

[KMW89] M. Krause, Ch. Meinel, S. Waack: Separating Complexity Classes Related to Certain Input Oblivious Logarithmic Space Bounded Turing Machines, Proc. 4th IEEE Structure in Complexity Theory, 1989, 240–259.

[KW89] M. Krause, S. Waack: On Oblivious Branching Programs of Linear Length, Proc. FCT'89, LNCS 380, 287–296.

[Lov89] L. Lovasz: Communication Complexity: A Survey. Technical Report CS-TR-204-89, Princeton University.

[Mei86] Ch. Meinel: P-Projection Reducibility and the Complexity Classes L(nonuniform) and NL(nonuniform), Proc. MFCS'86, LNCS 233, 527–535.

[Mei88] Ch. Meinel: Polynomial Size Ω-branching Programs and their Computational Power, Proc. STACS'88, LNCS 294, 81–90.

[Mei89] Ch. Meinel: Modified Branching Programs and Their Computational Power, LNCS 370, Springer Verlag, 1989.

[MW91] Ch. Meinel, S. Waack: Upper and Lower Bounds for Certain Graph-Accessibility-Problems on Bounded Alternating Branching Programs. Preprint No. 276/1991, TU Berlin, FB Mathematik, 1991.

[PŽ83] P. Pudlak, S. Žak: Space Complexity of Computations, Techn. Report Univ. of Prague, 1983.

[Sav70] W. Savitch: Relationships Between Nonderterministic and Deterministic Tape Complexities, J. Comp. Sys. Sci., Vol. 4, No. 2, 1970, 177–192.

CCS DYNAMIC BISIMULATION is PROGRESSING*

Ugo Montanari and Vladimiro Sassone
Dipartimento di Informatica – Università di Pisa
Corso Italia 40 - 56100 - Pisa - Italy
E-MAIL:{ugo,vladi}@dipisa.di.unipi.it

Abstract

Weak Observational Congruence (woc) defined on CCS agents is not a bisimulation since it does not require two states reached by bisimilar computations of woc agents to be still woc, e.g. $\alpha.\tau.\beta.nil$ *and* $\alpha.\beta.nil$ *are woc but* $\tau.\beta.nil$ *and* $\beta.nil$ *are not. This fact prevents us from characterizing CCS semantics (when* τ *is considered invisible) as a final algebra, since the semantic function would induce an equivalence over the agents that is both a congruence and a bisimulation.*

In the paper we introduce a new behavioural equivalence for CCS agents, which is the coarsest among those bisimulations which are also congruences. We call it Dynamic Observational Congruence because it expresses a natural notion of equivalence for concurrent systems required to simulate each other in the presence of dynamic, i.e. run time, (re)configurations. We provide an algebraic characterization of Dynamic Congruence in terms of a universal property of finality.

Furthermore we introduce Progressing Bisimulation, which forces processes to simulate each other performing explicit steps. We provide an algebraic characterization of it in terms of finality, two characterizations via modal logic in the style of HML, and a complete axiomatization for finite agents. Finally, we prove that Dynamic Congruence and Progressing Bisimulation coincide for CCS agents. Thus the title of the paper.

1 Introduction

Understanding concurrent systems is difficult, since many of our intuitions about sequential systems cannot be extended to concurrent and distributed systems. In particular, there is no prevalent, notion of system behaviour on which semantic constructions can be based.

Milner's *Calculus of Communicating Systems* (CCS) ([Mil80], [Mil89]) can be considered the touchstone of process description languages. Its semantics is given in two steps. First, a *Labelled Transition Systems* (LTS), which constitutes the abstract machine (the interpreter) of the language, is defined in the style of Plotkin's *Structured Operational Semantics* (SOS) ([Plo81]). Then behavioural equivalences are introduced.

A large number of such equivalences have been proposed. Several properties are interesting in the analysis of concurrent systems, and each definition stresses a particular aspect of systems behaviour. For instance, if we are interested only in the actions performed by a system, we consider a simple trace equivalence; otherwise, if we allow the possibility of replacing a subsystem by an equivalent one, we must define an equivalence which is a *congruence* with respect to language constructs. Moreover, if we follow the *interleaving* approach ([Mil80], [Mil89]), i.e. if we express concurrency of actions by saying that they may be executed in any order, then we will choose to observe *sequences* of actions. In a *truly concurrent* approach, on the other hand, ([Pet62], [NPW81], [Pra86], [DDM90]) we may want to observe *partial orderings* of actions. For an overview and a comparison of many behavioural equivalences see [DeN87] or [GvG89].

Among the equivalences proposed in the literature, we consider those based on *bisimulation* ([Mil80], [Par81], [vGW89]). Two processes are equivalent if they not only produce the same observations, but also reach equivalent states afterwards and, in the case of *Branching Bisimulation*, pass only through equivalent intermediate states. The advantages of those equivalences, besides their operational suggestiveness, are the existence of simple axiomatizations, the elegance of the proofs and their relationship with

*Research supported in part by **HEWLETT–PACKARD** Laboratories, Pisa Science Center.

equivalences based on logics ([Mil89]), on denotational semantics ([Ab88]) and on algebraic techniques ([BB88], [Acz87]).

Ferrari and Montanari ([FM90]) define *Strong Observational Congruence*, the simplest equivalence based on bisimulation, in an algebraic way. They see the CCS transition system as equipped with an algebraic structure on both states (the operations of the language) and transitions (an operation for every SOS inference rule). Furthermore they define a collection (in fact a subcategory) of such transition systems, where the operations are not necessarily free and where morphisms relating two transition systems are *transition preserving*, i.e. they define simplification mappings which respect, besides operations on both nodes and arcs and besides labels (including τ's) on arcs, the transitions outgoing from any state. This subcategory has an initial and a final element, and the unique morphism from the former to the latter defines the coarsest equivalence on agents that is both a *congruence* and a *strong bisimulation*, i.e. *Strong Observational Congruence*.

A similar construction can be repeated by mapping computations instead of transitions, and disregarding τ's. We obtain the coarsest equivalence that is both a *congruence* and a *weak bisimulation*, but this equivalence is *not* the *Weak Observational Congruence*, since the latter is not a weak bisimulation. Actually, Van Glabbeek ([vGl87]) shows that Weak Observational Congruence is a bisimulation, but the operational semantics of CCS he assumes is not the usual one, e.g. $\alpha.\beta \xrightarrow{\alpha} \tau.\beta$. From these facts originated the idea of the new behavioural equivalence introduced in this paper.

The basic idea of *dynamic bisimulation* is to allow at every step of bisimulation not only the execution of an action (or a sequence of actions), but also the embedding of the two agents under measurement within the same, but otherwise arbitrary, context.

Conceptually, bisimulation can be understood as a kind of *game*, where two programs try in turn to match each other's moves. When a move consists of performing some computational steps and matching a move means to produce the same *observable* behaviour, then we obtain the notion of *observational equivalence*. This definition is independent of the particular observable behaviour (τ observable or not, sequences or partial orderings of actions, etc.), and it can be proved under very mild conditions that the maximal bisimulation relation *always* exists and is an equivalence ([MSg89]).

Instead of programs just being able to perform computational steps, we might consider *modular* (i.e. compositional) *software systems* which are statically configured at time zero. In our functional setting, this means to start the game by applying an arbitrary context to both agents. The resulting semantic notion is Milner's *Observational Congruence*.

Finally we may allow *dynamic reconfiguration*: at any instant in time the structure of both software systems may be modified, but in the same way; i.e. a context can be applied to both agents. In this way we obtain our new notion of *dynamic bisimulation*, and the resulting semantic equivalence is called *Dynamic Observational Congruence*. Of course the *bisimulation game* is not just an academic fancy but is motivated by practical considerations: equivalent (in the various senses discussed above) modules can actually replace each other consistently in any real system, guaranteeing *software modularity* and *reusability*. In particular, the issue of dynamic reconfiguration is relevant for *system programming* and for applications like *software development*, where executing new programs is essential, and like *industrial automation*, where halting execution may be extremely costly.

In this paper we show that *Dynamic Observational Congruence* is the *coarsest* bisimulation which is also a congruence, and thus it is algebraically characterized by the finality construction in the style of [FM90] outlined above.

Furthermore we introduce a new observational equivalence, *Progressing Bisimulation*, between states of a labelled transition system with a distinct action τ. The basic idea underlying Progressing Bisimulation is to force programs in the bisimulation game to match moves with explicit moves. This also justifies its name.

For Progressing Bisimulation we give an *algebraic characterization* in the category of labelled transition systems and two *modal logics* in the style of HML, one in which the modal operators may include τ's, and their meaning is that at least those τ's must appear in the models, the other in which the satisfaction relation forces agents to move. Then we provide a *complete axiomatization* for states with finite computations, consisting of the axioms for Strong Observational Congruence and of two of the three Milner's τ-laws (of course $\alpha.\tau.P = \alpha.P$ is no longer true).

Finally, we show that on the CCS transition system Progressing Bisimulation coincides with Dynamic Congruence. That gives all the characterizations above to Dynamic Congruence and gives meaning to the paper's title.

This presentation stresses the fact that we are in presence of two distinct, general concepts, which, in the case of CCS, coincide: *Dynamic Congruence*, which makes sense on the LTS of every language and has a nice algebraic characterization and *Progressing Bisimulation*, which makes sense on every LTS with a distinct action τ and has algebraic, logical and axiomatic characterizations.

The paper is organized as follows.

In section 2 we recall the basic concepts of CCS ([Mil80], [Mil89]). Section 3 provides the *operational definition* and an *algebraic characterization* of Dynamic Observational Congruence. The Progressing Bisimulation and its *algebraic*, *logical* and *axiomatic* characterizations are introduced in section 4. Finally, in section 5 we show that Dynamic Congruence and Progressing Bisimulation *coincide* in the CCS transition system, thus obtaining a full characterization of Dynamic Congruence.

In the paper, we follow the notations and definitions in the references, thus the expert reader can safely skip section 2. For space saving, proofs are only sketched: the full version can be found in [MS90].

2 Calculus of Communicating Systems

In this section we recall the basic definitions of Milner's *Calculus of Communicating Systems* (CCS). For a full introduction however, the reader is referred to [Mil80] and [Mil89].

Let $\Delta = \{\alpha, \beta, \gamma, \dots\}$ be a fixed set of actions, and let $\overline{\Delta} = \{\overline{\alpha} | \alpha \in \Delta\}$ be the set of complementary actions ($\bar{\ }$ being the operation of complementation). $\Lambda = \Delta \cup \overline{\Delta}$ (ranged over by λ) is the set of *visible actions*. Let $\tau \notin \Lambda$ be the *invisible action*; $\Lambda \cup \{\tau\}$ is ranged over by μ.

Definition 2.1 *(CCS Expressions and Agents)*
The syntax of CCS expressions is defined as follows:

$$E ::= x \mid nil \mid \mu.E \mid E \setminus \alpha \mid E[\Phi] \mid E + E \mid E|E \mid recx.E$$

where x is a variable, and Φ is a permutation of $\Lambda \cup \{\tau\}$ preserving τ and $\bar{\ }$ CCS agents *(ranged over by $P, Q, \dots$) are closed CCS expressions, i.e. expressions without free variables.* □

The operational semantics of CCS is defined in terms of *labelled transition systems* ([Kel76]) in which the states are CCS expressions and the transition relation is defined by axioms and inference rules driven by the syntactic structure of expressions.

Definition 2.2 *(CCS Transition Relation)*
The CCS transition relation $\xrightarrow{\mu}$ is defined by the following inference rules.

Act) $\mu.E \xrightarrow{\mu} E$

Res) $\dfrac{E_1 \xrightarrow{\mu} E_2}{E_1 \setminus \alpha \xrightarrow{\mu} E_2 \setminus \alpha} \quad \mu \notin \{\alpha, \overline{\alpha}\}$

Rel) $\dfrac{E_1 \xrightarrow{\mu} E_2}{E_1[\Phi] \xrightarrow{\Phi(\mu)} E_2[\Phi]}$

Sum) $\dfrac{E_1 \xrightarrow{\mu} E_2}{E_1 + E \xrightarrow{\mu} E_2 \text{ and } E + E_1 \xrightarrow{\mu} E_2}$

Com1) $\dfrac{E_1 \xrightarrow{\mu} E_2}{E_1|E \xrightarrow{\mu} E_2|E \text{ and } E|E_1 \xrightarrow{\mu} E|E_2}$

Com2) $\dfrac{E_1 \xrightarrow{\lambda} F_1 \text{ and } E_2 \xrightarrow{\overline{\lambda}} F_2}{E_1|E_2 \xrightarrow{\tau} F_1|F_2}$

Rec) $\dfrac{E_1[recx.E_1/x] \xrightarrow{\mu} E_2}{recx.E_1 \xrightarrow{\mu} E_2}$ □

The transition $P \xrightarrow{\mu} Q$ expresses that the agent P may evolve to become the agent Q through the action μ, μ being either a stimulus from the environment or the internal action τ which is independent from the environment. Computations are usually described by *multistep derivation* relations derived from $\xrightarrow{\mu}$. The relations we will need in the following are: $\stackrel{\epsilon}{\Longrightarrow}$ defined as $(\xrightarrow{\tau})^*$, where ϵ is the null string and * the transitive closure of relations; $\stackrel{\mu}{\Longrightarrow} = \stackrel{\epsilon}{\Longrightarrow} \xrightarrow{\mu} \stackrel{\epsilon}{\Longrightarrow}$, where $\mu \in \Lambda \cup \{\tau\}$; $\stackrel{t}{\Longrightarrow} = \stackrel{\mu_1}{\Longrightarrow} \stackrel{\mu_2}{\Longrightarrow} \cdots \stackrel{\mu_n}{\Longrightarrow}$, where $t = \mu_1 \mu_2 \dots \mu_n \in (\Lambda \cup \{\tau\})^*$ and $\stackrel{s}{\Longrightarrow} = \stackrel{\lambda_1}{\Longrightarrow} \stackrel{\lambda_2}{\Longrightarrow} \cdots \stackrel{\lambda_n}{\Longrightarrow}$, where $s = \lambda_1 \lambda_2 \dots \lambda_n \in \Lambda^*$.

The semantics given by labelled transition systems is too concrete: the addition of *behavioural equivalence* equates those processes which cannot be distinguished by any external observer of their behaviour. Park's notion of *bisimulation* ([Par81]) has become the standard device for defining behavioural equivalences.

Definition 2.3 *(Strong Bisimulation)*
Let $\mathcal{R}$ be a binary relation over CCS agents. Then Ψ, a transformation of relations, is defined by

$(P,Q) \in \Psi(\mathcal{R})$ *if and only if* $\forall \mu \in \Lambda \cup \{\tau\}$:
- *whenever* $P \xrightarrow{\mu} P'$ *there exists* Q' *s.t.* $Q \xrightarrow{\mu} Q'$ *and* $(P',Q') \in \mathcal{R}$;
- *whenever* $Q \xrightarrow{\mu} Q'$ *there exists* P' *s.t.* $P \xrightarrow{\mu} P'$ *and* $(P',Q') \in \mathcal{R}$.

A relation $\mathcal{R}$ is called strong bisimulation *if and only if* $\mathcal{R} \subseteq \Psi(\mathcal{R})$.
The relation $\sim = \cup\{\mathcal{R} | \mathcal{R} \subseteq \Psi(\mathcal{R})\}$ *is called* Strong Observational Equivalence. □

Since Ψ is a monotone function, $\sim$ is the largest strong bisimulation. Moreover, it is an equivalence relation over CCS agents.

Strong Observational Equivalence can be extended to CCS expressions by saying that two expressions are strongly equivalent if all the agents obtained by binding their free variables to CCS agents are strongly equivalent.

However, definition 2.3 does not consider τ actions as special actions representing the occurrence of invisible internal moves. If we take into account the special status of τ actions, agents are equivalent if they can perform the same sequences of visible actions and then reach equivalent states. The notion of *Weak Observational Equivalence* implements this kind of abstraction.

Weak Equivalence, written $\approx$, is defined by introducing a transformation Φ, obtained from the definition of Ψ by replacing $\xrightarrow{\mu}$ by $\stackrel{s}{\Longrightarrow}$, and taking its greatest fixed point, i.e. $\approx = \cup\{\mathcal{R} | \mathcal{R} \subseteq \Phi(\mathcal{R})\}$. Relation $\approx$ is the largest weak bisimulation, and it is an equivalence relation. It is extended to CCS expressions in the same way Strong Equivalence is.

An equivalence ρ is called *congruence* with respect to an operator f, if it is respected by the operator, i.e. $x \rho y$ implies $f(x) \rho f(y)$. The equivalences which are congruences with respect to all the operators of the language are very important: they provide algebras in which equality is mantained in every context, a property that can be exploited by algebraic techniques.

Formally, in our framework, a context $\mathcal{C}[\,]$ is a CCS expression without free variables and with exactly one "hole" to be filled by a CCS agent.

Relation $\sim$ is a congruence with respect all CCS operators, that is $E \sim F$ implies $\mathcal{C}[E] \sim \mathcal{C}[F]$ for each context $\mathcal{C}[\,]$, but it is well known that Weak Observational Equivalence is not a congruence. Indeed, we have $\tau.E \approx E$ but in general it is false that $\tau.E + E' \approx E + E'$, e.g. $\tau.\alpha.nil \approx \alpha.nil$ but $\beta.nil + \alpha.nil \not\approx \beta.nil + \tau.\alpha.nil$ because the first agent provides α and β as alternative actions, while the second agent may autonomously discard the β alternative by simply performing a τ action.

The largest congruence contained in $\approx$ is Milner's Weak Observational Congruence, written $\approx^c$ and defined by $P \approx^c Q$ if and only if for any context $\mathcal{C}[\,]$, $\mathcal{C}[P] \approx \mathcal{C}[Q]$.

Weak Observational Congruence has a main drawback: it is *not* a bisimulation. As an example consider the weakly congruent agents $\alpha.\tau.nil$ and $\alpha.nil$. When $\alpha.\tau.nil$ performs an α action becoming the agent $\tau.nil$, $\alpha.nil$ can only perform an α action becoming nil: clearly $\tau.nil$ and nil are not weakly congruent but only weakly equivalent. Our definition of Dynamic Observational Congruence remedies this situation.

3 Dynamic Observational Congruence

In this section we introduce *Dynamic Bisimulation* by giving its *operational definition* and its *algebraic characterization* in the style of [FM90].

The definition is given for CCS, but it can be given for any labelled transition system in which the concept of context makes sense, in particular for the LTS corresponding to any language.

3.1 Operational definition

We want to find the coarsest bisimulation which is also a congruence. Let $\mathcal{B}$ be the set of (weak) bisimulations and $\mathcal{C}$ be the set of congruences.

Definition 3.1 *(Dynamic Bisimulation)*
Let $\mathcal{R}$ be a binary relation over CCS agents. Then Φ_d, a transformation of relations, is defined as follows:

$(P,Q) \in \Phi_d(\mathcal{R})$ *if and only if* $\forall s \in \Lambda^*$ *and* $\forall \mathcal{C}[\,]$:
- *whenever* $\mathcal{C}[P] \stackrel{s}{\Longrightarrow} P'$ *there exists* Q' *s.t.* $\mathcal{C}[Q] \stackrel{s}{\Longrightarrow} Q'$ *and* $(P',Q') \in \mathcal{R}$;
- *whenever* $\mathcal{C}[Q] \stackrel{s}{\Longrightarrow} Q'$ *there exists* P' *s.t.* $\mathcal{C}[P] \stackrel{s}{\Longrightarrow} P'$ *and* $(P',Q') \in \mathcal{R}$.

A relation $\mathcal{R}$ is called dynamic bisimulation *if and only if* $\mathcal{R} \subseteq \Phi_d(\mathcal{R})$.
The relation $\approx^d = \cup\{\mathcal{R} | \mathcal{R} \subseteq \Phi_d(\mathcal{R})\}$ *is called* Dynamic Observational Equivalence. □

Relation $\approx^d$ is a dynamic bisimulation and it is an equivalence relation. Just as Strong and Weak Equivalence, it is extended to CCS expressions. In the following, whenever it makes sense, we will consider only CCS agents: obviously, results hold also for CCS expressions, by definition of the equivalences on them.
We show now that $\approx^d$ is the coarsest bisimulation which is also a congruence.

Lemma 3.2 *(Dynamic Bisimulations are Weak Bisimulations)*
$\mathcal{R} \subseteq \Phi_d(\mathcal{R})$ *implies* $\mathcal{R} \subseteq \Phi(\mathcal{R})$, *where Φ is the function defining Weak Observational Equivalence.*
Proof. Directly from the definitions of Φ and Φ_d (fixing the context $\mathcal{C}[\,] \equiv x$). □

As a corollary to the previous lemma, we obtain that $\approx^d \subseteq \approx$, i.e. Dynamic Equivalence refines Weak Observational Equivalence. However, the reverse inclusion does not hold as $P \approx \tau.P$ while $P \not\approx^d \tau.P$.

Lemma 3.3 *(Dynamic Bisimulations are Congruences)*
Let $\mathcal{R} \subseteq \Phi_d(\mathcal{R})$. *Then* $P\mathcal{R}Q$ *if and only if* $\mathcal{C}[P]\mathcal{R}\mathcal{C}[Q]$ *for each context* $\mathcal{C}[\,]$.
Proof. ($\Rightarrow$) If there were $\mathcal{C}[\,]$ such that $(\mathcal{C}[P],\mathcal{C}[Q]) \notin \mathcal{R}$ then $\lambda.\mathcal{C}[\,]$ would be a context for which the definition of Φ_d does not hold. So $(P,Q) \notin \Phi_d(\mathcal{R})$ and $\mathcal{R} \not\subseteq \Phi_d(\mathcal{R})$. ($\Leftarrow$) Obvious. □

As a corollary, we have that Dynamic Equivalence is a Congruence, i.e. $P \approx^d Q$ if and only if $\mathcal{C}[P] \approx^d \mathcal{C}[Q]$ for each context $\mathcal{C}[\,]$. Since $\approx^c$ is the coarsest congruence contained in $\approx$ and $\approx^d \subseteq \approx$, it follows that $\approx^d \subseteq \approx^c$.

Proposition 3.4 *(Dynamic Bisimulation $\Leftrightarrow$ Bisimulation and Congruence)*
$\mathcal{R} \in \mathcal{B} \cap \mathcal{C}$ *if and only if* $\mathcal{R} \subseteq \Phi_d(\mathcal{R})$.
Proof. ($\Rightarrow$) $\mathcal{R} \in \mathcal{B}$ implies $\mathcal{R} \subseteq \Phi(\mathcal{R})$ and $\mathcal{R} \in \mathcal{C}$ gives that $P\mathcal{R}Q$ implies $\mathcal{C}[P]\mathcal{R}\mathcal{C}[Q]$ $\forall \mathcal{C}[\,]$. Then if $(P,Q) \in \mathcal{R}$, $\forall \mathcal{C}[\,]$ $(\mathcal{C}[P],\mathcal{C}[Q]) \in \mathcal{R}$ and so $(\mathcal{C}[P],\mathcal{C}[Q]) \in \Phi(\mathcal{R})$ which, by definition of Φ, implies $(P,Q) \in \Phi_d(\mathcal{R})$. Therefore, $\mathcal{R} \subseteq \Phi_d(\mathcal{R})$. ($\Leftarrow$) If $R \subseteq \Phi_d(\mathcal{R})$ then $\mathcal{R} \subseteq \Phi(\mathcal{R})$ by lemma 3.2, so $\mathcal{R} \in \mathcal{B}$. Moreover if $(P,Q) \in \mathcal{R}$ then by lemma 3.3 $(\mathcal{C}[P],\mathcal{C}[Q]) \in \mathcal{R}$, so $\mathcal{R} \in \mathcal{C}$. □

Therefore, $\approx^d = \bigcup\{\mathcal{R} | \mathcal{R} \in \mathcal{B} \cap \mathcal{C}\}$ is the coarsest bisimulation which is also a congruence.

3.2 Algebraic characterization

In this subsection we show that $\approx^d$ has a corresponding object in **CatLCCS**, the category of CCS computations whose construction is due to Ferrari and Montanari ([FM90]).

As we have seen in section 2, the operational semantics of CCS is defined by a deductive system. Now, we structure those systems as *typed algebras* ([MSS90]), i.e. algebras in which types are assigned to elements, and which contain, as special elements, the types themselves.

Types allow us to distinguish between elements which are agents (typed by *state* and denoted by $u, v, w \ldots$), elements which are transitions (typed by $\rightarrow$ and denoted by t) and elements which are computations (typed by $\Rightarrow$ and denoted by c).

In the following $x : type$ will indicate that x has type *type*. Operations *source*() and *target*() and a function *label*() which respectively give source state, target state and observation, are defined on elements typed by $\rightarrow$ or $\Rightarrow$. We write $t : u \xrightarrow{\mu} v$ to denote a transition with $source(t) = u$, $target(t) = v$ and $label(t) = \mu$. Similarly, we write $c : u \stackrel{s}{\Longrightarrow} v$. A computation with empty observation will be indicated by $c : u \stackrel{\epsilon}{\Longrightarrow} v$, while we will write $u \Rightarrow v$ when we are not interested in the observation.

The definition of CCS models should be given by listing an appropriate *presentation* and saying that CCS models are the models of that presentation. Since such a presentation would be rather long, we prefer to give the definition as follows. The interested reader can find the rigorous definition in [FMM91].

Definition 3.5 *(CCS Models and Morphisms,* **CatLCCS***)*
A CCS Model is a typed algebra (with multityping) where elements typed state *have the algebraic structure of guarded CCS agents. Moreover, there is an operation on transitions for each rule in the CCS transition system, an operation idle and an operation* $\_;\_$*. They satisfy the following:*

$$[\mu, v >: \mu.v \xrightarrow{\mu} v$$

$$\frac{t : u \xrightarrow{\mu} v}{t[\Phi] : u[\Phi] \xrightarrow{\Phi(\mu)} v[\Phi]} \qquad \frac{t : u \xrightarrow{\mu} v}{t \setminus \alpha : u \setminus \alpha \xrightarrow{\mu} v \setminus \alpha} \text{ if } \mu \notin \{\alpha, \overline{\alpha}\}$$

$$\frac{t : u \xrightarrow{\mu} v}{t <\!+\, w : u + w \xrightarrow{\mu} v} \qquad \frac{t : u \xrightarrow{\mu} v}{w \,+\!> t : w + u \xrightarrow{\mu} v}$$

$$\frac{t : u \xrightarrow{\mu} v}{t\rfloor w : u|w \xrightarrow{\mu} v|w} \qquad \frac{t : u \xrightarrow{\mu} v}{w\lfloor t : w|u \xrightarrow{\mu} w|v}$$

$$\frac{t_1 : u_1 \xrightarrow{\lambda} v_1 \text{ and } t_2 : u_2 \xrightarrow{\overline{\lambda}} v_2}{t_1|t_2 : u_1|u_2 \xrightarrow{\tau} v_1|v_2}$$

$$\frac{t : u \xrightarrow{\lambda} v}{t : u \stackrel{\lambda}{\Longrightarrow} v} \qquad \frac{t : u \xrightarrow{\tau} v}{t : u \stackrel{\epsilon}{\Longrightarrow} v} \qquad idle(v) : v \stackrel{\epsilon}{\Longrightarrow} v \qquad \frac{c_1 : u \stackrel{s_1}{\Longrightarrow} v \text{ and } c_2 : v \stackrel{s_2}{\Longrightarrow} w}{c_1; c_2 : u \stackrel{s_1 s_2}{\Longrightarrow} w}$$

Finally, a CCS Model satisfies the following equations:

$$recx.u = u[recx.u/x] \qquad c_1; (c_2; c_3) = (c_1; c_2); c_3 \qquad \frac{c : u \Rightarrow v}{idle(u); c = c = c; idle(v)}$$

A CCS morphism is an homomorphism of algebras that respects types. This defines **CatLCCS***, the category whose objects are CCS Models and whose morphisms are CCS morphisms.* □

Note that the way in which we defined the operations also defines *source*, *target* and *label*. Note also that there are no rules and operations for recursion which is instead handled by imposing the axiom above on states. Moreover, τ's are completely forgotten in the observations. Finally, note that a CCS morphism respects *source* and *target* since they are operations of the algebra. It is easy to prove that CCS morphisms respect the function *label*.

As a general result on typed algebras ([MSS90]), we state that **CatLCCS** has an initial object $\Im$.

Weak Observational Congruence cannot be characterized algebraically in **CatLCCS**, even though Ferrari and Montanari showed ([FM90]) that this is possible in a category constructed ad hoc from it. This is because the definition of morphism implies that, from congruent states, corresponding transitions lead to congruent states, and this is *not* the case for Weak Observational Congruence.

The situation is different for $\approx^d$. In the following, we shall use $[P]$ to denote the state to which agent P evaluates in a particular CCS model.

The following lemma derives directly from the fact that h respects the algebraic structure of elements.

Lemma 3.6 *(CatLCCS morphisms respect contexts)*
If h is a CatLCCS morphism then $h([P]) = h([Q])$ implies $h([C[P]]) = h([C[Q]])$ for each context $C[\,]$. □

Definition 3.7 *(Transition Preserving Homomorphism)*
A CatLCCS morphism $h : C \to C'$ is called a transition preserving homomorphism if and only if:

- *$h : C \to C'$ is a surjective CCS morphism;*
- *$t' \in C'$, $t' : h(u) \Rightarrow v'$ implies $\exists t \in C,\ t : u \Rightarrow v$ with $h(t) = t'$.* □

Example 3.8 *(no tp-homomorphism maps $\tau.\alpha + \beta$ to $\alpha + \beta$ or $\tau.\alpha$ to α)*

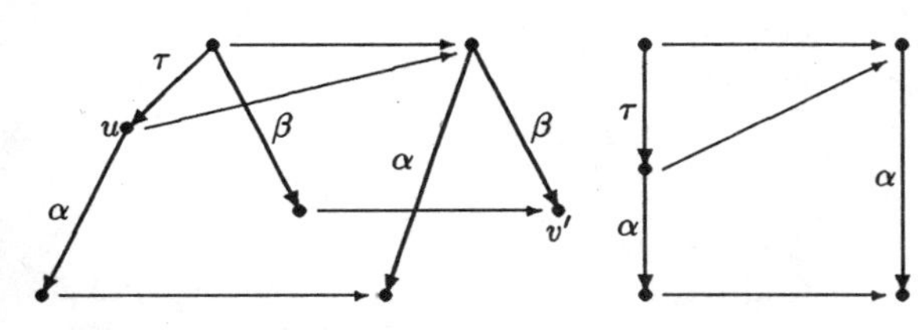

(A) **(B)**

The figures show two morphisms which are not tp-morphisms.

In case (A) we have $t : h(u) \stackrel{\beta}{\Longrightarrow} v'$ but $u \stackrel{\beta}{\not\Longrightarrow}$.

In case (B) the morphism cannot respect the algebra, for if it did we would have $h(\tau.\alpha) + h(\beta) = h(\alpha) + h(\beta)$ and so $h(\tau.\alpha + \beta) = h(\alpha + \beta)$ which is case (A). □

Proposition 3.9 *(tp-homomorphism ⇒ Dynamic Bisimulation)*
If $h : \Im \to C$ is a tp-homomorphism then $h([P]) = h([Q])$ implies $P \approx^d Q$.

Proof. We show that $\mathcal{R} = \{< P, Q > | h([P]) = h([Q])\}$ is a dynamic bisimulation. Let $(P, Q) \in \mathcal{R}$. If $\mathcal{C}[P] \stackrel{s}{\Longrightarrow} P'$ then there exists $t : [\mathcal{C}[P]] \stackrel{s}{\Longrightarrow} [P']$ in $\Im$. Then $h(t) : h([\mathcal{C}[P]]) \stackrel{s}{\Longrightarrow} h([P'])$. By lemma 3.6 we have $h(t) = t' : h([\mathcal{C}[Q]]) \stackrel{s}{\Longrightarrow} h([P'])$ and so by definition of tp-homomorphism there exists $t'' : [\mathcal{C}[Q]] \stackrel{s}{\Longrightarrow} [Q']$ with $h([Q']) = h([P'])$. Hence there exists $\mathcal{C}[Q] \stackrel{s}{\Longrightarrow} Q'$ and $(P', Q') \in \mathcal{R}$. Symmetrically if $\mathcal{C}[Q] \stackrel{s}{\Longrightarrow} Q'$. Therefore, by definition of Φ_d, $\mathcal{R} \subseteq \Phi_d(\mathcal{R})$. □

Proposition 3.10 *(Dynamic Congruence ⇒ tp-homomorphism)*
There exists an object $\Im / \approx^d$ of **CatLCCS** *such that the unique morphism $h_{\approx^d} : \Im \to \Im / \approx^d$ is a tp-homomorphism. Moreover, $P \approx^d Q$ implies $h_{\approx^d}([P]) = h_{\approx^d}([Q])$.*

Proof. Let $\mathcal{R}$ be the congruence over $\Im$ defined as follows:
$[P]\mathcal{R}[Q]$ iff $P \approx^d Q$ and $(t_1 : u_1 \stackrel{s}{\Longrightarrow} v_1)\mathcal{R}(t_2 : u_2 \stackrel{s}{\Longrightarrow} v_2)$ iff $u_1 \mathcal{R} u_2$ and $v_1 \mathcal{R} v_2$.
$\Im/\approx^d$ is obtained as the quotient of $\Im$ modulo $\mathcal{R}$ and $h_{\approx^d} : \Im \to \Im/\approx^d$ is the canonical map which sends each element to its equivalence class. □

Hence, as a corollary of the previous two propositions we have that $h_{\approx^d}$ coincides with $\approx^d$, i.e. $P \approx^d Q$ if and only if $h_{\approx^d}([P]) = h_{\approx^d}([Q])$.

Moreover, $\Im/\approx^d$ is the terminal object in the subcategory of the objects reachable from $\Im$ through tp-homomorphisms, that is the one corresponding to the coarsest dynamic bisimulation, i.e. $\approx^d$.

Proposition 3.11 *($\Im/\approx^d$ is terminal)*
The subcategory of **CatLCCS** *consisting of all objects reachable from $\Im$ through tp-homomorphisms and having morphisms which are tp-homomorphisms has $\Im/\approx^d$ as a terminal object.* □

4 Progressing Bisimulation

In this section we introduce a new bisimulation, *Progressing Bisimulation*, on the states of a labelled transition system with a distinct label τ. We will give an *algebraic characterization* of such a bisimulation and two *modal logical languages*, in the style of HML, adequate with respect to it. Furthermore we will provide a *complete axiomatization* of Progressing Equivalence for states with finite computations.

In the next section we will see that, when the transition system is the CCS transition system, Progressing Bisimulation coincides with Dynamic Congruence, thus completing its characterization for CCS.

We reiterate our two distinct results: the first, concerning Dynamic Congruence and guided by the concept of context, and the second concerning Progressing Bisimulation and its algebraic, logical and axiomatic characterizations. Both bisimulations are very general and go beyond CCS semantics, even though Dynamic Congruence is perhaps better justified in terms of practical considerations. Moreover, in the case of CCS they coincide, giving us plenty of characterizations of Dynamic Congruence.

4.1 Operational definition and Algebraic characterization

Definition 4.1 *(Progressing Bisimulation)*
Let $\mathcal{R}$ be a binary relation over the states of a labelled transition system $T =< S, \longrightarrow, \Lambda \cup \{\tau\} >$. Then Φ_p, a function from relations to relations, is defined as follows:

$(s, r) \in \Phi_p(\mathcal{R})$ *if and only if* $\forall \mu \in \Lambda \cup \{\tau\}$:

- *whenever $s \stackrel{\mu}{\longrightarrow} s'$ there exists r' s.t. $r \stackrel{\mu}{\Longrightarrow} r'$ and $(s', r') \in \mathcal{R}$;*
- *whenever $r \stackrel{\mu}{\longrightarrow} r'$ there exists s' s.t. $s \stackrel{\mu}{\Longrightarrow} s'$ and $(s', r') \in \mathcal{R}$.*

A relation $\mathcal{R}$ is called progressing bisimulation, *if and only if $\mathcal{R} \subseteq \Phi_p(\mathcal{R})$.*
The relation $\approx^p = \cup\{\mathcal{R} | \mathcal{R} \subseteq \Phi_p(\mathcal{R})\}$ is called Progressing Equivalence. □

Relation $\approx^p$ is a progressing bisimulation and an equivalence relation.

Now we introduce an algebraic model of a labelled transition system. As for CCS Models (definition 3.5) the definition of LTS Models could be given more formally. The notations used here are those defined in the previous section.

Definition 4.2 *(LTS Models and Morphisms,* **LTS***)*
An LTS Model is a typed algebra (with multityping) where elements typed state are a set, i.e. they do not have any particular algebraic structure. Partial operations idle and $\_;\_$ *are defined so that they satisfy:*

$$\frac{t : u \xrightarrow{\lambda} v}{t : u \stackrel{\lambda}{\Longrightarrow} v} \qquad \frac{t : u \xrightarrow{\tau} v}{t : u \stackrel{\epsilon}{\Longrightarrow} v} \qquad idle(v) : v \stackrel{\epsilon}{\Longrightarrow} v \qquad \frac{c_1 : u \stackrel{s_1}{\Longrightarrow} v \text{ and } c_2 : v \stackrel{s_2}{\Longrightarrow} w}{c_1; c_2 : u \stackrel{s_1 s_2}{\Longrightarrow} w}$$

Moreover, an LTS Model satisfies the following equations:

$$c_1; (c_2; c_3) = (c_1; c_2); c_3 \qquad \frac{c : u \Rightarrow v}{idle(u); c = c = c; idle(v)}$$

An LTS morphism is a morphism of algebras that respects types and labelling. This defines **LTS***, the category whose objects are LTS Models and whose morphisms are LTS morphisms.* □

Clearly, given an LTS Model, elements typed by $\rightarrow$ represent transitions, elements typed by $\Rightarrow$ represent sequences of transitions (computations) and elements typed by *state* represent states of the transition system. Note that an LTS morphism also respects *source* and *target*.

We introduce a new kind of morphism which, besides preserving transitions, prevents τ-transitions to be mapped to *idle*-transitions. We refer to them as *progressing transition preserving morphisms.*

Definition 4.3 *(Progressing Transition Preserving Morphism)*
An LTS morphism $h : T \rightarrow T'$ is called a progressing transition preserving morphism if and only if:

- *$h : T \rightarrow T'$ is a surjective LTS morphism;*
- *$t' \in T'$, $t' : h(s) \Rightarrow r'$ implies $\exists t \in T$, $t : s \Rightarrow r$ with $h(t) = t'$;*
- *$h(t) = idle(h(s))$ implies $t = idle(s)$.* □

Example 4.4 *(ptp-morphisms map $\tau.\mu + \mu$ to $\tau.\mu$ but not $\mu.\alpha + \mu.\beta$ to $\mu.(\alpha + \beta)$)*

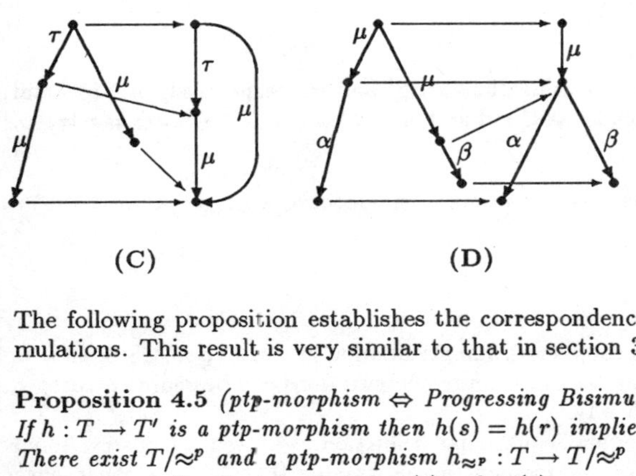

(C) (D)

Cases (A) and (B) of example 3.8 are not ptp-morphisms, the first because it does not preserve transitions, the second because it maps a not-idle to an idle transition.
In case (C) we have a ptp-morphism mapping a $\xrightarrow{\mu}$ transition to a computation $\xrightarrow{\tau};\xrightarrow{\mu} = \stackrel{\mu}{\Longrightarrow}$, while the morphism in case (D) is not a ptp, for it does not preserve transitions. □

The following proposition establishes the correspondence between ptp-morphisms and progressing bisimulations. This result is very similar to that in section 3.2.

Proposition 4.5 *(ptp-morphism $\Leftrightarrow$ Progressing Bisimulation)*
If $h : T \rightarrow T'$ is a ptp-morphism then $h(s) = h(r)$ implies $s \approx^p r$.
There exist $T/\approx^p$ and a ptp-morphism $h_{\approx^p} : T \rightarrow T/\approx^p$ such that $s \approx^p r$ implies $h_{\approx^p}(s) = h_{\approx^p}(r)$.
Therefore, $s \approx^p r$ if and only if $h_{\approx^p}(s) = h_{\approx^p}(r)$.
Finally, the subcategory of **LTS** *consisting of all objects reachable from T through ptp-morphisms, and having morphisms which are ptp-morphisms, has $T/\approx^p$ as a terminal object.* □

4.2 Logical characterization

In this subsection we design two modal logical languages in the style of HML which are adequate with respect to Progressing Bisimulation, i.e. two states are progressing equivalent if and only if they cannot be distinguished by any formula of the language.

The proofs in the rest of the section follow Milner's scheme in [Mil89], and so they are sketched very roughly. For the definitions of the modal languages adequate with respect to Strong Congruence and Weak Equivalence see [Mil89, chapter 10].

We introduce now a language whose modal operator may consider τ's, with the meaning that at least those τ's must be observed in the models. In the following, A^+ will mean $A^* \setminus \{\epsilon\}$.

Definition 4.6 *(The language : $PL_\tau^{\approx^p}$)*
$PL_\tau^{\approx^p}$ is the smallest class of formulae containing the following:

- *If $\varphi \in PL_\tau^{\approx^p}$ then $\forall t \in (\Lambda \cup \{\tau\})^+$ $\langle\!\langle t \rangle\!\rangle_\tau \varphi \in PL_\tau^{\approx^p}$*
- *If $\varphi \in PL_\tau^{\approx^p}$ then $\neg\varphi \in PL_\tau^{\approx^p}$*
- *If $\varphi_i \in PL_\tau^{\approx^p}$ $\forall i \in I$ then $\bigwedge_{i\in I} \varphi_i \in PL_\tau^{\approx^p}$, where I is an index set.* □

Definition 4.7 *(Satisfaction relation)*
The validity of a formula $\varphi \in PL_\tau^{\approx^p}$ at state r ($r \models_\tau \varphi$) is inductively defined as follows :

- *$r \models_\tau \langle\!\langle t \rangle\!\rangle_\tau \varphi$ if and only if $\exists r'$ s.t. $r \stackrel{t}{\Longrightarrow} r'$ and $r' \models_\tau \varphi$*
- *$r \models_\tau \neg\varphi$ if and only if* not *$r \models_\tau \varphi$*
- *$r \models_\tau \bigwedge_{i\in I} \varphi_i$ if and only if $\forall i \in I$ $r \models_\tau \varphi_i$.* □

There is another modal language we can naturally associate to $\approx^p$: $PL^{\approx^p}$. Its syntax is obtained from that of $PL_\tau^{\approx^p}$ by substituting the modal operator $\langle\!\langle t \rangle\!\rangle_\tau$ with the operator $\langle\!\langle s \rangle\!\rangle_p$ for $s \in \Lambda^*$.
The satisfaction relation $\models$ is obtained by replacing in definition 4.7 the clause for $r \models_\tau \langle\!\langle t \rangle\!\rangle_\tau \varphi$ with the clause $r \models \langle\!\langle s \rangle\!\rangle_p \varphi$ if and only if $\exists r'$ s.t. $r \stackrel{\bar{s}}{\Longrightarrow} r'$ and $r' \models \varphi$, where $\bar{s} = ($ if $s \neq \epsilon$ then s else $\tau)$.

The difference between the two languages is that $PL^{\approx^p}$ does not consider τ's in its modal operator, but takes care of them in the satisfaction relation, while the reverse holds for $PL_\tau^{\approx^p}$. The language used is a matter of taste.

Proposition 4.8 *($PL^{\approx^p}$ and $PL_\tau^{\approx^p}$ induce the same equivalence)*
$\forall \psi \in PL^{\approx^p}$ $s \models \psi \Leftrightarrow r \models \psi$ if and only if $\forall \varphi \in PL_\tau^{\approx^p}$ $s \models_\tau \varphi \Leftrightarrow r \models_\tau \varphi$. □

We show that the equivalences induced by $PL^{\approx^p}$ and $PL_\tau^{\approx^p}$ coincide with $\approx^p$.

Proposition 4.9 *($PL_\tau^{\approx^p}$ is adequate w.r.t. $\approx^p$)*
$s \approx^p r$ if and only if $\forall \varphi \in PL_\tau^{\approx^p}$ $s \models_\tau \varphi \Leftrightarrow r \models_\tau \varphi$.

Proof. Following Milner's scheme, we define stratifications $PL_{\tau\kappa}^{\approx^p}$ and $\approx_\kappa^p$, respectively of $PL_\tau^{\approx^p}$ and $\approx^p$, for κ an ordinal, and prove that for each κ, $s \approx_\kappa^p r$ if and only if $\forall \varphi \in PL_{\tau\kappa}^{\approx^p}$ $s \models_\tau \varphi \Leftrightarrow r \models_\tau \varphi$. The proposition follows easily from that. □

As a corollary to propositions 4.8 and 4.9, we have that also $PL^{\approx^p}$ is adequate w.r.t. $\approx^p$, that is $s \approx^p r$ if and only if $\forall \psi \in PL^{\approx^p}$ $s \models \psi \Leftrightarrow r \models \psi$.

4.3 Axiomatic characterization

Going back to CCS, in this subsection we give a complete axiomatization of $\approx^p$ for finite CCS agents. It is worth noticing that every labelled transition system with finite computations can be represented by a finite sequential CCS agent, in a straightforward way.

Obviously, to carry on a proof with axioms and equational deduction, we need an observational equivalence which is actually a congruence. For the moment let us assume that $\approx^p$ is a congruence with respect the CCS operators: in the next section, we will prove that this is indeed the case (see proposition 5.1).
For the axiomatizations of Strong and Weak Observational Congruence see [Mil89, pp. 160–165].
Let us begin relating $\approx^p$ on CCS agents to $\sim$ and $\approx^c$. The following is a direct consequence of the definitions of Ψ and Φ_p.

Proposition 4.10 *(Strong Congruence refines Progressing Bisimulation)*
$\sim \subseteq \approx^p$, where $\sim$ is Strong Observational Congruence. □

Thus $\approx^p$ inherits all the properties of $\sim$, in particular *monoid laws* and the *expansion theorem* ([Mil89, pp. 62, 67–76]) hold for $\approx^p$. Concerning the τ-laws ([Mil89, p. 62]) we have:

Proposition 4.11 *(Progressing Bisimulation and τ-laws)*
i. $P + \tau.P \approx^p \tau.P$ *ii.* $\alpha.(P + \tau.Q) + \alpha.Q \approx^p \alpha.(P + \tau.Q)$ *iii.* $\alpha.\tau.P \not\approx^p \alpha.P$ □

A CCS agent is *finite* if it does not contain recursion, and it is *serial* if it contains no parallel composition, restriction or relabelling. It is clear that with the use of the *expansion theorem* every finite agent can be equated to a finite serial agent. Therefore, a complete axiomatization for finite and serial agents can be considered an axiomatization for finite agents (considering the expansion theorem as an axiom scheme). In the rest of the subsection every CCS agent must be considered finite and serial.

We introduce a new set of axioms similar to the ones given for Strong and Weak Observational Congruence: it contains the monoid laws and two of the three τ-laws.

Definition 4.12 *(Axioms System $\mathcal{A}$)*

$\mathbf{A_1}: P+Q=Q+P$	$(\mathbf{T_1}: \mu.\tau.P=\mu.P)$
$\mathbf{A_2}: P+(Q+R)=(P+Q)+R$	$\mathbf{T_2}: P+\tau.P=\tau.P$
$\mathbf{A_3}: P+P=P$	$\mathbf{T_3}: \mu.(P+\tau.Q)+\mu.Q=\mu.(P+\tau.Q)$
$\mathbf{A_4}: P+nil=P$	

$$\mathcal{A}=\{\mathbf{A_1},\mathbf{A_2},\mathbf{A_3},\mathbf{A_4}\}\cup\{\mathbf{T_2},\mathbf{T_3}\}$$ □

Now, we prove that $\mathcal{A}$ is a complete set of axioms for $\approx^p$, i.e. two agents are progressing equivalent if and only if they can be proved equal by the axioms of $\mathcal{A}$ and equational deduction (denoted by $\vdash$).

Proposition 4.13 *($\mathcal{A}$ is a complete axiomatization of $\approx^p$)*
$P\approx^p Q$ if and only if $\mathcal{A}\vdash P=Q$.

Proof. ($\Leftarrow$) As previously noticed axioms $\mathcal{A}$ are true for $\approx^p$.
($\Rightarrow$) Following Milner's scheme, we define a standard form (SF) for CCS agents such that, using axioms $\mathbf{A_3}$, $\mathbf{A_4}$, $\mathbf{T_2}$ and $\mathbf{T_3}$, we can prove that each P is equal to a P' in SF with $P\approx^p P'$. By induction on the number of nested prefixes, it is easy to show that, if $P\approx^p Q$ and if P and Q are in SF, they can be proved equal using axioms $\mathbf{A_1}$ and $\mathbf{A_2}$. So, if $P\approx^p Q$ then $\mathcal{A}\vdash P=Q$. □

5 Dynamic Congruence and Progressing Bisimulation

In this section we show that Dynamic Congruence and Progressing Bisimulation *coincide* when $\approx^p$ is considered on CCS.

This gives many characterizations to $\approx^d$: in fact, we have two characterizations by finality through particular kinds of *abstraction morphisms* (the one encoding the CCS algebra into states and transitions, the other just considering the naked labelled transition system), two logical characterizations via HML-like modal languages and, finally, an axiomatization for finite agents, besides the two operational characterizations given by the *bisimulation game*.

Proposition 5.1 *($\approx^p$ is a congruence)*
Let E,F be CCS expressions and $E\approx^p F$. Then

i.. $\mu.E\approx^p\mu.F$	*ii..* $E+Q\approx^p F+Q$	*iii..* $E\mid Q\approx^p F\mid Q$
iv.. $E[\Phi]\approx^p F[\Phi]$	*v..* $E\setminus\alpha\approx^p F\setminus\alpha$	*vi..* $rec x.E\approx^p rec x.F$

Proof. The proof is standard. For *i–v* it is enough to exhibit appropriate progressing bisimulations. For instance, $\mathcal{R}=\{(P_1|Q,P_2|Q)|P_1\approx^p P_2\}$ shows *iii*. Point *vi* is less trivial and must be done by induction on the depth of the proofs by which actions are inferred, exploiting a concept analogous to Milner's bisimulation up to $\approx^c$. □

Proposition 5.2 *(Dynamic and Progressing Bisimulations coincide)*
$\approx^d=\approx^p$.

Proof. We shall prove that $\approx^d\subseteq\approx^p$ showing that $\approx^d\subseteq\Phi_p(\approx^d)$. Suppose $P\approx^d Q$ and $P\xrightarrow{\mu}P'$. If $\mu\neq\tau$ then we have that $\exists Q\overset{\mu}{\Longrightarrow}Q'$ and $P'\approx^d Q'$. Otherwise, if $\mu=\tau$ it must be that $\exists Q\overset{\epsilon}{\Longrightarrow}Q'$ in which at least one τ move is done and $P'\approx^d Q'$. Actually, if this were not the case, we could find a context for which the definition of $\approx^d$ does not hold: $\mathcal{C}[\;]\equiv x+\alpha.\overline{P}$, where $\overline{P}$ is not dynamically equivalent to each α-derivate of P', if there exists any, otherwise $\overline{P}\equiv nil$ will do.
Symmetrically if $Q\xrightarrow{\mu}Q'$. Then we have $(P,Q)\in\Phi_p(\approx^d)$ and so $\approx^d\subseteq\Phi_p(\approx^d)$.
By similar technique we show that $\approx^p\subseteq\approx^d$. Suppose that $P\approx^p Q$. Since $\approx^p$ is a congruence then $\forall\mathcal{C}[\;]\ \mathcal{C}[P]\approx^p\mathcal{C}[Q]$. If $\mathcal{C}[P]\overset{s}{\Longrightarrow}P'$ with $s\in\Lambda^*$ then, by repeated application of the definition of $\approx^p$, we have $\mathcal{C}[Q]\overset{s}{\Longrightarrow}Q'$ and $P'\approx^p Q'$. Symmetrically if $\mathcal{C}[Q]\overset{s}{\Longrightarrow}Q'$. So $\approx^p\subseteq\Phi_d(\approx^p)$. □

Acknowledgements. We wish to thank GianLuigi Ferrari for many interesting discussions, which have significantly contributed to our understanding of the subject. Special thanks to Rocco De Nicola for his suggestions on the organization of the paper.

References

[Ab88] S. Abramsky. *A Domain Equation for Bisimulation.* Technical report, Department of Computing, Imperial College, London, 1988.

[Acz87] P. Aczel. *Non-Well-Founded Sets.* CSLI Lecture Notes n. 14, Stanford University, 1987.

[BB88] D. Benson and O. Ben-Shachar. Bisimulations of Automata. *Information and Computation*, n. 79, 1988.

[DDM90] P. Degano, R. De Nicola, and U. Montanari. A Partial Ordering Semantics for CCS. *Theoretical Computer Science*, n. 75, 1990.

[DeN87] R. De Nicola. Extensional Equivalence for Transition Systems. *Acta Informatica*, n. 24, 1987.

[FM90] G. Ferrari and U. Montanari. Towards the Unification of Models for Concurrency. In *Proceedings of CAAP '90.* LNCS n. 431, Springer Verlag, 1990.

[FMM91] G. Ferrari, U. Montanari, and M. Mowbray. On Causality Observed Incrementally, Finally. In *Proceedings of TAPSOFT '91.* LNCS n. 493, Springer Verlag, 1991.

[GvG89] U. Goltz and R. van Glabbeek. Equivalence Notions for Concurrent System and Refinement of Actions. In *Proceedings of MFCS '89.* LNCS n. 379, Springer Verlag, 1989.

[Kel76] R. Keller. Formal Verification of Parallel Programs. *Communications of the ACM*, vol. 7, 1976.

[Mil80] R. Milner. *A Calculus of Communicating Systems.* Lecture Notes in Computer Science, n. 92. Springer Verlag, 1980.

[Mil89] R. Milner. *Concurrency and Communication.* Prentice Hall, 1989.

[MS90] U. Montanari and V. Sassone. *Dynamic Bisimulation.* Technical Report TR 13/90, Dipartimento di Informatica, Università di Pisa, 1990.

[MSg89] U. Montanari and M. Sgamma. Canonical Representatives for Observational Equivalences Classes. In *Proceedings of Colloquium on the Resolution of Equations in Algebraic Structures.* North Holland, 1989.

[MSS90] V. Manca, A. Salibra, and G. Scollo. Equational Type Logic. *Theoretical Computer Science*, n. 77, 1990.

[NPW81] M. Nielsen, G. Plotkin, and G. Winskel. Petri Nets, Event Structures and Domains, Part 1. *Theoretical Computer Science*, n. 13, 1981.

[Par81] D. Park. Concurrency and Automata on Infinite Sequences. In *Proceedings of GI.* LNCS n. 104, Springer Verlag, 1981.

[Pet62] C.A. Petri. *Kommunikation mit Automaten.* PhD thesis, Institut für Instrumentelle Mathematik, Bonn, FRG, 1962.

[Plo81] G. Plotkin. *A Structured Approach to Operational Semantics.* DAIMI FN-19, Computer Science Dept., Aarhus University, 1981.

[Pra86] V. Pratt. Modelling Concurrency with Partial Orders. *International Journal of Parallel Programming*, n. 15, 1986.

[vGl87] R. van Glabbeek. Bounded Nondeterminism and the Approximation Induction Principle in Process Algebra. In *Proceedings of STACS 87.* LNCS n. 247, Springer Verlag, 1987.

[vGW89] R. van Glabbeek and W. Weijland. Branching Time and Abstraction in Bisimulation Semantics. In *Proceedings of IFIP 11th World Computer Congress*, August 1989.

Syntax and semantics of a monotonic framework for non-monotonic reasoning

M.A. Nait Abdallah
Department of Computer Science
University of Western Ontario
London, Ontario, Canada

Abstract

We present the proof theory and the model theory of a monotonic framework for default reasoning, and we extend logic programming techniques to this framework. Standard formalizations of default reasoning do not syntactically separate between hard knowledge and conjectural knowledge. Such a separation is fundamental in our framework. To illustrate our approach, we show how it solves the Yale Shooting problem.

1 Introduction

To make this paper somewhat conceptually self-contained, we first recall very briefly some general notions on logic programming with ions. We then explain how we apply these ideas to *non-monotonic reasoning.*

Ions have been introduced in [3] for expressing *generalized SLD-derivations* in a framework where more than one logic program is relevant to the derivation. Essentially, an *ion* is a pair of the form (Q, g), i.e. a formula g that is given in some logical context Q where it is to be solved. In such a pair, the Q-part "qualifies" (or "puts in context") the g-part. The logic of ions [3, 4, 5] makes clear that every statement should be, whenever necessary, *localized* with respect to the *context* where this statement is made. It is well-known that overlooking the necessity of such a context may lead to serious problems; this is highlighted for example by Quine [11]. Various kinds of ions have been studied in some details: *open ions* [4], who inherit everything from their environment, *closed ions* [5], who inherit nothing from their environment, and *type-2 ions* [6, 8], who inherit some clauses from their environment under certain conditions.

In this paper, we extend this formal approach using ions as *syntactic objects* to perform "*commonsense reasoning*". The general motivation of this work is that we want to separate in our *formalization of commonsense reasoning* safe, irrefutable, hard knowledge—called *kernel knowledge* in our terminology— from tentative, plausible, conjectural knowledge—or *belt knowledge* [10]. This then allows to *factor out* the *monotonic component* of non-monotonic reasoning, and postpone the *non-monotonic component* to the last stages of the reasoning process. Ions are essential in this factorization, as they allow to show [9] that the hiding of some of their essential parameters is the sole cause of the apparent non-monotonicity of default reasoning. The basic ideas used in our approach are the notion of *ion*, Scott's *denotational semantics* and the *philosophical logic of scientific discovery* of Popper and Lakatos. The intuitions and motivations of our approach are discussed in more detail in [9, 10].

The framework we obtain has several interesting properties. First, it provides a denotational semantics for default reasoning. Second, it generalizes Reiter's framework. Third, it clearly separates syntax from semantics, and has a Tarskian model theory. Fourth, it solves the *Yale shooting problem* (which Reiter's logic cannot do) and several of its variants. Fifth, it has an easily implementable operational semantics.

In this paper, in contrast with the ionic framework of [3, 4, 5, 6, 8], the intended meaning of ion (P, g) is "*g if justification P is acceptable*". We shall give a model theory and a proof theory whose intent is to capture that notion. We show that, although the semantics of the logic programs built using these new ions ions are more complex than the semantics of classical logic programs, the very same methods (continuity and Knaster-Tarski theorem) and the very same tools (immediate consequence operator T used in van Emden-Kowalski theorem) can be used to solve problems Reiter's default logic cannot solve.

From a more general, *epistemological*, point of view, it also seems interesting that essentially the same "algebra of ions" applies to both the *provability ions* studied in [3, 4, 5, 6, 8] and the *justification ions* presented here. Thus concepts spawned from the necessity of modularizing logic programs [3] yield new tools for approaching and solving , from a computational point of view, problems discussed in the philosophical literature [11] and in the Artificial Intelligence literature.

To simplify the exposition, we concentrate on propositional logic, but our results generalize to first-order logic [9].

2 The syntax and proof theory of default ionic logic

The basic idea of our approach is the same as the one used by D. Scott when he made "*absence of information/termination*" into an element of the domain (of computation) called *bottom*. It was also used in [4] to solve the local definition problem in logic programming. Essentially, we make Reiter's default

$$\frac{:n}{p}$$

into an element of a set of generalized formulae, called *ionic logic formulae*, and denote it by the formula:

$$(n, p)_\star$$

The *default ion* $(n, p)_\star$ should be read: If n is consistent with our beliefs (i.e. current logical scope), then infer (or assert) p. For the sake of having a shorter notation, default ion $(n, n)_\star$ will be sometimes abbreviated as $[n]$. The intuitive meaning of $[n] = (n, n)_\star$ is *'n' is true if it is consistent with the current context.* In other words, we "weakly" assert n. From now on, statements of the form $[n]$ will be called *weak statements* and they shall be read "*weakly n*". Symmetrically, statements from "usual" logic will be called *strong statements.*

The default

$$\frac{m : n}{p}$$

will be translated by the ionic formula

$$m \to (n, p)_\star$$

In the terminology of Reiter's default logic, m is the prerequisite, n is the justification, and p is the conclusion. From the point of view adopted in this paper, a better name for p would be *conjecture*. In our approach, defaults $\frac{m:n}{n}$ and $\frac{: m \to n}{m \to n}$ will be *distinct* syntactic objects.

2.1 Syntax of propositional default ionic logic

We take a set P of propositional letters together with the single higher-order binary operator (we call *ionic operator*) $(\ . \ , \ . \ )_\star$, and the usual set of connectives.

The set of *propositional default ionic logic* formulae is the smallest set containing P, closed under connectives $\vee$, $\wedge$, $\neg$ and $\rightarrow$, and closed under the following *ionic operation*: for every *finite* set of formulae $\Phi = \{f_1 \ , \ \ldots \ , \ f_n\}$ and formula g, the expression $(\Phi, g)_\star$ is a formula, called *default (propositional) ion*. The formulae thus defined are called *(default) ionic formulae*.

The *default ion* $(\{f_1 \ , \ \ldots \ , \ f_n\}, g)_\star$ may be intuitively interpreted as follows: *If all of the f_i are simultaneously consistent with the current set of beliefs (logical context), then one can assert g.* In the case where $\Phi = \{f\}$ is a singleton, ion $(\Phi, g)_\star = (\{f\}, g)_\star$ will be simply denoted by $(f, g)_\star$. If, furthermore, f and g are identical, then the abbreviation $[g]$ shall be used for ion $(g, g)_\star$.

Examples of ionic formulae are : $p \wedge \neg(\Phi, p)_\star$, $(\Phi, p)_\star \rightarrow p$ and $bird \rightarrow (fly, fly)_\star$.

2.2 Axioms and proof rules for propositional default ionic logic (PDIL)

In addition to the *axioms inherited from propositional logic*, our logic will have the following specific axioms dealing with *default ions*. Our notations are as follows : a, b, f, g, h, k, j are arbitrary default ionic formulae, J, Γ and Δ are arbitrary finite sets of ionic formulae.

- justification elimination : $j \wedge (j \wedge k, g)_\star \rightarrow (k, g)_\star$
- elementary transformations :
  1. $(\Gamma, a \rightarrow b)_\star \rightarrow (a \rightarrow (\Gamma, b)_\star)$
  2. $(a \rightarrow (\Gamma, b)_\star) \rightarrow (\Gamma, a \rightarrow b)_\star$
  3. $(\Gamma, (\Gamma, a)_\star \rightarrow b)_\star \rightarrow (\Gamma, a \rightarrow b)_\star$
- $\vee$-intro in justification : $(\Gamma \cup \{j\}, g)_\star \wedge (\Gamma \cup \{k\}, g)_\star \rightarrow (\Gamma \cup \{j \vee k\}, g)_\star$
- $\wedge$-intro in conclusion : $(\Gamma, f)_\star \wedge (\Gamma, g)_\star \rightarrow (\Gamma, f \wedge g)_\star$
- Thinning : $(\Gamma, g)_\star \rightarrow (\Gamma \cup \Delta, g)_\star$
- Empty set of justifications : $(\emptyset, g)_\star \rightarrow g$, $(\mathbf{True}, g)_\star \rightarrow g$
- Justification introduction : $g \rightarrow (\Gamma, g)_\star$
- Abstraction :
  1. $(\Gamma, (\Delta, g)_\star)_\star \rightarrow (\Gamma \cup \Delta, g)_\star$
  2. $(\Gamma \cup \Delta, g)_\star \rightarrow (\Gamma, (\Delta, g)_\star)_\star$

We also have the following proof rules:

- J-modus ponens : $\dfrac{(\Gamma \cup \{j\}, g)_\star \quad k \rightarrow j}{(\Gamma \cup \{k\}, g)_\star}$
- C-modus ponens : $\dfrac{(\Gamma, a \rightarrow b)_\star \quad (\Gamma, a)_\star}{(\Gamma, b)_\star}$
- I-modus ponens : $\dfrac{a \quad a \rightarrow b}{b}$

It can be shown [9] that our logic is trivially *compact*, and thus *monotonic* and *continuous*. We also have the following deduction theorems for propositional default ionic logic.

Theorem 2.1 (Deduction theorem for default logic - weak form) *For any set A of default ionic formulae, and any default ionic formulae φ and ψ, if $A \cup \{\psi\} \vdash \varphi$ then $A \vdash (\psi \to \varphi)$.*

Theorem 2.2 (Deduction theorem—strong form) *For any set A of ionic formulae, and any default ions $(J,\psi)_\star$ and $(J,\varphi)_\star$, if $A \cup \{(J,\psi)_\star\} \vdash (J,\varphi)_\star$ then $A \vdash (J, \psi \to \varphi)_\star$.*

2.3 Syntax of first-order predicate default ionic logic

We use the following alphabets: a set R of relation symbols, a set F of function symbols, a set V of object variables, and a set Ξ of *justification variables.* Elements of Ξ will be 0-ary relation variables; they will be denoted by $\gamma, \varphi, \ldots \in \Xi$. We shall also use the usual connectives plus the singleton set $I = \{(\ . \ , \ . \ )_\star\}$ of ionic operators.

From the formal point of view, justifications play in default ionic logic a role similar to the role played by *procedures* in logic fields [5]. In logic field theory, procedures are named by procedure variables, and these procedure variables are used to *call* the associated procedures by means of the *copy rule.* In default ionic logic *justification variables* will allow us to *parameterize* over a set of possible justifications, and they will be used for *instantiation* purposes: a justification variable will be instantiated by some actual justification set.

This displays a striking dissimilarity between our approach and Reiter's approach. In Reiter's logic, justifications lead an *underground life* controlling the actual value of the extensions associated with the given default theory. The way this control is performed is not explicit in the syntax of the logic. In our approach, justifications are first-class objects, both syntactically and semantically, and the way they intervene in the reasoning is analyzed and displayed in detail.

The *terms* and *atomic formulae* of first-order default ionic logic are defined as usual. The set of *first-order default ionic logic* formulae is the smallest set (i) containing the atomic formulae, (ii) closed under connectives $\vee$, $\wedge$, $\neg$ and $\to$, (iii) closed under universal and existential quantification over object variables and justification variables, and (iv) closed under the following *ionic operations*: For every *enumerated* set of formulae $\Phi = \{f_1 \ , \ \ldots \ , \ f_n\}$ and formula g, the expression $(\Phi, g)_\star$ is a formula. For every justification variable $\gamma \in \Xi$, and formula g, the expression $(\gamma, g)_\star$ is a formula. Formulae of the form $(\Phi, g)_\star$ and $(\gamma, g)_\star$ thus constructed are called *default ions.*

For example, the intuitive meaning of default ionic formula $\exists\gamma(\gamma, g)_\star$ is that g *holds under some justification*, or g *has some justification.*

2.4 Axioms and proof rules for first-order predicate default ionic logic (FDIL)

As for the propositional case, axioms of our first-order logic are of two kinds: *axioms inherited from predicate logic*, and specific axioms dealing with *default ions.* The latter can be further decomposed into those axioms that are inherited from PDIL, and those axioms that are specific to first-order default ions. There will be one axiom that is specific to first-order default ions:

- $\forall x(\Gamma, g)_\star \to (\Gamma, \forall x g)_\star$, if variable x does not occur free in Γ.

The *proof rules* are the three modus ponens rules inherited from the propositional case, together with the *generalization rule*:

$$\frac{g}{(\forall x \ . \ g)}$$

Given Γ be a set of formulae, the set of all Γ-*theorems* of first-order default ionic logic is the smallest set Δ such that (*i*) $\Gamma \cup Axioms \subset \Delta$, and (*ii*) Δ is closed under the three modus ponens rules I, J and C, and the generalization rule.

Theorem 2.3 (Deduction theorem for first-order default ionic logic) *If* $A \cup \{\psi\} \vdash \varphi$, *with the free variables held constant for the last assumption formula* ψ, *then* $A \vdash (\psi \rightarrow \varphi)$.

3 Model theory for default ionic logic

We now consider the model theory of default ionic logic. The general principle underlying our approach is that we want to separate safe, irrefutable, hard knowledge (or *kernel knowledge*) from tentative, plausible, conjectural knowledge (or *belt knowledge*).

In his default logic [12], given a set S of first-order axioms and default rules, Reiter defines an *extension* as a set of possible beliefs about the world. Some of these beliefs may have to be abandoned and replaced by new ones, which gives rise to what has been called *non-monotonicity*. Other beliefs, however, should be held quite firmly, as they are not *revisable*: those are logical consequences of the predicate logic part of system S. In our view, there is, to some extent, a symmetry between *revisable beliefs* (belt knowledge) and *non-revisable beliefs* (kernel knowledge). Quite often this symmetry between the two kinds of beliefs, has been overlooked in favour of the revisable ones.

Our approach is inspired from Imre Lakatos' analysis of the nature of knowledge [2]. According to Lakatos, given a body of knowledge K, there is a dichotomy between the part of K that cannot be changed (the *'hard core'* or *kernel* of K, which we call *kernel knowledge*), and the part of K that can be revised and changed depending on incoming new information (the *protective belt* of K, which we call *belt knowledge*).

Due to the (possible) nesting of defaults, some of the semantic definitions we are about to give are (mutually) recursive. That these recursions always terminate is guaranteed by the constructive proof of theorem 3.1 below, which says that default interpretations exist.

3.1 Justifications, valuations and default interpretations

A *justification* is either a positive justification or a negative justification. A *positive justification* (resp. *negative justification*) is an expression of the form $+\Gamma$ (resp. $-\Gamma$) where Γ is a set of default ionic formulae. The intuitive meaning of $+\Gamma$, respectively $-\Gamma$, is "Γ *is an acceptable justification*", respectively "Γ *is not an acceptable justification*".

A set of justifications J is *stable* if and only if the following are verified:

(*i*) $-\Gamma \cup \{j\} \in J$ and $-\Gamma \cup \{k\} \in J$ imply $-\Gamma \cup \{j \vee k\} \in J$.

(*ii*) $+(\Gamma \cup \Delta) \in J$ implies $-\Gamma \notin J$.

(*iii*) $-\Gamma \in J$ implies $-(\Gamma \cup \Delta) \in J$.

(*iv*) $+\Gamma \in J$ and $+\Delta \in J$ if and only if $+(\Gamma \cup \Delta) \in J$.

(*v*) $-(\Gamma \cup \Delta) \in J$ implies $-\Gamma \in J$ or $-\Delta \in J$.

Notice that the empty set of justifications is trivially stable.

Throughout the rest of this section, we assume that some general interpretive structure $\mathcal{A}$ is given. We will only work with that structure.

A *valuation* is a *partial* mapping from the symbols of the alphabet to the interpretive structure $\mathcal{A}$. The nowhere defined (or undefined) valuation will be denoted by $\perp$. Two valuations (i_0 and i_1) are *compatible* if and only if they are equal on the intersection of their domains of definition. Two such valuations have a least defined common extension, denoted by $i_0 \sqcup i_1$.

As indicated earlier, intuitively, our interpretations will contain two parts: a "hard part" i_0, that corresponds to the usual notion of interpretation of classical logic, and a "soft part" (J, i_1), that corresponds to that part that is purely "default ionic" and is subject to the acceptability of the justifications being used. This is analogous to our earlier dichotomy between strong statements and weak statements.

A *default interpretation* is a triple $i = (i_0, J, i_1)$—which will be denoted by $i_0 \oplus (J, i_1)$—where i_0 and i_1 are valuations and J is a set of justifications, such that:

(*i*) Valuations i_0 and i_1 are compatible.

(*ii*) J is a stable set of justifications.

(*iii*) J is i_0-coherent.

(*iv*) $\forall\ \Gamma$, if $i_0 \models \Gamma$, then $+\Gamma \notin J$ and $-\Gamma \notin J$.

(*v*) $\forall \Gamma\ , \forall j\ ,\ k$, if $-\Gamma \cup \{j \wedge k\} \in J$ and $i(j) = 1$, then $-(\Gamma \cup \{k\}) \in J$.

(*vi*) $\forall \Gamma\ , \forall j\ ,\ k$, if $i(j) = 1$, then $+(\Gamma \cup \{k\}) \in J$ if and only if $+\Gamma \cup \{j \wedge k\} \in J$.

(*vii*) $\forall \Gamma\ , \forall j\ ,\ k$, if $+(\Gamma \cup \{j\}) \in J$ and $i_0 \not\models \Gamma \cup \{j \vee k\}$ then $+\Gamma \cup \{j \vee k\} \in J$.

In default interpretation $i_0 \oplus (J, i_1)$, valuation i_0 is called the *kernel* valuation of the interpretation, and i_1 is called the *belt* valuation of the interpretation. We shall abbreviate $\perp\ \oplus (J, i_1)$ as (J, i_1), and $i_0 \oplus (\emptyset, \perp\ )$ as i_0.

A default interpretation $i_0 \oplus (J, i_1)$ *extends* a valuation v if and only if kernel valuation i_0 extends v. Given some valuation v, a set of justifications K is *v-coherent* if and only if there exists some default interpretation $\mathbf{j} = i_0 \oplus (J, i_1)$ extending v such that $\mathbf{j}$ satisfies all positive justifications of K (i.e. $\mathbf{j} \models \Gamma$ for each $+\Gamma \in K$), and for any such extension $\mathbf{j}$, for any negative justification $\Delta \in K$, $\mathbf{j}$ does not satisfy Δ.

The existence of default interpretations is far from obvious. The proof of the following theorem uses a construction similar to the one used by D. Scott for giving models of the λ-calculus.

Theorem 3.1 *Default interpretations exist.*

Valuations are partial interpretations. Thus, they should be considered as possible values of partial models of the world. Since some of these models may be more precise than others, in the sense that they provide a more complete picture of the world, there is a natural *information ordering* that may be defined on valuations. We say that default interpretation $i_0 \oplus (J, i_1)$ is *less defined* than default interpretation $i'_0 \oplus (J', i'_1)$ (or that $i'_0 \oplus (J', i'_1)$ *extends* $i_0 \oplus (J, i_1)$) and we write $i_0 \oplus (J, i_1) \sqsubseteq i'_0 \oplus (J', i'_1)$, if and only if partial function i'_0 extends partial function i_0, partial function i'_1 extends partial function i_1 and set J' contains set J. The *undefined* default interpretation $\perp \oplus (\emptyset, \perp)$ is the smallest element of the information ordering.

We also define another partial ordering, called *default ordering* and denoted by $\prec$, whose aim is to formalize the principle "Apply as many defaults as you possibly can." This ordering has to do with maximality in Reiter's extension sets.

We first define preorder $\prec_1$ as follows: $i \prec_1 i'$, where $i = i_0 \oplus (J, i_1)$ and $i' = i'_0 \oplus (J', i'_1)$, if and only if the following three conditions hold: (*i*) $i_0 \sqcup i_1 = i'_0 \sqcup i'_1$, (*ii*) valuation i'_0 is a strict extension of valuation i_0 (we want the smallest possible kernel), and (*iii*) $J' \subset J$ and $J \neq J'$.

We then define preorder $\prec_2$ as follows: $i_0 \oplus (J, i_1) \prec_2 i'_0 \oplus (J', i'_1)$ if and only if the following three conditions hold: (i) $i_0 = i'_0$, (ii) valuation i_1 is an extension of valuation i'_1, and (iii) there exist H and Γ such that $J = H \cup \{+\Gamma\}$ and $J' = H \cup \{-\Gamma\}$. (We want to apply as many of the defaults as possible: J is obtained from J' by replacing some negative justification $-\Gamma$ by its positive counterpart $+\Gamma$.)

We define the *default ordering relation* $\prec$ as the reflexive and transitive closure of the union of the two above relations $\prec_1$ and $\prec_2$.

3.2 Satisfaction of PDIL default ionic formulae by default interpretations

Let $i = i_0 \oplus (J, i_1)$ be a default interpretation. We shall use the customary interpretation rules for the connectives. Truth is denoted by 1, and falsehood by 0. We define in the propositional case:

$$
\begin{array}{rl}
(i_0 \oplus (J, i_1))((\Gamma, g)_\star) = & 1 \text{ if } -\Gamma \in J \\
& (i_0 \oplus (J, i_1))[[g]] \text{ if } +\Gamma \in J \text{ and } i_0 \not\models \Gamma \\
& (i_0 \oplus (J, i_1))(g) \text{ if } +\Gamma, -\Gamma \notin J \text{ or } i_0 \models \Gamma \\
\forall p \in P \text{ such that } i \in \delta(p) \; , & i(p) = i_0(p) \\
(i_0 \oplus (J, i_1))[[(\Gamma, g)_\star]] = & 1 \text{ if } -\Gamma \in J \\
& (i_0 \oplus (J, i_1))[[g]] \text{ if } +\Gamma \in J \text{ or } i_0 \models \Gamma \\
& \text{undefined otherwise} \\
\forall p \in P \quad \text{such that } i \in \delta_w(p) \; , & i[[p]] = (i_0 \sqcup i_1)(p)
\end{array}
$$

We say that formula g is *kernel-satisfied* by interpretation i (i is a *kernel model* of g) if and only if $i(g) = 1$. Formula g is *belt-satisfied* by interpretation i (i is a *belt model* of g) if and only if $i[[g]] = 1$. By definition, g is *satisfied* by i if and only if i is a kernel model of g.

Theorem 3.2 (Consistency of IC-logic) *The axioms for PDIL are valid for default interpretations in the following sense: whenever a default interpretation i returns a truth value for an axiom, this truth value is equal to* true.

Furthermore, the I-modus ponens and the C-modus ponens inference rules are sound for default interpretations.

A default interpretation $i = i_0 \oplus (J, i_1)$ is *justified* if and only if, for any default formulae j and k, we have: (i) $i_0 \not\models j$, $i(j) = 1$ imply $+j \in J$, and (ii) $i(k) = 0$ implies $-k \in J$.

Theorem 3.3 (Consistency of J-logic) *The axioms of PDIL are valid for justified default interpretations in the following sense: whenever a justified default interpretation i returns a truth value for an axiom, this truth value is equal to* true.

Furthermore, the I-modus ponens, the C-modus ponens and the J-modus ponens inference rules are sound for justified default interpretations.

4 Default logic programs and their semantics

We discuss here only a restricted class of default logic programs. More general default logic programs are considered in [9].

A *default logic program* is a finite set of default ionic clauses. A *default ionic clause* is a formula of the form

$$A \leftarrow B_1 \wedge \ldots \wedge B_m$$

where $m \geq 0$, A is an atomic formula or a simple default ion, and the B_i's are atomic formulae. (In the propositional case, atomic formulae are simply proposition symbols). A ***simple default ion*** is an ion $(\Gamma, g)_*$ where g is an atomic formula.

4.1 Least fixpoint semantics of default logic programs

The *default Herbrand base* H_b is the set of all (ground) default ions and atomic formulae that can be built from symbols in the program. We define a *default Herbrand model* of P as being any subset of the default Herbrand base such that whenever the premisses of a clause of P are all in S, then the conclusion of P is also in S.

If P is a default logic program, then the associated T operator is defined as the one-step C-modus ponens or I-modus ponens operator $T : \mathcal{P}(H_b) \rightarrow \mathcal{P}(H_b)$:

$$\begin{aligned} T(S) = \; & \{A \in H_b : (A \leftarrow B_1 \wedge \ldots B_m) \in P \;,\; B_i \in S\} \cup \\ & \cup \{(a_{\sigma(1)}, (a_{\sigma(2)}, \ldots (a_{\sigma(m)}, A)_* \ldots)_*)_* \in H_b :) \, (A \leftarrow B_1 \wedge \ldots \wedge B_m) \in P \;,\; \ldots \\ & \ldots (a_i, B_i)_* \in S \;,\; \sigma \text{ transposition over } \{1, \ldots, m\}\} \end{aligned}$$

(By membership in P, we mean here being a variable-free rule variant of a clause of P.

Theorem 4.1 *Let P be a default logic program and let T be the operator associated with P. Then T is continuous for the Scott topology, i.e. $T(\cup_i S_i) = \cup_i T(S_i)$ for any increasing chain of subsets (S_i), and its least fixpoint is the least default Herbrand model of the default logic program.*

4.2 Continuous semantics of default logic programs

Let P be a default logic program. Let T be the associated operator. An *approximating system* for P is a sequence (m_n) of default interpretations such that: (i) sequence (m_n) is ascending for the extension ordering, i.e. for every n, m_{n+1} is an interpretation that extends m_n, (ii) for every natural number n, interpretation m_n is a model of $T^n(\emptyset)$, and (iii) the least upper bound m of (m_n) for the extension ordering $\sqsubseteq$ is a model of $lfp(T)$.

An approximating system $(m_n)_{n\in N}$ for P is *regular* if and only if for every approximating system $(m'_n)_{n\in N}$ for P, for every natural number n, whenever interpretations m_n and m'_n are comparable for the default ordering $\prec$, we have $m_n \prec m'_n$. A model m of $lfp(T)$ is *regular* if and only if it is the limit of some regular approximating system $(m_n)_{n\in N}$ for P.

An approximating system (m_n) for P is *continuous* if and only if $\forall n$, m_n is $\prec$-minimal among all the models of $T^n(\emptyset)$. A model m is *continuous* iff it is the limit of some continuous approximating system.

4.2.1 Constrained regular models

Let P be a default logic program. A set of *constraints* over the alphabet of P is intuitively a set C of pairs (a, b), where a and b are distinct atoms occurring in P.Given a set C of constraints, a *C-constrained valuation* is a valuation v such that for any $(a, b) \in C$, whenever one of the atoms a, b is in the domain of v, then the other atom is also in the domain of v, and we have $v(a) = -v(b)$, where $-$ denote the negation function over the boolean values. All of the above definitions can be generalized when we replace valuations by *constrained valuations.*

A given default logic program may have zero, one, or more than one constrained continuous model.

Example

Nixon is a quaker and a republican. Quakers are typically pacifists. Republicans are typically non-pacifists. Is Nixon pacifist? This yields the program: $\{q \ , \ r \ , \ q \rightarrow [p] \ , \ r \rightarrow [p']\}$ The *constraint* is (p, p'). Whence the following sequences of partial models: $m_1 = \{\mathbf{T}r, \mathbf{T}q\}$, $m_2 = \{\mathbf{T}r, \mathbf{T}q\} \oplus (\{+p, -\neg p\}, \{\mathbf{T}p\})$ and $m_2' = \{\mathbf{T}r, \mathbf{T}q\} \oplus (\{-p, +\neg p\}, \{\mathbf{F}p\})$.

We thus obtain two constrained continuous models, one where *Nixon is a pacifist*, and one where *he is a non-pacifist.*

4.3 Operational semantics of default logic programs

We have two kinds of derivation steps on default ions. First, we have the usual SLD-derivation steps, which are defined exactly as in classical logic programming. Second, we define *justified SLD-derivation steps* as follows: $g \rightarrow g'$ is a *justified SLD-derivation step* if and only if there exists a 4-tuple $(\lambda, r, \theta, \Gamma\theta)$ such that λ is an occurrence of an atom in g, r is a rule variant $(\Gamma, A)_\star \leftarrow B_1 \wedge \ldots \wedge B_m$ from the program with no free variable in common with g, and θ is a most general unifier for A and the atom of occurrence λ in g. Goal g' is then obtained from g by replacing the atom of occurrence λ in g by $B_1\theta \wedge \ldots B_m\theta$, and applying substitution θ to all the other atoms of g. Furthermore $\Gamma\theta$ is the *justification prefix* that should be attached to $A\theta$ in order to have $B_1\theta \wedge \ldots B_m\theta$ implies $(\Gamma\theta, A\theta)_\star$.

A *derivation* is a finite sequence of such steps. It is *successful* if the empty clause is the last goal in the derivation. From an algorithmic point of view, a successful derivation, will in general yield two kinds of bindings on our computation objects. First, we get the usual *answer substitution* θ computed by the derivation; θ will bind the object variables of the initial goal. Second, we also get a binding on the "justification contexts": this binding is obtained by putting together all the justifications prefixes encountered during the computation. Such a sequence of justifications provides a binding at the justification level, and will be called the *answer justification prefix* computed by the derivation. The justification context corresponding to the (answer) justification prefix is built inside-out during the computation; it essentially says which justifications should be acceptable in order for the success of the derivation to be meaningful.

4.4 Examples

Consider the following *frame problem for temporal projection* (Yale shooting problem): *After performing an action, things normally remain the same. After loading a gun, the gun is loaded. After a man is shot with a loaded gun, he dies. After loading the gun, waiting, and shooting Fred, is it the case that Fred will die?* The following default logic program formalizes this problem. We use default ion $(a, a)_\star$ to express the statement "*normally a*".

1. $result(s_1, load, s_0) \leftarrow$
2. $result(s_2, wait, s_1) \leftarrow result(s_1, load, s_0)$
3. $result(s_3, shoot, s_2) \leftarrow result(s_2, wait, s_1)$
4. $t(alive, s_0) \leftarrow$
5. $t(loaded, s) \leftarrow result(s, load, s')$
6. $ab(alive, shoot, s) \leftarrow t(loaded, s)$
7. $t(dead, s) \leftarrow t(loaded, s') \wedge result(s, shoot, s')$
8. $t(f, s) \leftarrow t(f, s') \wedge result(s, e, s') \wedge \neg ab(f, e, s')$
9. $(\neg ab(f, e, s), \neg ab(f, e, s))_\star \leftarrow result(s'', e, s) \wedge t(f, s)$

We obtain the following structure of models for the sets $T^n(\emptyset)$: $m_1 \sqsubseteq m_6' \sqsubseteq m_7'$ and $m_1 \sqsubseteq m_2 \sqsubseteq m_3 \sqsubseteq m_4 \sqsubseteq m_5 \sqsubseteq m_6 \sqsubseteq m_7$, where each m_n is a minimal model of $T^n(\emptyset)$, m_6' is a minimal model of $T^6(\emptyset)$ and m'_7 is a minimal model of $T^7(\emptyset)$. Thus $lfp(T) = T^7(\emptyset)$ has two $\prec$-minimal models, m_7 and m'_7. But only m_7 is continuous, i.e. can be obtained as a limit of "finite" approximations (in the sense of denotational semantics). Model m'_7 is not continuous, as there is no approximation process using $\prec$-minimal partial models leading from m_1 to m'_7.

In *regular, continuous* model m_7 Fred is *dead* in state s_3, after the shooting. In *non-regular* model m'_7, Fred is still alive after the shooting, and it is not the case that the gun is still loaded after the wait period.

From the operational point of view, query $t(dead, s_3)$ succeeds with *answer justification prefix* $\lambda\alpha.(\neg ab(loaded, wait, s_1), \alpha)_\star$. This says, intuitively, that Fred was dead after the shooting if the justification that the gun was not unloaded during the wait period is acceptable. Query $t(alive, s_3)$ also succeeds, with answer prefix justification:

$$\lambda\alpha.(\neg ab(alive, load, s_0), (\neg ab(alive, wait, s_1), (\neg ab(alive, shoot, s_2), \alpha)_\star)_\star$$

This essentially says that Fred will be alive after the shooting, if the justifications that he survived the loading of the gun, the waiting period, and the shooting, are all acceptable.

Let us now take a slightly more sophisticated version of the Yale shooting problem, say the Assassin problem [1], where *During the wait period, someone will attempt to unload the gun, and will succeed if he knows how. He may or may not know how to unload the gun.* Applying the same method, we obtain two symmetrical regular, continuous models. In the first model *The person knows how to unload the gun* is true, and Fred is alive after the shooting. This model is obtained as the limit of approximating system m_n. In the second model *The person knows how to unload the gun* is false, and Fred is dead after the shooting. This model is obtained as the limit of approximating system m'_n.

References

[1] Haugh B. *Simple causal minimizations for temporal persistence and projection*, Proc. AAAI-87, (1987), pp. 218-223

[2] Lakatos I. *Falsification and the methodology of scientific research programmes*, in Problems in the Philosophy of Science, Lakatos and Musgrave (eds.), North Holland, Amsterdam (1970), pp. 91-196

[3] Nait Abdallah M. A. *Procedures in logic programming*, Springer LNCS 225 (1986), pp. 433-447

[4] Nait Abdallah M. A. *Ions and local definitions in logic programming*, Springer LNCS 210 (1986), pp. 60-72

[5] Nait Abdallah M. A. *AL-KHOWARIZMI: A formal system for higher order logic programming*, Springer LNCS 233 (1986), pp. 545-553

[6] Nait Abdallah M. A. *Logic programming with ions*, Springer LNCS 267 (1987), pp. 11-20

[7] Nait Abdallah M. A. *An extended framework for default reasoning*, Springer LNCS 380 (1989), pp. 339-348

[8] Nait Abdallah M. A. *A logico-algebraic approach to the model theory of knowledge*, Theoretical Computer Science (1989)

[9] Nait Abdallah M. A. *Default ionic logic: A continuous framework for reasoning with partial information* (in preparation)

[10] Nait Abdallah M. A. *Kernel knowledge versus belt knowledge in default reasoning: a logical approach*, Proceedings of the Second International Conference on Computing and Information, Carleton, Canada, (May 1991)

[11] Quine W.V.O. *The problem of interpreting modal logic* , Journal of Symbolic Logic, (1947),

[12] Reiter R. *A logic for default reasoning* , Artificial Intelligence 13, (1980), pp. 81-132

On the cardinality of sets of infinite trees recognizable by finite automata

Damian Niwiński
Institute of Mathematics, University of Warsaw
ul.Banacha 2, 00-913 Warszawa 59, Poland

Abstract

We show that a Rabin recognizable set of trees is uncountable iff it is of the cardinality *continuum* iff it contains a non-regular tree. If a Rabin recognizable set L *is* countable, it can be represented as

$$L = M[t_1/x_1, \ldots, t_n/x_n]$$

where M is a regular set of finite terms and $t_1, \ldots, t_n$ are regular trees. We also design an algorithm which, given a Rabin automaton $\mathcal{A}$, computes the cardinality of the set of trees recognized by $\mathcal{A}$.

Introduction. RABIN(1969) introduced finite-state automata on infinite trees, now called *Rabin automata*, in order to prove decidability of the theory *S2S*, viz. the monadic second order theory of the binary tree. [1] This decidability result is often called *Rabin's Tree Theorem*. The result was actually achieved *via* a series of results on automata, namely decidability of their emptiness problem and effective closedness under finite union, projection and, the most difficult, complementation. Complexity of these construction has been studied much more recently in connection with the fact that a number of logics expressing properties of concurrent programs can be, in a sense, reduced to Rabin automata (see VARDI AND WOLPER(1986), EMERSON AND JUTLA(1988), SAFRA(1988)).

Not so much is known, however, about *decidability* of other natural problems concerning automata on infinite trees. For instance, is it decidable if a given Rabin automaton is equivalent to some *Büchi* automaton? (See a survey by THOMAS(1990) for more information on automata on infinite trees.) Or, more generally, can the *Rabin index* [2] of a recognizable set be effectively computed?

In the present paper, we make one step in the indicated direction by showing that, given a (finite presentation of) Rabin recognizable set L, the cardinality of L is effectively computable. In particular, it is decidable if a Rabin recognizable set of trees is countable

[1]The binary tree is considered as a logical structure, the universe of which consist of the set of words over the alphabet $\{\ell, r\}$ and the basic relations are those of the left and the right successors. The theory *S2S* can express many interesting properties of sequential and parallel programs, and can be viewed as a universal theory for most ot the known propositional logics of programs.

[2]The Rabin index of an automaton is the number of pairs in the acceptance condition (see below). The Rabin index of a set L is the minimum of indicies of automata recognizing L.

or not. We also show that, in the latter case, its cardinality is necessarily *continuum* (in other words Continuum Hypothesis holds for Rabin recognizable sets). Note that, because of the above-mentioned reductions of logics of programs to Rabin automata, our results induce a method to evaluate the number of (non-isomorphic) models of a given formula φ of, say *PDL, PAL*, $L\mu$, CTL^*,... (this application is not detailed in this paper).

The paper is organized as follows. After the preliminary Section 1, we prove in Section 2 that if a Rabin recognizable set L contains a non-regular tree then L is of cardinality continuum. In Section 3 we show that in the case when L is countable, it can be presented in the form

$$L = M[t_1/x_1, \ldots, t_k/x_k]$$

where M is a regular set of finite trees (terms) over the alphabet of L extended by $x_1, \ldots, x_k$ and $t_1, \ldots, t_k$ are regular trees; moreover the t_i's are the only trees t occurring (as subtrees of trees) in L which have the property that "t is a subtree of its own". Finally, in Section 4, we show an algorithm for computing the cardinality of L.

1 Trees and automata

The set of natural numbers $0, 1, \ldots$ is denoted by ω and identified with the first infinite cardinal number. We say that a set X is *countable* if its cardinality, $|X|$, is less or equal to ω. The cardinality of the powerset of ω, is denoted c (continuum).

For a set X, X^* is the set of finite words over X, including the empty word λ. The *length* of a word w is denoted by $|w|$, note that $|\lambda| = 0$. The (proper) *initial segment relation* is denoted $\leq$ ($<$). The same symbols are occasionally used for the standard inequality relation on natural numbers, but confusion should not arise. The following relation will be also useful:

$$u \lhd v \text{ iff } \textit{not } v \leq u$$

($\lhd$ is not an ordering). The concatenation of words u, $w \in X^*$ is presented by uw, this notation is also extended to sets of words $L, K \subseteq X^*$, $LK = \{uw : u \in L \text{ and } w \in K\}$.

A nonempty subset T of X^* closed under initial segments is called a *tree*. The elements of T are usually called *nodes* , the $\leq$-maximal nodes are *leaves* and λ is the *root* of T. If $u \in T$, $x \in X$ and $ux \in T$ then ux is an *immediate successor* of u in T. An infinite sequence $P = (w_0, w_1, \ldots)$ such that $w_0 = \lambda$ and, for each m, w_{m+1} is an immediate successor of w_m is called a *path* in T. We recall the celebrated *König's Lemma* (c.f.,e.g., KURATOWSKI-MOSTOWSKI(1976)) .

> If $T \subseteq X^*$ is an infinite tree and each $w \in T$ has only a finite number of immediate successors in T then T has an infinite path.

If S is an arbitrary set and T is a tree then a mapping $t : T \to S$ is called an *S-valued* tree or shortly an *S-tree*; in this context T is the *domain* of t denoted by $T = dom(t)$. We say "root of t", "path in t" etc., referring to the corresponding objects in $dom(t)$.

For an S-tree $t : dom(t) \to S$ and a node $v \in dom(t)$, the *subtree* of t induced by v is the S-tree denoted by $t.v$ and defined by $dom(t.v) = \{w : vw \in dom(t)\}$ and $t.v(w) = t(vw)$, for $w \in dom(t.v)$. Let

$$subtree(t) \;=\; \{t.w \;:\; w \in dom(t)\}.$$

A tree t is said to be *regular* if the set *subtree(t)* is *finite*. Notice that any subtree of a regular tree is also regular.

Now suppose that $A \subseteq dom(t)$ is an *antichain* with respect to $\leq$ (i.e. any two elements of A are incomparable) and let f be a function which associates an S-tree $f(w)$ with each $w \in A$. Then the ***substitution*** $t[f]$ is the S-tree defined by

- $dom(t[f]) = \{w \in dom(t) : \forall w' \in A, w \lhd w'\} \cup \bigcup_{w \in A} w\, dom(f(w))$,
- $t[f](u) = \begin{cases} f(w)(v) & \text{if } u = wv, w \in A \\ t(u) & \text{otherwise} \end{cases}$

In the case A is finite, say $A = \{w_1, \ldots, w_k\}$, we shall often express f explicitly, writing for example $t[w_1 : t_1, \ldots, w_k : t_k]$.

We also introduce a concept of *limit*. Suppose $t_0, t_1, \ldots$ is a sequence of S-trees such that $dom(t_0) \subseteq dom(t_1) \subseteq \ldots$, and, for each $w \in \bigcup_{n<\omega} dom(t_n)$ there is some $m(w)$ such that $\forall m \geq m(w), t_m(w) = t_{m(w)}(w)$. Then we define the limit of the sequence t_n, $t = lim\ t_n$ by

- $dom(t) = \bigcup_{n<\omega} dom(t_n)$,
- $t(w) = t_{m(w)}(w)$, for $w \in dom(t)$

We now fix a finite alphabet Σ. For notational convenience, we shall focus on full binary trees over Σ, i.e. the Σ-trees with $dom(t) = \{1,2\}^*$. Thus, any node $w \in dom(t)$ has exactly two successors $w1$ and $w2$. Let T_Σ be the collection of all such trees. The extension of our results to n-ary trees or ranked trees (where the number of successors of a node depends on the actual label) is routine.

Rabin automata on infinite trees. A Rabin automaton on Σ-trees is a tuple $\mathcal{A} = \langle Q, q_0, Tr, Acc\rangle$, where Q is a finite set of *states*, q_0 is an *initial state*, $Tr \subseteq Q \times \Sigma \times Q \times Q$ is a set of *transitions* and Acc is an acceptance condition, $Acc = \langle (L_1, U_1), \ldots, (L_n, U_n)\rangle$, where $U_i, L_i \subseteq Q$, for $i = 1, \ldots, n$.Number n is called the *Rabin index* of the automaton.

A *q-run* of the automaton $\mathcal{A}$ on a tree t is a Q-tree $r : dom(t) \to Q$ such that $r(\lambda) = q$ and for each node $w \in dom(t)$, $\langle r(w), t(w), r(w1), r(w2)\rangle \in Tr$. Now let $P = (w_0, w_1, \ldots)$ be a path in t. It is convenient to have the following notation. For $i \leq j$, let

$$r([w_i \ldots w_j]) = \{r(w_\ell) : i \leq \ell \leq j\},$$

also let $r([w_i \ldots \infty]) = \{r(w_\ell) : i \leq \ell\}$. Let

$$Inf(r, P) \;=\; \{q \in Q \;:\; r(w_m) = q \text{ for infinitely many } m\,\}.$$

(Clearly, $Inf(r,P) = \bigcap_{m<\omega} r([w_m \ldots \infty]) = r([w_{m_0} \ldots \infty])$, for some m_0) We say that a pair (L_i, U_i) is *appropriate* for a set of states Q' if

$$Q' \cap L_i = \emptyset \text{ and } Q' \cap U_i \neq \emptyset.$$

The q-run r is *accepting* if, for every path P there exist some $i \in \{1, \ldots, n\}$ s.t. the pair (L_i, U_i) is appropriate for $Inf(r, P)$. We say that the automaton $\mathcal{A}$ *accepts a tree* t, if there exists an accepting q_0-run of $\mathcal{A}$ on t. A set of trees $L \subseteq T_\Sigma$ is said to be *(Rabin) recognizable* if $L = L(\mathcal{A})$ for some automaton $\mathcal{A}$.

We now recall the notion of recognizability of sets of *finite* trees or *terms* (c.f., e.g., GECSEG AND STEINBY(1984)). Let $x_1, \ldots, x_n$ be symbols not in Σ. Consider the alphabet $\Sigma \cup \{x_1, \ldots, x_n\}$, also denoted $\Sigma(x_1, \ldots, x_n)$ or $\Sigma(\vec{x})$ for short . A $\Sigma(\vec{x})$-*term* is a $\Sigma(\vec{x})$-tree satisfying the following conditions:

1. $dom(t)$ is finite,
2. for $w \in dom(t)$, either w is a leaf (no successors) or w has two successors in $dom(t)$, $w1$ and $w2$,
3. for $w \in dom(t)$, $t(w) \in \{x_1, \ldots, x_n\}$ iff w is a leaf.

The set of all $\Sigma(\vec{x})$-terms will be denoted $finT_{\Sigma(\vec{x})}$. An automaton over $\Sigma(\vec{x})$-terms is a tuple $\mathcal{A} = \langle Q, q_0, Tr \rangle$, where $Tr \subseteq (Q \times \Sigma \times Q \times Q) \cup (Q \times \{x_1, \ldots, x_n\})$. An (accepting) run of $\mathcal{A}$ on a term t is a tree r with $dom(r) = dom(t)$ s.t. $r(\lambda) = q_0$ and, for each $v \in dom(t)$ different from a leaf, $\langle r(v), t(v), r(v1), r(v2) \rangle \in Tr$, whereas for v being a leaf, $\langle r(v), t(v) \rangle \in Tr$. The set of terms accepted by $\mathcal{A}$ is denoted $L_{fin}(\mathcal{A})$ and is said to be *regular*.

2 Non-regular trees

We start with observation that a non-regular tree satisfies in fact an apparently stronger condition.

Lemma 2.1 *If a tree $t \in T_\Sigma$ is not regular then there exists an infinite path in t, $P = (u_0, u_1, \ldots)$, such that $t.u_m \neq t.u_n$, for $m \neq n$.*

Proof. Consider a total, ω-type (strict) ordering on $dom(t)$:

$x \sqsubset y$ iff $|x| < |y|$ or $|x| = |y|$ and there exist $w \in dom(t)$ s.t. $w1 \leq x$ and $w2 \leq y$.

Let $V = \{v \in dom(t) : \text{for each } w \sqsubset v,\ t.w \neq t.v\}$. Since t is not regular, V is infinite. Let $V_{\leq} = \{v : v \leq u, \text{ for some } u \in V\}$. By definition, $V_{\leq}$ is a tree and then, by König's Lemma, there exists an infinite path contained in $V_{\leq}$, say, $P = (u_0, u_1, \ldots)$. We claim that

$t.u_i \neq t.u_j$, for $i \neq j$.

For suppose $i < j$. Let v be the $\sqsubset$-least node in V such that $u_j \leq v$, say $v = u_j w$ $(\lambda \leq w)$. Since $|u_i| < |u_j|$, we have $u_i w \sqsubset u_j w = v$. Hence, by definition of V, $t.u_j w \neq t.u_i w$. Consequently, $t.u_j \neq t.u_i$, as required. □

We now introduce a technical notion that will enable us to produce continuum trees by "pumping" a non-regular tree. Given a tree $t \in T_\Sigma$ and a node $v \in dom(t)$ *different from the root*, we call a pair $\langle t, v\rangle$ a *pointed tree*. Two pointed trees $\langle t, v\rangle$, $\langle t', v'\rangle$ are said to be *essentially different* if there exists a node w such that $(w \lhd v) \wedge (w \lhd v') \wedge (t(w) \neq t'(w))$. Given an automaton $\mathcal{A}$, $q \in Q$ and $i \in \{1, \ldots, n\}$, a *q-i-accepting* run of $\mathcal{A}$ on a pointed tree $\langle t, v\rangle$ is an accepting q-run, r say, of $\mathcal{A}$ on t such that moreover $r(v) = q$ and the pair (L_i, U_i) is appropriate for the set $r([\lambda \ldots v])$ (c.f.Section 1) A pointed tree $\langle t, v\rangle$ is *q-i-accepted* by $\mathcal{A}$ if it has a q-i-accepting run.

We say that a state $q \in Q$ is *accessible* if there exists a tree $t \in L(\mathcal{A})$ and an accepting q_0-run r of $\mathcal{A}$ on t such that, for some v, $r(v) = q$. We can state the following.

Theorem 2.2 *Let $\mathcal{A}$ be an automaton on Σ-trees with the Rabin index n. Then the following conditions are equivalent:*

1. *$L(\mathcal{A})$ is of the cardinality c,*
2. *$L(\mathcal{A})$ is uncountable,*
3. *$L(\mathcal{A})$ contains a non-regular tree,*
4. *there exist two essentially different special trees that are both q-i-accepted by $\mathcal{A}$, for some $i \in \{1, \ldots, n\}$ and some accessible state q.*

Proof "(1)⇒(2)" is trivial."(2)⇒(3)" follows immediately from counting agrument: there is only countably many regular trees in T_Σ.

"(3)⇒(4)" Let t be a non-regular tree in $L(\mathcal{A})$. By Lemma 2.1, there is an infinite path $P = (u_0, u_1, \ldots)$ in t such that all subtrees $t.u_i$ are parwise different. Let r be an accepting run of $\mathcal{A}$ on t and let $Q' = \mathit{Inf}(r, P)$. Suppose that a pair (L_i, U_i) is appropriate for Q'. Let $q \in Q'$. We choose some $m < \ell$ s.t. $r(u_m) = r(u_\ell) = q$ and $r([u_m \ldots \infty]) = Q'$. There must be some w s.t. $t.u_m(w) \neq t.u_\ell(w)$, thus $t(u_m w) \neq t(u_\ell w)$. Now, we can choose $k > \ell$ such that $r(u_k) = q$, $r([u_\ell \ldots u_k]) = Q'$ (hence, consequently, $r([u_m \ldots u_k]) = Q'$) and k is so large that $u_m w \lhd u_k$ and $u_\ell w \lhd u_k$. Let $u_k = u_m v = u_\ell v'$. Then also $w \lhd v$ and $w \lhd v'$ and so the pointed trees $\langle t.u_m, v\rangle$ and $\langle t.u_\ell, v'\rangle$ are essentially different and both are $q-i-$accepted by $\mathcal{A}$ (the required runs are obtained by suitable restrictions of r).

"(4)⇒(1)" Let $\langle t, v\rangle$ and $\langle t', v'\rangle$ be essentially different pointed trees *q-i*-accepted by $\mathcal{A}$ and let r and r' be corresponding runs. For each sequence $\alpha = (\alpha_n)_{n<\omega}$ of 0's and 1's, we construct a special tree t_α and a *q-i-accepting* run of $\mathcal{A}$ on t_α as follows. Intuitively, the sequence α induces a pattern of subsequent substitutions of t and t'. More specifically, we first define the sequences of trees $(s_n^\alpha)_n$ and $(r_n^\alpha)_n$ and additionally, a sequence of nodes $(v_n)_n$, as follows.

- If $\alpha_0 = 0$ then $s_0 = t$, $r_0 = r$, $v_0 = v$.
- otherwise $s_0 = t'$, $r_0 = r'$, $v_0 = v'$.

Next,

- if $\alpha_{n+1} = 0$ then $s_{n+1} = s_n[v_n : t]$, $r_{n+1} = r_n[v_n : r]$, $v_{n+1} = v_n v$.
- otherwise s_{n+1}, r_{n+1}, v_{n+1} are defined similarly with t', r', v' replacing t, r, v.

It follows by definition that r_n is an accepting q-i-run of $\mathcal{A}$ on the pointed tree $\langle s_n, v_n\rangle$. Also, it is clear that the sequences s_n and r_n satisfy the condition which is necessary for existence of a limit (Section 1). Let

- $s^\alpha = \lim_{n<\omega} s_n$
- $r^\alpha = \lim_{n<\omega} r_n$

We claim that r^α is an accepting q–run of $\mathcal{A}$ on s^α. Indeed, for a path P induced by the sequence v_n, the pair $\langle L_i, U_i\rangle$ is appropriate for $Inf(r^\alpha, P)$. Any other path is, from certain point on, included in a copy of t or t' and so is accepted by r or r' respectively. Now let t_0 be a tree in $L(\mathcal{A})$ s.t. there exists a run, say r_0 and $w \in dom(t_0)$ s.t. $r_0(w) = q_0$ (we have supposed that q is accessible) ; we fix such a t_0 independently from α. We set

$$t^\alpha = t_0[w : s^\alpha]$$

Clearly, t^α is accepted by $\mathcal{A}$ *via* the run $r_0[w : r^\alpha]$. Now, it follows from our construction that if two sequences α_n and β_n are different then $t^\alpha \neq t^\beta$. (Indeed, let m be the least index s.t. $\alpha(m) \neq \beta(m)$. Then $s^\alpha_m \neq s^\beta_m$ and consequently $s^\alpha \neq s^\beta$.)

The above argument implies that the cardinality of $L(\mathcal{A})$ is $\geq c$. The converse inequality is obvious.□

3 Normal form of countable sets

We say that a tree $t \in T_\Sigma$ is *alive* if there is some $v \in dom(t)$, $v \neq \lambda$, such that $t.v = t$. For a tree t, let

$$alive(t) = \{t' : t' \in subtree(t) \text{ and } t' \text{ is alive}\}$$

For a set of trees $L \subseteq T_\Sigma$, let

$$alive(L) = \bigcup_{t\in L} alive(t)$$

Now suppose $t.u = t$, $u \neq \lambda$. Let $u^0 = \lambda$, $u^{n+1} = u^n u$. Note that $t.u^n = t$. Let P be a completion of the sequence u^n to a path in t. Let $\mathcal{A}$ be an automaton and suppose that r is an accepting q-run of $\mathcal{A}$ on t. Let (L_i, U_i) be a pair appropriate for the set $Inf(r, P)$. There must be some q' s.t. $r(u^m) = q'$, for infinitely many m's. Choose m s.t. $r(u^m) = q'$ and moreover $r([u^m \ldots \infty]) = Inf(r, P)$. For each $k < \omega$, we can find $k' \geq k$ s.t. $r(u^{m+k'}) = q'$ and $r([u^m \ldots u^{m+k'}]) = Inf(r, P)$. Let $r' = r.u^m$, r' is an accepting q'-run of $\mathcal{A}$ on $t.u^m = t$ which have moreover the following property:

$$\forall k \exists w : |w| > k \wedge t.w = t \wedge r' \textit{ is a } q' - i\textit{-accepting run of } \mathcal{A} \textit{ on the pointed tree } \langle t, w\rangle$$

In this situation we say that r' is a q'*-lively-i-accepting* run of $\mathcal{A}$ on the tree t. So, we have just proved the following fact.

Lemma 3.1 *Each $t \in alive(L(\mathcal{A}))$ is q-lively-i-accepted by $\mathcal{A}$, for some q,i.* □

Now, let $x_1, \ldots, x_n$ be symbols not in Σ. For $t \in finT_{\Sigma(\bar{x})}$ and $t_1, \ldots, t_n \in T_\Sigma$, let $t[t_1/x_1, \ldots, t_n/x_n]$ be the tree in T_Σ obtained by replacing each occurrence of x_i in t by t_i, more specifically, this is the tree $t[f]$ (c.f. Section 1), where f is defined on the set $\{v \in dom(t) : t(v) \in \{x_1, \ldots, x_n\}\}$, by $f(v) = t_i$ iff $t(v) = x_i$. For a set of terms $M \subseteq finT_{\Sigma(\bar{x})}$, $M[t_1/x_1, \ldots, t_n/x_n] =_{df} \bigcup_{t\in M} t[t_1/x_1, \ldots, t_n/x_n]$.

The following fact can be easily observed using König's Lemma.

Lemma 3.2 *If t is a regular tree then there exist $m < \omega$, $s \in finT_{\Sigma(\vec{x})}$ and alive trees $t_1, \ldots, t_m$ s.t.*

$$t = s[t_1/x_1, \ldots, t_m/x_m].$$

□

Theorem 3.3 *If a set $L \subseteq T_\Sigma$ is Rabin recognizable and countable then L can be presented in the form*

$$L = M[t_1/x_1, \ldots, t_n/x_n]$$

where $M \subseteq finT_{\Sigma(\vec{x})}$ is a regular set of terms and

$$\{t_1, \ldots, t_n\} = \text{alive}(L).$$

Proof. Let $L = L(\mathcal{A})$, $\mathcal{A} = \langle Q, q_0, Tr, Acc\rangle$, where $Acc = \langle (L_1, U_1), \ldots, (L_p, U_p)\rangle$. We first show that, under assumption of countability of L, *alive*(L) is finite, more precisely

$$(*)\ |alive(L)| \leq |Q| \cdot p$$

For suppose that the inequality (*) does not hold. By this and Lemma 3.1, there are two *distinct* alive trees, say t and t' that are both q-lively-i-accepted by $\mathcal{A}$ for some pair q, i. But then we can clearly choose w and w' such that the pointed trees $\langle t, w\rangle$ and $\langle t', w'\rangle$ are essentially different (it is enough that $|w|, |w'| > |u|$, where $t(u) \neq t'(u)$) and both are q-i-accepted by $\mathcal{A}$. This, by Theorem 2.2, implies that $L(\mathcal{A})$ is uncountable, a contradiction. So, (*) is proven.

Let *alive*$(L) = \{t_1, \ldots, t_n\}$. Let

$$Q_i = \{q \in Q : \text{there is an accepting } q\text{-run of } \mathcal{A} \text{ on } t_i\}$$

for $i = 1, \ldots, n$. Let $\mathcal{B} = \langle Q^{\mathcal{B}}, q_0^{\mathcal{B}}, Tr^{\mathcal{B}}\rangle$ be an automaton on $\Sigma(\vec{x})$ terms defined by $Q^{\mathcal{B}} = Q$, $q_0^{\mathcal{B}} = q_0$, $Tr^{\mathcal{B}} = Tr \cup \{\langle q, x_i\rangle : i \in \{1, \ldots, n\}, q \in Q_i\}$. Let $M = L_{fin}(\mathcal{B})$.

Now, it is easy to show that

$$L = M[t_1/x_1, \ldots, t_n/x_n]\ .$$

□

4 Algorithm

We now apply the results of the preceding sections in order to design an algorithm which, given a Rabin automaton $\mathcal{A}$, computes the cardinality of $L(\mathcal{A})$. As we already know, this can be a natural number, ω or c.

Following an idea of EMERSON(1985), we introduce a concept of a *binary Σ-graph* (Σ-graph for short) which can be presented as a tuple

$$G = \langle V, root, succ_1, succ_2, lab\rangle$$

where V is a *finite* set of nodes, $root \in V$, for $i = 1, 2$, $succ_i$ is an (everywhere defined) unary operations on V and $lab : V \to \Sigma$ is a labelling function. We shall additionally require that the set V is *finitely generated* from $root$, viz., for every $v \in V$, there exists a path $root = v_0, v_1, \ldots, v_m = v$ such that, for $i < m$, $v_{i+1} = succ_j(v_i)$, for some $j \in \{1, 2\}$. Clearly, a Σ-graph can be unravelled into a regular tree into a natural way. More precisely, let a mapping $h_G : \{1,2\}^* \to V$ be defined by

- $h_G(\lambda) = root$,
- $h_G(wi) = succ_i(h_G(w))$, $i = 1, 2$.

The tree $t_G \in T_\Sigma$ defined by

$$t_G(v) = lab(h_G(v)),\ v \in \{1,2\}^*$$

is called the *unfolding* of G. Conversely, a regular tree t can clearly be obtained as the unfolding of infinitely many graphs.

A Σ-graph G will be called *alive* if there is a non-trivial path from $root$ to $root$, in other words, if $root = succ_i(v)$ for some $i \in \{1, 2\}$ and $v \in V$ (possibly $v = root$). Clearly, the unfolding of an alive graph is an alive tree (the converse need not be true, viz. an alive tree can be obtained by unfolding a graph which is not alive in our sense).

The notion of Rabin acceptance carries over Σ-graphs in a direct way. A q-run of an automaton $\mathcal{A} = \langle Q, q_0, Tr, Acc\rangle$ on a Σ-graph G is a Q-graph, of the form $R = \langle V, root, succ_1, succ_2, lab^R\rangle$, where $V, root$ and the successor mappings are as in G, $lab^R(root) = q$ and, for each $v \in V$, $\langle lab^R(v), lab(v), lab^R(succ_1(v)), lab^R(succ_2(v))\rangle \in Tr$. A run R is accepting if, for each infinite path P, there is an i such that a pair (L_i, U_i) is appropriate for Inf*(R,P)* (where the notions involved are defined analogically as for trees). A q-run R on an alive graph G is q-lively-i-accepting if it is accepting and moreover, there is a path, say $root = v_0, v_1, \ldots, v_m = root$ ($m \geq 1$), s.t. the pair (L_i, U_i) is appropriate for the set $\{lab_R(v_j) : j = 0, 1, \ldots, m\}$.

RABIN(1972) showed (essentially) that if an automaton $\mathcal{A}$ accepts some tree in T_Σ then it also accepts a finite Σ-graph. EMERSON(1985) further showed that this graph can be choosen with number of nodes linear in the size of $\mathcal{A}$'s transition diagram. We need a similar property for q-i-acceptance.

Lemma 4.1 *If an automaton $\mathcal{A}$ accepts some alive tree in T_Σ then it also accepts some alive Σ-graph G with number of nodes $O(n^2)$, where $n = |Tr|$.*

Sketch of proof. Let t be an alive tree and let r be a q-lively-i-accepting run of $\mathcal{A}$ on t. Let $u \neq \lambda$ be such that $t.u = t$, $r(u) = r(\lambda) = q$ and the pair (L_i, U_i) is appropriate for $r([\lambda \ldots u])$. Let $\lambda = v_0, v_1, \ldots v_m = u$ be the path leading from the root of t to u. We can further assume that all the pairs $\langle t(v_j), r(v_j)\rangle$ are different. Indeed, otherwise we could cut off the fragment between duplicated nodes and apply the same change all along the path induced by u^n. Next, by the above-mentioned results of RABIN and EMERSON, we can assume that, for $j < m$, if $v_j\ell$ is a successor different from v_{j+1} then the tree $t.v_j\ell$ is regular and can be obtained by unfolding of a graph with number of nodes linear in the size of $\mathcal{A}$. We define our graph G as follows:

- take the path $v_0, v_1, \ldots, v_{m-1}$ and set, for $j = 0, \ldots, m-2$, $v_{j+1} = succ_k(v_j)$, where $v_{j+1} = v_jk$;

- draw an edge from v_{m-1} to $v_0 = root$, by setting $succ_k(v_{m-1}) = v_0$, where $v_m = v_{m-1}k$;
- graft a graph suitable for $t.v_j\ell$ to the remaining successor of v_j, $j = 0, 1 \ldots, m-1$.

It is easy to see that G satisfies the required conditions. □

The proof of the following fact is routine.

Lemma 4.2 *Given a run R of $\mathcal{A}$ on an alive Σ-graph G, it can be decided if R is a q-lively-i-accepting (in time exponential in the size of G).* □

We can already outline an algorithm that, given a Rabin automaton $\mathcal{A} = \langle Q, q_0, Tr, Acc\rangle$, $Acc = \langle (L_1, U_1), \ldots, (L_p, U_p)\rangle$, computes $|L(\mathcal{A})|$. Let K be the constant induced by the Lemma 4.1 (viz. $|V| \leq K \cdot |Tr|^2$). Let $K_{\mathcal{A}} = K \cdot |Tr|^2$.

- Create a list ∇ as follows. For each $q \in Q$, $i \in \{1, \ldots, p\}$, check if there is a Σ-graph q-lively-i-accepted by $\mathcal{A}$ with number of nodes $\leq K_{\mathcal{A}}$. If it is the case, select one such graph and put it on the list ∇.
- If ∇ is empty then $L(\mathcal{A}) = \emptyset$ and so $|L(\mathcal{A})| = 0$. (Indeed, otherwise $L(\mathcal{A})$ contains a regular tree and then, by Lemmas 3.2 and 4.1, we must have put some graph on the list ∇.) Suppose $\nabla = (G_1, \ldots, G_n)$. Let t_i be the unfolding of G_i. For $i = 1, \ldots, n$, compute a set Q_i defined as in the proof of Theorem 3.3,

  $$Q_i = \{q \in Q : \text{there is an accepting } q\text{-run of } \mathcal{A} \text{ on } t_i\}$$

  Then define an automaton $\mathcal{B}$ exactly as in the proof of Theorem 3.3.
- Let $M = L_{fin}(\mathcal{B})$. For each t_i, design an automaton recognizing the singleton of t_i (easy), then construct an automaton, $\mathcal{C}$ say, recognizing $L = M[t_1/x_1, \ldots, t_n/x_n]$. Clearly, $L(\mathcal{C}) \subseteq L(\mathcal{A})$.
- Verify if $L(\mathcal{C}) = L(\mathcal{A})$. (Use Rabin Tree Theorem mentioned in the Introduction).
- If no, the answer is $|L(\mathcal{A})| = c$. (Indeed, $L(\mathcal{A}) - L(\mathcal{C})$ contains a regular tree t. By Lemma 3.2, $t = s[t'_1/z_1, \ldots, t'_m/z_m]$, where $s(z_1, \ldots, z_m)$ is a finite term and $t'_1, \ldots, t'_m$ are alive. By hypothesis, at least one t'_j is *not* on the list ∇. By Lemma 3.1, t'_j is q-lively-i-accepted by $\mathcal{A}$, for some q, i. Then, by Lemma 4.1, we have already some tree t_ℓ with the same property on the list ∇. But $t_\ell \neq t'_j$. Then, by Theorem 2.2, $|L(\mathcal{A})| = c$.)
- If $L(\mathcal{C}) = L(\mathcal{A})$ then $L(\mathcal{A})$ is countable. Now, it is routine to check if the set M (defined above) is infinite (in this case $|L(\mathcal{A})| = \omega$) or finite. In the latter case, we can also produce all elements of M and then using the list ∇, we can further produce the (finite representations of) elements of $L(\mathcal{A})$. However, in so-obtained list, some trees may duplicate. Fortunately, the equality between regular trees is decidable (it can be derived from Rabin Tree Theorem or proved directly, c.f.COURCELLE(1983)). So we are done.

Note. Clearly, the above algorithm is elementary. Evaluation of its complexity will follow in the full version of the paper.

References

B.COURCELLE(1983) Fundamental properties of infinite trees, *Theor.Comput.Sci.* 25, 95-169.

E.A.EMERSON(1985) Automata, tableaux and temporal logics, *in: Proc. Workshop on Logics of Programs*, LNCS 193, 79-88.

E.A.EMERSON AND C.JUTLA(1988) The complexity of tree automata and logics of programs, *Proc. 29th IEEE Symp. on Foundations of Computer Science*, , N.Y.,328-337.

F.GECSEG AND M.STEINBY(1984) Tree Automata, Akademiai Kiado, Budapest.

K.KURATOWSKI AND A.MOSTOWSKI(1976) Set Theory, *North-Holland.*

M.O.RABIN(1969) Decidability of second-order theories and automata on infinite trees, *Trans.Amer.Soc.*141, 1-35.

M.O.RABIN(1970) Weakly definable relations and special automata, *in:* Mathematical Logic in Foundations of Set Theory (Y.Bar-Hillel,ed.),1-23.

M.O.RABIN(1972) Automata on infinite objects and Church 's problem, *Amer.Math.Soc.*, 1-22.

M.O.RABIN(1977) Decidable theories, *in:* Handbook of Mathematical Logic (J.Barwise,ed

S.SAFRA(1988) On The Complexity of ω-Automata, *Proc. 29th IEEE Symp. on Foundations of Computer Science*, White Plains, N.Y., 319-327.

W.THOMAS(1990) Automata on infinite objects, *in:* Handbook of Theoretical Computer Science, vol.B (J.van Leeuven,ed.), 133-191.

M.Y.VARDI AND P.L.WOLPER(1986) Automata-theoretic techniques for modal logics of programs, *J. Comput.System Sci.* 32, 183-221.

Extending Temporal Logic by Explicit Concurrency

Barbara Paech
Institut für Informatik, Ludwig-Maximilians-Universität
Theresienstr.39, D-8000 München 2

Abstract

I define a logic with structural operators for sequential and parallel composition. It is shown to characterize the state spaces of occurrence nets. The logic has the finite model property and a sound and complete axiomatization. It is the first formalism to combine negation with structural operators.

1 Introduction

In the realm of program specification and verification *temporal logics* are a well established reasoning tool. The models of their basic form, *Linear Temporal Logic* (LTL for short, see e.g. [Kr87, MP89]), are sequences of states. Even with this basic model LTL has been applied successfully to *concurrent* systems. However, in that context the behaviour description with temporal logic is not fully satisfactory, since concurrency is only *indirectly* expressed. Consider the formula $\Box(A \wedge \alpha \rightarrow \circ B)$ which is typically used to describe the state changes induced by an atomic action α. From a set of such formulae it cannot be deduced, whether the specified program can exhibit any concurrency. This is only implicit in the partitioning of the set of atomic actions. As an example consider the two simple systems N_{conc} and N_{seq} depicted in figure 1. N_{conc} is specified by $\Box((A \wedge x \rightarrow \circ B) \wedge (C \wedge y \rightarrow \circ D))$ and, similarly, N_{seq} is specified by $\Box((A \wedge x \rightarrow \circ B) \wedge (B \wedge y \rightarrow \circ C))$. The two specifications are very similar, although N_{seq} does not exhibit any concurrency. So the important feature of concurrency is only captured in the atomic formulae. This example shows that the structure of the system is not easily gathered from its temporal logic description making temporal formulae often difficult to read. So it seems desirable to include structuring operators like ; for sequentialization and $\|$ for concurrency into the logic. Then N_{seq} could be described by the formula $(A \wedge x; B \wedge y; C)$, while N_{conc} is characterized by $(A \wedge x; B) \| (C \wedge y; D)$.
Obviously, the models of a logic with $\|$ should also have explicit concurrency. This motivates an extension of the usual sequences to partially ordered models. In a recent paper [PP90] Peled and Pnueli argue for a logic based on trace semantics. A *trace* is an equivalence class of sequences which correspond to the same conflict-free (but possibly concurrent) behaviour of a system. Such behaviour is often called a *process*. However, it is possible to give a much more intuitive extension of the linear semantics. The same way a sequence corresponds to a path in (and therefore a *linearly*

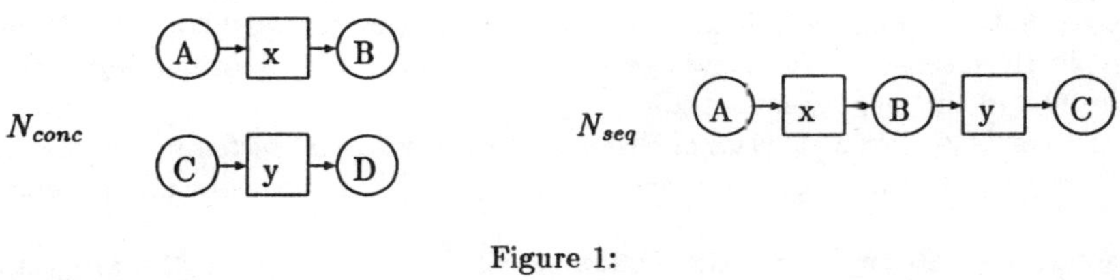

Figure 1:

ordered substrucure of) the state space of the system, one can characterize processes as a certain class of substructures of the state space. It turns out that these substructures are characterized by the *diamond property* and *concurrency closure*. They are called *occurrence graphs* in the following. Interestingly, although there are several partial order models (e.g. event structures, traces, occurrence nets as discussed in [NRT90]), occurrence graphs have not been studied before. In this paper I show that the state spaces of *occurrence nets* are occurrence graphs and vice versa. Having settled the question of the models I define *Net Logic*, whose main temporal operators are (iterated) sequential and parallel composition (denoted by $;,^+,||,^{||}$). These operators are very different from the usual temporal operators, like $\Box$, *unless*. $||$ is used to decompose models into two independent parallel components, while ; describes sequential composition such that all actions of the first component precede the ones of the second. It turns out that each finite occurrence graph has a unique decomposition according to these operators. The usual temporal operators seem not sufficient to characterize the structural properties particular to occurrence graphs. Note that in [RP86, Pae88] temporal logics with an operator ; (often called *chop*) have been investigated. In the context of linear models ; has turned out to be a natural extension of the usual temporal operators. This is no longer true for partially ordered models. $\Box$, *unless* are based on the *next-state*-relation, while ; has to be based on the *next-inevitable-state*-relation. Inevitability has also been studied in [MOP89].
In the third part of the paper I sketch a sound and complete axiomatization for Net Logic. It also satisfies the finite model property. The proofs are based on the observation that Net Logic can be reduced to some kind of dynamic logic, whose atomic programs are given by formulae describing occurrence graphs which are not sequentially decomposable. This class of formulae can also be characterized syntactically (the corresponding set of formulae is denoted as $\mathcal{AL}$). Then I derive a normal form for Net Logic formulae on *finite* occurrence graphs, where negation and concurrency only occur inside the $\mathcal{AL}$ formulae.
In the final section I discuss possible extensions and some related work.

2 Occurrence graphs

In this paper a concurrent system is modelled by a special kind of *elementary net system* (ENS for short). In contrast to e.g. [Thi87] I allow isolated conditions in the initial state and require events to be simple (that means characterized by their pre- and postset). The latter ensures that there is no difference between an event t and its *characteristic pair* $({}^\bullet t, t^\bullet)$. Examples of ENS are given in figure 1. If not otherwise stated, in the figures of this paper the initial states are always given by the minimal conditions.
A pair (x, y) of sets is called *enabled at* the set g (written as $g[(x,y)\rangle$), if $x \subseteq g$ and $y \cap g = \emptyset$. $g[(x,y)\rangle$ and $g' = g \setminus x \cup y$ is abbreviated as $g[(x,y)\rangle g'$, for an event t $g[({}^\bullet t, t^\bullet)\rangle$ and $g[({}^\bullet t, t^\bullet)\rangle g'$ are abbreviated as $g[t\rangle$ and $g[t\rangle g'$, respectively. Two pairs (x,y) and (u,v) of sets are called *concurrent*, if $(x \cup y) \cap (u \cup v) = \emptyset$. Again, two events t_1, t_2 are called *concurrent*, if $({}^\bullet t_1, t_1^\bullet)$ and $({}^\bullet t_2, t_2^\bullet)$ are. As usual, for the enabledness of an event loops are prohibited and therefore (and by simplicitity of the events) for each element $(g, g') \in R$ there exists exactly one event t such that ${}^\bullet t = g \setminus g'$ and $t^\bullet = g' \setminus g$. So one can omit the labelling of the edges in the *state space* $S(N)$ of an ENS N. Note, however, that one could include loops by defining $g[t\rangle \iff {}^\bullet t \subseteq g, t^\bullet \cap g \subseteq {}^\bullet t$. Then an edge has to be explicitly labelled by the events generating it. This makes notation more complicated, but all the results of this paper remain valid.
To describe processes of ENS a subclass of ENS called *occurrence net systems* (ONS for short) can be used. These consist of an occurrence net (see e.g. [R85]) together with a particular starting state.
Now what about the state spaces of ONS? Following [ER90] state spaces of ENS are just a special

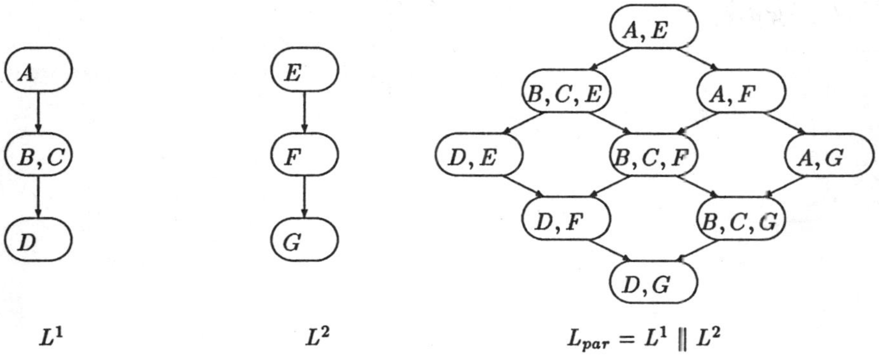

Figure 2:

case of so called *initialized labelled partial set 2- structures* (ilps2s for short) characterized by the following *closure property* : for all $g, g_1, g_2 \in G$: if $R(g_1, g_2)$, and $g[osd(g_1, g_2)\rangle$, then there exists $g' \in G$ with $R(g, g')$ and $osd(g_1, g_2) = osd(g, g')$, where $osd(h_1, h_2) = (h_1 \setminus h_2, h_2 \setminus h_1)$. However, in the context of ONS, it suffices to require only two special cases: *diamond property* and *concurrency closure*. The diamond property requires that different interleavings of concurrent events have the same overall effect and concurrency closure enforces that concurrency is made explicit in the branching structure. While the diamond property is often used in reasoning about concurrent systems, concurrency closure is new. In both cases only neighbouring elements are involved and therefore it is much easier to check for these two properties than for closure in general.

Definition 1 (occurrence graph) *An ilps2s* $L = (G, R, g_{in})$ *is called* an occurrence graph, *if*
- $G = R^*(g_{in})$ *and*
- $ENS(L) = (B, E, G, g_{in})$ *is an ONS, where* $B = loc(L) = \bigcup G$ *and* $E = \{osd(g, g') : (g, g') \in R\}$ *and* $F = \{(b, (x, y)) : b \in x, (x, y) \in E\} \cup \{((x, y), b) : b \in y, (x, y) \in E\}$ *and*
- *(diamond property) for all* $g, g_1, g_2 \in G$ *with* $R(g, g_i)$ *for* $i = 1, 2$ *and* $osd(g, g_1)$ *and* $osd(g, g_2)$ *concurrent, follows that there exists* $g_3 \in G$ *with* $R(g_1, g_3)$, $R(g_2, g_3)$ *and* $osd(g_1, g_3) = osd(g, g_2)$, $osd(g_2, g_3) = osd(g, g_1)$ *and*
- *(concurrency closure) for all* $g, g_1, g_2 \in G$ *with* $R(g, g_1), R(g_1, g_2)$ *such that* $osd(g, g_1)$ *and* $osd(g_1, g_2)$ *are concurrent, follows that there exists* $g_3 \in G$ *with* $R(g, g_3)$ *and* $osd(g, g_3) = osd(g_1, g_2)$.

Note that $G = R^*(g_{in})$ does not follow from $ENS(L)$ being an ONS. As an example of an occurrence graph consider L_{par} in figure 2. One can show that every occurrence graph is a distributive lattice.
Since $ENS(S(N))$ corresponds to a subnet of N for any ONS N, and therefore also is an ONS, it directly follows that every state space of an ONS is an occurrence graph. By case analysis on $R^*(g, g_1)$ or $R^*(g_1, g)$ one can show that every occurrence graph P satisfies the closure property and therefore is the state space of $ENS(P)$. This is captured in the following theorem.

Theorem 2
1. Let $N = (B, E, F, g_{in})$ *be an ONS. Then* $S(N)$ *is an occurrence graph and* $ENS(S(N)) = N$.
2. Let $L = (G, R, g_{in})$ *be an occurrence graph. Then* $ENS(L)$ *is an ONS with* $S(ENS(L)) = L$.

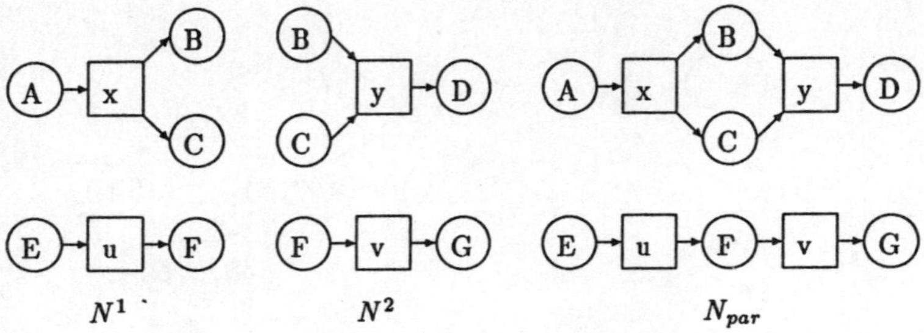

Figure 3:

3 Net logic

In this section I define a temporal logic over occurrence graphs called *Net Logic* (*NL* for short). First I define parallel and sequential composition on occurrence graphs. Two occurrence graphs are parallel composable, if they are independent of each other. Then the composed occurrence graph is given by the interleaving of all events.

Definition 3 (parallel composition of occurrence graphs)
Let $L_i = (G_i, R_i, g_{in}^i), i = 1,2$ *be two occurrence graphs such that* $loc(L_1) \cap loc(L_2) = \emptyset$. $L = (G, R, g_{in})$ *is called* the parallel composition of L_1 and L_2 *(* $L = L_1 \parallel L_2$*) iff*

- $g_{in} = g_{in}^1 \cup g_{in}^2$ *and*
- $G = \{g_1 \cup g_2 : g_i \in G_i, i = 1,2\}$ *and*
- $R = \{(g_1 \cup g_2, g_1' \cup g_2) : R_1(g_1, g_1'), g_2 \in G_2\} \cup \{(g_1 \cup g_2, g_1 \cup g_2') : R_2(g_2, g_2'), g_1 \in G_1\}$

It is trivial to check that the parallel composition of two occurrence graphs is again an occurrence graph. Note that a simple parallel composition of ONS can be defined by taking the disjoint union of their conditions and events, respectively. Then, obviously, $S(N_1 \parallel N_2) = S(N_1) \parallel S(N_2)$. As an example consider $L_{par} = L^1 \parallel L^2$ in figure 2.
The sequential composition is more complicated. Usually, e.g. [Win80, R85], a sequential composition operator $\bullet$ of ONS is defined by melting the maximal and minimal states. As an example consider the ONS N^1, N^2, N_{par} of figure 3, where $N_{par} = N^1 \bullet N^2$. A corresponding definition on the level of occurrence graphs yields $L_{wrong} = S(N^1) \bullet S(N^2)$ as given in figure 4. However, L_{wrong} violates the concurrency closure at the states $\{B,C,E\}, \{B,C,F\}, \{D,F\}$ and $\{A,F\}, \{B,C,F\}, \{B,C,G\}$. Therefore, also $L_{par} = S(N^1 \bullet N^2) \neq L_{wrong}$. The problem is that the state $\{B,C,F\}$ is not common to all paths in L_{par}. This is captured in the concept of *inevitability*.

Definition 4 (inevitable) *Let* $L = (G, R, g_{in})$ *be an occurrence graph. A state* $g \in G$ *is called* inevitable, *if for all maximal paths* $R(g_i, g_{i+1}), i \in I$ *starting in* g_{in} *holds* $g = g_i$ *for some* $i \in I$.

The above definition could also be rephrased by requiring that, for every $g' \in G$, $R^*(g, g')$ or $R^*(g', g)$ holds.
So to *decompose* an occurrence graph $L = (G, R, g_{in})$ sequentially it suffices to stipulate $L = L_1; L_2$ iff there exists an inevitable state $g \in G$ such that L_1 and L_2 correspond to the structure of predecessors and successors of g, respectively. For composition,however, one needs a way to ensure that the maximal state of the first component is inevitable in the composed occurrence graph. Here the basic idea is that the events executable at the end of the first component (denoted by $tcomp_b$)

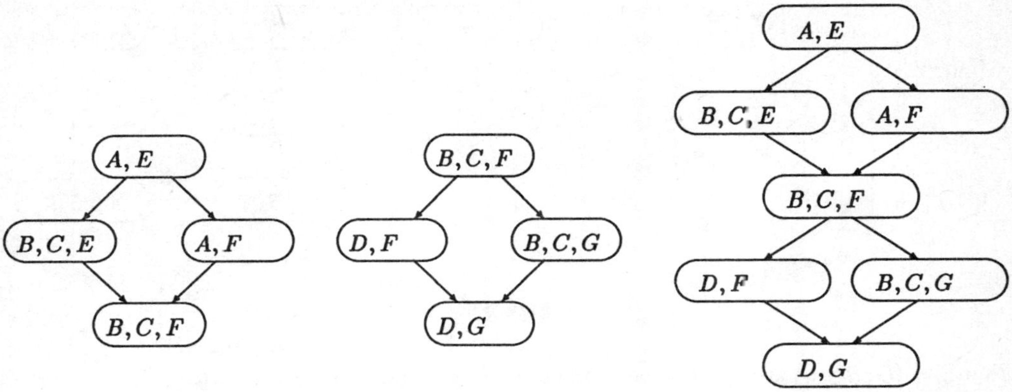

$S(N^1)$ $S(N^2)$ $L_{wrong} \neq S(N_{par})$

Figure 4:

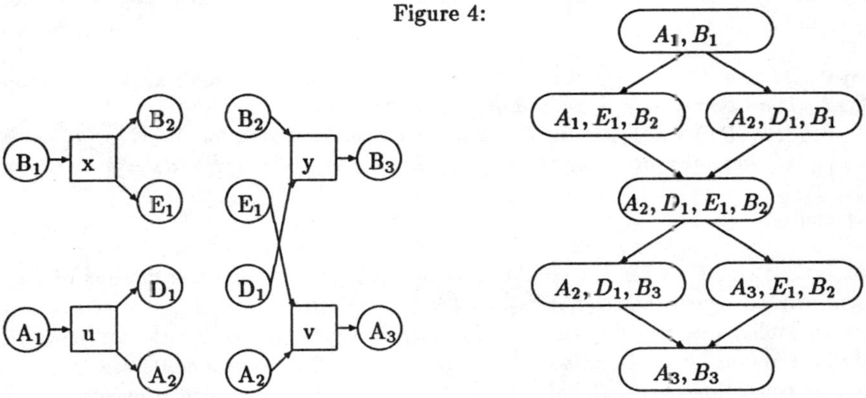

N^1_{syn} N^2_{syn} $S(N^1_{syn} \bullet N^2_{syn})$

Figure 5:

and the events executable at the beginning of the second component (denoted by $tcomp_f$) must not be concurrent. In figure 3 $tcomp_b(S(N^1)) = \{x, u\}, tcomp_f(S(N^2)) = \{y, v\}$ and $\{x, v\}$ and also $\{y, u\}$ are concurrent. On the other hand in figure 5 $tcomp_b(S(N^1_{syn})) = \{x, u\}, tcomp_f(S(N^2_{syn})) = \{y, v\}$ and none of the pairs $\{x, y\}\{x, v\}, \{u, y\}, \{u, v\}$ is concurrent. Looking at the corresponding state spaces one finds $S(N^1_{syn} \bullet N^2_{syn}) = S(N^1_{syn}) \bullet S(N^2_{syn})$.
For technical reasons I also stipulate that there should be no conditions neither involved in the $tcomp_b$-events of the first (denoted by $scomp_b$) nor in the $tcomp_f$- events of the second component (denoted by $scomp_f$), but common to both components.

Definition 5 (sequential composition)
Let $L_i = (G_i, R_i, g^i_{in}), i = 1, 2$ be two occurrence graphs such that

- $G_1 \cap G_2 = \{g^2_{in}\}$ *and* $loc(L_1) \cap loc(L_2) = g^2_{in}$ *and*
- L_1 *is finite with maximal state* g^2_{in} *and*
- *all* $(x_1, y_1) \in tcomp_b(L_1), (x_2, y_2) \in tcomp_f(L_2)$ *are not concurrent and*
- $scomp_b(L_1) \cap scomp_f(L_2) = \emptyset$.

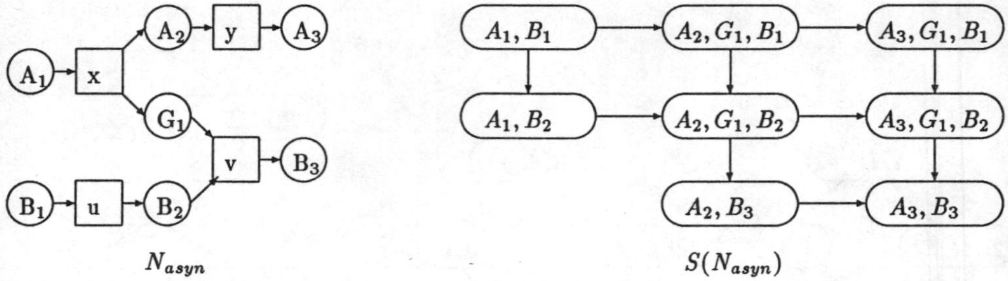

Figure 6:

Then $L = (G, R, g_{in})$ *is called* the sequential composition of L_1 and L_2 *(*$L = L_1; L_2$*) iff* $G = G_1 \cup G_2$ *and* $R = R_1 \cup R_2$

Two occurrence graphs satisfying the above definitions are called *sequentially composable.* The following facts hold:

Theorem 6 *Let* $L_i = (G_i, R_i, g^i_{in}), i = 1,2$ *be two sequentially composable occurrence graphs.*
1. $L = L_1; L_2$ *is an occurrence graph and* g^2_{in} *is inevitable in* L.
2. Let $L'_i = (G'_i, R'_i, h^i_{in}), i = 1,2$ *be two occurrence graphs such that* $loc(L'_1) \cap loc(L'_2) = \emptyset$. *Then* $L_1; L_2 = L'_1 \parallel L'_2$ *implies that either* $G_1 = \{g^1_{in}\}, R_1 = \emptyset$ *or* $G_2 = \{g^2_{in}\}, R_2 = \emptyset$ *or* $G'_1 = \{\emptyset\}$ *or* $G'_2 = \{\emptyset\}$.
3. ; is associative.

Note that again ; could be defined on the level of ONS (transferring the notions of *tcomp* and *scomp* appropriately) such that $S(N_1; N_2) = S(N_1); S(N_2)$.
By the orthogonality of ; and $\parallel$ (theorem 6.2) every finite occurrence graph can be decomposed recursively into unique parallel and sequential components. The atomic units are
• *state*-graphs consisting of one global state and an empty reachability relation or
• T_1-graphs consisting of two global states g, g' connected by an edge such that $osd(g, g') = (g, g')$ or
• $asyn_0$-graphs, which involve some kind of asynchrony as e.g. shown in figure 6.
Formulae describing the unit-graphs are taken as basic in Net Logic.
Now I have collected all the preliminaries for defining *Linear Net Logic* (LNL for short).

Definition 7 (syntax of LNL) *There are two kind of formulae:* state-formulae *and* process-formulae.
1. Let $\Phi_0 = \{P, Q, ..\}$ *be a set of atomic formulae. Every atomic formula is a state- and a process-formula.*
2. If A, B *are state- (process-) formulae , then* $\neg A, (A \vee B)$ *are also state- (process-) formulae.*
3. If A *is a process-formula and* B *is a state-formula, then* $(A; B)$ *is a state-formula.*
4. $T_1, asyn_0, state$ *are process-formulae.*
5. If A, B *are process-formulae, then* A^+, $(A; B), (A \parallel B)$ *and* $A^{\parallel}$ *are also process-formulae.*
The set of state-formulae is denoted by $\mathcal{F}_{LNL}$, *the set of process-formulae by* $\mathcal{F}_{proc}$.

Definition 8 (semantics of LNL) *Let* $L = (G, R, g_{in})$ *be an occurrence graph and* $\pi : loc(L) \rightarrow \Phi_0$ *be a proposition labelling such that* $\pi(s_1) \neq \pi(s_2)$ *holds for all* $g \in G$ *and* $s_1 \neq s_2 \in g$. *Then* $M = (L, \pi)$ *is called a* process-model. *It is a* finite process-model, *if* G *is finite.*

Note that the proposition labelling of process-models is local. Following the arguments of Parikh for deterministic distributed transition logic [P88] one can show that a logic on occurrence graphs with a *global* proposition labelling $\pi : G \rightarrow \Phi_0$ can characterize a grid and is therefore undecidable (for details consult [Pae90]).

The *truth of a (state-) process-formula* is determined in a process-model $M = (L, \pi)$ with $L = (G, R, g_{in})$ and $g \in G$ as follows (I write $min(M)$ for g_{in}. $succ(g, M)$ denotes the substructure of M given by the successors of g.):

$M, g \models P$	$\iff$	there exists $s \in g$ such that $P = \pi(s)$, where $P \in \Phi_0$
$M, g \models T_1$	$\iff$	$G = \{g, g'\}, R = \{(g, g')\}$ and $osd(g, g') = (g, g')$
$M, g \models state$	$\iff$	$R = \emptyset$
$M, g \models asyn_0$	$\iff$	$M, g \not\models T_1$ and $M, g \not\models state$ and ($M = M_1; M_2$ implies $M_1 \models state$ or $M_2 \models state$) and ($M = M_1 \parallel M_2$ implies $G_1 = \{\emptyset\}$ or $G_2 = \{\emptyset\}$)
$M, g \models \neg A$	$\iff$	$M, g \not\models A$
$M, g \models A \vee B$	$\iff$	$M, g \models A$ or $M, g \models B$
$M, g \models A op B$	$\iff$	there exist process-models M_1 and M_2 such that $succ(g, M) = M_1 op M_2$ and $M_1, min(M_1) \models A$ and $M_2, min(M_2) \models B$, where $op \in \{;, \parallel\}$
$M, g \models A^+$	$\iff$	there exist process-models $M_1, ..., M_n$ with $n \geq 1$ such that $succ(g, M) = M_1; ...; M_n$ and for all $i \in \{1, ..., n\}$ holds $M_i, min(M_i) \models A$
$M, g \models A^{\parallel}$	$\iff$	there exist process-models $M_1, ..., M_n$ with $n \geq 1$ such that $succ(g, M) = M_1 \parallel ... \parallel M_n$ and, for all $i \in \{1, ..., n\}$, holds $M_i, min(M_i)) \models A$ and $M_i = M_{i1} \parallel M_{i2}$ implies $M_{i1} \models state$ or $M_{i2} \models state$

A state- or process-formula A is called *LNL-satisfiable* , if there exists a process-model M and $g \in G$ such that $M, g \models A$. It is called *(finitely) valid* ($\models_{fpr} A$, resp. $\models_{LNL} A$), if for all (finite) process-models M and $g \in G$ holds $M, g \models A$.
Note that, since only state-formulae comprise the language $\mathcal{F}_{LNL}$ of LNL, the operator $\parallel$ is restricted to *finite* process-models. The evaluation of $\parallel$ on infinite models doesn't seem to be axiomatizable.
In the following I discuss some examples of LNL-formulae:

- For linear process-models not involving any concurrency (e.g. L^1 or L^2 of figure 2) LNL corresponds to *Linear Regular Logic* (LRL for short) as defined in [Pae88]. In that context a nexttime-formula $\circ A$ can be expressed as $\neg(T_1; \neg A)$ and, similarly, $\Box A$ corresponds to $A \wedge \neg(T_1^+; \neg A)$. *AunlessB* cannot be expressed uniformly, but as shown in [Pae88] because of the star operator the expressibility of LRL exceeds the one of LTL.
- The structure of occurrence graphs is apparent in their LNL-description. E.g. for L_{par} of figure 2 $L_{par}, \{A, E\} \models (A; (B \parallel C); D) \parallel (E; F; G)$ holds. Note that there are many other process-models satisfying this formula, since an atomic formula X is satisfied by any process-model whose initial state contains a label X. Substituting all atomic formulae X by $X \wedge state$ and ; by $;T_1;$ in the above formula, the model is forced to be isomorphic to L_{par}. However, it is not possible to restrict the number of local states in a global state (note that also the *empty* model satisfies *state* and therefore $\models_{LNL} X \rightarrow state \parallel X$ for all $X \in \mathcal{F}_{proc}$). Otherwise the normal form theorem of the next section would not be provable.
- The finite set $class = \{state, state \parallel T_1, state \parallel asyn_0, lin_0, (lin \wedge \neg lin_0), par\}$ classifies process-models according to their structure, where $lin_0 \equiv \neg state; \neg state$ and $lin \equiv state \parallel lin_0$ and $par \equiv \neg state \parallel \neg state$. It is easy to show that *class* is exclusive (that means that no two formulae of *class* have a common model) and $\models_{LNL} \bigvee class$. Note that one has to distinguish lin_0 and $lin \wedge \neg lin_0$, because only the latter can be true for units of sequential composition.
- $asyn_0$-graphs can only be characterized according to their first and last states. E.g. for $S(N_{asyn})$ of figure 6 $S(N_{asyn}), \{A_1, B_1\} \models (A \parallel B \wedge state); asyn_0; (A \parallel B \wedge state) \wedge \neg par \wedge \neg lin$ holds.

• The non-trivial units of a parallel composition are described by $atom_{par_0} \equiv T_1 \vee lin_0 \vee asyn_0$. So for $par_0 \equiv (atom_{par_0}) \parallel (atom_{par_0})^{\parallel}$ holds $\models_{LNL} par \leftrightarrow state \parallel par_0$. Also, because of the validity of $\bigvee class$ holds $\models_{LNL} state \vee state \parallel (atom_{par_0})^{\parallel}$.

• The non-trivial units of a sequential composition are described by $atom_{lin_0} \equiv (state \parallel T_1 \vee state \parallel asyn_0 \vee par \vee (lin \wedge \neg lin_0))$ (note that $\models_{LNL} X \rightarrow state; X$). Therefore $\models_{fpr} lin_0 \leftrightarrow atom_{lin_0}; (atom_{lin_0})^+$ and, again because of the validity of $\bigvee class$ follows $\models_{fpr} state \vee (atom_{lin_0})^+$. The sequential decomposition into unit-graphs defines the *next-inevitable-state-relation.*

Definition 9 (next-inevitable-state-relation, R_{inev}) *Let $L = (G, R, g_{in})$ be an occurrence graph. The* inevitable-state-set E *and the* next-inevitable-state-relation R_{inev} *are defined as the least sets $E \subseteq G, R_{inev} \subseteq E \times E$ such that $g_{in} \in E$ and for $g \in E$ also $g' \in E, (g, g') \in R_{inev}$, if $succ(g, L) = L_1; L_2$ with $L_1, g \models atom_{lin_0}$ and $min(L_2) = g'$.*

So e.g. for $L = S(N^1_{syn} \bullet N^2_{syn})$ of figure 5 $L, \{A_1, B_1\} \models (A \parallel B \wedge atom_{lin_0}); (D \parallel E \wedge atom_{lin_0}); (A \parallel B)$ holds.

4 On the axiomatizability of net logic

The formulae of LNL are based on the process-formulae the same way that formulae of *dynamic logic* (*PDL* for short, see e.g. [H84]) are based on program-formulae, where $A; B$ corresponds to $\langle\alpha\rangle B$. However, programs of PDL are given by the set $Reg(\Pi_0)$ of regular expressions over the set Π_0 of atomic programs. So in order to reduce LNL to some kind of PDL first $\mathcal{F}_{proc}$ has to be reduced to $Reg(\Sigma)$ for some suitable $\Sigma \subseteq \mathcal{F}_{proc}$. As noted before, $\models_{fpr} state \vee (atom_{lin_0})^+$. So the idea is to define an alphabet $\mathcal{AL}$ characterizing the $atom_{lin_0}$-models. However, since an $atom_{lin_0}$-model may be parallel decomposable into components that are in turn sequentially decomposable, one has to give a whole *hierarchy* of alphabets Σ^k with $\mathcal{AL}^k \subseteq \Sigma^k$. Note that the hierarchy starts with an alphabet which only characterizes the first and last states of its models. So infinitely many process-models are subsumed by some $X \in \Sigma^0$. One of the major points of the following normal form theorem is that for every formula the structure of its models needs only to be characterized up to a finite depth.
Now one can show that according to the complexity (given by the number of $\parallel$ and ;) of a process-formula X the formula $X \wedge h$ can be normalized, where $h \in class$. The essential point is to show that each alphabet Σ^k is closed under negation and conjunction. Then the proof proceeds by structural induction on X. Details can be found in [Pae90].

Theorem 10 *For every $h \in class$ and $m \in \mathbb{N}_0$ and $X \in \mathcal{F}_{proc}$ with $compl(X) \leq m$ there exists a finite subset $NF^m_h(X) \subseteq \Sigma^m$ such that $\models_{fpr} (X \wedge h) \leftrightarrow \bigvee NF^m_h(X)$.*

Using the validity of $\bigvee class$, the normal form theorem directly follows from the theorem above.

Theorem 11 (normal form theorem) *Let $m \in \mathbb{N}$ and $\mathcal{S} = \{\bigwedge H_1 \wedge \neg(\bigvee H_2) \wedge state : H_1 \cup H_2 \subseteq \Phi_0, finite, H_1 \cap H_2 = \emptyset\}$. Then for every formula $X \in \mathcal{F}_{proc}$ with $compl(X) < m$ there exists $H \subseteq \mathcal{S} \cup Reg(\mathcal{AL})$ such that $\models_{fpr} X \leftrightarrow \bigvee H$.*

So it is possible to restrict the use of $\neg$ to $\mathcal{S}$ and the use of $\parallel$ to $\mathcal{AL}^k$. Then it is straightforward to show that for $\Sigma_G = \mathcal{S}$ and $\Sigma_R = \bigcup_{k \in \mathbb{N}_0} \mathcal{AL}^k$ every formula $X \in \mathcal{F}_{LNL}$ can be equivalently substituted by $X' \in \mathcal{F}_{LNL}$ such that all process-subformulae of X' are in $\Sigma_G \cup Reg(\Sigma_R)$. The elimination of negation (up to atomic formulae) shows that properties of finite process-models can be described constructively. This is important, since constructive specifications are usually easier to understand. In particular, it is possible to give canonical models for such specifications. It is also possible to derive the normal form automatically. This is important, since the formulae used for the normal form get quite complex.

As mentioned before, it is an interesting consequence of the theorem that the structure of a model is only relevant up to a certain depth for the evaluation of a formula. While this is trivial for a constructive specification, it is not obvious for formulae involving negation.

The normal form theorem is the basic ingredient for the completeness proof. However, there are two further problems to be solved:

• PDL has to be generalized to a logic on *semantically labeled* transition systems, since - in contrast to atomic PDL-programs - $\mathcal{AL}^k$-formulae themselves have models.

• A correspondence between process-models and $\mathcal{AL}^k$-labeled transition systems has to be found, which preserves the validity of formulae.

The solutions of both of these problems are non-trivial. Details can be found again in [Pae90]. Altogether the following theorem holds:

Theorem 12 *LNL has a sound and complete axiomatization. It has the finite model property and is therefore decidable.*

5 Concluding remarks

LNL could be enhanced in several ways to make it a full-fledged specification logic.

• It is straightforward to extend LNL to a branching time logic (on ilps2s) the same way LRL has been extended to BRL in [Pae88]. First one defines the notion of processes *within* a ilps2s. Then R_{inev} has to be refined to ensure inevitability within a set of processes with a common prefix. The next-inevitable-state-relation R_{inev} may shed some light on the interplay of concurrency and conflict. In the context of concurrent systems conflict is usually taken as a local relation. However, in my view this makes it difficult to analyze the behaviour of a system. By allowing conflict only at the inevitable states one abstracts from the particular local states, where conflict arises. The information is only collected at points, where it is relevant to other parts of the system.

• LNL lacks an operator describing the usual reachability relation. The inclusion of e.g. the nexttime operator $\circ$ would allow a mixture of structured and unstructured specifications. Such a logic seems to be difficult to axiomatize, since it presupposes a general method to derive a structured specification from an unstructured one. On the other hand the always operator $\Box$ seems to be easy to include. So Net Logic could be taken as a framework for an investigation of the interplay of structural and non-structural operators.

• As pointed out before, LNL cannot say very much about asynchronous processes, since they have no parallel or sequential structure. However, it would be interesting to include an operator describing an asynchronous process as some kind of merge of other processes.

I conclude with a short survey of related work. The structural operators of LNL have similarities with CCS-based operations on nets, see e.g. [Tau89]. The $\|$-operator of LNL is much weaker, because the components have to be independent. On the other hand LNL is the first formalism to combine negation and structural operators. This also distinguishes it from CDL, an extension of PDL by concurrency [Pel87], since there parallel composition is only allowed on the level of programs. Penczek [Pen88] and Mukund et al. [MT89] have also given logics including explicit operators for concurrency. However, since these logics are defined on event structures, concurrency is viewed as a local relation between events instead of the structural $\|$ of LNL. The logic CCTL defined by Penczek [Pen89] uses the branching time operators of CTL to talk about processes within a state space. This way also some kind of inevitability can be described, but again the structure is not apparent in the formulae.

Finally, I want to note that because of the direct correspondence between occurrence nets and occurrence graphs, Net Logic can also be interpreted on occurrence nets. Therefore it is possible to specify nets without resorting to their state spaces.

References

[ER90] A.Ehrenfeucht, G.Rozenberg *Partial (set) 2-structures, Part I,II* (Acta Informatica 27,4, 1990)

[H84] D.Harel *Dynamic Logic* (D.Gabbay, F.Guenthner (eds.) Handbook of philosophical logic, Vol. II, pp 497-604, 1984)

[Kr87] F.Kröger *Temporal logics of programs* (EATCS Monographs on TCS, Vol. 8, Springer, 1987)

[MOP89] A.Mazurkiewicz, E. Ochmanski, W.Penczek *Concurrent systems and inevitability* (TCS 64, pp 281-304, 1989)

[MP89] Z.Manna, A.Pnueli *Completing the temporal picture* (ICALP, LNCS 372 , pp 534-558, Springer, 1989)

[MT89] M.Mukund, P.S.Thiagarajan *An axiomatization of event structures* (LNCS 405, pp 143 - 160, Springer, 1989)

[NRT90] M.Nielsen, G.Rozenberg, P.S.Thiagarajan *Behavioural notions for elementary net systems* (Distributed Computing 4, pp 45 -57, 1990)

[P88] R.Parikh *Decidability and Undecidability in distributed transition systems* ("A perspective in theoretical computer science", Vol.16, 1988)

[Pae88] B.Paech *Gentzen-Systems for propositional temporal logics* (CSL, LNCS 385,pp 240-253, Springer, 1988)

[Pae90] B.Paech *Concurrency as a modality* (LMU München, Ph.D.thesis, 1990)

[Pel87] D.Peleg *Communication in Concurrent Dynamic Logic* (JCSS 35, pp 23-58, 1987)

[Pen88] W. Penczek *A temporal logic for event structures* (Fundamenta Informaticae XI, pp 297-326, 1988)

[Pen89] W.Penczek *A concurrent branching time temporal logic* (CSL, LNCS 440, pp 337-354, 1989)

[PP90] D.Peled, A.Pnueli *Proving partial order liveness properties* (ICALP, LNCS 443, pp 553-571, 1990)

[R85] W.Reisig *Petri Nets - An introduction* (EATCS Monographs on TCS, Vol. 4, Springer, 1985)

[RP86] R.Rosner, A.Pnueli *A Choppy Logic* (Logic in Computer Science, pp 306-314, 1986)

[Tau89] D.Taubner *The finite representation of abstract programs* (LNCS 369, Springer, 1989)

[Thi87] P.S.Thiagarajan *Elementary net systems* (Advances in PN, LNCS 254, pp 26-59, Springer, 1987)

[Win80] J.Winkowski *Behaviour of concurrent systems* (TCS 12, pp39-60, 1980)

An Extensional Partial Combinatory Algebra based on λ-terms

Ramón PINO PEREZ

Université Lille I and LIFL U.A. 369 du CNRS
Cité Scientifique 59655 Villeneuve d'Ascq Cedex France
e-mail: pino@lifl.lifl.fr

Abstract

We build an extensional partial combinatory algebra on λ-terms. The main tool for our construction is the equality between two relations $\sim_c$ and $\sim$ on λ-terms.The first relation has been defined by Plotkin in [13]. The second relation looks like the bisimulation relation of Abramsky [1] but in the setting of eager evaluation. One consequence of this equality is that the algebra obtained is monotonic.

Introduction

Our main concern is the studying of the partial λ-calculus which is a variant of the λ-calculus introduced by E. Moggi [9, 10] to study equivalence of programs in the setting of call by value evaluation. This calculus is stronger than the λ_v-calculus of Plotkin [13] (see section 2.1). The partial λ-calculus is weaker than λ-calculus but more adapted for the study of the equivalence of programs in the setting of call by value evaluation.

As we know, Extensional Partial Combinatory Logic is a formalism equivalent to partial λ-calculus (see [10]) but with a simpler semantics. So, we are interested in finding models of Partial Combinatory Logic.

In a preliminary version of this work [12] we related a partial λ-calculus to Extensional Partial Combinatory Logic. For space reasons we had to cut off this part of the paper and we have concentrated our exposition on the construction of a combinatory algebra based on λ-terms.

Partial Combinatory Logic was studied by Bethke [6] who devised a method to construct extensional partial combinatory algebras. Recently Hoofman and Schellinx [7] generalized Bethke's method to give uniform constructions of extensional partial combinatory algebras in a category theoretical setting. Independently, the author [11] used Hoofman and Schellinx's method to construct functionals models of partial λ-calculus (so extensional partial combinatory algebras via the equivalence theorem above mentioned).

In this paper, our approach is different from the precedent methods (essentially based in graph models). We provide the set of λ-terms of a *partial* applicative structure (*i.e.* the application is not always defined) in such a way that it is extensional (essential property to obtain a model for partial λ-calculus). The first idea is to consider the λ_v-calculus as a partial applicative structure, *i.e.* the universe of this structure is the set of values (modulo the theory λ_v) and application of M to N is defined if the term MN is equivalent to a value. This idea does not work because the structure obtained in this way is a partial combinatory algebra which is not extensional (see the remark 2.4). So, we defined an equivalence relation on λ-terms which is a refinement of λ_v. This relation is like the bisimulation of Abramsky and Ong [1, 2] but in the setting of call by value evaluation. The clue to prove that this relation generates an extensional partial combinatory algebra is to see that the bisimulation relation is actually the contextual relation defined by Plotkin in [13].

We point out that a result very close to ours has been found by L. Egidi, F. Honsell and S. Ronchi della Rocca [8]. Their proof method uses domain theory. Our method is completely syntactical.

1 Partial Combinatory Logic

The purpose of this section is to define (Extensional) Partial Combinatory Logic.

Partial Combinatory Logic (PCL for short) is a theory in the setting of partial logic (see[4]). We recall briefly what is partial logic. The syntax is the same as in first order logic. But there is a unary predicate symbol , noted $\Downarrow$, with the following axioms and rules (we write $t \Downarrow$ instead of $\Downarrow(t)$):
For quantification

$$\frac{A \Rightarrow B \quad x \notin FV(A)}{A \Rightarrow \forall x B} \; (\mathsf{Q}_1) \qquad \frac{A \Rightarrow B \quad x \notin FV(B)}{\exists x A \Rightarrow xB} \; (\mathsf{Q}_2)$$

$$\forall x A \wedge t \Downarrow \Rightarrow A[t/x] \quad (\mathsf{Q}_3) \qquad A[t/x] \wedge t \Downarrow \Rightarrow \exists x A \quad (\mathsf{Q}_4)$$

For the equality, the abbreviation $t \simeq s$ for $t \Downarrow \vee s \Downarrow \Rightarrow t = s$ is introduced. The axioms are

$$x = x \wedge (x = y \Rightarrow y = x) \quad (\mathsf{E}_1) \qquad t \simeq s \wedge \phi(t) \Rightarrow \phi(s) \quad (\mathsf{E}_2)$$

Convergence axioms

$$R(t_1, \ldots t_n) \Rightarrow t_1 \Downarrow \wedge \ldots t_n \Downarrow \quad (\mathsf{C}_1) \qquad f(t_1, \ldots t_n) \Downarrow \Rightarrow t_1 \Downarrow \wedge \ldots t_n \Downarrow \quad (\mathsf{C}_2)$$

$$c \Downarrow \quad (\mathsf{C}_3) \qquad x \Downarrow \quad (\mathsf{C}_4)$$

where R is any n-ary predicate symbol, f is any n-ary function symbol, c is any constant symbol and x is any variable.

The propositional axioms and rules of inference are the same as usual.

The semantics of partial logic is the same as the semantics of first order logic but the only difference is that functional symbols are interpreted by *partial* functions.

Note that a particular case of the axiom (C_1) is $t = s \Rightarrow t \Downarrow \wedge s \Downarrow$ called axiom (E_3) .

Now let us return to PCL. Its language is $\{\cdot, \mathrm{K}, \mathrm{S}\}$ where $\cdot$ is a function symbol of arity 2, K and S are constant symbols. As usual $\cdot(t_1, t_2)$ is noted $t_1 t_2$ and $(\ldots((t_1 t_2)t_3)\ldots)t_n$ is noted $t_1 t_2 \ldots t_n$. The axioms of PCL are:

$$(\mathsf{ax}_1) \;\; \mathrm{K}xy = x \qquad (\mathsf{ax}_2) \;\; \mathrm{S}xyz \simeq xz(yz) \qquad (\mathsf{ax}_3) \;\; \mathrm{S}xy \Downarrow$$

Extensional Partial Combinatory Logic (EPCL for short) is PCL+ ext, where ext is the following axiom: $\forall z \, (xz \simeq yz) \Rightarrow x = y$

An applicative structure $\mathcal{A} = \langle A, \cdot, \mathrm{K}, \mathrm{S} \rangle$ is said to be a (extensional) partial combinatory algebra ((EPCA) PCA for short) if (EPCA $\models$ EPCL) PCA $\models$ PCL.

2 A EPCA: The eager λ-calculus

The aim of this section is to build a EPCA from a very familiar structure, namely the λ-calculus. We shall do it in such a way that the EPCA obtained is non trivial, *i.e* it is a extensional *partial* combinatory algebra and it is *not* a extensional (total) combinatory algebra (see [3] for the definition). In order to do it we revisited the Plotkin's paper about λ_v-calculus [13]. Actually we shall prove that the relation $\approx_V$ in [13] (see definition 2.6) can be defined in the style of Abramsky's bisimulation (see [1]). The advantages of these two ways of defining the relation $\approx_V$ are in one hand that they allow the construction of a EPCA and on the other hand that we obtain a model of Moggi's monotonic partial λ-calculus (see subsection 2.4).

2.1 The λ_v-calculus

Let us remind a few points related to the λ_v-calculus (for more details see [13]).

The terms of the λ_v-calculus are the terms of the λ-calculus. We noted by Λ the set of λ-terms.

A term is said to be a *value* iff it is not an application. So, the values are the constants, the variables and the abstractions.

The only axiom distinguishing the λ_v-calculus of the λ-calculus is the β axiom (there is no the η axiom because if not all the terms were values). Instead of β we have the following β_v axiom: $(\lambda x.M)N = M[N/x]$ only when N is a value.

All the others axiom and inference rules of λ-calculus are the same in the λ_v-calculus.

Definition 2.1 *Let $\triangleright$ be a binary relation on Λ. $\triangleright$ is said to be contextual if it is reflexive and the following conditions hold*

1. $M \triangleright N \wedge P \triangleright Q \Rightarrow MP \triangleright NQ.$
2. $M \triangleright N \Rightarrow \lambda x.M \triangleright \lambda x.N.$

Definition 2.2 1. *The relation $\xrightarrow{*}_v$ is the smallest relation reflexive, transitive and contextual on Λ such that $(\lambda x.M)N \longrightarrow_v M[N/x]$ if N is a value.*

2. *The left reduction , $\longrightarrow_{vl}$, is the smallest relation on Λ such that:*
   (a) $(\lambda x.M)N \longrightarrow_{vl} M[N/x]$ *if N is a value.*
   (b) $MN \longrightarrow_{vl} M'N$ *if* $M \longrightarrow_{vl} M'$
   (c) $(\lambda x.M)N \longrightarrow_{vl} (\lambda x.M)N'$ *if* $N \longrightarrow_{vl} N'$

3. $\xrightarrow{*}_{vl}$ *is the reflexive and transitive closure of* $\longrightarrow_{vl}$ *(note that there is no reductions* $\xrightarrow{*}_{vl}$ *under the abstractions,* i.e. $\lambda x.M \xrightarrow{*}_{vl} N \Rightarrow \lambda x.M \equiv N$ *in other words the abstractions are in normal form for* $\xrightarrow{*}_{vl}$*).*

The following results are proved, among others, in [13] :

1. The relation $\xrightarrow{*}_v$ possesses the Church-Rosser property.
2. If $M \xrightarrow{*}_v N$ and N is a value then there exists a value N' such that $M \xrightarrow{*}_{vl} N'$.

2.2 The eager λ-calculus

We define a relation $\sim$ on Λ like Abramsky's bisimulation [1]. The difference is in the convergence relation $\downarrow$ on Λ_0 used by us which corresponds to eager reduction (see lemma 2.1). Later we show that this relation is the relation $\approx_V$ defined by Plotkin in [13]. This equivalence will be the main point in this section. Here are the definitions.

Definition 2.3 *We define $\downarrow$, a binary relation on Λ_0 in the following way: $P \downarrow Q$ iff there is a proof of this fact with the two following rules*

$$\lambda x.M \downarrow \lambda x.M \qquad\qquad \frac{M \downarrow \lambda x.P \quad N \downarrow Q \quad P[Q/x] \downarrow R}{MN \downarrow R}$$

When $M \downarrow N$ we say that M converges (to N) and it is denoted by $M \downarrow$. When there is no N such that $M \downarrow N$ we say that M diverges and it is denoted by $M \uparrow$.

Let $\mathcal{V}_0$ be the set of convergent closed terms. Let $\mathcal{V}$ be the set of convergent terms.

Remark 2.1 The following properties result from the definition of $\downarrow$

1. If $M \downarrow$ then $M \downarrow \lambda x.M'$ for some M'.
2. If $MN \downarrow$ then $M \downarrow$ and $N \downarrow$.
3. If $(\lambda x.M)N \downarrow$ and $N \downarrow N'$ then $M[N'/x] \downarrow$.

Lemma 2.1 *$M \downarrow N \iff N$ is an abstraction and $M \xrightarrow{*}_{vl} N$.*

Proof: ($\Rightarrow$) by induction on the lenght of the proof of $M \downarrow N$.
($\Leftarrow$) by induction on the lenght of the path going from M to N. □

The previous lemma allows us to define the number of steps of the convergence of M into N as the number of steps in the left reduction (by value) from M to N. More precisely we have the following definition:

Definition 2.4 *When $M \downarrow N$ we say M converge to N in n steps, noted $M \downarrow_n N$, iff:*

- *either M is an abstraction and $n = 0$*
- *or there is a sequence $M_0, \ldots, M_n$ such that $M_0 \equiv M$, $M_n \equiv N$ and $M_i \longrightarrow_{vl} M_{i+1}$ for $i = 0, \ldots, n-1$*

We say that M converge in n steps, noted $M \downarrow_n$, if there exists N such that $M \downarrow_n N$.

Remark 2.2 When $M \downarrow_n N$, n and N are unique because $\longrightarrow_{vl}$ is determinist and abstraction are normal forms for $\longrightarrow_{vl}$.

Definition 2.5 *For every $i \in \omega$ we define a binary relation $\lesssim_i$ on Λ_0 by induction over ω in the following way:*

- $\lesssim_0 = \Lambda_0 \times \Lambda_0$
- $\lesssim_{i+1} = \{(M, N) \in \Lambda_0 \times \Lambda_0 : M \downarrow \lambda x.M' \Rightarrow (N \downarrow \lambda x.N' \wedge \forall P \in \mathcal{V}_0\ M'[P/x] \lesssim_i N'[P/x])\}$

Now we define a binary relation $\lesssim$ *on* Λ_0 *by* $\lesssim\, = \bigcap_{i \in \omega} \lesssim_i$.
The relation $\lesssim$ *is extended to* Λ *in the following manner*

$$M \lesssim N \iff \textit{for any substitution } \rho : \mathsf{Var} \to \mathcal{V}_0 \;\; M\rho \lesssim N\rho$$

Finally put $M \sim N \iff M \lesssim N \wedge N \lesssim M$.

Lemma 2.2 *For any* $n \geq 1$, $M \lesssim_n N$ *iff the following conditions hold:*

1. $M \downarrow \Rightarrow N \downarrow$.

2. *For any* i *such that* $1 \leq i < n$ *and for any sequence* $P_1, \ldots, P_i$ *the elements of which are in* $\mathcal{V}$ $(MP_1 \cdots P_i) \downarrow \Rightarrow (NP_1 \cdots P_i) \downarrow$.

Proof: By induction on n. □

As a straightforward consequence from lemma 2.2 we obtain the following:

Proposition 2.1 $M \lesssim N$ *iff the following conditions hold:*

1. $M \downarrow \Rightarrow N \downarrow$.

2. *For any* $i \in \omega$ *and for any sequence* $P_1, \ldots, P_i$ *the elements of which are in* $\mathcal{V}$ $(MP_1 \cdots P_i) \downarrow \Rightarrow (NP_1 \cdots P_i) \downarrow$.

Proposition 2.2 *Let be* $M, N \in \Lambda_0$. $M \lesssim N$ *if and only if*

$$M \downarrow \lambda x.M' \;\Rightarrow\; N \downarrow \lambda x.N' \wedge \forall P \in \mathcal{V}_0 \;\; M'[P/x] \lesssim N'[P/x]$$

Proof: Consider the operator $H : \mathcal{P}(\Lambda_0 \times \Lambda_0) \to \mathcal{P}(\Lambda_0 \times \Lambda_0)$ defined by

$$H(R) = \{(M, N) : M \downarrow \lambda x.M' \;\Rightarrow\; N \downarrow \lambda x.N' \wedge \forall P \in \mathcal{V}_0 \, (M'[P/x], N'[P/x]) \in R\}$$

It is clear that this operator is monotonous with respect to the inclusion (actually it commutes with arbitrary intersections), so by Tarski's theorem [14] it has a maximal fix point

$$\mathsf{R} = \bigcap_{i \in \omega} R_i$$

where $R_0 = H(\Lambda_0 \times \Lambda_0)$ and $R_{n+1} = H(R_n)$. But $R_i = \lesssim_{i+1}$ for any $i \in \omega$, then $\mathsf{R} = \lesssim$ what we wanted to prove. □

Corollary 2.1 *Let be* $M \in \Lambda_0$. *If* $M \uparrow$ *then for any* $N \in \Lambda$, $M \lesssim N$

Proof: It is straightforward from the characterization of $\lesssim$ given by the proposition 2.2. □

Lemma 2.3 *If* $M \xrightarrow{*}_{vl} P$, $N \xrightarrow{*}_{vl} Q$ *and* $P \lesssim Q$ *then* $M \lesssim N$

Proof: Straightforward by the lemma 2.1 and the proposition 2.2. □

Lemma 2.4 *For any* $M, N, P \in \Lambda$ *if* $M \lesssim N$ *then* $MP \lesssim NP$.

Proof: It is enough to prove the lemma under the assumption $M, N, P \in \Lambda_0$. If $MP \uparrow$ then, by the corollary 2.1, $MP \lesssim NP$. Now suppose $MP \downarrow$, then $M \downarrow \lambda x.M'$ and $P \downarrow P'$ (see remark 2.1). By the proposition 2.2 and the fact that $M \lesssim N$ and $M \downarrow$, $N \downarrow \lambda x.N'$. So (again by the proposition 2.2) $M'[P'/x] \lesssim N'[P'/x]$. But $MP \xrightarrow{*}_{vl} M'[P'/x]$ and $NP \xrightarrow{*}_{vl} N'[P'/x]$, so, by the lemma 2.3, $MP \lesssim NP$. □

Corollary 2.2 *For any $M, N, P_1, \ldots P_n \in \Lambda$ if $M \lesssim N$ then $MP_1 \cdots P_n \lesssim NP_1 \cdots P_n$*

Proof: By induction on n and the previous lemma. □

Definition 2.6 *We define the binary relations $\lesssim_c$ and $\sim_c$ on Λ_0 by the following*

- *$M \lesssim_c N$ iff for any closed context $C[\,]$, $C[M]\downarrow \Rightarrow C[N]\downarrow$*
- *$M \sim_c N$ iff $M \lesssim_c N$ and $N \lesssim_c M$*

Note that the relation $\sim_c$ is the relation $\approx_V$ defined by Plotkin in [13]. Moreover the following result is proved therein (theorem 5):

Theorem 2.1 *If $\lambda_v \vdash M = N$ then $M \sim_c N$*

The definition of $\lesssim_c$ is extended to Λ in the following way:

$$M \lesssim_c N \text{ iff } \forall \rho : \mathsf{Var} \to \mathcal{V}_0 \; M\rho \lesssim_c N\rho$$

The following theorem is analogous to the proposition 2.3.6 in [2] and our proof is inspired by theirs. Actually the technique used in the proof is a variation of a similar technique introduced by Berry in [5].

Theorem 2.2 $\lesssim \; = \; \lesssim_c$

Proof: The inclusion $\lesssim_c \subset \lesssim$ is straightforward by using the proposition 2.1. For the inclusion $\lesssim \subset \lesssim_c$ it is enough to prove the following statement: For all $M, N \in \Lambda_0$

$$M \lesssim N \;\Rightarrow\; \forall n \in \omega \; \forall C[\,] \in \Lambda_0 \; C[M]\downarrow_n \Rightarrow C[N]\downarrow$$

That is proved by induction on n.
If $n = 0$ then either $C[\,]$ is of the form $\lambda x.C'[\,]$ and the statement holds, or $C[\,] = [\,]$ and in this case $M\downarrow$ then $N\downarrow$ because of the proposition 2.2 and $M \lesssim N$. So the statement holds too.

Now suppose $C[M]\downarrow_{n+1}$. We reason according to leftmost redex. There are 5 cases:

1. The leftmost redex is in $C[\;]$ that is $C[\;] = (\lambda x.C_1[\;])(\lambda x.C_2[\;])\vec{P}[\;]$. Define $D[\;] = (C_1[\;])[\lambda x.C_2[\;]/x]\vec{P}[\;]$. Then $D[M]\downarrow_n$ because $C[M] \xrightarrow{1}_{vl} D[M]$. So, by induction hypothesis $D[N]\downarrow$. But $C[N] \xrightarrow{1}_{vl} D[N]$, so, by the lemma 2.1, $C[N]\downarrow$.

2. The leftmost redex in $C[M]$ is $(\lambda x.C_1[M])M$ that is $C[\,] = (\lambda x.C_1[\,])[\,]\vec{P}[\,]$. Put $D[\,] = (C_1[\;])[[\;]/x]\vec{P}[\;]$. Then $D[M]\downarrow_n$. Therefore $D[N]\downarrow$. Now note that $M\downarrow M$ because $(\lambda x.C_1[M])M$ is the leftmost redex, so $N\downarrow N'$ because $M \lesssim N$. Then, by the theorem 2.1, $D[N']\downarrow$ because $\lambda_v \vdash N = N'$ and $D[N]\downarrow$. Now note that $C[N'] \longrightarrow_{vl} D[N']$ so $C[N']\downarrow$ (lemma 2.1). But $\lambda_v \vdash C[N] = C[N']$ (because $\lambda_v \vdash N = N'$ and the inferences rules of the λ_v) therefore, by the theorem 2.1 again (applied to context $[\,]$), $C[N]\downarrow$.

3. The lefmost redex in $C[M]$ is MM and $C[\;] = [\;][\;]\vec{P}[\;]$. In this case $M = \lambda x.M'$. Put $D[\;] = M'[[\;]/x]\vec{P}[\;]$. Then $D[M]\downarrow_n$. So by induction hypothesis, $D[N]\downarrow$ *i.e.* $M'[N/x]\vec{P}[N]\downarrow$. But $M \lesssim N$ and $M\downarrow$ therefore $N\downarrow \lambda x.N'$. In particular $\lambda_v \vdash N = \lambda x.N'$. By the theorem 2.1, applied to context $M'[[\;]/x]$, $M'[N/x] \lesssim_c M'[(\lambda x.N')/x]$. Now by the inclusion $\lesssim_c \subset \lesssim$ proved above $M'[N/x] \lesssim M'[(\lambda x.N')/x]$. Then, by the corollary 2.2, $M'[N/x]\vec{P}[N] \lesssim M'[(\lambda x.N')/x]\vec{P}[N]$. But $M'[N/x]\vec{P}[N]\downarrow$ so $M'[(\lambda x.N')/x]\vec{P}[N]\downarrow$. Note that the proposition 2.2 implies $M'[(\lambda x.N')/x] \lesssim N'[(\lambda x.N')/x]$. Then, by the proposition 2.1 $N'[(\lambda x.N')/x]\vec{P}[N]\downarrow$. Finally note that $C[N] \xrightarrow{*}_{vl} N'[(\lambda x.N')/x]\vec{P}[N]$, so $C[N]\downarrow$.

4. The leftmost redex in $C[M]$ is $M(\lambda x.C_1[M])$ and $C[\,] = [\,](\lambda x.C_1[\,])\vec{P}[\,]$. In this case $M = \lambda x.M'$. Put $D[\,] = M'[\lambda x.C_1[\,]/x]\vec{P}[\,]$. Then $D[M]\downarrow_n$ so, by induction hypothesis, $D[N]\downarrow$. But $M \lesssim N$, so $N \downarrow \lambda x.N'$. Then, by the proposition 2.2, $M'[\lambda x.C_1[N]/x] \lesssim N'[\lambda x.C_1[N]/x]$. Applying the proposition 2.1 we obtain $N'[(\lambda x.C_1[N])/x]\vec{P}[N]\downarrow$. Finally note that $C[N] \xrightarrow{*}_{vl} N'[(\lambda x.C_1[N])/x]\vec{P}[N]$, so $C[N]\downarrow$.

5. The leftmost redex in $C[M]$ is in M and $C[\,] = [\,]\vec{P}[\,]$. If $\vec{P}[\,]$ is empty (*i.e.* $C[\,] = [\,]$) the result is trivial, otherwise there exists M' and $1 \le k < n+1$ such that $M \downarrow_k M'$. Put $D[\,] = M'\vec{P}[\,]$. $D[M]\downarrow_{n+1-k}$, so by induction hypothesis, $D[N]\downarrow$. Note that $N \downarrow N'$ because $M \lesssim N$ and $M\downarrow$. Then by the definition of $\lesssim$, $M' \lesssim N'$ because $M \lesssim N$. So by the proposition 2.1 $N'\vec{P}[N]\downarrow$. But $C[N] \xrightarrow{*}_{vl} N'\vec{P}[N]$ so $C[N]\downarrow$. □

Definition 2.7 *The eager λ-calculus (λ_e for short) is defined as the set $\{(M,N) : M \sim N\}$. $\lambda_e \vdash M = N$ and $\lambda_e \not\vdash M = N$ will denote the facts $(M,N) \in \lambda_e$ and $(M,N) \notin \lambda_e$ respectively.*

Theorem 2.3 *λ_e is a λ_v theory,* i.e.

- *There exists $M, N \in \Lambda$ such that $\lambda_e \not\vdash M = N$ (λ_e is consistent).*
- *The deductive closure of $\lambda_v + \lambda_e$ under the inference rules of λ_v is contained in λ_e*

Proof: For the first point note that $\Omega\uparrow$ and $\lambda x.x\downarrow$ so $\lambda_e \not\vdash \Omega = \lambda x.x$.

For the second point it is enough to prove that the axiom β_v is in λ_e and that λ_e is closed by the inference rules of λ_v. But, by virtue of the theorems 2.1 et 2.2, the axiom β_v is in λ_e. Concerning the inference rules, reflexivity and symmetry are trivials. Transitivity follows from the theorem 2.2.
Let us verify the ξ-rule *i.e.* $M \sim N$ implies $\lambda x.M \sim \lambda x.N$. Let $\rho : \mathsf{Var} \to \mathcal{V}_0$. We have to prove $(\lambda x.M)\rho \sim (\lambda x.N)\rho$. But by the proposition 2.2 this is equivalent to prove $M\rho' \sim N\rho'$ where $\rho' = \rho[v/x]$ for any $v \in \mathcal{V}_0$, and this is true because by hypothesis $M \sim N$.
Now we verify the congruence rules for the application, *i.e.* $M \sim N$ implies $MP \sim NP$ and $PM \sim PN$. We have to prove $(MP)\rho \sim (NP)\rho$ (the reasoning for right congruence of application is analogous). By virtue of the theorem 2.2 we have to prove $C[(MP)\rho] \sim C[(NP)\rho]$ for any closed context $C[\,]$. Put $D[\,] = C[[\,]P\rho]$. $D[\,]$ is a closed context and it is straightforward to see that $D[M\rho] = C[(MP)\rho]$ and $D[N\rho] = C[(NP)\rho]$. But $D[M\rho] \sim D[N\rho]$ because by hypothesis $M \sim N$. □

The theory λ_e has a weak η-rule. First we define convergence for open terms.

Definition 2.8 *Let M be in Λ, M is said convergent iff for all $\rho : \mathsf{Var} \to \mathcal{V}_0$, $M\rho\downarrow$.*

Proposition 2.3 (**η_e-rule**) *Let M be in Λ. If $M\downarrow$ then for all $x \notin FV(M)$*

$$\lambda_e \vdash \lambda x.Mx = M$$

Proof: Let $\rho : FV(M) \to \mathcal{V}_0$. We have to prove that $(\lambda x.Mx)\rho \sim M\rho$. Note that $(\lambda x.Mx)\rho = \lambda x.(M\rho)x$ so $(\lambda x.Mx)\rho \downarrow \lambda x.(M\rho)x$. By hypothesis $M\rho \downarrow \lambda x.M'$. So, we must prove that for any $P \in \mathcal{V}_0$ $(M\rho)P \sim M'[P/x]$. But $(M\rho)P \xrightarrow{*}_{vl} M'[P/y]$ so, by virtue of the theorem 2.1 and the theorem 2.2, $P \in \mathcal{V}_0$ $(M\rho)P \sim M'[P/x]$. □

Note that the general η-rule is not provable in λ_e because $\Omega\uparrow$ and $\lambda x.\Omega x\downarrow$ so $\lambda_e \not\vdash \Omega = \lambda x.\Omega x$.

Remark 2.3 The main consequence of proposition 2.2 is that it allows two points of view for the theory λ_e. So, to prove some properties of λ_e we can choose one of its characterizations. For instance, to prove the theorem 2.3 we used $\sim_c$ for the congruence rules for the application and we used $\sim$ to prove the ξ-rule and the η_e-rule (proposition 2.3).

2.3 λ_e is an EPCA

We are going to see that λ_e can be presented as a partial applicative structure and that we can choose two terms $\overline{\mathrm{K}}$ and $\overline{\mathrm{S}}$ and a relation $\downarrow$ in such a way that the structure is an EPCA.

Define $\mathcal{A}_{\lambda_e} = \langle A, \cdot, \overline{\mathrm{K}}, \overline{\mathrm{S}}, \downarrow \rangle$ where:

- $A = \{M \in \Lambda/\lambda_e : M \downarrow\}$ where Λ/λ_e is the quotient of Λ by λ_e.
- $\cdot$ is the application of terms of Λ modulo λ_e.
- $\overline{\mathrm{K}} = \lambda xy.x$
- $\overline{\mathrm{S}} = \lambda xyz.xz(yz)$
- $\downarrow$ is the convergence predicate of the definition 2.8.

Lemma 2.5 $\mathcal{A}_{\lambda_e}$ *is a structure for partial logic,* i.e. *rules of inference and axioms of partial logic are satisfied by* $\mathcal{A}_{\lambda_e}$.

Proof: It is a straightforward verification by using the definition of the semantics. □

Lemma 2.6 $\mathcal{A}_{\lambda_e}$ *satisfies* ext i.e. $\mathcal{A}_{\lambda_e} \models \forall z(xz.\simeq yz) \rightarrow x = y$

Proof: If $\lambda_e \vdash xz = yz$ then, by the ξ-rule, $\lambda_e \vdash \lambda z.xz = \lambda z.yz$. By the η_e-rule (proposition 2.3), $\lambda_e \vdash \lambda z.xz = x$ and $\lambda_e \vdash \lambda z.yz = y$ so by transitivity $\lambda_e \vdash x = y$. □

Proposition 2.4 $\mathcal{A}_{\lambda_e}$ *is an* EPCA.

Proof: We have seen that $\mathcal{A}_{\lambda_e}$ is an applicative structure for partial logic (lemma 2.5), so it is enough to see that $\mathcal{A}_{\lambda_e} \models$ EPCA:

- $\mathcal{A}_{\lambda_e} \models \mathrm{K}xy = x$ because, by β_v-conversion (theorem 2.3), $\lambda_e \vdash \overline{\mathrm{K}}xy = x$.
- $\mathcal{A}_{\lambda_e} \models \mathrm{S}xy \Downarrow$ because, by β_v-conversion, $\lambda_e \vdash \overline{\mathrm{S}}xy = \lambda z.xz(yz)$ and, by definition 2.8, $\lambda z.xz(yz) \downarrow$.
- $\mathcal{A}_{\lambda_e} \models \mathrm{S}xyz \simeq xz(yz)$ because, by β_v-conversion, $\lambda_e \vdash \overline{\mathrm{S}}xyz = xz(yz)$ so $\overline{\mathrm{S}}v_1v_2v_3 \downarrow$ iff $v_1v_3(v_2v_3) \downarrow$ for any v_1, v_2, v_3 in $\mathcal{V}_0$ and that case $\lambda_e \vdash \overline{\mathrm{S}}v_1v_2v_3 = v_1v_3(v_2v_3)$.
- $\mathcal{A}_{\lambda_e} \models$ ext because of the lemma 2.6. □

Remark 2.4 The λ_v-calculus is an exemple of non extensional PCA. Define

$$\mathcal{A}_{\lambda_v} = \langle \Lambda_0/\lambda_v, \cdot, \overline{\mathrm{K}}, \overline{\mathrm{S}}, \downarrow \rangle$$

where $\cdot, \overline{\mathrm{K}}, \overline{\mathrm{S}}, \downarrow$ are as defined above and Λ_0/λ_v is the quotient of Λ_0 by λ_v. It is easy to verify that $\mathcal{A}_{\lambda_v} \models$ PCL. But extensionality is not satisfied: on one hand $\mathcal{A}_{\lambda_v} \models (\lambda x.\Omega)z \simeq (\lambda x.\Omega\Omega)z$ because Ω and $\Omega\Omega$ are undefined in $\mathcal{A}_{\lambda_v}$. On the other hand $\mathcal{A}_{\lambda_v} \not\models \lambda x.\Omega = \lambda x.\Omega\Omega$ because, by confluence of the λ_v-calculus, $\lambda_v \not\vdash \lambda x.\Omega = \lambda x.\Omega\Omega$.

2.4 $\mathcal{A}_{\lambda_e}$ is monotonic

First define partial monotonic logic as partial logic plus a binary predicate symbol $\leq$ with axioms to express the reflexivity, asymmetry, transitivity for $\leq$ plus monotonicity of functions symbols with respect to $\leq$, *i.e.*

$$x_1 \leq y_1 \wedge \ldots \wedge x_n \leq y_n \wedge f(x_1, \ldots, x_n) \Downarrow \Rightarrow f(x_1, \ldots, x_n) \leq f(y_1, \ldots, y_n)$$

Then extensional monotonic partial combinatory logic is the theory axiomatized in partial logic by PCL plus the following axiom called monext

$$\forall z \; xz \lesssim yz \Rightarrow x \leq y$$

where the formula $t \lesssim s$ is an abbreviation for the formula $t \Downarrow \vee s \Downarrow \Rightarrow t \leq s$.

We are going to provide $\mathcal{A}_{\lambda_e}$ of an order $\lesssim$ to interpret $\leq$ in a such way that we obtain an extensional monotonic partial combinatory algebra. To do that, we interpreted the symbol $\leq$ by the relation $\lesssim$ of the definition 2.5.

Remark 2.5 *By the proof of theorem 2.2,* $\lesssim \subset \lesssim_c$ *and* $\lesssim_c \subset \lesssim$ *so the application of* $\mathcal{A}_{\lambda_e}$ *is monotonic (with respect to* $\lesssim$*).*

Lemma 2.7 $\mathcal{A}_{\lambda_e} \models$ monext

Proof: Similar to the proof of lemma 2.6. □

The remark 2.5 and the lemma 2.7 entail the following

Theorem 2.4 $\mathcal{A}_{\lambda_e}$ *is a extensional monotonic partial combinatory algebra.*

Acknowledgements

I am very grateful to Pierre-Louis Curien and the anonymous referees for their enlightening remarks. They helped me to improve this last version. In particular an anonymous referee and Curien show me how to simplify the theorem 2.3.

References

[1] ABRAMSKY, A., (1989), The Lazy Lambda Calculus. In "Declarative Programming". David Turner, editor. Addison-Wesley. To appear.

[2] ABRAMSKY, A., ONG C.-H. L. (1989), Full Abstraction in the Lazy Lambda Calculus. To appear.

[3] BARENDREGT, H. (1984), "The Lambda Calculus: Its Syntax and Semantics ". North-Holland. Amsterdam. 1984.

[4] BEESON, M. (1985), "Foundations of Constructive Mathematics". Springer-Verlag. 1985

[5] BERRY, G. (1981) "Some Syntactic and Categorical Constructions of Lambda Calculus Models". Rapport de Recherche de L'Institut National de Recherche en Informatique et Automatique (INRIA). Rocquancourt. 1981.

[6] BETHKE, I. (1988), "Notes on Partial Combinatory Algebras". Ph.D. Thesis. University of Amsterdam. 1988.

[7] HOOFMAN, R., SCHELLINX, H. (1991) Collapsing Graph Models by Preorders. ITLI Prepublication Series ML-91-04. Amsterdam. 1991.

[8] EGIDI, L., HONSELL, F., RONCHI DE LA ROCCA, S. The for Lazy call-by-value λ-calculus. Presented at Jumelage on typed lambda calculus, Paris, 1-5 February 1991.

[9] MOGGI, E. (1986), Categories of partial morphisms and the Partial Lambda Calculus. In "Proceedings of the Workshop on Category Theory and Computer Programming". Guilford 1985. *Lecture Notes in Computer Science* vol. 240. Springer-Verlag, 1986.

[10] MOGGI, E. (1988), The Partial Lambda Calculus, Ph.D. Thesis, University of Edinburgh. 1988.

[11] PINO PÉREZ, R. (1990), Contribution à l'étude du Lambda Calcul Partiel. Forthcoming thesis (draft). Université Paris 7.

[12] PINO PÉREZ, R. (1991), Semantics of the partial lambda calculus. Rapport de recherche LIFL. Université Lille I. 1991.

[13] PLOTKIN, G. (1975), Call-by-name, call-by-value and the λ-calculus. *Theoretical Computer Science* (1), 1975.

[14] TARSKI, A. (1955) A lattice-theoretical fixpoint theorem and its applications. *Pacific Journal of Mathematics*. Vol. 5, pp 285-309.

Once More on Order-Sorted Algebras

Axel Poigné *

Abstract A new definition of order-sorted algebras is given which unifies the approaches to be found in the literature. We claim that some conceptual clarification is obtained without losing out on technical simplicity. The corresponding theory is developed and related to the literature. Specifically, we prove completeness with regard to an appropriate deduction system.

0. Order-sorted algebras have a long genesis. Since its first manifestation in computer science [Goguen 78], a variety of interpretations of "order-sorting" have surfaced (such as [Gogolla 84], [Poigné 84], [Smolka 86], [Poigné 88]) which reflect different assumptions. Some ideas are shared, for instance that subsorting denotes inclusion and that overloading is allowed.

Opinions are split whether to accept "ad-hoc polymorphism" [Strachey 67] as a semantic concept. Ad-hoc polymorphism refers to an accidental overloading of operators as often found in many-sorted signatures, where the same symbol is used for semantically unrelated operators, e.g. + : nat nat → nat and + : bool bool → bool. Goguen and Meseguer (e.g. [Goguen, Meseguer 89], up to now the last in a series of papers by the same authors) argue in favour of ad-hoc polymorphism, observing that otherwise many-sorted algebra is not subsumed. Other authors [Gogolla 84], Poigné 84] commit themselves to the slogan of "sort-independent semantics", meaning that an operation applied to data should always give the same result whatever typing is involved.

This short note aims at a reconciliation of both schools of thought handling ad-hoc polymorphism, but retaining the technical simplicity of the sort-independent semantics.

I acknowledge inspiration by Goguen's and Meseguer's paper [Goguen, Meseguer 89], though the basic idea is already to be found in [Poigné 88] in another formal framework. This paper being a study of foundations, we rely on the references for motivation and applications of order-sorted algebra.

1. Let us recall the definition of a many-sorted signature and of many-sorted algebras.

1.1 DEFINITION • A ***many-sorted signature*** (S, Σ) consists of a set S (of *sorts*) and $S^* \times S$-sorted family $(\Sigma_{w,s} \mid w \in S^*, s \in S)$. Elements $\sigma \in \Sigma_{w,s}$ are called operation symbols, for short *operators*. We often use the notation $\sigma_{w,s} \in \Sigma$ instead of $\sigma \in \Sigma_{w,s}$.

• A ***many-sorted*** (S, Σ)***-algebra*** $\mathcal{A}$ consists of a set $s^{\mathcal{A}}$ for every sort $s \in S$, and an operation (function) $\sigma_{w,s}{}^{\mathcal{A}} : w^{\mathcal{A}} \to s^{\mathcal{A}}$ for every operator $\sigma_{w,s} \in \Sigma$ where $\lambda^{\mathcal{A}} = \{\emptyset\}$ and $ws^{\mathcal{A}} = w^{\mathcal{A}} \times s^{\mathcal{A}}$.

• a (S, Σ) ***-homomorphism*** $h : \mathcal{A} \to \mathcal{B}$ consists of a S-sorted family $(h_s : s^{\mathcal{A}} \to s^{\mathcal{B}} \mid s \in S)$ of functions such that $h_s(\sigma_{w,s}{}^{\mathcal{A}}(a_1,...,a_n)) = \sigma_{w,s}{}^{\mathcal{B}}(h_{s_1}(a_1),...,h_{s_n}(a_n))$.

These data form a category (S, Σ)-**Alg**.

* GMD F1P, Schloss Birlinghoven, Postfach 1240, 5205 St. Augustin 1, em: ap@gmdzi.uucp

NOTATION We use a as abbreviation for tuples $a_1,...,a_n$. The notation a_i refers to the i's component of a tuple. Functions, relations etc. are extended canonically to tuples, e.g. $a = b$ stands for $a_i = b_i$ for all $i = 1,...,n$.

Order-sorting adds a partial order on sorts and allows for overloading. We typically find that an *order-sorted signature* consists of a triple $(S, \leq, \Sigma)$ such that (S, Σ) is a many-sorted signature, and $(S, \leq)$ is a partially ordered set (poset). The *subsort relation* $s \leq s'$ is semantically interpreted as inclusion $s^{\mathcal{A}} \subseteq s'^{\mathcal{A}}$ abandoning many-sortedness in that overloading cannot be eliminated by sort information any more.

Goguen and Meseguer recover many-sortedness by introduction of the *monotonicity condition*

$$\sigma \in \Sigma_{w,s} \cap \Sigma_{w',s'},\ w \leq w' \quad \Rightarrow \quad s \leq s'$$

on signatures (where $w = s_1 ... s_n \leq s_1'...s_n' = w'$ if $s_i \leq s_i'$ for i = 1,...,n), and by the condition

$$\sigma \in \Sigma_{w,s} \cap \Sigma_{w',s'},\ w \leq w', a \in w^{\mathcal{A}} \quad \Rightarrow \quad \sigma_{w,s}{}^{\mathcal{A}}(a) = \sigma_{w',s'}{}^{\mathcal{A}}(a).$$

as a semantical counterpart. Homomorphisms are restricted by

$$s \leq s',\ a \in s^{\mathcal{A}} \ \Rightarrow \ h_s(a) = h_{s'}(a).$$

(We refer to such structures as *GM-signatures, SIG-GM-algebras* and *SIG-GM-homomorphisms*).

Overloading is ad-hoc since the operators with the same name may behave differently semantically. Specifically, if the order is discrete, many-sorted signatures and many-sorted algebras are obtained as a special case.

However, we notice a disturbing property: consider the signature with sorts $s1 \leq s3$, $s2 \leq s4$, $s5$ and operators a_{s1}, b_{s2}, $\sigma_{s1\ s4,\ s5}$, $\sigma_{s3\ s2,\ s5}$. Then an algebra $\mathcal{A}$ is well defined where $s1^{\mathcal{A}} = s2^{\mathcal{A}} = s3^{\mathcal{A}} = s4^{\mathcal{A}} = \{0\}$ and $s5^{\mathcal{A}} = \{1,2\}$ and $\sigma_{s1\ s4,\ s5}{}^{\mathcal{A}}(0, 0) = 1$, $\sigma_{s3\ s2,\ s5}{}^{\mathcal{A}}(0, 0) = 2$. But the standard term construction ($\sigma(t) \in T_{s'}$ whenever $\sigma \in \Sigma_{w,s}$, $s \leq s'$ and $t \in T_w$, see 3.1) generates $\sigma(a, b)$ as only element of sort $s5$, hence does not define an initial algebra. The problem disappears if the signature is *regular*, i.e. given $\sigma \in \Sigma_{w'',s''}$ and given $w \leq w''$ in S^* there is a least rank $(w', s') \in S^* \times S$ such that $w \leq w'$ and $\sigma \in S_{w',s'}$.

Gogolla [Gogolla 83] as well as [Poigné 84, 90] (and similarly [Smolka 86]) impose the following semantical condition

$$\sigma \in \Sigma_{w,s} \cap \Sigma_{w',s'}, a \in w^{\mathcal{A}} \cap w'^{\mathcal{A}} \quad \Rightarrow \quad \sigma_{w,s}{}^{\mathcal{A}}(a) = \sigma_{w',s'}{}^{\mathcal{A}}(a).$$

Here sorts should be thought of as to determine subsets of a "universal" set $A = \bigcup_{s \in S} A_s$, and operators should be thought of as (totally defined) components of a partial function $\sigma^{\mathcal{A}} : A^n \to A$. Data cannot be distinguished by sort information, hence the definition is (somewhat confusingly) one-sorted in spirit, the subsorts being only a means to introduce a certain degree of partiality.

In this paper, we propose to extend the latter approach to a many-sorted one: we consider a set $\{(S_{\mathbf{s}}, \leq) \mid \mathbf{s} \in \mathbf{S}\}$ of partial orders instead of only one, and recast all definitions appropriately. The "sorts" $s \in S_{\mathbf{s}}$ may be considered as *subsorts* of the *principal sort* $\mathbf{s} \in \mathbf{S}$ (for sake of a better name), the carrier $\mathbf{s}^{\mathcal{A}}$ being the union $\bigcup_{s \in \mathbf{s}} A_s$. Operators $\sigma_{w,s}$ are instances of a global operator σ on the respective principal sorts. The actual formalization is more or less to be a matter of taste. We propose the following definition which allows to rephrase the arguments of [Poigné 90] at little expense.

1.2 DEFINITION An ***order-sorted signature*** $(S, \leq, \mathbf{S}, \Sigma)$ consists of a many-sorted signature (S, Σ), a partial order $(S, \leq)$, and a partition[1] $\mathbf{S}$ of S such that

---

[1] A family $\{C_i \subseteq X \mid i \in I\}$ is called a *partition* of X if $X = \bigcup_{i \in I} C_i$ and if $C_i \cap C_j = \emptyset$ if $i \neq j$ for all $i, j \in I$.

(i) $s \leq s'$ implies $|s| = |s'|$, and

(ii) $\sigma \in \Sigma_{w,s} \cap \Sigma_{w',s'}$ and $|w| = |w'|$ implies $|s| = |s'|$

where $|s| \in \mathbf{S}$ is the *principal sort* such that $s \in |s|$, and where $|\lambda| = \lambda$ and $|s_1...s_n| := |s_1|...|s_n| \in \mathbf{S}^*$.

We say that sorts s and s' are ***related*** (notation $s \leftrightarrow s$) if $|s| = |s'|$ and that operators $\sigma_{w,s}$ and $\sigma_{w',s'}$ are related ($\sigma_{w,s} \leftrightarrow \sigma_{w',s'}$) if $|w| = |w'|$.

REMARKS • If the partition consists of singletons then we obtain a many-sorted signature, except for the condition (ii) which collapses to: $\sigma \in \Sigma_{w,s} \cap \Sigma$ implies $s = s'$.

• The paper [Poigné 88] considers many-sorted structures with specific unary sorted relation symbols $\_ \varepsilon \tau$, referred to as "types", which are used to model a subsorting structure in that the corresponding subsets are obtained as extension of "types". Operations are partial in this framework, and special provisions have to be taken that "types" really behave like subsorts.

The semantic condition of Gogolla is now restricted to related operators. We henceforth assume that $SIG = (S, \leq, \mathbf{S}, \Sigma)$ is a fixed order-sorted signature.

1.3 DEFINITION An ***(order-sorted) SIG-algebra*** $\mathcal{A}$ is a (S, Σ)-algebra such that

$$\sigma_{w,s} \leftrightarrow \sigma_{w',s'},\ a \in w^{\mathcal{A}} \cap w'^{\mathcal{A}} \quad \Rightarrow \quad \sigma_{w,s}{}^{\mathcal{A}}(a) = \sigma_{w',s'}{}^{\mathcal{A}}(a).$$

A ***homomorphism*** $h : \mathcal{A} \to \mathcal{B}$ of order sorted algebras is a (S, Σ)-homomorphism such that

$$h_s(a) = h_{s'}(a) \ \text{ if } \ s \leftrightarrow s' \ \text{ and } \ a \in s^{\mathcal{A}} \cap s'^{\mathcal{A}}.$$

Let ***SIG-OSAlg*** denote the category of order-sorted *SIG*-algebras.

REMARK We have ad-hoc polymorphism if operators are unrelated but have the same name.

NOTATION Let $\mathbf{s}^{\mathcal{A}} = \bigcup_{s \in \mathbf{s}} s^{\mathcal{A}}$ be the *carrier of principal sort* $\mathbf{s}$, and let $h_{\mathbf{s}} : \mathbf{s}^{\mathcal{A}} \to \mathbf{s}^{\mathcal{B}},\ a \mapsto h_s(a)$ where $a \in s^{\mathcal{A}}$. The latter is well defined due to the additional condition on order-sorted homomorphisms.

The definition subsumes those given in [Gogolla 84] and [Poigné 84, 90] (take the trivial partition with one partial order). Many-sorted algebra is obtained if the principal sorts consist of a one-element poset. The comparison to the definition of [Goguen, Meseguer 89] is more subtle and will be postponed till section 6.

2. The following term construction is a common feature to all approaches.

2.1 DEFINITION Let $X = (X_s \mid s \in S)$ be an *S-sorted set*. A *S*-sorted set $T_{SIG}(X)$ is defined inductively by

(i) $x_s \in T_{SIG}(X)_{s'}$ if $x \in X_s$ and $s \leq s'$,

(ii) $\sigma(t) \in T_{SIG}(X)_{s'}$ if $\sigma \in \Sigma_{w,s}$, $s \leq s'$, and $t \in T_{SIG}(X)_w$.

REMARK Note that we do not assume disjointness of the X_s's. We disambiguate variables by subscripts.

2.2 FACT $T_{SIG}(X)$ *with operations* $\sigma_{w,s} : w^{T_{SIG}(X)} \to s^{T_{SIG}(X)}$ *such that* $\sigma_{w,s}(t) = \sigma(t)$ *defines an order-sorted SIG-algebra we denote by* $T_{SIG}(X)$ *as well.*

2.3 LEMMA *If* $t \in T_{SIG}(X)_s \cap T_{SIG}(X)_{s'}$ *then* $|s| = |s'|$.

PROOF If $x_{s''} \in T_{SIG}(X)_s \cap T_{SIG}(X)_{s'}$ then $x_{s''} \in X_{s''}$ and $s'' \leq s$ by definition of the term algebra.

If $\sigma(t) \in T_{SIG}(X)_s \cap T_{SIG}(X)_{s'}$, then there exist $\sigma_{w1,s1}, \sigma_{w2,s2}$ such that $s1 \leq s$ and $s2 \leq s'$ and $t \in T_{SIG}(X)_{w1} \cap T_{SIG}(X)_{w2}$. Since $|w1| = |w2|$ by inductive assumption[1], we have $|s1| = |s2|$.

2.4 PROPOSITION $T_{SIG}(X)$ *is a free SIG-algebra generated by X meaning that for every SIG-algebra* $\mathcal{A}$ *and for every S-sorted mapping* $\Theta = (\Theta_s : X_s \to s^{\mathcal{A}} \mid s \in S)$ *there exists a unique homomorphism* $\Theta^{\#} : T_{SIG}(X) \to \mathcal{A}$ *such that* $\Theta^{\#}(x) = \Theta(x)$ *for* $x \in X_s$.

PROOF Define $\Theta^{\#}_{s'}(x) = \Theta_s(x)$ for $x \in X_s$, $s \leq s'$, and $\Theta^{\#}_{s'}(\sigma(t)) = \sigma_{w,s}{}^{\mathcal{A}}(\Theta^{\#}(t))$ for some $\sigma \in \Sigma_{w,s}$ such that $s \leq s'$ and $t \in T_{SIG}(X)_w$. We check that $\Theta^{\#}$ is well defined: Let $x_{s''} \in T_{SIG}(X)_s \cap T_{SIG}(X)_{s'}$. Then $\Theta^{\#}_s(x_{s''}) = \Theta_{s'}(x)$. If $\sigma(t) \in T_{SIG}(X)_s \cap T_{SIG}(X)_{s'}$ and $s \leftrightarrow s'$, then there exist $\sigma_{w1,s1}, \sigma_{w2,s2}$ such that $s1 \leq s$ and $s2 \leq s'$ and $t \in T_{SIG}(X)_{w1} \cap T_{SIG}(X)_{w2}$. Since $\Theta^{\#}_{w1}(t) = \Theta^{\#}_{w2}(t)$ by inductive assumption we have

$$\Theta^{\#}_s(\sigma(t)) = \sigma_{w1,s1}{}^{\mathcal{A}}(\Theta^{\#}_{w1}(t)) = \sigma_{w2,s2}{}^{\mathcal{A}}(\Theta^{\#}_{w2}(t)) = \Theta^{\#}_{s'}(\sigma(t)).$$

The middle equality uses 2.3 and (ii) of 1.2. We conclude that $\Theta^{\#}$ is well defined (use $s = s'$) and a *SIG*-homomorphism. It is straightforward to establish uniqueness.

Compared to [Goguen, Meseguer 89] the term construction works with a milder requirement than regularity of signatures. We pay in that the "connected components" of Goguen and Meseguer are specified explicitly. We take this rather as an advantage conceptually, since structure hidden in the conditions and the proofs of [Goguen, Meseguer 89] becomes explicit. Of course, regularity implies existence of a least sort for every term, which is a desirable feature on practical grounds. But we believe we can well live without it, typically at not too much expense. For instance, parsing of terms is achieved if we compute the set of all sorts of a term instead of the least sort.

TECHNICAL REMARK The monotonicity condition resp. condition (ii) of 1.2 is necessary even in the many-sorted case if we want ad-hoc polymorphism to work. Consider the *many-sorted* signature with operators $a_{s1}, f_{s1,s2}, f_{s1,s3}, g_{s2,s4}, g_{s3,s4}$. If we apply the standard term construction, then there exists only one term $g(f(a))$ of sort s_4. However $g_{s2,s4}{}^{\mathcal{A}}(f_{s1,s2}{}^{\mathcal{A}}(a_{s1}{}^{\mathcal{A}}))$ and $g_{s3,s4}{}^{\mathcal{A}}(f_{s1,s3}{}^{\mathcal{A}}(a_{s1}{}^{\mathcal{A}}))$ computes different values in the algebra with $s1^{\mathcal{A}} = s2^{\mathcal{A}} = s3^{\mathcal{A}} = \{0\}$ and $s4^{\mathcal{A}} = \{1, 2\}$, and $g_{s2,s4}{}^{\mathcal{A}}(0) = 1$, $g_{s3,s4}{}^{\mathcal{A}}(0) = 2$. Thus the standard term construction does not yield an initial algebra.

As everybody uses this term construction, I guess the implicit assumption is that operators can be disambiguated if necessary.

Even if we drop condition (ii) of 1.2 (resp. monotonicity), an appropriate term construction defines an initial algebra. Just annotate the operator symbols with enough sort information, e.g. use the rule

$\sigma_{|s|}(t) \in T_{SIG}(X)_{s'}$ if $\sigma \in \Sigma_{w,s}$, $s \leq s'$, and $t \in T_{SIG}(X)_w$.

Of course, we loose out on polymorphism.

3. We simplify order-sorted congruences as defined in [Gogolla 83], [Poigné 84,90].

3.1 DEFINITION An S-indexed relation $R = (R_s \subseteq s^{\mathcal{A}} \times s^{\mathcal{A}} \mid s \in S)$ is called a *SIG-congruence* if

(i) $R_s \subseteq s^{\mathcal{A}} \times s^{\mathcal{A}}$ is an equivalence relation for each $s \in S$,

(ii) $(\sigma_{w,s}{}^{\mathcal{A}}(a), \sigma_{w',s'}{}^{\mathcal{A}}(b)) \in R_{|s|}$ if $\sigma_{w,s} \leftrightarrow \sigma_{w',s'}$, $a \in w^{\mathcal{A}}$, $b \in w'^{\mathcal{A}}$ and $(a, b) \in R_{|w|}$, and

(iii) R_s is the transitive closure of $\{ (a, b) \in R_s \mid a, b \in s^{\mathcal{A}}, s \in s\} \subseteq s^{\mathcal{A}} \times s^{\mathcal{A}}$

The ***quotient*** $\mathcal{A}_{/R}$ is defined by

$s^{\mathcal{A}_{/R}} := \{ [a] \mid a \in s^{\mathcal{A}} \}$, where $[a] := \{b \mid (a, b) \in R_{|s|} \}$

---

[1] The induction principle used here is discussed in [Poigné 90]

with operations being defined on the congruences as usual.

REMARK It is necessary to define transitivity with regard to principal sorts. Consider the following example with the only principal sort $\{s1, s2\}$ and the algebra $\mathcal{A}$ with carriers $s1^{\mathcal{A}} = \{a, b\}$, $s2^{\mathcal{A}} = \{b, c\}$. The congruence generated by a relation $R = \{(a,b), (b,c)\}$ should comprise the element (a,c) due to transitivity since elements of the same principal sort are considered as equal, whatever the subsorts are, they belong to. This element is not obtained if transitive closure is only taken with regard to sorts. The same phenomenon is to be found in the proof system below where the euqivalence axioms are stated with regard to the principal sorts.

3.2 PROPOSITION (i) *$\mathcal{A}_{/R}$ is a SIG-algebra, and $\varepsilon : \mathcal{A} \to \mathcal{A}_{/R}$, $a \mapsto [a]$ is a homomorphism.*
(ii) *Given a homomorphism $h : \mathcal{A} \to \mathcal{B}$ such that*

$$R_s \subseteq Ker(h)_s := \{(a, b) \in s^{\mathcal{A}} \times s^{\mathcal{A}} \mid h_s(a) = h_s(b)\}$$

then there exists a unique homomorphism $[h] : \mathcal{A}_{/R} \to \mathcal{B}$ such that $h = [h] \circ \varepsilon$.

PROOF Straightforward or adapt 2.9 in [Poigné 90]

Congruences are specified by conditional equations.

3.3 DEFINITION An ***order-sorted presentation*** *PRES* consists of a signature *SIG* and a set Γ of conditional equations of the form

$$\forall X.t =_s t' \ \underline{\text{if}}\ t =_w t'$$

where X is a S-sorted set, and where $t, t' \in T_{SIG}(X)_s$ and $\mathbf{t}, \mathbf{t'} \in T_{SIG}(X)_w$. (If the set of conditions is empty, we use $\forall X.t =_s t$).

A ***PRES-algebra*** is a *SIG*-algebra which satisfies all the conditional equations in Γ. A *SIG*-algebra $\mathcal{A}$ *satisfies* a conditional equation $\forall X.t =_s t'$ $\underline{\text{if}}$ $\mathbf{t} =_w \mathbf{t'}$ if, for all homomorphisms $\Theta : T_{SIG}(X) \to \mathcal{A}$, the equality $\Theta_s(t) = \Theta_s(t')$ holds whenever $\Theta_w(\mathbf{t}) = \Theta_w(\mathbf{t'})$.

3.4 REMARK Equality is more restrictive than in [Goguen, Meseguer 89] where, in analogy, conditional equations are indexed by principal sorts, i.e. are of the form $\forall X.t =_{\mathbf{s}} t\ \underline{\text{if}}\ \mathbf{t} =_{\mathbf{w}} \mathbf{t'}$. Then the definition of satisfaction is to be modified: for all homomorphisms $\Theta : T_{SIG}(X) \to \mathcal{A}$, $\Theta_s(t) = \Theta_s(t')$ holds whenever $\Theta_{\mathbf{w}}(\mathbf{t}) = \Theta_{\mathbf{w}}(\mathbf{t'})$. We comment on this variation subsequently.

3.5 DEFINITION For $t, t' \in T_{SIG}(X)_s$ let $t \sim_s t'$ if $\Theta_s(t) = \Theta_s(t')$ holds for all *SIG*-homomorphisms $\Theta : T_{SIG}(X) \to \mathcal{A}$. Moreover, let $t \sim_{\mathbf{s}} t'$ be the transitive closure of $\{ (t, t') \mid t \sim_s t', s \in \mathbf{s}\}$.

3.6 LEMMA *$t \sim t'$ defines a SIG-congruence on $T_{SIG}(X)$.*

PROOF $t \sim_{\mathbf{s}} t'$ is an equivalence relation since $\{ (t, t') \mid t \sim_s t', s \in \mathbf{s}\}$ is reflexive and symmetric. Let $\sigma_{w,s} \leftrightarrow \sigma_{w',s'}$ and $\mathbf{t} \sim_{|w|} \mathbf{t'}$. Then $\mathbf{t} = \mathbf{t}_1 \sim_{w1} \mathbf{t}_2 \sim_{w2} \ldots \sim_{wn} \mathbf{t}_{n+1} = \mathbf{t'}$ for some $\mathbf{t}_i$'s and w_i's. As all sorts are related, we have $\Theta_w(\mathbf{t}) = \Theta_{w1}(\mathbf{t}_1) = \Theta_{w1}(\mathbf{t}_2) = \Theta_{w2}(\mathbf{t}_2) = \ldots = \Theta_{wn}(\mathbf{t}_{n+1}) = \Theta_{w'}(\mathbf{t'})$. Hence $\Theta_s(\sigma_{w,s}(\mathbf{t})) = \sigma_{w,s}^{\mathcal{A}}(\Theta_w(\mathbf{t})) = \sigma_{w',s'}^{\mathcal{A}}(\Theta_{w''}(\mathbf{t'})) = \Theta_{s''}(\sigma_{w',s'}(\mathbf{t'}))$.

These preliminaries allow us to construct a free *PRES*-algebra as usual.

3.7 PROPOSITION *The algebra* $T_{PRES}(X) = T_{SIG}(X)_{/\sim}$ *is a free SIG-algebra generated by X, meaning that for every PRES-algebra* $\mathcal{A}$ *and for every S-sorted mapping* $\Theta = (\Theta_s : X_s \to s^{\mathcal{A}} \mid s \in S)$ *there exists a unique homomorphism* $\Theta^{\#} : T_{PRES}(X) \to \mathcal{A}$ *such that* $\Theta^{\#}(x) = \Theta(x)$ *for* $x \in X_s$.

PROOF Combine 2.4 and 3.2 in the standard way.

REMARK In case of the modified definition of 3.3, the definition of congruence can be simplified by dropping condition (iii) of 3.1. The free algebra is then obtained by factorization using the relation $t \sim_s t'$ defined by $\Theta_s(t) = \Theta_s(t')$ holds for all *SIG*-homomorphisms $\Theta : T_{SIG}(X) \to \mathcal{A}$

4. Order-sorted logic corresponds to the logic of [Poigné 84,90]. The proof rules are presented in a sequent style with sequents of the form $\{\varphi_1,\dots,\varphi_n\} \vdash_X \varphi$ where the atomic formulas are either of the form $t =_s t'$ with $t, t' \in T_{SIG}(X)_s$ or of the form $\Phi \vdash_X t =_s t'$ with $t, t' \in T_{SIG}(X)_s$.

Monotonicity $\qquad \Phi \vdash_X \varphi \quad$ if $\varphi \in \Phi$

Equivalence $\qquad \Phi \vdash_X t =_s t$

$$\frac{\Phi \vdash_X t =_s t'}{\Phi \vdash_X t' =_s t} \qquad \frac{\Phi \vdash_X t =_s t' \quad \Phi \vdash_X t' =_s t''}{\Phi \vdash_X t =_s t''}$$

Subsorting

$$\frac{\Phi \vdash_X t =_s t'}{\Phi \vdash_X t' =_{|s|} t} \qquad \frac{\Phi \vdash_X t =_{|s|} t'}{\Phi \vdash_X t' =_s t} \quad \text{if } t, t' \in T_{SIG}(X)_s$$

Compatibility

$$\frac{\Phi \vdash_X t =_{|w|} t'}{\Phi \vdash_X \sigma(t) =_{|s|} \sigma(t')} \quad \text{where } \sigma_{w,s} \leftrightarrow \sigma_{w',s'}$$

Substitutivity

$$\frac{\Phi \vdash_Y \Theta(t) =_w \Theta(t')}{\Phi \vdash_Y \Theta(t) =_s \Theta(t')} \quad \text{where } \forall X.t =_s t' \text{ \underline{if} } t =_w t' \text{ and } \Theta : T_{SIG}(X) \to T_{SIG}(Y)$$

REMARK Note that the proof system depends on *PRES*.

4.1 PROPOSITION *Order-sorted (conditional equational) logic is sound and complete.*

I.e. for all presentations *PRES*, the following statements are equivalent.

(i) $t =_w t' \vdash_X t =_s t'$ is derivable by the proof rules above.

(ii) $\forall X.t =_s t$ <u>if</u> $t =_w t'$ is satisfied by every order-sorted *PRES*-algebra.

PROOF (i) ⇒ (ii) (*soundness*): Straightforward induction on proof trees.
(ii) ⇒ (i) (*completeness*): Define an algebra $T_{PRES,\Phi}(X)$ by $s^{T_\Phi(X)} = \{ [t] \mid t \in T_{SIG}(X)_s\}$ where $[t] = \{ t' \mid \Phi \vdash_X t =_s t'\}$ and $\sigma_{w,s}([t]) = [\sigma(t)]$. $T_{PRES,\Phi}(X)$ is a *SIG*-algebra since, by compatibility, $\sigma_{w,s}([t]) = [\sigma(t)] = \sigma_{w',s'}([t])$ if $\sigma_{w,s} \leftrightarrow \sigma_{w',s'}$ and it satisfies the equations of *PRES* by substitutivity and monotonicity. But $T_{PRES,t=_w t'}(X)$ satisfies $\forall X.t =_s t$ <u>if</u> $t =_w t'$ iff $t =_w t' \vdash_X t =_s t'$.

REMARK • As usual, $T_{PRES,\varnothing}(X) \cong T_{PRES}(X)$.
• With regard to the modified approach, we obtain a sound and complete proof system if sequents are restricted to deal with atomic formulas of the form $t =_s t'$ only, and if the subsorting rules are omitted.

5. Signature resp. presentation morphisms, if properly defined, induce reducts and free constructions.

5.1 DEFINITION A ***signature morphism*** $H : SIG \to SIG'$ consists of

- a monotone mapping $H : (S, \leq) \to (S', \leq')$ which *preserves principle types*, i.e. $|H(s)| = |H(s')|$ if $|s| = |s'|$, and
- a $S^*\times S$*-sorted function* $H : \Sigma \to \Sigma$ such that
  - $H_{w,s} : \Sigma_{w,s} \to \Sigma_{H(w),H(s)}$, and
  - $H_{w,s}(\sigma) = H_{w',s'}(\sigma)$ if $\sigma_{w,s} \leftrightarrow \sigma_{w',s'}$.

A ***presentation morphism*** $H : PRES \to PRES'$ is a signature morphism $H : SIG \to SIG'$ such that $\forall H(X).H(t) =_{H(s)} H(t')$ if $H(t) =_{H(w)} H(t') \in \Gamma$ if $\forall X.\ t =_s t'$ if $t =_w t' \in \Gamma$.

(As standard, H(_) replaces all sorts and operators as specified by the signature morphism including the subscripts of variables and operators. For a precise definition see [Poigné 90])

We state without proof:

5.2 PROPOSITION *Let $H : PRES \to PRES'$ be a presentation morphism, and $\mathcal{A}$ be a PRES'-algebra. Then $\mathcal{A}_H$ with carriers $s^{\mathcal{A}_H} = H(s)^{\mathcal{A}}$ and operators $\sigma_{w,s}{}^{\mathcal{A}_H}(a) = H(\sigma)_{H(w),H(s)}{}^{\mathcal{A}}(a)$ defines a PRES-algebra, the H-**reduct** of $\mathcal{A}$.*

5.3 PROPOSITION *Let $H : PRES \to PRES'$ be a presentation morphism, and $\mathcal{A}$ be a PRES-algebra. Then $H(\mathcal{A}) = T_{SIG'}(A)_{/\sim}$ is a free PRES'-algebra generated by $\mathcal{A}$,*

where - A is an S'-sorted set with $A_{s'} = \{(s, a) \mid H(s) = s', a \in s^{\mathcal{A}}\}$,

- $T_{SIG'}(A)$ is the term algebra generated by A,

and where

- the SIG'-congruence $t \sim_s t'$ is obtained as the transitive closure of $\{(t, t') \mid t \sim_s t', s \in s\}$,

where

$t \sim_s t'$ if $h_s(t) = h_s(t')$ for all $PRES$-homomorphisms $h : \mathcal{A} \to \mathcal{B}_H$

(We overload notation in that $h : T_{SIG'}(A) \to \mathcal{B}$ is the unique homomorphism such that $h_{s'}((s, a)) = h_s(a)$ if $s' = H(s)$).

($H(\mathcal{A})$ is free that for every $PRES'$-algebra $\mathcal{B}$ and for every homomorphism $h : \mathcal{A} \to \mathcal{B}_H$ there exists a unique homomorphisms $h^{\#} : H(\mathcal{A}) \to \mathcal{B}$ such that $h^{\#}{}_{H(s)}([(s, a)]) = h_s(a)$ for $a \in s^{\mathcal{A}}$.)

PROOF Combine 2.4 and 3.2 as standard.

6. Comparing our approach with the one of [Goguen, Meseguer 89], a natural candidate for a partition is given by the *connected sets*, i.e. subsets $[s] = \{s' \mid s \equiv s'\} \subseteq S$ where $\equiv$ is the symmetric and transitive closure of $\leq$. Unfortunately, regular order-sorted signature in the sense of [Goguen, Meseguer 89] fail to satisfy the condition (ii) of 1.2 : consider the regular signature SIG2 with sorts $s1 \leq s3 \geq s2, s4$ and operators $\sigma: s1 \to s4, \sigma; s2 \to s4$. Then a GM-model $\mathcal{B}$ is given by $s1^{\mathcal{B}} = s2^{\mathcal{B}} = s3^{\mathcal{B}} = \{a\}, s1^{\mathcal{B}} = \{b, c\}$ and

$\sigma_{s1,s4}{}^{\mathcal{B}}(a) = b$ and $\sigma_{s2,s4}{}^{\mathcal{B}}(a) = c$ which is not a model in our sense if $\{s1 \leq s3 \geq s2\}$ and $\{s4\}$ are the principal sorts. Hence we fall short of our original goal to unify the different approaches to order-sorting[1].

So we loose out on semantic representation of ad-hoc polymorphisms. Nevertheless, the order-sorted logic proposed here is still equivalent[2] to conditional equational many-sorted logic (and thus to Goguen's and Meseguer's order-sorted logic by [Goguen, Meseguer 89]): many-sorted logic is obviously obtained by restriction if the principal sorts are trivial. For the other direction one extends the proof of [Goguen, Meseguer 89, 5.7] which explicitly specifies the implicit intersections.

Remains the final question whether we gain on some points. Let us consider two examples:

- In the counter example above, application of the operator σ to the value a yields different results depending which sort is assigned to a. There should be little dispute whether this is wanted or not, but one may doubt if this setup occurs in real life, meaning with regard to a specification constructed by means of standard modularization operators such as parameter passing or reduction. We can speak only about values we can refer to, i.e. we have a term t the denotation of which is a. Then, of course, t and thus a should have a least sort as a consequence of regularity (cf. [Goguen, Meseguer 89], in contradiction to the example given. If such a least sort s would exist, it must be a subsort of $s1$ and $s2$, and the counter example would collapse since regularity then implies existence of a least operator $\sigma_{s,s4}$, and thus agreement of $\sigma_{s1,s4}$ and $\sigma_{s2,s4}$ on a.
- Assume that, within the GM-world, the signature above is an extension of the signature SIG1 comprising only the sorts $s1 \leq s3 \geq s2$ and that we have a corresponding GM-algebra $\mathcal{A}$ with carrier sets $s1^{\mathcal{A}} = s2^{\mathcal{A}} = s3^{\mathcal{A}} = \{a\}$. Starting from $\mathcal{A}$, order-sorted term construction generates only one term $f(a)$ of type $s4$ which is not good enough for a free algebra: we cannot extend the SIG1-homomorphism $h : \mathcal{A} \rightarrow \mathcal{B}$ with $h_{s1}(a) = b$ and $h_{s2}(a) = c$ to a SIG1-homomorphism $h^{\#} : \mathrm{T}(\mathcal{A}) \rightarrow \mathcal{B}$ since it cannot be the case that $h^{\#}(\sigma(a)) = b$ <u>and</u> $h^{\#}(\sigma(a)) = c$. As a loophole one might annotate the operators with sort informations, e.g. generate $\sigma_{s1}(a)$ and $\sigma_{s2}(a)$, but this is definitely against the spirit of order-sorting resp. overloading. Moreover, if we consider an extended SIG1 with an additional operator $\sigma : s3 \rightarrow s4$, we must factorize by $\sigma_{s1}(a) = \sigma_{s2}(a) = \sigma_{s3}(a)$ which adds the additional complication of subsorting inducing equalities. (REMARK This problem does not occur when constructing term algebras from variables as in [Goguen, Meseguer 89] since the variables of different sort are assumed to be disjoint.)

The examples are meant to demonstrate that the additional generality in semantic representation of ad-hoc polymorphism may be irrelevant for those models we are interested in and, secondly, that we cannot rely on order-sorted term construction in every context.

Of course, the arguments are not conclusive, and it may be a matter of taste whether to stress unrestricted use of term construction, as we do, or to look for a maximum of ad-hoc polymorphism. We hope to have found a reasonable compromise; our ad-hoc polymorphism exactly corresponds to that of many-sorted signatures but for principal sorts, while subsorting polymorphism is restricted to operators with the same principal sort. We can use order-sorted term construction without restrictions which allows to develop all

---

[1] In the version submitted (see also [Poigné 91]), I stated that my approach subsumes that of [Goguen, Meseguer 89]. Unfortunately, I did not notice the trivial counter example but produced a flaw in an essential lemma. However, parts of the original claim survive; many-sorted algebras is subsumed, and, as I believe, more insight in the mechanism of order-sorting is gained.

[2] Two logics are equivalent if for each theory in the one logic there exists a theory in the other logic such that the respective categories of models are equivalent.

the important infrastructure of categories of algebras very much standard lines. We pay a price in that we introduce principal sorts, which to consider as natural or just as an unnecessary complication is left to the reader.

ACKNOWLEDGEMENTS Many thanks to the careful referees who caused me to reexamine the submitted version.

7. REFERENCES

[Gogolla 83] Martin Gogolla, Algebraic specifications with subsorts and declarations, Forschungsbericht Nr. 169, Universität Dortmund, Abteilung Informatik, 1983, also in: Bruno Courcelle (editor), *Proceedings Ninth CAAP (Bordeaux)*, Cambridge University Press, 1984

[Goguen 78] Joseph Goguen, *Order Sorted Algebra*, Technical Report 14, UCLA Computer Science Department, 1978

[Goguen, Meseguer 89] Joseph Goguen, José Meseguer, *Order-sorted algebra I : Equational Deduction for Multiple Inheritance, Overloading, Exceptions and Partial Operations*, Technical Report SRI-CSL-89-10. Computer Science Laboratory, SRI International, July 1989

[Poigné 84] Axel Poigné, Another look at parameterization using algebraic specifications with subsorts, in : *Proceedings MFCS 84*, LNCS 176, Springer, 1984

[Poigné 88] Axel Poigné, Partial algebras, subsorting and dependent types, in: D.Sannella, A.Tarlecki (editors), *Recent Trends in Data Type Specification*, LNCS 332, 1988

[Poigné 90] Axel Poigné, Parameterization for order-sorted algebraic specification, JCSS Vol.40, No.2, 1990

[Smolka 86] Gert Smolka, *Order-Sorted Horn Logic : Semantics and Deduction*, Technical Report SEKI-Report Sr-86-17, Fachbereich Informatik, Universität Kaiserslautern, 1986

[Strachey 67] Christopher Strachey, Fundamental concepts in programming languages. *Lecture Notes from International Summer School in Programming*, Copenhagen, 1967

Composition of two semi commutations*

Y. Roos P.A. Wacrenier

CNRS URA 369, L.I.F.L.
Université de Lille 1
U.F.R. I.E.E.A. Informatique
59655 Villeneuve d'Ascq cedex

Abstract

We give a decidable necessary and sufficient condition on two semi commutations θ_1, θ_2 such that the composition of f_{θ_1} and f_{θ_2} is a semi commutation function. This characterization uses the commutation graphs of θ_1 and θ_2, and the non commutation graphs of θ_1 and θ_2^{-1}. Then we deduce a decidable graphic characterization of confluent semi commutations.

Introduction

The free partially commutative monoids study was initiated by Cartier and Foata [3] whose aim was to solve some combinatory problems. Trace languages, which are subsets of a free partially commutative monoid, were proposed by Mazurkiewicz [16] as tools for the description of concurrent program behaviour. Important results have been found and several syntheses have been written about this subject (see [1,2,12,13,15,17,18,19,20,22]).

A partially commutative alphabet is a couple (A, θ) in which A is an alphabet, and θ, the independance relation, is a symmetric and irreflexive binary relation over A. Associated with the commutation relation θ, an application $f_\theta : 2^{A^*} \rightarrow 2^{A^*}$ can be defined by: *for every language L over the alphabet A, $(L)f_\theta$ is the set of the words which are equivalent to some word of L for the congruence generated by θ.* Thus f_θ is a unary operation over languages which is named *partial commutation function associated with θ.*

M. Clerbout and M. Latteux [7] introduced the notion of semi commutation which generalizes the notion of commutation: a semi commutation is an irreflexive independance relation over A.

When new operators, as semi commutation functions, are defined a natural question is *what about their composition?* In [10] M. Clerbout and D. Gonzalez defined atomic semi commutation and proved that semi commutations can be decomposed into weaker semi commutations if and only if they are not atomic.

*This work is supported by the P.R.C. Mathématiques et Informatique, and by the EBRA project ALGEBRAIC and SYNTACTIC METHODS in COMPUTER SCIENCE.

Unfortunately the composition of two semi commutation functions , even atomic, is not always a semi commutation function. Our purpose is to give a decidable characterization of semi commutations θ_1, θ_2 such that the composition of f_{θ_1} and f_{θ_2} is a semi commutation function. The characterization is based on the commutation graphs of θ_1 and θ_2, and the non commutation graphs of θ_1 and θ_2^{-1}. As a consequence of this result we get a decidable characterization of confluent semi commutations.

Preliminaries

Notations

In the following text X is the used alphabet ; u, v and w are words in X^*.

u^R is the mirror of the word u i.e. if $u = x_1x_2 \ldots x_n$ then $u^R = x_n \ldots x_2x_1$.

$|w|$ is the length of the word w.

$|w|_x$ is the number of occurrences of the letter x that appear in the word w.

$(w)com = \{u \in X^* / \forall x \in X, |w|_x = |u|_x\}$ is the *commutative closure* of the word w.

$(w)\Pi_Y$ is the *projection* of the word w *over the subalphabet* Y, i.e. the image of w by the homomorphism Π_Y which is defined by:

$$\forall x \in X, \text{ if } x \in Y \text{ then } (x)\Pi_Y \ = \ x \ , \text{ else } \ (x)\Pi_Y = \varepsilon$$

We shall say that w' is a *subword* of w if there exist words of X^* $v_0, v_1, \ldots, v_n$ such that $w = v_0x_1v_1 \ldots x_iv_i \ldots x_nv_n$ where $w' = x_1x_2 \ldots x_n$.

Semi commutations

A *semi commutation relation* defined over an alphabet X is an irreflexive relation: it is a subset of $X \times X \backslash \{(x,x)/x \in X\}$.

To each semi commutation relation θ, we associate a rewriting system $S =< X, P >$ which is named *semi commutation system* in which P is the set $\{xy \to yx/(x,y) \in \theta\}$. We shall write $u \xrightarrow[\theta]{1} v$ if there is a rule $xy \to yx$ in P and two words w and w' such that $u = wxyw'$ and $v = wyxw'$. We shall write $u \xrightarrow[\theta]{*} v$ if there are words w_1, w_2, ..., w_n $(n \geq 1)$ such that $w_1 = u$, $w_n = v$, and for each $i < n$, $w_i \xrightarrow[\theta]{1} w_{i+1}$. Then we shall write that there is a *derivation* from u to v, and the integer $n-1$ is named the *derivation length.*

To each semi commutation θ we associate its *commutation graph* which is the directed graph defined by (X, θ) where X is the vertex-set and θ the edge-set.

We denote $\bar{\theta}$ the *non commutation relation* associated with the semi commutation θ. Hence: $\bar{\theta} = \{(x,y)/(x,y) \notin \theta\}$

To each semi commutation θ we associate its *non commutation graph* which is the directed graph defined by $(X, \{(x,y)/x \neq y \text{ and } (x,y) \notin \theta\})$ where X is the vertex-set set and $\{(x,y)/x \neq y \text{ and } (x,y) \notin \theta\}$ the edge-set.

We denote θ^{-1} the *converse semi commutation* associated with the semi commutation θ. Hence: $\theta^{-1} = \{(y,x)/(x,y) \in \theta\}$.

To each semi commutation relation θ, we associate a *semi commutation function* f_θ which is defined by: $\forall w \in X^*, (w)f_\theta = \{u \in X^*/w \xrightarrow[\theta]{*} u\}$.
By extension: if L is a language which is a subset of X^*, then : $(L)f_\theta = \bigcup_{w \in L}(w)f_\theta$.

If θ_1 and θ_2 are two semi commutation relations defined over X, we shall write $f_{\theta_1}f_{\theta_2}$ for the composition of f_{θ_1} and f_{θ_2}. Hence: $\forall w \in X^*(w)f_{\theta_1}f_{\theta_2} = [(w)f_{\theta_1}]f_{\theta_2}$

We shall write $f_{\theta_1}f_{\theta_2} \subset f_{\theta_2}f_{\theta_1}$ if: $\forall w \in X^*(w)f_{\theta_1}f_{\theta_2} \subset (w)f_{\theta_2}f_{\theta_1}$

A semi commutation θ over X is *confluent* if and only if for all words u, v, w of X^* such that $w \xrightarrow[\theta]{*} u$ and $w \xrightarrow[\theta]{*} v$ there exists a word w' of X^* such that $u \xrightarrow[\theta]{*} w'$ and $v \xrightarrow[\theta]{*} w'$.

Preliminary results

First we shall prove that if $\theta_1, \theta_2, \ldots, \theta_n$ are semi commutations over an alphabet X and the composition $f_{\theta_1}f_{\theta_2}\ldots f_{\theta_n}$ is equal to a semi commutation function f_θ then $f_\theta = f_{\theta_n}f_{\theta_{n-1}}\ldots f_{\theta_1}$. We shall use two trivial lemmas.

Lemma 1 *Let $\theta_1, \theta_2, \ldots, \theta_n$ be semi commutations over an alphabet X. Then*

$$(f_{\theta_1}f_{\theta_2}\ldots f_{\theta_n})^{-1} = f_{\theta_n^{-1}}f_{\theta_{n-1}^{-1}}\ldots f_{\theta_1^{-1}}$$

Lemma 2 *Let θ be a semi commutation over an alphabet X and u, v be two words of X^*. Then:*

$$u \xrightarrow[\theta]{*} v \Longrightarrow u^R \xrightarrow[\theta^{-1}]{*} v^R$$

Proposition 1 *Let $\theta_1, \theta_2, \ldots, \theta_n$ be semi commutations over an alphabet X. If $f_{\theta_1}f_{\theta_2}\ldots f_{\theta_n}$ is equal to a semi commutation function f_θ then $f_\theta = f_{\theta_n}f_{\theta_{n-1}}\ldots f_{\theta_1}$*

Proof

Let u and v be two words of X^* such that $u \xrightarrow[\theta]{*} v$.

From lemma 2 we have $u \xrightarrow[\theta]{*} v \Longleftrightarrow u^R \xrightarrow[\theta^{-1}]{*} v^R$ and from lemma 1 we know that $f_{\theta^{-1}} = f_{\theta_n^{-1}}f_{\theta_{n-1}^{-1}}\ldots f_{\theta_1^{-1}}$ therefore we have: $u^R = w_n \xrightarrow[\theta_n^{-1}]{*} w_{n-1} \xrightarrow[\theta_{n-1}^{-1}]{*} \ldots \xrightarrow[\theta_1^{-1}]{*} w_0 = v^R$.

So for each i in $[1,n]$ we have $w_i \xrightarrow[\theta_i^{-1}]{*} w_{i-1}$. From lemma 2 we have, for each i in $[1,n]$, $w_i^R \xrightarrow[\theta_i]{*} w_{i-1}^R$. So we have : $u = w_n^R \xrightarrow[\theta_n]{*} w_{n-1}^R \xrightarrow[\theta_{n-1}]{*} \dots \xrightarrow[\theta_1]{*} w_0^R = v$ □

Generally the converse of proposition 1 is false:
Let $X = \{a,b,c\}$, $\theta_1 = \{(a,c)\}$, $\theta_2 = \{(a,b)\}$ and $\theta_3 = \theta_1$. We have $f_{\theta_1} f_{\theta_2} f_{\theta_3} = f_{\theta_3} f_{\theta_2} f_{\theta_1}$ but $f_{\theta_1} f_{\theta_2} f_{\theta_3}$ is not a semi commutation function: $(abcb) f_{\theta_1} f_{\theta_2} f_{\theta_3} = \{abcb, bacb, bcab\}$ and $(abcb) f_{\theta_1} f_{\theta_2} f_{\theta_3} f_{\theta_1} f_{\theta_2} f_{\theta_3} = \{abcb, bacb, bcab, bcba\}$

Nevertheless when we work with only two semi commutations we easily get:

Proposition 2 *Let θ_1 and θ_2 be two semi commutations over an alphabet X. These following assertions are equivalent:*

1. $f_{\theta_1} f_{\theta_2}$ *is a semi commutation function*
2. $f_{\theta_1} f_{\theta_2} = f_{\theta_2} f_{\theta_1}$
3. $f_{\theta_2} f_{\theta_1} \subset f_{\theta_1} f_{\theta_2}$
4. $f_{\theta_1} f_{\theta_2} = f_{\theta_1 \cup \theta_2}$

From proposition 2 we can deduce that a semi commutation is confluent if and only if the composition of the semi commutation with its converse is a semi commutation.

Proposition 3 *A semi commutation θ over an alphabet X is confluent if and only if $f_\theta f_{\theta^{-1}}$ is a semi commutation function.*

Proof

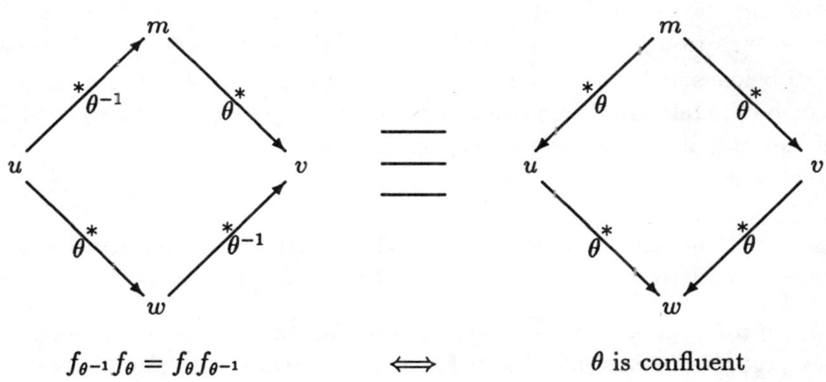

$f_{\theta^{-1}} f_\theta = f_\theta f_{\theta^{-1}}$ $\iff$ θ is confluent

Composition of two semi commutation functions

Thanks to proposition 2 we know that the composition of two semi commutation functions is a semi commutation function if and only if this composition is equal to the function associated with the union of the two semi commutations. Hence $f_{\theta_1} f_{\theta_2}$ is a semi commutation function if and only if there do not exist two words u and v such that $v \notin u\, f_{\theta_1} f_{\theta_2}$ and $v \in u f_{\theta_1 \cup \theta_2}$.

The following notations are used in this paper:

- f_{θ_1} and f_{θ_2} are two semi commutation functions over the alphabet X.
- if u and v are two words of X^* then the couple (u, v) satisfies the property P if and only if $v \notin u\, f_{\theta_1} f_{\theta_2}$ and $v \in u f_{\theta_1 \cup \theta_2}$.
- if u is a word of X^* and θ a semi commutation over X we shall denote $\pi(u, \theta)$ the set of the couples (x, y) such that xy is a subword of u and (x, y) does not belong to θ.

The aim of the next lemma is to propose a necessary and sufficient condition for a word, the letters of which are all different, to belong to the image of another word by the composition of the two semi commutation functions f_{θ_1} and f_{θ_2}.

Lemma 3 *Let u and v be two words of X^* such that v belongs to $(u)com$ and u does not contain two occurrences of the same letter. Then v belongs to $u f_{\theta_1} f_{\theta_2}$ if and only if $\pi(u, \theta_1) \cup \pi(v, \theta_2^{-1})$ does not contain any directed cycle.*

Sketch of proof
Let $\Pi = \pi(u, \theta_1) \cup \pi(v, \theta_2^{-1})$.
Clearly v belongs to $u f_{\theta_1} f_{\theta_2}$ if and only if there exists a word w such that w belongs to $u f_{\theta_1}$ and v belongs to $w f_{\theta_2}$. So the word w must belong to $u f_{\theta_1} \cap v f_{\theta_2^{-1}}$. It is easily seen that if (x, y) belongs to Π then the letter x must be before the letter y in w.
Let *is_before* be the relation defined on $X \times X$ by: x *is_before* $y \iff (x, y) \in \Pi$. The word w exists if and only if *is_before* is an order on X, hence, if and only if Π does not contain any directed cycle.

Thanks to the numerotation lemma[21], which permits to distinguish the different occurrences of each letter occurring in a word, and thanks to lemma 3 it is easy to get:

Lemma 4 *Let u and v be two words of X^* such that (u, v) satisfies the property P then there exists a couple of words (u', v') satisfying the property P such that u' does not contain two occurrences of the same letter.*

The next lemma proves that if there exists a couple of words satisfying the property P then there exists a couple of words (u, v) satisfying the property P such that the derivation of u in v uses at least one rule of θ_1 and only one rule of $\bar{\theta}_1 \cap \theta_2$.

Lemma 5 *If there exists a couple of words (u, v) satisfying the property P then there exists a couple of words (u', v') such that:*

- *the couple (u', v') satisfies the property P*
- $u' \xrightarrow[\bar{\theta}_1 \cap \theta_2]{1} w_1 \xrightarrow[\theta_1]{*} w_2 \xrightarrow[\theta_1 \cap \bar{\theta}_2]{1} v'$
- *u' belongs to $(u)com$*

Proof

The couple (u,v) satisfies the property P, let us write a derivation:

$$u = u_0 \xrightarrow[\theta_1 \cup \theta_2]{1} u_1 \xrightarrow[\theta_1 \cup \theta_2]{1} \dots \xrightarrow[\theta_1 \cup \theta_2]{1} u_i \xrightarrow[\theta_1 \cup \theta_2]{1} \dots \xrightarrow[\theta_1 \cup \theta_2]{1} u_{n-1} \xrightarrow[\theta_1 \cup \theta_2]{1} u_n = v$$

Let j be the greatest integer such that (u_j, u_n) satisfies the property P. Clearly $j \neq n$. Suppose that $u_j \xrightarrow[\theta_1]{*} u_{j+1}$. We know that the couple (u_{j+1}, u_n) does not satisfy the property P therefore $u_{j+1} \xrightarrow[\theta_1]{*} m \xrightarrow[\theta_2]{*} u_n$ this leads us to a contradiction: $u_j \xrightarrow[\theta_1]{*} u_{j+1} \xrightarrow[\theta_1]{*} m \xrightarrow[\theta_2]{*} u_n$ hence (u_j, u_n) does not satisfy the property P.

So we have:

$$u_j = w_0 \xrightarrow[\bar{\theta}_1 \cap \theta_2]{1} w_1 \xrightarrow[\theta_1]{1} \dots \xrightarrow[\theta_1]{1} w_l \xrightarrow[\theta_2]{1} \dots \xrightarrow[\theta_2]{1} w_{k-1} \xrightarrow[\theta_2]{1} w_k = u_n$$

Now let h be the smallest integer such that the couple (w_0, w_h) satisfies the property P. Clearly $h \neq 0$.

Suppose that $w_{h-1} \xrightarrow[\theta_2]{*} w_h$. we know that the couple (w_0, w_{h-1}) does not satisfy the property P, hence we have $w_0 \xrightarrow[\theta_1]{*} m' \xrightarrow[\theta_2]{*} w_{h-1}$. This leads us to a contradiction: $w_0 \xrightarrow[\theta_1]{*} m' \xrightarrow[\theta_2]{*} w_{h-1} \xrightarrow[\theta_2]{*} w_h$ but the couple (w_0, w_h) satisfies the property P.

It is easily seen that:

$$w_0 \xrightarrow[\bar{\theta}_1 \cap \theta_2]{1} w_1 \xrightarrow[\theta_1]{*} w_{h-1} \xrightarrow[\theta_1 \cap \bar{\theta}_2]{1} w_h$$

As $w_0 = u_j$ and u_j belongs to $(u)com$, w_0 belongs to $(u)com$. □

Due to lemma 4 we can decide if the composition of two semi commutations is a semi commutation: it is sufficient to compare the image by the composition and by the union of the two semi commutations for each word which contains exactly one occurrence of every letter of the alphabet. The next proposition proposes another characterization using the commutation graphs of θ_1 and of θ_2 and the non commutation graphs of θ_1 and of θ_2^{-1}.

Proposition 4 *The composition of two semi commutation functions is a semi commutation function if and only if for each set* $\{(x_0,x_1),(x_1,x_2),\ldots,(x_i,x_{i+1}),\ldots,(x_n,x_0)\}$ *included in* $\bar{\theta}_1 \cup \bar{\theta}_2^{-1}$ *such that:*

- $x_p = x_q \iff p = q$
- $(x_0, x_n) \in \theta_1 \cap \bar{\theta}_2$
- $(x_i, x_{i+1}) \in \bar{\theta}_1 \cap \theta_2$

there exist two integers $h \in [0,i]$ *and* $j \in [i+1,n]$ *such that* (x_h, x_j) *belongs to* $\bar{\theta}_1 \cap \bar{\theta}_2$.

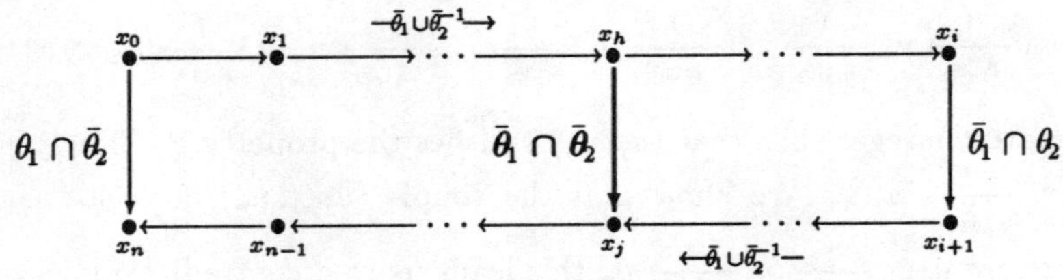

Sketch of proof

$\Longrightarrow$

Let us suppose that there exists a directed cycle $\{(x_0,x_1),(x_1,x_2),\ldots,(x_i,x_{i+1}),\ldots,(x_n,x_0)\}$ included in $\bar{\theta}_1 \cup \bar{\theta}_2^{-1}$, such that $x_p = x_q \iff p = q$, $(x_0,x_n) \in \theta_1 \cap \bar{\theta}_2$, $(x_i,x_{i+1}) \in \bar{\theta}_1 \cap \theta_2$ and there are no integer $h \in [0,i]$ and $j \in [i+1,n]$ such that $(x_h,x_j) \in \bar{\theta}_1 \cap \bar{\theta}_2$.
Let $w_1 = x_0x_1\ldots x_i$ and $w_2 = x_{i+1}x_{i+2}\ldots x_n$. It is easily seen that $w_1w_2 \xrightarrow[\theta_1 \cup \theta_2]{*} w_2w_1$. $\pi(w_1w_2,\theta_1) \cup \pi(w_2w_1,\theta_2^{-1})$ contains a directed cycle so, from lemma 3, $w_2w_1 \notin w_1w_2 f_{\theta_1} f_{\theta_2}$.

$\Longleftarrow$

We suppose that $f_{\theta_1} f_{\theta_2}$ is not a semi commutation function. Hence there exists a couple of words of minimun length (u,v) satifies property P and such that $u \xrightarrow[\bar{\theta}_1 \cap \theta_2]{1} u_1 \xrightarrow[\theta_1]{*} v_1 \xrightarrow[\theta_1 \cap \bar{\theta}_2]{1} v$. First we show that $\pi(u,\theta_1) \cup \pi(v,\theta_2^{-1})$ is a directed cycle equal to $\{(x_0,x_1),(x_1,x_2),\ldots,(x_i,x_{i+1}),\ldots,(x_n,x_0)\}$. As $u \xrightarrow[\bar{\theta}_1 \cap \theta_2]{1} u_1$ there exists a couple of letters (x_0,x_n) such that $(x_0,x_n) \in \theta_1 \cap \bar{\theta}_2$and as $v_1 \xrightarrow[\theta_1 \cap \bar{\theta}_2]{1} v$ there exists a couple of letters $(x_i,x_{i+1}) \in \bar{\theta}_1 \cap \theta_2$. Then, due to $u_1 \xrightarrow[\theta_1]{*} v$, $v\Pi_{xy} = xy$ for each couple of letters (x,y) of $\pi(u_1,\theta_1)$. From the definition of $\pi(v,\theta_2^{-1})$, we know that for each couple (x,y) of $\pi(v,\theta_2^{-1})$ we have $v\Pi_{xy} = xy$. Hence for each couple (x,y) of $\pi(v,\theta_2^{-1}) \cup \pi(u_1,\theta_1)$ we have $v\Pi_{xy} = xy$. It is easily seen that $\Pi/\{(x_i,x_{i+1})\}$ is included in $\pi(v,\theta_2^{-1}) \cup \pi(u_1,\theta_1)$. Hence we have $v = x_{i+1}\ldots x_nx_0\ldots x_i$. Let us suppose now that there exist h in $[0,i]$ and j in $[i+1,n]$ such that $(h,j) \neq (0,n)$ and (x_h,x_j) belongs to $\bar{\theta}_2$: this leads us to a contradiction: (x_j,x_h) belongs to $\bar{\theta}_2^{-1}$and $v\Pi_{x_hx_j} = x_jx_h$ so (x_j,x_h) belongs to $\pi(v,\theta_2^{-1})$ and $(h,j) \neq (0,n)$ so Π contains two directed cycles. $\square$

Applications

As a partial commutation is a symmetric semi commutation we can use the proposition 4 to decide whether the composition of two partial commutation functions is a partial commutation function:

Corollary 1 *The composition of two partial commutation functions f_{θ_1} and f_{θ_2} is a partial commutation function if and only if there exists no subalphabet A of X such that the graph $(A, \overline{\theta_1 \cap \theta_2})$ is a cycle which is neither in $(A, \bar{\theta}_1)$, nor in $(A, \bar{\theta}_2)$*

Characterization of confluent semi commutations

We proved (proposition 3) that the composition of a semi commutation function f_θ with $f_{\theta^{-1}}$ is a semi commutation function if and only if the semi commutation θ is confluent. Then we shall use the fourth proposition to give a graphic characterization of such semi commutations. As a matter of fact, we obtain a new proof of a result of V.Diekert, E.Ochmanski and K.Reinhardt [14].

Corollary 2 *A semi commutation θ over an alphabet X is confluent if and only if for each set $\{(x_0, x_1), (x_1, x_2), \ldots, (x_i, x_{i+1}), \ldots, (x_n, x_0)\}$ included in $\bar{\theta}$ such that:*

- $x_p = x_q \Longleftrightarrow p = q$
- $(x_0, x_n) \in \theta$
- $(x_{i+1}, x_i) \in \theta$

there exist two integers h in $[0, i]$ and j in $[i+1, n]$ such that (x_h, x_j) belongs to $\bar{\theta} \cap \bar{\theta}^{-1}$.

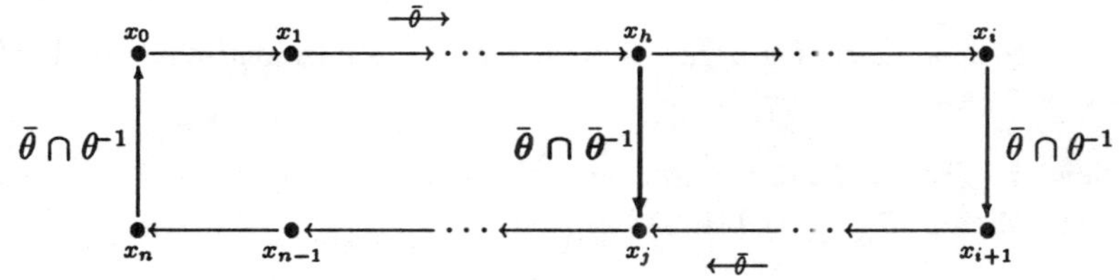

References

[1] IJ.J.Aabersberg & G.Rozenberg. *Theory of Traces.* The. Comp. Sci. 60 (1986) 1-83.

[2] J.Berstel & J.Sakarovich. *Recents results in the theory of rational sets.* Lect. Notes in Comp. Sci. 233 (1986) 15-28.

[3] P.Cartier & D.Foata. *Problèmes combinatoires de commutations et réarrangements.* Lect. Notes in Math. 85 (1969).

[4] M.Clerbout. *Commutations partielles et familles de langages.* Thèse Lille, 1984.

[5] M.Clerbout & M.Latteux. *Partial commutations and faithful rational transductions.* The. Comp. Sci. 34 (1984) 241-254.

[6] M.Clerbout & M.Latteux. *On a generalization of partial commutations.* in M.Arato, I.Katai, L.varga, eds, Prc.Fourth Hung. Computer Sci.Conf (1985) 15-24.

[7] M.Clerbout & M.Latteux. *Semi-commutations.* Information and Computation 73 (1987) 59-74.

[8] M.Clerbout. *Compositions de fonctions de commutation partielle.* RAIRO Inform.Theor. 23 (1986) 395-424.

[9] M.Clerbout, M.Latteux & Y.Roos. *Decomposition of partial commutations.* ICALP'90 Lect. Notes in Comp. Sci. 443 (1990) 501-511.

[10] M.Clerbout & D.Gonzalez. *Decomposition of semi commutations.* MFCS'90 Lect. Notes in Comp.Sci. 452 (1990) 209-216.

[11] R.Cori & D.Perrin. *Automates et commutations partielles.* RAIRO Inform. Theor. 19 (1985) 21-32.

[12] R.Cori. *Partially abelian monoids.* Invited lecture, STACS, Orsay (1986).

[13] V.Diekert *Combinatorics on Traces.* Lect. Notes in Comp. Sci. 454 (1990)

[14] V.Diekert, E.Ochmanski & K.Reinhardt. *On confluent semi commutation systems – decidability and complexity results.* To appear (ICALP'91).

[15] M.Latteux. *Rational cones and commutations. Machines, Languages and Complexity.* J.Dassow and J.Kelemen eds., Lect. Notes in Comp.Sci. (1989) 37-54.

[16] A.Mazurkiewicz. *Concurrent program schemes and their interpretations.* DAIMI PB 78, University of Aarhus (1977).

[17] A.Mazurkiewicz. *Traces, histories and graphs : instances of process monoids.* Lect. Notes in Comp.Sci. 176 (1984) 115-133.

[18] Y.Metivier. *On recognizable subsets of free partially commutative monoids.* Lect. Notes in Comp.Sci. 226 (1986) 254-264.

[19] E.Ochmanski. *Regular behaviour of concurrent systems.* Bulletin of EATCS 27 (1985) 56-67.

[20] D.Perrin. *Words over a partially commutative alphabet.* NATO ASI Series F12, Springer (1985) 329-340.

[21] Y.Roos. *Contribution à l'étude des fonctions de commutations partielles.* Thèse, Université de Lille (1989).

[22] W.Zielonka. *Notes on asynchronous automata.* RAIRO Inform. Theor. 21 (1987) 99-135.

An Efficient Decision Algorithm for the Uniform Semi-Unification Problem
Extended Abstract

Peter Ružička
VUSEI-AR
Dúbravská 3, 842 21 Bratislava, Czecho-Slovakia

Abstract: An algorithm to decide whether a pair of terms is semi-unifiable is presented. It is based on compact graph representation of terms and on efficient implementation of variable replacements. The time complexity of the decision algorithm is $O(n^2)$ in the worst case, where n is the size of the input terms.

1. Uniform Semi-Unification

Semi-unification problem is to decide whether for a given finite number of pairs of input terms s_i,t_i there are substitutions σ,ρ_i such that it holds $\rho_i(\sigma(s_i)) = \sigma(t_i)$ for i=1,...,n. A special case of equal "matching" substitutions $\rho_1 = \rho_2 = ... = \rho_n$ corresponds to the semi-unification for a pair of input terms. We define the uniform semi-unification as a problem to decide whether for two input terms s,t there are substitutions σ, ρ such that it holds $\rho(\sigma(s)) = \sigma(t)$.

The semi-unification problem has a number of applications in many fields of computer science, including logic programming, term rewriting, type inference, theorem proving and natural language processing.

The notion of semi-unification was introduced by Musser and Lankford (1978) and made popular by Dershowitz (1987). The general case of non-uniform semi-unification was proved to be undecidable by Kfoury, Urzyczyn, Tiuryn (1990). Fortunately, the uniform semi-unification is decidable. This was proved independently by Pudlák (1988) and Kapur, Musser, Narendran, Stillman (1988). Decision algorithms for uniform semi-unification problem have been also obtained by several other authors - Leiss (1989) and Henglein (1989).

This paper is concerned with the development of an efficient decision algorithm for the uniform semi-unification problem. For this purpose we introduce a new class of expressions called τ-terms and a new type of substitution rules called τ-variable replacement rules. We show that these notions are sufficient to perform the semi-unification. A symbol τ plays the role of a "binding-symbol to a variable" (called τ-variable) to which a variable replacement as the "matching" substitution will be applied. We show how compact graph representation for τ-terms and τ-variable replacements can be used. The result is a decision algorithm for the uniform semi-unification with $O(n^2)$ time complexity where n is the sum of the input terms sizes. This result asymptotically improves the efficiency of the only known polynomial time uniform semi-unification decision algorithm obtained by Kapur, Musser, Narendran and Stillman (1988).

2. Terms, substitutions and their graph representation

In unification algorithms a computation on terms is performed by means of applying a sequence of simple variable replacement steps. Hence, all what is needed to do in order to perform an efficient unification computation is to design a suitable data structure for a term on which one can implement a variable replacement rule efficiently.

In this introductory section we first introduce notions of a term and a variable replacement rule in a string form which is simple enough to explain basic concepts. Later on we build up a formalism based on a graph representation which is more suitable for time and space efficient implementation of unification computation. Finaly the equivalency of both formalisms is certified.

Let V be a countable set of variable symbols, F be a countable set of function symbols ($V \cap F = \emptyset$) and N be a set of integers. Let arity be a function $V \cup F \mapsto N$ such that for $x \in V$ it holds $arity(x) = 0$. Constants are function symbols with arity equal to 0. Denote F_o to be a set of constants, $F_o \subseteq F$.

Let $T(V, F)$ be a set of well-formed terms in the string form, inductively defined using the following rules:

1. $V \subseteq T(V, F)$, $F_o \subseteq T(V, F)$;
2. if $f \in F$, arity(f)=n, $t_1, ..., t_n \in T(V, F)$, then $f(t_1, ..., f_n) \in T(V, F)$.

The size of an arbitrary term $t \in T(V, F)$ is defined as an integer equal to the number of occurrences of symbols from $V \cup F$ in the term t. A subterm s of a term t (denoted as t[s]) is a well-formed term which contiguously occurs in t as a substring.

The substitution σ is a mapping $V \mapsto T(V, F)$. Each substitution can be naturally extended to a morphism $T(V, F) \mapsto T(V, F)$. In this paper we shall use only substitutions σ such that $\sigma(x) = x$ holds except for a finite subset of V. The substitution σ can therefore be described in the form $[x_1 \mapsto t_1, ..., x_n \mapsto t_n]$ where $x_i \in V$, $t_i \in T(V, F)$ and $x_i \mapsto t_i$ is a variable replacement rule. The identity substitution is denoted as []. To apply a substitution σ to a term t one has to replace all occurrences of each variable x_i in t by the corresponding term t_i. A term s is called a σ-instance of a term t, if it holds $s = \sigma(t)$. By a computation (related to a substitution σ applied to a term t) we mean a sequence of variable replacement steps, where each variable replacement rule applied in computation is in σ and the result of the computation is a σ-instance of a term t.

The unifier of the two given terms s and t is a substitution σ such that $\sigma(s) = \sigma(t)$ holds. The goal of the unification problem is to decide whether there is a unifier of a given pair of input terms. The semi-unifier of the two given terms s and t is a substitution σ for which there is a substitution ρ such that a term $\sigma(t)$ is a ρ-instance of a term $\sigma(s)$. The goal of the uniform semi-unification problem is to decide whether there is a semi-unifier of a given pair of input terms.

Unification algorithms seldom use the string form to represent terms. In efficient unification algorithms the terms can be implemented as directed acyclic graphs (term-dags) and a variable replacement rule can be implemented as a simple local operation on vertices and edges of those graphs.

A term can be represented in the form of a term-dag. The term-dag is a directed acyclic graph with vertices labelled by symbols from the set $V \cup F$. The term-dags are multigraphs (they can have multiple edges, i.e. there can be more than one edge connecting a given pair of vertices) and the outgoing edges of each vertex of a term-dag are ordered. The label of an arbitrary vertex v is denoted as label(v). The out-degree of a vertex v with $label(v) \in V$ is 0. These vertices are further called "variable vertices". The variable vertices are shared, i.e. for each variable occurring in input terms there is exactly one vertex with this variable symbol in the corresponding term-dag. The out-degree of a vertex labelled by a symbol $f \in F$ is exactly arity(f). These vertices are further called "functional vertices". To reference the k-th successor of a vertex v in the term-dag G we use the notation successor(G,v,k).

One can easily translate a term in the string form into a term-dag and vice versa. The case from a string into a dag is trivial. Hence, we only mention here the opposite case. To each vertex v of a term-dag G one can associate a term of the form $term(G, v)$, recursively defined as follows:

if v is a variable vertex
then term(G,v) = label(v)
else (* let f = label(v) ; n = arity(f);
v_k = successor(G,v,k) for k=1,2,...,n *)
$term(G, v) \;=\; f(term(G, v_1), ..., term(G, v_n))$.

The variable replacement rule $x \mapsto t$ can be implemented on the term-dag G by means of the operation $replace(G, v, v')$, where v,v' are two distinct vertices in the term-dag G and x=label(v) , t=term(G,v'). The result of the operation $replace(G, v, v')$, in the case that there is no path from v' to v in G, is the graph G' satisfying the following properties:

1. G' has the same vertices as G with the same labels;
2. G' has the same edges as G except for those edges in G which lead to the vertex v ; these edges are replaced in G' by edges going to the vertex v' (the order of the outgoing edges is of course preserved).

The correctness of the variable replacement rule implementation on term-dags can be obtained from the following properties:

Lemma 1. *Let G be a term-dag containing two distinct vertices v,v' and there is no path from v' to v in G.*

1. *Then G' is the term-dag.*
2. *If v is a variable vertex and σ is a variable replacement in a form $label(v) \mapsto term(G, v')$, then for each vertex w it holds*

$$term(replace(G, v, v'), w) = \textbf{if } w = v \textbf{ then } label(v) \textbf{ else } \sigma(term(G, w))$$

Fact (2) in Lemma 1 states that, if there is no path in G from v' to v, the operation $replace(G, v, v')$ simulates the variable replacement (substitution) $label(v) \mapsto term(G, v')$. In the term-dag G' the vertex v is isolated. This property corresponds to the fact that the variable $label(v)$ does not occur in the term after the variable replacement.

3. A simple algorithm for the uniform semi-unification

The main result of this section is to establish the formal certification of the observation that the uniform semi-unification on terms can be expressed as a unification on τ-terms.

In the uniform semi-unification problem one can deal syntactically with expressions called τ-terms. Let τ be a special symbol, $\tau \notin V \cup F$. A set of τ-terms (denoted as τ-T(V,F)) is recursively defined in the following way:

1. if $x \in V, i \in N$, then $\tau^i x \in \tau - T(V,F)$;
2. if $c \in F_o, i \in N$, then $\tau^i c \in \tau - T(V,F)$;
3. if $f \in F, arity(f) = n$ for $n \geq 1, t_1, ..., t_n \in \tau - T(V,F), i \in N$, then $\tau^i f(t_1, ..., t_n) \in \tau - T(V,F)$.

By τ-variable we mean a τ-term in the form $\tau^i x$ for $x \in V, i \in N$. The τ-substitution σ can be viewed as a finite set of τ-variable replacements $[\tau^{c_i} x_i \mapsto t_i]$ for $x_i \in V, c_i \in N, t_i \in \tau - T(V,F), 1 \leq i \leq n$. A skeleton of τ-term t is a term t' obtained from the τ-term t by omitting all occurrences of the symbol τ in it. The skeleton of τ-term $\tau f(h(\tau x), y)$ is the term $f(h(x), y)$. A τ-term is in a simple form when either it does not contain the symbol τ or it contains symbols τ only in subterms of the form $\tau^i x$ for $x \in V, i \in N$.

On the set of all τ-terms τ-T(V,F) we define an equivalence relation $\simeq$ (to express the movement of symbol τ) which fulfils the following conditions:

1. if $x \in V, i \in N$, then $\tau^i x \simeq \tau^i x$;
2. if $c \in F_o, i, j \in N$, then $\tau^i c \simeq \tau^j c$;
3. if $f(t_1, ..., t_n) \in \tau$-T(V,F),i,j $\in N$, then $\tau^{i+j} f(t_1, ..., t_n) \simeq \tau^i f(\tau^j t_1, ..., \tau^j t_n)$;
4. if $f \in F, t_1, ..., t_n, s_1, ..., s_n \in \tau - T(V,F)$ and $t_1 \simeq s_1, ..., t_n \simeq s_n$, then $f(t_1, ..., t_n) \simeq f(s_1, ..., s_n)$.

A class of equivalent τ-terms according to $\simeq$ can be uniquely represented by the τ-term representative in a simple form. For example $\tau f(h(\tau y), x)$ is equivalent to $f(\tau h(\tau y), \tau x)$ and to $f(h(\tau^2 y), \tau x)$ and only the last τ-term is in a simple form.

We shall need a relation $\succ$ between two terms from τ-T(V,F) in a special case when at least one of τ-terms is a τ-variable (i.e. it is in the form $\tau^i x$ for $x \in V, i \in N$) as follows:

a. a total ordering $\succ$ is defined on the set V;
b. if t contains a symbol from F, $x \in V, i \in N$, then $\tau^i x \succ t$;
c. if $x, y \in V, i, j \in N$, then $\tau^i x \succ \tau^j y$ for $x \succ y$ or $x = y$ and $i \geq j$.

We present an algorithm which transforms an input set of τ-term equations into an equivalent set of equations in solved form in a step-by-step manner. An input set of τ-term equations is $\{\tau s = t\}$, where $s, t \in T(V,F)$. A set of τ-term equations is said to be in solved form iff the following three conditions are fulfilled:

a. all equations are of the form $\tau^i x = u$ for $x \in V, i \in N, u \in \tau - T(V,F)$, $\tau^i x \succ u$;
b. all variables x in left-hand sides of equations are pairwise disjoint;
c. there do not exist $\tau^i x$ and $f(t_1, ..., t_n)$ such that $f(t_1, ..., t_n)[\tau^i x]$, where $\tau^i x$ is a left-hand side and $f(t_1, ..., t_n)$ is a right-hand side of an equation.

Later on we show how the step-by-step transformation of input set of equations $\{\tau s = t\}$ simulates a sequence of computation steps of semi-unification for input terms s,t.

A transformation is denoted as

$$E \equiv \{s = t\} \cup A \;\Rightarrow\; E := \{...\} \cup A \quad \textbf{if}\ cond$$

which means that the equation s=t is a candidate for transformation in the case when the condition cond is fulfilled. In the case the equation s=t is selected, it is replaced by equations denoted as {...}.

Five basic transformations are the following:

(SU1) Exchange of sides in equation:
$E \equiv \{u = \tau^i x\} \cup A \;\Rightarrow\; E := \{\tau^i x = u\} \cup A$
$if\ x \in V, i \in N, \tau^i x \succ u$

(SU2) Elimination of equation:
$E \equiv \{u = u\} \cup A \;\Rightarrow\; E := A$

(SU3) Decomposition of equation (Elimination of function symbol):
$E \equiv \{f(t_1, ..., t_n) = f(s_1, ..., s_n)\} \cup A \Rightarrow E := \{t_1 = s_1, ..., t_n = s_n\} \cup A$

(SU4) τ-variable replacement:
$E \equiv \{\tau^i x = u\} \cup A \;\Rightarrow\; E := \{\tau^i x = u\} \cup A_{\{\tau^i x \mapsto u\}}$
$if\ x \in V, u \neq \tau^i x, \tau^i x \succ u, \tau^i x\ occurs\ in\ A$

(SU5) Movement of symbol τ:
$E \equiv \{u = t[\tau f(t_1, ..., t_n)]\} \cup A \;\Rightarrow\; E := \{u = t[f(\tau t_1, ..., \tau t_n]\} \cup A$
$E \equiv \{u = t[\tau c]\} \cup A \;\Rightarrow\; E := \{u = t[c]\} \cup A$
$E \equiv \{t[\tau f(t_1, ..., t_n)] = u\} \cup A \;\Rightarrow\; E := \{t[f(\tau t_1, ..., \tau t_n] = u\} \cup A$
$E \equiv \{t[\tau c] = u\} \cup A \;\Rightarrow\; E := \{t[c] = u\} \cup A$

Two inconsistency conditions are the following:

(IC1) Non-homogeneous terms:

$$E \equiv \{f(t_1, ..., t_n) = g(s_1, ..., s_n)\} \cup A \quad and \quad f \neq g$$

(IC2) Occur-check:

$$E \equiv \{\tau^i x = f(t_1, ..., t_n)\} \cup A \quad and\ f(t_1, ..., t_n)\ contains\ a\ subterm\ \tau^i x$$

An algorithm SSUNIF (for symbolic semi-unification) is presented such that it transforms an equation $\{\tau s = t\}$ into a set of equations in solved form by means of basic transformations SU1 - SU5.

algorithm $SSUNIF(s,t)$
Input : $An\ equation\ E \equiv \{\tau s = t\};$
Output : $A\ set\ of\ equations\ E\ in\ solved\ form,\ if\ s,t\ are\ semi-unifiable;$
$otherwise\ failure$
begin
repeat
select any equation (in the form $u = v$) from the set E ;
if *for the equation $u = v$ inconsistency condition $IC1$ or $IC2$ holds*
then *stop with failure*
else *apply basic transformation $SU1 - SU5$ to the equation $u = v$*
until *basic transformations are not applicable to any equation in E*
end

The fundamental assertions about the algorithm SSUNIF, concerning termination, correctness and solution properties, are formulated in the next theorem.

Theorem 1. *Algorithm SSUNIF has the following properties:*

1. *the algorithm terminates, no matter which choices are made;*
2. *if the algorithm stops with failure, then input terms are not semi-unifiable;*
3. *if the algorithm terminates with success, then the set of input equations has been transformed into an equivalent set of equations in solved form;*
4. *a set of equations in solved form represents a substitution σ which is*
   a. *an idempotent substitution;*
   b. *a most general semi-unifier of two input terms.*

The time complexity of SSUNIF is exponential in the size of the input equation, because even the number of basic transformation applications performed by SSUNIF can be exponential in the worst case.

SSUNIF is the nondeterministic algorithm because it selects any equation from the set E in a nondeterministic way. If the set of equations E is organized following LIFO (last-in-first-out) discipline and if in transformation SU3 "new" equations $t_1 = s_1, ..., t_n = s_n$ are added into the set E in the order starting with equation $t_n = s_n$ and ending with equation $t_1 = s_1$, then a step "select any equation from the set E" in the algorithm SSUNIF is performed by choosing the last added equation into E. In this way a nondeterministic selection step can be replaced by a deterministic one.

The deterministic version of the algorithm SSUNIF for the input $\{\tau s = t\}$ can be interpreted as a deterministic algorithm performing on input τ-terms τs,t (or equivalently on their tree representations). The only transformation which changes τ-terms is SU4 (so called τ-variable replacement transformation). It follows directly that deterministic version of the algorithm SSUNIF searches through term-trees corresponding to input τ-terms τs,t using DFS (depth-first search) strategy and performs variable replacements according to the transformation SU4. In Section 5 we show in detail how such an algorithm can be implemented on compact graph structure.

4. τ-terms, τ-substitutions and their graph representation

In this section we show how semi-unification viewed as unification on τ-terms can be transformed from the string representation to a more compact but equivalent graph representation analogically as terms have been represented by graphs in Section 2.

In efficient uniform semi-unification algorithms τ-terms are represented as τ-term-digraphs. The τ-term-digraph is a directed graph with vertices labelled by symbols from the set $V \cup F$ or by pairs of symbols from $V \times F$ and with edges labelled by whole numbers. For technical reasons we add an "input" edge (leading to a root of the digraph but "not outgoing" from any vertex of the digraph). τ-term-digraphs are multigraphs in which the outgoing edges are ordered. In τ-term-digraphs all paths from the root with the property that each prefix of the path is of non-negative cost (where the cost of a path is defined as the sum of labels of all edges in the path) are of finite length. A simple τ-term-digraph is a τ-term-digraph in which only variable vertices are shared.

There is a direct relationship between τ-terms and τ-term-digraphs in the following sense. For each τ-term there is a simple τ-term-digraph representing this τ-term and reversely to each τ-term-digraph a unique τ-term is associated in a simple form.

1. τ-term-digraph associated to τ-term
   a. construct a skeletal of τ-term;
   b. for a skeletal of τ-term construct a corresponding simple term-dag;
   c. if there is a subterm $\tau^i f(t_1, ..., t_n)$ of τ-term, then the edge, leading to the vertex of term-dag representing the skeletal of subterm $f(t_1, ..., t_n)$, has the label i.

2. τ-term associated to τ-term-digraph

   We show how to each vertex v of a τ-term-digraph G a τ-term term(G,v) in a simple form is associated. Let the cost of the current path from the root of G to vertex v be $i \geq 0$.

```
if  v is a variable vertex
then  term(G,v) = τ^i label(v)
else if  label(v) is a constant
    then  term(G,v) = label(v)
    else  {* f = label(v), n = arity(f), v_k = successor(G,v,k),
           c_k = label(v,v_k)  for k = 1,2,...,n *}
          if  i + c_1 ≥ 0,...,i + c_n ≥ 0
          then  term(G,v) = f(term(G,v_1),...,term(G,v_n))
          else
             if  i + c_1 < 0,...,i + c_n < 0
             then  term(G,v) = τ^i label(v)
             else  does not exist a corresponding τ − term
             fi
          fi
    fi
fi
```

A substitution $\tau^i x \mapsto \tau^j t$ is implemented on τ-term-digraph G by means of the operation replace(G,v,i,v',j), where v,v' are two distinct vertices in the τ-term-digraph G and x=label(v) for $x \in V$ and t=term(G,v'). Moreover, it is assumed that t is in τ-reduced form, i.e. there is no term s such that it holds $t \simeq \tau s$. The result of the operation replace(G,v,i,v',j), in the case that there is no path of non-negative cost from v' to v in G, is the graph G' satisfying the following properties:

1. G' has the same vertices as G with the same labels;
2. G' has the same edges as G except for those edges in G which lead to the vertex v; these edges are replaced in G' by edges going to the vertex v' (the order of the outgoing edges is of course preserved);
3. G' has the same edge-labels as G except for the following differences:
   a. in case $i \geq j$ labels of all edges leading to v' in G are increased by i-j and labels of all edges outgoing from v' in G are decreased by i-j ;
   b. in case $i < j$ labels of all edges leading to v in G are increased by j-i.

The correctness of the implementation of a substitution on τ-term-digraphs can be obtained from the following properties:

Lemma 2. *Let G be a τ-term-digraph, v and v' be two distinct vertices with the path from the root of G to v and v' of costs i and j, respectively, and in the case $i \geq j$ there is no path of cost at least i-j from v' to v in G.*

1. *Then G' is the τ-term-digraph.*
2. *If v is a variable vertex and σ is a τ-variable replacement in the form $x \mapsto \tau^{j-i} term(G, v')$ in case $i < j$ and in the form $\tau^{i-j} x \mapsto term(G, v')$ in case $i \geq j$, then for each vertex w with path of cost k from the root of G to w it holds:*
   a. $\tau^k term(replace(G, v, i, v', j), w) = \sigma(\tau^k term(G, w))$ *for* $w \neq v', i \geq j$ *or* $w \neq v\ ,\ i < j$;
   b. $\tau^{k+i-j} term(replace(G, v, i, v', j), v') = \sigma(\tau^k term(G, v'))$ *for* $i \geq j$;
   c. $\tau^{k+j-i} term(replace(G, v, i, v', j), v') = \sigma(\tau^k term(G, v))$ *for* $i < j$.

Using Lemma 2 it follows by induction that for an arbitrary τ-term-digraph G and an arbitrary τ-substitution σ it holds:

$$term(G', root(G')) = \sigma(term(G, root(G)))$$

where G' is obtained from G by applying replace operations corresponding to the τ-substitution σ.

5. An efficient algorithm for the uniform semi-unification

We present a recursive decision algorithm SEMI-UNIF.

Recursive algorithm $SEMI-UNIF(u,i,v,j) : bool$

Input : A pair of distinct vertices u, v in the τ − term − digraph G such that $\tau^i term(u)$ and $\tau^j term(v)$ are τ − terms to be semi − unified.

Output : Bool such that $bool = true$ iff two given τ − terms $\tau^i term(u)$ and $\tau^j term(v)$ are semi − unifiable, otherwise $bool = false$.

Method :

```
begin
  if one of the input vertices, say u, is a variable vertex
  then if occur(u, i, v, j)
       then bool := false
       else replace(u, i, v, j);
            bool := true
       fi
  else if label(u) ≠ label(v)
       then bool := false
       else {* n is the number of successors of u and v *}
          bool := true;
          for k := 1 to n
          do w1 := successor(u, k); {* cost of the edge (u, w1) is c1 *}
             w2 := successor(v, k); {* cost of the edge (v, w2) is c2 *}
             if w1 ≠ w2
             then if i + c1 ≥ 0 and j + c2 ≥ 0
                  then bool := SEMI − UNIF(w1, i + c1, w2, j + c2)
                  else bool := false
                  fi
             else case
                    w1, w2 are identical variable vertices : bool := true
                      {* w1 "remembers" value gcd(i + c1, j + c2, m)
                      for j + c2 > 0 or value 0 for j + c2 = 0, where m is
                      previously "remembered" value of w1, w2 *};
                    w1, w2 are identical constant vertices : bool := true;
                    w1, w2 are identical functional vertices : bool := true
                      for i + c1 = j + c2, otherwise bool := false
                  endcase
             fi
          od;
          if bool
          then equal − cost(u, v, n);
               replace(u, 0, v, 0)
          fi
       fi
  fi
end
```

SEMI-UNIF(u,i,v,j) is computing on a τ-term-digraph G, where u,v are two vertices of G and i,j are costs of paths from roots of G to u,v, respectively. Two τ-terms $\tau s, t$ given for semi-unification are at first represented as a simple τ-term-dag with roots r_1, r_2, respectively. Distinct input terms s,t are semi-unifiable iff it holds SEMI-UNIF(r_1,1,r_2,0)=true.

The goal of $occur(u, i, v, j)$ is to determine whether in case $i \geq j$ there is a path from v to u of the cost greater than or equal to $i - j$. In a positive answer it returns the value $true$, otherwise the value $false$. The boolean procedure $occur$ can use the marking strategy to avoid cycling. In the marking strategy a path from the starting vertex v is marked as visited. A cycle is detected when a new vertex in the search through a subgraph with the root v has already been visited. It is sufficient to go through each cycle only once in order to find out whether it contains a vertex u or not. The cost of a path from the root v is continuously updated.

The test whether an input vertex u is a variable one is positive iff either there are no outgoing edges from the vertex u or for all outgoing edges the cost of the path through them falls down into negative value. The case when an input vertex v is a variable vertex and an input vertex u is not a variable vertex is solved analogically as the reverse case. In the case when both u, v are variable vertices, it holds $\tau^i term(u) \succ \tau^j term(v)$.

The procedure $equal - cost(u, v, n)$ makes the costs of all n appropriate edges outgoing from u, v equal. When corresponding successors w_1, w_2 are identical variable vertices, with "remembered" value m, then procedure equal-cost regards the costs of path into w_1, w_2, respectively, as equal according to modulo m. This case can be illustrated on terms $h(f(x, y), z, z)$ and $h(z, f(y, y), f(x, y))$.

The main properties of the SEMI-UNIF algorithm, i.e. correctness and asymptotical time complexity, are summarized in two next results.

Theorem 2. *The algorithm SEMI-UNIF is correct in the sense of the described input-output relation.*

We sketch an idea of the proof. It is based on the identical behaviour of the algorithms SEMI-UNIF and deterministic version of SSUNIF on the input τ-terms $\tau s, t$. The correctness of the replace operation in SEMI-UNIF algorithm follows directly from Lemma 2 in case of distinct vertices. Each replace operation in SEMI-UNIF algorithm (representing a τ-variable replacement rule) is simulated in deterministic version of SSUNIF by repeated application of this "τ-variable replacement" transformation (until it is applicable). Each successor operation from the functional vertex in SEMI-UNIF algorithm is simulated in the deterministic version of SSUNIF by the "decomposition of equation" transformation. The correctness of SEMI-UNIF algorithm then follows from the step-by-step simulation of SEMI-UNIF by the deterministic version of SSUNIF and from the correctness of SSUNIF algorithm.

Theorem 3. *The worst case time complexity of the SEMI-UNIF algorithm is $O(n^2)$, where n is the sum of the input terms sizes.*

Proof :
Let n be a total size of the input terms. During semi-unification no new vertices are created (the number of vertices is upper bounded by n) and the number of edges is not increased (the upper bound is $n-2$). Each call of the replace operation "isolates" one variable vertex or functional vertex in the τ-term-digraph, that means that replace is called at most n times. The number of successor operations performed by the SEMI-UNIF algorithm is upper bounded by $O(n^2)$. This follows from the fact that there are at most n vertices in the τ-term-digraph in any stage of semi-unification process and that the length of the cycle cannot be longer as n. Hence, for an arbitrary semi-unification path from the root of the τ-term-digraph of length at least $2n$ one can compute the effect of replace operation in these cycles. As for each vertex of the τ-term-digraph one can prolong the path of the length at most O(n) to compute the replace operation or to come to identical vertices, there are needed at most $O(n^2)$ successor operations to be performed by the SEMI-UNIF algorithm. The number of occur operations is not greater than the number of replace operations, that means that occur is called at most n times. In an occur operation at most n vertices have to be visited in the τ-term-digraph with O(n) edges and thus again at most $O(n^2)$ edges have to be passed to perform occur operations in SEMI-UNIF algorithm. Since the sequence of O(n) occur and replace operations and $O(n^2)$ successor operations can be realized in the time $O(n^2)$, the worst case time complexity of the SEMI-UNIF algorithm is $O(n^2)$ when an arithmetical operation performed in replace and successor operations can be realized in a unit of time. This completes the proof of Theorem 3.

6. References

1. Dershowitz,N.: Termination of Rewriting. Journal of Symbolic Computation 3, No 1,2, 1987, pp. 69–116
2. Henglein,F.: Polymorphic Type Inference and Semi-Unification. Technical Report 443, Department of Computer Science, Courant Institute of Mathematical Sciences, New York University, 1989, p. 109
3. Kapur,D.-Musser,D.-Narendran,P.-Stillman,J.: Semi-unification. In: Proceedings of the 8th Conference on Foundations of Software Technology and Theoretical Computer Science, LNCS 338, Springer-Verlag, 1988, pp. 435–454
4. Kfoury,A.J.-Tiuryn,J.-Urzyczyn,P.: The undecidability of the semi- unification. In: Proceedings of the 22nd ACM Symposium on Theory of Computing, 1990, pp. 468–476
5. Lankford,D.S.-Musser,D.R.: A finite termination criterion. Unpublished draft. USC Information Sciences Institute, Marina Del Rey, California, 1978
6. Leiss,H.: Semi-unification and type inference for polymorphic recursion. Bericht INF2-ASE-5-89, Siemens AG, München, 1989, p. 46
7. Pudlák,P.: On a unification problem related to Kreisel's conjecture. Commentationes Mathematicae Universitatis Caroline 29, No 3, 1988, pp.551–556

Different Modifications of Pointer Machines and their Computational Power

Extended Abstract

KONSTANTIN V.SHVACHKO

6, Polevaya Str., Apt. 52
Pereslavl-Zalessky, 152140,
USSR

Abstract. *Kolmogorov - Uspensky machines* (KUM) and *storage modification machines* (SMM) are the variants of machines with graph storage, called here *pointer machines* (PM). It is an open problem whether SMMs are more powerful than KUMs. An approach to solution of this problem is suggested here. It is shown that with some restrictions weak variants of PMs, called tree pointer machines, and Turing machines with tree memory can not recognize in real time some language real time recognizible by SMMs.

INTRODUCTION. A.N.Kolmogorov and V.A.Uspensky formulated in [K53] and [KU58] a general "approach to the definition of an algorithm". From this Kolmogorov - Uspensky algorithms became one of the main computational models that is considered in the complexity theory. Actually it is a class of computational models in sence of [US87], where these models were called *the Kolmogorov type models*, i.e. models with local transformation of an information. It is very wide class. Turing machines (TM) and Post machines belong to it, but not random access machines (RAM) and Markov's normal algorithms.

In this work we consider a subclass of Kolmogorov type models, which we call *Pointer machines* (PM). The main common thing of these models is their graph storage. V.A.Uspensky and A.L.Semenov in [US87] gave the most general definition for the representative model of this class called Kolmogorov - Uspensky machines. In this paper we mean by Kolmogorov - Uspensky machines the model, which is described in [KU58]. This model was the first example of PM appeared in literature. The distinctive feature of Kolmogorov - Uspensky machines is that they are equipped with storage device organized as an undirected graph. We follow [US87] giving an informal description of PMs.

Fix a finite set of "directions". Directions correspond *uniquely* to edges of PM's storage graph in such a way that the edges coming out from any one node have different directions. Storage graph has one distinguished node called the center. Other nodes are attainable from the center through paths defined uniquely by sequences of directions. In a computational process machine transforms its graph storage (memory configuration) according to some rules. Transformations depend on the input and memory configuration. Namely, at each step machine separates an r-neighbourhood of the center and changes it to some new graph. A "sticking function" is defined to determine how to connect nodes out of the removed r-neighbourhood with nodes of new graph.

Storage graph of *Kolmogorov - Uspensky machines* (KUM) introduced in [KU58] is a non-directed graph, called Kolmogorov Б-complex. Directions in it are determined by labels of neighbouring nodes. Another example of pointer machines called *storage modification machines* (SMM) was introduced by A.Shonhage in [S70], [S80]. Memory configurations of SMMs are directed graphs, called Δ-structures. Directions in Δ-structures are defined by labeling edges. KUM and SMM are the basic variants of PMs. One can define also other classes of PMs by fixing any class of graphs for machines' storage configurations. It will be considered here for example *tree pointer machines* (TPM) with storage graphs being directed trees.

Natural problem is to compare computational powers of models as within the class of pointer machines, so as with respect to other known computational models. It is easy to see that SMMs are not weaker than KUMs, i.e. KUMs can be real time simulated by SMMs. (About real time simulations see in [S80].) The reverse problem is hard and open for today ([S80],[US87],[Gu88]). The difficulty here is that Δ-structures of SMMs can have unbounded in-degree, while for Kolmogorov Б-complexes of KUMs this is impossible.

With respect to other models computational power of PMs is characterized by following three results. D.Yu.Grigoriev in [G76] proved that Turing machines (even with multidimensional tapes) are weaker than PMs. On the other hand A.Shonhage ([S73], [S80]) showed that SMMs can simulate TMs in real time. It is shown also in [S80] that SMMs are equivalent to the most weak modification of random access machines - successor RAM.

In this paper I tried to approach the solution of the problem "KUM vs. SMM". It is shown here that with some restrictions week variants of PMs, which we call TPMs, can not recognize in real time some language, which is real time recognizible on SMMs. I hope that this language and our method would be used for separating the computational powers of KUMs and SMMs.

The rest of the text is organized as follows. In Section 1 we describe three computational models that are examined in the paper. They are two modifications of TPM – WTPM and STPM – and Turing machines with tree memory (TTM). As a matter of fact the result of Grigoriev is fulfilled for TTMs, i.e. TMs are weaker than TTMs. But it is not known whether TTMs are weaker than Kolmogorov - Uspensky machines or even SMMs. In Section 2 we introduce the language L. It is shown here that there exists SMM recognizing L in real time. We present in the paper a new technique based on the Main (Section 3) and the Combinatorial (Section 5) lemmas for proving three identical theorems: language L cannot be recognized in real time in each of the three computational models with some additional restriction (Section 4) on working steps of a machine recognizing L. This restriction is that after reading a certain part of an input string a machine begins to work in a special (simple) way. In Section 6 the theorem is proved for WTPM, in Section 8 for STPM, and in Section 9 for TTM.

1. COMPUTATIONAL MODELS. We consider two variants of tree pointer machines. Storage configurations in both models are trees. Each node of a tree has no more than k outgoing edges labeled by natural numbers from $\{0, 1, \dots, k-1\}$ in such a way, that all edges outgoing from one node have different labels.

In the first model nodes of a tree are not marked. These trees are called k-trees, and corresponding (weak) computational model is denoted by WTPM. In the second model nodes of a tree are labeled by letters from a finite alphabet Б. These trees are called (Б, k)-trees and corresponding (strong) computational model is denoted by STPM. The computational process of WTPM and STPM is defined in a usual for Kolmogorov - Uspensky machines way.

We also will consider Turing machines, that are equipped with an ordinary one-way input tape and a storage device ("tape"), that is a tree. Each node of this tree is a storage cell. It can contain a symbol from a finite alphabet Σ. Everything else for these machines, denoted below by TTM, is defined similar to usual Turing machines.

2. STATEMENT OF THE PROBLEM. Let us consider the language:

$$L = \bigcup_n \{b_{0^n}@\ldots@b_{1^n}\#x\#y\# \mid b_i \in \{0,1\}^{[n/2]}, x, y \in \{0,1\}^n, b_x = b_y\}$$

a

Sequence $B = b_{0^n}@\ldots@b_{1^n}$ is called *base segment*. It is indexed by the set of words $\{0,1\}^n$ with lexicogr fic order on it. Sequence B determines a partition of the set $\{0,1\}^n$ into $2^{[n/2]}$ equivalence classes so that x and y are equivalent (denoted by $x \overset{B}{\sim} y$) iff $b_x = b_y$. For a given base segment B and representatives x and y of two classes the string $B\#x\#y\# \in L$ iff $x \overset{B}{\sim} y$.

Theorem 1. There exists SMM that recognizes L in real time (RT).

Proof. While reading B the SMM should build two full binary trees T_1 and T_2 of heights n and $[n/2]$, resp. Leaves of T_1 represent strings of length n, and leaves of T_2 represent different codes b_i of equivalence classes. The equivalence relation is represented by edges directed from leaves of T_1 to leaves of T_2. So, if a leaf t_1 of T_1 is connected with a leaf t_2 of T_2, then the word, that corresponds to t_1, belongs to the class represented by t_2. It is easy to build this graph in RT. Obviously that the equivalence of x and y can be established by this graph in RT. □

Open problem. Prove that KUMs can not recognize L in RT.

This problem will be solved for TPMs with some additional restriction on the behaviour of a machine recognizing L. An important note is that SMMs with the same additional restriction can nevertheless recognize L in RT.

3. THE MAIN LEMMA. Let pointer machine M recognize L. $S_M(Bw)$ denotes the storage configuration (graph) of machine M, that it constructs after reading part Bw of the input word, for $|w| \leqslant 2n$ (symbols $\#$ are not counted in the length – denoted by $|\cdot|$).

Definition.

$$\xi_B(x\#y\#) = \begin{cases} 1, & \text{if } x \overset{B}{\sim} y \\ 0, & \text{otherwise} \end{cases}$$

Let us extend the domain of ξ_B to shorter sequences. If ξ_B is defined for all words of length m ($m \leqslant 2n$), then for a word w of length $|w| = m-1$ let $\xi_B(w) = \xi_B(w0)\xi_B(w1)$. So $\xi_B(w)$ is a sequence that charaterizes the restriction of equivalence relation, defined by B, on the set of words from $\{0,1\}^n$ with common prefix w.

Lemma (Main). Let pointer machine M recognize L. B is a base segment, w_1, w_2 are words, $|w_1|, |w_2| \leqslant 2n$ and $S_M(Bw_1) = S_M(Bw_2)$. **Then**

1) If w_1 and w_2 have acceptable extensions, i.e. each sequence $\xi_B(w_1)$ and $\xi_B(w_2)$ contains at least one "1", then $|w_1| = |w_2|$,

2) $\xi_B(w_1) = \xi_B(w_2)$.

Proof. (1) Let $|w_1| < |w_2|$ and w_1u_1, w_2u_2 be extensions such that $\xi_B(w_1u_1) = \xi_B(w_2u_2) = 1$. Since memory configurations after reading inputs Bw_1 and Bw_2 are equal: $S_M(Bw_1) = S_M(Bw_2)$, so the result of M on the input Bw_1u_2 will coincide with that on Bw_2u_2, i.e. "1". And M accepts the word Bw_1u_2 that is not in L (because its length is less than $2^n[n/2] + 2n$).

(2) Starting from the same state $S_M(Bw_1) = S_M(Bw_2)$ and getting equal input sequences machine will produce equal output values. □

4. COMPUTATIONAL RESTRICTIONS.

Definition. Let storage configuration of M at a moment t be a graph $S_M(t) =< V(t), E(t) >$, where $V(t)$ and $E(t)$ denote sets of vertices and edges, resp. A step t of M is said to be *not essential* if

$$V(t+1) \subseteq V(t) \ \& \ E(t+1) \subseteq E(t),$$

otherwise step t is called *essential*. A step t of M is said to be a *move* if it changes a location of the center of the graph.

So a step t is not essential if machine does not construct new vertices or edges on it. Non-essential step is a move if inclusions above are proper. We restrict the behaviour of TPM recognizing L in the following way.

Restriction. Machine must not make essential steps while reading last part $x\#y\#$ of any input.

Let us denote this interval by I. Informally the Restriction means that WTPM can change its memory configurations during I only by moves, and STPM can move and change labels of vertices.

It is easy to see that after a little modification of the algorithm from theorem 1 it can be implemented for constructing the SMM, recognizing L in RT, that makes a unique essential step at time interval I. It will be shown below that WTPMs cannot recognize L in RT, when the Restriction is fulfilled.

5. THE COMBINATORIAL LEMMA.

Lemma (Combinatorial). There exists a partition B of the set of words $\{0,1\}^n$ into $2^{[n/2]}$ equivalence classes such that

1) $\forall w_1w_2(|w_i| < n \ \& \ w_1 \neq w_2 \Rightarrow \xi_B(w_1) \neq \xi_B(w_2))$,

2) $\forall w(|w| = n \Rightarrow \forall u_1u_2(|u_i| \leqslant n/2 \ \& \ u_1 \neq u_2 \Rightarrow \xi_B(wu_1) \neq \xi_B(wu_2)))$.

Proof. We use counting arguments to prove the existence of a partition B that satisfies the first lemma's propositions. For proving proposition 2 we consider some one-person "chess" game and turn our problem into it. The details of the proof will appear in the full paper. □

Note. The combinatorial lemma (CL) states the existence of a partition B such that, firstly, for all words w of lengths less than n characteristic sequences $\xi_B(w)$ are different and, secondly, for all words u of lengths $n \leqslant |u| \leqslant 3n/2$ with common prefix w ($|w| = n$) their characteristic sequences $\xi_B(u)$ are different, i.e. sequences $\xi_B(u_1)$ and $\xi_B(u_2)$ can be equal only if u_1 and u_2 does not have a common prefix of length n.

6. THE MAIN RESULT FOR WTPM.

Theorem 2. WTPM that recognizes L and satisfies the Restriction cannot work in RT.

Consider WTPM M that recognizes L and satisfies the Restriction. Suppose also that M works on I in RT.

Definition. $\gamma_B(w)$ denotes a subgraph of the graph $S_M(B)$, which M examines while reading part w of an input $x\#y\#$. For a WTPM that makes only moves $\gamma_B(w)$ is some path in $S_M(B)$. Let $E_B(w)$ be the end-node of this path. $E_B(t)$ denotes a section of $S_M(B)$ at level t, which is a set of end-nodes $E_B(w)$ for all w of length t.

Corollary of the main lemma.

$\gamma_B(w_1) = \gamma_B(w_2) \Rightarrow \xi_B(w_1) = \xi_B(w_2)$

$E_B(w_1) = E_B(w_2) \Rightarrow \xi_B(w_1) = \xi_B(w_2)$

Corollary 1 of the combinatorial lemma.

$$\exists B(\forall t < n)(Card(E_B(t)) \geqslant 2^t \ \& \ \forall w_1 w_2(|w_i| = t \Rightarrow E_B(w_1) \neq E_B(w_2)))$$

Proof. Consider a partition B that satisfies CL(1). We will prove by induction on t the right part of conjunction, which implies the left one. For $t = 1$ from CrML we have $\xi_B(0) \neq \xi_B(1) \Rightarrow E_B(0) \not\leftarrow E_B\ 1)$. Assume the proposition is fulfilled for $t-1$. Remember that storage structure of WTPM is a tree. That is why it is necessary to prove only for each w of length $t-1$ that for its extensions $w0$ and $w1$ endnodes $E_B(w0)$, $E_B(w1)$ are different. It appears again from CrML. □

Corollary 2 of the combinatorial lemma.

$$\exists B(\forall t < 3n/2)(Card(E_B(t)) \geqslant 2^t \ \& \ \forall w_1 w_2(|w_i| = t \Rightarrow E_B(w_1) \neq E_B(w_2)))$$

Proof. Consider a partition B that satisfies CL(1,2). For each w of length $n-1$ consider a section $E_{Bw}(s)$ of the graph $S_M(Bw)$ at its level $s \leqslant n/2$, i.e. $E_{Bw}(s) = \{E_B(wu) \mid |u| = s\}$. By proposition CL(2) and CrML and reasoning as in Cr1CL we obtain $Card(E_{Bw}(s)) \geqslant 2^s$. By Cr1CL for each pair of words $w_1 \neq w_2$ of lengths $n-1$ we have $E_B(w_1) \neq E_B(w_2)$. According to the storage's tree structure and the Restriction we conclude that $E_{Bw_1}(s) \bigcap E_{Bw_2}(s) = \varnothing$ for any $s < n/2$. □

Proof of theorem 2. We have shown that the number of vertices in graph $S_M(B)$, constructed by M after reading partition's code B of length $|B| = 2^n[n/2]$, is at least $2^{3n/2}$. Surely such graph cannot be built in time $O(n2^n)$. So WTPM M does not work in RT. □

7. DISCUSSION. The corollary 1 of CL is fulfilled for an arbitrary SMM that does not make essential steps on time interval I. The corollary 2 of CL does not hold already for KUM with the Restriction mentioned above. This follows from the fact that sequences $\xi_B(w)$ for many w of length n are equal and SMMs and KUMs have an opportunity to "stick together" subgraphs $S_M(Bw_1)$ and $S_M(Bw_2)$ if $\xi_B(w_1) = \xi_B(w_2)$, where $|w_1| = |w_2| \geqslant n$.

As we mentioned above there exists SMM recognizing L in RT that makes a unique essential step on time interval I. This unique step consists of constructing exactly one new node in the graph $S_M(B)$.

Consider a WTPM that makes constantly many essential steps on I. The result (Theorem 2) would be true for it. As it follows from lemmas WTPM while recognizing L on time interval I must change memory configuration (move over the graph or reconstruct it) at every step. I.e. constantly many essential steps can keep off moving over the graph only for some times. After a machine resumes motion all the information (part of a graph) constructed at the essential steps will be lost and will not be able to be used in further computations. Thus the estimate on the number of vertices in $S_M(B)$ holds true and that is why recognition of L in RT by these machines is impossible.

8. THE MAIN RESULT FOR STPM. In this part we shall prove that Theorem 2 holds for STPMs in the case when the Restriction is understood in the sense of Section 4. I.e. machine recognizing L is permitted either to move center of its storage or change labels of vertices in the graph. Suppose for simplicity that a machine can change label only of the central node. We shall see later that this limitation is not substantial.

Let vertices of a storage graph be labeled by symbols of finite alphabet Б and have out-degree at most k, let r be the local parameter of STPM, i.e. the radius of a neighbourhood in a graph that machine observes at any time.

Consider STPM M recognizing L and satisfying the Restriction. Suppose also that M observes neighbourhood $\mathcal{O}$ of a (Б, k)-tree $S_M(Bw)$ after reading an input Bw. Consider a set $W(w) \subseteq \{0,1\}^*$ that consists of words u after reading which M makes a move:

$$W(w) = \{u \mid \gamma_B(wu) \neq \gamma_B(w) \ \ \& \ \ (\forall v \subset u)(\gamma_B(wv) = \gamma_B(w))\}$$

Reading a word $u \in W(w)$ machine M changes root labels of $S_M(Bw)$ and after reading all the u it moves. The M's behaviour on inputs u from $W(w)$ depends on the r-neighbourhood $\mathcal{O}$ that it observes. Since the number of r-neighbourhoods is limited then the number of different sets $W(w)$ for various w is bounded by a constant that does not depend on the input's length.

It will be convenient for us to represent a set $W(w)$ as a binary tree, the number of leaves in which equals to $Card(W(w))$. Edges of the tree are marked with 0(left) and 1(right). And each path in it from the root to a leaf corresponds to some $u \in W(w)$. Square of $W(w)$ is the number of its internal (not leaves) vertices. We shall not distinguish below notations of a set $W(w)$ and its representation with binary tree. We say that a tree is *regular* if its nodes have out-degree either 0 or 2.

Lemma 1. Let M recognize L and satisfy the Restriction. Consider a base segment B, for which the CL propositions are held. Let a word w have length $|w| < 3n/2 - |\text{Б}|$. **Then** the following properties are fulfilled.

1. A height of the tree $W(w)$ is at most $|\text{Б}|$.
2. $W(w)$ is regular tree.
3. A square of $W(w)$ is at most $|\text{Б}|$.
4. The number of leaves in $W(w)$ is at most k: $Card(W(w)) \leqslant k$.
5. If the square of $W(w)$ is s and the number of its leaves is q, then $q = s + 1$ and $s \leqslant min(|\text{Б}|, k - 1)$.

Proof is based on applications of ML and CL.

Consider a tree $T(2n)$ of M's inputs on time interval I, i.e. a full binary tree of height $2n$ with edges labeled by 0(left) and 1(right). Each path in this tree corresponds to some input for M on time interval I.

Let's represent a process of M's computation on various inputs as a computation on the tree. We identify initially the tree $W(\lambda)$ with a subtree of $T(2n)$ in such a way that roots of $T(2n)$ and $W(\lambda)$ coincide, and the identified edges have equal labels. Nodes of $T(2n)$ that are identified with leaves of $W(\lambda)$ we mark with crosses. Each of this node becomes a root of a subtree that is identified in the same manner with the tree $W(u)$ for the corresponding $u \in W(\lambda)$.

Nodes of obtained subtrees we mark again with crosses. They become roots of new subtrees that in their turn are identified with corresponding trees ... and so on. After all we shall get a covering of $T(2n)$ by trees of form $W(w)$. Nodes of $T(2n)$ marked with crosses correspond to M's moves. As we are interested in the whole number of various moves, so we are trying to estimate the number of crosses in the tree $T(2n)$, which is called below the cardinality of $T(2n)$ covering.

Lemma (Covering). Let a finite set of regular trees W_i $(1 \leqslant i \leqslant p)$ such that their squares do not exceed s be given. Then for he full binary tree $T(h)$ of height h any its covering by trees of form W_i has cardinality at least $(2^{h+1} - 1)/s$.

Proof. As each internal node of W_i has outdegree 2, so the covering algorithm mentioned above guarantees that all $T(h)$ nodes will be covered. Square of W_i does not exceed s. Thus each subtree in the covering of $T(h)$ identified with some W_i covers at most s vertices of $T(h)$. A whole number of vertices in $T(h)$ is $2^{h+1}-1$. So the covering contains at least $(2^{h+1} - 1)/s$ subtrees. □

Theorem 3. STPM that recognizes L and satisfies the Restriction cannot work in RT.

Proof. Number of moves made by M with various inputs on time interval I is greater than the cardinality of some covering $T(3n/2)$ by trees of form $W(w)$. Using the Covering lemma and the lemma 1 we conclude that the number of M's moves on time interval I for various inputs is at least

$$\frac{1}{min(|\text{Б}|, k-1)} 2^{3n/2} = \Omega(2^{3n/2})$$

From CL and according to the storage tree structure almost all of these moves are different. Only those moves can coincide that M makes near the n-th level of the tree of inputs. But for each such move there is constantly many inputs on which M makes exactly this move. So it cannot affect the asymptotic behavior of the above bound.

Thus while reading base segment B of length $2^n[n/2]$ machine M working in RT must build a tree with $\Omega(2^{3n/2})$ nodes, which is impossible. □

9. THE MAIN RESULT FOR TTM. In Section 8 it was supposed that STPM recognizing L and satisfying the Restriction can make moves on time interval I and change labels of the central node only. It was promised also to show that this assumption is not essential, i.e. the result (Theorem 3) holds if STPM is permited to change labels of any node from the observing r-neighbourhood. Difficulties here are the same as one meets while trying to prove the result analogous to the Theorem 2 for TTMs.

Definition. Each TTM's step associated with the work memory is either a *move*, i.e. a passing to a neighbouring memory cell, or putting new symbol into the current cell, or changing of a machine state. A machine is said to make an *essential* step on time interval J if it visits some memory cell during J at least twice.

Restriction. TTM recognizing L must not make essential steps on time interval I, that corresponds to the part $x\#y\#$ of any input.

The Restriction for a TTM means that it is forbidden to move in the direction of the root – the memory cell from which machine starts on time interval I.

A restricted TTM has an opportunity to "remember" with the help of its states Q some finite information about previous computations and to "keep" it for a long time. This distinguishes TTMs from machine models considered above. WTPMs and STPMs that can change labels of the central node exclusively are forgetting all information about the past after each move. STPMs with an opportunity of unbounded changing all the available nodes have the same capability as TTMs. Further reasoning about TTMs can be easily reformulated for such (general case) STPMs.

Let some TTM M recognize L and satisfy the Restriction. $S_M(Bw)$ denotes the following tree. Its nodes are M's memory cells. Its root is a cell at which M occurs after reading input string Bw. Other nodes are cells accessible from the root by paths not going in the direction of the root and that were visited at least once during previous computations. Edges i $S_M(Bw)$ connect neighbouring memory cells. Nodes in $S_M(Bw)$ are marked with symbols that are written in the corresponding cells.

Let M's state after reading input word Bw be $q \in Q$ and let it observe a cell μ, which contains a symbol $a \in \Sigma$. M's configuration at this moment is denoted by a triple $K = < \mu, a, q >$. The Main lemma's analogue for TTMs is

Lemma (main). Let TTM M recognize L and satisfy the Restriction. Let M's configurations after reading Bw_1 and Bw_2 be $K_1 = < \mu_1, a_1, q_1 >$ and $K_2 = < \mu_2, a_2, q_2 >$, resp.

Then $S_M(Bw_1) = S_M(Bw_2)$ & $q_1 = q_2 \Rightarrow \xi_B(w_1) = \xi_B(w_2)$.

A scheme of the Main lemma proof for TTMs is the same of that we use above. Now we want to estimate using ML and CL a number of different moves that machine makes on various inputs. An extraordinary thing here is that TTM can make the same move on different inputs w_1 and w_2 while having not equal states $q_1 \neq q_2$. Thus the following situation is possible

$$\xi_B(w_1) \neq \xi_B(w_2),\ S_M(Bw_1) = S_M(Bw_2) \text{ and } q_1 \neq q_2$$

Let M's configurations after reading words $w_1, \ldots, w_p$ differ only in there states

$$K_i = < \mu, a, q_i >,\ q_i \neq q_j \text{ if } i \neq j, \text{ and } i, j = 1, \ldots, p.$$

Number p should not exceed $|Q|$.

For each w_i consider the set $W(w_i) \subseteq \{0,1\}^*$ of words after reading which M moves (see s.8). Let $V(\mu, a) = \bigcup_i \{q_i\} \times W(w_i)$. We will represent $V(\mu, a)$ by a group of trees $W(w_i)$ whose roots are marked with the corresponding states q_i. The height of such group is a maximal height of its trees. And group's square is a total sum of trees' squares.

Lemma 2.

1. A height of the group $V(\mu, a)$ does not exceed $|\Sigma|$.
2. A square of the group $V(\mu, a)$ does not exceed $|\Sigma| \cdot |Q|$.

Proof is based on applications of ML and CL.

Groups of trees $V(\mu, a)$ cover the tree $T(2n)$ of M's inputs on time interval I in the following way. Let $K_0 = < \mu_0, a_0, q_0 >$ denotes M's initial configuration on I. We identify $V(\mu_0, a_0) = \{q_0\} \times W(\lambda)$ with a subtree of $T(2n)$ in such a way that roots of $V(\mu_0, a_0)$ and $T(2n)$ coincide, and the identified edges have identical labels. Nodes of $T(2n)$ that are identified with leaves of $V(\mu_0, a_0)$ we mark with crosses. Choose words from $W(\lambda)$, after reading which M makes the same move to a cell μ_1. Crossed nodes that correspond to these words in $T(2n)$ are the ground of the next group $V(\mu_1, a_1)$ of trees. If M after reading w comes to μ_1 at a state q then a node that corresponds to w in $T(2n)$ becomes a root of the tree $\{q\} \times W(w)$ from $V(\mu_1, a_1)$. Continuing this process we get a covering of $T(2n)$ by tree groups $V(\mu, a)$.

Lemma (Covering). Let $T(h)$ be covered by groups of trees and a square of each group is at most s. Then cardinality of the covering of $T(2n)$ is at least $(2^{h+1}-1)/s$.

Theorem 4. TTM M that recognizes L and satisfies the Restriction cannot work in RT.

Proof. The Covering lemma, ML, CL and lemma 2 imply the following lower bound on the number of cells in $S_M(B)$

$$\frac{1}{|\Sigma|\cdot|Q|}2^{3n/2}=\Omega(2^{3n/2}).$$

M must visit all these cells while reading base segment B within a time of $O(n2^n)$. It is impossible. □

Note. The same result for STPMs (general case) follows from the lower bound on the number of moves that machine makes on time interval I for various inputs and the tree structure of STPM's storage.

ACKNOWLEDGEMENTS. I would like to thank A.L.Semenov for statement of the problem and permanent attention to the work, S.K.Lando and V.A.Shvachko for help.

REFERENCES.

1. [AHU] A.V.Aho, J.E.Hopcroft, J.D.Ullman, "The Design and Analysis of Computer Algorithms," Addison-Wesley, Reading, Mass., 1974.
2. Ya.M.Barzdin', *The complexity of simmetry recognition on Turing machines*, Problems of Cybernetics **15** (1965), 245–248, Nauka, Moscow. (Russian).
3. [G76] D.Yu.Grigoriev, *Kolmogorov algorithms are stronger than Turing machines*, ICMML VII, Notes of Scientific Seminars Steklov Mathematical Institute, Leningrad Department **60** (1976), 29–37, Nauka, Leningrad. (Russian).
4. [Gu88] Yu.Gurevich, *Kolmogorov machines and related issues*, The column on Logic in Computer Science, Bulletin EATCS **35** (1988), 71–82.
5. D.E.Knuth, "The Art of Computer Programming," vol.2, Seminumerical Algorithms, Addison-Wesley, Reading, MA, 1971.
6. [K53] A.N.Kolmogorov, *On the Notion of Algorithm*, Uspekhi Mat. Nauk **8** 4 (1953), 175–176. (Russian).
7. [KU58] A.N.Kolmogorov, V.A.Uspensky, *On the Definition of an Algorithm*, Uspekhi Mat. Nauk **13** 4 (1958), 3–28. (Russian); *AMS Translations*, **29** 2 (1963), 217–245.
8. K.Mehlhorn, S.Naher, H.Alt, *A lower bound on the compexity of the union-split-find problem*, SIAM J. Computing **17** 6 (1988), 1092–1102.
9. [S73] A.Schönhage, *Real-time simulation of multidimensional Turing machines by storage modification machines*, Technical memorandum **37** (1973), M.I.T. Project MAC, Cambridge, MA.
10. [S80] A.Schönhage, *Storage Modification Machines*, SIAM J. Computing **9** 3 (1980), 490–508.
11. A.Schönhage, *Tapes versus poi ters, a study in implementing fast algorithms*, EATCS Bulletin **30** (1986), 23–32.
12. J.Tromp, P. van Emde Boas, *Associative storage modification machines*, Cent. Math. and Comput. sci., NCS **R9014** (1990), 1–13.

13. [US87] V.A.Uspensky, A.L.Semenov, "The Theory of Algorithms: Basic Developments and Applications," Nauka, Moscow, 1987. (Russian); LN in computer science **122** (1981), 100–234.
14. P. van Emde Boas, *Space mesures for storage modification machines*, Inf. Proc. Lett. **30** (1989), 103–110.

Lecture Notes in Computer Science

For information about Vols. 1–429
please contact your bookseller or Springer-Verlag

Vol. 430: J.W. de Bakker, W.-P. de Roever, G. Rozenberg (Eds.), Stepwise Refinement of Distributed Systems. Models, Formalisms, Correctness. Proceedings, 1989. X, 808 pages. 1990.

Vol. 431: A. Arnold (Ed.), CAAP '90. Proceedings, 1990. Vl, 285 pages. 1990.

Vol. 432: N. Jones (Ed.), ESOP '90. Proceedings, 1990. IX, 436 pages. 1990.

Vol. 433: W. Schröder-Preikschat, W. Zimmer (Eds.), Progress in Distributed Operating Systems and Distributed Systems Management. Proceedings, 1989. V, 206 pages. 1990.

Vol. 434: J.-J. Quisquater, J. Vandewalle (Eds.), Advances in Cryptology – EUROCRYPT '89. Proceedings, 1989. X, 710 pages. 1990.

Vol. 435: G. Brassard (Ed.), Advances in Cryptology – CRYPTO '89. Proceedings, 1989. Xlll, 634 pages. 1990.

Vol. 436: B. Steinholtz, A. Sølvberg, L. Bergman (Eds.), Advanced lnformation Systems Engineering. Proceedings, 1990. X, 392 pages. 1990.

Vol. 437: D. Kumar (Ed.), Current Trends in SNePS – Semantic Network Processing System. Proceedings, 1989. Vll, 162 pages. 1990. (Subseries LNAI).

Vol. 438: D.H. Norrie, H.W. Six (Eds.), Computer Assisted Learning – ICCAL '90. Proceedings, 1990. Vll, 467 pages. 1990.

Vol. 439: P. Gorny, M. Tauber (Eds.), Visualization in Human-Computer Interaction. Proceedings, 1988. Vl, 274 pages. 1990.

Vol. 440: E. Börger, H. Kleine Büning, M.M. Richter (Eds.), CSL '89. Proceedings, 1989. Vl, 437 pages. 1990.

Vol. 441: T. Ito, R.H. Halstead, Jr. (Eds.), Parallel Lisp: Languages and Systems. Proceedings, 1989. Xll, 364 pages. 1990.

Vol. 442: M. Main, A. Melton, M. Mislove, D. Schmidt (Eds.), Mathematical Foundations of Programming Semantics. Proceedings, 1989. Vl, 439 pages. 1990.

Vol. 443: M.S. Paterson (Ed.), Automata, Languages and Programming. Proceedings, 1990. IX, 781 pages. 1990.

Vol. 444: S. Ramani, R. Chandrasekar, K.S.R. Anjaneyulu (Eds.), Knowledge Based Computer Systems. Proceedings, 1989. X, 546 pages. 1990. (Subseries LNAI).

Vol. 445: A.J.M. van Gasteren, On the Shape of Mathematical Arguments. Vlll, 181 pages. 1990.

Vol. 446: L. Plümer, Termination Proofs for Logic Programs. Vlll ,142 pages. 1990. (Subseries LNAI).

Vol. 447: J.R. Gilbert, R. Karlsson (Eds.), SWAT '90. 2nd Scandinavian Workshop on Algorithm Theory. Proceedings, 1990. Vl, 417 pages. 1990.

Vol. 448: B. Simons, A. Spector (Eds.), Fault Tolerant Distributed Computing. Vl, 298 pages. 1990.

Vol. 449: M.E. Stickel (Ed.), 10th International Conference on Automated Deduction. Proceedings, 1990. XVI, 688 pages. 1990. (Subseries LNAI).

Vol. 450: T. Asano, T. Ibaraki, H. Imai, T. Nishizeki (Eds.), Algorithms. Proceedings, 1990. Vlll, 479 pages. 1990.

Vol. 451: V. Marík, O. Stepánková, Z. Zdráhal (Eds.), Artificial Intelligence in Higher Education. Proceedings, 1989. IX, 247 pages. 1990. (Subseries LNAI).

Vol. 452: B. Rovan (Ed.), Mathematical Foundations of Computer Science 1990. Proceedings, 1990. VIII, 544 pages. 1990.

Vol. 453: J. Seberry, J. Pieprzyk (Eds.), Advances in Cryptology – AUSCRYPT '90 Proceedings, 1990. IX. 462 pages. 1990.

Vol. 454: V. Diekert, Combinatorics on Traces. Xll, 165 pages. 1990.

Vol. 455: C.A. Floudas, P.M. Pardalos, A Collection of Test Problems for Constrained Global Optimization Algorithms. XIV, 180 pages. 1990.

Vol. 456: P. Deransart, J. Maluszyn´ski (Eds.), Programming Language Implementation and Logic Programming. Proceedings, 1990. Vlll, 401 pages. 1990.

Vol. 457: H. Burkhart (Ed.), CONPAR '90 – VAPP IV. Proceedings, 1990. XIV, 900 pages. 1990.

Vol. 458: J.C.M. Baeten, J.W. Klop (Eds.), CONCUR '90. Proceedings, 1990. Vll, 537 pages. 1990.

Vol. 459: R. Studer (Ed.), Natural Language and Logic. Proceedings, 1989. Vll, 252 pages. 1990. (Subseries LNAI).

Vol. 460: J. Uhl, H.A. Schmid, A Systematic Catalogue of Reusable Abstract Data Types. Xll, 344 pages. 1990.

Vol. 461: P. Deransart, M. Jourdan (Eds.), Attribute Grammars and their Applications. Proceedings, 1990. Vlll, 358 pages. 1990.

Vol. 462: G. Gottlob, W. Nejdl (Eds.), Expert Systems in Engineering. Proceedings, 1990. IX, 260 pages. 1990. (Subseries LNAI).

Vol. 463: H. Kirchner, W. Wechler (Eds.), Algebraic and Logic Programming. Proceedings, 1990. Vll, 386 pages. 1990.

Vol. 464: J. Dassow, J. Kelemen (Eds.), Aspects and Prospects of Theoretical Computer Science. Proceedings, 1990. Vl, 298 pages. 1990.

Vol. 465: A. Fuhrmann, M. Morreau (Eds.), The Logic of Theory Change. Proceedings, 1989. X, 334 pages. 1991. (Subseries LNAI).

Vol. 466: A. Blaser (Ed.), Database Systems of the 90s. Proceedings, 1990. Vlll, 334 pages. 1990.

Vol. 467: F. Long (Ed.), Software Engineering Environments. Proceedings, 1969. Vl, 313 pages. 1990.

Vol. 468: S.G. Akl, F. Fiala, W.W. Koczkodaj (Eds.), Advances in Computing and lnformation – ICCI '90. Proceedings, 1990. VII, 529 pages. 1990.

Vol. 469: I. Guessarian (Ed.), Semantics of Systeme of Concurrent Processes. Proceedings, 1990. V, 456 pages. 1990.

Vol. 470: S. Abiteboul, P.C. Kanellakis (Eds.), ICDT '90. Proceedings, 1990. Vll, 528 pages. 1990.

Vol. 471: B.C. Ooi, Efficient Query Processing in Geographic Information Systems. Vlll, 208 pages. 1990.

Vol. 472: K.V. Nori, C.E. Veni Madhavan (Eds.), Foundations of Software Technology and Theoretical Computer Science. Proceedings, 1990. X, 420 pages. 1990.

Vol. 473: I.B. Damgård (Ed.), Advances in Cryptology – EUROCRYPT ’90. Proceedings, 1990. VIII, 500 pages. 1991.

Vol. 474: D. Karagiannis (Ed.), Information Syetems and Artificial Intelligence: Integration Aspects. Proceedings, 1990. X, 293 pages. 1991. (Subseries LNAI).

Vol. 475: P. Schroeder-Heister (Ed.), Extensions of Logic Programming. Proceedings, 1989. VIII, 364 pages. 1991. (Subseries LNAI).

Vol. 476: M. Filgueiras, L. Damas, N. Moreira, A.P. Tomás (Eds.), Natural Language Processing. Proceedings, 1990. VII, 253 pages. 1991. (Subseries LNAI).

Vol. 477: D. Hammer (Ed.), Compiler Compilers. Proceedings, 1990. VI, 227 pages. 1991.

Vol. 478: J. van Eijck (Ed.), Logics in AI. Proceedings, 1990. IX, 562 pages. 1991. (Subseries in LNAI).

Vol. 480: C. Choffrut, M. Jantzen (Eds.), STACS 91. Proceedings, 1991. X, 549 pages. 1991.

Vol. 481: E. Lang, K.-U. Carstensen, G. Simmons, Modelling Spatial Knowledge on a Linguistic Basis. IX, 138 pages. 1991. (Subseries LNAI).

Vol. 482: Y. Kodratoff (Ed.), Machine Learning – EWSL-91. Proceedings, 1991. XI, 537 pages. 1991. (Subseries LNAI).

Vol. 483: G. Rozenberg (Ed.), Advances In Petri Nets 1990. VI, 515 pages. 1991.

Vol. 484: R. H. Möhring (Ed.), Graph-Theoretic Concepts In Computer Science. Proceedings, 1990. IX, 360 pages. 1991.

Vol. 485: K. Furukawa, H. Tanaka, T. Fullsaki (Eds.), Logic Programming ’89. Proceedings, 1989. IX, 183 pages. 1991. (Subseries LNAI).

Vol. 486: J. van Leeuwen, N. Santoro (Eds.), Distributed Algorithms. Proceedings, 1990. VI, 433 pages. 1991.

Vol. 487: A. Bode (Ed.), Distributed Memory Computing. Proceedings, 1991. XI, 506 pages. 1991.

Vol. 488: R. V. Book (Ed.), Rewriting Techniques and Applications. Proceedings, 1991. VII, 458 pages. 1991.

Vol. 489: J. W. de Bakker, W. P. de Roever, G. Rozenberg (Eds.), Foundations of Object-Oriented Languages. Proceedings, 1990. VIII, 442 pages. 1991.

Vol. 490: J. A. Bergstra, L. M. G. Feljs (Eds.), Algebraic Methods 11: Theory, Tools and Applicatlons. VI, 434 pages. 1991.

Vol. 491: A. Yonezawa, T. Ito (Eds.), Concurrency: Theory, Language, and Architecture. Proceedings, 1989. VIII, 339 pages. 1991.

Vol. 492: D. Sriram, R. Logcher, S. Fukuda (Eds.), Computer-Aided Cooperative Product Development. Proceedings, 1989 VII, 630 pages. 1991.

Vol. 493: S. Abramsky, T. S. E. Maibaum (Eds.), TAPSOFT ’91. Volume 1. Proceedings, 1991. VIII, 455 pages. 1991.

Vol. 494: S. Abramsky, T. S. E. Maibaum (Eds.), TAPSOFT ’91. Volume 2. Proceedings, 1991. VIII, 482 pages. 1991.

Vol. 495: 9. Thalheim, J. Demetrovics, H.-D. Gerhardt (Eds.), MFDBS ’91. Proceedings, 1991. VI, 395 pages. 1991.

Vol. 496: H.-P. Schwefel, R. Männer (Eds.), Parallel Problem Solving from Nature. Proceedings, 1991. XI, 485 pages. 1991.

Vol. 497: F. Dehne, F. Fiala. W.W. Koczkodaj (Eds.), Advances in Computing and Intormation - ICCI ’91 Proceedings, 1991. VIII, 745 pages. 1991.

Vol. 498: R. Andersen, J. A. Bubenko jr., A. Sølvberg (Eds.), Advanced Information Systems Engineering. Proceedings, 1991. VI, 579 pages. 1991.

Vol. 499: D. Christodoulakis (Ed.), Ada: The Choice for ’92. Proceedings, 1991. VI, 411 pages. 1991.

Vol. 500: M. Held, On the Computational Geometry of Pocket Machining. XII, 179 pages. 1991.

Vol. 501: M. Bidoit, H.-J. Kreowski, P. Lescanne, F. Orejas, D. Sannella (Eds.), Algebraic System Specification and Development. VIII, 98 pages. 1991.

Vol. 502: J. Bārzdiņš, D. Bjørner (Eds.), Baltic Computer Science. X, 619 pages. 1991.

Vol. 503: P. America (Ed.), Parallel Database Systems. Proceedings, 1990. VIII, 433 pages. 1991.

Vol. 504: J. W. Schmidt, A. A. Stogny (Eds.), Next Generation Information System Technology. Proceedings, 1990. IX, 450 pages. 1991.

Vol. 505: E. H. L. Aarts, J. van Leeuwen, M. Rem (Eds.), PARLE ’91. Parallel Architectures and Languages Europe, Volume I. Proceedings, 1991. XV, 423 pages. 1991.

Vol. 506: E. H. L. Aarts, J. van Leeuwen, M. Rem (Eds.), PARLE ’91. Parallel Architectures and Languages Europe, Volume II. Proceedings, 1991. XV, 489 pages. 1991.

Vol. 507: N. A. Sherwani, E. de Doncker, J. A. Kapenga (Eds.), Computing in the 90’s. Proceedings, 1989. XIII, 441 pages. 1991.

Vol. 508: S. Sakata (Ed.), Applied Algebra, Algebraic Algorithms and Error-Correcting Codes. Proceedings, 1990. IX, 390 pages. 1991.

Vol. 509: A. Endres, H. Weber (Eds.), Software Development Environments and CASE Technology. Proceedings, 1991. VIII, 286 pages. 1991.

Vol. 510: J. Leach Albert, B. Monien, M. Rodríguez (Eds.), Automata, Languages and Programming. Proceedings, 1991. XII, 763 pages. 1991.

Vol. 511: A. C. F. Colchester, D.J. Hawkes (Eds.), Information Processing in Medical Imaging. Proceedings, 1991. XI, 512 pages. 1991.

Vol. 512: P. America (Ed.), ECOOP ’91. European Conference on Object-Oriented Programming. Proceedings, 1991. X, 396 pages. 1991.

Vol. 513: N. M. Mattos, An Approach to Knowledge Base Management. IX, 247 pages. 1991. (Subseries LNAI).

Vol. 514: G. Cohen, P. Charpin (Eds.), EUROCODE ’90. Proceedings, 1990. XI, 392 pages. 1991.

Vol. 515: J. P. Martins, M. Reinfrank (Eds.), Truth Maintenance Systems. Proceedings, 1990. VII, 177 pages. 1991. (Subseries LNAI).

Vol. 516: S. Kaplan, M. Okada (Eds.), Conditional and Typed Rewriting Systems. Proceedings, 1990. IX, 461 pages. 1991.

Vol. 517: K. Nökel, Temporally Distributed Symptoms in Technical Diagnosis. IX, 164 pages. 1991. (Subseries LNAI).

Vol. 518: J. G. Williams, Instantiation Theory. VIII, 133 pages. 1991. (Subseries LNAI).

Vol. 519: F. Dehne, J.-R. Sack, N. Santoro (Eds.), Algorithms and Data Structures. Proceedings, 1991. X, 496 pages. 1991.

Vol. 520: A. Tarlecki (Ed.), Mathematical Foundations of Computer Science 1991. Proceedings, 1991. XI, 435 pages. 1991.